Lecture Notes in Computer Science 3467

Commenced Publication in 1973
Founding and Former Series Editors:
Gerhard Goos, Juris Hartmanis, and Jan van Leeuwen

T0134754

Jürgen Giesl (Ed.)

Term Rewriting and Applications

16th International Conference, RTA 2005
Nara, Japan, April 19-21, 2005
Proceedings

 Springer

Volume Editor

Jürgen Giesl
RWTH Aachen
Lehr- und Forschungsgebiet für Informatik II
Ahornstr. 55, 52074 Aachen, Germany
E-mail: giesl@informatik.rwth-aachen.de

Library of Congress Control Number: 2005923620

CR Subject Classification (1998): F.4, F.3.2, D.3, I.2.2-3, I.1

ISSN   0302-9743
ISBN-10   3-540-25596-6 Springer Berlin Heidelberg New York
ISBN-13   978-3-540-25596-3 Springer Berlin Heidelberg New York

Springer is a part of Springer Science+Business Media

springeronline.com

© Springer-Verlag Berlin Heidelberg 2005
Printed in Germany

Typesetting: Camera-ready by author, data conversion by Olgun Computergrafik
Printed on acid-free paper   SPIN: 11416999   06/3142   5 4 3 2 1 0

V

Preface

This volume contains the proceedings of the *16th International Conference on Rewriting Techniques and Applications* (RTA 2005), which was held on April 19–21, 2005, at the Nara-Ken New Public Hall in the center of the Nara National Park in Nara, Japan.

RTA is the major forum for the presentation of research on all aspects of rewriting. Previous RTA conferences were held in Dijon (1985), Bordeaux (1987), Chapel Hill (1989), Como (1991), Montreal (1993), Kaiserslautern (1995), Rutgers (1996), Sitges (1997), Tsukuba (1998), Trento (1999), Norwich (2000), Utrecht (2001), Copenhagen (2002), Valencia (2003), and Aachen (2004).

This year, there were 79 submissions from 20 countries, of which 31 papers were accepted for publication (29 regular papers and 2 system descriptions). The submissions came from France (10 accepted papers of the 23.1 submitted papers), USA (5.6 of 11.7), Japan (4 of 9), Spain (2.7 of 6.5), UK (2.7 of 4.7), The Netherlands (1.7 of 3.8), Germany (1.3 of 2.3), Austria (1 of 1), Poland (1 of 1), Israel (0.5 of 0.8), Denmark (0.5 of 0.5), China (0 of 4), Korea (0 of 4), Taiwan (0 of 1.3), Australia (0 of 1), Brazil (0 of 1), Russia (0 of 1), Switzerland (0 of 1), Sweden (0 of 1), and Italy (0 of 0.3).

Each submission was assigned to at least three Program Committee members, who carefully reviewed the papers, with the help of 112 external referees. Afterwards, the submissions were discussed by the Program Committee during one week through the Internet by means of Andrei Voronkov's *EasyChair* system. I want to thank Andrei very much for providing his system which was very helpful for the management of the submissions and reviews and for the discussion of the Program Committee.

The Program Committee decided to award a prize of 100,000 Yen for the best paper to the article *Extending the Explicit Substitution Paradigm* by Delia Kesner and Stéphane Lengrand. Moreover, student travel grants were awarded to Jérôme Rocheteau (author of the paper *λμ-Calculus and Duality: Call-by-Name and Call-by-Value*) and Wojciech Moczydłowski (co-author of the paper *Termination of Single-Threaded One-Rule Semi-Thue Systems*).

RTA 2005 had three invited talks, by Yoshihito Toyama (*Confluent Term Rewriting Systems*), Philip Wadler (*Call-by-Value is Dual to Call-by-Name – Reloaded*), and Amy Felty (*A Tutorial Example of the Semantic Approach to Foundational Proof-Carrying Code*). The talk by Amy Felty was a joint invited talk of RTA and the collocating *7th International Conference on Typed Lambda Calculi and Applications* (TLCA 2005).

Since RTA 2005 marked the 20th anniversary of RTA, this anniversary was celebrated with a special session of invited talks on the history and future of RTA and rewriting. For this session, we invited Gérard Huet (*Before RTA: Early Days in Rewriting Research*), Jean-Pierre Jouannaud (*Twenty Years Later*), and Nachum Dershowitz (*Open. Closed. Open.*). I want to thank both the invited speakers of RTA and of the anniversary session for their interesting and inspiring talks.

RTA 2005 was held as part of the *Federated Conference on Rewriting, Deduction, and Programming* (RDP), together with the following events. I wish to thank the organizers of these events for making the conference even more attractive:

- *7th International Conference on Typed Lambda Calculi and Applications*, TLCA 2005 (Program Chair: Paweł Urzyczyn, Conference Chair: Masahito Hasegawa)
- *6th International Workshop on Rule-Based Programming*, RULE 2005 (Horatiu Cirstea, Narciso Martí-Oliet)
- *19th International Workshop on Unification*, UNIF 2005 (Laurent Vigneron)
- *5th International Workshop on Reduction Strategies in Rewriting and Programming*, WRS 2005 (Roberto Di Cosmo, Yoshihito Toyama)
- IFIP Working Group 1.6 on Term Rewriting (Claude Kirchner)

Many people helped to make RTA 2005 a success. In particular, I want to thank Hitoshi Ohsaki, the conference chair of RTA 2005, and the other members of the Organizing Committee, who organized the conference in a very careful and completely perfect way. I am also very grateful to the members of the Program Committee, to the external reviewers, to the former and current publicity chairs of RTA (Femke van Raamsdonk and Ralf Treinen), to the sponsors of the conference, and to René Thiemann and Peter Schneider-Kamp for helping with many technical problems.

February 2005   Jürgen Giesl

Conference Organization

Program Chair

Jürgen Giesl   *RWTH Aachen*

Conference Chair

Hitoshi Ohsaki   *AIST*

Program Committee

Franz Baader   *Dresden*
Mariangiola Dezani   *Torino*
Jürgen Giesl   *Aachen*
Bernhard Gramlich   *Vienna*
Florent Jacquemard   *Cachan*
Claude Kirchner   *Nancy*
Pierre Lescanne   *Lyon*
Aart Middeldorp   *Innsbruck*
Hitoshi Ohsaki   *Amagasaki*
Vincent van Oostrom   *Utrecht*
Christine Paulin-Mohring   *Orsay*
Frank Pfenning   *Pittsburgh*
Femke van Raamsdonk   *Amsterdam*
Mark-Oliver Stehr   *Urbana*
Rakesh Verma   *Houston*
Andrei Voronkov   *Manchester*

Organizing Committee

Kokichi Futatsugi   *Tatsunokuchi*
Masami Hagiya   *Tokyo*
Teruo Higashino   *Osaka*
Maki Ishida   *Amagasaki*   (Secretary)
Hitoshi Ohsaki   *Amagasaki*
Toshiki Sakabe   *Nagoya*
Hiroyuki Seki   *Nara*
Yoshihito Toyama   *Tohoku*

RTA Steering Committee

Jürgen Giesl	*Aachen*	
Delia Kesner	*Paris*	
Robert Nieuwenhuis	*Barcelona*	
Ralf Treinen	*Cachan*	(Publicity Chair)
Vincent van Oostrom	*Utrecht*	
Femke van Raamsdonk	*Amsterdam*	(Chair)

Sponsors

Foundation for Nara Institute of Science and Technology
Information Processing Society of Japan (IPSJ)
Japan Society for the Promotion of Science (JSPS)
Kayamori Foundation of Informational Science Advancement
Nara Convention Bureau (NCB)
The Institute of Electronics, Information and Communication Engineers (IEICE)
The Telecommunications Advancement Foundation (TAF)

Referees

Takahito Aoto
Philippe Audebaud
Jürgen Avenhaus
Steffen van Bakel
Franco Barbanera
Clara Bertolissi
Stefan Blom
Viviana Bono
Olivier Bournez
Roberto Bruni
Pierre Castéran
Kaustuv Chaudhuri
Horatiu Cirstea
Evelyne Contejean
Veronique Cortier
Ferruccio Damiani
Rowan Davies
Giorgio Delzanno
Catherine Dubois
Rachid Echahed
Santiago Escobar
Moreno Falaschi
Azadeh Farzan
Germain Faure
Jean-Christophe Filliatre
Thomas Forster
Laurent Fribourg
Gerhard Friedrich
Fabio Gadducci
Florent Garnier
Alfons Geser
Herman Geuvers
Silvio Ghilardi
Robert Glück
Isabelle Gnaedig
Jean Goubault-Larrecq
Makoto Hamana
Michael Hanus
Joe Hendrix
Fritz Henglein
Hugo Herbelin
Nao Hirokawa
Yuichi Kaji
Delia Kesner
Jeroen Ketema
Assaf Kfoury
Florent Kirchner
Francis Klay
Konstantin Korovin
Keiichirou Kusakari
Ralf Küsters
Jordi Levy
Zhiyao Liang
Ugo de Liguoro
Sébastien Limet
Salvador Lucas
Denis Lugiez
Bas Luttik
Chris Lynch
Sebastian Maneth
Mircea Marin
Ursula Martin
Jean-François Monin
Pierre-Etienne Moreau
Georg Moser
Fabrice Nahon
Aleks Nanevski
Paliath Narendran
Naoki Nishida

Albert Oliveras
Peter Ölveczky
Vincent Padovani
Miguel Palomino
Luca Paolini
David Plaisted
Jeff Polakow
Emmanuel Polonovski
Silvio Ranise
Christophe Ringeissen
Simona R. Della Rocca
Michael Rusinowitch
Claudio Russo
Tatiana Rybina
Claudio Sacerdoti

Masahiko Sakai
Sylvain Salvati
Gernot Salzer
Anderson Santana
Manfred Schmidt-Schauß
Peter Schneider-Kamp
Takahiro Seino
Hiroyuki Seki
Vitaly Shmatikov
Jakob Grue Simonsen
Toshinori Takai
Prasanna Thati
René Thiemann
Cesare Tinelli
Sophie Tison

Ashish Tiwari
Yoshihito Toyama
Ralf Treinen
Xavier Urbain
Kumar Neeraj Verma
Paliath Narendran
Fer-Jan de Vries
Christoph Walther
Geoffrey Washburn
Herbert Wiklicky
Toshiyuki Yamada
Hans Zantema
Pascal Zimmer

Table of Contents

XII   Table of Contents

Confluent Term Rewriting Systems

Yoshihito Toyama

RIEC, Tohoku University,
Katahira 2-1-1, Aoba-ku, Sendai 980-8577, Japan
toyama@nue.riec.tohoku.ac.jp

Abstract. The confluence property is one of the most important properties of term rewriting systems, and various sufficient criteria for proving this property have been widely investigated. A necessary and sufficient criterion for confluence of terminating term rewriting systems, in which every reduction must terminate, was demonstrated by Knuth and Bendix (1970). For non-terminating term rewriting systems, Rosen (1973) proved that left-linear and non-overlapping term rewriting systems (i.e., no variable occurs twice or more in the left-hand side of a rewriting rule and two left-hand sides of rewriting rules must not overlap) are confluent, and the non-overlapping limitation was somewhat relaxed by Huet (1980), Toyama (1988), and van Oostrom (1997). However, few criteria have been proposed for confluence of term rewriting systems that are non-left-linear and non-terminating. Thus, it is still worth while extending criteria for these systems.

A powerful technique for showing confluence of a non-left-linear non-terminating term rewriting system is a divide-and-conquer method based on modularity by Toyama (1987) or persistency by Zantema (1994), Aoto and Toyama (1997). The method guarantees that if the system is decomposed into small subsystems and each of them is confluent then this system has the confluence property. Another useful technique is a transformational method based on conditional-linearization by Klop and de Vrijer (1991), Toyama and Oyamaguchi (1994), or a labelling technique. In this method we apply a non-confluence preserving transformation on a term rewriting system. Then the term rewriting system is confluent if the transformed system is confluent, because of non-confluence preserving. In this talk we will illustrate these techniques through various examples and discuss the relation among them.

J. Giesl (Ed.): RTA 2005, LNCS 3467, p. 1, 2005.

Generalized Innermost Rewriting

Jaco van de Pol[1,2] and Hans Zantema[2]

[1] Department of Software Engineering, CWI, P.O. Box 94.079,
1090 GB Amsterdam, The Netherlands
Jaco.van.de.Pol@cwi.nl
[2] Department of Computer Science, TU Eindhoven, P.O. Box 513,
5600 MB Eindhoven, The Netherlands
H.Zantema@tue.nl

Abstract. We propose two generalizations of innermost rewriting for which we prove that termination of innermost rewriting is equivalent to termination of generalized innermost rewriting. As a consequence, by rewriting in an arbitrary TRS certain non-innermost steps may be allowed by which the termination behavior and efficiency is often much better, but never worse than by only doing innermost rewriting.

1 Introduction

In term rewriting one can rewrite according various reduction strategies, for instance innermost or outermost. It may occur that by one strategy a normal form is reached while rewriting according another strategy may go on forever. For instance, innermost rewriting of $f(a)$ by the TRS consisting of the two rules $a \to b, f(a) \to f(a)$ yields a normal form in one step while outermost rewriting goes on forever. By the TRS consisting of the two rules $a \to a, f(x) \to b$ the behavior of $f(a)$ is opposite: now innermost rewriting goes on forever while outermost rewriting yields a normal form in one step. We say that a particular strategy is not worse than another strategy if for every term for which the latter yields a normal form, the same holds for the former. The above examples show that innermost and outermost are incomparable. For orthogonal TRSs it is known that no strategy is worse than innermost ([8]); later it was proved that the same result holds more generally for non-overlapping TRSs ([5]). In practice, many implementations use innermost rewriting since it is easy to implement and often efficient. On the other hand, for implementing lazy evaluation for functional programming it is essential to do non-innermost rewriting since otherwise computations will not terminate. The main idea of lazy rewriting ([3]) is that doing a computation is postponed until the result is required for continuation.

In this paper we consider strategies for arbitrary TRSs allowing overlaps and even non-confluence, as often occurs in applications in theorem proving. For instance, the natural rules to obtain conjunctive normal forms are not confluent. We present two generalizations of innermost rewriting that are always provably not worse than innermost, and allow non-innermost steps as they are preferred in lazy rewriting. In the usual definition of innermost rewriting it is required that all proper subterms of a redex are in normal form; in our definition of generalized innermost rewriting we require this only for particular proper subterms.

J. Giesl (Ed.): RTA 2005, LNCS 3467, pp. 2–16, 2005.
© Springer-Verlag Berlin Heidelberg 2005

As an example, consider the following TRS for computation of factorials:

$$\mathsf{fac}(x) \to \mathsf{if}(\mathsf{eq}(x,0), \mathsf{succ}(0), x * \mathsf{fac}(\mathsf{pred}(x)))$$
$$\mathsf{if}(\mathsf{true}, x, y) \to x$$
$$\mathsf{if}(\mathsf{false}, x, y) \to y$$

completed by standard rules for eq, pred and *. Now $\mathsf{fac}(\mathsf{succ}^n(0))$ is not weakly innermost normalizing, for any $n \geq 0$, but it is weakly generalized innermost normalizing. Moreover, straightforward implementations of generalized innermost rewriting easily find the corresponding normal form.

In case of constructor systems that are either right-linear or non-root-overlapping, our generalization corresponds to arbitrary rewriting. But our results also apply for TRSs not being constructor systems or having overlaps. For instance, in the above example we may have rules like $(x * y) * z \to x * (y * z)$.

Apart from only considering whether a reduction will terminate or not, also lengths of reductions to normal form may be considered. Then it is natural for calling one strategy not worse than another strategy to require additionally that if both strategies yield a normal form then the former strategy does not take more steps. For this extra requirement it is obvious that duplicating rules should be avoided. By doing so we prove that indeed this stronger result of being not worse than innermost is obtained for generalized innermost rewriting.

Many implementations of rewriting including OBJ ([4, 7]) make use of similar modifications of the innermost strategy to achieve better termination behavior or efficiency. The basis of our work was in [10, 11], where JITty strategy annotations and a corresponding implementation was proposed to postpone computation steps in a slightly more general way than eager annotations in OBJ. All these implementations are essentially deterministic and apply particular cases of generalized innermost rewriting, while often non-innermost steps are done. Typically, the termination behavior and efficiency of these implementations is much better, but by our results never worse than by only doing innermost rewriting.

The outline of the paper is as follows. After recalling some standard notation in Section 2, we introduce our basic generalization of innermost rewriting in Section 3. Unfortunately, some further restrictions have to be given in order to have the desired properties, as we show by an example. This is done in two ways. In Section 4 this is done by avoiding duplication, by which also results are obtained involving the number of rewrite steps. In Section 5 this is done by avoiding root overlaps and applying priority of rules. In particular an open problem from [10] is solved.

2   Basic Notation

As usual, for binary relations $\to, \to'$ on a set T we define relation composition by $\to \cdot \to' := \{(t, t') \mid \exists t'' : t \to t'' \wedge t'' \to' t'\}$. With $\to^n$ we denote the n-fold relation composition, and $\to^+$ and $\to^*$ denote the transitive, and the transitive reflexive closure of $\to$, respectively. Finally, $\leftarrow$ denotes the inverse of $\to$. Using these notations many properties can be expressed shortly, for instance, local confluence of $\to$ is expressed by $\leftarrow \cdot \to \subseteq \to^* \cdot \leftarrow^*$.

An object t is in *normal form* in $\rightarrow$, if no s exists satisfying $t \rightarrow s$. The set $\mathsf{NF}(\rightarrow)$ denotes the set of all $\rightarrow$-normal forms. An object $t \in T$ is *weakly normalizing* in $\rightarrow$, denoted by $\mathsf{WN}(t, \rightarrow)$, if there exists a reduction $t \rightarrow^* s$ and $s \in \mathsf{NF}(\rightarrow)$. An object $t \in T$ is *terminating* in $\rightarrow$, denoted by $\mathsf{SN}(t, \rightarrow)$, if there is no infinite $\rightarrow$-reduction sequence starting in t.

In the sequel, we consider a fixed set of function symbols and a fixed set of variables $\mathcal{V}$, from which terms are built as usual. By $\mathsf{arity}(f)$ we denote the number of arguments expected by f. The set of variables occurring in a term t is denoted by $\mathsf{Var}(t)$; we will use $\mathsf{LinVar}(t)$ to denote the set of variables that occur exactly once in t. A substitution is a mapping from variables to terms, and application of substitution σ to a term t is denoted by $t\sigma$, or by t^σ when σ is complex expression.

A *position* is defined to be a list of positive integers, as in [1]; ε denotes the empty list, corresponding to the root position. The set of valid positions in a term t is denoted by $\mathsf{Pos}(t)$; the subterm at position p in a term t is denoted by $t|_p$; its root symbol is denoted by $(t)_p$. Replacing the subterm at position p in C by s is denoted by $C[s]_p$, or simply $C[s]$, where $C[]$ is called the context. Two positions that are not comparable in the prefix order are called *disjoint*.

We consider rewriting w.r.t. a fixed TRS, consisting of a set of rules of the form $\ell \rightarrow r$, such that $\ell \notin \mathcal{V}$ and $\mathsf{Var}(r) \subseteq \mathsf{Var}(\ell)$. Rules $\ell_1 \rightarrow r_1$ and $\ell_2 \rightarrow r_2$ are *root overlapping* if for some σ and τ, $\ell_1\sigma = \ell_2\tau$.

Let Def denote the set of *defined symbols*, that is, symbols that occur as the root symbol of the left hand side of a rule. A TRS is called a *constructor system* if for all left hand sides ℓ the root symbol is the only occurrence of a symbol in Def.

3   Generalized Innermost Rewriting

Recall that we consider a fixed TRS.

Definition 1. *A term t rewrites to a term u, written $t \rightarrow u$, if*

- $t = C[\ell\sigma]$ *and* $u = C[r\sigma]$ *for a rule* $\ell \rightarrow r$, *context C and substitution σ.*

The subterm $\ell\sigma$ of t is called the corresponding redex. *In the sequel, the set* NF *always denotes the set of normal forms* $\mathsf{NF}(\rightarrow)$.

Definition 2. *A term t rewrites* innermost *to a term u, notation $t \rightarrow_i u$, if*

- $t = C[\ell\sigma]$ *and* $u = C[r\sigma]$ *for a rule* $\ell \rightarrow r$, *context C and substitution σ; and*
- $\ell\sigma|_p \in \mathsf{NF}$ *for all* $p \in \mathsf{Pos}(\ell) \setminus \{\varepsilon\}$.

It is easy to see that this corresponds to the usual notion of innermost rewriting. It is also easy to see that $\mathsf{NF}(\rightarrow_i) = \mathsf{NF}(\rightarrow)$. We now introduce a generalization of innermost rewriting in which the second condition is slightly weakened.

Definition 3. *A term t rewrites* generalized innermost *to a term u, notation $t \rightarrow_g u$, if*

- $t = C[\ell\sigma]$ *and* $u = C[r\sigma]$ *for a rule* $\ell \rightarrow r$, *context C and substitution σ; and*
- $\ell\sigma|_p \in \mathsf{NF}$, *for all* $p \in \mathsf{Pos}(\ell) \setminus \{\varepsilon\}$, *such that* $(\ell)_p \in \mathsf{Def}$.

Clearly, $\rightarrow_i \subseteq \rightarrow_g \subseteq \rightarrow$, hence $\mathsf{NF}(\rightarrow_g) = \mathsf{NF}(\rightarrow)$. Note that for constructor systems, $\rightarrow_g = \rightarrow$, i.e. generalized innermost rewriting corresponds to general rewriting.

We hope that generalized innermost rewriting has better normalization properties than innermost rewriting. This means first of all that if a term t is $\rightarrow_i$-terminating, then it should be $\rightarrow_g$-terminating. However, this is not always the case, as is witnessed by the following example inspired by [9]:

Example 4. Consider the TRS consisting of the three rules

$$f(a, b, x) \rightarrow f(x, x, x), \ \ c \rightarrow a, \ \ c \rightarrow b$$

$\rightarrow_i$ is terminating on $f(a, b, c)$, while $\rightarrow_g$ is not terminating on $f(a, b, c)$.

We see in this example that redexes can be copied by the first rule, and that the copies can behave differently, because the last rules are root-overlapping. Both ingredients are essential for this counter example. In Section 4, we restrict the generalized innermost strategy to avoid duplication of redexes. In Section 5 we restrict the generalized innermost strategy to avoid root-overlaps by assuming a priority on the rules. In both cases we prove that innermost termination coincides with termination of the restricted generalized innermost reduction relation. The proofs of these results are quite different, and the sections can be read independently.

Moreover, if redex duplication is avoided, the number of generalized innermost steps is bounded by the number of innermost steps. In case the TRS is non-root-overlapping, we prove equivalence of $\rightarrow_i$-termination and $\rightarrow_g$-termination. The latter section also solves an open problem in [10].

4   Avoiding Duplication

In case one wants short reductions to normal form it is clear that duplicating rules should be avoided. For instance, in rewriting the term $f(a)$ by the TRS consisting of the two rules $f(x) \rightarrow g(x, x), a \rightarrow b$ we prefer innermost reduction since otherwise a will be duplicated before it is rewritten, requiring one more rewrite step afterward. Therefore we now adjust our definition of generalized innermost rewriting to a non-duplicating variant as follows.

Definition 5. *A term t rewrites non-dup-generalized innermost to a term u, notation $t \rightarrow_{ndg} u$, if*

- $t = C[\ell\sigma]$ and $u = C[r\sigma]$ for a rule $\ell \rightarrow r$, context C and substitution σ; and
- $\ell\sigma|_p \in \mathsf{NF}$ for all $p \in \mathsf{Pos}(\ell) \setminus \{\epsilon\}$ for which either
  - $(\ell)_p \in \mathsf{Def}$, or
  - $(\ell)_p$ is a variable occurring more than once in r.

Clearly, $\rightarrow_i \subseteq \rightarrow_{ndg} \subseteq \rightarrow_g \subseteq \rightarrow$, hence $\mathsf{NF}(\rightarrow_{ndg}) = \mathsf{NF}(\rightarrow)$. Compared to generalized innermost rewriting the extra condition is that subterms on variable positions with multiple occurrences in r should be in normal form. One can

argue this is more related to non-right-linearity than to duplication, but since
the intension is to avoid duplicating steps we chose to call this 'non-dup'. The
desired theorem (Theorem 9) does not hold if we weaken the extra condition to
non-duplication in the sense that the variable occurs not more often in r than
in ℓ: in the TRS

$$f(x,x) \rightarrow g(x,x), \; a \rightarrow b, \; a \rightarrow c, \; g(b,c) \rightarrow g(b,c)$$

the term $f(a,a)$ is innermost terminating while it admits an infinite generalized
innermost reduction.

Note that non-dup-generalized innermost rewriting applies to all TRSs, the
only extra restriction is that a subterm in a particular position in a redex should
be in normal form. For non-dup-generalized innermost rewriting we will show
that it is not worse than innermost without any restriction on the order of the
rules, and moreover, the lengths of reductions to normal form by this strategy
are not worse than by the innermost strategy. First we need some lemmas.

Lemma 6. *Let t' be an innermost redex of $t = C[t']$, and $t \rightarrow_i^n u$ for some term
u and $n \geq 0$. Then either*

- *$u = C'[t']$ for some context C' and $C[t''] \rightarrow_i^n C'[t'']$ for any term t'', or*
- *there exists v such that $t' \rightarrow_i v$ and $C[t'] \rightarrow_i C[v] \rightarrow_i^{n-1} u$.*

Proof. Induction on n. For $n = 0$ we choose $C' = C$ and we are in the first case.
If $n > 0$ and t' is rewritten in the first step then we are in the second case. In
the remaining case another redex is rewritten in the first step of $t \rightarrow_i^n u$, say
t_1 rewriting to t_2. Assume t' is left from t_1 in t; if it is right the argument is
similar. So we have

$$t = D[t',t_1] \rightarrow_i D[t',t_2] \rightarrow_i^{n-1} u.$$

We apply the induction hypothesis to $D[t',t_2] \rightarrow_i^{n-1} u$. If the first case holds
then we obtain $u = C'[t']$ and $D[t'',t_2] \rightarrow_i^{n-1} C'[t'']$, hence $C[t''] = D[t'',t_1] \rightarrow_i$
$D[t'',t_2] \rightarrow_i^{n-1} C'[t'']$, and we are in the first case. If the second case holds for
$D[t',t_2] \rightarrow_i^{n-1} u$ then we have v satisfying $t' \rightarrow_i v$ and $D[t',t_2] \rightarrow_i D[v,t_2] \rightarrow_i^{n-2}$
u, yielding $C[t'] = D[t',t_1] \rightarrow_i D[v,t_1] \rightarrow_i D[v,t_2] \rightarrow_i^{n-2} u$ by which we
are in the second case, concluding the proof.   □

Lemma 7. *Let $t \rightarrow_{ndg} u \rightarrow_i^n w$ for terms t,u,w and $n \geq 0$, and $t \not\rightarrow_i u$. Then*

$$t \rightarrow_i^+ \cdot \rightarrow_{ndg} \cdot \rightarrow_i^k \cdot \leftarrow_i^m w$$

for some $k \geq n-1$ and m being either 0 or 1.

Proof. Write $t = \overline{C}[\ell\sigma]$ and $u = \overline{C}[r\sigma]$ for a rule $\ell \rightarrow r$ and a substitution σ,
satisfying $\ell\sigma \rightarrow_{ndg} r\sigma$ and $\ell\sigma \not\rightarrow_i r\sigma$. Hence $\ell\sigma$ admits a non-root reduction step,
say on position $q \neq \epsilon$. If $q \in \mathsf{Pos}(\ell)$ and $\ell|_q$ is not a variable then $\ell|_q \in \mathsf{Def}$ since
$\ell\sigma$ admits a reduction step on position q, but then $\ell\sigma|_q \in \mathsf{NF}$ by the definition
of $\rightarrow_{ndg}$, contradiction. Hence either $\ell|_q$ is a variable or $q \notin \mathsf{Pos}(\ell)$. In the latter

case there is a position $p \in \mathsf{Pos}(\ell)$ such that $\ell|_p$ is a variable and $\ell\sigma|_p$ admits a reduction step. So in both cases we have such a variable $x = \ell|_p$ for which $x\sigma \notin \mathrm{NF}$. Since $\ell\sigma \to_{ndg} r\sigma$ and $x\sigma \notin \mathrm{NF}$ we conclude that x occurs at most once in r. Write $\ell = C[x, \ldots, x]$ where x does not occur in C. Choose u' such that $x\sigma \to_i u'$. Define τ by $x\tau = u'$ and $y\tau = y\sigma$ for $y \neq x$. Then

$$t = \overline{C}[\ell\sigma] = \overline{C}[C[x\sigma, \ldots, x\sigma]] \to_i^+ \overline{C}[C[u', \ldots, u']] = \overline{C}[\ell\tau] \to_{ndg} \overline{C}[r\tau].$$

In case x does not occur in r we have $\overline{C}[r\tau] = \overline{C}[r\sigma] = u$ and we are done, choosing $m = 0, k = n$.

In the remaining case x occurs exactly once in r, so $r = D[x]$ where x does not occur in D. Let t' be the redex corresponding to $x\sigma \to_i u'$, so $x\sigma = E[t'] \to_i E[t''] = u'$ for some context E and some term t''. Now we apply Lemma 6 to $u = \overline{C}[r\sigma] = \overline{C}[D\sigma[E[t']]] \to_i^n w$, yielding two cases.

In the first case we obtain C' satisfying $w = C'[t']$ and $D\sigma[E[t'']] \to_i^n C'[t'']$. Now we obtain $k = n$ and $m = 1$ in $w = C'[t'] \to_i C'[t'']$ and

$$\overline{C}[r\tau] = \overline{C}[D\sigma[x\tau]] = \overline{C}[D\sigma[u']] = \overline{C}[D\sigma[E[t'']]] \to_i^n C'[t'']$$

and we are done.

In the remaining second case of applying Lemma 6 we obtain a term v satisfying $t' \to_i v$ and $u = \overline{C}[D\sigma[E[t']]] \to_i \overline{C}[D\sigma[E[v]]] \to_i^{n-1} w$. Note that it may be the case that $v \neq t''$. Define ρ by $x\rho = E[v]$ and $y\rho = y\sigma$ for $y \neq x$. Then

$$t = \overline{C}[\ell\sigma] \to_i^+ \overline{C}[\ell\rho] \to_{ndg} \overline{C}[r\rho] = \overline{C}[D\rho[x\rho]] = \overline{C}[D\sigma[E[v]]] \to_i^{n-1} w,$$

by which we are done choosing $m = 0, k = n - 1$.   □

Lemma 7 is the key lemma for our Theorems 9 and 12; the rest of the proofs of these theorems and corresponding lemmas hold for arbitrary finitely branching ARSs $\to_i$ and $\to_{ndg}$ satisfying the property of Lemma 7.

Lemma 8. *Let t be a term and $n \geq 0$. Assume $t \to_{ndg}^n u$ for some term u. Then there is a term v satisfying $t \to_i^n v$.*

Proof. Induction on n. For $n = 0$ it is trivial, for $n > 0$ assume $t \to_{ndg} t' \to_{ndg}^{n-1} u$. Applying the induction hypothesis on $t' \to_{ndg}^{n-1} u$ yields a term u' satisfying $t' \to_i^{n-1} u'$. If $t \to_i t'$ we are done; in the remaining case we apply Lemma 7 on $t \to_{ndg} t' \to_i^{n-1} u'$ yielding $t \to_i^+ t'' \to_{ndg} \cdot \to_i^k u''$ for some u'' and $k \geq n - 2$. For $n = 1$ we are done. For $n > 1$ we apply the induction hypothesis on the first $n - 1$ steps of $t'' \to_{ndg} \cdot \to_i^k u''$, note that all $\to_i$ steps are $\to_{ndg}$ steps too. This yields a term v' satisfying $t \to_i^+ t'' \to_i^{n-1} v'$; now we can choose v to be the term obtained after n steps in this reduction.   □

Theorem 9. *Let the given TRS be finite, and let t be a term. Then $\to_i$ is terminating on t if and only if $\to_{ndg}$ is terminating on t.*

Proof. The 'if'-part trivially holds since every $\to_i$ step is a $\to_{ndg}$ step. For the 'only if'-part assume t admits an infinite $\to_{ndg}$-reduction while $\to_i$ is terminating on t. So the reduction graph of t with respect to $\to_i$ does not contain infinite paths. Since the TRS is finite, this graph is finitely branching. Hence by König's lemma this graph is finite and acyclic. Hence a number N exists such that all $\to_i$ reductions of t have length $\leq N$. Since t admits an infinite $\to_{ndg}$-reduction there is a term u satisfying $t \to_{ndg}^{N+1} u$, contradicting Lemma 8.   □

Theorem 9 does not hold if $\to_{ndg}$ is replaced by $\to_g$, as witnessed by the TRS of Example 4. Hence the restriction in the definition of $\to_{ndg}$ that $\ell\sigma|_p \in$ NF if $\ell|_p$ is a variable occurring more than once in r, is essential for Theorem 9.

Next we consider lengths of reductions: we will show that if t reduces by $\to_{ndg}$ in n steps to a normal form, then t admits either an infinite innermost reduction or an innermost reduction of at least n steps to the same normal form.

Lemma 10. *Let t, w be terms, let $\to_i$ be terminating on t and let $t \to_{ndg} \cdot \to_i^n w$ for $n \geq 0$. Then $t \to_i^{n'} \cdot \leftarrow_i^* w$ for some $n' > n$.*

Proof. We apply induction on $\to_i$, i.e., in proving that $t \to_{ndg} \cdot \to_i^n w$ implies $t \to_i^{n'} \cdot \leftarrow_i^* w$ we assume the induction hypothesis that a similar property replacing t by t' holds for all n and all t' satisfying $t \to_i^+ t'$. So assume $t \to_{ndg} u \to_i^n w$. If $t \to_i u$ we are done, otherwise we apply Lemma 7 yielding t', t'' satisfying $t \to_i^+ t' \to_{ndg} \cdot \to_i^k t'' \leftarrow_i^* w$ for $k \geq n-1$. Now applying the induction hypothesis on $t' \to_{ndg} \cdot \to_i^k t''$ yields $t \to_i^+ t' \to_i^{k'} \cdot \leftarrow_i^* t'' \leftarrow_i^* w$ for $k' > k$. Since $1 + k' > k + 1 \geq n$ we are done.   □

Lemma 11. *Let $\to_{ndg}$ be terminating on a term t and $t \to_{ndg}^n v$ for some $n \geq 0$. Then $t \to_i^k \cdot \leftarrow_i^* v$ for some $k \geq n$.*

Proof. Induction on n. For $n = 0$ it is trivial, for $n > 0$ assume $t \to_{ndg} u \to_{ndg}^{n-1} v$. Observe that $\to_{ndg}$ is terminating on u. Applying the induction hypothesis on $u \to_{ndg}^{n-1} v$ yields $k' \geq n-1$ and w satisfying $u \to_i^{k'} w$ and $v \to_i^* w$. Applying Lemma 10 on $t \to_{ndg} u \to_i^{k'} w$ yields $t \to_i^k \cdot \leftarrow_i^* w$ for $k > k'$. Since $v \to_i^* w$ and $k \geq k' + 1 \geq n$ we are done.   □

Theorem 12. *Let the given TRS be finite, let t be a term and let u be a normal form such that $t \to_{ndg}^n u$. Then either t admits an infinite $\to_i$-reduction, or $t \to_i^k u$ for $k \geq n$.*

Proof. Assume t does not admit an infinite $\to_i$-reduction. Then $\to_i$ terminates on t, and by Theorem 9 also $\to_{ndg}$ terminates on t. Then by Lemma 11 we have $t \to_i^k \cdot \leftarrow_i^* u$ for some $k \geq n$. Since u is a normal form we have $t \to_i^k u$.   □

Hence a $\to_{ndg}$-reduction to normal form is never worse (counted in number of steps) than an innermost reduction to the same normal form.

Theorem 12 does not hold if $\to_{ndg}$ is replaced by $\to_g$: in the TRS consisting of the two rules

$$f(x) \to g(x, x), \quad a \to b,$$

the term $f(a)$ admits a three step $\rightarrow_g$-reduction

$$f(a) \rightarrow_g g(a,a) \rightarrow_g g(b,a) \rightarrow_g g(b,b),$$

while the only innermost reduction $f(a) \rightarrow_i f(b) \rightarrow_i g(b,b)$ of $f(a)$ contains only two steps. Clearly for obtaining short reduction sequences in duplicating positions in left hand sides the argument first should be in normal form, as is required by the restriction in the definition of $\rightarrow_{ndg}$ that $\ell\sigma|_p \in \mathsf{NF}$ if $\ell|_p$ is a variable occurring more than once in r.

On can wonder whether the finiteness condition in Theorems 9 and 12 are essential. It is claimed by Vincent van Oostrom that it is not. However, for proving so a different approach will be required: the abstract reduction property described in Lemma 7 is not sufficient to conclude the properties claimed in Theorems 9 and 12 in case of infinite branching. For instance, by defining the abstract reduction systems on natural numbers

$$\rightarrow_i = \{(0,n) \mid n > 0\} \cup \{(n+1,n) \mid n > 0\}, \quad \rightarrow_{ndg} = \rightarrow_i \cup \{(0,0)\}$$

the property described in Lemma 7 holds but the properties claimed in Theorems 9 and 12 do not. This example is due to Vincent van Oostrom.

5   Avoiding Root Overlaps

In this section we will deal with the case that top rewrite steps are deterministic. This holds if the TRS is non-root-overlapping. For general TRSs this can be forced by fixing a priority on the rules. In this case, we can show that generalized innermost reduction steps commute in a proper way with parallel innermost reduction steps. We will first identify the required commutation diagram in an abstract setting (Lemma 13).

In many cases, the non-root overlapping criterion is too restrictive. In order to apply our theory to any TRS, we will assume that overlapping rules will be applied in a fixed order. For implementations, this is a natural restriction. This idea is implemented by a partial order on the rules, following *priority rewrite systems* [2]. For a fixed TRS, any partial order gives a particular strategy $\rightarrow_{g>}$ (Definition 15). Finally, we show that the commutation diagram holds for these strategies (Lemma 20). As a conclusion, we obtain that if innermost rewriting is terminating, then generalized innermost rewriting with ordered rules is terminating, Theorem 21.

The following lemma holds for all ARSs. One can think of $\rightarrow$ and $\twoheadrightarrow$ as innermost rewriting and parallel innermost rewriting, respectively, and $\Rightarrow$ as an extension of it, such as generalized innermost rewriting. Later we will use another instance.

Lemma 13. *Let binary relations* $\Rightarrow$, $\rightarrow$ *and* $\twoheadrightarrow_n$ $(n \geq 0)$ *be given. We write* $\twoheadrightarrow$ *for* $\bigcup_{n \geq 0} \twoheadrightarrow_n$. *Assume* $\rightarrow \subseteq \Rightarrow$, $\rightarrow \subseteq \twoheadrightarrow$ *and* $\twoheadrightarrow_n \cdot \Rightarrow \subseteq (\rightarrow^* \cdot \Rightarrow \cdot \twoheadrightarrow) \cup \twoheadrightarrow_{n-1}$.

1. *Assume that* $x \twoheadrightarrow y$ *and* $\mathsf{SN}(y, \Rightarrow)$. *Then* $\mathsf{SN}(x, \Rightarrow)$.
2. *Assume that* $\mathsf{WN}(x, \rightarrow)$, *and for all* y, *if* $y \in \mathsf{NF}(\rightarrow)$ *then* $\mathsf{SN}(y, \Rightarrow)$. *Then* $\mathsf{SN}(x, \Rightarrow)$.

Proof. Part 1. Assume $\mathsf{SN}(y, \Rightarrow)$. Then $\Rightarrow^+$ is well-founded on successors of y. By induction on y_0 (for all y_0 with $y \Rightarrow^* y_0$) ordered by $\Rightarrow^+$, we will prove: $\forall n \forall x_0 : x_0 \twoheadrightarrow_n y_0 \Rightarrow \mathsf{SN}(x_0, \Rightarrow)$. The latter is proved by induction on n (inner induction). So assume $x_0 \twoheadrightarrow_n y_0$. It suffices to prove $\mathsf{SN}(z, \Rightarrow)$ for all z with $x_0 \Rightarrow z$. The main assumption gives two cases. In the first case, $y_0 \to^* \cdot \Rightarrow \cdot \twoheadleftarrow z$. In particular, as $\to \subseteq \Rightarrow$, we find y_1, such that $y_0 \Rightarrow^+ y_1$ and $z \twoheadrightarrow y_1$. By the outer induction hypothesis, $\mathsf{SN}(z, \Rightarrow)$. In the second case, $z \twoheadrightarrow_{n-1} y_0$. Then, by the inner induction hypothesis, $\mathsf{SN}(z, \Rightarrow)$.

Part 2. By $\mathsf{WN}(x, \to)$, we find a reduction $x \to^* y$, for some $y \in \mathsf{NF}(\to)$. Hence by using the assumptions, $x \twoheadrightarrow^* y$ and $\mathsf{SN}(y, \Rightarrow)$. By induction on the length of this reduction sequence and by Part 1, we obtain $\mathsf{SN}(x, \Rightarrow)$. $\square$

From now on, we assume a fixed TRS with a partial-order $>$ on rules, such that at least any two root-overlapping rules are comparable. Note in particular that for non-root-overlapping systems, taking the empty partial order is allowed. We now first define innermost rewriting with priority, which in case of overlaps gives priority to the smallest rule.

Definition 14. *A term t rewrites* innermost with priority *to a term u, notation $t \to_{i>} u$, if*

- $t = C[\ell\sigma]$ and $u = C[r\sigma]$ for a rule $\ell \to r$, context C and substitution σ; and
- $\ell\sigma|_p \in \mathsf{NF}$ for all $p \in \mathsf{Pos}(\ell) \setminus \{\epsilon\}$; and
- there is no rule $(\ell' \to r')$ with $(\ell' \to r') < (\ell \to r)$ and substitution τ, such that $\ell'\tau = \ell\sigma$.

Innermost reductions with priority in which $C[]$ is the empty context are called top innermost reductions, *notation $t \mapsto_{i>} u$.*

The first two conditions ensure innermost behavior and the third condition enforces the priority restrictions. Next, we define the generalized innermost rewriting with priority:

Definition 15. *A term t rewrites* generalized innermost with priority *to a term u, notation $t \to_{g>} u$, if*

- $t = C[\ell\sigma]$ and $u = C[r\sigma]$ for a rule $\ell \to r$, context C and substitution σ; and
- $\ell\sigma|_p \in \mathsf{NF}$ for all $p \in \mathsf{Pos}(\ell) \setminus \{\epsilon\}$ such that $(\ell)_p \in \mathsf{Def}$; and
- there is no rule $(\ell' \to r')$ with $(\ell' \to r') < (\ell \to r)$ and substitution τ, such that $\ell'\tau = \ell\sigma$; and
- for all rules $(\ell' \to r')$ with $(\ell' \to r') < (\ell \to r)$ with the same top symbol f, and for all $1 \le i \le \mathsf{arity}(f)$ if $\ell\sigma|_i \notin \mathsf{NF}$ then $\ell'|_i \in \mathsf{LinVar}(\ell')$.

Top generalized innermost reductions with priority in which $C[]$ is the empty context are denoted by $t \mapsto_{g>} u$.

The first two clauses ensure generalized innermost behavior, the third clause enforces priority restrictions, and the last technical clause ensures that doing some (innermost) steps doesn't influence which rule is chosen. (Adding a similar last clause to the definition of $\to_{i>}$ would not give a different relation.)

Note that $\to_{i>} \subseteq \to_{g>} \subseteq \to$. Moreover, if $>$ is well-founded, then $\mathsf{NF}(\to_{i>}) = \mathsf{NF}(\to_{g>}) = \mathsf{NF}(\to)$. Note that for non-root-overlapping systems, by choosing $>= \emptyset$, the last two conditions can be removed. So in this case, $\to_{g>} = \to_g$.

Top generalized innermost reduction with priority is deterministic in the following sense:

Lemma 16. *Assume that root-overlapping rules are comparable. If $t \mapsto_{g>} u$ and $t \mapsto_{g>} v$, then $u = v$.*

Proof. By definition of $\mapsto_{g>}$, we find rules $\ell_1 \to r_1$ and $\ell_2 \to r_2$ and substitutions σ and ρ, such that $t = \ell_1\sigma$ and $t = \ell_2\rho$. Then these rules have root-overlap, so they are comparable. By the third condition of Definition 15, $(\ell_1 \to r_1) \not> (\ell_2 \to r_2)$ and $(\ell_2 \to r_2) \not> (\ell_1 \to r_1)$, so $\ell_1 = \ell_2$ and $r_1 = r_2$. So σ and τ coincide on the variables in ℓ_1, so in particular on variables in r_1, so $u = r_1\sigma = r_2\tau = v$. $\square$

Note that $\mapsto_{i>} \subseteq \mapsto_{g>}$, so the lemma also holds when one or both reduction steps are replaced by $\mapsto_{i>}$. One can prove that $\to_{i>}$ is confluent, because it satisfies the diamond property, and $\to_{g>}$ is weakly confluent, but not confluent, due to possible non-left-linearity.

We will also need parallel $\to_{i>}$-reduction:

Definition 17. *t rewrites to u with n-step parallel innermost reduction with priority (notation $t \twoheadrightarrow_{i>}^n u$) if $t = C[t_1, \ldots, t_n]$ and $u = C[t_1', \ldots, t_n']$ and for each j with $1 \leq j \leq n$, $t_j \mapsto_{i>} t_j'$. With $\twoheadrightarrow_{i>}$ we denote $\bigcup_{n \geq 0} \twoheadrightarrow_{i>}^n$.*

Note that $\twoheadrightarrow_{i>}^n$ doesn't coincide with the n-fold composition of $\twoheadrightarrow_{i>}$, as in $(\twoheadrightarrow_{i>})^n$.

Next we need an operation for simultaneous replacement of subterms.

Definition 18. *Let $t_1, \ldots, t_n, t_1', \ldots, t_n'$ be given, such that $t_j \mapsto_{i>} t_j'$ (for all $1 \leq j \leq n$). We define the operation α on terms and extend it to substitutions as follows:*

 - *for a term t the term $\alpha(t)$ is obtained from t by simultaneously replacing all occurrences of t_j by t_j' (for all j);*
 - *$\alpha(\sigma)(x) = \alpha(x\sigma)$ for all variables x and all substitutions σ.*

Note that this is uniquely defined due to Lemma 16

We use the following facts on α and $\twoheadrightarrow_{i>}$:

Lemma 19. *Let $t_1, \ldots, t_n, t_1', \ldots, t_n'$ be given, such that $t_j \mapsto_{i>} t_j'$ (for $1 \leq j \leq n$).*

1. *$t\sigma \twoheadrightarrow_{i>} t^{\alpha(\sigma)}$; in particular, if $t\sigma \in \mathsf{NF}$ then $t^{\alpha(\sigma)} = t^\sigma$.*
2. *Assume that for all $p \in \mathsf{Pos}(\ell)$ with $(\ell)|_p \notin \mathcal{V}$, we have $\ell\sigma|_p \notin \{t_1, \ldots, t_n\}$. Then $\alpha(\ell^\sigma) = \ell^{\alpha(\sigma)}$.*

Proof. 1: All t_j are innermost redexes, so they occur at disjoint positions.
  2: Induction on ℓ. $\square$

We now prove the key lemma of this section:

Lemma 20. *Assume that any root-overlapping rules in the TRS are comparable in the partial order $>$. Then*

$$\twoheadleftarrow_{i>}^{n} \cdot \rightarrow_{g>} \subseteq (\rightarrow_{i>}^{*} \cdot \rightarrow_{g>} \cdot \twoheadleftarrow_{i>}) \cup \twoheadleftarrow_{i>}^{n-1} .$$

Proof. Let $x \twoheadleftarrow_{i>}^{n} y$ and $x \rightarrow_{g>} z$. Then $x|_p = \ell\sigma$ and $z = x[r\sigma]_p$ for certain $\ell \rightarrow r$, σ, and p; and $\ell\sigma \mapsto_{g>} r\sigma$, so, writing f for the top symbol of ℓ, we have the conditions of Definition 15:

C2 $\ell\sigma|_q \in$ NF for all $q \in \text{Pos}(\ell) \setminus \{\epsilon\}$ such that $(\ell)_q \in$ Def; and
C3 there is no rule $(\ell' \rightarrow r')$ with $(\ell' \rightarrow r') < (\ell \rightarrow r)$ and substitution τ, such that $\ell'\tau = \ell\sigma$; and
C4 for all rules $(\ell' \rightarrow r')$ with $(\ell' \rightarrow r') < (\ell \rightarrow r)$ with the same top symbol f, and for all $1 \leq i \leq \text{arity}(f)$, either $\ell'|_i \in \text{LinVar}(\ell')$, or $\ell\sigma|_i \in$ NF.

Moreover, we find pairwise disjoint positions $p_1, \ldots, p_n$, terms $t_1, \ldots, t_n, t'_1, \ldots, t'_n$, such that $x|_{p_j} = t_j$ and $t_j \mapsto_{i>} t'_j$, for all $1 \leq j \leq n$. Note that p cannot be strictly below any p_j, because t_j are innermost redexes.

We distinguish cases. Case 1: $\ell\sigma = t_j$ for some $1 \leq j \leq n$. By Lemma 16, $r\sigma = t'_j$. Now if $p = p_j$, for some $1 \leq j \leq n$, we can write x as $C[t_1, \ldots, t_n]$, and prove the lemma as follows:

$$z = x[r\sigma]_p = C[t_1, \ldots, t'_j, \ldots, t_n] \twoheadleftarrow_{i>}^{n-1} C[t'_1, \ldots, t'_n] = y$$

Otherwise, p is disjoint from all p_j, so we can write x as $C[\ell\sigma, t_1, \ldots, t_n]$, and prove the lemma as follows:

$$y = C[\ell\sigma, t'_1, \ldots, t'_n] \rightarrow_{g>} C[r\sigma, t'_1, \ldots, t'_n] \twoheadleftarrow_{i>}^{n} C'[r\sigma, t_1, \ldots, t_n] = z$$

Case 2: for all j, $\ell\sigma \neq t_j$. Then all p_j are either disjoint from p, or strictly below p. Assume (w.l.o.g.) that positions $p_1, \ldots, p_k$ are strictly below p, and $p_{k+1}, \ldots, p_n$ are disjoint from p (for some $0 \leq k \leq n+1$). So we find context $C[]$ and $D[]$ such that $x = C[\ell\sigma, t_{k+1}, \ldots, t_n]$ and $\ell\sigma = D[t_1, \ldots, t_k]$. We can write $y = C[D[t'_1, \ldots, t'_k], t'_{k+1}, \ldots, t'_n]$ and $z = C[r\sigma, t_{k+1}, \ldots, t_n]$. In order to apply $\ell \rightarrow r$ to y, we first have to reduce any remaining copies of t_j in $D[t'_1, \ldots, t'_k]$ (in case ℓ is non-left-linear). This is done by the simultaneous replacement α, defined in Definition 18 for terms $t_1, \ldots, t_k$.

Define $v := C[\alpha(\ell\sigma), t'_{k+1}, \ldots, t'_n]$. We next show that $\alpha(\ell\sigma)$ is an instance of ℓ, using Lemma 19.2. So let $q \in \text{Pos}(\ell)$, $(\ell)|_q \notin \mathcal{V}$, and assume $\ell\sigma|_q = t_j$ (for some $0 \leq j \leq k$). Then $(\ell\sigma)_q \in$ Def, so $(\ell)_q \in$ Def (it is not in $\mathcal{V}$ by assumption). Note that $q \neq \varepsilon$ (by assumption of Case 2). Hence, by C2, $\ell\sigma|_q \in$ NF, in contradiction with t_j being an innermost redex. So by Lemma 19.2, we indeed get that $\alpha(\ell\sigma) = \ell\alpha(\sigma)$.

So we proved $v = C[\ell\alpha(\sigma), t'_{k+1}, \ldots, t'_n]$. Define $w := C[r\alpha(\sigma), t'_{k+1}, \ldots, t'_n]$. Then by Lemma 19.1, $\ell\sigma \twoheadleftarrow_{i>} \ell\alpha(\sigma) = \alpha(\ell\sigma)$. Then also $D[t'_1, \ldots, t'_k] \twoheadleftarrow_{i>} \alpha(\ell\sigma)$ (this contracts a subset of the redexes from $\ell\sigma$). Then also $y \twoheadleftarrow_{i>} v$, as $\twoheadleftarrow_{i>}$ is closed under context. Similarly, $r\sigma \twoheadleftarrow_{i} r\alpha(\sigma)$ by Lemma 19.1, so $z \twoheadleftarrow_{i} w$, because $p_{k+1}, \ldots, p_n$ are all disjoint from p.

Finally, we must check that $\ell^{\alpha(\sigma)} \mapsto_{g>} r^{\alpha(\sigma)}$, in order to conclude that $v \to_{g>} w$. This boils down to checking conditions 2–4 of Definition 15, assuming that these conditions (called C2–C4) hold for $\ell^{\sigma} \mapsto_{g>} r^{\sigma}$.

Cond 2) Let $q \in \mathsf{Pos}(\ell) \setminus \{\varepsilon\}$ with $(l)_q \in \mathsf{Def}$. Then (using equations of Lemma 19.1, 19.2 and usual commutation of substitutions and positions), we have $\ell^{\alpha(\sigma)}|_q = \alpha(\ell^{\sigma}|_q) = \ell^{\sigma}|_q \in \mathsf{NF}$ by C2.

Cond 3) Let $(\ell' \to r') < (\ell \to r)$ and τ be given, such that $\ell'\tau = l^{\alpha(\sigma)}$. We will construct τ', such that $\ell'\tau' = l^{\sigma}$, contradicting C3, as follows: $\tau'(x) := \mathbf{if}$ $x = \ell'|_i \in \mathsf{LinVar}(\ell')$ for some $0 \le i \le \mathsf{arity}(f)$, $\mathbf{then}$ $\ell|_i\sigma$; $\mathbf{else}$ $\tau(x)$. Then for each $0 \le i \le \mathsf{arity}(f)$, by C4, either $\ell'|_i \in \mathsf{LinVar}(\ell')$, so $\ell'|_i\tau' = \ell|_i\sigma$ by definition of τ', or $\ell\sigma|_i \in \mathsf{NF}$, so $\ell|_i\sigma = \ell|_i^{\alpha(\sigma)} = \ell'|_i\tau = \ell'|_i\tau'$. This holds for all arguments i, hence $\ell'\tau' = l^{\sigma}$, contradicting C3.

Cond 4) Let $(\ell' \to r') < (\ell \to r)$. By C4, either $\ell\sigma|_i \in \mathsf{NF}$, hence also $\ell^{\alpha(\sigma)}|_i = \ell\sigma|_i \in \mathsf{NF}$, or $\ell'|_i \in \mathsf{LinVar}(\ell')$.

Summarizing, we obtain:

$$\begin{aligned}
y &= C[D[t'_1, \ldots, t'_k], t'_{k+1}, \ldots, t'_n] \\
\rightarrowtail_{i>} v &= C[\alpha(l^{\sigma}), t'_{k+1}, \ldots, t'_n] \\
&= C[l^{\alpha(\sigma)}, t'_{k+1}, \ldots, t'_n] \\
\to_{g>} w &= C[r^{\alpha(\sigma)}, t'_{k+1}, \ldots, t'_n] \\
\leftarrowtail_{i>} z &= C[r^{\sigma}, t_{k+1}, \ldots, t_n]
\end{aligned}$$

which proves the lemma, by observing that $\rightarrowtail_{i>} \subseteq \to^*_{i>}$. $\qquad\square$

Theorem 21. *Assume that any root-overlapping rules in the TRS are comparable by the partial order $>$. Then for any term t, if $\mathsf{WN}(t, \to_{i>})$, then $\mathsf{SN}(t, \to_{g>})$.*

Proof. Assume $\mathsf{WN}(t, \to_{i>})$. We check the conditions of Lemma 13.2, with $\Rightarrow = \to_{g>}$, $\to = \to_{i>}$ and $\rightarrowtail_n = \rightarrowtail_{i>}^n$. Clearly, $\to_{i>} \subseteq \to_{g>}$ and $\to_{i>} = \rightarrowtail_{i>}^1 \subseteq \rightarrowtail_{i>}$. The next condition is obtained from Lemma 20. Finally, $\to_{i>}$-normal forms are $\to_{g>}$-normal forms, so in particular $\to_{g>}$-terminating. Hence we can apply Lemma 13.2, and obtain $\mathsf{SN}(t, \to_{g>})$ $\qquad\square$

Note that if we drop the third condition of Definition 15, the previous theorem would not hold. This is witnessed by Example 4. Also the fourth condition is essential, as witnessed by the following example. Consider the TRS consisting of the rules

$$e \to a, \quad d \to d, \quad \alpha : f(x, x) \to c, \quad \beta : f(a, y) \to d$$

with $\alpha < \beta$. Then $f(a, e) \to_{i>} f(a, a) \to_{i>} c$ is the only $\to_{i>}$-reduction from $f(a, e)$, so $\mathsf{SN}(f(a, e), \to_{i>})$. However, without the fourth condition of Definition 15 we would have $f(a, e) \to_{g>} d \to_{g>} d$, leading to an infinite reduction. A similar left-linear example exists:

$$e \to b, \quad d \to d, \quad \alpha : f(x, b) \to c, \quad \beta : f(a, y) \to d$$

with again $\alpha < \beta$. Now $\mathsf{SN}(f(a, e), \to_{i>})$, but without the fourth condition of Definition 15, we would obtain $f(a, e) \to_{g>} d \to_{g>} d$, leading to an infinite reduction. The following corollary shows that generalized innermost rewriting with priority is not worse than usual innermost rewriting.

Corollary 22. *Assume that any root-overlapping rules in the TRS are comparable by the partial order* $>$. *Let* t *be a term.*

- $\mathsf{SN}(t, \to_{i>})$ *if and only if* $\mathsf{SN}(t, \to_{g>})$.
- *If* $\mathsf{SN}(t, \to_i)$, *then* $\mathsf{SN}(t, \to_{g>})$.

Proof. This follows from Theorem 21 using $\to_{i>} \subseteq \to_i$ and $\to_{i>} \subseteq \to_{g>}$.   □

The reverse of the second doesn't hold, as witnessed by the two rules $(a \to b) < (a \to a)$. The infinite reduction $a \to_i a$ is disabled in $\to_{g>}$ by the terminating smaller rule.

Corollary 23. *Let the TRS be non-root-overlapping and let* t *be a term. Then* $\mathsf{WN}(t, \to_i) \Leftrightarrow \mathsf{SN}(t, \to_g) \Leftrightarrow \mathsf{SN}(t, \to_i)$.

Proof. Assume $\mathsf{WN}(t, \to_i)$. The TRS is non-root overlapping, so we can take $>= \emptyset$, then $\to_i = \to_{i>}$ and $\to_g = \to_{g>}$. By Theorem 21, we obtain $\mathsf{SN}(t, \to_g)$. The implication $\mathsf{SN}(t, \to_g) \Rightarrow \mathsf{SN}(t, \to_i)$ follows from $\to_i \subseteq \to_g$; the implication $\mathsf{SN}(t, \to_i) \Rightarrow \mathsf{WN}(t, \to_i)$ is universal.   □

In [10] the conjecture was stated that if a TRS is innermost terminating, then any in-time JITty annotation induces a terminating strategy. A JITty annotation for a function symbol is a list consisting of argument positions and rules for that symbol, which deterministically describes in which order to evaluate the arguments or apply the rules. An annotation induces a rewrite relation $\to_{strat}$. E.g., given rules α : if(true, x, y) $\to x$ and β : if(false, x, y) $\to y$, the annotation if : $[1, \alpha, \beta, 2, 3]$ denotes that if(s, t, u) is evaluated by first evaluating s, then trying rule α, then β, and if this failed, normalize t and u, respectively. A strategy annotation is *in-time* if for every rule $\ell \to r$ in it, all argument positions in ℓ distinct from $\mathsf{LinVar}(\ell)$ occur before it. We can now solve this conjecture.

Corollary 24. *Let the TRS be finite, and let strat be an in-time strategy annotation in the sense of [10]. For all terms* t, *if* $\mathsf{SN}(t, \to_i)$ *then* $\mathsf{SN}(t, \to_{strat})$.

Proof. Define $(\ell \to r) < (\ell' \to r')$ if and only if ℓ and ℓ' have the same top symbol, and $\ell \to r$ occurs before $(\ell' \to r')$ in the strategy annotation. Then $\to_{strat} \subseteq \to_{g>}$; conditions 2 and 4 of Definition 15 are enforced by the in-time requirement, and condition 3 is enforced because the order coincides with the order in the annotation. Assume $\mathsf{SN}(t, \to_i)$. Note that $\to_{i>} \subseteq \to_i$ and they have the same normal forms, so $\mathsf{WN}(t, \to_{i>})$. By Theorem 21, $\mathsf{SN}(t, \to_{g>})$, hence $\mathsf{SN}(t, \to_{strat})$.   □

The last result doesn't hold for all (eager) OBJ annotations, in which 0 is used to denote application of any rule. Consider again the example TRS consisting of the following three rules α : $f(a, b, x) \to f(x, x, x)$, β : $c \to a$, γ : $c \to b$. The system is innermost terminating, so any JITty annotation gives a terminating strategy, including f : $[1, 2, \alpha, 3]$, and either c : $[\beta, \gamma]$ or c : $[\gamma, \beta]$. However, the OBJ-annotation f : $[1, 2, 0, 3]$ and c : $[0]$ (where 0 stands for the application of any rule) admits an infinite sequence.

6 Conclusions

We introduced two generalizations $\to_{ndg}$ and $\to_{g>}$ of innermost rewriting $\to_i$ and $\to_{i>}$, respectively, for which we proved that for every term t the properties $\mathsf{SN}(t, \to_i)$ and $\mathsf{SN}(t, \to_{ndg})$ are equivalent, and the properties $\mathsf{SN}(t, \to_{i>})$ and $\mathsf{SN}(t, \to_{g>})$ are equivalent. As a main application of these results we see that for particular strategies as they are applied in implementations ([4, 11]) we may conclude that they are not worse than innermost rewriting as long as they are contained in $\to_{ndg}$ or $\to_{g>}$. This comparison describes worst case behavior; in typical applications we observe that the particular strategies terminate where innermost rewriting does not. Roughly speaking we can say that these strategies allow a kind of lazy rewriting without loss of efficiency or termination behavior.

We want to emphasize that these strategies apply for all TRSs without any restriction, and have the same set of normal forms as the full general rewrite relation, in contrast to other approaches like context-sensitive rewriting ([6]). Moreover, our strategies do not depend on user-defined options, except for the order of root-overlapping rules in Section 5.

In implementations typically strategies are deterministic. For a proper order $>$ on the rules and $\to \in \{\to_i, \to_{i>}, \to_{g>}, \to_{ndg}\}$ let $\overset{d}{\to}$ be a deterministic instance of $\to$, i.e., $\overset{d}{\to} \subseteq \to$ and for every term t not being a normal form there is exactly one u satisfying $t \overset{d}{\to} u$. Then for every term t we have the following properties:

$$
\begin{array}{cccc}
\mathsf{SN}(t, \to_{ndg}) & \Leftrightarrow\ \mathsf{SN}(t, \to_i) & \Rightarrow\ \mathsf{SN}(t, \to_{i>}) & \Leftrightarrow\ \mathsf{SN}(t, \to_{g>}) \\
\Downarrow & \Downarrow & \Updownarrow & \Downarrow \\
\mathsf{SN}(t, \overset{d}{\to}_{ndg}) & \mathsf{SN}(t, \overset{d}{\to}_i) & \mathsf{SN}(t, \overset{d}{\to}_{i>}) & \mathsf{SN}(t, \overset{d}{\to}_{g>}) \\
\Updownarrow & \Updownarrow & \Updownarrow & \Updownarrow \\
\mathsf{WN}(t, \overset{d}{\to}_{ndg}) & \mathsf{WN}(t, \overset{d}{\to}_i) & \mathsf{WN}(t, \overset{d}{\to}_{i>}) & \mathsf{WN}(t, \overset{d}{\to}_{g>}) \\
\Downarrow & \Downarrow & \Updownarrow & \Downarrow \\
\mathsf{WN}(t, \to_{ndg}) & \Leftarrow\ \mathsf{WN}(t, \to_i) & \Leftarrow\ \mathsf{WN}(t, \to_{i>}) & \Rightarrow\ \mathsf{WN}(t, \to_{g>})
\end{array}
$$

In this diagram the equivalences in the first line are Theorem 9 and Corollary 22; the vertical equivalences involving $\to_{i>}$ follow from Theorem 21. All other implications and equivalences are immediate from the definitions. For none of the implications the converse holds as is easily checked by considering the term a w.r.t. the two rules $a \to a, a \to b$ and the term $f(a)$ w.r.t. the two rules $a \to a, f(x) \to b$, for various orders of the rules and deterministic instances. A remaining question is whether $\mathsf{WN}(t, \to_{ndg})$ and $\mathsf{WN}(t, \to_{g>})$ are comparable. They are not, as follows from the following examples. Let the TRS consist of the rules $a \to a$, $f(x) \to g(x, x)$ and $g(x, y) \to b$ and let $>$ be empty. Then we have $\mathsf{WN}(f(a), \to_{g>})$ but not $\mathsf{WN}(f(a), \to_{ndg})$. Conversely, let the TRS consist of the rules $f(x) \to g(x)$, $f(x) \to h(x)$, $g(g(x)) \to g(g(x))$ and $h(h(x)) \to h(h(x))$. Then $f(f(x)) \to_i^+ g(h(x))$ and $f(f(x)) \not\to_{g>}^+ g(h(x))$ for any order $>$. Hence both $\mathsf{WN}(f(f(x)), \to_i)$ and $\mathsf{WN}(f(f(x)), \to_{ndg})$ hold, while for no order $>$ on the rules $\mathsf{WN}(f(f(x)), \to_{g>})$ holds.

For non-root-overlapping TRSs we proved that termination of $\to_g$ and $\to_i$ are equivalent. As an open question we leave whether in this claim the condition of being non-root-overlapping can be weakened to confluence.

Acknowledgment

We like to thank Vincent van Oostrom for fruitful discussions and suggestions for improvement.

References

1. F. Baader and T. Nipkow. *Term Rewriting and All That*. Cambr. Univ. Pr., 1998.
2. J.C.M. Baeten, J.A. Bergstra, J.W. Klop, and W.P. Weijland. Term rewriting systems with rule priorities. *Theoretical Computer Science*, 67:283–301, 1989.
3. W.J. Fokkink, J.F.Th. Kamperman, and H.R. Walters. Lazy rewriting and eager machinery. *ACM Transactions on Programming Languages and Systems*, 22(1):45–86, 2000.
4. J. Goguen, T. Winkler, J. Meseguer, K. Futatsugi, and J.-P. Jouannaud. Introducing OBJ. In J. Goguen and G. Malcolm, editors, *Software Engineering with OBJ: algebraic specification in action*. Kluwer, 2000.
5. B. Gramlich. Abstract relations between restricted termination and confluence properties of rewrite systems. *Fundamenta Informaticae*, 24:3–23, 1995.
6. S. Lucas. Context-sensitive rewriting strategies. *Information and Computation*, 178:294–343, 2002.
7. M. Nakamura and K. Ogata. The evaluation strategy for head normal form with and without on-demand flags. In K. Futatsugi, editor, *The 3rd Int. W. on Rewriting / Logic and its Applications (WRLA2000)*, volume 36 of *Electronic Notes in Theoretical Computer Science*. Elsevier, 2001.
8. M. J. O'Donnell. *Computing in systems described by equations*, volume 58 of *Lecture Notes in Computer Science*. Springer, 1977.
9. Y. Toyama. Counterexamples to the termination for the direct sum of term rewriting systems. *Information Processing Letters*, 25:141–143, 1987.
10. J. C. van de Pol. Just-in-time: on strategy annotations. In *Int. Workshop on Reduction Strategies in Rewriting and Programming (WRS 2001)*, volume 57 of *Electronic Notes in Theoretical Computer Science*, 2001.
11. J. C. van de Pol. JITty: a rewriter with strategy annotations. In *Proc. 13th RTA*, volume 2378 of *Lecture Notes in Computer Science*, pages 367–370, 2002.

Orderings for Innermost Termination

Mirtha-Lina Fernández[1,*], Guillem Godoy[2], and Albert Rubio[2]

[1] Universidad de Oriente, Santiago de Cuba, Cuba
mirtha@csd.uo.edu.cu
[2] Universitat Politècnica de Catalunya, Barcelona, España
{ggodoy,rubio}@lsi.upc.es

Abstract. This paper shows that the suitable orderings for proving innermost termination are characterized by the *innermost parallel monotonicity, IP-monotonicity* for short. This property may lead to several innermost-specific orderings. Here, an IP-monotonic version of the *Recursive Path Ordering* is presented. This variant can be used (directly or as ingredient of the *Dependency Pairs* method) for proving innermost termination of non-terminating term rewrite systems.

1 Introduction

Rewrite systems are sets of rules used to compute by replacing an instance of the left-hand side of a rule (*redex*) by the corresponding instance of the right-hand side. The replacements are repeated until a term with no redex (*normal form*) is eventually reached. Every replacement (*rewriting step*) involves a non-deterministic choice of both, the redex and the rewriting rule to be applied. Hence, in general one can produce an infinite number of rewriting step sequences started on the same term. A term rewrite system (TRS) is terminating if it has no infinite rewriting sequence.

A common way of restricting the number of rewriting sequences to be inspected when searching for a normal form is to use a rewriting strategy. A TRS can be terminating under a specific strategy whereas not in general. The termination proof for a strategy may be easier and weaker conditions for modularity can be applied. Moreover, for some classes of TRS, proving termination under a particular rewriting strategy suffices for ensuring general termination. Therefore, it turns out to be very important to develop techniques for proving termination of rewriting under strategies.

One of the most commonly used rewriting strategies is the *innermost* one, in which only innermost redexes are reduced. This strategy corresponds to the *"call by value"* computation rule of programming languages and enjoys all the aforementioned advantages. Therefore, studies on properties of innermost rewriting are useful for program verification.

The first and most successful technique for proving innermost termination of rewriting was the *Dependency Pairs* method (DP) [1]. In [20], the *size-change*

* Supported by the project CORDIAL-2, Ref. AML/B7-311-97/0666/II-0021F1A.

principle for functional programming [14] was adapted in order to prove innermost termination of rewriting. Moreover, it was combined with DP, obtaining the best of both methods. Other approaches are described in [4, 8, 19]. All these methods are used with general purpose orderings as ingredient, like the *Recursive Path Ordering* (RPO) [6, 12], the Knuth-Bendix Ordering and polynomial interpretations over the reals [3, 15].

In this paper, we study the relationship between innermost termination and well-founded orderings. Stability and monotonicity (which are always required for termination proofs) can be relaxed for termination of this strategy. In innermost rewriting only normalized substitutions are considered. Moreover, very recently it was shown that for innermost termination some monotonicity requirements can be discarded for some function symbols [7]. Here we provide a different approach for relaxing the monotonicity. Our approach was obtained by noting that innermost normalization and termination of the innermost parallel rewriting strategy are equivalent [18]. The latter strategy reduces all innermost redexes of a term at the same time. Therefore, for innermost termination we need to demand monotonicity only after each maximal parallel innermost rewriting step. We call this property *IP-monotonicity* and we show that the suitable orderings for direct innermost termination proofs are IP-monotonic. Another characterization for innermost termination is obtained by combining the innermost parallel relation and DP. As consequence, an innermost termination criterion relying on IP-monotonic quasi-orderings instead of IP-monotonic orderings is also obtained.

IP-monotonicity may lead to new, practical and innermost-specific orderings. In particular, we present an IP-monotonic version of the RPO, called the innermost RPO (iRPO). Its practical application is shown by means of examples. We also show that, for non-overlaying TRSs, the non-strict version of iRPO is an IP-monotonic quasi-ordering. Thus, it can be used as ingredient of DP and effectively combined with the argument filtering method [1, 13].

The rest of the paper is organized as follows. In Section 2 we introduce basic notions and notations. In Section 3 we characterize innermost termination in terms of IP-monotonic (quasi-) orderings. Section 4 is devoted to iRPO and the stability issue.

2   Preliminaries

We assume familiarity with the basics of term rewriting termination (see e.g. [2]).

The set of terms over a signature $\mathcal{F}$ is denoted as $\mathcal{T}(\mathcal{F}, \mathcal{X})$, where $\mathcal{X}$ represents a set of variables. Variables are denoted with the letters x, y, z while s, t, u (possibly with subscripts and apostrophes) denote terms. The arity of a function symbol f is denoted as $ar(f)$. The symbol labelling the root of a term t is denoted as $root(t)$. The notation $\bar{t}$ will be ambiguously used to denote either the tuple $(t_1, \ldots, t_n)$ or the multiset $\{t_1, \ldots, t_n\}$.

We assume positions within terms represented by sequences of positive integers, ordered by the prefix ordering. Positions are denoted with the letters p, q (possibly with apostrophes) while for integers we use i, j, k. The *root* position is denoted by λ and $p.q$ denotes the *concatenation* of p and q. The set of positions of a term t is $\mathcal{P}os(t)$. The *subterm* of t at position p is denoted as $t|_p$. The *subterm relation* denoted as $t \rhd t|_p$ in case of $p > \lambda$. The term t with the subterm at position p replaced by s is denoted as $t[s]_p$. Occasionally, we use $t[s]$ to indicate that s is subterm of t.

We say that binary relation $\succ$ is compatible with another binary relation $\succsim$ if $e_1 \succsim e_1' \succ e_2' \succsim e_2$ implies $e_1 \succ e_2$. We call $(\succsim, \succ)$ a *compatible pair* if $\succ$ is well-founded and $\succsim \circ \succ \subseteq \succ$ or $\succ \circ \succsim \subseteq \succ$ [13]. The syntactic equality is denoted as $\equiv$.

Let $\succ$ be an ordering on terms and let $\approx$ be an equivalence relation compatible with $\succ$. The lexicographic extension $\succ^{lex}$ of $\succ$ wrt. $\approx$ for n-tuples is defined as $(s_1,\ldots,s_n) \succ^{lex} (t_1,\ldots,t_n)$ iff $s_1 \approx t_1,\ldots,s_{k-1} \approx t_{k-1}$ and $s_k \succ t_k$ for some $k \in \{1\ldots n\}$. The extension of $\approx$ to multisets, denoted as $\approx^{mul}$, is the smallest relation s.t. $\emptyset \approx^{mul} \emptyset$ and $S \cup \{s\} \approx^{mul} S' \cup \{t\}$ if $s \approx t \wedge S \approx^{mul} S'$. The extension of $\succ$ to multisets w.r.t. $\approx$ is defined as the smallest ordering $\succ^{mul}$ s.t. $M \cup \{s\} \succ^{mul} N \cup \{t_1,\ldots,t_n\}$ if $M \approx^{mul} N$ and $s \succ t_i$ for all $i \in \{1\ldots n\}$.

A TRS over $\mathcal{F}$ is denoted as $\mathcal{R}$. The *defined* symbols of $\mathcal{R}$ are $\mathcal{D} = \{root(l) \mid l \rightarrow r \in \mathcal{R}\}$. A *rewriting step* with $\mathcal{R}$ is written as $s \rightarrow_{\mathcal{R}} t$. The notation $\rightarrow_{\mathcal{R},>\lambda}$ is used for a rewriting step at position $p \neq \lambda$. We omit the subscript $\mathcal{R}$ whenever is clear from the context.

A TRS $\mathcal{R}$ is *terminating* if $\rightarrow$ is well-founded, i.e. there is no infinite sequence $s_1 \rightarrow s_2 \rightarrow \ldots$ (sometimes denoted as $s_1 \rightarrow^\infty$). Alternatively, $\mathcal{R}$ is terminating iff all its rules are included in a reduction ordering [16]. One of the most popular reduction orderings for proving termination is the *Recursive Path Ordering* [6] which is defined below. RPO uses a precedence and can be adapted for dealing with statuses as proposed in [12].

Definition 1. *A precedence $\succ_{\mathcal{F}}$ is an ordering on $\mathcal{F}$ compatible with an equivalence relation $\approx_{\mathcal{F}}$. Let $\{\mathcal{L}ex, \mathcal{M}ul\}$ be a partition of $\mathcal{F}$ called statuses of $\mathcal{F}$. The precedence $\succ_{\mathcal{F}}$ is compatible with the statuses of $\mathcal{F}$ if $f \approx_{\mathcal{F}} g$ implies that both f and g belong to the same part, either $\mathcal{L}ex$ or $\mathcal{M}ul$.*

Definition 2. *Let $\succ_{\mathcal{F}}$ be a precedence over $\mathcal{F}$ compatible with the statuses $\{\mathcal{L}ex, \mathcal{M}ul\}$. Then $s = f(\bar{s}) \succ_{rpo} t$ if one of the following conditions holds:*

1. $s' \succ_{rpo} t$ or $s' \approx_{rpo} t$, for some $s' \in \bar{s}$
2. $t = g(\bar{t})$, $f \succ_{\mathcal{F}} g$ and $s \succ_{rpo} t'$ for all $t' \in \bar{t}$
3. $t = g(\bar{t})$, $f \approx_{\mathcal{F}} g$, $f \in \mathcal{F}_{\mathcal{L}ex}$, $\bar{s} (\succ_{rpo})^{lex} \bar{t}$ and $s \succ t'$, for all $t' \in \bar{t}$,
4. $t = g(\bar{t})$, $f \approx_{\mathcal{F}} g$, $f \in \mathcal{F}_{\mathcal{M}ul}$ and $\bar{s} (\succ_{rpo})^{mul} \bar{t}$,

where $s \approx_{rpo} t$ iff $s \equiv t$ or one of the following conditions holds:

(a) $root(s) \approx_{\mathcal{F}} root(t)$, $root(s) \in \mathcal{F}_{\mathcal{L}ex}$ and $s_1 \approx_{rpo} t_1,\ldots,s_n \approx_{rpo} t_n$,
(b) $root(s) \approx_{\mathcal{F}} root(t)$, $root(s) \in \mathcal{F}_{\mathcal{M}ul}$ and $\bar{s}(\approx_{rpo})^{mul}\bar{t}$,

Theorem 1. *[12] $\succ_{rpo}$ is a reduction ordering compatible with the congruence relation $\approx_{rpo}$.*

Given a TRS $\mathcal{R}$, $f(t,\ldots,t_n)$ is said to be *argument normalized* w.r.t. $\mathcal{R}$ if for all $k = 1\ldots n$, t_k is in normal form w.r.t. $\mathcal{R}$. A pair (s,t) is said to be *argument normalized* if s is so. A *normalized substitution* σ is s.t. $x\sigma$ is in normal form w.r.t.

$\mathcal{R}$ for all $x \in \mathcal{D}om(\sigma)$. An *innermost redex* is an argument normalized redex. A term s *rewrites innermost* to t w.r.t. $\mathcal{R}$, written $s \rightarrow_i t$, iff $s \rightarrow t$ at position p and $s|_p$ is an innermost redex. It is said that $\mathcal{R}$ is *innermost terminating* if $\rightarrow_i$ is well-founded.

Example 1. The system $\mathcal{R} = \{g(x, y) \rightarrow x, g(x, y) \rightarrow y, f(0, 1, x) \rightarrow f(x, x, x)\}$ was given by Toyama for proving that termination is not modular for disjoint unions of TRS [21]. This illustrative example has the infinite rewriting sequence: $f(0, 1, g(0, 1)) \rightarrow f(g(0, 1), g(0, 1), g(0, 1)) \xrightarrow{+} f(0, 1, g(0, 1)) \ldots$ However, every innermost rewriting sequence is terminating.

A TRS $\mathcal{R}$ is *innermost confluent* if $\rightarrow_i$ is confluent. We say that $\mathcal{R}$ is *non-overlaying* if there are no two different rules (after renaming variables so that both rules have distinct variables) having unifiable left-hand sides. If $\mathcal{R}$ is non-overlaying, then it is innermost confluent.

3   Characterizing Innermost Termination of Rewriting

In this section we focus on innermost termination, trying to characterize it by means of orderings. The basic idea to achieve this is the fact that all innermost redexes of a term t are in pairwise disjoint positions and moreover, all must be rewritten before reaching a normal form. Hence, if t can be normalized using the innermost strategy, all its innermost redexes can be reduced simultaneously by the parallel innermost strategy [17].

Definition 3. *A term s is reduced innermost in parallel to t w.r.t. $\mathcal{R}$, written $s \dashrightarrow_i t$, iff $s \xrightarrow{+}_i t$ and either $s \rightarrow_i t$ at position λ or $s = f(\bar{s})$, $t = f(\bar{t})$ and for all $k = 1 \ldots |\bar{s}|$ either $s_k \dashrightarrow_i t_k$ or $s_k = t_k$ is a normal form.*

It is easy to see that when $s \dashrightarrow_i t$, t can be obtained by consecutive one-step reductions of all innermost redexes in s. For instance, using the TRS of Example 1 we have $f(g(0, 1), g(0, 1), g(0, 1)) \dashrightarrow_i f(0, 1, 0)$. The innermost parallel rewrite relation is not only included in the transitive-closure of the innermost rewrite relation but it also characterizes innermost termination. The latter follows from Krishna Rao's contribution concerning the selection invariance for innermost normalization [18]. That is, the choice of innermost redex to be reduced at any step is irrelevant for innermost termination. Thereby, if a TRS is innermost normalizing under a particular strategy then it is innermost normalizing under any other strategy. In order to prove this fact, an oracle based reasoning was used. The following theorem provides a simpler proof for the same result.

Theorem 2. *A TRS $\mathcal{R}$ is innermost terminating iff $\dashrightarrow_i$ is terminating.*

Proof. The *left-to-right* implication is trivial. For the other direction, it is enough to prove that, for any infinite rewriting sequence $s \rightarrow_i^\infty$ there exists an alternate derivation $s \dashrightarrow_i s' \rightarrow_i^\infty$.

First, we show that given a derivation $s \xrightarrow{+}_i t$ where t is argument normalized there exists an alternate derivation $s \dashrightarrow_i s' \xrightarrow{*}_i t$, and we do it by structural induction. If the first rewrite step in $s \xrightarrow{+}_i t$ is at position λ, then this derivation is already of the form $s \;+\!\!\!+\!\!\!\rightarrow_i s' \xrightarrow{*}_i t$. Otherwise, either there is no rewrite step at λ or the first step at λ is on an argument normalized term obtained from s by at least one rewriting step. In any case, the original derivation is of the form $s = f(\bar{s}) \xrightarrow{+}_{i,>\lambda} f(\bar{t}) \xrightarrow{*}_i t$, where $f(\bar{t})$ is argument normalized and every $s_k \in \bar{s}$ is either a normal form and we call $s'_k = s_k$, or $s_k \xrightarrow{+}_i t_k$ and by induction hypothesis $s_k \;+\!\!\!+\!\!\!\rightarrow_i s'_k \xrightarrow{*}_i t_k$. Therefore, $s = f(\bar{s}) \;+\!\!\!+\!\!\!\rightarrow_i f(\bar{s'}) \xrightarrow{*}_i t$, as desired.

Now, given a derivation $s \rightarrow_i^\infty$, we show that there exists an alternate derivation $s \;+\!\!\!+\!\!\!\rightarrow_i s' \rightarrow_i^\infty$ by structural induction. If the first rewrite step is at λ position, the result trivially holds. Otherwise, if there is some rewrite step at λ, then this derivation is of the form $s \xrightarrow{+}_i t \rightarrow_i^\infty$ where t is argument normalized, and by our previous statement, there exists an alternate derivation $s \;+\!\!\!+\!\!\!\rightarrow_i s' \rightarrow_i^* t \rightarrow_i^\infty$, and the result holds. If there is no rewrite step at λ in $s \rightarrow_i^\infty$, then s is of the form $f(\bar{s})$ and for some $s_k \in \bar{s}$, say s_1, there exists an infinite rewriting sequence $s_1 \rightarrow_i^\infty$. By induction hypothesis, there exists an alternate derivation $s_1 \;+\!\!\!+\!\!\!\rightarrow_i s'_1 \rightarrow_i^\infty$. For the rest of s_k's, either s_k is a normal form and we call $s'_k = s_k$, or a parallel innermost rewriting step can be applied on s_k, i.e. $s_k \;+\!\!\!+\!\!\!\rightarrow_i s'_k$ for some s'_k. Therefore, there exists an alternate derivation $s = f(\bar{s}) \;+\!\!\!+\!\!\!\rightarrow_i f(\bar{s'}) \rightarrow_i^\infty$, and the result follows. $\square$

This theorem leads us to define the *innermost parallel monotonicity, IP-monotonicity* for short, directly from $\;+\!\!\!+\!\!\!\rightarrow_i$.

Definition 4. *A binary relation $\succ$ is IP-monotonic w.r.t. a TRS $\mathcal{R}$ iff $\;+\!\!\!+\!\!\!\rightarrow_i \;\subseteq\; \succ$.*

The IP-monotonicity hides a weak kind of stability and monotonicity. This can be seen in the next lemma, which is a straightforward conclusion from Definition 4.

Lemma 1. *A binary relation $\succ$ is IP-monotonic w.r.t. $\mathcal{R}$ iff*

  - *$l\sigma \succ r\sigma$ for all $l \rightarrow r \in \mathcal{R}$ and substitution σ s.t. $l\sigma$ is argument normalized and*
  - *$\bar{s} \;+\!\!\!+\!\!\!\rightarrow_i \bar{t}$ implies $f(\bar{s}) \succ f(\bar{t})$ for all $f \in \mathcal{F}$.*

Using this lemma is easy to see that any transitive, monotonic and stable binary relation including $\mathcal{R}$ is also in-monotonic w.r.t. $\mathcal{R}$. Therefore, reduction orderings suffices for innermost termination. However, termination of this strategy is indeed characterized by IP-monotonic and well-founded orderings.

Theorem 3. *A TRS $\mathcal{R}$ is innermost terminating iff there is a well-founded relation $\succ$ which is IP-monotonic w.r.t. $\mathcal{R}$.*

Proof. The left-to-right implication can be easily shown by taking $\;+\!\!\!+\!\!\!\rightarrow_i^+$. For the converse, if $\mathcal{R}$ is not innermost terminating, by Theorem 2, there exists an infinite rewriting sequence $s_1 \;+\!\!\!+\!\!\!\rightarrow_i s_2 \;+\!\!\!+\!\!\!\rightarrow_i \ldots$. By IP-monotonicity of $\succ$ w.r.t. $\mathcal{R}$, $s_1 \succ s_2 \succ \ldots$, contradicting the well-foundedness of $\succ$. $\square$

In the context of DP, innermost termination was characterized through the use of chains. Given a TRS $\mathcal{R}$, $\langle f(\bar{s}), g(\bar{t}) \rangle$ is a *dependency pair* of $\mathcal{R}$ if $f(\bar{s}) \rightarrow$

$u[g(\bar{t})] \in \mathcal{R}$ and $g \in \mathcal{D}$.[1] The set of all dependency pairs of $\mathcal{R}$ is denoted as $\mathcal{DP}(\mathcal{R})$. A sequence of dependency pairs $S = \langle s_1, t_1 \rangle \langle s_2, t_2 \rangle \langle s_3, t_3 \rangle \ldots$ of $\mathcal{R}$ is an innermost $\mathcal{R}$-chain if there is a substitution σ s.t. for all $j > 0$, $s_j\sigma$ is argument normalized and $t_j\sigma \xrightarrow{*}_i s_{j+1}\sigma$ holds. A TRS $\mathcal{R}$ is innermost terminating iff there is no infinite innermost $\mathcal{R}$-chain [1].

Since in every innermost $\mathcal{R}$-chain $s_j\sigma$ is argument normalized , by the proof of Theorem 2, we have $t_j\sigma \overset{*}{\Vdash}_i s_{j+1}\sigma$, for all $j > 0$. Therefore, the parallel innermost relation can also be used for characterizing innermost termination by means of chains. Furthermore, we can use a compatible pair $(\succsim, \succ)$ s.t. $\succsim$ is IP-monotonic w.r.t. $\mathcal{R}$.

Theorem 4. *A TRS $\mathcal{R}$ is innermost terminating iff there is a compatible pair $(\succsim, \succ)$ s.t. $\succsim$ is IP-monotonic w.r.t. $\mathcal{R}$ and $s\sigma \succ rt\sigma$ for all $\langle s, t \rangle \in \mathcal{DP}(\mathcal{R})$ and substitution σ s.t. $s\sigma$ is argument normalized.*

Proof. For the *right-to-left* direction suppose $\mathcal{R}$ is not innermost terminating. Then, there is an infinite innermost $\mathcal{R}$-chain $\langle s_1, t_1 \rangle \langle s_2, t_2 \rangle \langle s_3, t_3 \rangle \ldots$ and a substitution σ s.t. for all $j > 0$, $s_j\sigma$ is argument normalized and $t_j\sigma \xrightarrow{*}_i s_{j+1}\sigma$.

Since $t_j\sigma \overset{*}{\Vdash}_i s_{j+1}\sigma$ and $\succsim$ is IP-monotonic, we have $t_j\sigma \ (\succsim \cup \equiv) \ s_{j+1}\sigma$. Besides $s_j\sigma \succ t_j\sigma$ holds by assumption. Hence, seeing that $\succsim \circ \succ \subseteq \succ$ or $\succ \circ \succsim \subseteq \succ$, we obtain the infinite sequence $s_1\sigma \succ s_2\sigma \succ s_3\sigma \succ \ldots$ which contradicts the well-foundedness of $\succ$.

For the *left-to-right* direction we take $\succsim = \succ = (\rightarrow_i \cup \rhd)^+$. Clearly, $\succ \circ \succ \subseteq \succ$, $\succ$ is IP-monotonic w.r.t. $\mathcal{R}$ and orients $\mathcal{DP}(\mathcal{R})$. Finally, when $\mathcal{R}$ is innermost terminating, $\overset{+}{\rightarrow}_i$ is a monotonic and well-founded ordering and thereby $(\rightarrow_i \cup \rhd)^+$ is also well-founded. $\square$

4   An Example of IP-Monotonic Ordering

Multiset extensions have been used for defining successful reduction orderings like RPO and MSPO [5]. This is because they preserve suitable properties like irreflexivity, transitivity, stability and well-foundedness. Besides, every ordering $\succ$ is monotonic on $\succ^{mul}$ in the sense that $s \succ t$ implies $\{s_1, \ldots, s, \ldots, s_n\} \succ^{mul} \{s_1, \ldots, t, \ldots, s_n\}$. Once the terms of a multiset are rewritten with $\Vdash_i$ w.r.t. a TRS $\mathcal{R}$, all reducible terms decrease w.r.t. every IP-monotonic ordering $\succ$ whereas normal forms remain untouched. Therefore, the original multiset also decreases w.r.t. $\succ^{mul}$. Even more, the comparison with $\succ^{mul}$ still holds if we remove all multiple occurrences from the original and the reduced multisets.

Based on this fact we adapt (actually extend) RPO for proving innermost termination. This is achieved just by adding a new status $\mathcal{F}_{Set}$ which allows certain

---

[1] The original notion of dependency pair is $\langle \widehat{f}(\bar{s}), \widehat{g}(\bar{t}) \rangle$ where $\widehat{f}$ and $\widehat{g}$ are *marked* (or *tuple*) symbols associated to f and g resp. This renaming allows to apply a different treatment to function symbols when they appear on top of dependency pairs. We have chosen the unmarked version for simplicity but using the marked version does not affect our results.

terms to be compared using the *set* (instead of the multiset) of their arguments. The new ordering is the first which is innermost-specific; therefore we call it the *innermost Recursive Path Ordering*. Its definition can be formulated either by cases like RPO or by transformation, i.e. first we eliminate repetitions and then compare with RPO. The latter alternative provides an elegant definition and straightforward proofs for iRPO's properties.

Definition 5. *Given $\mathcal{F}_{Set} \subseteq \mathcal{F}$, the transformation ϕ over $\mathcal{T}(\mathcal{F}, \mathcal{X})$ is defined as*

- $\phi(x) = x$, if $x \in \mathcal{X}$
- $\phi(f(s_1, \ldots, s_n)) = f(\phi(s_1), \ldots, \phi(s_n))$, if $f \notin \mathcal{F}_{Set}$
- *otherwise* $\phi(f(s_1, \ldots, s_n)) = f(\phi(s_{j_1}), \ldots, \phi(s_{j_m}))$ *where* $j_1 < \ldots < j_m$ *are the j's in $\{1 \ldots n\}$ s.t. $s_k \not\equiv s_j$ for all $k < j$. In other words, the tuple $(s_{j_1} \ldots s_{j_m})$ is just $(s_1 \ldots s_n)$ after removing repetitions from left to right.*

Given an RPO ordering $\succ_{rpo}$ and $\mathcal{F}_{Set} \subseteq \mathcal{F}_{Mul}$, the corresponding $\succ_{irpo}$ ordering is defined as $s \succ_{irpo} t$ iff $\phi(s) \succ_{rpo} \phi(t)$. If $\approx_{rpo}$ is the equivalence relation corresponding to $\succ_{rpo}$, then $\approx_{irpo}$ is defined as $s \approx_{irpo} t$ iff $\phi(s) \approx_{rpo} \phi(t)$. The union of $\succ_{irpo}$ and $\approx_{irpo}$ is denoted as $\succsim_{irpo}$.

Note that after applying ϕ, some symbols in $\mathcal{F}_{Set}$ may become varyadic. Besides, repetitions are removed before applying ϕ, not later. For example, if $\mathcal{F}_{Set} = \{h, g\}$, $\phi(h(g(a, b, b, a), g(a, b, c, b), g(a, b, b, c))) = h(g(a, b), g(a, b, c), g(a, b, c))$. Although ϕ removes repetitions from left to right, any other fixed order would give the same definition of $\succ_{irpo}$ and $\approx_{rpo}$ above. Clearly, the transformed terms might be different (for instance, choosing the right-to-left order $\phi(g(a, b, c, b)) = g(a, c, b)$). But this is irrelevant since the multiset comparison is used for comparing the affected arguments.

The following proposition is a direct consequence of the definition of $\approx_{irpo}$, $\succ_{irpo}$ and Theorem 1.

Proposition 1. *$\succ_{irpo}$ is a well-founded ordering compatible with the equivalence relation $\approx_{irpo}$.*

Now, we show that if the set of argument normalized instances of a TRS $\mathcal{R}$ can be oriented using iRPO then the ordering is IP-monotonic w.r.t. $\mathcal{R}$. Therefore, by Theorem 3 and Proposition 1, it can be used for proving innermost termination.

Theorem 5. *$\succ_{irpo}$ is IP-monotonic w.r.t. a TRS $\mathcal{R}$ iff $l\sigma \succ_{irpo} r\sigma$, for every rule $l \to r \in \mathcal{R}$ and substitution σ s.t. $l\sigma$ is argument normalized.*

Proof. The *left-to-right* implication follows by definition of IP-monotonicity. For the other direction we need to show that $s \dashrightarrow_i t$ implies $s \succ_{irpo} t$. By assumption, $s \succ_{irpo} t$ whenever $s \dashrightarrow_i t$ at position λ. Otherwise $s = f(\bar{s})$, $t = f(\bar{t})$ and for all $k \in \{1 \ldots |\bar{s}|\}$ either $s_k = t_k$ is a normal form, or $s_k \dashrightarrow_i t_k$ and using structural induction we have $s_j \succ_{irpo} t_j$, and hence $\phi(s_j) \succ_{rpo} \phi(t_j)$. Moreover, for some $k \in \{1 \ldots |\bar{s}|\}$, s_k is not a normal form, and consequently $s_k \succ_{irpo} t_k$ and $\phi(s_k) \succ_{rpo} \phi(t_k)$. Now, if $f \notin \mathcal{F}_{Set}$, then $s \succ_{irpo} t$ by monotonicity and transitivity of $\succ_{rpo}$. Otherwise, let $\phi(s) = f(s'_1, \ldots, s'_m)$, $\phi(t) = f(t'_1, \ldots, t'_n)$.

Moreover, let S, T and S' be the multisets $\{s'_1, \ldots, s'_m\}$, $\{t'_1, \ldots, t'_n\}$ and $S' = \{\phi(s_k) \mid k \in \{1 \ldots |\bar{s}|\}, s_k \text{ is a normal form w.r.t } \mathcal{R}\}^2$ respectively. Then $S' \subset S$, $S' \subseteq T$ and for all $v \in T - S'$ there is some $u \in S - S'$ s.t. $u \succ_{rpo} v$ holds by induction. Therefore, $\{s'_1, \ldots, s'_m\} \succ^{mul}_{rpo} \{t'_1, \ldots, t'_n\}$ by definition of the multiset extension, and $\phi(s) \succ_{rpo} \phi(t)$ holds.   □

Example 2. Toyama's TRS (see Example 1) can be shown innermost terminating using iRPO. For the first two rules, $g(x, y)\sigma \succ_{rpo} x\sigma$ and $g(x, y)\sigma \succ_{rpo} y\sigma$ hold by case 1 for every substitution σ. Moreover, every instance of the last rule can be oriented by defining $\mathcal{F}_{Set} = \mathcal{F}_{Mul} = \{f\}$. Note that depending on the value of $x\sigma$ we have the following situations, all of them holding by case 3.

1. if $x\sigma = 0$ then $\phi(f(0, 1, x)\sigma) = f(0, 1) \succ_{rpo} f(0) = \phi(f(x, x, x)\sigma)$,
2. if $x\sigma = 1$ then $\phi(f(0, 1, x)\sigma) = f(0, 1) \succ_{rpo} f(1) = \phi(f(x, x, x)\sigma)$,
3. otherwise $\phi(f(0, 1, x)\sigma) = f(0, 1, x\sigma) \succ_{rpo} f(x\sigma) = \phi(f(x, x, x)\sigma)$.

Since the transformation ϕ unites duplicated arguments, other Toyama-like examples can be included in iRPO (e.g. [1, Examples 5.2.3,5.2.13,5.2.14]). When such multiple occurrences appear at top level, the techniques for cancelling cycles in the estimated innermost dependency graph [1, 4, 9] also handle many of these systems. However, as the next example shows, the latter does not hold in general.

Example 3. The next TRS is a more complex variant of Toyama's example.

$$\mathcal{R}_1 = \begin{cases} f(x, x, y) \to h(y) \\ h(x) \to x \\ h(f(x, y, z)) \to f(z, z, y) \\ h(f(x, y, z)) \to f(y, y, x) \\ c(f(0, 1, x), x) \to c(f(x, x, x), h(x)) \end{cases}$$

This system has the following infinite sequence (the redex used in each rewriting step appears underlined):

$c(f(0, 1, \underline{h(f(0, 1, 0))}), h(f(0, 1, 0))) \to$
$c(f(\underline{h(f(0, 1, 0))}, h(f(0, 1, 0))), h(h(f(0, 1, 0)))) \to$
$c(f(f(1, 1, 0), \underline{h(f(0, 1, 0))}), h(f(0, 1, 0))), h(h(f(0, 1, 0)))) \to$
$c(f(f(1, 1, 0), \underline{f(0, 0, 1)}, h(f(0, 1, 0))), h(h(f(0, 1, 0)))) \to$
$c(f(\underline{f(1, 1, 0)}, f(0, 0, 1), h(f(0, 1, 0))), h(h(f(0, 1, 0)))) \to$
$c(f(\underline{h(0)}, f(0, 0, 1), h(f(0, 1, 0))), h(f(0, 1, 0))) \to$
$c(f(0, \underline{f(0, 0, 1)}, h(f(0, 1, 0))), h(f(0, 1, 0))) \to$
$c(f(0, \underline{h(1)}, h(f(0, 1, 0))), h(f(0, 1, 0))) \to$
$c(f(0, 1, h(f(0, 1, 0))), h(f(0, 1, 0))) \to \ldots$

However, $\mathcal{R}_1$ is indeed innermost terminating. Proving this fact automatically is hard to obtain with the existing methods. Obviously, no termination technique can be used in this case. Furthermore, the estimations for the innermost

---

[2] We construct S' by selecting just one occurrence of every normal form in $\bar{s}$. Note that S, T and even S' may have repeated elements, because ϕ is not injective.

dependency graph do not cancel the problematic cycle corresponding to the last rule. The use of polynomials with negative coefficients has been proposed for innermost termination proofs [1, 7]. But the practical results concerning the automated generation of such polynomials are still few, and for instance the method described in [11] cannot be applied to this system.

Nevertheless, innermost termination of $\mathcal{R}_1$ can be proved using iRPO with $\mathcal{F}_{Set} = \mathcal{F}_{Mul} = \{f\}$, $\mathcal{F}_{Lex} = \{c\}$ and the precedence $c \succ_{\mathcal{F}} h$ and $f \succ_{\mathcal{F}} h$. Every instance of the first rule decreases by case 2. For the next three rules, $l\sigma \succ_{irpo} r\sigma$ holds by case 1, for every substitution σ (note that $f(x, y, z)\sigma \succsim_{irpo} f(z, z, y)\sigma$). Finally, considering the situations of the previous example the last rule is easily oriented using case 4.

4.1   Innermost Stability for iRPO

Theorem 5 is not suitable for automation since one has to check infinitely many instantiation of the rules. Hence, though stability is not necessary for innermost termination, it is always a desirable property when proving termination.

Unlike RPO, iRPO is not stable. The problem comes from the fact that two different terms can be equal after applying a substitution. Thereby, in general $\phi(s\sigma) \neq \phi(s)\sigma$ when $s \rhd t$ and $root(t) \in \mathcal{F}_{Set}$. For example, for $\mathcal{F}_{Set} = \{f\}$ and $\sigma = \{y \mapsto x\}$ we have $\phi(f(c(x), c(y))\sigma) = f(c(x)) \neq \phi(f(c(x), c(y)))\sigma$ $f(c(x), c(x))$. Due to this $\succ_{irpo}$ is not stable. For example, $f(c(x), c(y)) \succ_{irpo}$ $f(c(x), c(x))$ and $f(c(x), c(y)) \succ_{irpo} f(x, c(x))$ hold but do not after applying the former substitution. Note that $\phi(f(c(x), c(y))\sigma) = f(c(x)) = \phi(f(c(x), c(x))\sigma)$ and $\phi(f(c(x), c(y))\sigma) = f(c(x)) \not\succ_{rpo} f(x, c(x)) = \phi(x, f(c(x))\sigma)$.

Definition 6. *The problem of* iRPO *stability is defined as follows.*

Instance: *two terms s and t, a term rewrite system $\mathcal{R}$, and an iRPO ordering* $\succ_{irpo}$.

Question: *Is $s\sigma \succ_{irpo} t\sigma$ for any substitution σ such that $s\sigma$ is argument normalized?*

As we will see, this problem is co-NP-complete. The following algorithm non-deterministically decides the complement of the iRPO stability problem, i.e., if there exists a substitution σ such that $s\sigma$ is argument normalized and $s\sigma \not\succ_{irpo} t\sigma$ for given terms s and t.

Algorithm 1

1. Let $E_p \subseteq \{(i, j) \mid 1 \leq i < j \leq ar(root(s|_p))\}$ be an selection of pairs for every position $p \in Pos_{\mathcal{F}_{Set}}(s)$.
2. Let σ be the m.g.u. of the set of equations $\{s|_{p.i} = s|_{p.j} \mid p \in Pos_{\mathcal{F}_{Set}}(s), (i, j) \in E_p\}$. Check if $s\sigma$ is argument normalized w.r.t. $\mathcal{R}$ and $s\sigma \not\succ_{irpo} t\sigma$, and give this result as output.

Lemma 2. *The Algorithm 1 non-deterministically decides the complement of the iRPO stability problem.*

Before giving the proof of the previous lemma, we will need the following two technical results.

Lemma 3. *Let σ be the m.g.u. of a set of equations S. For every term s occurring in S and for every position $p \in \mathcal{P}os_{\mathcal{F}}(s\sigma)$, there exists a term s' occurring in S and a position $p' \in \mathcal{P}os_{\mathcal{F}}(s')$ such that $s\sigma|_p \equiv s'\sigma|_{p'}$.*

Proof. This can easily be proved by induction on the number of steps of many known unification algorithms. In those algorithms, the m.g.u. σ is incrementally obtained by, first, making σ_0 to be the identity substitution. Then, at some step, it is modified by an assignment of the form $\sigma_{i+1} := \sigma_i\{x \mapsto t\sigma_i\}$, where t is a term occurring in S, $x\sigma_i = x$, and any variable y occurring in $t\sigma_i$ satisfies that x does not occur in $y\sigma_i$. It is not difficult to see that, if σ_i satisfies the condition of the lemma, then σ_{i+1} does.   □

Lemma 4. *For every term s and substitution σ we have $\phi(s)\phi(\sigma) \succsim_{rpo} \phi(s\sigma)$.[3] Moreover, if for every pair of positions $p.i$ and $p.j$ of s with $p \in \mathcal{P}os_{\mathcal{F}_{Set}}(s)$ it holds that $(s|_{p.i} \equiv s|_{p.j}) \Leftrightarrow (s|_{p.i}\sigma \equiv s|_{p.j}\sigma)$, then $\phi(s)\phi(\sigma) \equiv \phi(s\sigma)$*

Proof. Clearly $\phi(s)\phi(\sigma) \succsim_{rpo} \phi(s\sigma)$ holds since $\phi(s\sigma)$ can be obtained from $\phi(s)\phi(\sigma)$ by eventually removing some subterms at positions below a symbol with multiset status. Now, assume that for every pair of positions $p.i$ and $p.j$ of $\mathcal{P}os(s)$, it holds that $(s|_{p.i} \equiv s|_{p.j}) \Leftrightarrow (s|_{p.i}\sigma \equiv s|_{p.j}\sigma)$. Proving $\phi(s)\phi(\sigma) \equiv \phi(s\sigma)$, is equivalent to see that any position $p.i$ with $p \in \mathcal{P}os_{\mathcal{F}_{Set}}(s)$ satisfies that for all j in $1\ldots i-1$, $s|_{p.j} \equiv s|_{p.i}$ if and only if $s|_{p.j}\sigma \equiv s|_{p.i}\sigma$ (i.e. the removing action of ϕ coincides on s and $s\sigma$ at positions in $\mathcal{P}os(s)$). But this is trivial by our assumption.   □

Now, we are ready to prove Lemma 2.

Proof. (Of Lemma 2) If the algorithm gives a positive answer, then it is clear that there exists a substitution σ (the one obtained by the algorithm) satisfying that $s\sigma$ is argument normalized and $s\sigma \not\succ_{irpo} t\sigma$.

Hence, it remains to see that, if for some substitution σ, $s\sigma$ is argument normalized and $s\sigma \not\succ_{irpo} t\sigma$, then there is a selection E_p for every $p \in \mathcal{P}os_{\mathcal{F}_{Set}}(s)$ that produces a positive answer. The selection we need for every of such p's is $E_p = \{(i,j) \mid s\sigma|_{p.i} \equiv s\sigma|_{p.j}\}$. Let σ' be the m.g.u. of the corresponding set of equations S in the algorithm. We have that $\sigma = \sigma'\sigma''$ for some σ'' (since σ is an unifier of S), and that for all $p \in \mathcal{P}os_{\mathcal{F}_{Set}}(s)$, $s\sigma'|_{p.i} \equiv s\sigma'|_{p.j}$ if and only if $s\sigma'|_{p.i}\sigma'' \equiv s\sigma'|_{p.j}\sigma''$. Moreover, $s\sigma'$ is argument normalized since $s\sigma'\sigma'' = s\sigma$ is. It remains to see that $s\sigma' \not\succ_{irpo} t\sigma'$, or, equivalently, that $\phi(s\sigma') \not\succ_{rpo} \phi(t\sigma')$. We do it by contradiction, i.e. assume that $\phi(s\sigma') \succ_{rpo} \phi(t\sigma')$. By stability of $\succ_{rpo}$, it holds that $\phi(s\sigma')\phi(\sigma'') \succ_{rpo} \phi(t\sigma')\phi(\sigma'')$. By the first part of Lemma 4, $\phi(t\sigma')\phi(\sigma'') \succsim_{rpo} \phi(t\sigma'\sigma'') \equiv \phi(t\sigma)$. If we could prove $\phi(s\sigma')\phi(\sigma'') \equiv \phi(s\sigma'\sigma'')$ then we would obtain $\phi(s\sigma) \succ_{rpo} \phi(t\sigma)$, and hence, $s\sigma \succ_{irpo} t\sigma$, which is a contradiction with our assumption.

In order to prove $\phi(s\sigma')\phi(\sigma'') \equiv \phi(s\sigma'\sigma'')$, we want to apply the second part of Lemma 4. We already know that for all $p \in \mathcal{P}os_{\mathcal{F}_{Set}}(s)$, $s\sigma'|_{p.i} \equiv s\sigma'|_{p.j}$ if and only if $s\sigma'|_{p.i}\sigma'' \equiv s\sigma'|_{p.j}\sigma''$. It remains to see that this property extends to $s\sigma'$,

---

[3] Here, ϕ is adapted to substitutions in a natural way, i.e. $x\phi(\sigma) = \phi(x\sigma)$.

i.e., for all $p \in \mathcal{P}os_{\mathcal{F}_{Set}}(s\sigma')$, $s\sigma'|_{p.i} \equiv s\sigma'|_{p.j}$ if and only if $s\sigma'|_{p.i}\sigma'' \equiv s\sigma'|_{p.j}\sigma''$. For this goal, it is enough to see that for any position $p \in \mathcal{P}os_{\mathcal{F}_{Set}}(s\sigma')$, there exists a position $p' \in \mathcal{P}os_{\mathcal{F}_{Set}}(s)$ such that $s\sigma'|_{p} \equiv s\sigma'|_{p'}$. But this is easy by means of Lemma 3 as follows. First, note that if instead of considering the set of equations S we consider $S \cup \{s = s\}$, then σ' continues being the m.g.u. of this set. Now, let p be a position in $\mathcal{P}os_{\mathcal{F}_{Set}}(s\sigma')$. By Lemma 3, there exists a term s' in $S \cup \{s = s\}$ and a non-variable position p' in s' such that $s\sigma'|_{p} \equiv s'\sigma'|_{p'}$. But this term s' can be considered to be s, since all terms occurring in S are subterms of s. □

Theorem 6. *The iRPO stability problem is co-NP-complete*

Proof. Since we have proved the correctness of the Algorithm 1, for seeing that the complement of this problem belongs to NP, it only remains to see that such an algorithm takes polynomial time. The selection E_p for every p and the corresponding set of equations need polynomial time. A most general unifier σ can be represented in polynomial space on the given set of equations by means of DAG's, and computed in polynomial time. Checking the irreducibility of $s\sigma$, obtaining the DAG's representing $\phi(s\sigma)$ and $\phi(t\sigma)$, and checking if $\phi(s\sigma) \not\succ_{rpo} \phi(t\sigma)$ takes polynomial time as well.

For proving that the complement is an NP-hard problem we give a reduction from 3-SAT. Given an instance of 3-SAT with variables $x_1 \ldots x_n$ and clauses $c_1 \ldots c_m$, we construct the following terms s and t based on the signature $\{h, f, g, g', 0, 1\}$ where h and f have lexicographic status and arity 2, g and g' have set status and arities 4 and 5 respectively, and 0 and 1 are constants. In s and t appear the (term) variables $x_1, \ldots, x_n, \overline{x_1}, \ldots, \overline{x_n}$.

$$s = h(\, f(v_1, f(v_2 \ldots, f(v_{n-1}, v_n) \ldots)) \, , \, f(u_1, f(u_2 \ldots, f(u_{m-1}, u_m) \ldots)) \,)$$
$$t = h(\, f(v'_1, f(v'_2 \ldots, f(v'_{n-1}, v'_n) \ldots)) \, , \, f(u'_1, f(u'_2 \ldots, f(u'_{m-1}, u'_m) \ldots)) \,)$$

where $v_i = g(x_i, \overline{x_i}, 0, 1)$, $v'_i = g(x_i, \overline{x_i}, x_i, \overline{x_i})$ and if c_i is a clause with literals l_j, l_k, l_o, then $u_i = g'(l_j, l_k, l_o, 0, 1)$ and $u'_i = g'(l_j, l_k, l_o, 0, 0)$.

Regardless the precedence, it is easy to see that there exists σ satisfying $s\sigma \not\succ_{irpo} t\sigma$ if and only if the original 3-SAT problem is satisfiable. Note that, since $g \in \mathcal{F}_{Set}$, the term $g(x_i, \overline{x_i}, 0, 1)\sigma$ is not greater than $g(x_i, \overline{x_i}, x_i, \overline{x_i})\sigma$ only if σ assigns 0 and 1, or 1 and 0, to the variables x_i and $\overline{x_i}$, respectively. Besides, the term $g'(l_j, l_k, l_o, 0, 1)\sigma$ is not not greater than $g'(l_j, l_k, l_o, 0, 0)\sigma$ only if σ satisfies every clause c_i with literals l_j, l_k and l_o. By considering an empty $\mathcal{R}$ the result follows. □

4.2   Using iRPO for DP

In general, the compatible pair $(\succsim_{irpo}, \succ_{irpo})$ cannot be used for proving innermost termination with DP. This is because $\succsim_{irpo}$ is not IP-monotonic w.r.t. an arbitrary TRS $\mathcal{R}$. Unlike iRPO, the latter holds even if $\succsim_{irpo}$ orients every rule instance whose left-hand side is argument normalized. Note that after an innermost parallel step, it may happen that an occurrence of a duplicated argument of a symbol in $\mathcal{F}_{Set}$ decreases w.r.t. $\succ_{irpo}$ while another occurrence remains equal w.r.t. $\approx_{irpo}$. Hence, the corresponding set of arguments may neither decrease w.r.t $(\succ_{irpo})^{mul}$ nor remain equal w.r.t. $(\approx_{irpo})^{mul}$.

As an alternative, we may combine the argument filtering technique [1, 13] with $\succ_{irpo}$ in order to obtain a compatible pair. An argument filtering over a signature $\mathcal{F}$ is a function π s.t. for all $f \in \mathcal{F}$, either $\pi(f) \in \{1 \ldots ar(f)\}$ or $\pi(f) \subseteq \{1 \ldots ar(f)\}$. It induces a mapping from $\mathcal{T}(\mathcal{F}, \mathcal{X})$ to $\mathcal{T}(\mathcal{F}_\pi, \mathcal{X})$ as follows:

$$
\begin{cases}
\pi(x) = x & if \ x \in \mathcal{X} \\
\pi(f(t_1, \ldots, t_n)) = \pi(t_i) & if \ \pi(f) = i, \\
\pi(f(t_1, \ldots, t_n)) = f(\pi(t_{i_1}), \ldots, \pi(t_{i_m})) & if \ \pi(f) = [i_1, \ldots, i_m],
\end{cases}
$$

where $[i_1, \ldots, i_m]$ denotes an ordered set and $\mathcal{F}_\pi$ consists of all symbols f s.t. $\pi(f)$ is a set (the arity of every $f \in \mathcal{F}_\pi$ is $|\pi(f)|$). Given a set of pairs $\mathcal{P}$, $\pi(\mathcal{P})$ denotes $\{(\pi(s), \pi(t)) \mid (s, t) \in \mathcal{P}\}$.

Given a binary relation $\succ$, the relation $\succ_\pi$ is defined as $s \succ_\pi t$ iff $\pi(s) \succ \pi(t)$. It not difficult to see that when $\succ$ is monotonic (resp. stable) we have that $s \succ_\pi t$ implies $u[s] \succeq_\pi u[t]$ (resp. $s\sigma \succ_\pi t\sigma$). Therefore, this method has been used for obtaining a monotonic quasi-ordering from a monotonic ordering while preserving stability and well-foundedness.

The iRPO ordering is already defined via a transformation: $s \succ_{irpo} t$ iff $\phi(s) \succ_{rpo} \phi(t)$. Hence, two possibilities seem natural to be considered for combining it with an argument filtering π: we can compare two terms s and t by either $\phi(\pi(s)) \succ_{rpo} \phi(\pi(t))$ or $\pi(\phi(s)) \succ_{rpo} \pi(\phi(t))$. Applying π before ϕ does not work well. An argument filtering might transform a redex into a filtered normal form. Hence, since the transformation ϕ removes duplicated arguments, some innermost parallel reductions might be lost, i.e. it might happen that $s +\!\!\!\rightarrow_i t$ but $\pi(s) \not\succeq \pi(t)$. The next example illustrates this situation.

Example 4. The following non-overlapping system is not innermost terminating.

$$
\mathcal{R}_2 = \begin{cases}
h(0) \rightarrow 0 \\
h(1) \rightarrow 1 \\
f(0, 1, h(2)) \rightarrow f(h(0), h(1), h(2))
\end{cases}
$$

If we remove the argument of h then we obtain the ordering constraints $h \succ 0, h \succ 1, f(0, 1, h) \succ f(h, h, h)$. These constraints are satisfied by $\succ_{irpo}$ with $\mathcal{F}_{Set} = \{f\}$, $h \succ_\mathcal{F} 0$ and $h \succ_\mathcal{F} 1$. Therefore, one could falsely prove (innermost) termination of $\mathcal{R}_2$.

Note that $f(h(0), h(1), h(2)) = s +\!\!\!\rightarrow_i t = f(0, 1, h(2))$ and even $s \succ_{irpo} t$ but since $\{h\} \not\succ_{irpo}^{mul} \{0, 1, h\}$, we have $f(h, h, h) = \pi(s) \not\succ_{irpo} \pi(t) = f(0, 1, h)$.

Hence, we consider the other possibility, i.e. to apply the transformation ϕ before the filtering π. In this case, it is natural to demand that π does not affect the symbols in $\mathcal{F}_{Set}$ (i.e. $\pi(f(t_1, \ldots, t_n)) = f(\pi(t_1), \ldots, \pi(t_n))$, for all $f \in \mathcal{F}_{Set}$), since some arguments might be previously removed by ϕ. In general, this approach does not work either.

Example 5. The TRS $\mathcal{R}_3 = \{a \rightarrow b, a \rightarrow c, g(c) \rightarrow d, f(g(b), d) \rightarrow f(g(a), g(a))\}$ is not innermost terminating. But taking $\mathcal{F}_{Set} = \{f\}$, $\pi(g) = \emptyset$ and the precedence $a \succ_{\mathcal{F}} b, a \succ_{\mathcal{F}} c, g \succ_{\mathcal{F}} d$, we have $\pi(\phi(\mathcal{R}_3)) \subset \succ_{rpo}$. Hence, one may erroneously conclude $\mathcal{R}_3$ is innermost terminating.

Nevertheless, for non-overlaying TRSs this approach indeed yields the desired result. Non-overlayingness is not a very restrictive condition for a TRS in the context of innermost rewriting. This strategy corresponds to the usual behavior of programming languages, where arguments are fully evaluated before applying a function. If a program is deterministic, which is the usual situation, then it corresponds to a non-overlaying system. Besides, this family of TRSs includes non-overlapping ones for which termination and innermost termination coincide [10].

Definition 7. *Let* $\succ_{rpo}$ *be an RPO ordering with* $\mathcal{F}_{Set} \subseteq \mathcal{F}_{Mul}$ *and* π *be an argument filtering over* $\mathcal{F} - \mathcal{F}_{Set}$. *The corresponding* $\succ_{irpo,\pi}$ *ordering is defined as* $s \succ_{irpo,\pi} t$ *iff* $\pi(\phi(s)) \succ_{rpo} \pi(\phi(t))$. *If* $\approx_{rpo}$ *is the equivalence relation corresponding to* $\succ_{rpo}$, *then* $\approx_{irpo,\pi}$ *is defined as* $s \approx_{irpo,\pi} t$ *iff* $\pi(\phi(s)) \approx_{rpo} \pi(\phi(t))$. *The union of* $\succ_{irpo,\pi}$ *and* $\approx_{irpo,\pi}$ *is denoted as* $\succsim_{irpo,\pi}$.

The next proposition follows directly from Proposition 1 and Definition 7.

Proposition 2. $\succ_{irpo,\pi}$ *is a well-founded ordering compatible with the equivalence relation* $\approx_{irpo,\pi}$.

Now, we prove the IP-monotonicity of $\succsim_{irpo,\pi}$.

Theorem 7. *Let* $\mathcal{R}$ *be a non-overlaying TRS. If* $l\sigma \succsim_{irpo,\pi} r\sigma$, *for every rule* $l \rightarrow r \in \mathcal{R}$ *and substitution* σ *s.t.* $l\sigma$ *is argument normalized then* $\succsim_{irpo,\pi}$ *is IP-monotonic w.r.t.* $\mathcal{R}$.

Proof. We need to show that $s \dashv\!\!\mid\!\!\rightarrow_i t$ implies $s \succsim_{irpo,\pi} t$, and we prove it by induction on the size of s. If this rewrite step is at position λ the result trivially follows. Otherwise, $s = f(\bar{s})$, $t = f(\bar{t})$ and for all $k = 1 \ldots |\bar{s}|$, either s_k is a normal form and $t_k = s_k$, or $s_k \dashv\!\!\mid\!\!\rightarrow_i t_k$ and by induction hypothesis $s_k \succsim_{irpo,\pi} t_k$. Now, when $f \notin \mathcal{F}_{Set}$, if $\pi(f)$ is either the empty set or a natural number the result is trivial; otherwise $s \succsim_{irpo,\pi} t$ is obtained using monotonicity and transitivity of $\succsim_{rpo}$. In case of $f \in \mathcal{F}_{Set}$, first note that, by non-overlayingness, if $s_i \equiv s_j$, then $t_i \equiv t_j$. Therefore, if for some t_i, all the t_j's with $j < i$ are different from t_i (and hence t_i is not removed by the transformation ϕ) then, all the s_j's with $j < i$ are different from s_i. As consequence, to every element in $\phi(\bar{t})$ we can associate a distinct element in $\phi(\bar{s})$ that is greater w.r.t. $\succsim_{irpo,\pi}$, and hence, $s \succsim_{irpo,\pi} t$. $\qquad\square$

Combining Theorems 4, 7 and Proposition 5 we have that, for non-overlaying TRSs, the compatible pair $(\succsim_{irpo,\pi}, \succ_{irpo,\pi})$ can be effectively used for innermost termination proofs with DP.

Corollary 1. *A non-overlaying TRS* $\mathcal{R}$ *is innermost terminating if*

- $l\sigma \succsim_{irpo,\pi} r\sigma$ *for all* $l \rightarrow r \in \mathcal{R}$ *and substitution* σ *s.t.* $l\sigma$ *is argument normalized and*
- $s\sigma \succ_{irpo,\pi} t\sigma$ *for all* $\langle s, t \rangle \in DP(\mathcal{R})$ *and substitution* σ *s.t.* $s\sigma$ *is argument normalized.*

Finally we point out that Algorithm 1 for the iRPO stability problem can be easily adapted for checking if $l\sigma \succsim_{irpo,\pi} r\sigma$, for every substitution σ s.t. $l\sigma$ is argument normalized.

5 Conclusions

In this paper we introduce the first syntactical ordering which can be used for proving innermost termination of non-terminating TRSs. The ordering is a variant of the most popular reduction ordering, RPO, and we call it the *innermost RPO*. The iRPO was obtained by considering, for some function symbols, sets instead of multisets of arguments. Hence, it is specially recommended for dealing with duplicated arguments in right-hand sides. The use of sets entails non-stability as drawback. However, for the (quasi-) orderings presented here, the problem of checking stability is decidable and co-NP-complete. The algorithm for doing this checking considers those m.g.u. which duplicate arguments in left-hand sides. But usually there are not many of such arguments. Therefore, we think that in many practical situations the stability of iRPO can be computed efficiently.

The iRPO enjoys a property, called IP-monotonicity, which is essential for innermost termination. This property demands monotonicity just after each (maximal) parallel innermost rewriting step. We believe that this weaker condition might be useful for defining other innermost-specific orderings.

References

1. T. Arts and J. Giesl. Termination of term rewriting using dependency pairs. *Theoretical Computer Science*, 236(1-2):133–178, 2000.
2. F. Baader and T. Nipkow. *Term Rewriting and All That*. Cambridge University Press, 1998.
3. A. Ben-Cherifa and P. Lescanne. Termination of rewriting systems by polynomial interpretations and its implementation. *Science of Computer Programming*, (9):137–160, 1987.
4. C. Borralleras. *Ordering-based methods for proving termination automatically*. PhD thesis, Dpto. LSI, Universitat Politècnica de Catalunya, España, 2003.
5. C. Borralleras, M. Ferreira, and A. Rubio. Complete monotonic semantic path orderings. In *Proc. CADE*, volume 1831 of *LNAI*, pages 346–364, 2000.
6. N. Dershowitz. Orderings for term rewriting systems. *Theoretical Computer Science*, 17(3):279–301, 1982.
7. M.L. Fernández. Relaxing monotonicity for innermost termination. *Information Processing Letters*, 93(3):117–123, 2005.
8. O. Fissore, I. Gnaedig, and H. Kirchner. Induction for innermost and outermost ground termination. Technical Report A01-R-178, LORIA, Nancy, France, 2001.
9. J. Giesl, R. Thiemann, P. Schneider-Kamp, and S. Falke. Improving dependency pairs. In *Proc. LPAR*, volume 2850 of *LNAI*, pages 165–179, 2003.
10. B. Gramlich. On proving termination by innermost termination. In *Proc. RTA*, volume 1103 of *LNCS*, pages 93–107, 1996.
11. Nao Hirokawa and Aart Middeldorp. Polynomial interpretations with negative coefficients. In *Proc. AISC*, volume 3249 of *LNAI*, pages 185–198, 2004.

12. S. Kamin and J. J. Levy. Two generalizations of the recursive path ordering. Dept. of Computer Science, Univ. of Illinois, Urbana, IL, 1980.
13. K. Kusakari, M. Nakamura, and Y. Toyama. Argument filtering transformation. In *Proc. PPDP*, volume 1702 of *LNCS*, pages 47–61, 1999.
14. C. S. Lee, N. D. Jones, and A. M. Ben-Amram. The size-change principle for program termination. In *Proc. POPL*, pages 81–92, 2001.
15. S. Lucas. Polynomials for proving termination of context-sensitive rewriting. In *Proc. FOSSACS*, volume 2987 of *LNCS*, pages 318–332, 2004.
16. Z. Manna and S. Ness. On the termination of Markov algorithms. In *Proc. Int. Conf. on System Science*, pages 789–792, 1970.
17. M. J. O'Donnell. *Computing in Systems Described by Equations*, volume 58 of *LNCS*. 1977.
18. M.R.K. Krishna Rao. Some characteristic of strong innermost normalization. *Theoretical Computer Science*, (239):141–164, 2000.
19. J. Steinbach and H. Xi. Freezing - termination proofs for classical, context-sensitive and innermost rewriting. Technical report, Institut für Informatik, T.U. München, Germany, 1998.
20. R. Thiemann and J. Giesl. Size-change termination for term rewriting. In *Proc. RTA*, volume 2706 of *LNCS*, pages 264–278, 2003.
21. Y. Toyama. Counterexamples to termination for the direct sum of term rewriting systems. *Information Processing Letters*, 25:141–143, 1987.

Leanest Quasi-orderings
Preliminary Version

Nachum Dershowitz[1] and E. Castedo Ellerman[2]

[1] School of Computer Science, Tel Aviv University,
Ramat Aviv, Tel Aviv 69978, Israel
nachumd@tau.ac.il
[2] 36 Lancaster St,
Cambridge, MA 02140, USA
castedo@castedo.com

> *In my poor, lean lank face*
> *nobody has ever seen*
> *that any cabbages were sprouting.*
>
> *—Abraham Lincoln*

Abstract. A convenient method for defining a quasi-ordering, such as those used for proving termination of rewriting, is to choose the minimum of a set of quasi-orderings satisfying some desired traits. Unfortunately, a minimum in terms of set inclusion can be non-existent even when an intuitive "minimum" exists. We suggest an alternative to set inclusion, called "leanness", show that leanness is a partial ordering of quasi-orderings, and provide sufficient conditions for the existence of a "leanest" ordering.

1 Introduction

Well-founded partial orderings (admitting no infinite strictly decreasing sequences) are the standard tool for proving algorithm termination. States of the program are assigned values in the underlying set, such that program steps always result in a decrease in the ordering, thereby establishing termination. Quasi-orderings (reflexive-transitive binary relations) are often more convenient for this purpose than partial or total orderings: the ordering on states induced by a partial ordering of values is in fact a quasi-ordering. In this paper, the unqualified term "ordering" will always refer to a *quasi-ordering*.

A non-empty set of quasi-orderings can be defined by a set of conditions (such as weak-monotonicity and weak-subterm for *quasi-simplification orderings*); then we can identify a particular, ideal ordering by choosing the minimum ordering in the set. Unfortunately, at times, a set of orderings will have no minimum in the usual set-theoretic sense of minimum. (One example where there is a meaningful such minimum may be found in [7].) Accordingly, this paper suggests a more general definition of "minimum" which often leads to a unique ordering that is intuitively the desired minimum ordering.

J. Giesl (Ed.): RTA 2005, LNCS 3467, pp. 32–45, 2005.

The notion of "leanness" defined here embodies a preference for thinness of quasi-orderings near their bottom. By "thinness" we mean that equivalence classes are smaller. Our definition is especially useful when defining orderings by incrementally adding constraints. Investigations of alternate choices of partial orderings for rewriting include [3–5], which are regarding multiset orderings. A classification of some string orderings appears in [6].

We begin with a motivating example. Then, in Section 3, we define the leanness relation. This is followed by a section devoted to conditions guaranteeing the existence of a leanest ordering. Section 5 illustrates the ideas with an example of a leanest (*lexicographic-path-ordering*-like) tree ordering. We conclude with a brief discussion.

2   A String Example

As usual, a quasi-ordering A may be viewed as a set of ordered pairs, where each ordered pair is a comparison. We use $x \precsim_A y$ to denote $\langle x, y \rangle \in A$, a comparison according to ordering A. As usual, $x \prec_A y$ will denote $x \precsim_A y$ but not $y \precsim_A x$. We will also have recourse to denote comparable, *but unequal* elements by $x \lozenge_A y$, as short for $x \precsim_A y$ or $y \precsim_A x$ but $x \neq y$.

Consider a simple example of a set of conditions defining a set of quasi-orderings. Let Ω denote the set of all quasi-orderings A of strings over $\Sigma = \{\mathbf{a}, \mathbf{b}, \mathbf{c}\}$ that satisfy all three of the following conditions:

1. $\varepsilon \precsim_A \mathbf{a} \precsim_A \mathbf{b} \precsim_A \mathbf{c}$;
2. if $v \precsim_A w$ and $x \precsim_A y$, then $vx \precsim_A wy$;
3. if $v \prec_A w$, then $vx \precsim_A wy$,

for all strings $v, w \in \Sigma^*$ and symbols $x, y \in \Sigma$.

Intuitively it might seem that there should be a minimum ordering that satisfies these conditions. In it, the empty string ε would be the smallest element followed by $\mathbf{a}$, $\mathbf{b}$ and $\mathbf{c}$ in strictly increasing order. Following this pattern we can enumerate a total "length-first lexicographic" ordering in the following fashion:

$$\varepsilon \prec \mathbf{a} \prec \mathbf{b} \prec \mathbf{c} \prec \mathbf{aa} \prec \mathbf{ab} \prec \mathbf{ac} \prec \mathbf{ba} \prec \cdots$$

Example 1. Let M be the above quasi-ordering, which may be defined as follows:

$$v \precsim_M w := m(v) \leq m(w),$$

where m is the homomorphism:

$$m(\varepsilon) = 1 \,,$$
$$m(w\mathbf{a}) = m(w)3 \,,$$
$$m(w\mathbf{b}) = m(w)3 + 1 \,,$$
$$m(w\mathbf{c}) = m(w)3 + 2$$

for any string w.   □

A natural definition for the minimum (or, "least defined") ordering is the minimum in terms of the subset relation: the ordering that, as a set of comparisons, is a subset of all other orderings in Ω. Surprisingly, perhaps, M is not a minimum of Ω in this sense. Furthermore, no such minimum in terms of subset exists.

To see this, we consider another ordering which intuitively is greater than M, but of which M is not a subset.

Example 2. We make an intuitively less minimal ordering N by forcing **a** and **b** to be equivalent. Like M, let ε be strictly less than **a** and **c** be strictly greater than **b**. The next equivalence classes in N are

$$\{\mathbf{aa}, \mathbf{ab}, \mathbf{ba}, \mathbf{bb}\}$$

followed (strictly) by

$$\{\mathbf{ac}, \mathbf{bc}\} \ .$$

Like M, we define the entirety of N with a mapping:

$$v \precsim_N w := n(v) \leq n(w),$$

where n is the string-homomorphism:

$$\begin{aligned} n(\varepsilon) &= 1 \ , \\ n(w\mathbf{a}) &= n(w)2 \ , \\ n(w\mathbf{b}) &= n(w)2 \ , \\ n(w\mathbf{c}) &= n(w)2 + 1 \end{aligned}$$

for any string w.

This ordering N also satisfies all the conditions for bona fide membership in Ω.   □

With **a** and **b** equivalent in N but strictly increasing in M, a more striking difference between N and M is made possible. In N the string **ac** is strictly greater than **ba**, since $n(\mathbf{ac}) = 5$ and $n(\mathbf{ba}) = 4$. However in M, **ac** is strictly less than **ba**, since $m(\mathbf{ac}) = 11$ and $m(\mathbf{ba}) = 12$.

The following diagram displays comparisons for M, N, and any relation S that is a subset of both M and N.

	M	N	S
$\mathbf{a} \precsim \mathbf{b}$	√	√	√
$\mathbf{b} \precsim \mathbf{a}$	×	√	×
$\mathbf{ac} \precsim \mathbf{ba}$	√	×	×

We see that $\mathbf{ac} \precsim_S \mathbf{ba}$ must not hold even though $\mathbf{a} \prec_S \mathbf{b}$ holds. This means that any ordering that is a subset of both M and N cannot satisfy the third condition for membership in Ω. Thus Ω cannot have an ordering that is the subset-minimum (the minimum in terms of the subset relation).

Nevertheless, intuitively, M is "more minimal" than N, since it omits the inequality $\mathbf{b} \precsim \mathbf{a}$. So, instead of comparing quasi-orderings in terms of the subset relation, we propose an alternative: "leanness" of orderings. In this alternative relation of quasi-orderings, M is in fact the "leaner" of the two. Both M and N "start off" the same, with $\varepsilon \prec \mathbf{a}$, but then diverge with the comparison of $\mathbf{a}$ and $\mathbf{b}$. Whereas N has an equivalence class of $\mathbf{a} \simeq \mathbf{b}$, M has only $\mathbf{a}$. This is why M is to be preferred.

In the next section, we formalize these observations to obtain a general definition of a "leanness" relation on quasi-orderings.

3 Leanness

The comparison of $\mathbf{ac}$ and $\mathbf{ba}$ in examples M and N proved problematic because viewing comparisons outside the context of the comparisons around it results in a "subset tie". By taking into account what happens lower down in an ordering, such ties can be avoided. The rationale is that the constraints that characterize the family of orderings in question are typically inductive, for which reason the ordering imposed on smaller elements ought to be more significant.

Instead of looking at comparisons by themselves, we want to work with a construct that takes into account the position of the comparisons. For that purpose, we need to extend the notion of an initial segment of a linear order to quasi-orderings.

Definition 1 (Initial segments and super-segments).

1. *A quasi-ordering A is an* initial segment *of a quasi-ordering B if B extends A and everything smaller than an element of B that is non-reflexively ordered in A is also smaller in A. In symbols:*

$$A \trianglelefteq B := A \upharpoonright \mathrm{Dom}^* A = B \upharpoonright \mathrm{Dom}^* A,$$

   where $\mathrm{Dom}^* A := \{x : \exists y.\ y \Diamond_A x\}$ and $A \upharpoonright D := \{\langle x, y \rangle \in A :\ y \in D\}$.
2. *An initial segment of B is a* strict *initial segment of B if it is not equal to B, denoted $A \triangleleft B$.*
3. *If $A \trianglelefteq B$, then B is a* super-segment *of A; it is a* strict *super-segment if $A \neq B$.*
4. *We use $\mathcal{I}(B)$ to denote the set of all initial segments of B, and $\mathcal{I}(\Omega)$ for all the initial segments of members of Ω.*

Proposition 1. *For any quasi-orderings A and B,*

1. $\mathrm{Dom}^*(A \cup B) = (\mathrm{Dom}^* A) \cup (\mathrm{Dom}^* B)$,
2. $\mathrm{Dom}^*(A \cap B) \subseteq (\mathrm{Dom}^* A) \cap (\mathrm{Dom}^* B)$,
3. $A \subseteq B$ implies $\mathrm{Dom}^* A \subseteq \mathrm{Dom}^* B$.

Proposition 2. *For any quasi-orderings A and B and sets D and E,*

1. $(A \cup B) \upharpoonright D = (A \upharpoonright D) \cup (B \upharpoonright D)$,
2. $(A \cap B) \upharpoonright D = (A \upharpoonright D) \cap (B \upharpoonright D)$,

3. $A \upharpoonright (D \cup E) = (A \upharpoonright D) \cup (A \upharpoonright E)$,
4. $A \upharpoonright (D \cap E) = (A \upharpoonright D) \cap (A \upharpoonright E)$,
5. $A \subseteq B$ implies $A \upharpoonright D \subseteq B \upharpoonright D$,
6. $D \subseteq E$ implies $A \upharpoonright D \subseteq A \upharpoonright E$,
7. $D \subseteq E$ and $A \upharpoonright E = B \upharpoonright E$ implies $A \upharpoonright D = B \upharpoonright D$,
8. $A \upharpoonright (\text{Dom}^* B \setminus \text{Dom}^* A) \subseteq B \upharpoonright (\text{Dom}^* B \setminus \text{Dom}^* A)$,
9. $\mathcal{I}(A) \subseteq \mathcal{I}(B)$ implies $A \subseteq B$.

Proposition 3. *For all orderings A, $\mathcal{I}(A)$ is closed under union and intersection.*

Proposition 4. *The initial-segment relation $\trianglelefteq$ is a partial ordering of quasi-orderings.*

All orderings have the identity relation as a trivial initial segment. In the case of M and N the first non-trivial initial segments differ: M has an initial segment ordering ε strictly below **a**:

$$\varepsilon \prec_M \mathbf{a} ,$$

whereas N has an initial segment ordering ε strictly below the equivalence class $\{\mathbf{a}, \mathbf{b}\}$:

$$\varepsilon \prec_N \mathbf{a} \simeq_N \mathbf{b} .$$

That M's initial segment is a subset of N's initial segment is the first indication that M is leaner than N.

In the general case of arbitrary quasi-orderings A and B, there may be no single "next" initial segment that marks the divergence between orderings A and B. The key property however is of initial segments that are found in one ordering but not the other.

We now have the building blocks necessary to define a "leaner" relation for quasi-orderings in general. In the simple case of examples M and N, there was one initial segment from M that was a subset of one initial segment from N. In the general case, *any* initial segments of N will be considered as long as they are not initial segments of M. Similarly, more than just one initial segment from M can be a subset of initial segments from N, just as long as the initial segment from M is not an initial segment of N.

Definition 2 (Leanness). *Quasi-ordering A is* leaner *than quasi-ordering B, symbolized $A \sqsubseteq B$, iff for every initial segment B_0 of B and not of A there is an initial segment A_0 of A and not of B that is a subset of B_0:*

$$\forall B_0 \in \mathcal{I}(B) \setminus \mathcal{I}(A). \; \exists A_0 \in \mathcal{I}(A) \setminus \mathcal{I}(B). \; (A_0 \subseteq B_0) .$$

Leanness is a partial ordering, as we will see below. In the case of M and N, we do have $M \sqsubseteq N$. It is also the case that M is the leanest ordering in Ω.

Remark 1. The definition of leanness resembles a Smyth powerdomain construction [8] on initial segments (but removes common elements from comparison) and the multiset extension [1] of proper superset (but applies to infinite sets).

The initial segment relation, $\unlhd$, and the leanness relation, $\sqsubseteq$, play complementary roles. For any two distinct quasi-orderings (ordering the same set of elements), $\unlhd$ will always leave the two orderings incomparable, whereas $\sqsubseteq$ may make them comparable. When quasi-orderings are partial orderings (i.e. they are anti-symmetric) the leanness relation does not compare any two distinct orderings with the same set of non-reflexively ordered elements.

Possibly counter-intuitive at first is the following result.

Lemma 1. *For any two quasi-orderings A and B,*

$$A \unlhd B \text{ implies } B \sqsubseteq A .$$

Proof. For any two quasi-orderings A and B with $A \unlhd B$, we have $\mathcal{I}(A) \backslash \mathcal{I}(B) = \emptyset$, so the definition of $B \sqsubseteq A$ is vacuously true. $\quad\square$

The reverse direction is not generally true for quasi-orderings, but is true for (antisymmetric) well-orderings.

Lemma 2. *For any two well-orderings A and B,*

$$A \sqsubseteq B \text{ implies } B \unlhd A .$$

Proof. Suppose B is not an initial segment of A. Then $\mathcal{I}(B) \setminus \mathcal{I}(A)$ is nonempty. Let S be the set $\mathrm{Dom}^* B \setminus \mathrm{Dom}^* (A \sqcap B)$. Since B is well-ordered (and antisymmetric), there must exist a minimum $x \in S$ under the ordering B. Let B_0 be

$$B \upharpoonright \mathrm{Dom}^* (\{x\} \cup \mathrm{Dom}^* (A \sqcap B)) .$$

Then $B_0 \in \mathcal{I}(B) \backslash \mathcal{I}(A)$. Any $A_0 \in \mathcal{I}(A)$ that includes x must also include some $y \notin \mathrm{Dom}^* B_0$, since $A_0 \neq B_0$. For this reason, any $A_0 \in \mathcal{I}(A) \setminus \mathcal{I}(B)$ cannot be a subset of B_0. Thus $A \not\sqsubseteq B$. $\quad\square$

In general, leanness is always a partial ordering (of quasi-orderings). The proof proceeds as follows:

Lemma 3. *For any quasi-orderings A, A_0 and B,*

$$A_0 \unlhd A \text{ and } A_0 \subseteq B \subseteq A \text{ implies } A_0 \unlhd B .$$

In particular, if $A_0 \subseteq B_0$ are both initial segments of some ordering, then $A_0 \unlhd B_0$.

Proof. Assume $A_0 \unlhd A$. By definition, $A_0 \upharpoonright \mathrm{Dom}^* A_0 = A \upharpoonright \mathrm{Dom}^* A_0$.

$$A_0 \subseteq B \subseteq A \text{ implies } A_0 \upharpoonright \mathrm{Dom}^* A_0 \subseteq B \upharpoonright \mathrm{Dom}^* A_0 \subseteq A \upharpoonright \mathrm{Dom}^* A_0$$
$$\text{implies } A_0 \upharpoonright \mathrm{Dom}^* A_0 = B \upharpoonright \mathrm{Dom}^* A_0$$
$$\text{implies } A_0 \unlhd B .$$

$\quad\square$

Theorem 1. *Leanness is a partial ordering of quasi-orderings.*

Proof. For any quasi-ordering A, $A \sqsubseteq A$ is trivially true since $\mathcal{I}(A) \setminus \mathcal{I}(A) = \emptyset$. Thus leanness is reflexive.

Consider any quasi-orderings A and B, with $A \sqsubseteq B \sqsubseteq A$. Suppose there exists $A_1 \in \mathcal{I}(A) \setminus \mathcal{I}(B)$. Then there exists $B_0 \in \mathcal{I}(B) \setminus \mathcal{I}(A)$ with $B_0 \subseteq A_1$ and there must also exist $A_0 \in \mathcal{I}(A) \setminus \mathcal{I}(B)$ with $A_0 \subseteq B_0$. By Lemma 3, A_0 must be a member of $\mathcal{I}(B)$, a contradiction. Thus $\mathcal{I}(A) \setminus \mathcal{I}(B)$ must be empty. Likewise $\mathcal{I}(B) \setminus \mathcal{I}(A)$ must also be empty. Thus $\mathcal{I}(A) = \mathcal{I}(B)$, and hence $A = B$, implying anti-symmetry.

For transitivity, suppose
$$A \sqsubseteq B \sqsubseteq C.$$
Consider any $C_0 \in \mathcal{I}(C) \setminus \mathcal{I}(A)$. We show there must exist $A_0 \in \mathcal{I}(A) \setminus \mathcal{I}(C)$ such that $A_0 \subseteq C_0$.

Case 1: $C_0 \trianglelefteq B$. Since $A \sqsubseteq B$, there must be an $A_0 \in \mathcal{I}(A) \setminus \mathcal{I}(B)$ such that $A_0 \subseteq C_0$. If $A_0 \trianglelefteq C$ then $A_0 \trianglelefteq B$ since $A_0 \subseteq C_0 \trianglelefteq B$. Since $A_0 \ntrianglelefteq B$ we must have $A_0 \ntrianglelefteq C$. Thus there exists $A_0 \in \mathcal{I}(A) \setminus \mathcal{I}(C)$ with $A_0 \subseteq C_0$.

Case 2: $C_0 \ntrianglelefteq B$. Since $B \sqsubseteq C$, there must exist $B_0 \in \mathcal{I}(B) \setminus \mathcal{I}(C)$ such that $B_0 \subseteq C_0$.

Case 2a: $B_0 \trianglelefteq A$. Thus there exists $A_0 = B_0 \in \mathcal{I}(A) \setminus \mathcal{I}(C)$ with $A_0 = B_0 \subseteq C_0$.

Case 2b: $B_0 \ntrianglelefteq A$. Since $A \sqsubseteq B$, there must exist $A_0 \in \mathcal{I}(A) \setminus \mathcal{I}(B)$ such that $A_0 \subseteq B_0$. By Lemma 3, if A_0 were in $\mathcal{I}(C)$ then A_0 must be in $\mathcal{I}(B)$, thus $A_0 \ntrianglelefteq C$. Thus we have $A_0 \in \mathcal{I}(A) \setminus \mathcal{I}(C)$ with $A_0 \subseteq B_0 \subseteq C_0$.

Thus $A \sqsubseteq C$. □

4 Leanest

The set Ω from Section 2 provided a good example of orderings that include a leanest ordering, but have no subset-minimum. In general, for any set of orderings Ψ, the two properties presented below are sufficient to know that a leanest ordering exists in Ψ. The structure these properties depend on is not from the elements ordered or from the way the set of orderings is defined, but rather from the set of all initial segments of members of Ψ (denoted $\mathcal{I}(\Psi)$).

The first property is a closure property: any sequence of initial segments has an upper bound. The second property is a kind of "tie breaker". Intuitively, it ensures that, for any initial segment in $\mathcal{I}(\Psi)$, there is a "winning" minimum super-segment to follow, which we will call the "successor segment".

Definition 3 (Successor segment). *Given a set of quasi-orderings Ψ and quasi-ordering $A \in \Psi$, a strict super-segment in Ψ of A is a successor segment in Ψ of A if it is a subset of all strict super-segments in Ψ of A.*

Theorem 2 (Existence). *A set of quasi-orderings Ψ has a unique leanest ordering if*

1. *every ascending sequence $C_0 \trianglelefteq C_1 \trianglelefteq C_2 \trianglelefteq \ldots$ in poset $(\mathcal{I}(\Psi), \trianglelefteq)$ has an upper bound in $\mathcal{I}(\Psi)$; and*
2. *every non-maximal member of $(\mathcal{I}(\Psi), \trianglelefteq)$ has a successor segment in $\mathcal{I}(\Psi)$.*

If chains in $(\mathcal{I}(\Psi), \trianglelefteq)$ can be uncountable, then the term "sequences" in the first condition needs to be interpreted to mean *transfinite* sequences, which are functions from ordinals instead of just non-negative integers. In this paper, sequences can mean transfinite sequences or the usual kind of sequences. Typically the usual kind of sequences suffice.

The proof of existence of leanest orderings requires introducing some fundamental constructs and identifying some intermediate results. We start with the analogues to union and intersection for initial segments.

Definition 4 (Greatest common initial segment). *For any two quasi-orderings A and B, define the* greatest common initial segment *to be*

$$A \sqcap B := \bigcup (\mathcal{I}(A) \cap \mathcal{I}(B)) .$$

Define the greatest common initial segment *of a set of quasi-orderings S to be*

$$\bigsqcap_{A \in S} A := \bigcup \left(\bigcap_{A \in S} \mathcal{I}(A) \right) .$$

Because $\mathcal{I}(A)$ is closed under union, $A \sqcap B$ must always be an initial segment of both A and B.

Proposition 5. *The greatest common initial segment operation $\sqcap$ is associative, commutative and idempotent.*

Definition 5 (Least common super-segment). *For any two quasi-orderings A and B, let the* least common super-segment *be the intersection of all quasi-orderings for which A and B are both initial segments. Symbolically:*

$$A \sqcup B := \bigcap \{C : A, B \trianglelefteq C\} .$$

With $\sqcap$ and $\sqcup$ defined, we move on to an alternative definition of leanness which at times is more convenient to use that the original definition.

Lemma 4. *For any quasi-orderings A, B, and $B_0 \in \mathcal{I}(B) \setminus \mathcal{I}(A)$, if $A \subseteq B_0 \sqcup (A \sqcap B)$ then there exists $A_0 \in \mathcal{I}(A) \setminus \mathcal{I}(B)$ such that $A_0 \subseteq B_0$.*

Proof. Consider any quasi-orderings A and B. We select a set of initial segments of A that intuitively "have what makes A not an initial segment of B":

$$S := \{R \trianglelefteq A : R \sqcup (A \sqcap B) = A\} .$$

Each member of S includes all of the comparisons of A that are "above and beyond" what A and B have in common. The intersection of all members of S gives us "just what makes A not an initial segment of B":

$$A_0 := \bigcap_{R \in S} R .$$

Since all members of $\mathcal{S}$ are in $\mathcal{I}(A)$, A_0 must also be in $\mathcal{I}(A)$. Since all members of $\mathcal{S}$ have or lack the same comparisons of A that prevent membership in $\mathcal{I}(B)$, A_0 must also not be in $\mathcal{I}(B)$. We are left with $A_0 \in \mathcal{I}(A) \setminus \mathcal{I}(B)$.

Consider any $B_0 \in \mathcal{I}(B) \setminus \mathcal{I}(A)$ with $A \subseteq B_0 \sqcup (A \sqcap B)$. With

$$A_0 \sqcup (A \sqcap B) = \bigcap_{R \in \mathcal{S}} (R \sqcup (A \sqcap B)) = \bigcap_{R \in \mathcal{S}} A = A$$

we get

$$A_0 \sqcup (A \sqcap B) \subseteq B_0 \sqcup (A \sqcap B)$$

and conclude that for any comparison $\langle x, y \rangle \notin A \sqcap B$, $\langle x, y \rangle \in A_0$ must imply $\langle x, y \rangle \in B_0$. If $\langle x, y \rangle \in A \sqcap B$ and $\langle x, y \rangle \in A_0$, then $\langle x, y \rangle \in B_0$ must hold for A_0 to be the smallest member of $\mathcal{S}$. Thus, $A_0 \subseteq B_0$. $\qquad\square$

Theorem 3. *For any quasi-orderings A and B, $A \sqsubseteq B$ iff for every strict super-segment B_0 of $A \sqcap B$ in B there exists a strict super-segment A_0 of $A \sqcap B$ in A such that $A_0 \subseteq B_0$.*

Proof. Assume $A \sqsubseteq B$. Consider any $B_0 \trianglelefteq B$ with $A \sqcap B \triangleleft B_0$. Since $B_0 \in \mathcal{I}(B) \setminus \mathcal{I}(A)$ there exists $A_1 \in \mathcal{I}(A) \setminus \mathcal{I}(B)$ such that $A_1 \subseteq B_0$. Let $A_0 = A_1 \sqcup (A \sqcap B)$. Since B_0 is a super-segment of $A \sqcap B$, $A_0 \subseteq B_0$.

Assume

$$\forall B_0 \in B. \ (A \sqcap B \triangleleft B_0 \rightarrow \exists A_0 \in A. \ (A \sqcap B \triangleleft A_0 \text{ and } A_0 \subseteq B_0)) \ .$$

Consider any $B_1 \in \mathcal{I}(B) \setminus \mathcal{I}(A)$. Let $B_0 = B_1 \sqcup (A \sqcap B)$. Since $B_0 \trianglelefteq B$ with $A \sqcap B \triangleleft B_0$, there must exists $A_0 \trianglelefteq A$ with $A \sqcap B \triangleleft A_0$ and $A_0 \subseteq B_0$. By Lemma 4, there exists $A_1 \in \mathcal{I}(A) \setminus \mathcal{I}(B)$ with $A_1 \subseteq B_1$. Thus $A \sqsubseteq B$. $\qquad\square$

Corollary 1. *For any quasi-orderings A and B with $A \not\trianglelefteq B$, if $A \subseteq B_0$ for every B_0 with $A \sqcap B \triangleleft B_0 \trianglelefteq B$, then $A \sqsubseteq B$.*

Proof. For every $B_0 \trianglelefteq B$ with $A \sqcap B \triangleleft B_0$, there exists $A \trianglelefteq A$ with $A \sqcap B \triangleleft A$ and $A \subseteq B_0$. $\qquad\square$

Next we introduce constructions and results for working with sequences of orderings.

Definition 6 (Dual-Chain). *For any sequence C with $C_0 \sqsupseteq C_1 \sqsupseteq C_2 \sqsupseteq \cdots$ descending in poset $(\Psi, \sqsubseteq)$ let the dual-chain of C be the sequence of common initial segments*

$$\Delta_\alpha (C) := \bigcap_{\beta \geq \alpha} C_\beta \ .$$

Let the dual-chain limit be their union:

$$\lim{}^\Delta C := \bigcup_\alpha \Delta_\alpha (C) \ .$$

Some results about dual-chains follow.

Proposition 6

1. *The dual-chain of a descending sequence in poset $(\Psi, \sqsubseteq)$ is an ascending sequence in $(\mathcal{I}(\Psi), \trianglelefteq)$ and a descending sequence in poset $(\mathcal{I}(\Psi), \sqsubseteq)$.*
2. *The least upper bound of a dual-chain in terms of $\trianglelefteq$ is the dual-chain limit.*
3. *For any descending sequence C in $(\Psi, \sqsubseteq)$,*

$$C_\alpha \sqcap \lim^\Delta C = \Delta_\alpha(C).$$

Lemma 5. *For any set of quasi-orderings Ψ and any sequence C with $C_0 \sqsupseteq C_1 \sqsupseteq C_2 \sqsupseteq \ldots$ descending in poset $(\Psi, \sqsubseteq)$, such that every non-maximal member of $(\mathcal{I}(\Psi), \trianglelefteq)$ has a successor segment in $\mathcal{I}(\Psi)$, if the dual-chain limit of C is not in $\mathcal{I}(C)$, then it is leaner than all members of C.*

Proof. Let L be the dual-chain limit of C. Consider any $C_\alpha \in C$.

Since $L \ntrianglelefteq C_\alpha$, $\Delta_\alpha(C) \triangleleft L$ and for some $\beta' > \alpha$ we have $\Delta_\alpha(C) \triangleleft \Delta_\beta(C)$ for any $\beta > \beta'$.

Consider any $A_0 \trianglelefteq C_\alpha$ with $L \triangleleft A_0$. Since $A_0 \ntrianglelefteq L$, there must exist some $\beta > \beta' > \alpha$ and $A_0 \ntrianglelefteq C_\beta$. Since $C_\beta \sqsubseteq C_\alpha$ and $A_0 \in \mathcal{I}(C_\alpha) \setminus \mathcal{I}(C_\beta)$ there must exist some $B_0 \in \mathcal{I}(C_\beta) \setminus \mathcal{I}(C_\alpha)$ such that $B_0 \sqsubseteq A_0$.

Both $\Delta_\beta(C)$ and $\Delta_\alpha(C) \sqcup B_0$ are initial segments of C_β and strict super-segments of $\Delta_\alpha(C)$. The intersection of $\Delta_\beta(C)$ and $\Delta_\alpha(C) \sqcup B_0$ and must also be an initial segment of B and a strict super-segment of $\Delta_\alpha(C)$. Thus we have

$$\Delta_\alpha(C) \triangleleft \Delta_\alpha(C) \sqcap (\Delta_\alpha(C) \sqcup B_0) \trianglelefteq L.$$

By Theorem 3, $L \sqsubseteq C_\alpha$. $\qquad\qquad\qquad\qquad\qquad\qquad\qquad\qquad\qquad\qquad\qquad\quad\square$

Lemma 6. *For any set of quasi-orderings Ψ and any sequence C with $C_0 \sqsupseteq C_1 \sqsupseteq C_2 \sqsupseteq \ldots$ descending in poset $(\Psi, \sqsubseteq)$, $\mathcal{I}(C)$ cannot include a successor segment in $\mathcal{I}(\Psi)$ of the dual-chain limit of C.*

Proof. Let S be a successor segment in $\mathcal{I}(\Psi)$ of $\lim^\Delta C$. Suppose there did exist α such that $S \trianglelefteq C_\alpha$. Since $\lim^\Delta C \triangleleft S$ there must exist some $\beta > \alpha$ such that S is not an initial segment of C_β. Thus C_β must be strictly leaner than C_α and $C_\alpha \sqcap C_\beta = \lim^\Delta C$ and for any B_0 with $\lim^\Delta C \triangleleft B_0 \trianglelefteq C_\beta$ it must be the case that S is a subset of B_0. By Corollary 1, $C_\alpha \sqsubseteq C_\beta$, a contradiction. $\qquad\square$

Lemma 7. *For any set of quasi-orderings Ψ, if every non-maximal member of $(\mathcal{I}(\Psi), \trianglelefteq)$ has a successor segment in $\mathcal{I}(\Psi)$, then Ψ is down-directed (for all $A, B \in \Psi$ there must exist a $C \in \Psi$ such that $C \sqsubseteq A, B$).*

Proof. If either A or B is an initial segment of the other then, by Lemma 1, the proof is trivial.

Assume neither A nor B is an initial segment of the other. Thus $A \sqcap B$ must be a proper initial segment of both A and B. If every non-maximal member of $(\mathcal{I}(\Psi), \trianglelefteq)$ has a successor segment in $\mathcal{I}(\Psi)$, then $A \sqcap B$ has a successor segment S in $\mathcal{I}(\Psi)$ where $S \subseteq A_0$ and $S \subseteq B_0$ whenever $(A \sqcap B) \triangleleft A_0 \trianglelefteq A$ and $(A \sqcap B) \triangleleft B_0 \trianglelefteq B$.

If $S \trianglelefteq A$ then S is not an initial segment of B and $A \sqcap B = S \sqcap B$ and thus $A \sqsubseteq S \sqsubseteq B$ by Lemma 1 and Corollary 1.

Similarly, if $S \trianglelefteq B$ then $B \sqsubseteq A$.

If S is not an initial segment of A nor of B then $A \sqcap B = S \sqcap A$ and $A \sqcap B = S \sqcap B$ and thus by Corollary 1, $S \sqsubseteq A$. Since $S \trianglelefteq \Psi$ there exists some $C \in \Psi$ with $S \trianglelefteq C$. By Lemma 1, $C \sqsubseteq A$. $\qquad\square$

The proof of the leanest-ordering existence theorem follows next:

Proof (of Existence Theorem). Consider any sequence C with $C_0 \sqsupseteq C_1 \sqsupseteq C_2 \sqsupseteq \ldots$ descending in poset $(\Psi, \sqsubseteq)$. The dual-chain of C is an ascending sequence in poset $(\mathcal{I}(\Psi), \sqsubseteq)$. By the first condition on $\mathcal{I}(\Psi)$, the dual-chain must have an upper bound U in poset $(\mathcal{I}(\Psi), \trianglelefteq)$. Since $\mathcal{I}(U)$ is closed under union, the dual-chain limit V must be in $\mathcal{I}(U)$, and thus also in $\mathcal{I}(\Psi)$. We seek $L \trianglelefteq \Psi$ leaner than every member of sequence C.

Case 1: The dual-chain limit V is not in $\mathcal{I}(C)$. Choose $L = V$. By Lemma 5, $L \sqsubseteq C_\alpha$ for all α.

Case 2: $V \in C$. Choose $L = V$. For some α, $C_\alpha = L$. For every $\beta > \alpha$, $C_\alpha \trianglelefteq C_\beta$ and $C_\beta \sqsubseteq C_\alpha$ thus $C_\beta = C_\alpha$ by Lemma 1 and Theorem 1. Again, $L \sqsubseteq C_\alpha$ for all α.

Case 3: $V \in \mathcal{I}(C) \backslash C$. For some α, $\lim^\Delta C \triangleleft C_\alpha$. Using the second condition on $\mathcal{I}(\Psi)$, we choose L to be the successor segment in $\mathcal{I}(\Psi)$ of $\lim^\Delta C$. Consider any $\alpha' > \alpha$. By Lemma 6, L is not an initial segment of $C_{\alpha'}$ and hence $C_{\alpha'} \sqcap L \triangleleft L$. Since $\lim^\Delta C$ is an initial segment of both $C_{\alpha'}$ and L, $\lim^\Delta C \trianglelefteq C_{\alpha'} \sqcap L$. Since L is the successor segment of $\lim^\Delta C$, it must be the case that $\lim^\Delta C = C_{\alpha'} \sqcap L$. Thus, for all A_0 with $C_{\alpha'} \sqcap L \triangleleft A_0 \trianglelefteq C_{\alpha'}$, we have $L \sqsubseteq A_0$. By Corollary 1, $L \sqsubseteq C_{\alpha'}$. It follows that $L \sqsubseteq C_\alpha$ for all α.

In all three cases, we have $L \trianglelefteq \Psi$ leaner than every member of C. There must exist some $M \in \Psi$ such that $L \trianglelefteq M$. By Lemma 1, $M \sqsubseteq L$, thus M is a lower bound to C in poset $(\Psi, \sqsubseteq)$.

If chains can be uncountable, then "sequences" must be interpreted to mean *transfinite* sequences. Otherwise, the usual kind of sequences suffices. Since every descending sequence in $(\Psi, \sqsubseteq)$ has a lower bound, every chain must have a lower bound, because every chain can be countably enumerated (or well-ordered if uncountable) and a descending subsequence extracted. By Zorn's Lemma, Ψ must have at least one minimally lean member. By Lemma 7, Ψ is directed, so there can be only one. Thus, there must exist a unique leanest member of Ψ. $\qquad\square$

5   Application to Binary Trees

Earlier, we described a very simple leanest string ordering. With the Existence Theorem, leanest orderings of greater complexity can be found. In the example to follow, binary trees serve as elements rather than strings.

The most basic and trivial tree is the empty tree denoted $\square$. This tree has no nodes or branches. From the empty tree $\square$, more interesting trees can be built using the operation of $(x \wedge y)$, which places tree x to the left of a root node and tree y to the right. For instance,

$$(\square \curlywedge \square) = \quad \bullet$$

$$(\square \curlywedge (\square \curlywedge \square)) =$$

$$((\square \curlywedge \square) \curlywedge \square) =$$

$$((\square \curlywedge \square) \curlywedge (\square \curlywedge \square)) =$$

The following set of conditions for quasi-orderings comes from [2]:

Definition 7 ([2]). *Let Ω be the set of all quasi-orderings A of finite binary trees that satisfy the following three tree-ordering conditions:*

- **Growth:** $(x \curlywedge y) \succsim_A x, y$;
- **Monotonicity:** *if* $y \precsim_A z$ *then* $(x \curlywedge y) \precsim_A (x \curlywedge z)$ *and* $(y \curlywedge x) \precsim_A (z \curlywedge x)$;
- **Lexicography:** $(x_1 \curlywedge x_0) \precsim_A (y_1 \curlywedge y_0)$ *if* $x_1 \prec_A y_1$ *and* $x_0 \precsim_A (y_1 \curlywedge y_0)$.

We establish that the first condition of the Existence Theorem holds for Ω.

Lemma 8. *Every chain in $(\mathcal{I}(\Omega), \trianglelefteq)$ has an upper bound.*

Proof. Consider any chain $\mathcal{C}$ in poset $(\mathcal{I}(\Omega), \trianglelefteq)$. Let $L = \cup_{A \in \mathcal{C}} A$ and D be the set of all trees that are not ordered by L (the complement of $\mathrm{Dom}^* L$). Let K be the ordering defined as

$$x \precsim_K y := \begin{cases} x \precsim_L y, & \text{or} \\ x \in \mathrm{Dom}^* L \text{ and } y \in D, \text{ or} \\ x, y \in D \end{cases}$$

The ordering K places D as an equivalence class ordered strictly above $\mathrm{Dom}^* L$. It must satisfy all of the conditions to be a member of Ω and thus $L \trianglelefteq \Omega$ and is an upper bound of $\mathcal{C}$. □

Next we establish that the second condition of the Existence Theorem holds for Ω.

Lemma 9. *For every $A \trianglelefteq \Omega$, the set of strict super-segments of A has a subset-minimum.*

Proof. Consider the set S of strict super-segments in $\mathcal{I}(\Omega)$ of A. Because Ω consists of total *well-founded* orderings (see [2]), every super-segment in S must have an equivalence class ordered as less than all other elements outside of A. Let T be the set of all these equivalence classes and let B be the intersection of all these equivalence classes.

No two members of T can be disjoint, since otherwise one could construct an ordering of trees that satisfies the conditions of membership in Ω, but the ordering would not be total – a contradiction.

Furthermore, there can be no subset descending sequence of members of T, for otherwise one could construct an ordering of trees that satisfies the conditions of membership in Ω, but again the ordering would not be well-founded, which is a contradiction.

Were B empty, then either two members of T would be disjoint or there would be a subset descending sequence of members of T. Since neither can be the case, B must be non-empty.

Let C be the ordering of A with B placed as an equivalence class strictly above A. Let D be the ordering of C with an equivalence class strictly greater than C consisting of all binary trees not in C. If D is not a member of Ω then one of the members of S cannot be the initial segment of a member of Ω; one of the conditions for membership in Ω must be violated.

Thus, C is the subset-minimum of all strict super-segment in $\mathcal{I}(\Omega)$ of A.   □

It follows from Theorem 2 that Ω has a leanest (lpo-like) ordering.

6   Conclusion

The minimum quasi-ordering satisfying certain conditions is a convenient definition for a well-quasi-ordering used in proving termination of rewriting. A set of quasi-orderings – rather than partial orderings – is easier to define since anti-symmetry is not required. But we have seen that this definition technique comes with a possible snag: There may be no subset-minimum. In particular, conditions that involve a strict comparison ($\prec$) can preclude the existence of a subset-minimum.

To compensate for this problem, we described an alternative to a subset-minimum ordering, namely, the "leanest ordering", building on fundamental notions for quasi-orderings. By establishing two properties on a set of quasi-orderings, a leanest ordering is guaranteed to exist. These properties are defined independent of what kind of elements are ordered and what conditions define a set of quasi-orderings, which should help the results of this paper be applicable in a wide range of situations.

Acknowledgement

We thank Ed Reingold for his interest and a referee for his comments.

References

1. Nachum Dershowitz and Zohar Manna. Proving termination with multiset orderings. *Communications of the ACM*, 22(8):465–476, August 1979.
2. Nachum Dershowitz and Edward M. Reingold. Ordinal arithmetic with list expressions. In A. Nerode and M. Taitslin, editors, *Proceedings of the Symposium on Logical Foundations of Computer Science (Tver, Russia)*, volume 620 of *Lecture Notes in Computer Science*, pages 117–126, Berlin, July 1992. Springer-Verlag.

3. Jean-Pierre Jouannaud and Pierre Lescanne. On multiset orderings. *Information Processing Letters*, 15(2):57–63, September 1982.
4. Ursula Martin. Extension functions for multiset orderings. *Information Processing Letters*, 26:181–186, December 1987.
5. Ursula Martin. A geometrical approach to multiset orderings. *Information Processing Letters*, 67:37–54, May 1989.
6. Ursula Martin and Elizabeth Scott. The order types of termination orderings on monadic terms, strings and multisets. *J. Symbolic Logic*, 62(2):624–635, 1997.
7. Alexander Serebrenik and Danny De Schreye. On termination of meta-programs. *Theory and Practice of Logic Programming*. To appear; available at `http://arxiv.org/abs/cs.PL/0110035`.
8. Michael B. Smyth. Powerdomains. *J. of Computer and System Sciences*, 16(1):23–36, February 1978.

Abstract Modularity

Michael Abbott[1], Neil Ghani[1], and Christoph Lüth[2]

[1] Department of Mathematics and Computer Science, University of Leicester
michael@araneidae.co.uk, ng13@mcs.le.ac.uk
[2] FB 3 – Mathematik und Informatik, Universität Bremen
cxl@informatik.uni-bremen.de

Abstract. Modular rewriting seeks criteria under which rewrite systems inherit properties from their smaller subsystems. This divide and conquer methodology is particularly useful for reasoning about large systems where other techniques fail to scale adequately. Research has typically focused on reasoning about the modularity of specific properties for specific ways of combining specific forms of rewriting.

This paper is, we believe, the first to ask a much more general question. Namely, what can be said about modularity independently of the specific form of rewriting, combination and property at hand. A priori there is no reason to believe that anything can actually be said about modularity without reference to the specifics of the particular systems etc. However, this paper shows that, quite surprisingly, much can indeed be said.

1 Introduction

The key properties of term rewriting systems (TRSs) are *confluence* and *strong normalisation*. One technique for establishing these properties is *modularity* which seeks criteria under which TRSs inherit properties from their smaller (and hence easier to reason about) subsystems. This divide and conquer methodology is particularly useful for reasoning about large systems where other techniques fail to scale adequately.

Research originally focused on *disjoint unions* of term rewrite systems where the systems do not share any operators. Here, confluence is modular [23] and strong normalisation is modular for *non-collapsing* TRSs and for *non-duplicating* TRSs [22]. Subsequently, a variety of alternative proof techniques have been developed [9, 10, 16, 19]. Modularity for *conditional term rewriting systems* (CTRSs) was first studied by Middeldorp [17] who showed that confluence is modular for certain types of CTRS while strong normalisation is again only modular in the presence of extra syntactic restrictions. Several unions permitting the sharing of term constructors have been proposed but, for each of these, confluence and strong normalisation are only modular again in the presence of various syntactic restrictions [18, 20].

These examples demonstrate how modularity is typically studied for specific combinations of rewrite systems, or specific notions of rewriting, or specific properties. This paper, we believe, is the first to ask what can be said about

J. Giesl (Ed.): RTA 2005, LNCS 3467, pp. 46–60, 2005.

modularity independently of these specifics. That is we do not ask about the modularity of confluence, or termination, or weak termination, but rather seek a general modularity theorem applicable to all of these properties. Similarly, we wish to get away from modularity for specific notions of rewriting such as term rewriting, or graph rewriting, or equational rewriting and instead prove general modularity results applicable to as many forms of rewriting as possible. And finally, we wish to avoid commitment to modularity for specific ways of combining rewrite systems but rather extract conditions that are uniformly applicable to a variety of different such mechanisms.

A priori, there is no reason to believe that such an abstract theory of modularity should exist. Certainly it is hard to see how the conditions on the modularity of confluence, strong normalisation *etc.* are instances of the same general theme. This paper demonstrates that such a theory is indeed possible. Of course, it will not be able to magically prove the most general results for any specific situation. Rather, its contribution is to provide a platform of general results which can be instantiated for a specific situation as the need arises. In order to develop such a theory of abstract modularity, we have to build it upon a theme which unifies key features of specific modularity results. We believe the key concept in modularity is the notion of *layer structure* on the terms of the combined TRS which describes how rewrites in the combined TRS decompose into rewrites from the component TRSs. If rewrites do not preserve this layer structure (i.e. if there are collapsing rewrites), then non-trivial interactions between the layers may occur and modularity may fail.

Our results show that, providing rewrite systems preserve the layer structure, properties are inherently modular. We were quite surprised to find such a powerful result by using the techniques we have developed in our previous work [11–13]. To this end, we generalised our approach by treating not just term rewriting systems, but all rewrite systems that arise as *monads*, and by abstracting from specific properties to properties in general, given by a subcategory of the base.

Our general modularity result requires two conditions: i) that the rewrite systems do not collapse layers which is reflected in a condition on the monad representing the rewrite system; and ii) that the semantic and syntactic treatment of properties coincide. The latter condition ought to be automatic in the sense that it should hold for any reasonable property; it does for all well-known ones, such as confluence, termination and weak termination. Overall, we believe that this paper delivers on the promise of clean and simple results in rewriting based upon the categorical methodology.

The paper is structured as follows: In Sect. 2, we explain our abstract notion of the data structures we rewrite. In Sect. 3, we show how to model the actual rewriting by monads. In Sect. 4, we develop our semantic notion of properties of rewriting systems, and show they coincide with the well-known syntactic properties. Sect. 5 introduces an abstract notion of combining systems modelled by monads and shows the general modularity results, the key result being Thm. 27.

For this paper, we assume a very basic knowledge of category theory (comprising concepts such as categories, functors, push-outs and adjoints), but will explain all more sophisticated concepts as they are needed, and concentrate on examples and intuition rather than technical categorical proofs. For an introduction to category theory, see [14].

2   Abstract Data Structures

To enable our modularity results to as many forms of rewriting as possible, we need to extract their common feature. One possibility are abstract reduction systems (ARSs) which model a rewrite system by the one step reduction relation it induces. This semantics therefore throws away the structure of the data being rewritten, including the key concepts of substitution and layer which are central to modularity. Thus, it is unlikely that the ARS semantics of rewriting can be used as the basis of an abstract theory of modularity.

For us, rewriting consists of a data structure where subterms can be replaced with other terms and, as such, substitution is the fundamental property. Thus, we propose to use monads as they take as primitive an abstract notion of data endowed with a well behaved notion of substitution.

Definition 1 (Monad). *A monad* $\mathsf{T} = \langle T, \eta, \mu \rangle$ *on a category* $\mathcal{C}$ *is given by a functor* $T : \mathcal{C} \to \mathcal{C}$, *called the* action, *and two natural transformations,* $\eta : \mathsf{Id} \to T$, *called the* unit, *and* $\mu : TT \to T$, *called the* multiplication *of the monad, satisfying the* monad laws: $\mu \cdot T\eta = \mathsf{Id} = \mu \cdot \eta_T$, *and* $\mu \cdot T\mu = \mu \cdot \mu_T$.

Good introductions to the theory of monads in our sense are [2, 15, 21]. The canonical example of a monad is the one arising from the term algebra over a signature:

Definition 2 (Signature). *A (single-sorted) signature consists of a function* $\Sigma : \mathbb{N} \to \mathsf{Set}$. *The set of n-ary operators of* Σ *is defined* $\Sigma_n \stackrel{def}{=} \Sigma(n)$

Definition 3 (Term Algebra). *Given a signature* Σ *and a set of variables* X, *the terms* $T_\Sigma(X)$ *built over* X *are defined inductively:*

$$\frac{x \in X}{x \in T_\Sigma(X)} \qquad \frac{f \in \Sigma_n \quad t_1, \ldots t_n \in T_\Sigma(X)}{f(t_1, \ldots, t_n) \in T_\Sigma(X)}$$

Lemma 4. *The map* $X \mapsto T_\Sigma(X)$ *defines a monad* T_Σ *on* Set.

Proof. Given a function $f : X \to Y$, renaming of variables defines a function $T_\Sigma(f) : T_\Sigma(X) \to T_\Sigma(Y)$. Every variable is a term, which gives us a family $\eta_X : X \to T_\Sigma(X)$ while substitution defines a family $\mu_X : T_\Sigma T_\Sigma(X) \to T_\Sigma(X)$. The monad laws state that substitution behaves correctly, i.e. is associative and has variables as left and right units, which is easily checked by induction.   □

Our interest in monads is that they describe a number of other computationally interesting data structures possessing well behaved notions of substitutions, as the following examples show.

Example 5 (Strings). The map sending an alphabet X to the set X^* of words over X extends to a monad T^* : Set $\to$ Set. Substitution here takes a word consisting of words and flattens it into one word.

Example 6 (Groups and Rings). The map sending X to the free group $G(X)$ over X extends to a monad. Similarly, the map sending X to the set of free polynomials over X extends to a monad as well. In both cases, substitution is defined structurally, as for the term algebra above.

These examples can be generalised to any algebraic theories:

Example 7 (Algebraic theories). Given an algebraic theory $\langle \Sigma, E \rangle$ where Σ is a signature and E a set of equations, let $\sim_E$ be the congruence generated from E, and $T_{\langle \Sigma, E \rangle}(X) = T_\Sigma(X)/ \sim_E$ be the term algebra quotiented by this congruence, then the map $X \mapsto T_{\langle \Sigma, E \rangle}(X)$ extends to a monad.

Furthermore, monads have another key advantage when applied to modularity, in that the interleaving of monads models the layer structure, e.g. $T_\Sigma(T_\Omega(X))$ consists of terms with a Σ-layer over a Ω-layer with variables built from X.

   As a mild technical condition, we require these monads to be finitary which corresponds to the fact that the data structure under question is built inductively. Formally, a monad is finitary iff it preserves filtered colimits [1] (i.e. $T(X)$ is built from a finite subset of X). Finitary monads on a category $\mathcal{C}$ and monad morphisms form a category $\mathsf{Mon}(\mathcal{C})$. Motivated by all of this, we make our first definition of a rewrite structure, which is the structure containing the data over which rewriting takes place.

Definition 8. *A rewrite structure is a finitary monad* T : Set $\to$ Set.

   To summarise, this section observed that in order to do rewriting, the fundamental properties required were the construction of some form of term calculus and a notion of substitution for that calculus. These concepts are perfectly captured by a rewrite structure.

3   Abstract Rewriting

A monad on Set builds a set of terms from a set of variables. Incorporating rewrites into this framework means that we are actually building a relation of terms and rewrites between them from a relation consisting of a set of variables and (what we consider to be) rewrites between these variables, called *variable rewrites*. That variables rewrite to other variables may seem odd from a rewriting perspective but in modularity these variables represent terms from sublayers, and terms in a sublayer certainly can rewrite to others. Further, as we shall see, adding variable rewrites does not affect properties such as confluence.

   The exact nature of these relations depends upon what we are interested in studying. If we are interested in one-step reduction or one-step completion, we take relations, if we are interested in many-step reduction, we take preorders, or if we are interested in labelled rewriting, graphs or categories. For termination, we want to preserve reduction sequence, so we take transitive relations or well-orders. Here, the base category is the category Pre of preorders and monotone morphisms between them, but the reader should be aware that our general treatment can, and will be used with other base categories. We could make the following definitions parametric over the choice of Pre, and work with an arbitrary category $\mathcal{V}$ such that there is an adjunction as in Lemma 9 below, but we prefer a more concrete definition here.

Lemma 9. *The functor* D : Set → Pre, *which maps a set to the discrete preorder over it, is left adjoint to the functor* V : Pre → Set, *which maps a preorder to its underlying set.*

Proof. The adjunction is established by the isomorphism $\mathsf{Pre}(\mathsf{D}X,P) \cong \mathsf{Set}(X, \mathsf{V}P)$ for any set X and preorder P.   □

In fact, D also has a left adjoint C : Pre → Set, which maps preorder to its set of connected components, and V has a further left adjoint, which maps a set to the total order on it. Now, there are a number of different ways of adding rewrites to a rewrite structure, that is to turn a monad on Set into a monad on Pre:

1. We can define the monad M_R : Pre → Pre to send a preorder X to the pre-order defined as the abstract reduction semantics where there are no variables, but constants from X with associated variable rewrites. However, we still need to define the ARS semantics for each form of rewriting.
2. We can define the action of the monad concretely as in our previous work on term rewriting systems, e.g. [11, 12]. The advantage of this is that it gives a precise description of the rewrite monad, but at the cost of having to repeat the exercise every time we change the data structure.
3. We can define a rewrite presentation to be a parallel pair in Mon(Set), lift to Mon(Pre) and take the *coinserter*. This was the approach in [3]. The advantage of this approach is that it gives a precise and abstract formation of the rewrite system associated to any presentation, but at the cost of the technical overhead of coinserters.

In this paper, we choose an axiomatic approach which allows us to derive as many results as possible on a general level, and then instantiate them.

Definition 10. *Let M be a rewrite structure. An M-rewrite system is a finitary monad M_R : Pre → Pre such that M_R is a lifting of M, i.e. the following diagram commutes:*

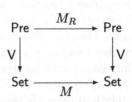

The condition says that the monad M_R which calculates terms and rewrites agrees with the monad M on the terms. Thus one can think of M_R as acting as M on terms, but adding in extra rewrites.

We have been speaking informally of the ARS semantics for rewrite systems, but now we make this precise. If R is a rewrite system, one usually fixes a countable infinite set of variables X, and considers the resulting ARS (which would be $M_R(X)$ here). But we can be more specific, as all countable infinite sets are isomorphic to the set $\mathbb{N}$ of natural numbers, so we take the discrete preorder $D\mathbb{N}$ as the canonical representation of all variables, and call $M_R(D\mathbb{N})$ the *representing ARS*. Thus, the difference between the monadic semantics and the ARS semantics is that the monadic semantics builds terms and rewrites over an arbitrary, not fixed, preorder of variables and variable rewrites. This extra flexibility is precisely what is required by modular rewriting as we can instantiate the variables and variable rewrites to be the terms and rewrites from a sublayer.

Given a rewrite structure M, there is always an empty (or discrete) M-rewrite system $M_\emptyset$ with no rewrite rules.

Lemma 11. *For a rewrite structure M, there is a free M-rewrite system $M_\emptyset$.*

Proof. The functor $\mathsf{V\_D} : \mathsf{Mon(Pre)} \to \mathsf{Mon(Set)}$ has a left adjoint, denoted L, as shown in [3]. This computes the free lifting $M_\emptyset = L(M)$. ☐

If M_R is an M-rewrite system, then by definition $M\mathsf{V} = \mathsf{V}M_R$. Precomposing with D, and noting $\mathsf{VD} = 1$, we get $M = \mathsf{V}M_R\mathsf{D}$, and hence a canonical embedding $\kappa : M_\emptyset \to M_R$ which embeds the empty M-rewrite system in any other M-rewrite system. Given a M-rewrite system, we will often want to abstractly use the idea that rewrites created by M_R are either created by an underlying rewrite system or by the variable rewrites. This is captured by asking that the diagram (1) be a push-out, where ε is the counit of the adjunction of Lemma 9.

$$
\begin{array}{ccc}
M_\emptyset(\mathsf{DV}X) & \xrightarrow{\ M_\emptyset\varepsilon\ } & M_\emptyset X \\
{\scriptstyle \kappa_{\mathsf{DV}X}}\big\downarrow & & \big\downarrow{\scriptstyle \kappa_X} \\
M_R(\mathsf{DV}X) & \xrightarrow[\ M_R\varepsilon\ ]{} & M_R X
\end{array}
\qquad (1)
$$

We say an M-rewrite system M_R is *cocartesian* iff $\kappa : M_\emptyset \to M_R$ is a cocartesian natural transformation, i.e. all components form push-out squares. Most M-rewrite systems are cocartesian, because $M_R(X)$ is the coproduct of the monads representing R and representing the rewrites of X.

We finish this section with some examples.

Example 12 (Term Rewriting). A term rewriting system $\langle \Sigma, R \rangle$ has as a rewrite structure the term algebra monad T_Σ and as a T_Σ-rewrite system the monad $\mathsf{T}_{\langle \Sigma, R \rangle}$ which sends a preorder X to the smallest ordered Σ-algebra $\mathsf{T}_{\langle \Sigma, R \rangle}(X)$ containing X for which R is sound [3]. Cocartesianness follows from the inductive construction of $\mathsf{T}_{\langle \Sigma, R \rangle}(X)$ [11].

Example 13 (String Rewriting and Gröbner Bases). String rewriting can be regarded as rewriting over the free monoid, i.e. words, while Gröbner bases can be regarded as rewriting over free rings, i.e. polynomials. The rewrite structure here is given by Example 5 and 6, with M_R adding in a reduction structure between the words and polynomials respectively.

Thus, for example, Gröbner bases have as a rewrite system the monad that computes for a preorder X, the smallest preorder on the free ring over X containing X for which the ring operations are monotone and for which R is sound.

Example 14 (Equational Rewriting). The rewrite structure for an equational term rewriting system $\langle \Sigma, E, R \rangle$ is the monad $T_{\langle \Sigma, E \rangle}$ from Example 7, and as a $T_{\langle \Sigma, E \rangle}$-rewrite system the monad $T_{\langle \Sigma, E, R \rangle}$ which sends a preorder X to the smallest preorder on $T_{\langle \Sigma, E \rangle}(X)$ for which the Σ-constructors are monotone and for which R is sound. Cocartesianness follows from cocartesianness of $T_{\langle \Sigma, R \rangle}$.

Further examples could be developed, e.g. the rational monad suffices as a rewrite structure to consider rational rewriting. Recent work on abstract syntax shows that structures with variable binding are monads. This monadic approach to higher order rewriting has been developed by Hamana [8].

These examples follow the general pattern. Given a rewrite structure M, a M-rewrite system is given by triples (Y, l, r) where Y is a set and l, r are elements of MY. The associated M-rewrite system M_R maps a preorder X to the smallest preorder on $M(X)$ for which the operations in M are monotone and for which the interpretations of l is greater than that of r. When this order relation is defined inductively, the cocartesianness of $\kappa : M_\emptyset \to M_R$ follows.

To summarise, M-rewrite systems provide a model of rewriting covering a large variety of different forms of rewriting. In fact, given any underlying data structure for rewriting which forms a monad, i.e. possesses a well behaved notion of substitution, we can model rewriting over that data structure by a monad which sends a preorder to the smallest preorder over $MV(X)$ containing X, which forms an M-algebra which validates the rewrites.

4   Abstract Properties

Properties of rewrite systems are often given via properties of the associated abstract reduction system, e.g. a TRS is confluent iff the rewrites built from the TRS using a countably infinite set of variables form a confluent preorder. If we are going to reason about rewriting using M-rewrite systems, we need a definition of properties in terms of the representing monad M_R. The direct translation is that M_R satisfies P if the representing ARS $M(\mathbb{DN})$ does. However, given the need for variable rewrites in modularity, it is only reasonable to ask the relation $M_R(X)$ to satisfy a property if the relation X does, and thus an alternative definition would be that M_R satisfies P iff it preserves P. We say that property is *monadic* if these two notions coincide:

Definition 15. *Let P be a property of preorders, characterising a subcategory $\mathcal{K}$ of* Pre. *We say P is* monadic *if the following holds: $M(\mathbb{DN}) \in \mathcal{K}$ iff whenever $X \in \mathcal{K}$ then $M_R(X) \in \mathcal{K}$.*

If this definition is sensible it must be satisfied by the standard properties such as confluence and strong normalisation, so we first check if these two are monadic.

4.1   Abstract Confluence

In this section we prove that confluence is monadic according to Def. 15, i.e. a finitary monad $M_R : \mathsf{Pre} \to \mathsf{Pre}$ preserves confluence iff its representing ARS $M_R(\mathrm{DN})$ is confluent. One direction is easy: if M_R preserves confluence, then since DN is discrete and hence trivially confluent, $M_R(\mathrm{DN})$ is also confluent.

Note that in the following cocartesianness is not required , and that previous results [11, 12] restricted to the case where M_R is the representing monad for a TRS and used an explicit inductive construction of this monad.

To prove our result, we use a characterisation of confluence in terms of maps.

Lemma 16. *If X is a finite preorder then it is confluent iff the map $f : X \to \mathrm{D}CX$ has a right adjoint $g : \mathrm{D}CX \to X$, denoted as $f \dashv g$.*

Proof. Let X be confluent. To each connected component of X assign an upper bound of the connected component which exists by confluence and finiteness. This defines a monotone function $g : \mathrm{D}CX \to X$ which satisfies $fg = 1$ and $1 \to gf$, establishing $f \dashv g$. Conversely, given such a right adjoint, it is obvious that each connected component has a minimal element, making X confluent. □

Adjoints like the above allow us to reflect confluence.

Lemma 17. *Let X, Y be preorders, and maps $f : X \to Y$, $g : Y \to X$ such that $1 \to gf$. Then if Y is confluent so is X.*

Proof. Let $b \leftarrow a \to c$ be a span in X with completion $fb \to d \leftarrow fc$ in Y, which has an image $gfb \to gd \leftarrow gfc$ in X. Since $b \to gfb$, and $c \to gfc$ in X, gd is a completion of $b \leftarrow a \to c$, hence X is confluent. □

Lemma 18. *Let $\mathsf{M} = \langle M, \eta, \mu \rangle$ be a monad such that $M\mathrm{DN}$ is confluent, then $M\mathrm{D}X$ is confluent for every finite X.*

Proof. We proceed by assuming that $M\mathrm{D}X$ is inhabited, e.g. by $* \in M\mathrm{D}X$. We can assume this without loss of generality, as given any span $b \leftarrow a \to c$ in $M\mathrm{D}X$ which needs completing, we can take $* = a$.

With X finite, we have $f : X \hookrightarrow \mathbb{N}$ and hence $\mathrm{D}f : \mathrm{D}X \hookrightarrow \mathrm{DN}$, and we can define a map $g : \mathrm{DN} \to M\mathrm{D}X$ by cases so that the following commutes:

$$
\begin{array}{ccc}
\mathrm{D}X & \xrightarrow{\ \mathrm{D}f\ } & \mathrm{DN} \\
& \searrow{\scriptstyle \eta_{\mathrm{D}X}} & \downarrow{\scriptstyle g} \\
& & M\mathrm{D}X
\end{array}
\qquad
g(x) = \begin{cases} \eta_{\mathrm{D}X}(x) & \text{for } x \in \mathrm{D}X \\ * & \text{for } x \notin \mathrm{D}X \end{cases}
$$

Now $g^\flat : M\mathrm{DN} \to M\mathrm{D}X$ (the Kleisli extension of g) is defined as $g^\flat = \mu_{\mathrm{D}X} \cdot Mg$, and hence satisfies $g^\flat \cdot M\mathrm{D}f = \mu_{\mathrm{D}X} \cdot Mg \cdot M\mathrm{D}f = \mu_{\mathrm{D}X} \cdot M\eta_{\mathrm{D}X} = 1$. This allows us to reflect confluence of $M\mathrm{DN}$ along $M\mathrm{D}f : M\mathrm{D}X \to M\mathrm{DN}$ using Lemma 17, making $M\mathrm{D}X$ confluent. □

Lemma 19. *Let M be a functor $M : \mathsf{Pre} \to \mathsf{Pre}$ taking finite discrete preorders to confluent preorders. Then M also takes finite confluent preorders to confluent preorders.*

Proof. Let X be a finite confluent preorder. By Lemma 16 there are two adjoint maps $f \dashv g : X \to \mathrm{D}\mathrm{C}X$, lifting to two adjoint maps $Mf \dashv Mg : MX \to M(\mathrm{D}\mathrm{C}X)$. By Lemma 18, the preorder $M(\mathrm{D}\mathrm{C}X)$ is confluent and hence by Lemma 17, MX is confluent. $\qquad\qquad\square$

The following lemma uses the finite accessibility of Conf, which is a technical property and means that all confluent preorders are finitely generated. This allows us to deduce confluence of infinite preorders from the confluence of their finite suborders. We can then establish the monadicity of confluence as follows:

Lemma 20. *If M is a finitary monad and $M\mathrm{D}\mathrm{N}$ is confluent then MX is confluent whenever X is.*

Proof. By Lemma 18, $M\mathrm{D}X$ is confluent for finite X if $M\mathrm{D}\mathrm{N}$ be confluent. By Lemma 19, MP is confluent for every finite confluent preorder P. Finally, use finite accessibility of Conf to write any confluent P as a filtered colimit $P \cong \operatorname{colim} P_i$ of finite confluent preorders. We can now write

$$MP \cong M \operatorname{colim} P_i \cong \operatorname{colim} M P_i$$

concluding that MP can be written as a filtered colimit of confluent preorders and is therefore confluent. $\qquad\qquad\square$

4.2 Abstract Strong Normalisation

Strong normalisation can be treated in a similar way. We do need a different base category though, as we need to exclude identity rewrites. First a few preliminaries. Let Trans be the category of transitive, but not necessarily reflexive, orders and monotone functions between them, and let WO_f be the full subcategory of well-founded, finitely branching orders. These are the strongly normalising orders that we are interested in. For any $X \in \mathsf{Set}$, we have the discrete order on X which is transitive but not reflexive (and of course in WO_f), which by abuse of language we call $\mathrm{D}X$; and similarly, for the underlying set of a transitive order Y we use $\mathrm{V}Y$. This overloading of notation makes sense, as we are now using a different instance of Definition 10 (with Trans for Pre).

To characterise WO_f algebraically, we use maps into and from ω, the natural numbers ordered by the strictly-greater relation $>$ (or strict reverse inclusion), and their dual ω^{op} with the reversed order, as follows.

Lemma 21. *If $X \in \mathsf{Trans}$, then $X \in \mathsf{WO}_f$ iff there is a map $X \to \omega$. If X is not in WO_f then there is a map $\omega^{op} \to X$.*

Note the finite branching is required to ensure that to each element of a well-founded order we can assign an element of ω (since each element only reduces to a finite number of direct successors), and that this assignment is monotone.

We now show that strong normalisation is monadic. According to Def. 15, we need to show that given an M-rewriting system M_R, $M_R(\text{DN})$ is SN iff the monad M_R preserves strong normalisation, or equivalently, restricts to WO_f. One direction is easy: if M_R preserves SN, then since DN is discrete and hence trivially well-founded, $M_R(\text{DN})$ is also SN. Our aim is to show the converse, and as with confluence, we build up to our result systematically. Note that opposed to confluence, we now need M_R to be cocartesian.

The first result shows that we can show strong normalisation by replacing all variables with a canonical element. For this, let $\mathbf{1}$ be the one-element set. Then, there is exactly one map $!_X : X \to \text{D1}$ in Trans if and only if X is discrete as the map needs to be monotone and D1 has an empty order structure.

Lemma 22. *Let $T : \text{Trans} \to \text{Trans}$ and $T\text{D1} \in \text{WO}_f$, then $TDX \in \text{WO}_f$ for all sets X.*

Proof. By Lemma 21, there is a map $T\text{D1} \to \omega$. Since DX is discrete, there is a map $D!_X : DX \to \text{D1}$. Applying T and composing with the first map gives a map $TDX \to \omega$, hence by Lemma 21, $TDX \in \text{WO}_f$. $\qquad\square$

Now observe that the free lifting $M_\emptyset$ of the monad M is strongly normalising because the only rewrites in $M_\emptyset$ are variable rewrites. The second main step in showing that normal SN implies monadic SN is to show that adding an order structure to the variables does not affect SN:

Lemma 23. *Let M_R be a cocartesian M-rewrite system with $\kappa : M_\emptyset \to M_R$ a cocartesian transformation. If $M_R(\text{D1})$ is SN and X is SN, then $M_R X$ is SN.*

Proof. We know that $M_\emptyset X$ and by Lemma 22 $M_R(DVX)$ are well-founded and hence there are maps $\alpha_1 : M_R(DVX) \to \omega$ and $\alpha_2 : M_\emptyset X \to \omega$. α_1 and α_2 do not form a cone over the square in Diagram (2), i.e. $\alpha_1 \cdot \kappa \neq \alpha_2 \cdot M^\dagger \varepsilon$, so we define new maps $\beta_1 : M_R(DVX) \to \omega$ and $\beta_2 : M_\emptyset X \to \omega$ by

$$\beta_1(t) = \max\{\alpha_1(t), \alpha_2(t)\} \qquad \beta_2(t) = \max\{\alpha_1(t), \alpha_2(t)\}$$

This can be done as all the orders mentioned above have the same carrier. That the β_i are monotone is easily checked and, since they have the same underlying function, we have a cone over the square in diagram (2), and since this is a pushout square (because of the cocartesianness of M_R), we have a map $M_R X \to \omega$ as required.

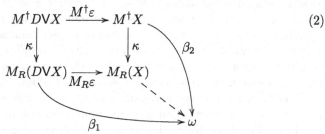

$$\begin{array}{ccc} M^\dagger DVX & \xrightarrow{M^\dagger \varepsilon} & M^\dagger X \\ {\scriptstyle \kappa} \downarrow & & \downarrow {\scriptstyle \kappa} \\ M_R(DVX) & \xrightarrow[M_R\varepsilon]{} & M_R(X) \end{array} \qquad (2)$$

$\qquad\qquad\qquad\qquad\qquad\qquad\qquad\qquad\qquad\qquad\square$

Putting all the pieces together, we get the main result:

Lemma 24. *Let M_R be a cocartesian M-rewrite system. The ARS $M_R(\text{DN})$ is SN iff the monad M_R is SN.*

Summing up this section, we have shown that confluence and SN are monadic, i.e. their usual definition in terms of the representing ARS coincides with them being preserved by the monad representing an M-rewrite system.

5   Abstract Combinators

Modularity deals with combinations of systems, so we are now going to consider the combination of M-rewrite systems. We do so by defining combinators for putting together the representing monads. The appropriate categorical construction here is the *colimit*, but computing the colimit of monads in full generality is a technically involved exercise. Even if we restrict ourselves to the coproduct, corresponding to the disjoint union M-rewrite systems, the construction is very unwieldy, and hence much research has recently focused on developing simpler algorithms which are correct in specific situations. Ideal monads are one such situation which correspond to the idea of layers being non-collapsing.

5.1   Ideal Monads

Intuitively, ideal monads are monads whose variable part can be separated from the non-variable part. Formally:

Definition 25 (Ideal Monad). *A monad $\mathsf{T} = \langle T, \eta, \mu \rangle$ is ideal iff there is a functor T_0 such that $T = \mathsf{Id} + T_0$, the unit is the left injection and there is a natural transformation $\mu_0 : T_0 T \rightarrow T_0$ such that*

$$
\begin{array}{ccc}
T_0 T & \xrightarrow{\ \text{inr}_T\ } & TT \\
{\scriptstyle \mu_0}\big\downarrow & & \big\downarrow{\scriptstyle \mu} \\
T_0 & \xrightarrow[\ \text{inr}\ ]{} & T
\end{array}
$$

where $\text{inr} : T_0 \rightarrow \mathsf{Id} + T_0$ is the right injection into the coproduct.

We write ideal monads in the form $\mathsf{Id} + T_0$ for simplicity (where Id is the identity functor) and leave the restricted form of multiplication μ_0 implicit. A monad morphism $f : \mathsf{Id} + T_0 \rightarrow R$ whose source is an ideal monad has its action on Id forced by the monad laws and is hence of the form $[\eta^R, f_0]$ where $f_0 : T_0 \rightarrow R$. Examples of ideal monads over Set include the term monads T_Σ, the string and ring monads from Examples 5 and 6, and in general any algebraic theory $\mathsf{T}_{\langle \Sigma, E \rangle}$ where both sides of every equation are either variable terms or non-variable terms; hence, a counter-example is the group monad from Example 6.

The fundamental observation behind the construction of the coproduct $R+S$ of two ideal monads $R = \mathsf{Id} + R_0$ and $S = \mathsf{Id} + S_0$ is that $R + S$ should contain

R and S as submonads, and further, that $R + S$ should be closed under the application of R_0 and S_0. Hence, $R + S$ should consist of alternating sequences beginning from R_0 or S_0, and we ask for the least functors satisfying the following mutually recursive equations:

$$T_1 \cong R_0(\mathsf{Id} + T_2) \qquad\qquad T_2 \cong S_0(\mathsf{Id} + T_1).$$

The solution is computed as the least fixpoint of a functor Φ on the product category $(\mathsf{Pre} \to \mathsf{Pre}) \times (\mathsf{Pre} \to \mathsf{Pre})$ (so Φ takes pairs of functors as arguments):

$$\langle T1, T2 \rangle = \mu\Phi \qquad \Phi\langle F, G \rangle = \langle R_0 \cdot (\mathsf{Id} + G), S_0 \cdot (\mathsf{Id} + F) \rangle \qquad (3)$$

To solve the fixpoint equation, note that the functor $c_0 : \mathsf{Pre} \to \mathsf{Pre}$ which constantly returns the initial object is initial in the functor category $\mathsf{Pre} \to \mathsf{Pre}$. We can then use the following standard construction: for a finitary functor $F : \mathcal{C} \to \mathcal{C}$, the least fixpoint μF is given by the colimit of the following chain (if it exists and there is an initial object 0, with $! : 0 \to X$ the unique map out of the initial object):

$$0 \xrightarrow{\;!\;} F0 \xrightarrow{\;F!\;} F^2 0 \xrightarrow{\;F^2!\;} F^3 0 \quad \cdots \qquad (4)$$

Now, intuitively T_1 consists of elements in $R + S$ whose top layer is a non-variable R-layer (captured by the use of R_0) and whose next layers are either variables or a non-variable S layer, etc. In our opinion, this is a very elegant way of capturing the layer structure in the disjoint union of two systems. The following result proves our intuition correct and can be found in [7].

Theorem 26. *The action of the coproduct of ideal monads $\mathsf{Id} + R_0$ and $\mathsf{Id} + S_0$ is the functor $T = \mathsf{Id} + (T_1 + T_2)$, where T_1 and T_2 are defined as in (3).*

The central result of this paper is that those rewrite systems whose representing monad is ideal have good modularity properties and, further, that these are actually rather easy to derive. Note that Theorem 26 holds for all ideal monads, i.e. all ideal M-rewrite systems, not only term-generated ones. We now prove the central theorem from which all our modularity results can be uniformly derived.

Theorem 27. *Let P be a monadic property represented by a subcategory $\mathcal{K}$ of Pre. If $\mathcal{K}$ has coproducts, an initial object and ω-colimits, then P is modular for the disjoint union of ideal M-rewriting systems.*

Proof. Let R and S be ideal M-rewriting systems satisfying P. To show that their disjoint union has the property P, we have to show that, given $X \in \mathcal{K}$, $R + S(X) = X + T_1(X) + T_2(X)$ is in $\mathcal{K}$.

By Theorem 26, $T_1(X)$ and $T_2 X$ are given by the initial fixpoint of Φ in (3) at X; i.e. the colimit of the chain (5). We know that both R_0 and S_0 preserve

$$\langle 0, 0 \rangle \xrightarrow{\;!\;} \langle R_0 X, S_0 X \rangle \to \langle R_0(X + S_0 X), S_0(X + R_0 X) \rangle \to \quad \cdots \qquad (5)$$

$\mathcal{K}$, and since $\mathcal{K}$ has coproducts and an initial object, all objects of the chain (5) are in $\mathcal{K}$, and since $\mathcal{K}$ has ω-colimits, so is the fixpoint, i.e. $T_1(X)$ and $T_2(X)$. With $\mathcal{K}$ having coproducts, we get that $X + T_1(X) + T_2(X) \in \mathcal{K}$. $\qquad\square$

We finish this section by using Theorem 27 to uniformly derive a number of modularity results.

Example 28 (Confluence of Non-Collapsing TRSs). Take $\mathcal{K}$ to be the full subcategory Conf of Pre whose objects are confluent orders. A TRS Θ is confluent iff $T_\Theta(DN)$ is confluent, where T_Θ is its representing monad (Ex. 12). Using Lemma 20, this is equivalent to T_Θ being confluent.

It remains to show that confluence satisfies the preconditions of Theorem 27. Clearly, the empty preorder (the initial object) is confluent, and the disjoint union of two confluent preorders is confluent. Further, given an ω-chain of confluent preorders, their colimit (i.e. the least upper bound) will be confluent as well (this is the finite accessibility of Conf mentioned above), allowing us to conclude the result.

Example 29 (Strong Normalisation for Non-Collapsing TRS). In this example, we have to change the base category from Pre to Trans (as in Sect. 4.2), and let $\mathcal{K}$ to be the category WO_f of finitely branching well-founded orders. As in the previous example, we can use Lemma 24 to show that termination of a TRS Θ and termination of the monad T_Θ coincide.

It remains to show that strong normalisation satisfies the preconditions of Theorem 27. The empty relation is SN. The disjoint union preserves SN. WO_f actually fails to have all filtered colimits, but fortunately it does have colimits of chains which preserve the normalisation rank, as is the case for the chain in (5). Thus, Theorem 27 applies.

Example 30 (Adding Equations; Modularity for Equational TRSs). Let R be a confluent non-collapsing TRS. Assume we want to add to R a fresh associative operator $\otimes$ and prove the resulting system remains confluent.

The monad $T_{\langle\otimes,E\rangle}$ given by the algebraic theory with one operation $\otimes$ and the equation E stating associativity of $\otimes$ is confluent (trivially, as it contains no rewrites); note that the base category of this monad is Pre, not Set (in fact, we treat the algebraic theory as an equational rewrite system without rewrites). We have already established that confluent preorders satisfy the preconditions of Theorem 27. With T_R the monad representing the TRS R, can easily deduce that $T_R + T_{\langle\otimes,E\rangle}$ satisfies confluence, hence $R + \langle\otimes, E\rangle$ is confluent as well.

This can be generalised to two arbitrary, non-collapsing equational term rewriting systems: if both are confluent or SN, so will be their disjoint union.

6   Conclusion and Future Work

We have demonstrated that there is indeed a theory of modularity which abstracts from the specific notion of rewriting, property and combination under consideration. Moreover, we believe that our use of monads has helped to establish these results in an elegant and straightforward way. Underlying this is the simple representation of the layer structure as the interleaving of monads and the use of variable rewrites to model rewrites in sublayers. As mentioned in the introduction, the point about these examples is not that they are the most general results for a specific modularity problem, but rather that we have a uniform principle that works in a variety of different situations.

In the above, we have used **Pre** as the base category, and switched to **Trans** when considering strong normalisation. The exact way to model this would have been define a rewrite system as parameterised over a base category $\mathcal{C}$, which has to satisfy certain properties, but we felt this would make the exposition more categorical and less rewriting. We have also omitted rewriting of infinite terms as the corresponding monads are not finitary. Our methodology still works, but requires us to work at a higher rank (with transfinite constructions), the technicalities of which we felt would distract from the concrete term rewriting contribution of the present paper.

The applications to graph rewriting need to be examined more closely. Graph rewriting has the dual modularity results then normal term rewriting (i.e. confluence is not modular but SN is). We can model term graphs with monads [4, 5], but the precise relation of the monadic properties to the properties of term graph rewriting systems is not clear.

We would like to comment on the limitations of this work. Higher-order systems with variable binding are essentially not covered at all, because although this can be modelled in the monad framework [6], higher-order systems are not ideal monads (the reason is that free variables can be captured when building a new layer).

In future work we wish to make these ideas accessible to a wider audience by developing many more different examples and applications. We also plan to extend the methodology to other methods of combining rewrite systems than the disjoint union, in particular modelling constructor sharing systems, where first tentative steps have already been taken.

References

1. J. Adámek and J. Rosický. *Locally Presentable and Accessible Categories.* London Mathematical Society Lecture Note Series 189. Cambridge University Press, 1994.
2. M. Barr and C. Wells. *Toposes, Triples and Theories.* Grundlehren der mathematischen Wissenschaften 278. Springer Verlag, 1985.
3. N. Ghani and C. Lüth. Rewriting via coinserters. *Nordic Journal of Computing*, 10:290– 312, 2004.
4. N. Ghani, C. Lüth, and F. de Marchi. Solving algebraic equations using coalgebra. *Journal of Theoretical Informatics and Applications*, 37:301–314, 2003.
5. N. Ghani, C. Lüth, and F. de Marchi. Monads of coalgebras: Rational terms and term graphs. To appear in *Mathematical Structures in Computer Science*.
6. N. Ghani and T. Uustalu. Explicit substitutions and higher-order syntax (extended abstract). In F. Honsell, M. Miculan, and A. Momigliano, editors, *Proc. of 2nd ACM SIGPLAN Wksh. on Mechanized Reasoning about Languages with Variable Binding, MERLIN'03*, pages 1–7. ACM Press, New York, 2003.
7. N. Ghani and T. Uustalu. Coproducts of ideal monads. *Journal of Theoretical Informatics and Applications*, 38:321–342, 2004.
8. M. Hamana. Term rewriting with variable binding: an initial algebra approach. In *Proceedings of the 5th ACM SIGPLAN international conference on Principles and practice of declaritive programming*, pages 148–159. ACM Press, 2003.

9. J. W. Klop, A. Middeldorp, Y. Toyama, and R. de Vrijer. A simplified proof of Toyama's theorem. *Information Processing Letters*, 49:101–109, 1994.
10. M. Kurihara and I. Kaji. Modular term rewriting systems and the termination. *Information Processing Letters*, 34(1):1–4, Feb. 1990.
11. C. Lüth. *Categorical Term Rewriting: Monads and Modularity*. PhD thesis, University of Edinburgh, 1998.
12. C. Lüth and N. Ghani. Monads and modular term rewriting. In *Category Theory in Computer Science CTCS'97*, LNCS 1290, pages 69–86. Springer Verlag, 1997.
13. C. Lüth and N. Ghani. Monads and modularity. In A. Armando, editor, *Frontiers of Combining Systems FroCos 2002, 4th International Workshop*, LNAI 2309, pages 18–32. Springer Verlag, 2002.
14. S. MacLane. *Categories for the Working Mathematician, Graduate Texts in Mathematics* 5. Springer Verlag, 1971.
15. E. G. Manes. *Algebraic Theories, Graduate Texts in Mathematics* 26. Springer Verlag, 1976.
16. M. Marchiori. Modularity of completeness revisited. In J. Hsiang, editor, *Proceedings of the 6th International Conference on Rewriting Techniques and Applications RTA'95*, LNCS 914, pages 2–10, Springer Verlag, 1995.
17. A. Middeldorp. *Modular Properties of Term Rewriting Systems*. PhD thesis, Vrije Universiteit te Amsterdam, 1990.
18. A. Middeldorp and Y. Toyama. Completeness of constructor systems. Technical Report CS-R9058, Centrum voor Wiskunde en Informatica, Amsterdam, 1990.
19. E. Ohlebusch. On the modularity of termination of term rewriting systems. *Theoretical Computer Science*, 136:333–360, 1994.
20. E. Ohlenbusch. On the modularity of confluence of constructor-sharing term rewriting systems. In S. Tison, editor, *Trees in Algebra and Programming – CAAP 94*, LNCS 787. Springer Verlag, 1994.
21. E. Robinson. Variations on algebra: monadicity and generalisations of equational theories. Manuscript. Available at
http://www.dcs.qmw.ac.uk/~edmundr/pubs/algebras/algebras.ps, 1994.
22. M. Rusinowitch. On the termination of the direct sum of term-rewriting systems. *Information Processing Letters*, 26(2):65–70, 1987.
23. Y. Toyama. On the Church-Rosser property for the direct sum of term rewriting systems. *Journal of the ACM*, 34(1):128–143, 1987.

Union of Equational Theories:
An Algebraic Approach[*]

Piotr Hoffman

Warsaw University, Institute of Informatics,
Banacha 2, 02-097 Warszawa, Poland
piotrek@mimuw.edu.pl

Abstract. We consider the well-known problem of deciding the union of decidable equational theories. We focus on monadic theories, i.e., theories over signatures with unary function symbols only. The equivalence of the category of monadic equational theories and the category of monoids is used. This equivalence facilitates a translation of the considered decidability problem into the word problem in the pushout of monoids which themselves have decidable word problems. Using monoids, existing results on the union of theories are then restated and proved in a succint way. The idea is then analyzed of first guaranteeing that the union is a "jointly conservative" extension and then using this property to show decidability of the union. It is shown that "joint conservativity" is equivalent to the corresponding monoid amalgam being embeddable; this allows one to apply results from amalgamation theory to this problem. Then we prove that using this property to show decidability is a more difficult matter: it turns out that even if this property and some additional conditions hold, the problem remains undecidable.

Introduction

Assume two equational theories E_1 and E_2 are given. If one has procedures for deciding these theories, then the natural question arises of whether they can be extended to a procedure deciding the union theory, $E_1 \cup E_2$. This question is practically important, since in principle it is equivalent to the question whether theorem provers for E_1 and E_2 can be joined into a theorem prover for $E_1 \cup E_2$.

The general answer is, quite obviously, in the negative. For a long time it has been known that if the signatures Σ_1 and Σ_2, over which E_1 and E_2 are built, are disjoint, then the answer is in the affirmative [Pig74]. The problem is usually only discussed in the case of E_1 and E_2 being conservative extensions of an "intersection theory" E over the signature $\Sigma = \Sigma_1 \cap \Sigma_2$. Recently it has been proven that in this case, if one additionally assumes that E_1 and E_2 are in a certain sense *effectively constructible* over Σ, then the answer is positive. This has been proven by Fiorentini and Ghilardi [FG03] and by Baader and

---

[*] This work has been partially supported by European Union project AGILE IST-2001-32747.

J. Giesl (Ed.): RTA 2005, LNCS 3467, pp. 61–73, 2005.

Tinelli [BT02]. Both papers are quite complex, which is to a large extent due to them dealing with signatures in which symbols of arbitrary arity may exist.

In this paper we only deal with *monadic theories*, i.e., theories in which all function symbols are unary. The category of such equational theories turns out to be equivalent to the category of monoids (Theorem 1). Showing this equivalence and transposing the considered problem from the level of theories to the level of monoids is the topic of Sections 1 and 2. It should be noted that the approach of Fiorentini and Ghilardi is, in fact, very similar – however, since they do not restrict themselves to monadic theories, they have to use the much more complex notion of category with products and product-preserving functors.

Moving to the category of monoids allows us to define the property of effective constructability is a concise way (see the definition of *base* in Section 3). The proof of the result of Fiorentini/Ghilardi and Baader/Tinelli for the monadic case also turns out very simple (Theorem 2). The main benefit of Section 3 is thus that known results are presented in such a way that they are easy to understand; this should facilitate future research on this topic. As an example, a simple extension of the known result to *cobases* is proposed (Corollary 1).

We then consider an idea stemming from [BT02], namely of dividing the considered problem into two stages. In the first stage, two properties are enforced: first, that the union theory conservatively extends E_1 and E_2 (conservativity property); second, that if a Σ_1-term t_1 is equivalent in the union theory to a Σ_2-term t_2, then there exists a Σ-term u such that $t_1 = u$ in E_1 and $u = t_2$ in E_2 (interpolation property). In the second stage, one would use the conservativity and interpolation properties to decide the union theory (we call this "lifting").

In Section 4 the above idea is formalized and it is shown that, in the monoid case, the conservativity and interpolation properties are very useful and natural. In particular, it is shown that the theories E_1 and E_2 conservatively extending E are really equivalent to a *monoid amalgam*, an object well known in semigroup theory [How96a, Hig92]. Moreover, the conservativity property is equivalent to the amalgam being *weakly embeddable*, while adding the interpolation property is equivalent to it being *embeddable*. These equivalences make it possible to use results from amalgamation theory concerning the embeddability of amalgams to guarantee or prove that conservativity and interpolation hold. In particular, we show that if the theory E, or rather the corresponding monoid, is unitary or quasiunitary in E_1 and E_2, then both properties hold. We also show that in certain cases the monoid corresponding to E can, by itself, guarantee that these poperties hold, in particular if it is a group or inverse semigroup.

Section 5 is concerned with stage two of the program presented above, i.e., with lifting the conservativity and interpolation to a solution to our main problem. The results of this section are negative, showing that conservativity and interpolation, even if augmented by one of the theories (monoids) having a base, may still fail to guarantee the decidability of the union theory (Theorem 3).

This negative result suggests that one should search for some additional condition which would guarantee that the lifting phase is indeed correct. The result is also interesting even if one abstracts away from the proposed two-stage program, since it shows that even a very slight weakening of the assumption of both theories having bases moves the problem into the realm of undecidability.

Finally, Section 6 contains conclusions and suggestions on directions for future work. In particular, the problem of extending the techniques and results of this paper to arbitrary arities and to multi-sorted logic is considered.

1 Monadic Theories vs Monoids

A *signature* Σ defines a set of *function symbols* together with their *arities*. By $T_\Sigma(X)$ the set of *terms* over Σ with variables X is denoted. A Σ-*equation* has the form $t_1 = t_2$, where $t_1, t_2 \in T_\Sigma(X)$. A Σ-*theory* is any set of Σ-equations; a *theory* is a pair (Σ, E), where E is a Σ-theory. A Σ-*model* consists of a non-empty set M and, for any function symbol f with arity n in Σ, of an interpretation of f, that is, of a function $M(f) : M^n \to M$. If $X = \{x_1, \ldots, x_n\}$ is an n-element set and t is a Σ-term over X, then any Σ-model M naturally defines a function $M(t) : M^n \to M$, called the *interpretation* of t. An equation $t_1 = t_2$ *holds* in M if $M(t_1) = M(t_2)$. A Σ-model M is a *model* of a Σ-theory E if all equations in E hold in M. An equation is a *consequence* of E if it holds in all models of E.

A *signature morphism* $\sigma : \Sigma \to \Delta$ is a map taking any function symbol f of arity n in Σ to a Δ-term t over $\{x_1, \ldots, x_n\}$. Signature morphisms compose naturally via term substitution. Moreover, the morphism $\sigma : \Sigma \to \Sigma$ taking any f with arity n to the term $f(x_1, \ldots, x_n)$ is an identity. Thus, signatures and signature morphisms form a category. Note that any term and, hence, any equation can be translated via a signature morphism. A *theory morphism* $\sigma :$ $(\Sigma, E) \to (\Delta, F)$ is a signature morphism $\sigma : \Sigma \to \Delta$ such that the translated equations in $\sigma(E)$ are all consequences of F. The composition of two theory morphisms is a theory morphism and identity morphisms are theory morphisms. Thus, theories and theory morphisms form a category.

The map taking any signature or any theory to the class of its models is, actually, functorial, since if $\sigma : (\Sigma, E) \to (\Delta, F)$ is a theory morphism and N is a Δ-model of F, then one can define the Σ-model $M = N|_\sigma$, called the σ-*reduct* of N, and M will be a model of E; this map is functorial, since the reduct along identity is identity and since reducts compose (contravariantly).

We focus on *monadic* theories. A signature is monadic if all its function symbols are unary. A term is monadic if it is built solely of unary symbols and uses only the variable "x". An equation is monadic if it is built of monadic terms. A theory is monadic if it is a theory over a monadic signature and consists only of monadic equations.

Let (Σ, E) be a monadic theory. Consider the set of all monadic Σ-terms and let $t_1 \equiv_E t_2$ hold iff $t_1 = t_2$ is a consequence of E. The set of equivalence classes w.r.t. $\equiv_E$ forms a monoid if one defines 1 to be the equivalence class of x and $[t_1]_{\equiv_E}[t_2]_{\equiv_E} = [t_2[t_1/x]]_{\equiv_E}$. This monoid is denoted by $\alpha(\Sigma, E)$. In fact, this operation is functorial, since if $\sigma : (\Sigma, E) \to (\Delta, F)$ is a theory morphism, then the map defined by $\alpha(\sigma)([t]_{\equiv_E}) = [\sigma(t)]_{\equiv_F}$ is a homomorphism and, moreover, this correspondence preserves identities and composition. Thus α is a functor from the category of monadic theories to the category of monoids.

Similarly, any monoid S defines a monadic theory $\beta(S) = (\Sigma, E)$, where function symbols in Σ are all elements of $S \setminus \{1\}$ and, for any $s_1, s_2 \in S \setminus \{1\}$

and $s \in S$ with $s_1 s_2 = s$ holding in S, E contains the equation $s_2(s_1(x)) = s(x)$ if $s \neq 1$, or the equation $s_2(s_1(x)) = x$ if $s = 1$. This map again extends to a functor, since any homomorphism $h : S \to T$ defines a theory morphism $\beta(h)$ taking any function symbol in $\beta(S)$, i.e., any element $s \in S \setminus \{1\}$, to the term x if $h(s) = 1$ or to the term consisting of a single function symbol $h(s)$ in $\beta(T)$ if $h(s) \neq 1$. Thus β is a functor from the category of monoids to the category of monadic theories.

We now have:

Theorem 1. *The functors α and β form an equivalence of the category of monadic theories and the category of monoids.*

Proof. We only define the natural isomorphisms $\gamma : \alpha; \beta \to \mathrm{id}$ and $\delta : \beta; \alpha \to \mathrm{id}$.

For any theory $T = (\Sigma, E)$, γ_T takes any function symbol K over $\beta(\alpha(T))$, where K is a term t over Σ, to the term t over Σ. The inverse transformation γ_T^{-1} takes any function symbol f in Σ to the term $K(x)$ built of a single function symbol K in $\beta(\alpha(T))$, where K is the term $f(x)$ over Σ.

Now take any monoid S and let $(\Sigma, E) = \beta(S)$. Then δ_S takes any element K of $\alpha(\beta(S))$, i.e., any term $K = s_n(\cdots s_1(x) \cdots)$, where $s_1, \ldots, s_n \in S \setminus \{1\}$, $n \geq 0$, to the element $s_1 \ldots s_n$ of S. The inverse transformation δ_S^{-1} takes any $s \in S$ to an element K of $\alpha(\beta(S))$, namely to the term $s(x)$ if $s \neq 1$ or to the term x if $s = 1$.

It is easy to check that γ and δ are indeed natural isomorphisms. $\square$

This equivalence allows one to translate notions from one category to notions from the other.

A signature morphism $\gamma : (\Sigma, E) \to (\Delta, F)$ is said to be *conservative*, if $\gamma(t_1) = \gamma(t_2)$ being a consequence of F implies that $t_1 = t_2$ is a consequence of E, for all Σ-terms t_1 and t_2; the equivalent notion is a homomorphism $g : S \to T$ which is one-one, i.e., a monomorphism.

The notion equivalent to checking that an equation $t_1 = t_2$ is a consequence of E is solving the *word problem* in S, i.e., checking whether $s_1 \ldots s_n = s_1' \ldots s_k'$, for $s_1, \ldots, s_n, s_1', \ldots, s_k' \in S$. Thus the decidability of E is equivalent to the word problem in S being decidable.

It should be noted that Fiorentini and Ghilardi also use an equivalence similar to the one defined in this section, but because they cover all, not only monadic, theories, instead of monoids they use categories with products and product-preserving functors.

2 Union of Monadic Theories vs Amalgam of Monoids

Let (Σ_1, E_1) and (Σ_2, E_2) be decidable monadic theories conservatively extending a theory (Σ, E), where $\Sigma = \Sigma_1 \cap \Sigma_2$. The equivalence constructed in the previous section gives us what, in monoid theory, is called an *amalgam*, namely monoids S and T with decidable word problems and monomorphisms $i : U \to S$ and $j : U \to T$; here S corresponds to E_1, T to E_2 and U to E. Because monomorphisms in the category of monoids are simply one-to-one homomorphisms, one can without loss of generality assume that i and j are inclusions, i.e., that $U = S \cap T$. In the sequel, an amalgam is written as $[U \subseteq_{i,j} S, T]$ or, if i and j are inclusions and $U = S \cap T$, just as $[U \subseteq S, T]$.

We are interested in deciding the theory $(\Sigma_1 \cup \Sigma_2, E_1 \cup E_2)$. Note that this theory is, in fact, the pushout of the inclusions of E into E_1 and E_2 in the category of monadic theories. Therefore, due to the equivalence of categories, our problem turns into the problem of solving the word problem in the pushout P of the span of monomorphisms $i : U \to S$ and $j : U \to T$; this pushout is given by certain homomorphisms $\mu : S \to P$ and $\nu : T \to P$ satisfying $i; \mu = j; \nu$. We call P, μ and ν the *pushout of the amalgam*.

The above considerations show that deciding the union of decidable monadic theories is equivalent to the following problem: given a monoid amalgam $[U \subseteq_{i,j} S, T]$ with S and T having decidable word problems, solve the word problem in the pushout P of the amalgam.

From now on we therefore assume that we have monoids S and T with decidable word problems. We write $U = S \cap T$, and $i : U \to S$ and $j : U \to T$ are the inclusions. Finally, $\mu : S \to P$ and $\nu : T \to P$ form the pushout; note that the pushout always exists (the category of monoids is cocomplete). We concentrate on solving the word problem in P. Note that, obviously, μ and ν are jointly epimorphic; more specifically, P may be seen as built of all sequences of elements of S and T, quotiented by an appropriate equivalence $\sim$. This equivalence is the least congruence satisfying $\langle 1 \rangle \sim \langle \rangle$ and, for all $s_1, s_2 \in S$ and $t_1, t_2 \in T$, $\langle s_1, s_2 \rangle \sim \langle s_1 s_2 \rangle$ and $\langle t_1, t_2 \rangle \sim \langle t_1 t_2 \rangle$.

3 The Solution via Bases

Solutions to the considered problem have been presented independently by Fiorentini/Ghilardi [FG03] and Baader/Tinelli [BT02]. In both solutions, virtually the same assumptions are made, cf. Prop. 7.10 of [BT02]. The main assumption is what Fiorentini and Ghilardi call *(effective) constructibility* of E_1 over E and of E_2 over E. An algebraic characterization of this notion may be found in Prop. 10.4 of [FG03]. The corresponding notion in the world of monoids is the notion of *(computable) base* (this terminology is borrowed from [BT02]).

Assume U is a submonoid of a monoid S. A *U-base* of S is a set $G \subseteq S$ such that $1 \in G$ and, for any $s \in S$, there exists a unique pair $(g, u) \in G \times U$ such that $s = gu$. The base is *computable* if the pair (g, u) may be computed by a recursive function.

It should be noted that, in fact, if S has a decidable word problem and G is a U-base of S, then the base G is computable; this is because one can simply enumerate all pairs, knowing that exactly one must have the needed property.

Observe that if G is U-base, then $G \cap U = \{1\}$. For if $g \in G \cap U$, then the pairs $(g, 1)$ and $(1, g)$ must be equal and so $g = 1$.

Note also that any monoid S has a computable $\{1\}$-base, where $\{1\}$ is the trivial monoid; one has to take $G = S$. If the original signatures Σ_1 and Σ_2 were disjoint, then U is indeed the trivial monoid. Hence, in this case both bases always exist.

For monadic theories, the results of Fiorentini/Ghilardi and Baader/Tinelli may thus, in algebraic terms, be expressed as follows:

Theorem 2. *Assume $[U \subseteq S, T]$ is an amalgam and the monoids S and T both have U-bases and decidable word problems. Then the pushout P of the amalgam has a decidable word problem.*

Proof. Assume G_S and G_T are the respective bases. Consider the set Q of all non-empty sequences of elements of G_S, G_T and U. Let $\rightarrow$ be a reduction relation on these sequences defined by:

- $x, g \rightarrow g', u'$ if $g \neq 1$, $xg = g'u'$ and $g' \neq 1$, where $x \in S$, $g, g' \in G_S$, $u' \in U$, or symmetrically for T,
- $x, g \rightarrow u'$ if $g \neq 1$, $xg = u'$, where $x \in S$, $g \in G_S$ and $u' \in U$, or symmetrically for T,
- $u, u' \rightarrow uu'$ if $u, u' \in U$,
- $g, 1 \rightarrow 1$ if $g \neq 1$, $g \in G_S$, or symmetrically for T.

It is obvious that $\rightarrow$ is terminating. It is also confluent: if two reductions overlap, then we must have elements x, x', x'' such that x, x' reduces and x', x'' reduces. The following cases thus arise:

1. x, x', x'' all come from S or all come from T; then both reductions may be performed in any order,
2. otherwise, assume $x' \neq 1$; since x, x' reduces and since it cannot be the case that $x, x' \in U$ (because we would then be in case 1), we must have that $x' \in G_S \setminus U$ or symmetrically for T, because if $x' \in U \setminus \{1\}$ and $x \notin U$, then x, x' is not a redex; but since x', x'' reduces, we must then have $x'' \in S$, or symmetrically for T, which would mean that we are actually in case 1; therefore $x' = 1$, so both reductions may be performed in any order.

Consider two sequences q, q' from Q to be equivalent, written $q \sim q'$, if they have the same normal formal form with respect to $\rightarrow$. Let P be the quotient $Q/\sim$; then P with concatenation modulo $\sim$ and with $[1]_\sim$ as identity is a monoid. Define $\mu : S \rightarrow P$ by $\mu(s) = [g, u]_\sim$, where $gu = s$, $(g, u) \in G_S \times U$; define $\nu : T \rightarrow P$ analogically. We claim that μ must then be a homomorphism. Of course, $\mu(1) = [1]_\sim$. Also, if $s_1 = g_1 u_1$ and $s_2 = g_2 u_2$, with $g_1, g_2 \in G_S$, $u_1, u_2 \in U$, and if $s_1 s_2 = gu$ with $g \in G_S$, $u \in U$, then $\mu(s_1)\mu(s_2) = [g_1, u_1, g_2, u_2]_\sim = [g, u]_\sim = \mu(s_1 s_2)$. This is due to there existing $g_0 \in G_S$, $u_0 \in U$ such that $u_1 g_2 = g_0 u_0$; then $g_1, u_1, g_2, u_2 \rightarrow g_1, g_0, u_0, u_2$ (unless $g_0 = 1$); there further exist $g_0' \in G_S$ and $u_0' \in U$ such that $g_1 g_0 = g_0' u_0'$; then $g_1, g_0, u_0, u_2 \rightarrow g_0', u_0', u_0, u_2 \rightarrow g_0', u_0' u_0, u_2 \rightarrow g_0', u_0' u_0 u_2$ (unless $g_0' = 1$); but since $g_0' u_0' u_0 u_2 = g_1 u_1 g_2 u_2 = s_1 s_2 = gu$, we must have $g = g_0'$ and $u = u_0' u_0 u_2$, which completes the proof of the claim. An analogical argument shows that $\nu : T \rightarrow P$ is a homomorphism as well. Moreover, it is clear that $\mu(u) = [1, u]_\sim = \nu(u)$ and so $i; \mu = j; \nu$. Thus P, μ and ν form a cocone. Moreover, for any cocone $\mu' : S \rightarrow P'$, $\nu' : T \rightarrow P'$, there exists a unique factorizing $\eta : P \rightarrow P'$ given by $\eta([x_1, \ldots, x_n]_\sim) = h(x_1) \cdot h(x_2) \cdots h(x_n)$, where $h(x_i) = \mu'(x_i)$ for $x_i \in S$ and $h(x_i) = \nu'(x_i)$ for $x_i \in T$ and where "$\cdot$" denotes multiplication in P'. This proves that P is a pushout.

What remains to be shown is that P has a decidable word problem. But this is obvious, since one may simply compute the normal forms of both sides of an equation and then check whether they are equal.   □

The algorithm for deciding the word problem in the pushout is thus straightforward: one simply has to repeatedly apply the relation $\rightarrow$, obtaining a normal form.

In the monadic case, a trivial generalization of this algorithm may be considered. If U is a submonoid of S, then a U-*cobase* of S is a set $G \subseteq S$ such that $1 \in G$ and, for any $s \in S$, there exists a unique pair $(u, g) \in U \times G$ such that $s = ug$. It is obvious that the following corollary holds:

Corollary 1. *Assume* $[U \subseteq S, T]$ *is an amalgam and the monoids S and T both have U-cobases and decidable word problems. Then the pushout P of the amalgam has a decidable word problem.*   □

It is an interesting question whether the notion of cobase can be generalized to non-monadic theories, or whether the use of cobases is excluded by function symbols with arities other than 1.

4   Conservativity and Interpolation vs Embeddability of Amalgams

In the paper [BT02], before the correctness of the whole proposed procedure is considered, two important lemmas are proved (Prop. 4.14 and Lemma 4.18). These lemmas show that under certain assumptions the union theory has the conservativity and interpolation properties.

The union theory $E_1 \cup E_2$ is said to have the *conservativity* property if it is a conservative extension of E_1 and a conservative extension of E_2. In other words, this property means that if t_1 and t_2 are terms over Σ_1, then the equality $t_1 = t_2$ is a consequence of $E_1 \cup E_2$ iff it is a consequence of E_1 alone, and that an analogous fact holds for Σ_2 and E_2.

The union theory is said to have the *interpolation property* if, for any term t_1 over Σ_1 and t_2 over Σ_2, the equation $t_1 = t_2$ is a consequence of $E_1 \cup E_2$ iff there exists a term t over $\Sigma_1 \cap \Sigma_2$, ·called the *interpolant*, such that $t_1 = t$ is a consequence of E_1 and $t = t_2$ is a consequence of E_2.

Together, the conservativity and interpolation properties mean that the equivalence relation defining the pushout P is trivial on sequences of length 1.

It is a natural idea to try to find a solution of the word problem in the pushout P in two stages: first to guarantee the conservativity and interpolation properties, and then, exploiting them, solve the problem. In this section the first stage is dealt with. In particular, we show that these properties correspond to classical notions of monoid amalgamation theory. This gives us many tools for checking whether these properties hold for a given union of theories. In the next section, the problem of "lifting" the solution for length 1 sequences to sequences of arbitrary length is considered, i.e., the problem of whether these properties can help us solve the word problem in P.

Let $\mu : S \to P$ and $\nu : T \to P$ be the pushout of the amalgam $[U \subseteq_{i,j} S, T]$. The word problem in P for sequences of length one is simply the problem of checking whether:

- $\mu(s) = \mu(s')$, for $s, s' \in S$,
- $\mu(s) = \nu(t)$, for $s \in S$, $t \in T$,
- $\nu(t) = \nu(t')$, for $t, t' \in T$.

It is clear that on the level of terms these problems are equivalent to checking whether the following equalities are consequences of the union theory $E_1 \cup E_2$:

- $s = s'$, for Σ_1-terms s and s',
- $s = t$, for a Σ_1-term s and a Σ_2-term t,
- $t = t'$, for Σ_2-terms t and t'.

In monoid theory, the problem of an amalgam $[U \subseteq_{i,j} S, T]$ being *embeddable* is considered, i.e., the question whether there exists a monoid Z such that $S, T \subseteq Z$, perhaps with the inclusions replaced by some homomorphisms. It can be easily shown that an amalgam is embeddable iff it is embeddable in its pushout P. This leads to the following definition:

- an amalgam is *weakly embeddable* if $\mu : S \to P$ and $\nu : T \to P$ are monomorphisms,
- an amalgam is *embeddable* if it is weakly embeddable and $\mu(S) \cap \nu(T) = \mu(i(U))$.

Note that, of course, $\mu(i(U)) = \nu(j(U))$.

Now, it is clear that an amalgam is weakly embeddable if and only if equalities of the form $\mu(s) = \mu(s')$ or $\nu(t) = \nu(t')$ hold in P iff $s = s'$ in S or $t = t'$ in T. On the level of terms this means that to check such an equality with respect to the theory $E_1 \cup E_2$ it suffices to check it with respect to the theories E_1 or E_2, respectively. In other words, the inclusions $E_1 \subseteq E_1 \cup E_2$ and $E_2 \subseteq E_1 \cup E_2$ are conservative. Thus:

Proposition 1. *The union theory $E_1 \cup E_2$ has the conservativity property iff the corresponding monoid amalgam is weakly embeddable.* $\square$

It is also clear that an amalgam is embeddable if and only if it is weakly embeddable and, additionally, any equality $\mu(s) = \nu(t)$ holds in P iff $s, t \in U$ and $s = t$. On the level of terms embeddablity thus means that, in addition to weak embeddability, for all terms s over Σ_1 and t over Σ_2, the equation $s = t$ is a consequence of $E_1 \cup E_2$ iff there exists a term u over $\Sigma_1 \cap \Sigma_2$ such that $s = u$ is a consequence of E_1 and $u = t$ is a consequence of E_2. Thus:

Proposition 2. *The union theory $E_1 \cup E_2$ has the conservativity and interpolation properties iff the corresponding monoid amalgam is embeddable.* $\square$

The following proposition simplifies checking that the equations considered at the beginning of this section hold. In particular, it may be used to prove the embeddability of an amalgam.

Proposition 3. *For any amalgam* $[U \subseteq_{i,j} S, T]$*, if* $\sigma : S \to T$ *is a homomorphism satisfying* $i; \sigma = j$*, then:*

- $\mu(s) = \nu(t)$ *implies* $\sigma(s) = t$,
- $\nu(t) = \nu(t')$ *iff* $t = t'$. $\square$

In particular, if $\sigma : S \to T$ is an isomorphism, then the amalgam is weakly embeddable. Also, $\mu(s) = \nu(t)$ then implies that s and t are connected via the isomorphism, i.e., that $\sigma(s) = t$ and $\sigma^{-1}(t) = s$. It should be noted that in this case, that is if the pairs $U \subseteq S$ and $U \subseteq T$ are isomorphic, checking $\mu(s) = \nu(t)$ may be performed by considering, in the pushout P, only sequences of length 2; this is a consequence of Isbell's Zigzag Theorem (see [Isb66, How96b]). In this case, the object in which equality should be checked is called the *tensor* and denoted by $S \otimes_U T$.

One can try to guarantee the embeddability of an amalgam by enforcing certain restrictions on the pairs $U \subseteq S$ and $U \subseteq T$. For instance, $U \subseteq S$ is said to be *unitary* if, for all $u \in U$ and $s \in S$, $us \in U$ implies $s \in U$ and $su \in U$ implies $s \in U$. On the theory level, this means that composing a "new" term with an "old" one must give something provably equal to an "old" term. The connection with bases used in Section 3 is as follows:

Proposition 4. *If S has a U-base and a U-cobase, then $U \subseteq S$ is unitary.*

Proof. Assume S has a U-base G and $su \in U$, with $s \in S$, $u \in U$. Then $s = gu'$, for some $g \in G$, $u' \in U$. Thus $su = (gu')u = g(u'u)$ and $su = 1(su)$. Hence, by the property of the base, $g = 1$ and so $s = u' \in U$. The second part of the proof is analogous. $\square$

It has been shown that if $U \subseteq S$ and $U \subseteq T$ are unitary, then the amalgam is embeddable (see [How96b]). This gives one method for guaranteeing that the amalgam is embeddable.

Actually, even more has been shown (we report these results after [How96b], where [Ren86a, Ren86b] are cited). The pair $U \subseteq S$ is said to be *quasiunitary* if there exists a map $\phi : S \to S$ such that:

- $\phi(\phi(s)) = \phi(s)$, for all $s \in S$,
- $\phi(us) = u\phi(s)$ and $\phi(su) = s\phi(u)$, for all $u \in U$, $s \in S$,
- $us \in U$ or $su \in U$ implies $\phi(s) \in U$, for all $u \in U$, $s \in S$.

It is obvious that if $U \subseteq S$ is unitary, then it is quasiunitary, since then as ϕ one may take the identity mapping. It turns out that if $U \subseteq S$ and $U \subseteq T$ are quasiunitary, then the amalgam is embeddable.

In some cases one can be sure that an amalgam is (weakly) embeddable by simply inspecting the monoid U (called the "core" of the amalgam). A monoid U is said to be a *(weak) amalgamation base* if any amalgam which has it as core is (weakly) embeddable. For example, if U is a group, then it is unitary in any S, since $us \in U$ implies that $s = u^{-1}us \in u^{-1}U \subseteq U$. Thus, any group is an amalgamation base. Note that a theory corresponds to a group if, for any term t, there is a term t^* such that $t(t^*(x)) = x$ and $t^*(t(x)) = x$ are consequences of the theory.

Actually, a much stronger result is know. A monoid U is said to be *inverse* if one can define an operation $(\_)^* : U \to U$ such that $(u^*)^* = u$, $uu^*u = u$ and $(u_1u_2)^* = u_2^*u_1^*$ for all $u, u_1, u_2 \in U$. It can be shown that any inverse monoid is an amalgamation base (again, see [Hal78]).

The above considerations give us a variety of methods for guaranteeing that the considered union theory has the conservativity and interpolation properties. A natural question would be whether it is decidable to check, given decidable theories E_1 and E_2 conservatively extending E, whether $E_1 \cup E_2$ has the conservativity and interpolation properties. It turns out that the answer is negative: this problem is undecidable. This is an easy consequence of the result of Sapir [Sap00], who proves (Theorem 1.2) that the problem of whether an amalgam of two finite semigroups is embeddable in a semigroup is undecidable.

5   Lifting

In the previous sections we have defined the conservativity and interpolation properties of the union theory $E_1 \cup E_2$, and we showed how to guarantee that these properties indeed hold. These properties allow us to solve the word problem in the pushout P for words of length 1 (i.e., on the term level, for so-called *pure* terms); what remains to be done is to lift this solution to words of arbitrary length.

Unfortunately, without additional assumptions this turns out to be impossible in general. That is, the word problem in the pushout P of an embeddable amalgam may be undecidable, even if the word problems in S and T are decidable. More precisely, we construct finitely presentable monoids S and T with decidable word problems and such that the amalgam $[U \subseteq S, T]$ is embeddable, but the word problem in its pushout P is undecidable.

Let MT be a deterministic, universal Turing machine. We may assume that there is exactly one accepting configuration and that in this configuration the machine is blocked. The problem of checking whether there is a computation in MT from a given configuration to the accepting configuration is undecidable. Now consider the problem of checking whether there is a two-way computation between those configurations, that is, a computation in which the machine may go in "reverse mode". Observe that since M is deterministic, if such a computation exists, then one may assume that it consists of a block of forward moves followed by a block of reverse moves. But since the last reverse move would have to end in the (blocked) accepting configuration, there cannot be any reverse moves. Therefore the two-way computation is actually a usual one-way computation in MT. Hence, the problem of checking whether a two-way computation exists is undecidable.

Let M be any Turing machine. Let Q be its set of states and Γ its set of tape symbols, with the symbol $0 \in \Gamma$ serving as blank; we assume that Q and Γ are disjoint. Also, let $\#$ be a special, distinct symbol which we will use to mark the end of the tape. Let U be the free monoid 0^* and let S be the monoid generated by 0 and $\#$ and satisfying the relation $\# = 0\#$. Note that U is a submonoid of S and that S has a decidable word problem. Also, the submonoid $G_S = \#^*$ is a U-base of S.

Let T be a monoid generated by $Q \cup \Gamma$ and satisfying the following relations: if in the state $q \in Q$ with the head over $a \in \Gamma$ the machine goes right and replaces a by $b \in \Gamma$ and changes its state to $q' \in Q$, then $aqx = bxq'$ for all $x \in \Gamma$; if it goes left then $aq = q'b$. Note that U is a submonoid of T, since if two elements are equal in T, then they must have the same length. Also, T has a decidable word problem, again because the relations do not change the length of elements of T.

Now let P be the pushout of the amalgam $[U \subseteq S, T]$. A *configuration* is a sequence of the form $a_1 \ldots a_n q b_1 \ldots b_k \#$, where $n > 0$, $k \geq 0$, $a_1, \ldots, a_n \in \Gamma$, $b_1, \ldots, b_k \in \Gamma \setminus \{0\}$, $q \in Q$. The problem of checking whether two configurations are equal in the pushout is equivalent to checking whether there is a two-way computation between the corresponding configurations of the Turing machine M. If M is the previously considered deterministic, universal Turing machine MT, then the latter problem is undecidable. Thus, if the machine is appropriately chosen, the word problem in P is undecidable.

It is easy to see that $U \subseteq S$ is unitary, since there is no way of removing the symbol $\#$ once it has been inserted, and an element of S not containing $\#$ at all must, in fact, be an element of U. Similarly, $U \subseteq T$ is unitary, since if $ut \in U$ for some $u \in U$ and $t \in T$, then t cannot possibly contain an element of Q, since this element could not be disposed of. But if t does not contain any element of Q, then no relation can be applied to ut, because all relations require at least one $q \in Q$ to appear. Hence, $ut \in U$ implies that t is built of 0s only, i.e., $t \in U$.

Thus, by the fact mentioned in the previous section, the amalgam $[U \subseteq S, T]$ is embeddable.

The above results may be summarized as follows:

Theorem 3. *There exists a monoid amalgam $[U \subseteq S, T]$ such that:*

- *S and T have decidable word problems,*
- *there exists a U-base of S,*
- *the pairs $U \subseteq S$ and $U \subseteq T$ are unitary and so the amalgam is embeddable,*

and such that the pushout P has an undecidable word problem.   □

It may be noted that the monoid U is of course isomorphic to $\langle \mathbb{N}, + \rangle$, while S is isomorphic to the monoid $\langle \mathbb{N} \times \mathbb{N}, \cdot \rangle$ with multiplication defined by $(n, m) \cdot (n', m') = (n + n', m')$; finally, the inclusion of U in S takes any number n to the pair $(0, n)$.

The above theorem lets us state the following corollary for theories:

Corollary 2. *There exist decidable theories E_1 and E_2 conservatively extending a theory E and such that the union $E_1 \cup E_2$ has the conservativity and interpolation properties, but is undecidable.*   □

This negative result should not be understood as meaning that the idea of exploiting conservativity and interpolation is flawed. However, what it means is that an appropriate property is still to be found that would facilitate the lifting of the conservativity and interpolation properties to a full solution of our problem. The fact that the monoid S has a U-base suggests that perhaps this property should not involve the existence of bases and that a new idea is needed here. We believe that, even from a purely algebraic point of view, the following question is natural and interesting: to what extent is the decidability of the word problem in the pushout of an *embeddable* amalgam a property stronger than the decidability of the word problem in S and T? That it is a stronger property at all has been proved above.

6   Conclusions

In the paper we investigate the problem of deciding the union of two decidable theories, but restricted to the monadic case. This problem is shown, via an equivalence of categories, to be equivalent to the decidability of the word problem in the pushout of an amalgam of monoids with decidable word problems.

Moving to the category of monoids leads to a simplified view of the problem, allowing us to produce a simple and easily comprehensible statement and proof of the known fact that the existence of bases guarantees decidability of the union theory. This simple proof should help in the search for yet more general theorems, in which assumptions are made that are weaker than the, admittedly rather strong, assumption on the existence of bases.

In the sections which follow, we make use of the equivalence of the considered problem to problems concerning monoid amalgams. We show that the conservativity and interpolation properties of the union theory, properties used, e.g., in [BT02], actually correspond to certain classical properties of amalgams, namely to weak embeddability and embeddability. This allows one to use the various ideas stemming from amalgamation theory to guarantee and prove that the union theory has the conservativity and interpolation properties. In particular, notions of unitary and quasiunitary extensions may be used here, as well as the property of the "core" theory E corresponding to an inverse monoid.

We then propose a program of proving decidability of the union theory by first guaranteeing conservativity and interpolation, and then exploiting these properties. While, as mentioned above, it turns out that the first part of the program may indeed be carried out, implementing the second part is shown to require some new approach. For, as stated in Corollary 2, the decidability of E_1 and E_2 plus conservativity and interpolation of the union theory $E_1 \cup E_2$ do not guarantee its decidability.

The latter result gives a clear suggestion on possible future work: one should look at properties of E_1 and E_2 that would guarantee that an implication of the above form holds. An additional argument in favor of this idea is that, in general, it is hardly reasonable to expect that a union theory without the conservativity and interpolation properties will be guaranteed decidable. Therefore we do not really lose much if we assume that these properties hold. The above task, in monoid terms, amounts to looking for such properties of embeddable amalgams of monoids with decidable word problems that would be strong enough to guarantee that the word problem in the pushout of the amalgam is decidable.

Another obvious direction of future work is a generalization to non-monadic theories. Of course, with respect to Theorem 2, this could simply lead back to the solution found in [FG03] and to categories with products and product-preserving functors. What is needed, however, is the development of an appropriate amalgamation theory, which would generalize known results for monoids to objects representing arbitrary equational theories. It remains to be seen whether categories with products and product-preserving functors turn out to be the ideal notion of such an object. It could also be fruitful to perform a limited generalization, e.g., to use operads instead of monoids: operads correspond to equational theories which are *strongly regular*, that is, defined by equations in which, on either side, the same sequence of variables appears (see [Lei04]).

Finally, a rather natural generalization, and one which should not involve too much difficulties, is switching from single-sorted to multi-sorted equational logic. For monadic theories this change simply means that, instead of monoids, one has to use categories. It seems that results concerning monoids transfer to categories, in particular, Theorem 2 holds for categories as well; from this an appropriate theorem on multi-sorted equational theories may be deduced. It may be noted that using multi-sorted logic and the operad approach proposed above would lead to the use of yet another categorical idea, namely of multicategories [Lei04], since it is precisely multicategories that may be used to represent strongly regular multi-sorted equational theories.

References

[BT02]   Franz Baader and Cesare Tinelli. Deciding the word problem in the union of equational theories. *Information and Computation*, 178(2):346–390, 2002.

[FG03]   Camillo Fiorentini and Silvio Ghilardi. Combining word problems through rewriting in categories with products. *Theoretical Computer Science*, 294:103–149, 2003.

[Hal78]   Tom E. Hall. Amalgamation and inverse and regular semigroups. *Transactions of the AMS*, 246:395–406, 1978.

[Hig92]   Peter M. Higgins. *Techniques of Semigroup Theory*. Oxford University Press, 1992.

[How96a]   John M. Howie. *Fundamentals of Semigroup Theory*. London Mathematical Society Monographs, New Series, No. 12. Oxford University Press, 1996.

[How96b]   John M. Howie. Isbell's zigzag theorem and its consequences. In K. H. Hofmann and M. W. Mislove, editors, *Semigroup Theory and its Applications, Proceedings of the 1994 Conference Commemorating the Work of Alfred H. Clifford, New Orleans, LA*, London Mathematical Society Lecture Notes Series Vol. 231, pages 81–91. Cambridge University Press, 1996.

[Isb66]   John R. Isbell. Epimorphisms and dominions. In *Conference on Categorical Algebra, La Jolla, California, 1965, Proceedings*, pages 232–246. Springer, 1966.

[Lei04]   Tom Leinster. *Higher Operads, Higher Categories*. London Mathematical Society Lecture Notes Series Vol. 298. Cambridge University Press, 2004.

[Pig74]   Don Pigozzi. The join of equational theories. *Colloquium Mathematicum*, 30(1):15–25, 1974.

[Ren86a]   James Renshaw. Extension and amalgamation in monoids and semigroups. *Proceedings of the London Mathematical Society (3)*, 52:119–141, 1986.

[Ren86b]   James Renshaw. Flatness and amalgamation in monoids. *Journal of the London Mathematical Society (2)*, 33:73–88, 1986.

[Sap00]   Mark V. Sapir. Algorithmic problems for amalgams of finite semigroups. *Journal of Algebra*, 229(2):514–531, 2000.

Equivariant Unification

James Cheney

University of Edinburgh

Abstract. Nominal logic is a variant of first-order logic with special facilities for reasoning about names and binding based on the underlying concepts of swapping and freshness. It serves as the basis of logic programming and term rewriting techniques that provide similar advantages to, but remain simpler than, higher-order logic programming or term rewriting systems. Previous work on nominal rewriting and logic programming has relied on nominal unification, that is, unification up to equality in nominal logic. However, because of nominal logic's equivariance property, these applications require a stronger form of unification, which we call *equivariant unification*. Unfortunately, equivariant unification and matching are **NP**-hard decision problems. This paper presents an algorithm for equivariant unification that produces a complete set of finitely many solutions, as well as **NP** decision procedure and a version that enumerates solutions one at a time. In addition, we present a polynomial time algorithm for *swapping-free* equivariant matching, that is, for matching problems in which the swapping operation does not appear.

1   Introduction

Gabbay and Pitts [6] introduced a novel approach to formalizing and reasoning about abstract syntax involving bound names, based on the fundamental ideas of name-swapping and freshness. We call this approach *nominal abstract syntax* (NAS). Initially, this approach was based on FM-set theory, a variant of standard ZF-set theory originally developed to prove the independence of the Axiom of Choice. However, Pitts [7] showed that this radical step can be avoided by incorporating the ideas of nominal abstract syntax into a logic (called *nominal logic*) whose intended semantics is based on FM set theory but rests on standard mathematical foundations.

The key elements of nominal logic are: a collection of infinitely many term symbols $\mathsf{a}, \mathsf{b}, \ldots \in Name$ called *names*; a binary relation $\#$ called *freshness* that can hold between a name and a value; a *swapping function* $(a\ b){\cdot}t$ that exchanges the values of names a and b in t; and an *abstraction function* $\langle a \rangle x$ that takes a name and value. Abstractions are considered equal up to α-equivalence; for example $\langle \mathsf{a} \rangle f(\mathsf{a}, \mathsf{c}) \approx \langle \mathsf{b} \rangle f(\mathsf{b}, \mathsf{c})$.

Nominal logic has been used as a basis for logic programming [1, 3] and term rewriting systems [4]. So far, these techniques have relied upon the (efficiently implementable) *nominal unification* algorithm of Urban, Pitts, and Gabbay [11] as a fundamental tool, just as first-order unification is used in ordinary logic

J. Giesl (Ed.): RTA 2005, LNCS 3467, pp. 74–89, 2005.

programming and term rewriting. However, as shown by Cheney [2, 3], nominal unification is not the right tool for this job: proof search and term rewriting using nominal unification is incomplete.

First-order unification is complete for first-order resolution and rewriting because ground atomic formulas are logically equivalent if and only if they are equal as terms. But due to nominal logic's *equivariance* property, this is not the case for nominal logic. The equivariance property states that validity is preserved by applying name-swappings uniformly: that is, $p(\bar{t}) \iff p((a\ b)\cdot\bar{t})$. As a result, atomic formulas (such as $p(\mathsf{a})$ and $p(\mathsf{b})$) may be equivalent without being equal nominal terms. Similarly, if a collection of rewriting rules $t \rightarrow u$ is used to define a relation $\rightarrow$ in nominal logic, then $t \rightarrow u$ is equivalent to $(a\ b)\cdot t \rightarrow (a\ b)\cdot u$.

Consider the following logic program clauses and rewriting rules:

$$spec(mono(T), [], T). \quad spec(all(\langle\mathsf{a}\rangle T), [\mathsf{a}|Vs], U) :- spec(T, Vs, U).$$
$$subst(var(\mathsf{a}), T, \mathsf{a}) \rightarrow T \quad subst(var(\mathsf{b}), T, \mathsf{a}) \rightarrow var(\mathsf{b})$$
$$subst(app(E1, E2), T, \mathsf{a}) \rightarrow app(subst(E1, T, \mathsf{a}), subst(E2, T, \mathsf{a}))$$
$$\mathsf{b}\ \#\ T \vdash subst(lam(\langle\mathsf{b}\rangle E), T, \mathsf{a}) \rightarrow lam(\langle\mathsf{b}\rangle subst(E, T, \mathsf{a}))$$

The *spec* predicate is taken from an αProlog [1] program that performs ML type inference. It relates a polymorphic type to a list of bound variables and a monomorphic type, and can be used in the forward direction to instantiate the bound variables of a polymorphic type to fresh names, or backwards to quantify the free type variables of an inferred type. The *subst* rewriting rules perform capture-avoiding substitution on λ-terms encoded using nominal abstract syntax. (Note that nominal rewriting rules can have freshness "guards", e.g. $\mathsf{a}\ \#\ X \vdash l \rightarrow r$ applies only when $\mathsf{a}\ \#\ X$.)

Nominal unification and matching do not (and should not) take equivariance into account. As a result, logic programs or rewriting systems may not work as desired when nominal unification is used for backchaining or nominal matching is used for term rewriting, respectively. The goal $spec(all(\langle\mathsf{a}\rangle mono(tvar(\mathsf{a}))), [\mathsf{b}], U)$ has solution $[U = tvar(\mathsf{b})]$ in nominal logic, but this solution cannot be found using nominal unification. As another example, in nominal logic the first rewriting rule for *subst* implies that $subst(var(\mathsf{b}), var(\mathsf{a}), \mathsf{b})$ rewrites to $var(\mathsf{a})$, but there is no substitution for T making $subst(var(\mathsf{a}), T, \mathsf{a}) \approx subst(var(\mathsf{b}), var(\mathsf{a}), \mathsf{b})$.

Therefore, it is necessary to unify or match modulo a stronger equational theory that takes equivariance into account. We call these problems *equivariant matching* and *equivariant unification*, respectively. Equivariant unification is of both practical and theoretical interest. On the theoretical side, Cheney [2] showed that equivariant unification is **NP**-hard. On the practical side, there are some interesting programs (such as *spec*) that only appear to be expressible using equivariant unification. In addition, equivariant matching seems desirable in nominal rewriting systems for clarity and simplicity. For example, in the nominal rewriting approach advocated by Fernandez et al. [4], the *subst* rewrite rules above will not work properly. Instead, the following rewrite system was used for capture-avoiding substitution:

$$subst'(\langle\mathsf{a}\rangle var(\mathsf{a}), T) \to T \qquad \mathsf{a} \# B \vdash subst'(\langle\mathsf{a}\rangle var(B), T) \to var(B)$$
$$subst'(\langle\mathsf{a}\rangle app(E1, E2), T) \to app(subst'(\langle\mathsf{a}\rangle E1, T), subst'(\langle\mathsf{a}\rangle E2, T))$$
$$\mathsf{b} \# T \vdash subst'(\langle\mathsf{a}\rangle lam(\langle\mathsf{b}\rangle E), T) \to lam(\langle\mathsf{b}\rangle subst'(\langle\mathsf{a}\rangle E, T))$$

In this paper we make two significant contributions:

- We present a **NP** algorithm for equivariant unification that produces at most finitely many different solutions. This is the first (terminating) algorithm to be developed for general equivariant unification[1]. Besides taking equivariance into account, our algorithm solves a more general form of nominal unification problems than those considered by [11]. This algorithm can be used to run arbitrary nominal logic programs and rewriting systems and may also be useful in analyzing such systems.
- We present a polynomial-time algorithm for *swapping-free* equivariant matching problems, that is, problems in which the swapping function symbol is not present. This is significant because typical nominal rewriting systems that require equivariance (including *subst*) are swapping-free. This algorithm can be used as the basis of efficient nominal term rewriting for a larger class of programs than considered by Fernandez, Gabbay, and Mackie [4].

The remainder of this paper is structured as follows. In the next section, we review nominal equational logic. In Section 3, we introduce *permutation graphs*, an important tool for solving basic equivariant unification and matching problems that is used in the rest of the paper. In Section 4, we present the equivariant unification algorithm and sketch proofs of its important properties. Likewise, in Section 5 we present the swapping-free equivariant matching algorithm and prove its properties. Section 6 discusses additional related work and future directions, and Section 7 concludes.

2   Background

We first consider the set $Term$ of ground nominal terms, given by the grammar

$$t ::= \langle\rangle \mid \langle t, u \rangle \mid f(t) \mid \mathsf{a} \mid \langle\mathsf{a}\rangle t$$

The first three cases denote units, pairing, and function symbols; we represent constant symbols c as functions applied to unit $f(\langle\rangle)$ and represent n-ary function applications $f(t_1, \ldots, t_n)$ using iterated pairing $f(\langle t_1, \langle t_2, \cdots \rangle\rangle)$. Names a, a' are drawn from a countably infinite set $Name$, and abstractions $\langle\mathsf{a}\rangle t$ represent terms with bound names.

Let $Perm$ be the set of (finite) permutations of names. We write $\pi{\cdot}t$ for the *action* of $\pi \in Perm$ on t, or the result of applying π to rename the names of t. The permutation action function and equality $\approx: Term \times Term$ and freshness

---

[1] Cheney [2] only established that the equivariant matching and unification problems **for terms involving only names, variables, and swappings** are **NP**-complete, but did not present algorithms or upper bounds for problems involving general nominal terms.

$$\pi \cdot \mathsf{a} = \pi(\mathsf{a}) \quad \pi \cdot \langle \rangle = \langle \rangle \quad \pi \cdot \langle t, u \rangle = \langle \pi \cdot t, \pi \cdot u \rangle \quad \pi \cdot f(t) = f(\pi \cdot t) \quad \pi \cdot \langle \mathsf{b} \rangle t = \langle \pi \cdot \mathsf{b} \rangle \pi \cdot t$$

$$\frac{(\mathsf{a} \neq \mathsf{b})}{\mathsf{a} \mathbin{\#} \mathsf{b}} \quad \frac{}{\mathsf{a} \mathbin{\#} \langle \rangle} \quad \frac{\mathsf{a} \mathbin{\#} t}{\mathsf{a} \mathbin{\#} f(t)} \quad \frac{\mathsf{a} \mathbin{\#} t \quad \mathsf{a} \mathbin{\#} u}{\mathsf{a} \mathbin{\#} \langle t, u \rangle} \quad \frac{}{\mathsf{a} \mathbin{\#} \langle \mathsf{a} \rangle t} \quad \frac{(\mathsf{a} \neq \mathsf{b}) \quad \mathsf{a} \mathbin{\#} t}{\mathsf{a} \mathbin{\#} \langle \mathsf{b} \rangle t} \quad \frac{}{\mathsf{a} \approx \mathsf{a}} \quad \frac{}{\langle \rangle \approx \langle \rangle}$$

$$\frac{t_1 \approx u_1 \quad t_2 \approx u_2}{\langle t_1, t_2 \rangle \approx \langle u_1, u_2 \rangle} \quad \frac{t \approx u}{f(t) \approx f(u)} \quad \frac{t \approx u}{\langle \mathsf{a} \rangle t \approx \langle \mathsf{a} \rangle u} \quad \frac{(\mathsf{a} \neq \mathsf{b}) \quad \mathsf{a} \mathbin{\#} u \quad t \approx (\mathsf{a}\,\mathsf{b}) \cdot u}{\langle \mathsf{a} \rangle t \approx \langle \mathsf{b} \rangle u}$$

Fig. 1. Swapping, equality, and freshness for ground terms.

$\#: Name \times Term$ relations are defined in Figure 1. Equality is syntactic equality except for abstractions, which are considered equal modulo renaming of the bound names to a fresh name. The freshness theory spells out when a name is fresh for (not free in) a term. In particular, $\mathsf{a} \mathbin{\#} \langle \mathsf{a} \rangle t$ holds unconditionally, while $\mathsf{a} \mathbin{\#} \langle \mathsf{b} \rangle t$ holds for $\mathsf{a} \neq \mathsf{b}$ if $\mathsf{a} \mathbin{\#} t$.

Our definitions freshness and equality are superficially different from those used by Urban et al., but they are equivalent for ground terms. Urban et al. unified nominal terms modulo an equational theory axiomatizing equality and freshness judgments $\nabla \vdash A$ in the presence of some assumptions ∇ of the form $\mathsf{a} \mathbin{\#} X$, for names a and variables X. We instead axiomatize equality for ground terms only. Note that both freshness and equality are *equivariant*, that is, $t \approx u \supset \pi \cdot t \approx \pi \cdot u$ and $\mathsf{a} \mathbin{\#} t \supset \pi \cdot \mathsf{a} \mathbin{\#} \pi \cdot t$ for any a, t, u, π.

We now generalize to non-ground nominal terms so that *name-variables* $A, B, \ldots \in NVar$ and *term-variables* $X, Y, \ldots \in Var$ are permitted. In addition, we add explicit syntax for *permutation terms* Π applied to nominal terms, including swappings, composition, inversion, and permutation variables $Q, R, \ldots \in PVar$. Consider terms of the form:

$$v, w ::= \mathsf{a} \mid A \qquad \Pi, \Pi' ::= Q \mid \mathsf{id} \mid (\mathsf{a}\,\mathsf{b}) \mid \Pi \circ \Pi' \mid \Pi^{-1}$$
$$a, b ::= \Pi \cdot v \qquad t, u ::= \Pi \cdot X \mid a \mid \langle \rangle \mid \langle t, u \rangle \mid f(t) \mid \langle a \rangle t$$

We write $FN(t)$, $FV(t)$, $FNV(t)$ and $FPV(t)$ for the sets of names, term variables, name variables, and permutation variables of t. This grammar forbids permutation terms except immediately around names or variables. We define $\Pi \cdot t$ for arbitrary terms t as follows:

$$\Pi \cdot \langle \rangle = \langle \rangle \quad \Pi \cdot \langle t, u \rangle = \langle \Pi \cdot t, \Pi \cdot u \rangle \quad \Pi \cdot f(t) = f(\Pi \cdot t)$$
$$\Pi \cdot \langle a \rangle t = \langle \Pi \cdot a \rangle \Pi \cdot t \quad \Pi \cdot (\Pi' \cdot t) = (\Pi \circ \Pi') \cdot t$$

Urban et al. considered a more restrictive language of nominal terms in which permutation variables were not present, and required a and b to be ground names in terms of the forms $a \mathbin{\#} t$, $(a\,b) \cdot t$, and $\langle a \rangle t$. These restrictions were crucial for obtaining an efficient, deterministic unification algorithm. To avoid confusion, we refer to such terms as *grounded terms*, and to Urban et al.'s algorithm as *grounded nominal unification*. There are several important differences between our nominal terms and grounded terms. For our nominal terms, permutations applied to names cannot always be simplified: for example, $Q \cdot \mathsf{a}$ cannot be simplified without knowing something about Q. Another difference is that name

$$\theta(\langle\rangle) = \langle\rangle \quad \theta(\langle t, u\rangle) = \langle\theta(t), \theta(u)\rangle \quad \theta(f(t)) = f(\theta(t))$$
$$\theta(\Pi \cdot v) = \theta(\Pi) \cdot \theta(v) \quad \theta(\mathsf{a}) = \mathsf{a} \quad \theta(\langle\mathsf{a}\rangle t) = \langle\theta(\mathsf{a})\rangle\theta(t)$$
$$\theta(\mathsf{id}) = \mathsf{id} \quad \theta(\Pi \circ \Pi') = \theta(\Pi) \circ \theta(\Pi') \quad \theta(\Pi^{-1}) = \theta(\Pi)^{-1} \quad \theta((\mathsf{a}\ \mathsf{b})) = (\theta(\mathsf{a})\ \theta(\mathsf{b}))$$

Fig. 2. Valuations.

variables are permitted in any place where a name would be permitted. Nominal unification is **NP**-complete for arbitrary terms [3, Ch. 7], but tractable for grounded terms [11]. General equivariant unification and matching are **NP**-complete even for grounded terms (see [2] and Section 4). However, equivariant matching is tractable for grounded, swapping-free terms (see Section 5).

We refer to atomic formulas $t \approx u$, $\mathsf{a} \# u$ as *constraints* C, conjunctions and $\exists$-quantifications of constraints as *problems* S, and disjunctions of problems as *extended problems* $\mathcal{M} = \left\{\begin{smallmatrix} S_1; \\ \cdots \\ S_n \end{smallmatrix}\right\}$. An arbitrary problem involving terms that may have permutation variables is called an *equivariant unification* problem. A problem involving no permutation variables is called a *nominal unification* problem. A problem in which all equations involving permutation variables are of the form $Q \cdot t \approx u$, where u is ground, is called an *equivariant matching* problem. Problems are *grounded*, *name–name*, or *swapping-free* if all terms are grounded, if only names, swappings, and name variables are present, or if the swapping operation is not present, respectively.

A *valuation* is a function θ mapping term variables to ground terms, name variables to ground names, and permutation variables to ground permutations. Valuations are extended to terms as shown in Figure 2. We say that $\theta \vDash t \approx u$ if $\theta(t) \approx \theta(u)$; similarly, $\theta \vDash \mathsf{a} \# t$ if $\theta(\mathsf{a}) \# \theta(t)$. If S is a set of constraints, then we write $\theta \vDash S$ if $\theta \vDash A$ for each $A \in S$, and $\theta \vDash \exists X.S$ if $\theta[X := t] \vDash S$ for some t.

We write $Solv(S)$ for $\{\theta \mid \theta \vDash S\}$ and $Solv(\mathcal{M})$ for $\bigcup_{S \in \mathcal{M}} Solv(S)$. A problem S is a *pre-solution* to $\mathcal{M}$ if $Solv(S) \subseteq Solv(\mathcal{M})$, and a *solution* if in addition $Solv(S) \neq \varnothing$. A solution S to $\mathcal{M}$ is *more general* than another solution T if $Solv(T) \subseteq Solv(S)$, and *most general* if no strictly more general solution exists. A set $\mathcal{M}'$ of (pre-)solutions to $\mathcal{M}$ is a *complete* for $\mathcal{M}$ if $Solv(\mathcal{M}) = Solv(\mathcal{M}')$ and *minimal* if each $S \in \mathcal{M}'$ is a most general solution.

Example 1. A complete minimal set of solutions to the problem $(A\ B) \cdot C \approx C$ is $\{\{C \approx A, A \approx B\}, \{A \# C, C \# B\}\}$. A complete minimal solution set to the problem $Q \cdot \mathsf{a} \# \langle \mathsf{b}\rangle C$ is $\{\{Q \cdot \mathsf{a} \approx \mathsf{b}\}, \{Q \cdot \mathsf{a} \# \mathsf{b}, Q \cdot \mathsf{a} \# C\}\}$. The equivariant matching problem $Q \cdot (A, (\mathsf{a}\ \mathsf{b}) \cdot A, B, (\mathsf{a}\ \mathsf{b}) \cdot B) \approx (\mathsf{a}, \mathsf{b}, \mathsf{c}, \mathsf{d})$ has no solutions. The problem $f(\langle\mathsf{a}\rangle A, \mathsf{b}) \approx f(\langle\mathsf{c}\rangle\mathsf{d}, \mathsf{d})$ has a unique most general solution $A \approx \mathsf{b}$.

3   Permutation Graphs

In this section we consider an important data structure for representing information about permutations, names, and freshness, called a *permutation graph* (or *p-graph*).

Definition 1. *A p-graph* $G = (N, V, PV, E_\approx, E_\#, E_Q, \ldots)$ *is a structure such that* $N \subseteq Name$, $V \subseteq Var$ *and* $PV \subseteq PVar$ *are finite,* $E_\approx$ *and* $E_\#$ *are undirected graphs on* $W = N \cup V$, *and* E_Q *is a directed graph on* W *for each* $Q \in PV$.

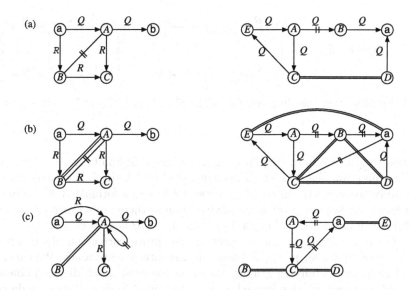

Fig. 3. (a) Example p-graphs. (b) Simplified versions. (c) Solved forms.

Note that the vertices v of a p-graph may be either names $a, b, \ldots$ or name variables $A, B, \ldots$. There are three kinds of edges: undirected equality edges (written using a double line $v == w$), undirected freshness edges (written as a broken line $v \dashrightarrow\!\!+\!\!\dashleftarrow w$), and directed permutation edges (written $v \xrightarrow{Q} w$). We consider the edge $v == w$ equivalent to the formula $v \approx w$, $v \dashrightarrow\!\!+\!\!\dashleftarrow w$ equivalent to $v \# w$, and $v \xrightarrow{Q} w$ equivalent to $Q \cdot v \approx w$. We write S_G for the problem corresponding to the edges of G and G_S for the p-graph corresponding the problem S.

For example, two small p-graphs are shown in Figure 3(a). Freshness and permutation edges are sometimes superimposed in these diagrams. These graphs correspond to the problems

$$\{Q \cdot a \approx A, Q \cdot A \approx b, R \cdot a \approx B, R \cdot B \approx C, R \cdot A \approx C, A \# B\}$$
$$\{Q \cdot E \approx A, Q \cdot A \approx B, A \# B, Q \cdot B \approx a, Q \cdot A \approx C, Q \cdot C \approx E, C \approx D, Q \cdot D \approx a\}$$

Testing the satisfiability of such problems is not straightforward, because there are hidden consequences. For example, the first set of constraints implies $R \cdot A \approx R \cdot B$, so $A \approx B$ since R is invertible. Similarly, in the second problem, since $A \# B$, we know $Q \cdot A \# Q \cdot B$, so $C \approx Q \cdot B \# Q \cdot A \approx a$. As a result of such observations, additional edges can be added to the graph to obtain a "simpler" graph with fewer hidden consequences. Our example graphs can be simplified in this way as shown in Figure 3(b).

In addition, when there is a variable equality edge involving a variable, such as $A \approx v \in G$, the graph can be simplified by collapsing A and v. This process is exactly analogous to substituting for A in the corresponding problem S_G. The results of collapsing the simplified example graphs are shown in Figure 3(c). The

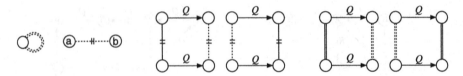

Fig. 4. Simplification rule diagrams for $(\approx_r^p), (\#_r^p), (\#_{\rightarrow}^p), (\#_{\leftarrow}^p), (\approx_{\rightarrow}^p), (\approx_{\leftarrow}^p)$ (respectively).

resulting graphs are fully simplified, and testing satisfiability is trivial because there are no remaining hidden consequences. The first graph is clearly unsatisfiable since there is a freshness edge corresponding to a formula $A \# A$. On the other hand, the second graph is satisfiable because there are no such edges. One satisfying valuation is $Q = (\mathsf{a}\ \mathsf{c})(\mathsf{a}\ \mathsf{b}), E = \mathsf{a}, A = \mathsf{b}, B = D = C = \mathsf{c}$.

In the rest of this section, we present and prove correct an algorithm for testing satisfiability for p-graphs based on this intuitive approach. We consider several rules for simplifying graphs, shown in Figure 4. Each diagram consists of a solid part and an edge formed with dotted lines. Such a diagram indicates that if G has a subgraph of the form described by the solid part, then the dotted edge should be added. These rules correspond to the following transformations on sets of formulas:

$$
\begin{aligned}
&(\approx_{ref}^p)\ S \rightarrow_p S, v \approx v &&(E_{\rightarrow}^p)\ S[v\ E\ v', Q{\cdot}v \approx w, Q{\cdot}v' \approx w'] \rightarrow_p S, w\ E\ w' \\
&(\#_{irr}^p)\ S \rightarrow_p S, \mathsf{a} \# \mathsf{b} &&(E_{\leftarrow}^p)\ S[w\ E\ w', Q{\cdot}v \approx w, Q{\cdot}v' \approx w'] \rightarrow_p S, v\ E\ v'
\end{aligned}
$$

where in the $(\approx_{ref}^p)$ and $(\#_{irr}^p)$ rules, v or $\mathsf{a} \neq \mathsf{b}$ must be in S, respectively; in the $(E_{\rightarrow}^p)$ and $(E_{\leftarrow}^p)$ rules, $E \in \{\approx, \#\}$; and $S[S_1] \rightarrow_p S, S_2$ means "If S contains the formulas S_1, then add S_2 to S," .

Once the simplification rules above have been applied we can "collapse" equality edges involving variables, as outlined in the informal example. We say that a variable in G is solved if it appears in just one equality edge in G, otherwise it is unsolved; if $A \approx v \in G$ and A is unsolved then we can solve A in G by replacing A with v in all the other edges of G. This transformation on the graph corresponds to a variable elimination step on its corresponding constraint set:

$$(\approx_{var}^p)\quad S, A \approx v \rightarrow_p S[A := v], A \approx v \quad (\text{if } A \in FV(S), A \neq v)$$

We define $G[A := v]$ as the result of removing A from G and replacing A with v in all edges of G; a collapsing step on $A \approx v$ transforms G to $G[A := v], A \approx v$. We write $G \rightarrow_p G'$ if G can be transformed to G' via a simplification or collapsing step.

When considering satisfiability, solved variables can be ignored. The *collapsing* $c(G)$ of a graph G is the graph formed by eliminating all solved vertices. If $c(G) = G$, we say that G is fully collapsed.

Lemma 1. *If $A \approx v \in G$ then $Solv(G[A := v], A \approx v) = Solv(G)$. Moreover, $c(G)$ is satisfiable if and only if G is.*

Proof. If $\theta \vDash G$, then $\theta \vDash A \approx v$, hence $\theta \vDash A[A := v]$ for each $A \in G$ besides $A \approx v$. Conversely, if $\theta \vDash G[A := v], A \approx v$, then $\theta \vDash A \approx v$ and so $\theta \vDash A$ for each $A \in G$. The second part follows by induction on the number of solved variables in G.

We say that a p-graph is in *normal form* if none of the simplification or collapsing rules apply. A normalized graph is *solved* if in addition, $E_\approx$ and $E_\#$ are disjoint. We consider the possible forms of fully collapsed solved forms.

Proposition 1. *In a fully collapsed solved form, $E_\approx = Id_W$, and each E_Q is a partial injective function on W.*

Proof. Clearly $Id_W \subseteq E_\approx$ since otherwise $(\approx^p_{ref})$ would apply. Suppose $v \approx w \in G$. If v is a variable A, then w must also be A because otherwise v and w could be collapsed. The case in which w is a variable is symmetric. If v and w are names, then since G is normalized we must have $\mathsf{a} \# \mathsf{b} \in G$ and $\mathsf{a} \approx \mathsf{b} \notin G$ for any distinct names $\mathsf{a} \neq \mathsf{b}$, so it must be the case that $v = \mathsf{a} = w$.

For the second part, let $Q \in PV$ be given and consider $(v, w), (v, w') \in E_Q$. Since G is normalized, we must have $v \approx v \in G$ and $w \approx w' \in G$. By the first part, $w = w'$. Hence E_Q is a function. Moreover, by a similar argument if $(v, w), (v', w) \in E_Q$ then $v = v'$, so E_Q is injective.

Proposition 2. *A normalized p-graph is satisfiable if and only if it is solved.*

Proof. For the forward direction, we prove the contrapositive. If G is normalized but not solved, then there exists $(v, w) \in E_\# \cap E_\approx$. No valuation satisfies both $v \approx w$ and $v \# w$, so G is unsatisfiable.

For the reverse direction, it suffices to consider only normalized, fully collapsed graphs. By Proposition 1, G must satisfy $E_\approx = Id_W$, and E_Q must be an injective function on W for each Q.

Recall that $W = V \cup N$, where N is the set of names and $V = \{A_1, \ldots, A_k\}$ the set of variables of G. Let $\mathsf{b}_1, \ldots, \mathsf{b}_k$ be k names fresh for each other and not appearing in N. Define $\theta(A_i) = \mathsf{b}_i$ for $A_i \in V$. Note that θ is a bijection between V and $B = \{\mathsf{b}_1, \ldots, \mathsf{b}_k\}$. It extends to a bijection $\theta : W \to B \cup N$. If $v \approx w \in G$ then $v = w$ so clearly $\theta(v) \approx \theta(w)$. On the other hand, if $v \# w \in G$, we must have $v \neq w$, so $\theta(v) \# \theta(w)$ since θ is bijective. This shows that any valuation based on θ satisfies the edges $E_\approx$ and $E_\#$ of G.

Since each E_Q is injective it can be completed to a bijection $\pi_Q : W \to W$. Define $\theta(Q) = \theta \circ \pi_Q \circ \theta^{-1}$ for each Q. Suppose $(v, w) \in E_Q$. By construction, $\pi_Q(v) = w$, so

$$\theta(Q \cdot v) = \theta(Q)(\theta(v)) = \theta(\pi_Q(\theta^{-1}(\theta(v)))) = \theta(\pi_Q(v)) = \theta(w)$$

as desired. This completes the proof that the valuation θ satisfies G.

Theorem 1 (Soundness). *If $G \to_p G'$ then $Solv(G') = Solv(G)$.*

Proof. Suppose $G \to_p G'$. For a simplification step, $G' = G \cup A$ where A is either $v \approx w$, $v \# w$, or $v \xrightarrow{Q} w$. Trivially, $Solv(G') \subseteq Solv(G)$ since $G \subseteq G'$. To show that $Solv(G) \subseteq Solv(G')$, we need only verify that $G \vDash A$ in each case.

For the $(\approx^p_{ref})$ rule, $A = v \approx v$, and $G \vDash v \approx w$. For the $(\#^p_{irr})$ rule, $A = $ a $\#$ b for names a $\neq$ b, and clearly $G \vDash$ a $\#$ b. For the $(E^p_\rightarrow)$ rule, we have $A = x'$ E y' and $v \xrightarrow{Q} v', w \xrightarrow{Q} w', v$ E $w \in G$ for $E \in \{\approx, \#\}$. Then $\theta(v)$ E $\theta(w)$, $\theta(Q)(\theta(w)) = \theta(w')$ and $\theta(Q)(\theta(v)) = \theta(v')$, so

$$\theta(v') = \theta(Q)(\theta(v))\ E\ \theta(Q)\theta(w) \approx \theta(w')$$

The cases for the $(E^p_\leftarrow)$ rules are symmetric, since permutations are invertible and $\approx$ and $\#$ are equivariant. The case for a $(\approx^p_{var})$ step is shown in Lemma 1.

Theorem 2 (Termination). *There are no infinite sequences of simplification steps. Moreover, p-graph normalization can be performed in polynomial time.*

Proof. Each reduction step either adds an edge or solves a vertex in G, so the maximum number of steps is bounded above by $(2+|PV|)\cdot|W|^2|V|$. Each reduction step can be identified and performed in polynomial time, and normalized and solved graphs can be recognized in polynomial time.

Corollary 1 (Completeness). *The relation $\rightarrow_p$ reduces any p-graph G to a normal form G' which is solved if and only if G is satisfiable; moreover, $Solv(G) = Solv(G')$.*

4   Equivariant Unification

In the previous section, we considered a very limited case of equivariant unification, namely solving systems of formulas of the form $v \approx w$, $v \# w$, and $Q \cdot v \approx w$. We showed that this problem can be solved in polynomial time using permutation graphs. In this section, we give an algorithm for reducing equivariant unification for arbitrary nominal terms to the problem of testing the satisfiability of a finite (but possibly exponential) number of permutation graphs. This algorithm can be easily be modified to obtain a nondeterministic polynomial time procedure for testing the satisfiability of such a problem, or as a procedure for enumerating the solutions one at a time.

We break the process into two phases. In the first phase, we simplify all problems involving subterms of the form $\langle\rangle, \langle t, u \rangle, \langle a \rangle t, f(t)$. After the first phase, the remaining satisfiable subproblems are of the form $a \# b, a \approx b$, where a, b are formed using only names, variables, and permutations. In the second phase, we convert these subproblems into p-graphs by eliminating permutations. Once each p-graph is constructed, we can test its satisfiability as shown in the previous section.

4.1   First Phase

The first phase of the algorithm (defined as a relation $\rightarrow_1$) is presented as a collection of multiset rewriting rules in Figure 5. Each rule is of the form $S \rightarrow_1 M$, and indicates that an extended problem $M'; S$ should be rewritten to the problem $M'; M$.

A problem S is in *solved form* if it consists only of constraints of the form $a \# b, a \approx b$, or $X \approx t$ where X does not appear in t or elsewhere in S; an

$$(\approx_1) \qquad\qquad S, \langle\rangle \approx \langle\rangle \rightarrow_1 S$$

$$(\approx_\times) \quad S, \langle t_1, t_2\rangle \approx \langle u_1, u_2\rangle \rightarrow_1 S, t_1 \approx u_1, t_2 \approx u_2$$

$$(\approx_f) \qquad\quad S, f(t) \approx f(u) \rightarrow_1 S, t \approx u$$

$$(\approx_{abs}) \qquad S, \langle a\rangle t \approx \langle b\rangle u \rightarrow_1 \left\{ \begin{array}{l} S, a \approx b, t \approx u; \\ S, a \# u, t \approx (a\ b)\cdot u \end{array} \right\}$$

$$(\approx_{var}) \qquad S, \Pi\cdot X \approx t \rightarrow_1 S[X := \Pi^{-1}\cdot t], X \approx \Pi^{-1}\cdot t$$
$$\text{(where } X \notin FV(t), X \in FV(S))$$

$$(\#_1) \qquad\qquad S, a \# \langle\rangle \rightarrow_1 S$$

$$(\#_\times) \qquad S, a \# \langle u_1, u_2\rangle \rightarrow_1 S, a \# u_1, a \# u_2$$

$$(\#_f) \qquad\quad S, a \# f(u) \rightarrow_1 S, a \# u$$

$$(\#_{abs}) \qquad S, a \# \langle b\rangle u \rightarrow_1 \left\{ \begin{array}{l} S, a \approx b; \\ S, a \# u \end{array} \right\}$$

Fig. 5. Equivariant unification: phase one.

extended problem is solved if all its problems are solved. "Stuck" subproblems S that are unsolved and can take no transition can always be removed from an extended problem.

Example 2. The problem $Q\cdot(\langle a\rangle\langle A, B\rangle) \approx \langle b\rangle\langle b, c\rangle$ reduces to a solved form as follows:

$$\langle Q\cdot a\rangle\langle Q\cdot A, Q\cdot B\rangle \approx \langle b\rangle\langle b, c\rangle \rightarrow_1 \left\{ \begin{array}{l} Q\cdot a \approx b, \langle Q\cdot A, Q\cdot B\rangle \approx \langle b, c\rangle; \\ Q\cdot a \# \langle b, c\rangle, \langle Q\cdot A, Q\cdot B\rangle \approx \langle(Q\cdot a)\cdot b, (Q\cdot a)\cdot c\rangle \end{array} \right\}$$
$$\rightarrow_1^* \left\{ \begin{array}{l} Q\cdot a \approx b, Q\cdot A \approx b, Q\cdot B \approx c; \\ Q\cdot a \# b, Q\cdot a \# c, Q\cdot A \approx (Q\cdot a)\cdot b, Q\cdot B \approx (Q\cdot a)\cdot c \end{array} \right\}$$

Some constraints in a solved form may be of the form $\Pi\cdot X \approx \Pi'\cdot X$ where X is not a name-variable so cannot be substituted with names. These constraints are always satisfiable so can be set aside. This leaves *name–name* constraints $a \# b, a \approx b$ involving only permutations, names, and name variables.

Theorem 3 (Soundness). *If $\mathcal{M} \rightarrow_1 \mathcal{M}'$ then $Solv(\mathcal{M}) = Solv(\mathcal{M}')$.*

Proof. The cases for $(\approx_1)$, $(\approx_\times)$, $(\approx_f)$, and $(\approx_{var})$ are straightforward. For $(\approx_{abs})$, it suffices to show that $Solv(\langle a\rangle t \approx \langle b\rangle u) = Solv(a \approx b, t \approx u) \cup Solv(a \# u, t \approx (a\ b)\cdot u)$. Clearly, if $\theta \vDash a \approx b, t \approx u$, or $\theta \vDash a \# u, t \approx (a\ b)\cdot u$, then $\theta \vDash \langle a\rangle t \approx \langle b\rangle u$ using the rules in Figure 1. If $\theta \vDash \langle a\rangle t \approx \langle b\rangle u$, then there are two cases. If $\theta(a) = \theta(b)$, then $\theta \vDash a \approx b, t \approx u$, so $\theta \vDash \mathcal{M} \uplus \{a \approx b, t \approx u\}$. Otherwise, we must have $\theta \vDash a \# u, t \approx (a\ b)\cdot u$, so $\theta \vDash \mathcal{M} \uplus \{a \# u, t \approx (a\ b)\cdot u\}$.

The cases involving freshness are straightforward, with the reasoning for $(\#_{abs})$ similar to that for $(\approx_{abs})$.

Theorem 4 (Termination). *The relation $\rightarrow_1$ terminates.*

Proof. We define a measure on terms as follows: $\mu(\langle\rangle) \approx 1$, $\mu(\langle t, u\rangle) = \mu(t) + \mu(u) + 1$, $\mu(f(t)) = \mu(f) + 1$, $\mu(\langle a\rangle t) = \mu(t) + 1$, $\mu(\Pi\cdot X) = \mu(\Pi\cdot a) = 0$. Let $\mu(t\ E\ u) = \mu(t) + \mu(u)$ and $\mu(S) = \sum_{A \in S} \mu(A)$. Let $\mu'(S)$ be the number of unsolved variables in S. Define $\nu(S) = (\mu'(S), \mu(S))$ and $\nu(\mathcal{M}) = \{\nu(S) \mid S \in \mathcal{M}\}$. It is straightforward to verify that if $P \rightarrow_1 P'$ then $\nu(\mathcal{M}) > \nu(\mathcal{M}')$ in the multiset order generated by the lexicographic order on $\mathbb{N} \times \mathbb{N}$.

$$(id) \qquad S[\mathit{id}\cdot v] \to_2 S[v]$$
$$(inv) \qquad S[\Pi^{-1}\cdot v] \to_2 \exists X.S[X], \Pi\cdot X \approx v$$
$$(comp)\ S[\Pi \circ \Pi'\cdot v] \to_2 \exists X.S[\Pi\cdot X], \Pi'\cdot v \approx X)$$
$$(swap)\ \ S[(a\ a')\cdot v] \to_2 \left\{ \begin{array}{r} S[a], a' \approx v; \\ S[a'], a \approx v; \\ \exists X.S[X], v \approx X, a \mathbin{\#} X, a' \mathbin{\#} X \end{array} \right\}$$
$$(\#_Q) \qquad S, Q\cdot v \mathbin{\#} w \to_2 \exists X.S, Q\cdot v \approx X, X \mathbin{\#} w$$

Fig. 6. Equivariant unification: phase two.

Lemma 2. *If $\mathcal{M}$ is satisfiable and $\to_1$-normalized, then $\mathcal{M}$ is in solved form.*

Proof. We prove that if $\mathcal{M}$ is unsolved and satisfiable, it is not normalized. Suppose $\mathcal{M}$ is satisfiable but not solved. Then there must be some constraint in $\mathcal{M}$ which is not of the form $a \mathbin{\#} b, a \approx b$, or $X \approx t$ where X is solved in $\mathcal{M}$. If the constraint is of the form $\Pi\cdot X \approx t$ where X is a term variable and t starts with a term symbol, then we must have $\theta(\Pi\cdot X) \approx \theta(t)$, which can only be the case if X does not appear in t, so ($\approx_{var}$) applies. Otherwise, the constraint must be of the form $t \approx u$ or $a \mathbin{\#} u$, where t, u start with term symbols. For the case of $a \mathbin{\#} u$, a step can be taken no matter which term symbol is at the head of u. For $t \approx u$, since $\theta \models t \approx u$, the head symbols of t and u must match, so that we can take a step. In any case, $\mathcal{M} \to \mathcal{M}'$ for some $\mathcal{M}'$.

Corollary 2 (Completeness). *The relation $\to_1$ reduces any finite equivariant unification problem to a finite complete set of pre-solutions.*

4.2   Second Phase

In the second phase, we reduce name–name constraints to p-graphs whose satisfiability can be checked easily. As a preprocessing step, we assume that all constraints of the form $\Pi_1\cdot v \approx \Pi_2\cdot w$ or $\Pi_1\cdot v \mathbin{\#} \Pi_2\cdot w$ are normalized to $(\Pi_2^{-1} \circ \Pi_1)\cdot v \approx w$ or $(\Pi_2^{-1} \circ \Pi_1)\cdot v \mathbin{\#} w$ respectively. This is without loss of generality because $\#$ and $\approx$ are preserved by applying permutations to both sides. The rules for the second phase of equivariant unification shown in Figure 6 reduce the results of the first phase to a form suitable for satisfiability checking via p-graphs. In several rules, we introduce fresh existentially-quantified variables; these are required not to already appear in the problem.

Example 3. We continue Example 2. The first subproblem, $Q\cdot a \approx b, Q\cdot A \approx b, Q\cdot B \approx c$, is already in solved form (and is satisfiable provided $A \approx a$). The second problem reduces as follows:

$$Q\cdot A \approx (Q\cdot a\ b)\cdot b, Q\cdot B \approx (Q\cdot a\ b)\cdot c, Q\cdot a \mathbin{\#} b, Q\cdot a \mathbin{\#} c$$
$$\to_2 (Q\cdot a\ b) \circ Q\cdot A \approx b, (Q\cdot a\ b) \circ Q\cdot B \approx c, S$$
$$\to_2 (Q\cdot a\ b)\cdot C_1 \approx b, Q\cdot A \approx C_1, (Q\cdot a\ b)\cdot C_2 \approx c, Q\cdot B \approx C_2, S$$
$$\to_2 \left\{ \begin{array}{ll} & Q\cdot a \approx C_1, b \approx b; \\ (*) & Q\cdot a \approx b, a \approx C_1; \\ (*)\ Q\cdot a \mathbin{\#} C_1, b \mathbin{\#} C_1, C_1 \approx b \end{array} \right\} \otimes \left\{ \begin{array}{ll} (*) & Q\cdot a \approx C_2, b \approx c; \\ (*) & a \approx C_2, Q\cdot a \approx c; \\ & Q\cdot a \mathbin{\#} C_2, b \mathbin{\#} C_2, C_2 \approx c \end{array} \right\} \otimes \{S'\}$$

where $S = Q \cdot a \mathbin{\#} b, Q \cdot a \mathbin{\#} c$ and $S' = Q \cdot A \approx C_1, Q \cdot B \approx C_2, S$, and $\mathcal{M} \otimes \mathcal{M}'$ denotes $\{T \wedge T' \mid T \in \mathcal{M}, T' \in \mathcal{M}'\}$. There are a total of nine cases; however, the starred subproblems are unsatisfiable, so there is only one solution.

Theorem 5 (Soundness). *If $\mathcal{M} \to_2 \mathcal{M}'$ then $Solv(\mathcal{M}) = Solv(\mathcal{M}')$.*

Proof. There are several cases, one for each rule replacing $\mathcal{M}, S$ with $\mathcal{M}, \mathcal{M}'$ for $S \to_2 \mathcal{M}'$. The cases for (id), (inv), $(comp)$, and $(\#_Q)$ are straightforward. For the $(swap)$ rule, it suffices to show that $Solv(S[(a\ b) \cdot v]) = T = Solv(S[b], a \approx v) \cup Solv(S[a], b \approx v) \cup Solv(\exists X.S[X], v \approx X, a \mathbin{\#} X, b \mathbin{\#} X)$. If $\theta \in Solv(S[(a\ b) \cdot v])$, then there are three cases. If $\theta \vDash a \approx v$, then $\theta \vDash (a\ b) \cdot v \approx b$ so $\theta \in Solv(S[b], a \approx v)$. The case for $\theta \vDash b \approx v$ is symmetric. If $\theta \vDash a \mathbin{\#} v, b \mathbin{\#} v$, then $\theta \vDash (a\ b) \cdot v \approx v$ so $\theta[X := \theta(v)] \vDash X \approx v, a \mathbin{\#} X, b \mathbin{\#} X, S[X]$, and $\theta \in Solv(\exists X.S[X], v \approx X, a \mathbin{\#} X, b \mathbin{\#} X)$. So in any case $\theta \in T$. The reverse direction, $T \subseteq Solv(S[(a\ b) \cdot v])$, is straightforward.

Theorem 6 (Termination). *The relation $\to_2$ terminates.*

Proof. We employ a measure μ that measures the complexity of the permutation terms remaining in $\mathcal{M}$. We define $\mu(v) \approx 0$, $\mu(\Pi \cdot v) = \mu(\Pi)$, $\mu((a\ b)) = 1 + \mu(a) + \mu(b)$, $\mu(\Pi \circ \Pi') = \mu(\Pi) + \mu(\Pi')$, $\mu(\Pi^{-1}) = \mu(\Pi) + 1$, and $\mu(id) = 1$. In addition, $\mu(a \mathbin{\#} v) = \mu(a \approx v) = \mu(a)$, $\mu(S) = \sum_{A \in S} \mu(A)$, and $\mu(\mathcal{M}) = \{\mu(S) \mid S \in \mathcal{M}\}$. If $\mathcal{M} \to_2 \mathcal{M}'$, then $\mu(\mathcal{M})$ is decreasing in the multiset ordering generated by $>_{\mathbb{N}}$.

Lemma 3. *If $\mathcal{M}$ is $\to_2$-normalized problem, then it is in solved form.*

Proof. Since $\mathcal{M}$ is normalized, it cannot contain any constraints of the form $\Pi \cdot v\ E\ w$ where $E \in \{\#, \approx\}$ and Π is not a variable, since otherwise one of the rules (id), $(comp)$, (inv), $(swap)$ can be applied. Similarly, $\mathcal{M}$ cannot contain a constraint of the form $Q \cdot v \mathbin{\#} w$, since otherwise $(\#_Q)$ applies. Because only constraints of the form $v \approx w$, $v \mathbin{\#} w$, and $Q \cdot v \approx w$ remain, $\mathcal{M}$ is in solved form.

Corollary 3 (Completeness). *The relation $\to_2$ reduces any finite name–name problem to a finite complete set of pre-solutions.*

Example 4. Consider the query $?\text{-}\ spec(all(\langle a \rangle tvar(a)), [b], U)$. Equivariant unification against a suitably renamed/permuted head clause $P \cdot spec(all(\langle a' \rangle T'), a' :: L', U')$ yields a single unifier $P \cdot a' \approx b, T' := tvar(P^{-1} \circ (a\ P \cdot a') \cdot a), L' := [], U := P \cdot U'$. The resulting subgoal $spec(tvar(P^{-1} \circ (a\ P \cdot a') \cdot a), [], U')$ produces the unique solution $U' := tvar(P^{-1} \circ (a\ P \cdot a') \cdot a)$. This gives the overall solution $U := tvar(P \circ P^{-1} \circ (a\ P \cdot a') \cdot a)$, which can be simplified to $U := tvar(b)$ since $P \cdot a' \approx b$.

5   Swapping-Free Equivariant Matching

In equivariant unification, only the abstraction and swapping operations cause branching. This implies (perhaps surprisingly) that equivariant unification is tractable for problems involving names, term symbols, and freshness but not

$$(\leq_1) \qquad S, l.t \leq l'.\langle\rangle \rightarrow_m S, t \approx \langle\rangle$$
$$(\leq_f) \qquad S, l.t \leq l'.f(u) \rightarrow_m \exists X.S, l.X \leq l'.u, t \approx f(X)$$
$$(\leq_\times) \quad S, l.t \leq l'.\langle u_1, u_2\rangle \rightarrow_m \exists X_1, X_2.S, l.X_1 \leq l'.u_1, l.X_2 \leq l'.u_2, t \approx \langle X_1, X_2\rangle$$
$$(\leq_{abs}) \qquad S, l.t \leq l'.\langle a\rangle u \rightarrow_m \exists X.S, lb.X \leq l'a.u, t \approx \langle b\rangle X \quad (b \notin FN(S))$$
$$(\leq_\approx) \qquad S, la.v \leq l'b.b \rightarrow_m S, v \approx a$$
$$(\leq_\#) \qquad S, la.v \leq l'b.c \rightarrow_m S, l.v \leq l'.c, a \# v \quad (b \neq c)$$

Fig. 7. Swapping-free equivariant matching.

abstraction and swapping. If we restrict attention to equivariant matching of grounded terms, however, we can get a stronger result: swapping-free grounded terms can be matched efficiently. We consider grounded problems of the form $t \leq u$, where u is ground; a solution is a ground substitution θ and ground permutation π such that $\theta(t) \approx \pi \cdot u$.

When one side of an equation is ground, the structure of the bound names on that side must be mirrored exactly on the other side. For example, consider the problem $\langle a\rangle\langle b\rangle X \leq \langle c\rangle\langle d\rangle e$, where X is a name-variable and e is a ground name. If $e = d$, then we must have $X = b$; if $e = c$, then we must have $X = a$; and if e is some name other than c, d then we must have $a, b \# X$ and $Q \cdot X = e$.

Also, in a problem of the form $\langle a_1\rangle \cdots \langle a_n\rangle X \leq \langle b_1\rangle \cdots \langle b_n\rangle t$, if t starts with unit, pairing, or a function symbol f, then X must also start with unit, pairing, or f, so we can proceed by *simulating* the head symbol of t by making an appropriate substitution of $X \approx \langle\rangle$, $X \approx \langle X_1, X_2\rangle$, or $X \approx f(X')$, where X_1, X_2, X' are new variables. More generally, if the problem is of the form $\langle a_1\rangle \cdots \langle a_n\rangle t \leq \langle b_1\rangle \cdots \langle b_n\rangle u$, then we can proceed by unifying t with $\langle\rangle$, $\langle X_1, X_2\rangle$, or $f(X')$, as appropriate.

Based on this intuition, we propose the following algorithm for matching swapping-free grounded terms with ground nominal terms. We write l, l' for lists of names $a_1 \cdots a_n$ and consider problems of the form $l.t \leq l'.u$ where u is ground. This problem is equivalent to the problem $\langle a_1\rangle \cdots \langle a_n\rangle t \leq \langle b_1\rangle \cdots \langle b_n\rangle u$.

The rules in Figure 7 define a relation $\rightarrow_m$ that reduces equivariant matching problems to the form $S_\leq \cup S_{NP}$, where $S_\leq$ is a collection of inequalities of the form $[].v \leq [].a$, and S_{NP} is a collection of equality and freshness constraints among grounded terms. The satisfiability of S_{NP} can be tested using grounded nominal unification; if successful, this results in a unifier $\langle \nabla, \sigma\rangle$, where ∇ is a set of freshness constraints and σ is a substitution. Now let Q be a permutation variable, let $S_Q = \nabla \cup \{Q \cdot \sigma(v) \approx a \mid [].v \leq [].a \in S_\leq\}$, and test the satisfiability of the p-graph G_{S_Q}.

We now state the important properties of $\rightarrow_m$. The proofs are complicated without being particularly enlightening so are omitted.

Theorem 7 (Soundness). *If* $S \rightarrow_m S'$, *then* $Solv(S) = Solv(S')$.

Theorem 8 (Termination). *The relation* $\rightarrow_m$ *is terminating, and normal forms can be computed in polynomial time.*

Lemma 4. *If* S *is a normalized swapping-free equivariant matching problem, then* S *is in solved form.*

Theorem 9 (Completeness). *The relation* $\to_m$ *reduces any equivariant matching problem S to a pre-solution S' such that $Solv(S) = Solv(S')$.*

Example 5. Recall the problem $subst(var(\mathsf{a}), T, \mathsf{a}) \leq subst(var(\mathsf{b}), var(\mathsf{a}), \mathsf{b})$ mentioned in the Introduction. The above algorithm reduces to the solved form $\mathsf{a} \leq \mathsf{b}, T := var(B), B \leq \mathsf{a}$. Since $subst(var(\mathsf{a}), T, \mathsf{a})$ rewrites to $T = var(B)$ and $B \leq \mathsf{a}$, we rewrite $subst(var(\mathsf{b}), var(\mathsf{a}), \mathsf{b})$ to $var(\mathsf{a})$.

6   Related and Future Work

Many researchers (see for example [9]) have studied the problem of E-unification, or unification with respect to a general equational theory E. However, nominal logic poses some unique challenges to standard E-unification techniques based on confluent rewrite systems. One reason is that it appears that the equivariance principle $p(\bar{x}) \approx p((a\ b)\cdot\bar{x})$ cannot be directed so as to obtain a confluent rewrite system for nominal terms.

There also may be an interesting connection between equivariant unification and unification modulo equational laws having to do with name-restriction in the π-calculus, such as $\nu a.p \equiv p$ (where $a \notin FN(p)$) and $\nu a.\nu b.p \equiv \nu b.\nu a.p$. More generally, it may be interesting to study E-unification for equational theories specified in nominal logic.

Cheney [2], Fernandez, Gabbay and Mackie [4], and Urban and Cheney [10] have developed increasingly general tests for identifying rewriting systems or logic programs for which nominal unification is adequate. These results demonstrate that nominal unification can often be used instead of equivariant unification to execute programs efficiently. Such special cases should be recognized and exploited whenever possible.

FreshML [8] is a ML-like functional programming language based on nominal abstract syntax. In FreshML, programs can perform pattern matching against terms involving abstraction and name-variables but not constant names, swappings; also, such pattern matching is not modulo equivariance. However, as usual in ML, variables may appear at most once in patterns, so the matching problems involved in FreshML can be solved efficiently and without backtracking: for example, to match u against an abstraction $\langle A \rangle t$, it suffices to generate a fresh name c and match u against $\langle \mathsf{c} \rangle t$. It would be interesting to see whether constant names could be incorporated into FreshML-style functional programming.

In logic programming, nondeterminism is often a bigger performance problem than exponential worst-case complexity, so it would be worthwhile to find ways of avoiding duplicate answers, delaying nondeterministic search, and detecting failure early. One possibility is to look for and factor out symmetries in subproblems as soon as they appear. Another step in this direction is to replace the $(\approx_{abs})$ and $(\#_{abs})$ rules with

$$(\approx_{abs})\ \ S, \langle a \rangle t \approx \langle b \rangle u \to_1 S, \mathsf{N}c.(a\ c)\cdot t \approx (b\ c)\cdot u$$
$$(\#_{abs})\ \ \ \ S, a \mathbin{\#} \langle b \rangle u \to_1 S, \mathsf{N}c.a \mathbin{\#} (b\ c)\cdot u$$

where $c \notin FN(a, b, t, u)$ and N is nominal logic's "new" or "fresh name" quantifier. This is correct because in nominal logic, $\langle a \rangle t \approx \langle b \rangle u \iff \mathsf{N}c.(a\ c) \cdot t \approx (b\ c) \cdot u$ and $a \mathbin{\#} \langle b \rangle u \iff \mathsf{N}c.a \mathbin{\#} (b\ c) \cdot u$. This approach concentrates the non-determinism in name–name constraints, which suggest that a practical approach may be to delay attempts to solve such constraints as long as possible.

We expressed equivariant unification in terms of permutation terms and variables. In contrast, in nominal logic, only the swapping operator is present; general permutations are not. It is not clear how solutions involving permutation variables produced by our algorithm relate to nominal logic. Thus, it would be an advantage if permutation variables could be eliminated from the results of logic programming queries. This issue needs to be investigated.

We have developed a prototype implementation of equivariant unification using Constraint Handling Rules [5], which are available in many Prolog implementations. This helped identify some subtle issues and is a first step towards incorporating nominal abstract syntax into standard logic programming languages.

7   Conclusions

Equivariant unification and matching are computationally hard problems requiring subtle algorithmic techniques. Solutions to these problems are necessary for complete implementations of nominal rewriting and logic programming. This paper makes two contributions building upon an important technical device called *permutation graphs*. We present an equivariant unification algorithm, the first terminating algorithm for this problem. This algorithm can be viewed as a nondeterministic polynomial time algorithm for reducing equivariant unification problems to finite complete sets of solutions. It is evident from the structure of the algorithm that the only sources of nondeterminism in equivariant unification are swappings and abstractions. Based on this observation, we developed an algorithm for efficient matching of swapping-free grounded terms. This algorithm can be used to run interesting nominal rewrite systems that do not work properly using nominal unification alone. However, there are several potential efficiency problems which will need to be addressed for equivariant unification to be practical.

Acknowledgments

This work was supported by EPSRC grant R37476. The author wishes to thank the anonymous reviewers for their comments.

References

1. J. Cheney and C. Urban. Alpha-Prolog: A logic programming language with names, binding and alpha-equivalence. In *Proc. 20th Int. Conf. on Logic Programming (ICLP 2004)*, number 3132 in LNCS, pages 269–283, 2004.
2. James Cheney. The complexity of equivariant unification. In *Proceedings of the 31st International Colloquium on Automata, Languages and Programming (ICALP 2004)*, volume 3142 of *LNCS*, pages 332–344. Springer-Verlag, 2004.

3. James R. Cheney. *Nominal Logic Programming*. PhD thesis, Cornell University, Ithaca, NY, August 2004.
4. Maribel Fernández, Murdoch Gabbay, and Ian Mackie. Nominal rewriting systems. In *Proceedings of the 6th Conference on Principles and Practice of Declarative Programming (PPDP 2004)*, 2004. To appear.
5. Thom Frühwirth. Theory and practice of constraint handling rules. *Journal of Logic Programming*, 37(1–3):95–138, October 1998.
6. M. J. Gabbay and A. M. Pitts. A new approach to abstract syntax with variable binding. *Formal Aspects of Computing*, 13:341–363, 2002.
7. A. M. Pitts. Nominal logic, a first order theory of names and binding. *Information and Computation*, 183:165–193, 2003.
8. M. R. Shinwell, A. M. Pitts, and M. J. Gabbay. FreshML: Programmming with binders made simple. In *Proc. 8th ACM SIGPLAN Int. Conf. on Functional Programming (ICFP 2003)*, pages 263–274, Uppsala, Sweden, 2003. ACM Press.
9. Wayne Snyder. *A Proof Theory for General Unification*, volume 11 of *Progress in Computer Science and Applied Logic*. Birkhäuser, 1991.
10. C. Urban and J. Cheney. Avoiding equivariance in alpha-Prolog. To appear, 2004.
11. C. Urban, A. M. Pitts, and M. J. Gabbay. Nominal unification. *Theoretical Computer Science*, 323(1–3):473–497, 2004.

Faster *Basic Syntactic Mutation* with Sorts for Some Separable Equational Theories

Christopher Lynch[1] and Barbara Morawska[2]

[1] Department of Computer Science, Box 5815, Clarkson University,
Potsdam, NY 13699-5815, USA
clynch@clarkson.edu
[2] Chair for Automata Theory, Institute for Theoretical Computer Science,
Dresden University of Technology, Germany
morawska@tcs.inf.tu-dresden.de

Abstract. Sorting information arises naturally in E-unification problems. This information is used to rule out invalid solutions. We show how to use sorting information to make E-unification procedures more efficient. We illustrate our ideas using Basic Syntactic Mutation. We give classes of problems where E-unification becomes polynomial. We show how E-unification can be separated into a polynomial part and a more complicated part using a specialized algorithm. Our approach is motivated by a real problem arising from Cryptographic Protocol Verification.

1 Introduction

The problem of Equational Unification is the problem of finding a substitution that makes two terms equivalent modulo an equational theory [1]. This problem arises in automated theorem proving and formal methods, such as analysis of cryptographic protocols [9]. In traditional formal methods tools for cryptographic protocol analysis, cryptographic algorithms are modeled as a black box. But it is possible to represent properties of algorithms using equational theories. Such approaches are becoming more and more common [2]. Therefore, E-unification becomes necessary in the analysis of cryptographic protocols. This paper is inspired by a problem in [12] from cryptographic protocol analysis.

The problem of E-unification is undecidable in general. So it is important to find ways to restrict the class of theories to make E-unification more efficient. For example, given an equational theory that is saturated under Paramodulation, E-unification is in NP using Basic Narrowing [10] and Basic Syntactic Mutation (BSM) [7]. If the equational theory is further restricted, BSM runs in polynomial time and gives a most general unifier. This has been used to show that E-unification for an approximation of the commutativity of exponents in the Diffie Hellman protocol runs in polynomial time [6]. This approximation is used in practice in the NRL Protocol Analyzer [8].

This class of theories is restrictive. It would be helpful if we could use real-life properties of the theory under consideration in order to reduce the search space further. Here, we use sorts to define whether a term makes sense. And then terms

J. Giesl (Ed.): RTA 2005, LNCS 3467, pp. 90–104, 2005.

that are not well-sorted do not have to be considered in the unification procedure. Sorts are also used to instantiate some variables. For example, a delete operation only makes sense when applied to an item and a set containing that item.

Specifically, we reconsider BSM, and give conditions for which BSM with sorts runs in polynomial time. We also give conditions for which a theory can be separated into two parts such that the E-unification procedure is divided into a polynomial time part and a part that runs a specialized E-unification procedure. The polynomial part can be performed first followed by the specialized procedure, with no interaction between them.

We illustrate our ideas with an example of the key hierarchy theory from cryptographic protocol analysis. This models a protocol where member keys are stored in a tree, along with an associated group key. Member keys can be added to or deleted from the tree. It is possible to access the group key if a member key is known. We can view the tree as a set. This theory is closed under paramodulation, so BSM runs in NP. We show how to use sorts to separate it into a polynomial time procedure yielding a most general unifier, followed by a separate procedure for E-unification modulo the theory of sets.

The paper is organized as follows. In Section 2, we introduce the key hierarchy problem that motivates our idea, and give some observations to show why sorts are needed. In Section 3, we introduce the kind of sorts used in this paper, and the properties they should have. Then we show how certain violations of the properties can be rectified, so that theories that do not obey all the properties can be forced to obey them. In Section 4, we give a sorting procedure, which we use later to provide sorts to unsorted variables. Then we give Sort Expansion inference rules that use that Sort procedure to give sorts to the variables in the equational theory and the goal, and sometimes expand the terms in the goal. In Section 5, we define a separable equational theory as one where the E-unification problem for different theories can be solved separately. In Section 6, we show how the Basic Syntactic Mutation procedure is modified for sorts. In Section 7, we show how sorts are used to make E-unification polynomial for certain theories. Finally, in Section 8 we relate our work to previous work on sorts. Proofs appear in the full version at `http://www.clarkson.edu/~clynch/papers/bsmsort_full.ps`.

2   Problem

In [7] we have shown that for some equational theories, called *deterministic*, we have a deterministic E-unification algorithm which runs in $O(n^2)$ time. E is *subterm collapsing* if there are terms s and t with s a proper subterm of t, and s equal to t modulo E. We call E *deterministic* if E is not subterm-collapsing, no two equations in E have the same root symbols at their sides, neither t nor s is a variable and if $s \approx t \in E$, then the root symbol of s is not the same as the root symbol of t.

Here we present another set of theories, which can allow subterm collapsing axioms, but also have a very fast E-unification solver. We first explain our ideas on an example of an equational theory of key hierarchy in security protocols[1]:

---

[1] In this example, x, y and M are variables.

$$E = \{pick(x, tree(y, f(x, M))) \approx y$$
$$add(x, tree(y, M)) \approx tree(y, f(x, M))$$
$$delete(x, tree(y, f(x, M))) \approx tree(y, M)\}$$

Here we want to deal with the member keys in a group of users and a group key. A group key is a root of a tree of member keys, and the member keys are leaves of that tree. $tree(y, M)$ means that y is a root of a tree with leaves M. $f(x, M)$ is just a set constructor like $cons(x, M)$ but for sets, and not for lists, hence x is an element and M is a set of elements, i.e. member keys. $pick(x, tree(y, f(x, M)))$ allows us to retrieve a group key if a member key is known. add and $delete$ adds and deletes member keys in a tree.

The theory is closed under Paramodulation[2], hence *Basic Syntactic Mutation* (*BSM*) provides an NP-algorithm for E-unification in E. The first step in the BSM-algorithm is to close a given E under the Right-Hand-Side Critical Pair rule[3], which in our example yields the following equational theory:

RHS(E)

$$= E \cup \{delete(x', tree(y, f(x', f(x, M)))) \approx add(x, tree(y, M)),$$
$$delete(x, tree(y, f(x, M))) \approx delete(x', tree(y, f(x', M))),$$
$$pick(x, tree(y, f(x, M))) \approx pick(x', tree(y, f(x', M'))),$$
$$pick(x, tree(tree(y', f(x', M')), f(x, M))) \approx add(x', tree(y', M'),$$
$$pick(x, tree(tree(y', M'), f(x, M))) \approx delete(x', tree(y', f(x', M'))) \}$$

We can see that $RHS(E)$ is not a deterministic equational theory. It is subterm-collapsing (e.g. $pick(x, tree(y, f(x, M))) \approx y$), the first condition is satisfied, but the second is not $(pick(x, tree(y, f(x, M))) \approx y)$ and neither is the third (e.g. $pick(x, tree(y, f(x, M))) \approx pick(x', tree(y, f(x', M'))))$.

The situation is still worse if we add axioms defining sets of member keys in order to obtain the full theory for key hierarchy.

$$E^* = E \cup \{f(x, f(y, M)) \approx f(y, f(x, M)),$$
$$f(x, f(x, M)) \approx f(x, M) \}$$

E^* is not finitely saturated under Paramodulation.
That leads us to some observations.

1. Considering intended interpretation, we can restrict subterms appearing in *pick*-, *add*-, *delete*-, *tree*- and f-terms[4]. A term $pick(u, v)$ makes no sense if u is not a member-key, and v a tree, $tree(s, t)$, where s is a group-key, and t a term representing a set of member-keys with u an element of t. Similar restrictions can be imposed on *add*-, *delete*-, *tree*- and f-expressions.
2. A *pick*-expression represents an operation on trees which returns a group key; *add*- and *delete*- expressions operate on trees and return a tree.

---

[2] See section 6 for definitions of Paramodulation and $RHS(E)$.
[3] The rule appears again on page 100.
[4] An f-term is a term with root symbol f.

3. f-expressions are troublesome, because they represent sets, and in order to enable them to properly unify, we have to add identities to E, which destroy its closure under Paramodulation, and inevitably add to the complexity of E-unification[5]. On the other hand we can postpone unification of f-terms until all *pick-*, *add-*, *delete-*, and *tree-* terms are eliminated from the goal. Hence our procedure may have two phases: first – polynomial, and the second one, nondeterministic polynomial, dealing with sets. At any rate, if the goal is ground, this phase can also be made polynomial.

We can look at it semantically: there is a universe of objects (in our example the objects are group-keys, member-keys, trees and sets). Terms in our goal (if it is unifiable) should map to some objects in the universe. But terms which are not well-sorted, do not map to anything in the universe.

The second observation is also justified by this semantic point of view. There are no special objects in our universe such as *pick*, *add* and *delete*. *pick* represents a group-key, which was accessed in a given way, and *add* and *delete* refer to trees. Assuming that our goal is E-unifiable, all its variables map eventually to objects: trees, group-keys, member-keys or sets of member-keys.

3 A Sort Theory Associated with an Equational Theory

The previous section suggests that we should use some sort theory to help us in the process of E-unification by restricting the search space for a solution. Which sort theory to use for this purpose is perhaps mostly decided by semantic information, outside of the possibilities of purely syntactic analysis of equations in E. Nevertheless we can have some requirements which make a sort theory suitable for our purposes. See [13] for the definition of a sort theory.

- **Simplicity.** The first such requirement is that the sort theory is simple, i.e. for every function symbol there is at most one sort declaration in it. For example, for our equational theory we could use the following sort theory:

$$L_E = \{G(pick(x_M, tree(y_G, f(x_M, z_S)))), T(add(x_M, y_T)),$$
$$T(delete(x_M, tree(y_G, f(x_M, z_S)))), T(tree(x_G, y_S)), S(f(x_M, y_S))\}$$

   where G represents group-keys, M represents member-keys, T represents trees and S represents sets. Notice that this sort theory is simple. The notation x_M means that x is of sort M.
- **Consistency with E.** If $u \approx v \in E$, then there are terms s, t and sort declarations in the sort theory $S(s)$, $S(t)$, such that there is a matching substitution σ, with $s\sigma = u$ and $t\sigma = v$. Obviously, L_E is consistent with E or $RHS(E)$.

   The next requirement ensures that in each sort there is at least one term E-unifying with all the other terms in this sort.

---

[5] We could also apply the inferences modulo the theory of sets, but this also adds to the complexity.

- **Property (*)**

  In each sort in a sort theory L, there must be at least one sort declaration with a term u, such that for each other term v in the sort declaration of the same sort, there is a substitution σ with $E \models u\sigma \approx v\sigma$.

  Our sort theory L_E satisfies this requirement. It is enough to inspect the equations in E or $RHS(E)$ to see that e.g. any term in the sort T can be chosen to be the term u in the description of the above requirement.

In order to define the fourth requirement, we first define a partial order on the terms appearing in sort declarations. If $S(f(u_1, \ldots, u_n)), S(g(v_1, \ldots, v_m))$ are sort declarations in L, we say that $f(u_1, \ldots, u_n) \geq_L g(v_1, \ldots, v_m)$, if g appears at a root position in a proper subterm of $f(u_1, \ldots, u_n)$.

Definition 1. *Given a sort theory L satisfying Property (*), for each sort S in L, we call a term u appearing in a sort declaration $S(u)$, such that u is minimal with respect to $\geq_L$ and for each other term v in the same sort, there is a substitution σ with $E \models u\sigma \approx v\sigma$ a **minimal term in the sort** S.*

- **Normal Term Requirement.** We require that for each sort S in L, there is a minimal term in S which does not contain any proper subterm of the sort S, or is of the form $g(x_S)$ and $E \models g(x_S) \approx x_S$. We call such a term, a **normal term in the sort** S.

In L_E, the normal term for the sort T can only be $tree(x_G, y_S)$, because it is the smallest term with the required property in this sort. However, L_E does not satisfy the normal term requirement, because it contains the sort declaration: $G(pick(x_M, tree(y_G, f(x_M, z_S))))$, which is the only sort declaration for the sort G, but the term $pick(x_M, tree(y_G, f(x_M, z_S)))$ does contain y_G which is not the only argument in this term. A similar problem appears with the sort declaration for sets, S. The term in the sort declaration $S(f(x_M, y_S))$ contains a subterm of the same sort.

Definition 2. *Given a sort theory L and an equational theory E, if L satisfies all the above requirements, then L is called a sort theory associated with E.*

Now we show how to refine a simple sort theory with Property (*) but not satisfying the normal term requirement for a sort theory associated with E, in such a way that we can have a sort theory satisfying all the requirements.

The normal term requirement is natural, because if we think about sorted terms as terms mapping into a domain of objects, the sort of such terms must be well-defined. In any case we can define a function $g(x)$ such that each term of a given sort is a fix point for this function (hence $g(v) = v$). Notice that if we add the sort declaration $S(g(x_S))$ to L and an equation $g(x) \approx x$ to E, $g(x_S)$ will satisfy requirements for the normal term in the sort S.

So we can modify a simple sort theory L with Property (*) which does not satisfy the normal term condition by adding to it a new function symbol and a sort declaration. At the same time we have to modify the equational theory E in such a way that the new sort theory is associated with the modified E.

Definition 3. *1. Given a sort theory L and an equational theory E, if L has Property (*), but for a sort S there is no term which can be normal in this sort, we add to L a new sort declaration $S(g(x))$, where g is a fresh function symbol, and set $g(x)$ as normal term in the sort S.*

2. If L' is a sort theory obtained in 1, and g is a new function symbol added to L in the process of the refinement, we add the equation $g(x) \approx x$ to E.

We call L' obtained in such a way, a **refinement** *of L, and E', which is obtained from E in 2, an equational theory* **modified for** *L'.*

If L is a simple sort theory, then the refined L' is also a simple sort theory. In our example, from L_E we obtain the following refined sort theory:

$$L'_E = \{G(pick(x_M, tree(y_G, f(x_M, z_S)))), G(group\text{-}key(x_G)), T(add(x_M, y_T)),$$
$$T(delete(x_M, y_T)), T(tree(x_G, y_S)), S(f(x_M, y_S)), S(set(x_S))\}$$

The equational theory E' modified for L' is the following:

$$E' = \{pick(x, tree(y, f(x, M))) \approx y$$
$$add(x, tree(y, M)) \approx tree(y, f(x, M))$$
$$delete(x, tree(y, f(x, M))) \approx tree(y, M)\} \cup$$

fixpoint equations: $\{group\text{-}key(x) \approx x, set(x) \approx x\}$

As a consequence of Definition 3 we know that if L is a simple sort theory such that L has Property (*) with respect to some equational theory E, and if L' is a refined sort theory obtained from L, and E' is an equational theory modified for L', then L' is a sort theory associated with E'.

4   *Sort* – A Procedure to Sort a Goal

We assume we have a sort theory L associated with an equational theory E. We want to use the sort theory for two tasks: to check whether the terms in the goal are not ill-sorted and to assign appropriate sorted terms to the variables in the goal in such a way, that the terms appearing there are sorted. For these purposes we use a unification procedure *Sort* which is defined by the following rules and an arbitrary selection rule[6].

1. If the procedure is called on a set of equations, we require that on one side of each equation is a renamed subterm from a sort theory.
2. A solved equation is an equation of the form
   - $x \approx v_S$, where x is an unsorted variable from the goal or equational theory, and v is a renamed subterm from a sort declaration of a sort theory, where x doesn't appear anywhere else in the goal, or
   - an equation of the form $u_S \approx v_S$, where both terms are renamed subterms from a sort theory.
3. Apart from the usual syntactic unification rules (Tautology, Orientation, Decomposition, Clash, Cycle), we have the following:

---

[6] This procedure was inspired by the one from [13].

(a) **Weakening**

$$\frac{v_S \approx f(v_1,\ldots,v_n)}{v_S \approx f(s_1,\ldots,s_n), f(s_1,\ldots,s_n) \approx f(v_1,\ldots,v_n)}$$

where $f(v_1,\ldots,v_n)$ is an unsorted term, $S(f(s_1,\ldots,s_n))$ is a renamed sort declaration for f-terms, and v_S is a subterm from a renamed sort declaration from the sort theory, of the sort S (possibly a variable of the sort S) which doesn't have f as its root symbol.

(b) **Sorted Fail**

$$\frac{v_S \approx f(v_1,\ldots,v_n)}{FAIL}$$

if v_S is a subterm (possibly a variable) from a sort declaration from the sort theory with sort S, $G(f(s_1,...,s_n))$ is a sort declaration for f-terms, and $G \neq S$ or there is no sort declaration for an f-term in L.

(c) **Variable Elimination** is applied only to unsorted variables.

We will call this procedure on equations of the form $s \approx t$, where s is a term from a sort theory, and t is an unsorted term. The procedure will return a set of substitutions for variables of s and equations of the form $u \approx v$ where both u and v are sorted. $Sortsub(s \approx t)$ is the substitution determined for the variables of s by the procedure. $Sorteqs(s \approx t)$ is the set of equations with sorted terms on both sides that result from the procedure.

By inspection of the rules of $Sort$, we can make sure that the rules preserve the form of the goal passed to the procedure, as expressed in the following lemma:

Lemma 1. *Given a goal equation where one side is a subterm from a sort declaration then the rules of* Sort *apply deterministically, and the conclusion of each rule is either* FAIL *or a set of equations, each of which has a subterm from the sort declaration on one side.*

The next lemma states that *Sort* always halts on a goal of the required form.

Lemma 2. *Given a sort theory L and a goal G such that each equation in G has a renamed subterm of a sort declaration on one side then sort G halts with a set of solved equations or Fails, in only a linear number of steps.*

Sort-Expansion of E

For each equation $f(u_1,\ldots,u_n) \approx g(v_1,\ldots,v_m)$ in E, we apply the following:

$$\frac{f(u_1,\ldots,u_n) \approx g(v_1,\ldots,v_m)}{f(u_1,\ldots,u_n)\sigma \approx g(v_1,\ldots,v_m)\sigma}$$

where there are sort declarations $S(f(s_1,\ldots,s_n))$ and $S(g(t_1,\ldots,t_m))$ in L, such that $\sigma = Sortsub(f(s_1,\ldots,s_n) \approx f(u_1,\ldots,u_n), g(t_1,\ldots,t_n)) \approx g(v_1,\ldots,v_m))$.

We are guaranteed that such σ exists by the fact that L is associated with E. Moreover, we know that for the variables in $f(u_1,\ldots,u_n) \approx g(v_1,\ldots,v_m)$, σ is a renaming assigning sorts to variables, since L is associated with E and therefore there is a matcher τ such that $f(s_1,\ldots,s_n)\tau = f(u_1,\ldots,u_n)$. The original equation in E is replaced by the equation obtained in the conclusion.

For every equation $f(u_1,\ldots,u_n) \approx x$ in E, which is not a *fixpoint equation*, we apply the following rule:

$$\frac{f(u_1,\ldots,u_n) \approx x}{f(u_1,\ldots,u_n)\sigma[x\sigma \mapsto g(v_1,\ldots,v_n)] \approx g(v_1,\ldots,v_n)}$$

where $S(f(s_1,\ldots,s_n))$ is a sort declaration in L such that $\sigma{=}Sortsub(f(s_1,\ldots,s_n) \approx f(u_1,\ldots,u_n))$, obtained by Sort procedure, $x\sigma$ is of sort S, $g(v_1,\ldots,v_m)$ is the normal term in sort S and $[x\sigma \mapsto g(v_1,\ldots,v_n)]$ is a substitution of a renamed variable $x\sigma$ the normal term in the sort S. (Again, for variables in $f(u_1,\ldots,u_n) \approx x$, σ is just a renaming assigning sorts to these variables.)

Notice that in this way we get rid of collapsing equations in E, i.e. equations of the form $x \approx t$ in E, except for the *fixpoint equations*. We will prove that the fixpoint equations do not need to be used in the E-unification inferences.

Sort-Expansion of the Goal

For each equation in the goal, apply one of the following rules:

$$\frac{\{f(u_1,\ldots,u_n) \approx g(v_1,\ldots,v_m)\} \cup G}{\{f(u_1,\ldots,u_n)\sigma \approx g(v_1,\ldots,v_m)\sigma\} \cup G\sigma \cup eqs}$$

where there are sort declarations $S(f(s_1,\ldots,s_n))$ and $S(g(t_1,\ldots,t_m))$ in L, $\sigma = Sortsub(f(u_1,\ldots,u_n) \approx f(s_1,\ldots,s_n), g(v_1,\ldots,v_m) \approx g(t_1,\ldots,t_m))$, and $eqs = Sorteqs(f(u_1,\ldots,u_n) \approx f(s_1,\ldots,s_n), g(v_1,\ldots,v_m) \approx g(t_1,\ldots,t_m))$, obtained by Sort procedure.

$$\frac{\{f(u_1,\ldots,u_n) \approx x\} \cup G}{\{f(u_1,\ldots,u_n)\sigma \approx x\sigma\} \cup G\sigma \cup eqs}$$

where there is a sort declaration $S(f(s_1,\ldots,s_n))$ in L, and $\sigma = Sortsub(f(u_1,\ldots,u_n) \approx f(s_1,\ldots,s_n))$, $x\sigma$ is of the sort S and $eqs = Sorteqs(f(u_1,\ldots,u_n) \approx f(s_1,\ldots,s_n))$, obtained by Sort procedure.

If σ does not exist, we fail, and the goal does not E-unify. If it does exist, we replace the goal with the new one, after application of each rule. We do not apply any expansion rule to an equation which was already expanded or added in the sort-expansion.

We cannot get rid of collapsing equations in the goal, as we did in E, since inference rules may create new collapsing equations. Hence we will have to deal with such equations in the inference rules.

The following definition states the conditions for *Sort* to succeed, i.e. the conditions for *Sort* not to return *Fail*.

Definition 4. *Given a sort theory L, and an equational theory E, we call an equation $u \approx v$ **sortable**, if there are ground substitutions σ and σ' obeying sort restrictions such that all of the following hold*

- *if u is not a variable, there is a sort declaration $S(u')$, such that the root symbol of u is the same as the root symbol of u' and if v is not a variable, there is a sort declaration $S(v')$, such that the root symbol of v is the same as the root of v',*
- *$u'\sigma \approx u\sigma$ and $v'\sigma' \approx v\sigma'$, and*
- *if $x \in dom(\sigma) \cap dom(\sigma')$, then $x\sigma$ and $x\sigma'$ are of the same sort.*

Theorem 1. *Given a sort theory L associated with an equational theory E, and a sortable equation $f(u_1,\ldots,u_n) \approx g(v_1,\ldots,v_m)$ (or $f(u_1,\ldots,u_n) \approx x$), there exists a set of equations, σ, containing no Fail, which is computed by $Sort(f(s_1,\ldots,s_n) \approx f(u_1,\ldots,u_n), g(t_1,\ldots,t_m) \approx g(v_1,\ldots,v_m))$ (or $Sort(f(s_1,\ldots,s_n) \approx f(u_1,\ldots,u_n))$ respectively).*

We have to prove that we do not change an equational theory with respect to a sorted ground domain by the sorted expansion.

Theorem 2. *Given equational theory E and a sort theory L, if E' is obtained by Sort-expansion from E and L is associated with E', then $E \models u \approx v$, u,v are sorted terms and there is a proof of $u \approx v$ obeying sort restrictions, iff $E' \models u \approx v$.*

As an example, if we use the sort theory:

$$L'_E = \{G(pick(x_M, tree(y_G, f(x_M, z_S)))), G(group\text{-}key(x_G)),$$
$$T(add(x_M, y_T)), T(delete(x_M, tree(y_G, f(x_M, z_S)))), T(tree(x_G, y_S)),$$
$$S(f(x_M, y_S)), S(set(x_S))\}$$

to sort the following equational theory:

$$E' = \{pick(x, tree(y, f(x, M))) \approx y$$
$$add(x, tree(y, M)) \approx tree(y, f(x, M))$$
$$delete(x, tree(y, f(x, M))) \approx tree(y, M)\} \cup$$
fixpoint equations: $\{group\text{-}key(x) \approx x, set(x) \approx x\}$

we get the following sorted equational theory:

$$Sorted(E') = \{pick(x_M, tree(group\text{-}key(y_G), f(x_M, z_S))) \approx group\text{-}key(y_G)$$
$$add(x_M, tree(y_G, z_S)) \approx tree(y_G, f(x_M, z_S))$$
$$delete(x_M, tree(y_G, f(x_M, z_S))) \approx tree(y_G, z_S)\} \cup$$
fixpoint equations: $\{group\text{-}key(x_G) \approx x_G, set(x_S) \approx x_S\}$

5   A Separable Equational Theory

At the end of the previous section, we have shown how to sort an equational theory E', but in the beginning of this paper we started with an equational theory E^*, which contains the troublesome equations defining unification between sets of member-keys.

Hence we have the following:

$$Sorted(E^*) = \{pick(x_M, tree(group\text{-}key(y_G), f(x_M, z_S))) \approx group\text{-}key(y)$$
$$add(x_M, tree(y_G, z_S)) \approx tree(y_G, f(x_M, z_S))$$
$$delete(x_M, tree(y_G, f(x_M, z_S))) \approx tree(y_G, z_S)\} \cup$$
$$\{f(x_M, f(y_M, z_S)) \approx f(y_M, f(x_M, z_S)),$$
$$f(x_M, f(x_M, y_S)) \approx f(x_M, y_S)\} \cup$$

fixpoint equations: $\{group\text{-}key(x_G) \approx x_G, set(x_S) \approx x_S\}$

As we have noticed already, the additional equations make the theory not be saturated under Paramodulation, hence prevent us from applying the techniques described in the next section.

But we can notice that the "good" equational theory E' is *separable* from the "bad" part of E^*, so that we can solve an E^*-unification problem in two steps: first by using the E'-unification procedure and then tackling the unification in sets. The general criterion for this is stated in the following definition.

Definition 5. *Given a sort theory L and an equational theory E, sorted by L, if $E' \subset E$ such that if $u \approx v$ is in E', and u and v are of sort S, then there is no subterm in $E \backslash E'$ of sort S, then E' is called separable in E.*

With a goal-directed E-unification procedure, like the one we use in the next section, we can try to solve a goal first with the procedure for E', where E' is separable. This procedure will return an $E \backslash E'$-unification problem.

This is obvious, since if u is a sorted subterm, and its sort is in $E \backslash E'$, then no subterms in u can be changed by equations from E'. Especially, no variables in u can be assigned terms of the sorts of the terms from E'.

6   Basic Mutation Refined with Sorts

We now look again into the rules from [7] in order to show that the above analysis can help us to get better E-unification time. Recall that we can apply the Basic Syntactic Mutation technique only in the case when an equational theory is saturated under Paramodulation. In the presence of sorts, we could require that the theory is closed under Paramodulation obeying sorts from a given sort theory L.

Paramodulation

$$\frac{u[s'] \approx v \quad s \approx t}{u[t]\sigma \approx v\sigma}$$

where $\sigma = mgu(s, s')$, $s\sigma \not\prec t\sigma$, $\prec$ is a reduction ordering on terms.

σ in the above rule should obey sort restrictions, and neither of the premises are fixpoint equations.

Since we assume our sort theory to be simple and $\sigma = mgu(s, s')$, σ not obeying sort restrictions, would mean that at least one of the terms is not well-sorted. But we assume that E is well-sorted, hence σ can be just an mgu of s and s', with no additional regard for sorts.

In Basic Mutation without sorts, $RHS(E)$ was constructed by the RHS Critical Pair rule. In the context of sorts, we also consider sort restrictions:

Right-Hand-Side Critical Pair

$$s \approx t \quad u \approx v$$
$$\overline{s\sigma \approx u\sigma}$$

where $s\sigma \not\prec t\sigma$, $u\sigma \not\prec v\sigma$, $\sigma \approx mgu(v, t)$
and $s\sigma \neq u\sigma$.

σ in the above rule should obey sort restrictions. As in the case of Paramodulation, we should not perform RHS Critical Pair inferences with *fixpoint equations*, which were added to E in sort-expansion. However, if $g(x_S) \approx x_S$ is a fixpoint equation, and a is a constant of sort S, then we must consider the equation $g(a) \approx a$ as a premise in Right-Hand-Side Critical Pair.

In our example (page 98), if E' is sort-expanded and closed under Paramodulation, sorted $RHS(E')$ is the following.

Sorted-$RHS(E')$
$$= E' \cup \{delete(x'_M, tree(y_G, f(x'_M, f(x_M, z_S)))) \approx add(x_M, tree(y_G, z_S)),$$
$$delete(x_M, tree(y_G, f(x_M, z_S))) \approx delete(x'_M, tree(y_G, f(x'_M, z_S))),$$
$$pick(x_M, tree(y_G, f(x_M, z_S))) \approx pick(x'_M, tree(y_G, f(x'_M, z'_S)))\}$$

The Basic Syntactic Mutation (BSM) rules with sorts are presented in [7]. The following is true in general for this set of inference rules:

Theorem 3. *Given a simple sort theory, associated with an equational theory E and a well-sorted goal, the inference rules of Sorted BSM preserve well-sortedness of the terms in the goal.*

As an immediate corollary to the definition of sort-expanded equational theory, we show that any rule involving an equation from E with a variable on one side, such as Variable Mutate, is not needed.

Corollary 1. *Given an equational theory sorted by an associated sort theory, and a goal, the inference rule Variable Mutate never applies to a selected goal equation.*

Now we can state formally completeness of BSM rules with sorts for a sorted equational theory and an equational goal.

Theorem 4. *Given an equational theory E, a sort theory L, E' which is E modified for L so that L is associated with E', such that E is closed under sorted*

Paramodulation, if $E \models G\gamma$ and $G\gamma$ is not solved, then there is a rule in the sorted BSM-rules based on sort-expanded E', such that $G \longrightarrow G'$ and there is $\gamma' \leq_E \gamma|_{Var(G)}$ and $E \models G'\gamma'$.

7   Redundancy of Rules and Determinism

We will show that sort restrictions can make some rules in the Sorted BSM procedure redundant. By deleting those rules from the set of our inference rules we obtain a Sorted BSM E-unification procedure, which in some cases is deterministic and polynomial.

BSM solves E-unification in non-deterministic polynomial time for theories saturated under Paramodulation. We will show that Sorted Deterministic BSM will solve it in polynomial time for our key hierarchy example.

Definition 6. *A rule R :*

$$\frac{s \approx t}{s_1 \approx t_1, \ldots, s_n \approx t_n}$$

is redundant in the set of rules S, such that S does not contain R, iff there is a rule R' in S applicable to $s \approx t$, such that $s_1 \approx t_1, \ldots, s_n \approx t_n$ is an instance of the conclusion of R' applied to $s \approx t$.

It is easy to check if some rules in the set of rules for BSM with sorts, for each function symbol in a given signature and a given equational theory are redundant. Deleting redundant rules may result in eliminating all choices between applications of different rules or choices of different equations from the equational theory for inferences, and hence in a deterministic procedure. By Corollary 1, we already know that there is no conflict between Mutate and Variable Mutate in Basic Mutation with Sorts, assuming that we have a sort theory associated with a sorted equational theory saturated under Paramodulation.

7.1   Collapsing Goal Equations

First we define a rule called Sort Imitation that is applied eagerly. This rule replaces all rules in BSM, which apply to equations of the form $x \approx v$. This is possible, because of the normal term requirement for a sort theory L associated with E.

Sort Imitation

$$\frac{\{x_S \approx v_S\} \cup G}{\{\boxed{g(v_1, \ldots, v_n)_S} \approx v_S[x \mapsto \boxed{g(v_1, \ldots, v_n)}]\} \cup G[x \mapsto \boxed{g(v_1, \ldots, v_n)}]}$$

where $x_S \approx v_S$ is selected (hence not solved) and $g(v_1, \ldots, v_n)_S$ is a normal term in the sort S in a sort theory L associated with E. This rule does not destroy the argument for termination for an E-unifiable sorted goal.

7.2   When Decomposition Is Redundant

We will show that if $f(u_1, \ldots, u_n) \approx f(v_1, \ldots, v_n)$ is in an equational theory, then Decomposition is often redundant. This is stated formally, with all needed assumptions, in the following theorem:

Theorem 5. *Let E be an equational theory. If L is a sort theory associated with E, $f(u_1, \ldots, u_n) \approx f(v_1, \ldots, v_n) \in E$, and all shared variables in $f(u_1, \ldots, u_n)$ and $f(v_1, \ldots, v_n)$ are at the same positions in these two terms, then the Decomposition rule for f-terms is redundant in the set of rules of Sorted-BSM.*

7.3   Deterministic BSM with Sorts

We are now ready to state the conditions which make our E-unification procedure with sorts terminating and deterministic.

Definition 7. *Given a sort theory L, an equational theory E is called L-deterministic if:*

1. *no two equations in E have the same root symbols at their sides,*
2. *L is associated with E.*

We show that if $E'' = RHS(E')$, E' is finite and saturated by Paramodulation, and there is a sort theory L associated with E, which is E'' modified for L, then if E is also L-deterministic, then BSM can be turned into a deterministic algorithm, which means that it will halt in a linear number of steps.

We define *Sorted-BSMd* as a deterministic version of *Sorted-BSM*, based on any selection rule and the rules in Figure 1.

Theorem 6. *Let $E'' = RHS(E')$, such that E' is finite and saturated by Paramodulation, and let L be a sort theory associated with E, which is E'' modified for L. Then if E is L-deterministic, the algorithm Sort-BSMd solves a goal G deterministically in polynomial time. E with sorts is then unitary.*

8   Conclusion

We have shown how sorts are necessary in real problems of E-unification. The example motivating our work is the key hierarchy theory from Cryptographic Protocol Analysis. Reasoning modulo equational theories is currently an important topic in Cryptographic Protocol Analysis, and so our results are applicable there. But the characteristics of our example problem are general, and apply to many E-unification problems from verification, such as modeling data structures with equational theories. In our example, we showed that the E-unification problem can be divided into two stages: the first stage runs in polynomial time, and the second stage can be solved solely by reasoning modulo the theory of sets. We believe that many natural problems have this same structure.

Our use of sorts has some differences with other uses of sorts. We deal with sorts in the context of E-unification. Much of the previous work considers sorts with syntactic unification, although there is some work dealing with equational

Decomposition:

$$\frac{\{f(u_1, \cdots, u_n) \approx f(v_1, \cdots, v_n)\} \cup G}{\{u_1 \approx v_1, \cdots, u_n \approx v_n\} \cup G}$$

where $f(u_1, \cdots, u_n) \approx f(v_1, \cdots, v_n)$ is selected and Decomposition is not redundant for f-terms.

Mutate:

$$\frac{\{f(u_1, \cdots, u_n) \approx g(v_1, \cdots, v_m)\} \cup G}{\bigcup_i \{u_i \approx \boxed{s_i}\} \cup \bigcup_i \{\boxed{t_i} \approx v_i\} \cup G}$$

where $f(u_1, \cdots, u_n) \approx g(v_1, \cdots, v_m)$ is selected, $f(u_1, \cdots, u_n)$ is not boxed, $f(s_1, \cdots, s_n) \approx g(t_1, \cdots, t_m) \in E$.

Sort Imitation:

$$\frac{\{x_S \approx v_S\} \cup G}{\{\boxed{g(v_1, \ldots, v_n)_S} \approx v_S[x \mapsto \boxed{g(v_1, \ldots, v_n)}]\} \cup G[x \mapsto \boxed{g(v_1, \ldots, v_n)}]}$$

where $x_S \approx v_S$ is selected and $g(v_1, \ldots, v_n)_S$ is a normal term in the sort S in a sort theory L associated with E.

Variable-Variable Elimination:

$$\frac{x_S \approx y_S, \quad G}{x \approx y, \quad G[x \mapsto y]} \qquad \left| \quad \frac{x \approx x \cup G}{G} \right.$$

where both x and y appear in G

Fig. 1. Sorted *BSMd* Inference Rules.

theories[3–5, 11]. However, there are other ways where our work differs. We do not consider that the equational theory and the goal are sorted. Instead, we have procedures that give sorts to the variables in the equational theory and the goal. We require that the sort theory is simple. Much other work only requires that the sort theory is order-sorted. Deletion requires that an item can only be deleted from a set if it is in the set. We could not capture this property in an order-sorted theory. However, we do not try to deal with subsorts. We also require certain properties of the equational theory, such that the terms of the equations are instances of the sorts, and each sort must have some normal form that is E-unifiable with everything else in the sort.

The properties we require naturally occur in many real life problems. We showed in this paper that these properties imply a more efficient algorithm for Sorted E-unification. Although our approach differs from other approaches on sorts, we also have some similarities. For example, our Sort algorithm is inspired by the Syntactic Sorted Unification procedure of [13], however we use it mainly to give sorts to unsorted variables. Since we do not need to unify sorted terms, it makes the algorithm more efficient.

104   Christopher Lynch and Barbara Morawska

The results of this paper should lead to more efficient E-unification procedures, because it bases the procedure more closely on the meaning of the terms.
We have also begun applying these ideas to Narrowing.

References

1. F. Baader and W. Snyder. Unification Theory. In J.A. Robinson and A. Voronkov,
   editors, Handbook of Automated Reasoning, volume I, pages 447–533. Elsevier
   Science Publishers, 2001.
2. H. Comon Intruder Theories (Ongoing Work). In Proceedings of the 7th International Conference on Foundations of Software Science and Computation Structures
   (FoSSaCS'04), Barcelona, Spain, March 2004, LNCS 2987, pages 1-4. Springer.
3. J. A. Goguen and J. Meseguer. Order-Sorted algebra I: equational deduction
   for multiple inheritance, overload, exceptions and partial operations. *Theoretical
   Computer Science*, 105, 217–273, 1992.
4. C. Hintermeier, C. Kirchner and H. Kirchner. Dynamically-Typed Computations
   for Order-Sorted Equational Presentations. In S. Abiteboul and E. Shamir (eds.):
   Automata Language and Programming: 21st International Colloquium, ICALP'94,
   Vol. 820 of LNCS, pp. 450–461, Springer, 1994.
5. J. P. Jouannaud and C. Kirchner. Solving Equations in Abstract Algebras: A Rule-
   Based Survey of Unification. In J. Lassez and G. Plotkin (eds.): *Computational
   Logic, Essays in Honor of Alan Robinson*. MIT Press, Chap. 8, pp. 257–321, 1991.
6. C. Lynch and C. Meadows Sound Approximations to Diffie-Hellman Using Rewrite
   Rules. In Proceedings of the Sixth International Conference on Information and
   Communications Security. Malaga, Spain. October, 2004.
7. C. Lynch and B. Morawska. Basic Syntactic Mutation. In Proceedings of Confer-
   ence on Automated Deduction (CADE), Vol. 2392 of LNAI, 471–485, 2002.
8. C. Meadows Analysis of the Internet Key Exchange Protocol Using the NRL
   Protocol Analyzer, Proceedings of the 1999 IEEE Symposium on Security and
   Privacy, IEEE Computer Society Press, May 1999.
9. C. Meadows. Formal Methods for Cryptographic Protocol Analysis: Emerging
   Issues and Trends, IEEE Journal on Selected Areas in Communication, Vol. 21,
   No. 1, pp. 44-54, January 2003.
10. R. Nieuwenhuis. Decidability and Complexity Analysis by Basic Paramodulation.
    Information and Computation, 147:1-21, 1998.
11. G. Smolka, W. Nutt, J. A. Goguen and J. Meseguer. Order-Sorted Equational
    Computation. In H. Ait-Kaci and M. Nivat (eds.): *Resolution of Equations in
    Algebraic Structures*, Vol. 2 of *Rewriting Techniques*. Academic Press, Chap. 10,
    pp. 297–267, 1989.
12. D. Wallner, E. Harder, and Ryan C. Agee. Key Management for Multicast: Issues
    and Architectures, RFC 2627, June 1999.
13. C. Weidenbach. Sorted Unification and Tree Automata in Bibel W. and Schmitt
    P. H. , editors, Automated Deduction – A Basis for Applications, Volume 1 of
    Applied Logic, Chapter 9, Kluwer, pp. 291-320, 1998

Unification in a Class of Permutative Theories

Thierry Boy de la Tour and Mnacho Echenim

LEIBNIZ laboratory, IMAG – CNRS,
INPG, 46 avenue Félix Viallet F-38031 Grenoble Cedex
{Thierry.Boy-de-la-Tour,Mnacho.Echenim}@imag.fr

Abstract. It has been proposed in [1] to perform deduction modulo leaf permutative theories, which are notoriously hard to handle directly in equational theorem proving. But unification modulo such theories is a difficult task, not tackled in [1]; a subclass of flat equations has been considered only recently, in [2]. Our emphasis on group theoretic structures led us in [6] to the definition of a more general subclass of leaf permutative theories, the *unify-stable* theories. They have good semantic and algorithmic properties, which we use here to design a complete unification algorithm.

1 Introduction

An equation $t = t'$, joining two linear terms t and t', is *leaf permutative* if t is a variant of t', i.e. one term is obtained from the other by a permutation of its variables. As the axiom of commutativity, the simplest non-trivial example, these equations cannot be oriented, an obvious obstacle to termination. The usual way to handle this is to embed these equations directly in the deductive mechanisms, including unification.

These lp-equations (lp is short for leaf permutative) are significantly more difficult to handle than commutativity, especially if we consider a set E of lp-equations, or lp-theory, because of possible interactions. If we consider for example the two lp-equations

$$f(x, y, z, u) = f(y, x, z, u),$$
$$f(x, y, z, u) = f(y, z, u, x),$$

then we can deduce the lp-equation $f(x, y, z, u) = f(x\sigma, y\sigma, z\sigma, u\sigma)$ for any permutation σ of $\{x, y, z, u\}$. These permutations are of course the elements of the symmetric group $\mathrm{Sym}(\{x, y, z, u\})$. Lp-theories naturally lead to permutation groups, which opens the perspective of using group theoretic tools in specialized deductive mechanisms. This was originally illustrated in [1].

Unfortunately, lp-equations cannot be considered simply as generators in a group. Consider for example the two equations $f(a, x, y) = f(a, y, x)$ and $f(x, y, c) = f(y, x, c)$, and the term $t = f(a, b, c)$, where a, b and c are constants. We may deduce $t = f(a, c, b)$, i.e. apply the permutation $(b\ c)$ to t, and we may also deduce $t = f(b, a, c)$, with permutation $(a\ b)$, but we cannot deduce

J. Giesl (Ed.): RTA 2005, LNCS 3467, pp. 105–119, 2005.

$t = f(c, b, a)$, i.e. we cannot apply the permutation $(a\ b)(b\ c) = (a\ c\ b)$ to t. Permutations cannot be freely composed, they are context dependent, which fundamentally departs from group theory. Here, the context of the first equation, $f(a, \circ, \circ)$, differs from the context of the second equation, $f(\circ, \circ, c)$.

These conflicts can be avoided by deciding that, at any possible position, one context applies at the exclusion of other contexts. Thus the notion of *stratified terms* from [1], which are terms T where contexts and their associated groups are attached to occurrences of function symbols in a non-overlapping way. Applying any permutation in one of these groups is safe since no context is affected. The so-called *stratified set* of terms thus obtained, noted S[T], which can be seen as a group theoretic construction (see [5]), is included in a congruence class modulo E, but may not be complete.

Of course, there is still the possibility of covering an E-congruence class as a finite union of stratified sets. However, even under this hypothesis, or if we only consider complete stratified sets, the unification algorithm given in [1] is not sufficient. Since it returns a set of most general substitutions θ such that S[T]θ and S[T']θ have a non-empty intersection, it basically performs unification of stratified sets, not E-unification (as is duly acknowledged in [1, p. 261]).

The difference can be illustrated on an example. We consider an lp-theory E with only one axiom:

$$f(g(x, y), z) = f(g(x, z), y),$$

so that complete stratified terms always exist. Since no subterm of $t = f(x, a)$ and $t' = f(y, b)$ is an instance of the context $f(g(\circ, \circ), \circ)$, then t and t' are complete stratified terms, and S[t] $= \{t\}$, S[t'] $= \{t'\}$. Since t and t' are not syntactically unifiable, then neither are their stratified sets. However, these two terms are E-unifiable, and, as we will see later, the substitution $\{x \leftarrow g(x_1, b), y \leftarrow g(x_1, a)\}$ is a most general E-unifier. The clauses $P(t)$ and $\neg P(t')$ are E-unsatisfiable, which would not be recognized by unifying stratified sets (unless the lp-axiom is added to the clauses, which is clearly unacceptable).

In [6], we introduced a class of lp-theories, so-called *unify-stable*, where every term t can be lifted in a consistent way to a complete stratified term lift(t). This is non-trivial in the sense that the contexts used as labels in lift(t) may not appear explicitly in the axioms of E. In the present paper we give a E-unification algorithm that is complete for all unify-stable theories E. It is more complex than C-unification (see e.g. [3]) because we have to consider possible expansions to the relevant contexts, as in the previous example.

In Section 2 we give a motivated definition of unify-stable theories, and prove a basic property of such theories. This is proved using a result from [6], which allows to keep the group-theoretic formalization to a minimum. Section 3 presents a rule-based unification algorithm, as well as its basic properties. Most efforts are devoted to proving termination. Compared to commutativity, a source of inefficiency arises from the necessity to expand with many different contexts; we give in Section 4 a way to reduce the number of contexts to be considered. A conclusion follows.

2   Unify-Stable Theories

We are not going to develop the full technical details necessary to give a clean group-theoretic formalization of stratified terms and sets. We just mention that this treatment uses the notion of *term trees*, and can be found in [5]. Of course, term trees are not unknown in unification theory, since they are used to improve efficiency by sharing subterms. It should however be noted that subterm sharing cannot be used in stratified terms, because occurrences may be freely permuted, e.g. if f is commutative then $f(f(a,b), f(a,b)) = f(f(a,b), f(b,a))$.

We will use the standard notions of terms and substitutions. We will note $s \preceq t$ when s is a subterm of t. We will also need the notion of contexts, which we may consider as ground terms over a signature extended by a special constant $\circ$, called the hole. This constant is special since each occurrence of $\circ$ in a context sometimes behaves like a new variable. More precisely, all holes in a context can be simultaneously substituted by terms (yielding a term) or by contexts (yielding a context). For example, if $c = f(\circ, \circ)$, then $c[s_1, s_2]$ is exactly $f(x_1, x_2)\theta$, where $\theta = \{x_1 \leftarrow s_1, x_2 \leftarrow s_2\}$.

We therefore have on contexts a notion of substitution very similar to the standard one. A context c' is an *instance* of a context c, noted $c \sqsubseteq c'$, if there is a context substitution $[c_1, \ldots, c_n]$ such that $c' = c[c_1, \ldots, c_n]$. We also say that c *subsumes* c'. All properties of the usual subsumption relation on terms are true for $\sqsubseteq$, which is an ordering relation, and not just a quasi-order. In particular, a nonempty set of contexts C with an upper bound (i.e. a common instance) has a least upper bound noted $\sqcup C$.

An lp-equation can be represented by a context c and a permutation σ of the occurrences of $\circ$ in c. More precisely, σ is a permutation of vertices in a term tree for c, but defining these notions is out of the scope of the present paper. The important point is that, in any term $t = c[t_1, \ldots, t_n]$, the permutation σ translates (through a group isomorphism) into a permutation μ on the integers $\{1, \ldots, n\}$, so that the action of σ on t, noted t^σ, is given by $c[t_{1\mu}, \ldots, t_{n\mu}]$ (where the trivial action i^μ is $\mu(i)$). In the sequel we simply identify σ with μ. For example, if τ swaps the two holes of $f(\circ, \circ)$, then $f(a,b)^\tau = (f(\circ, \circ)[a,b])^\tau = f(\circ, \circ)[b,a] = f(b,a)$, so we simply write $\tau = (1\ 2)$.

We use the notation $[\![c, \sigma]\!]$ to represent an lp-equation; in the previous example, the axiom of commutativity of f is $[\![f(\circ, \circ), \tau]\!]$. When the elements of a set of lp-equations $F = \{[\![c, \sigma_1]\!], \ldots, [\![c, \sigma_m]\!]\}$ bear the same context, the group G generated by $\{\sigma_1, \ldots, \sigma_n\}$ contains exactly the consequences of F of the form $[\![c, \pi]\!]$ (G is complete for c), and we directly represent F by $[\![c, G]\!]$. Every lp-theory E can then be uniquely decomposed as $\biguplus_{i=1}^{k} F_i$, where F_i is a set of lp-equations on a context c_i, and the c_i's are all different. Thus E can be represented by $\biguplus_{i=1}^{k} [\![c_i, G_i]\!]$, where G_i is defined from F_i as above. But this set may not be complete for all the contexts c_i's, as illustrated on the following example.

Example 1. We consider the commutativity axiom $[\![c, \tau]\!]$ of f, together with the lp-equation

$$g(f(h(x,y), h(z,u))) = g(f(h(y,x), h(u,z))),$$

which we write $[\![c', \tau']\!]$. This last equation is therefore represented by $[\![c', G']\!]$, where G' is the group generated by τ'. By applying the commutativity of f on the term $g(f(h(x, y), h(z, u)))$, we get

$$g(f(h(x, y), h(z, u))) \; = \; g(f(h(z, u), h(x, y))),$$

hence an equation $[\![c', \pi]\!]$ where $\pi \notin G'$.

The problem here is of course that we have a position p inside c' such that $c \sqsubseteq c'|p$. However, in this case, we also benefit from the fact that the context substitution applied to c is invariant under τ. In other words, the holes of c which may be permuted by its group G are matched to identical contexts in $c'|p$ (the context $h(\circ, \circ)$ in Example 1). We note this condition $c \sqsubseteq_G c'|p$, it permits to easily identify induced permutations, like π in Example 1.

If an lp-theory E, decomposed as $\biguplus_{i=1}^{k}[\![c_i, G_i]\!]$, verifies for all i, j the property $c_i \sqsubseteq c_j|p \Rightarrow c_i \sqsubseteq_{G_i} c_j|p$, we say it is *stable*. It is then possible to build a group $\Gamma_E(c_j)$ (by amalgamating the induced permutations, like π in Example 1) which is complete for c_j. See [6] for details.

This of course does not imply that stratified sets are complete. It is obvious that the lp-theory $A = \{[\![f(a, \circ, \circ), \tau]\!], \ [\![f(\circ, \circ, c), \tau]\!]\}$, given in the introduction, is stable. Yet there is no stratified term T such that $S[T]$ is the A-congruence class of $f(a, b, c)$.

We therefore need a stronger notion. The idea is to build stratified terms (which essentially means choosing once and for all which $[\![c, G]\!]$ to apply at some position p) by unifying conflicting contexts. In order to ensure that this operation does not introduce new conflicts, we came in [6] to the following definition.

Definition 1. *The lp-theory E is* unify-stable *if and only if for all i, j*

1. *if c_i and c_j are unifiable, then $c_i \sqsubseteq_{G_i} c_i \sqcup c_j$,*
2. *if there is a non-variable[1] position $p \neq \varepsilon$ in c_j such that c_i and $c_j|p$ are unifiable, then $c_i \sqsubseteq_{G_i} c_j|p$.*

We now briefly describe how a stratified term $\text{lift}(t)$ is built from a term tree t. We first consider the set C of contexts c_i of which t is an instance. If C is empty, we recursively lift the direct subterms of t. Otherwise, let $c = \sqcup C$, we label the root of t with $[\![c, \Gamma_E(c)]\!]$, and recursively lift the subterms t_i's such that $t = c[t_1, \ldots, t_n]$.

We have proved in [6] that the stratified set $S[\text{lift}(t)]$, i.e. the set of terms obtained from $\text{lift}(t)$ by applying all permutations in the groups labeling $\text{lift}(t)$ (and then removing all labels), is exactly the congruence class of t modulo the unify-stable theory E.

We therefore assume throughout the rest of the paper that E is unify-stable. In order to simplify the treatment, we further suppose that E contains the equations of the form $f(x_1, \ldots, x_n) = f(x_1, \ldots, x_n)$, for all f in the signature Σ.

---

[1] i.e. it does not correspond to an occurrence of $\circ$. Note that this condition is mistakenly missing in [6], but it is easy to see that adding it preserves the results of [6], especially the proof of Theorem 5.

There is no loss of generality, since adding these equations (and contexts) preserves unify-stability, and of course has no semantic influence. The only difference is that, in the lifting algorithm, the set C is never empty.

Definition 2. *Let C be the set of $\sqcup C$ for all nonempty subsets C of $\{c_1, \ldots, c_k\}$ such that all contexts in C have a common instance. Then we define*

$$\mathcal{L} = \{[\![c, \Gamma_E(c)]\!] \mid c \in \mathcal{C}\}.$$

We will then mostly use the following property, for which we give a slightly informal proof.

Theorem 1. *For all terms t and t', we have $t =_E t'$ if and only if there is a $[\![c, G]\!] \in \mathcal{L}$ such that $t = c[s_1, \ldots, s_m]$, $t' = c[s'_1, \ldots, s'_m]$, and there is a σ in G such that for all $i = 1, \ldots, m$, $s_{i^\sigma} =_E s'_i$.*

Proof. The if part is trivial. Conversely, suppose $t =_E t'$. Then by [6] Theorem 5, we have $T = \operatorname{lift}(t) \bowtie \operatorname{lift}(t') = T'$, which means that T and T' are isomorphic trees up to an admissible permutation. Since there must be a label at the root of each tree (because E contains the necessary trivial axioms), then it must be the same label $[\![c, G]\!] \in \mathcal{L}$ for both, and there must be an admissible permutation in $\sigma \in G$ such that $T_{i^\sigma} \bowtie T'_i$, where $T = c[T_1, \ldots, T_m]$ and $T' = c[T'_1, \ldots, T'_m]$. It is clear that s_i is obtained from T_i (and s'_i from T'_i) by removing all labels, and thus $s_{i^\sigma} =_E s'_i$.

3 Unification in Unify-Stable Theories

3.1 Notations and Generalities

E-unification problems will be noted $S = \{t_1 =^?_E t'_1, \ldots, t_n =^?_E t'_n\}$, and their set of E-unifiers $\mathcal{U}_E(S)$. If S is in solved form, it induces a substitution θ_S, which of course belongs to $\mathcal{U}_E(S)$.

Given two E-unifiers θ and θ' of S, we say that θ *E-subsumes* θ' if there is a substitution δ such that $x\theta\delta =_E x\theta'$ for all variables x occurring in S. A set U of E-unifiers of S is *complete* if every E-unifier of S is E-subsumed by an element of U. We also say that U is a *CSU* for S.

If there is a complete set of E-unifiers for S with only one element, this element is a *most general E-unifier* for S (or *mgu*). Since it is generally not the case that a solvable E-unification problem admits a most general E-unifier, as in commutative unification, we will use finite sets M of E-unification problems (or *extended* unification problems), with of course

$$\mathcal{U}_E(M) = \bigcup_{S \in M} \mathcal{U}_E(S).$$

We will also consider *syntactic* unification problems, i.e. with $E = \emptyset$, and note $t =^? t'$ for $t =^?_\emptyset t'$. Solvable unification problems always have a mgu θ such that $\theta\theta = \theta$ (it is *idempotent*); from now on, *the* mgu of a syntactic problem S is the mgu computed by the standard algorithm given in [3].

Example 2. Let E be the unify-stable theory whose associated set of labels is $\mathcal{L} = \{[\![f(\circ,\ldots,\circ), \operatorname{Sym}(n)]\!]\}$. Take the unification problem $f(x_1,\ldots,x_n) =^?_E f(a_1,\ldots,a_n)$, where the x_i's are variables, and the a_i's constants. For $\sigma \in \operatorname{Sym}(n)$, define $\theta_\sigma = \{x_1 \leftarrow a_{1^\sigma},\ldots,x_n \leftarrow a_{n^\sigma}\}$. Then $U = \{\theta_\sigma \mid \sigma \in \operatorname{Sym}(n)\}$ is a minimal CSU for the problem, of cardinality $n!$.

Example 3. We consider the unify-stable theory E with the equation $[\![f(h(\circ),\circ,\circ),$ $(2\ 3)]\!]$ (and the trivial equations for f and h). Let $S = \{f(x,y,a) =^?_E f(x,a,z)\}$, it is easy to see that the three substitutions

$$\theta_1 = \{y \leftarrow a,\ z \leftarrow a\}$$
$$\theta_2 = \{x \leftarrow h(x_1),\ y \leftarrow a,\ z \leftarrow a\}$$
$$\theta_3 = \{x \leftarrow h(x_1),\ y \leftarrow z\}$$

are E-unifiers. Moreover θ_2 is subsumed by θ_1, and E-subsumed by θ_3.

We now need to develop special tools in order to handle contexts, which are not, strictly speaking, Σ-terms.

Definition 3. *A E-unifier θ of $t =^?_E t'$ is under a context c if $t\theta$ and $t'\theta$ are both instances of c.*

The (syntactic) unification problem $t =^? c$ stands for the problem $t =^? c[x_1(c,t),\ldots,x_n(c,t)]$, where the $x_i(c,t)$'s are new variables, which do not occur in t, nor in any other problem $t' =^? c$ or $t =^? c'$. When no confusion is possible, we may write these simply x_i, or y_i...

Given two terms t and t' and a context c, if the syntactic unification problem $\{t =^? c;\ t' =^? c\}$ has a solution, then we note $\gamma_c(t,t')$ the restriction of its mgu to the set of variables appearing in t or t'.

Example 4. Let $t = f(x,h(a,b))$ and $c = f(g(\circ,\circ),\circ)$, the unification problem $t =^? c$ stands for $t =^? c[x_1,x_2,x_3]$, which is solvable, with mgu $\theta = \{x \leftarrow g(x_1,x_2), x_3 \leftarrow h(a,b)\}$.

The fact that $c[x_1,\ldots,x_n]$ is linear, with variables disjoint from any term appearing in a unification problem induces the following properties.

Lemma 1. *For non-variable terms t, t' and context c, consider the syntactic unification problem $\{t =^? c[x_1,\ldots,x_n]; t' =^? c[y_1,\ldots,y_n]\}$, and suppose that θ is a mgu for this problem. If x is a variable of t or t' that appears in $\operatorname{Dom}(\theta)$, then any non-variable subterm of $x\theta$ is an instance of a non-variable subcontext of c.*

Proof. Let x be such a variable, and consider the set P of non-variable subterms of $c[x_1,\ldots,x_n]$ and $c[y_1,\ldots,y_n]$ that correspond to occurences of x in t or t'. If P is empty, then $x\theta$ is a x_i or a y_i. Otherwise, say $P = \{c_1,\ldots,c_m\}$, where the c_i's are linear terms over disjoint sets of variables. If $m = 1$, then $x\theta = c_1$, and the result is immediate. If $m \geq 2$, then (a restriction of) θ is also a mgu of the linear unification problem $\{c_1 =^? c_i \mid i = 2,\ldots,m\}$, and since for all i, $x\theta = c_i\theta$, any non-variable subterm of $x\theta$ is an instance of a subterm of each c_i, and an instance of a non-variable subterm of at least one c_i. This proves the result.

Lemma 2. *Any unifier θ of $t =^?_E t'$ under c is subsumed by $\gamma_c(t, t')$.*

Proof. Let $S = \{t =^? c; \ t' =^? c\}$. Since $t\theta$ and $t'\theta$ are instances of c, there is a substitution δ such that $t\theta = c[x_1, \ldots, x_n]\delta$ and $t'\theta = c[y_1, \ldots, y_n]\delta$. The variables x_i (resp. y_i) do not appear in $t\theta$ (resp. $t'\theta$), hence $t\theta\delta = t\theta$ and $t'\theta\delta = t'\theta$. We also have $\mathrm{Dom}(\theta) \cap \{x_1, \ldots, x_n, y_1, \ldots, y_n\} = \emptyset$, so that

$$c[x_1, \ldots, x_n]\theta\delta = c[x_1, \ldots, x_n]\delta = t\theta = t\theta\delta,$$

and similarly for t'. This entails that $\theta\delta$ is a unifier of S, and therefore is subsumed by its mgu τ. The restriction of $\theta\delta$ to the variables in t, t' is equal to θ, and is also subsumed by the same restriction of τ, i.e. by $\gamma_c(t, t')$.

We finally state an easy property, that we provide without a proof.

Property 1. For t, t' terms unifiable under c', and $c \sqsubseteq c'$, let $\tau = \gamma_c(t, t')$. If $t\tau$ and $t'\tau$ are both instances of c', then $\tau = \gamma_{c'}(t, t')$.

3.2 Transformation Rules

We now show how to perform E-unification. The principles applied here are basically the same as those applied to commutative theories in [3], except for the decomposition rule. To present it in a simple way, we start by defining a function that, given a unification problem, returns an extended unification problem.

Definition 4. *Let t and t' be two terms, none of which is a variable, let S be a unification problem and $[\![c, G]\!]$ be an element of $\mathcal{L}$. If t and t' have no unifier under c, then we define $\mathrm{D}(t, t', [\![c, G]\!], S) = \emptyset$. Otherwise, let $\tau = \gamma_c(t, t')$, let $t\tau = c[s_1, \ldots, s_m]$, $t'\tau = c[s'_1, \ldots, s'_m]$, and for any permutation σ of $\{1, \ldots, m\}$ we extend S to the unification problem*

$$S_\sigma = \{s_{i\sigma} =^?_E s'_i \mid i = 1, \ldots, m\} \ \cup \ \{x =^?_E x\tau \mid x \in \mathrm{Dom}(\tau)\} \ \cup \ S,$$

and we define $\mathrm{D}(t, t', [\![c, G]\!], S) = \{S_\sigma \mid \sigma \in G\}$. The transformation rules for solving extended E-unification problems are given in Figure 1.

Example 5. Take the lp-theory E with the unique non-trivial equation

$$f(g(x, y), z) = f(g(x, z), y).$$

E is unify-stable and induces the set $\mathcal{L} = \{[\![f(\circ, \circ), I]\!], [\![g(\circ, \circ), I]\!], [\![c, G]\!]\}$, where I is the trivial group, $c = f(g(\circ, \circ), \circ)$, and $G = \mathrm{Sym}(\{2, 3\})$. Let $t = f(x, a)$ and $t' = f(y, b)$, and suppose we want to solve the unification problem $S = \{t =^?_E t'\}$. We start by computing:

$$\mathrm{D}(t, t', [\![f(\circ, \circ), I]\!], \emptyset) = \{ \ \{x =^?_E y; \ a =^?_E b\}\},$$
$$\mathrm{D}(t, t', [\![g(\circ, \circ), I]\!], \emptyset) = \emptyset,$$
$$\mathrm{D}(t, t', [\![c, G]\!], \emptyset) \quad = \{ \ \{ \ x_1 =^?_E y_1, \ x_2 =^?_E y_2, \ a =^?_E b,$$
$$x =^?_E g(x_1, x_2), \ y =^?_E g(y_1, y_2)\},$$
$$\{ \ x_1 =^?_E y_1, \ a =^?_E y_2, \ x_2 =^?_E b,$$
$$x =^?_E g(x_1, x_2), \ y =^?_E g(y_1, y_2)\}\}.$$

1. **trivial:**
   $$\{\{s =^?_E s\} \cup S\} \cup M \;\rightarrow\; \{S\} \cup M$$
2. **orient:**
   $$\{\{t =^?_E x\} \cup S\} \cup M \;\rightarrow\; \{\{x =^?_E t\} \cup S\} \cup M \text{ if } t \text{ is not a variable}$$
3. **clash:**
   $$\{\{f(t_1, \dots, t_n) =^?_E g(t'_1, \dots, t'_m)\} \cup S\} \cup M \;\rightarrow\; M$$
4. **occurrence test:**
   $$\{\{s =^?_E t\} \cup S\} \cup M \;\rightarrow\; M \text{ if } s \text{ is a proper subterm of } t$$
5. **replacement:**
   $$\{\{x =^?_E u\} \cup S\} \cup M \;\rightarrow\; \{\{x =^?_E u\} \cup S[x \leftarrow u]\} \cup M \text{ if } x \text{ appears in } S$$
6. **P-decomposition:**
   $$\{\{t =^?_E t'\} \cup S\} \cup M \;\rightarrow\; [\textstyle\bigcup_{L \in \mathcal{L}} D(t, t', L, S)] \cup M \text{ if neither } t \text{ nor } t' \text{ is a variable}$$

Fig. 1. The transformation rules.

Respectively note these three unification problems S_1, S_2 and S_3. Then the P-decomposition rule yields $\{S\} \rightarrow \{S_1, S_2, S_3\}$.

Applying the clash rule on S_1 and S_2, and the orient and replacement rules on S_3 yields:

$$\{S\} \rightarrow^* \{\{x_1 =^?_E y_1, y_2 =^?_E a, x_2 =^?_E b, x =^?_E g(y_1, b), y =^?_E g(y_1, a)\}\}.$$

This new unification problem is in solved form. Therefore, the unification problem has a mgu, which as expected maps x to $g(y_1, b)$ and y to $g(y_1, a)$.

The rest of the section is dedicated to the proof that these rules terminate and compute a CSU for any extended unification problem.

3.3   Termination

We define a complexity measure on extended unification problems, and show that each transformation step strictly decreases the measures of the associated extended unification problems. The main problem that arises is that the P-decomposition rule produces new variables and symbols, which makes it difficult to determine a standard complexity measure that strictly decreases after its application. The solution is to carefully consider the way new symbols and variables are created. Thanks to the unify-stable hypothesis, we prove that each P-decomposition rule strictly decreases the number of candidates on which the rule, when applied, will create new symbols.

Definition 5. *Given a term t and a context c, we note $t \curlywedge c$ when t and c have a common instance, but t is not an instance of c, and we define the set*

$$V_{t,c} = \{\langle s, c \rangle \mid s \preceq t, s \text{ is not a variable, and } s \curlywedge c\}.$$

To any unification problem S, we associate

$$\mathcal{V}(S) = \bigcup_{c \in \mathcal{C}, \; t =^?_E t' \in S} (V_{t,c} \cup V_{t',c}).$$

Intuitively, these sets will represent the positions of the terms on which the application of the P-decomposition rule would produce new symbols or variables. Of course, these sets are finite, and the main result we show is that the applications of our transformation rules decrease their cardinalities.

We first prove some easy properties.

Lemma 3. *Let t be a term, c a context, and θ a substitution. Then:*

1. *if $s \preceq t$, then $V_{s,c} \subseteq V_{t,c}$,*
2. *if $s \preceq t$ and $\langle s\theta, c \rangle \in V_{t\theta,c}$, then $\langle s, c \rangle \in V_{t,c}$.*

Proof. 1. This is an obvious consequence of the definition of $V_{t,c}$.
    2. Since $s\theta$ is unifiable with c, then so is s, and since $s\theta$ is not an instance of c, then neither is s.

In the following lemma we apply a substitution θ to a set $V_{t,c}$, which simply means that each element $\langle s, c \rangle$ is transformed into $\langle s\theta, c \rangle$.

Lemma 4. *Given two terms t, u and a context c, let $\theta = \{x \leftarrow u\}$. We have:*

$$V_{t\theta,c} \subseteq V_{t,c}\theta \cup V_{u,c}.$$

Proof. Let $\langle s, c \rangle \in V_{t\theta,c}$. Then either $s \preceq u$, or there is a non-variable $s' \preceq t$ such that $s = s'\theta$. If $s \preceq u$, then by definition $\langle s, c \rangle \in V_{u,c}$. Otherwise, by Lemma 3 we have $\langle s', c \rangle \in V_{t,c}$, so that , $\langle s, c \rangle = \langle s'\theta, c \rangle \in V_{t,c}\theta$.

We now use the definition of unify-stability to prove that the expanded variables introduced by the P-decomposition rule do not introduce new ways of applying this rule.

Lemma 5. *For any context $c \in \mathcal{C}$, and non-variable terms t, t' which are unifiable under c, let $\tau = \gamma_c(t, t')$ and $x \in \mathrm{Dom}(\tau)$. Then $\mathcal{V}(x =_E^? x\tau) = \emptyset$.*

Proof. By Definition 5 we have

$$\mathcal{V}(x =_E^? x\tau) = \biguplus_{c' \in \mathcal{C}} V_{x,c'} \cup V_{x\tau,c'},$$

so let c' be any context in $\mathcal{C}$, we trivially have $V_{x,c'} = \emptyset$. Suppose that there is an element $\langle s, c' \rangle$ in $V_{x\tau,c'}$. Then $x\tau$ cannot be a variable, neither can s, and we have $s \preceq x\tau$ and $s \wedge c'$.

By Lemma 1, there is a non-variable position $p \neq \varepsilon$ of c such that s is an instance of the subterm $c[x_1, \dots, x_n]|p$. By Definition 2, there is a subset $C \subseteq \{c_1, \dots, c_k\}$ such that $c = \sqcup C$, hence there must be a $c_i \in C$ such that p is a non-variable position of c_i. We then have $c_i|p \sqsubseteq c|p$, so that s is an instance of $c_i|p$.

Since s and c' have a common instance, then $c_i|p$ and c' are unifiable. As above, there is a $C' \subseteq \{c_1, \dots, c_k\}$ such that $c' = \sqcup C'$. For all $c_j \in C'$, obviously c_j and $c_i|p$ are also unifiable, and by Definition 1 we have $c_j \sqsubseteq c_i|p$. This means that $c_i|p$ is an upper bound of C', and therefore that $c' \sqsubseteq c_i|p$. This proves that s is also an instance of c', contradicting $s \wedge c'$.

Lemma 6. *Let t, t' be two non-variable terms unifiable under a context $c \in C$, and let $\tau = \gamma_c(t, t')$. Then for any term s and any $c' \in C$, we have $V_{s\tau,c'} \subseteq V_{s,c'}\tau$.*

Proof. We prove that for any subset θ of τ, and any term s, that $V_{s\theta,c'} \subseteq V_{s,c'}\theta$, by induction on the cardinality of θ. It is it is 1, then θ is a substitution $\{x \leftarrow u\}$, and by Lemma 4, we know that $V_{s\theta,c'} \subseteq V_{s,c'}\theta \cup V_{u,c'}$. But $u = x\tau$, hence by Lemma 5 we have $V_{u,c'} = \emptyset$.

We now suppose that $\theta' = \theta\{x \leftarrow u\}$ and that the inclusion holds for θ. By using the base case on $s\theta$ we get

$$V_{s\theta',c'} = V_{s\theta\{x \leftarrow u\},c'} \subseteq V_{s\theta,c'}\{x \leftarrow u\} \subseteq V_{s,c'}\theta\{x \leftarrow u\} = V_{s,c'}\theta'.$$

We can now prove the main result concerning the P-decomposition rule.

Theorem 2. *In the context of Definition 4, we have $|\mathcal{V}(S_\sigma)| \leq |\mathcal{V}(\{t =^?_E t'\} \cup S)|$. Moreover, if $t \prec c$ or $t' \prec c$, then $|\mathcal{V}(S_\sigma)| < |\mathcal{V}(\{t =^?_E t'\} \cup S)|$.*

Proof. By Lemma 5, $\mathcal{V}(\{x =^?_E x\tau\}) = \emptyset$ for all $x \in \mathrm{Dom}(\tau)$. We thus have

$$\mathcal{V}(S_\sigma) = \mathcal{V}(S) \cup \bigcup_{i=1}^{m} \mathcal{V}(s_{i\sigma} =^?_E s'_i) \quad \text{and} \quad \mathcal{V}(s_j =^?_E s'_i) = \biguplus_{c' \in C} V_{s_j,c'} \cup V_{s'_i,c'}.$$

For any $c' \in C$, since s_j is a subterm of $t\tau$, by Lemma 3 we have $V_{s_j,c'} \subseteq V_{t\tau,c'}$, and by Lemma 6 we get $V_{t\tau,c'} \subseteq V_{t,c'}\tau$. Similarly we have $V_{s'_i,c'} \subseteq V_{t',c'}\tau$, so that

$$\mathcal{V}(s_j =^?_E s'_i) \subseteq \biguplus_{c' \in C} V_{t,c'}\tau \cup V_{t',c'}\tau = \mathcal{V}(t =^?_E t')\tau.$$

This is true for all i, j, hence we obtain $\mathcal{V}(S_\sigma) \subseteq \mathcal{V}(S) \cup \mathcal{V}(t =^?_E t')\tau$. The result is then obvious.

Suppose now that, say, $t \prec c$. Then by definition, we have $\langle t, c \rangle \in V_{t,c}$, hence $\langle t\tau, c \rangle \in V_{t,c}\tau$. However, $t\tau$ is an instance of c, so $\langle t\tau, c \rangle$ cannot belong to any set $V_{s,c}$, nor of course to any set $V_{s,c'}$ with $c' \neq c$, and therefore $\langle t\tau, c \rangle \notin \mathcal{V}(S_\sigma)$. But we have proved that $\langle t\tau, c \rangle \in \mathcal{V}(S) \cup \mathcal{V}(t =^?_E t')\tau$, so the inequality is strict.

We also need to prove that this measure cannot increase by applying the other rules.

Lemma 7. *If $\{S\} \cup M \to \{S'\} \cup M$ by another rule than P-decomposition, we have $|\mathcal{V}(S')| \leq |\mathcal{V}(S)|$.*

Proof. This is obvious for the trivial and orient rules. For the replacement rule, suppose that $S = \{x =^?_E u\} \cup A$, and let $\theta = \{x \leftarrow u\}$, so that $S' = \{x =^?_E u\} \cup A\theta$. By Lemma 4 we have $V_{t\theta,c} \subseteq V_{t,c}\theta \cup V_{u,c}$ for any term t and $c \in C$, hence

$$\mathcal{V}(A\theta) = \bigcup_{c \in C, t =^?_E t' \in A} V_{t\theta,c} \cup V_{t'\theta,c}$$

$$\subseteq \bigcup_{c \in C, t =^?_E t' \in A} V_{t,c}\theta \cup V_{t',c}\theta \cup V_{u,c} = \mathcal{V}(A)\theta \cup \mathcal{V}(x =^?_E u),$$

so obviously $\mathcal{V}(S') \subseteq \mathcal{V}(A)\theta \cup \mathcal{V}(x =^?_E u)$. But the variable x does not occur in u, so that $V_{u,c}\theta = V_{u,c}$, and hence $\mathcal{V}(x =^?_E u)\theta = \mathcal{V}(x =^?_E u)$. This proves that $\mathcal{V}(S') \subseteq (\mathcal{V}(A) \cup \mathcal{V}(x =^?_E u))\theta = \mathcal{V}(S)\theta$, and we have the result.

In order to prove termination we will also use classical measures.

Definition 6. *We map each E-unification problem S to a tuple of integers $\langle n_0, n_1, n_2, n_3 \rangle$, where $n_0 = |\mathcal{V}(S)|$, and n_1, n_2, n_3 are defined as in [3], i.e. n_1 is the number of variables in S that are not solved in S (x is solved in S if $x =^?_E t \in S$ is the only occurrence of x in S), n_2 is the size of S (i.e. the total number of symbols), and n_3 is the number of equations of the form $t =^?_E x$ in S.*

We then map each extended E-unification problem M to the multiset $\dot M$ of the tuples the elements $S \in M$ are mapped to. The well-founded strict order $M < M'$ is then defined as $\dot M \ll \dot M'$, where $\ll$ is the multiset order based on the lexicographic order $\prec$ on tuples. We recall that $A \gg B$ if B is obtained from A by removing a multiset X and adding a multiset Y, such that for any tuple $y \in Y$ there is a tuple $x \in X$ such that $y \prec x$.

Theorem 3. *The set of rules given in Figure 1 terminates.*

Proof. We prove that if $M \to M'$ then $M' < M$. This is obvious for the clash and occurrence test rules, since then $M' \subsetneq M$.

For the trivial and orient rules, we have $\{S\} \cup M \to \{S'\} \cup M$. Let $\langle n_0, n_1, n_2, n_3 \rangle$ (resp. $\langle n_0', n_1', n_2', n_3' \rangle$) be the tuple corresponding to S (resp. S'), by Lemma 7 we have $n_0 \leq n_0'$. But it is known that, for these rules, we have $\langle n_1, n_2, n_3 \rangle \prec \langle n_1', n_2', n_3' \rangle$, hence we obtain $\{S'\} \cup M < \{S\} \cup M$.

We now consider P-decomposition, applied to $\{\{t =^?_E t'\} \cup S\} \cup M$, where t, t' are non-variable terms. Let $\langle n_0, n_1, n_2, n_3 \rangle$ be the tuple corresponding to $\{t =^?_E t'\} \cup S$. For any $[\![c, G]\!] \in \mathcal{L}$ such that t and t' are unifiable under c, and any $\sigma \in G$, let $\langle n_0', n_1', n_2', n_3' \rangle$ be the tuple corresponding to S_σ.

By Theorem 2 we have $n_0' \leq n_0$. If $t \wedge c$ or $t' \wedge c$ we even have $n_0' < n_0$, and we are done. Otherwise, t and t' are instances of c, and it is easy to see that $\gamma_c(t, t')$ is then empty. Hence, for all i the term s_i (resp. s_i') is a subterm of t (resp. t'). This proves that no variable is introduced in S_σ compared to $\{t =^?_E t'\} \cup S$, so that $n_1' \leq n_1$, and that at least two occurrences of the head symbol of t have disappeared, so that $n_2' < n_2$, and once again we obtain $\langle n_0', n_1', n_2', n_3' \rangle \prec \langle n_0, n_1, n_2, n_3 \rangle$.

3.4   Completeness

Once again, the main result we have to prove is that the P-decomposition rule preserves the set of solutions of a unification problem.

Lemma 8. *Let $S' = \{t =^?_E t'\} \cup S$ be a unification problem, and $[\![c, G]\!] \in \mathcal{L}$. Suppose $\theta \in \mathcal{U}_E(S)$ is such that $t\theta = c[t_1, \ldots, t_m]$, $t'\theta = c[t_1', \ldots, t_m']$, and that there exists a σ in G such that for $i = 1, \ldots, m$, $t_{i^\sigma} =_E t_i'$. Then θ is also in $\mathcal{U}_E(D(t, t', [\![c, G]\!], S))$.*

Furthermore, if there is a $[\![c', G']\!] \in \mathcal{L}$ such that $c \sqsubseteq c'$ and $t\theta$ and $t'\theta$ are both instances of c', then $\theta \in \mathcal{U}_E(D(t, t', [\![c', G']\!], S))$.

Proof. We will actually prove that $\theta \in \mathcal{U}_E(S_\sigma)$. Since $t\theta$ and $t'\theta$ are both instances of c, $\tau = \gamma_c(t, t')$ is guaranteed to exist, and by Lemma 2, there is a substitution δ such that $\theta = \tau\delta$. Let $t\tau = c[s_1, \ldots, s_m]$, and $t'\tau = c[s'_1, \ldots, s'_m]$. τ is idempotent, so $\tau\theta = \tau\tau\delta = \tau\delta = \theta$; therefore, $t\theta = t\tau\theta = c[s_1\theta, \ldots, s_m\theta]$, and $t'\theta = c[s'_1\theta, \ldots, s'_m\theta]$. This shows that for all $i = 1, \ldots, m$, $t_i = s_i\theta$ and $t'_i = s'_i\theta$; therefore, by hypothesis, that $s_{i\sigma}\theta =_E s'_i\theta$. Hence, θ is a solution of the unification problem $\{s_{i\sigma} =_E^? s'_i \mid i = 1, \ldots, m\}$. Since $\tau\theta = \theta$, we also have $x\theta = x\tau\theta$ for any variable x, and therefore θ is also a solution of the unification problem $\{x =_E^? x\tau \mid x \in \mathrm{Dom}(\tau)\}$. Finally, θ is a solution of S', and since $S \subseteq S'$, θ is also a solution of S. This proves that θ is a solution of S_σ, and therefore of $\mathrm{D}(t, t', [\![c, G]\!], S)$.

Now suppose that $[\![c', G']\!]$ exists. Let $\tau' = \gamma_{c'}(t, t')$, $t = c'[u_1, \ldots, u_n]$, and $t' = c'[u'_1, \ldots, u'_n]$. For the same reasons as previously, we have: $\tau'\theta = \theta$, and $c'[u_1\theta, \ldots, u_n\theta] =_E c'[u'_1\theta, \ldots, u'_n\theta]$. Since G' is complete for c' there is a μ in G' such that for all i, $u_{i\mu}\theta =_E u'_i\theta$. So, θ is a solution of $\{u_{i\mu} =_E^? u'_i \mid i = 1, \ldots, n\}$. By Property 1, $\tau = \tau'$, so θ is also a solution of $\{x =_E^? x\tau' \mid x \in \mathrm{Dom}(\tau')\} \cup S$, hence the result.

Lemma 9. $\mathcal{U}_E(\mathrm{D}(t, t', [\![c, G]\!], S)) \subseteq \mathcal{U}_E(\{t =_E^? t'\} \cup S)$.

Proof. Let $\theta \in \mathrm{D}(t, t', [\![c, G]\!], S)$, and $\tau = \gamma_c(t, t')$, then $t\tau = c[s_1, \ldots, s_m]$, and $t'\tau = c[s'_1, \ldots, s'_m]$. Let $S_\sigma \in \mathrm{D}(t, t', [\![c, G]\!], S)$ be the unification problem such that θ is a solution of S_σ. Then since $S \subseteq S_\sigma$, all we have to prove is that θ is a solution of $t =_E^? t'$.

θ is a solution of $\{x =_E^? x\tau \mid x \in \mathrm{Dom}(\tau)\}$, so, for every variable in $\mathrm{Dom}(\tau)$, $x\theta =_E x\tau\theta$, which proves that $t\theta =_E t\tau\theta$, and $t'\theta =_E t'\tau\theta$.

θ is also a solution of $\{s_{i\sigma} =_E^? s'_i \mid i = 1, \ldots, m\}$, so, for all i, $s_{i\sigma}\theta =_E s'_i\theta$, and by Theorem 1, $c[s_1, \ldots, s_m]\theta =_E c[s'_1, \ldots, s'_m]\theta$. Therefore, $t\theta =_E t\tau\theta =_E t'\tau\theta =_E t'\theta$, and we have the result by transitivity.

Theorem 4. $\mathcal{U}_E(\{t =_E^? t'\} \cup S) = \mathcal{U}_E(\bigcup_{L \in \mathcal{L}} \mathrm{D}(t, t', L, S))$.

Proof. One inclusion is direct from Lemma 9. To prove the converse, let θ be a solution of $\{t =_E^? t'\} \cup S$. Then we have $t\theta =_E t'\theta$, and by Theorem 1 there is a $[\![c, G]\!] \in \mathcal{L}$ such that the conditions of Lemma 8 are fulfilled, so we may conclude $\theta \in \mathcal{U}_E(\mathrm{D}(t, t', [\![c, G]\!], S))$, and the result is obvious.

Theorem 5. *If $M \to M'$ then $\mathcal{U}_E(M) = \mathcal{U}_E(M')$.*

Proof. The proof of the result for the first five transformation rules is exactly identical to the proof for syntactic unification, and Theorem 4 proves the result for P-decomposition.

So, now, we know that the system terminates, and that the set of solutions of the extended unification problem is preserved by each transformation rule. Since the only extended unification problems on which no transformation rule can be applied (i.e. the normal forms for $\to$) are those in solved form, we deduce that:

Corollary 1. *For any E-unification problem S, and any solved form M such that $\{S\} \to^* M$, the set $\{\theta_{S'} \mid S' \in M\}$ is a CSU for S.*

4 Optimization of the Algorithm

Several improvements can be made to this algorithm. First, consider the trivial rule, which only tests for syntactic equality. It can be replaced by the following rule, call it the E-trivial rule:

$$\{\{t =^?_E t'\} \cup S\} \cup M \rightarrow \{S\} \cup M \text{ if } t =_E t'.$$

This rule prevents from decomposing t and t' when they are equal modulo E, and may therefore save a significant amount of space. Testing whether $t =_E t'$ is **NP**-complete in general. However, for unify-stable theories, this test is in the Luks complexity class, and conjectured not to be **NP**-complete (see [6]). Thus, although no polynomial test exists to decide whether $t =_E t'$, it can be performed efficiently in general. As in [2], it is possible to carry out a pre-treatment on the initial unification problem, but we would not gain much efficiency. Indeed, most of the terms to be tested are obtained after application of the P-decomposition rule, and those that were not E-equal may become so after instantiation.

A more manageable source of improvement is to reduce the number of contexts considered for P-decomposition. This is possible thanks to the following result.

Lemma 10. Let $S' = \{t =^?_E t'\} \cup S$, and let $[\![c, G]\!]$ and $[\![c', G']\!]$ be two elements of $\mathcal{L}$, such that $c \sqsubseteq c'$ and $\tau = \gamma_c(t, t')$ exists.

If both $t\tau$ and $t'\tau$ are instances of c', then every solution of $\mathrm{D}(t, t', [\![c, G]\!], S)$ is a solution of $\mathrm{D}(t, t', [\![c', G']\!], S)$, and for any context d such that $\tau' = \gamma_{c \sqcup d}(t, t')$ exists, both $t\tau'$ and $t'\tau'$ are instances of $c' \sqcup d$.

Proof. Let θ be a solution of $\mathrm{D}(t, t', [\![c, G]\!], S)$. Then $t\theta$ and $t'\theta$ are instances of c, so by Lemma 2, there is a substitution δ such that $\theta = \tau\delta$. Therefore, $t\theta$ and $t'\theta$ are also instances of c', and θ is a solution of $\mathrm{D}(t, t', [\![c', G']\!], S)$ by Lemma 8.

Since $t\tau'$ and $t'\tau'$ are instances of $c \sqcup d$, they also are instances of c, so τ' is the restriction to the variables of t and t' of a solution of the syntactic unification problem $\{t =^? c; t' =^? c\}$. This proves that there is a δ such that $\tau' = \tau\delta$ by Lemma 2, hence that $t\tau'$ and $t'\tau'$ are instances of c'. Since they are also instances of d, they are instances of $c' \sqcup d$, which proves the result.

The first point proves that, in this case, if L is the element of $\mathcal{L}$ with context c, and $L' \in \mathcal{L}$ with context c', then all solutions of $\mathrm{D}(t, t', L, S')$ are also solutions of $\mathrm{D}(t, t', L', S')$, so that L may be discarded. The second one proves that this is also the case for the other elements of $\mathcal{L}$ whose contexts are instances of c but not of c'.

Example 6. Let E be the unify-stable theory from which we construct the set $\{L_i \mid i = 1, \ldots, 3\} = \{[\![c_i, G_i]\!] \mid i = 1, \ldots, 3\}$, along with the trivial equations for f and g, where:

$$c_1 = f(g(\circ, \circ), \circ, \circ) \text{ and } G_1 = \mathrm{Sym}(2),$$
$$c_2 = f(g(\circ, \circ), g(\circ, \circ), \circ) \text{ and } G_2 = \mathrm{Sym}(4),$$
$$c_3 = f(g(\circ, \circ), g(\circ, \circ), g(\circ, \circ)) \text{ and } G_3 = \mathrm{Sym}(4) \times \mathrm{Sym}(\{5, 6\}).$$

Let $t = f(x, g(a, b), y)$, $t' = f(y, x, g(c, d))$, and suppose we want to solve the unification problem $t =^?_E t'$. Since $\tau = \gamma_{c_1}(t, t') = \{x \leftarrow g(x_1, x_2), y \leftarrow g(y_1, y_2)\}$, then both $t\tau$ and $t'\tau$ are instances of c_3, and there is no need to search for the solutions for the other possible contexts, i.e. $f(\circ, \circ)$, c_1 and c_2. So we can restrict the search of a solution to an extended unification problem with $|G_3| = 4! \, 2! = 48$ elements, instead of one with $48 + 4! + 2! + 1 = 75$ elements.

In Figure 2, we present an algorithm written in pseudo-CAML that, given a unification problem $S' = \{t =^?_E t'\} \cup S$, uses the set $C = \{c_1, \ldots, c_k\}$ of contexts that appear in E to compute a subset $\mathcal{F}$ of $\mathcal{L}$. This subset is such that the associated extended unification problem generates the same set of E-unifiers as the original P-decomposition rule. $\mathcal{F}$ can thus safely replace $\mathcal{L}$ in the P-decomposition rule.

This algorithm starts by determining the contexts of C that t and t' are instances of, and computes their upper bound c. Note that these sets are guaranteed not to be empty: they each contain at least one element (the flat context associated to the head symbol of the term). Since any E-unifier of $t =^?_E t'$ is under c, all the contexts that subsume c (i.e. those in $Q \cup Q'$) can be discarded.

The recursive procedure subsearch(d, C) is meant to consider the contexts of the form $c' = \sqcup(C' \cup \{d\})$, for all $C' \subseteq C$. If t and t' have no common instance under d, then obviously they do not have a common instance under any context subsumed by d. Hence the contexts c' can only produce empty sets $D(t, t', [\![c', G']\!], S)$, and can therefore be skipped.

```
Pdec(t, t', S) =
  let F = ∅ in
  let rec subsearch(d, C) =
    if τ = γ_d(t, t') exists then
      let Q = {c ∈ C | tτ and t'τ are instances of c ⊔ d} in
      let d' = ⊔(Q ∪ {d}) in
      let L = [[d', G]] ∈ C in
      F := F ∪ {L}
      C := C \ Q
      for c in C do
        C := C \ {c}
        if d' ⊔ c exists then
          subsearch(d' ⊔ c, C)
      done
  in
  let Q = {c ∈ C | t is an instance of c} in
  let Q' = {c ∈ C | t' is an instance of c} in
  if c = (⊔Q) ⊔ (⊔Q') exists then
    C := C \ (Q ∪ Q')
    subsearch(c, C)
  return ⋃_{L∈F} D(t, t', L, S)
```

Fig. 2. Optimization of the P-decomposition rule.

Otherwise, Q contains all the elements c of C such that $t\tau$ and $t'\tau$ are instances of $c \sqcup d$, and is used to construct the context d'. So, $t\tau$ and $t'\tau$ are both instances of d'. By Lemma 10, there is no need to consider any element of $\mathcal{L}$ whose context c' is an instance of d but not of d', so, only the element of $\mathcal{L}$ whose context is d' is added to $\mathcal{F}$ before the next call to the procedure.

5   Conclusion

In this paper we have presented a simple E-unification algorithm for unify-stable theories. Since commutativity is a unify-stable theory, all negative results concerning C-unification are preserved, among which **NP**-completeness of the associated decision problem (see [7]; membership in **NP** should however be considered as a positive result, see [4]). Example 2 shows that simple unification problems can have a huge number of unifiers; and another problem is that minimality of the CSU is not guaranteed. The optimizations of Section 4 of course reduce the number of redundant solutions, but they would be useless on a theory of commutativity, because it has only one context in $\mathcal{L}$. Other redundancies come from the fact that all permutations of a context are considered. Some optimizations are possible at this level, but guaranteeing minimality would be very difficult, and probably a waste of time compared to a simple post-processing.

Acknowledgments

We thank the anonymous referee who found an error in the first version of Lemma 1.

References

1. J. Avenhaus and D. Plaisted. General algorithms for permutations in equational inference. *Journal of Automated Reasoning*, 26:223–268, April 2001.
2. Jürgen Avenhaus. Efficient algorithms for computing modulo permutation theories. In David Basin and Michaël Rusinowitch, editors, *Second International Joint Conference IJCAR*, volume 3097 of *Lecture Notes in Artificial Intelligence*, pages 415–429, Cork, Ireland, july 2004. Springer Verlag.
3. F. Baader and T. Nipkow. *Term Rewriting and All That*. Cambridge University Press, 1998.
4. T. Boy de la Tour and M. Echenim. **NP**-completeness results for deductive problems on stratified terms. In Moshe Vardi and Andrei Voronkov, editors, *LPAR*, LNAI 2850, pages 315–329. Springer Verlag, 2003.
5. Thierry Boy de la Tour and Mnacho Echenim. On the complexity of deduction modulo leaf permutative equations. To appear in Journal of Automated Reasoning, 2004.
6. Thierry Boy de la Tour and Mnacho Echenim. Overlapping leaf permutative equations. In David Basin and Michaël Rusinowitch, editors, *Second International Joint Conference IJCAR*, volume 3097 of *Lecture Notes in Artificial Intelligence*, pages 430–444, Cork, Ireland, july 2004. Springer Verlag.
7. M. Garey and D. S. Johnson. *Computers and intractability: a guide to the theory of* **NP**-*completeness*. Freeman, San Francisco, California, 1979.

Dependency Pairs
for Simply Typed Term Rewriting

Takahito Aoto[1] and Toshiyuki Yamada[2]

[1] Research Institute of Electrical Communication, Tohoku University, Japan
aoto@nue.riec.tohoku.ac.jp
[2] Faculty of Engineering, Mie University, Japan
toshi@cs.info.mie-u.ac.jp

Abstract. Simply typed term rewriting proposed by Yamada (RTA, 2001) is a framework of higher-order term rewriting without bound variables. In this paper, the dependency pair method of first-order term rewriting introduced by Arts and Giesl (TCS, 2000) is extended in order to show termination of simply typed term rewriting systems. Basic concepts such as dependency pairs and estimated dependency graph in the simply typed term rewriting framework are clarified. The subterm criterion introduced by Hirokawa and Middeldorp (RTA, 2004) is successfully extended to the case where terms of function type are allowed. Finally, an experimental result for a collection of simply typed term rewriting systems is presented. Our method is compared with the direct application of the first-order dependency pair method to a first-order encoding of simply typed term rewriting systems.

1 Introduction

Simply typed term rewriting, proposed by Yamada [20], is a framework of higher-order term rewriting. Equational specifications using higher-order functions, like functional programs, are naturally expressed in this framework. In contrast to the usual higher-order term rewriting [11, 12, 15], simply typed term rewriting dispenses with bound variables. In this respect, this framework reflects limited higher-order features. On the other hand, it is succinct and theoretically much easier to deal with.

The dependency pair method, introduced by Arts and Giesl [3], is a technique to prove the termination of (first-order) term rewriting systems. In contrast to simplification orders, such as the recursive path ordering, this method can also handle termination proofs of non-simply terminating term rewriting systems. Various improvements of the technique have been proposed [4–6, 10, 16, 19]. Implementation techniques as well as experimental results of termination provers based on the dependency pair method are reported in [6–9]. Extensions of the method to the higher-order case have been considered in [13, 17, 18].

In this paper, the dependency pair method of first-order term rewriting is extended to the case of simply typed term rewriting. We first study properties of rewrite sequences over minimal non-terminating simply typed terms on which

J. Giesl (Ed.): RTA 2005, LNCS 3467, pp. 120–134, 2005.

all the dependency pair techniques are based (Section 3). Then basic concepts such as the dependency pairs (Section 4) and the estimated dependency graph (Section 5) in the simply typed term rewriting framework are clarified.

Simply typed terms may have variables as head symbols because variables of function types are allowed. Hence we need to introduce a technique to deal with dependency pairs with head variables. It turns out that head variables in dependency pairs can always be instantiated for the termination proof. As a result, the subterm criterion, introduced by Hirokawa and Middeldorp [10], is successfully incorporated into the simply typed term rewriting framework (Section 6).

Finally, an experimental result for a collection of simply typed term rewriting systems is presented. The effectiveness of our method is supported by a comparison with the direct application of the first-order dependency pair method to a first-order encoding of simply typed term rewriting systems (Section 7).

2   Preliminaries

In this section, we briefly recall the terminology and the notations of simply typed term rewriting which were introduced in [20]. Some additional definitions needed in this paper are also given here.

A *simple type* is either the *base type* o or a *function type* $\tau_1 \times \cdots \times \tau_n \to \tau_0$. When clear, simple type is abbreviated as *type*. For the sake of simplicity, we consider only the single base type, although all the results in this paper can be extended to the case of multiple base types. The sets of *constants* and *variables* of type τ are denoted by Σ^τ and V^τ, respectively. Let $\Sigma = \bigcup_\tau \Sigma^\tau$ and $V = \bigcup_\tau V^\tau$. The set of *simply typed terms* of type τ over Σ and V is denoted by $T(\Sigma, V)^\tau$. We define $T(\Sigma, V) = \bigcup_\tau T(\Sigma, V)^\tau$. The *head symbol* of a simply typed term t is written as head(t).

A *context* of type τ is a simply typed term that contains one special symbol $\Box^\tau$, called the *hole*, of type τ. The term obtained by replacing the hole in a context C of type τ with a term t of the same type τ is denoted by $C[t]$. A context of the form $\Box^\tau$ is said to be *empty*. We omit the type of a hole when it is not important. A term s is a *subterm* of a term t (denoted by $s \trianglelefteq t$) if $C[s] = t$ for some context C, and is a *proper* subterm (denoted by $s \triangleleft t$) when $s \neq t$ holds in addition. For a *substitution* $\sigma : V \to T(\Sigma, V)$, an *instance* $\sigma(t)$ of a term t is also written as $t\sigma$. We say s *matches* t and write $s \precsim t$ if t is an instance of s. Clearly, the relation $\precsim$ is a quasi-order. We note that when s and t do not have variables in common, the existence of a term u such that $s \precsim u \succsim t$ is equivalent to the unifiability of s and t. We use mgu(s, t) to denote an arbitrary partial function to compute a most general unifier of s and t.

Every *simply typed rewrite rule* $l \to r$ must satisfy the following conditions: (1) l and r have the same type, (2) head(l) $\in \Sigma$, and (3) $V(r) \subseteq V(l)$. The type of a rewrite rule $l \to r$ is defined as that of l and r. Let $\mathcal{R} = \langle \Sigma, R \rangle$ be a *simply typed term rewriting system* (STTRS, for short). The *rewrite relation* induced by $\mathcal{R}$ is denoted by $\to_\mathcal{R}$. The reflexive closure, transitive closure, reflexive transitive closure of $\to_\mathcal{R}$ are written as $\to_\mathcal{R}^=$, $\to_\mathcal{R}^+$, $\to_\mathcal{R}^*$, respectively. When $\mathcal{R}$ is obvious from the context, we omit the subscript. A constant $f \in \Sigma$ is a *defined symbol* of an STTRS $\mathcal{R} = \langle \Sigma, R \rangle$ if $f = \text{head}(l)$ for some $l \rightarrow r \in R$, otherwise f is a *constructor symbol*. The sets of defined and constructor symbols are denoted by Σ_d and Σ_c, respectively.

Example 1 (simply typed term rewriting). Let $\mathcal{R} = \langle \Sigma, R \rangle$ be an STTRS where $\Sigma = \{\ 0^o,\ s^{o\rightarrow o},\ []^o,\ :^{o\times o\rightarrow o},\ +^{o\rightarrow o\rightarrow o},\ \text{fold}^{(o\rightarrow o\rightarrow o)\times o\rightarrow o\rightarrow o},\ \text{sum}^{o\rightarrow o}\ \}$, and

$$
R = \left\{
\begin{array}{lcl}
(+\ 0)\ y & \rightarrow & y \\
(+\ (s\ x))\ y & \rightarrow & s\ ((+\ x)\ y) \\
(\text{fold}\ F\ x)\ [] & \rightarrow & x \\
(\text{fold}\ F\ x)\ (:\ y\ ys) & \rightarrow & (F\ y)\ ((\text{fold}\ F\ x)\ ys) \\
\text{sum} & \rightarrow & \text{fold} + 0
\end{array}
\right\}.
$$

The function sum computes the sum of all elements of a list. Here is an example of a rewrite sequence of $\mathcal{R}$:

$$
\begin{array}{rl}
\text{sum}\ (:\ (s\ 0)\ [])\ \ \rightarrow_{\mathcal{R}} & (\text{fold} + 0)\ (:\ (s\ 0)\ []) \\
\rightarrow_{\mathcal{R}} & (+\ (s\ 0))\ ((\text{fold} + 0)\ []) \\
\rightarrow_{\mathcal{R}} & (+\ (s\ 0))\ 0 \\
\rightarrow_{\mathcal{R}} & s\ ((+\ 0)\ 0) \\
\rightarrow_{\mathcal{R}} & s\ 0.
\end{array}
$$

We have $\Sigma_d = \{\ +,\ \text{fold},\ \text{sum}\ \}$ and $\Sigma_c = \{\ 0,\ s,\ [],\ :\ \}$.

3   Chains of Minimal Non-terminating Terms

In this section, we study properties of minimal non-terminating terms, on which all the dependency pair techniques are based. Basic notions and properties of infinite rewrite sequences are lifted to the case of simply typed term rewriting. Since root symbols in first-order term rewriting correspond to head symbols in simply typed term rewriting, we now regard head rewrite steps instead of root rewrite steps.

Definition 1 (root and head rewrite steps). We write $s \xrightarrow{r} t$ when $s = l\sigma$ and $t = r\sigma$ for some rewrite rule $l \rightarrow r$ and some substitution σ. The relation $\xrightarrow{r}$ is called the *root rewrite step*. The *head rewrite step* $\xrightarrow{h}$ is defined recursively as follows: $s \xrightarrow{h} t$ if (1) $s \xrightarrow{r} t$, or (2) $s = (s_0\ u_1\ \cdots\ u_n)$, $t = (t_0\ u_1\ \cdots\ u_n)$, and $s_0 \xrightarrow{h} t_0$. The non-root and non-head rewrite steps are defined by $\xrightarrow{nr} = \rightarrow \setminus \xrightarrow{r}$ and $\xrightarrow{nh} = \rightarrow \setminus \xrightarrow{h}$, respectively.

By definition, $\xrightarrow{r} \subseteq \xrightarrow{h}$ and thus $\xrightarrow{nh} \subseteq \xrightarrow{nr}$ holds. If there are no terms of function types which can be rewriten, then the reverse inclusions also hold.

Example 2 (root and head rewrite step). Let $\mathcal{R}$ be the STTRS in Example 1. The following rewrite sequence exemplifies the use of the relations defined above:

$$\underline{\mathsf{sum}}\ (:\ (\mathsf{s}\ 0)\ [])\ \xrightarrow{\mathrm{h}}\ (\mathsf{fold} + 0)\ (:\ (\mathsf{s}\ 0)\ [])$$
$$\xrightarrow{\mathrm{r}}\ (+ (\mathsf{s}\ 0))\ ((\mathsf{fold} + 0)\ [])$$
$$\xrightarrow{\mathrm{nh}}\ (+ (\mathsf{s}\ 0))\ 0$$
$$\xrightarrow{\mathrm{r}}\ \mathsf{s}\ ((+ 0)\ 0)$$
$$\xrightarrow{\mathrm{nh}}\ \mathsf{s}\ 0.$$

We note that $\mathsf{sum}\ (:\ (\mathsf{s}\ 0)\ [])\ \xrightarrow{\mathrm{nr}}\ (\mathsf{fold} + 0)\ (:\ (\mathsf{s}\ 0)\ [])$.

We say a term is *terminating* if there is no infinite rewrite sequence starting from that term, otherwise *non-terminating*. We denote the set of non-terminating terms by $\mathrm{NT}(\mathcal{R})$. The set of minimal (with respect to the subterm relation $\unlhd$) non-terminating terms is denoted by $\mathrm{NT}_{\min}(\mathcal{R})$. We may omit the parameter $\mathcal{R}$ of $\mathrm{NT}(\mathcal{R})$ and $\mathrm{NT}_{\min}(\mathcal{R})$ when it is clear from the context. We say an STTRS $\mathcal{R}$ is terminating if all terms are terminating. Since the subterm relation is well-founded, non-termination implies the existence of a minimal non-terminating term.

Lemma 1 (chain of minimal non-terminating terms). Let $s_0 \in \mathrm{NT}_{\min}$. (1) There exist $s_1,\ldots,s_n,t \in \mathrm{NT}_{\min}$ $(n \geq 0)$ and $s_{n+1} \in \mathrm{NT}$ such that $s_0 \xrightarrow{\mathrm{nr}} s_1 \xrightarrow{\mathrm{nr}} \cdots \xrightarrow{\mathrm{nr}} s_n \xrightarrow{\mathrm{r}} s_{n+1} \unrhd t$. (2) There exist $s_1,\ldots,s_n,t \in \mathrm{NT}_{\min}$ $(n \geq 0)$ and $s_{n+1} \in \mathrm{NT}$ such that $s_0 \xrightarrow{\mathrm{nh}} s_1 \xrightarrow{\mathrm{nh}} \cdots \xrightarrow{\mathrm{nh}} s_n \xrightarrow{\mathrm{h}} s_{n+1} \unrhd t$, and $s_{n+1} = t$ whenever $s_n \xrightarrow{\mathrm{nr}} s_{n+1}$.

Proof. (1) The proof proceeds in the same way as the first-order case [3]. (2) By (1), there exist $s_1,\ldots,s_m,u \in \mathrm{NT}_{\min}$ $(m \geq 0)$ and $s_{m+1} \in \mathrm{NT}$ such that $s_0 \xrightarrow{\mathrm{nr}} s_1 \xrightarrow{\mathrm{nr}} \cdots \xrightarrow{\mathrm{nr}} s_m \xrightarrow{\mathrm{r}} s_{m+1} \unrhd u$. Since $\xrightarrow{\mathrm{r}} \subseteq \xrightarrow{\mathrm{h}}$, there exists at least one $i \in \{0,\ldots,m\}$ such that $s_i \xrightarrow{\mathrm{h}} s_{i+1}$. Let n be the smallest such i. We take $t = u$ if $n = m$, and take $t = s_{n+1}$ otherwise. $\square$

We are now going to give a more detailed description of Lemma 1 (2). For that purpose, we need to know some properties of head and non-head rewrite steps. We first introduce a notion of argument context in order to characterize a head rewrite step as a rewrite step accompanied by an argument context. We next introduce a notion of argument sequence in order to deal with properties of non-head rewrite steps.

Definition 2 (argument context). An *argument context*[1] is a context whose head symbol is the hole, more precisely, κ is an argument context of type τ if (1) $\kappa = \square^\tau$, or (2) $\kappa = (\kappa'\ t_1 \cdots t_n)$ for some argument context κ' of a function type and some terms $t_1,\ldots,t_n$ of appropriate types.

Example 3 (argument context). $\kappa = ((\square^{o\to(o\to o)}\ 0)\ (\mathsf{s}\ 0))$ is an argument context. Then we have $\kappa[+] = ((+ 0)\ (\mathsf{s}\ 0))$. Neither $((+ \square^o)\ (\mathsf{s}\ 0))$ nor $((+ 0)\ \square^o)$ is an argument context.

---

[1] A similar notion is called a *suffix context* in [13].

Lemma 2 (characterization of head rewrite steps). $s \xrightarrow{\text{h}} t$ if and only if there exist a rewrite rule $l \to r$, a substitution σ, and an argument context κ such that $s = \kappa[l\sigma]$ and $t = \kappa[r\sigma]$.

Proof. ($\Rightarrow$) By induction on the definition of $\xrightarrow{\text{h}}$. ($\Leftarrow$) By induction on κ. $\square$

Lemma 3 (preservation of head symbols). (1) For any argument context κ, $\text{head}(\kappa[t]) = \text{head}(t)$. (2) If $s \xrightarrow{\text{h}} t$ then $\text{head}(s) \in \Sigma_\text{d}$. (3) If $s \xrightarrow{\text{nh}} t$ then $\text{head}(s) = \text{head}(t)$. (4) If $s \to t$ and $\text{head}(s) \notin \Sigma_\text{d}$ then $s \xrightarrow{\text{nh}} t$ and $\text{head}(t) \notin \Sigma_\text{d}$.

Proof. (1) By induction on κ. (2) By (1) and Lemma 2. (3) By induction on s. (4) Follows immediately from (2) and (3). $\square$

Definition 3 (argument sequence). An *argument sequence* $\text{Arg}(t)$ of a simply typed term t is the empty sequence when $t \in V \cup \Sigma$; and is $\text{Arg}(t_0), t_1, \ldots, t_n$ when $t = (t_0\ t_1 \cdots t_n)$.

We write the length and the i-th element of a sequence X by $|X|$ and $X_{|i}$.

Example 4 (argument sequence). Let $t = ((\text{fold} + (\text{s}\ 0))\ (:\ (\text{s}\ 0)\ [\,]))$. The argument sequence of t is $+, (\text{s}\ 0), (:\ (\text{s}\ 0)\ [\,])$. We also have $|\text{Arg}(t)| = 3$ and $\text{Arg}(t)_{|2} = (\text{s}\ 0)$.

Lemma 4 (argument sequence of non-head rewrite steps). Suppose $s \xrightarrow{\text{nh}} t$. (1) $|\text{Arg}(s)| = |\text{Arg}(t)| > 0$. (2) Let $|\text{Arg}(s)| = |\text{Arg}(t)| = n$. Then there exists some $i \in \{1, \ldots, n\}$ such that $\text{Arg}(s)_{|i} \to \text{Arg}(t)_{|i}$ and $\text{Arg}(s)_{|j} = \text{Arg}(t)_{|j}$ for all $j \in \{1, \ldots, n\} \setminus \{i\}$.

Proof. (1) The contraposition can be shown by induction on s. (2) By induction on s using (1). $\square$

Lemma 5 (head symbols of minimal non-terminating terms). If $s \in \text{NT}_\text{min}$ then $\text{head}(s) \in \Sigma_\text{d}$.

Proof. Suppose $s \in \text{NT}_\text{min}$. Then there exists an infinite rewrite sequence $s = s_0 \to s_1 \to \cdots$. For a proof by contradiction, suppose $\text{head}(s) \notin \Sigma_\text{d}$. By repeatedly applying Lemma 3 (4), we know $s_0 \xrightarrow{\text{nh}} s_1 \xrightarrow{\text{nh}} s_2 \xrightarrow{\text{nh}} \cdots$. By Lemma 4 (1), all $\text{Arg}(s_k)$ have the same length for $k \geq 0$; so, let $n = |\text{Arg}(s_k)|$. Then by Lemma 4 (2), for each $k \geq 0$ there exists $i \in \{1, \ldots, n\}$ such that $\text{Arg}(s_k)_{|i} \to \text{Arg}(s_{k+1})_{|i}$ and $\text{Arg}(s_k)_{|j} = \text{Arg}(s_{k+1})_{|j}$ for all $j \in \{1, \ldots, n\} \setminus \{i\}$. Thus there exists an index i such that $\text{Arg}(s_0)_{|i} \to^= \text{Arg}(s_1)_{|i} \to^= \cdots$ is an infinite rewrite sequence, contradicting the minimality of s. $\square$

Lemma 6 (chain of minimal non-terminating terms in detail). If $s \in \text{NT}_\text{min}$ then there exist a rewrite rule $l \to r$, an argument context κ, a substitution σ, and simply typed terms $r' \trianglelefteq r$ and $t \in \text{NT}_\text{min}$ satisfying the following properties: (1) $s \xrightarrow{\text{nh}}{}^* \kappa[l\sigma] \xrightarrow{\text{h}} \kappa[r\sigma] \trianglerighteq t$, (2) if κ is empty then $l \not\trianglerighteq r'$ and $t = r'\sigma$ hold, otherwise $r = r'$ and $t = \kappa[r\sigma]$ hold, (3) $\text{head}(r'\sigma) \in \Sigma_\text{d}$.

Proof. By Lemma 1, there exist terms $s', t \in \mathrm{NT}_{\min}$ and $u \in \mathrm{NT}$ such that $s \xrightarrow{\mathrm{nh}}^* s' \xrightarrow{\mathrm{h}} u \trianglerighteq t$, and $u = t$ whenever $s' \xrightarrow{\mathrm{nr}} u$. In order to prove (1) and (2), first suppose $s' \xrightarrow{\mathrm{nr}} u$. By Lemma 2, there exist a rewrite rule $l \to r$, an argument context κ, and a substitution σ such that $s' = \kappa[l\sigma]$ and $u = \kappa[r\sigma]$. Thus (1) holds. By the definition of $\xrightarrow{\mathrm{nr}}$, κ is nonempty. Since $u = t$, (2) is satisfied by taking $r' = r$. Next, suppose $s' \xrightarrow{\mathrm{r}} u$. Then, there exist a rewrite rule $l \to r$, a substitution σ such that $s' = l\sigma$ and $u = r\sigma$. Hence (1) is satisfied by taking κ to be an empty context. No subterm of $\sigma(x)$ can be t for any $x \in \mathrm{V}(r)$, for otherwise, t would be a proper subterm of $l\sigma$ by $\mathrm{V}(r) \subseteq \mathrm{V}(l)$ and $l \notin V$, which contradicts the minimality of $l\sigma(= s')$. Hence $t = r'\sigma$ for some $r' \trianglelefteq r$. Moreover $l \ntrianglerighteq r'$ holds, for otherwise, we would obtain $l\sigma \triangleright r'\sigma = t$, contradicting again the minimality of $l\sigma$. This completes the proof of (2). Finally, (3) is obtained from (2) by using Lemma 5 and Lemma 3 (1). $\qquad\qquad\square$

4 Dependency Pairs

As shown in Lemma 6, every minimal non-terminating sequence contains infinitely many rewrite steps each accompanied by an argument context. To treat instances of a rewrite rule and instances of a rewrite rule within an argument context in a uniform way, we introduce a notion of argument expansion of a rewrite rule.

Definition 4 (argument expansion). The *argument expansion* $\mathrm{Exp}(l \to r)$ of a rewrite rule $l \to r$ is defined recursively as follows: if $l \to r$ is of base type then $\mathrm{Exp}(l \to r) = \emptyset$, otherwise $\mathrm{Exp}(l \to r) = \{l' \to r'\} \cup \mathrm{Exp}(l' \to r')$ where $l' = (l \; x_1 \; \cdots \; x_n)$, $r' = (r \; x_1 \; \cdots \; x_n)$, and $x_1, \cdots, x_n$ are distinct fresh variables of appropriate types.

We are now ready to give an extended definition of dependency pairs.

Definition 5 (dependency pairs). The set $\mathrm{DP}(\mathcal{R})$ of *dependency pairs* of an STTRS $\mathcal{R} = \langle \Sigma, R \rangle$ is defined as follows:

$$\mathrm{DP}(\mathcal{R}) = \{\langle l, r' \rangle \mid l \to r \in R, \; r' \trianglelefteq r, \; r' \ntrianglelefteq l, \; \mathrm{head}(r') \in \Sigma_{\mathrm{d}} \cup V\}$$
$$\cup \{\langle l', r' \rangle \in \mathrm{Exp}(l \to r) \mid l \to r \in R, \; \mathrm{head}(r) \in \Sigma_{\mathrm{d}} \cup V\}$$

A dependency pair $\langle l, r \rangle$ is also written as $l \mapsto r$. We abbreviate $\mathrm{DP}(\mathcal{R})$ as DP when $\mathcal{R}$ is clear from the context.

Note that head variables are allowed in the right-hand sides of dependency pairs. If no right-hand side of the rewrite rules in $\mathcal{R}$ has a head variable of a function type, then the first component of the definition of $\mathrm{DP}(\mathcal{R})$ is the same as the first-order dependency pairs. Moreover, if all the rewrite rules are of base type, then the second component is empty.

Example 5 (dependency pairs). Let $\mathcal{R}$ be the STTRS of Example 1. The set of dependency pairs of $\mathcal{R}$ is as follows:

$$\mathrm{DP}(\mathcal{R}) = \left\{ \begin{array}{lll} (1) & (+\ (\mathsf{s}\ x))\ y & \rightarrowtail (+\ x)\ y \\ (2) & (+\ (\mathsf{s}\ x))\ y & \rightarrowtail +\ x \\ (3) & (\mathsf{fold}\ F\ x)\ (:\ y\ ys) & \rightarrowtail (F\ y)\ ((\mathsf{fold}\ F\ x)\ ys) \\ (4) & (\mathsf{fold}\ F\ x)\ (:\ y\ ys) & \rightarrowtail F\ y \\ (5) & (\mathsf{fold}\ F\ x)\ (:\ y\ ys) & \rightarrowtail (\mathsf{fold}\ F\ x)\ ys \\ (6) & \mathsf{sum} & \rightarrowtail \mathsf{fold} + 0 \\ (7) & \mathsf{sum} & \rightarrowtail \mathsf{fold} \\ (8) & \mathsf{sum} & \rightarrowtail + \\ (9) & \mathsf{sum}\ x & \rightarrowtail (\mathsf{fold} + 0)\ x \end{array} \right\}.$$

Neither $+\ 0\ y \rightarrowtail y$ nor $(+\ (\mathsf{s}\ x))\ y \rightarrowtail +$ is a dependency pair of $\mathcal{R}$ because each right-hand side is a proper subterm of the corresponding left-hand side.

We introduce a new relation which expresses a root rewrite step using a dependency pair, because it is not in general type-preserving.

Definition 6 (dependency relation). The *dependency relation* $\rightarrowtail_D$ of a set of dependency pairs D is defined as follows: $s \rightarrowtail_D t$ if there exist a dependency pair $l \rightarrow r \in D$ and a substitution σ such that $s = l\sigma$, $t = r\sigma$ and $\mathrm{head}(t) \in \Sigma_d$.

Lemma 7 (chain of minimal non-terminating terms via dependency relation). If $s \in \mathrm{NT}_{\min}$ then $s \xrightarrow{\mathrm{nh}}^* \cdot \rightarrowtail_{\mathrm{DP}} t$ for some $t \in \mathrm{NT}_{\min}$.

Proof. By Lemma 6 and the definition of dependency pairs. $\square$

The next lemma gives a characterization of termination in terms of the dependency relation.

Lemma 8 (termination by dependency pairs). An STTRS is terminating if and only if the relation $\xrightarrow{\mathrm{nh}}^* \cdot \rightarrowtail_{\mathrm{DP}}$ on $\mathrm{NT}_{\min}$ is terminating.

Proof. ($\Rightarrow$) The contraposition can be shown by using the inclusions $\xrightarrow{\mathrm{nh}}^* \cdot \rightarrowtail_{\mathrm{DP}} \subseteq \rightarrow^+ \cdot \trianglerighteq$ and $\trianglerighteq \cdot \rightarrow^+ \subseteq \rightarrow^+ \cdot \trianglerighteq$. ($\Leftarrow$) The contraposition can be shown by using Lemma 7 and the existence of a (minimal) non-terminating term. $\square$

Definition 7 (dependency chain). Let D be a set of dependency pairs of an STTRS $\mathcal{R}$. A *dependency chain* of D is an infinite sequence $t_0, t_1, \ldots$ on $\mathrm{NT}_{\min}(\mathcal{R})$ such that $t_i \xrightarrow{\mathrm{nh}}^* \cdot \rightarrowtail_D t_{i+1}$ for all $i \geq 0$. The family of all sets of dependency pairs that admit dependency chain is denoted by $\mathbf{DC}(\mathcal{R})$. We define $\mathbf{DC}_{\min}(\mathcal{R}) = \{D \in \mathbf{DC}(\mathcal{R}) \mid D' \notin \mathbf{DC}(\mathcal{R}) \text{ for any } D' \subsetneq D\}$. As usual, we omit the parameter $\mathcal{R}$ when it is clear from the context.

We now come to the simply typed version of Theorem 6 of [3].

Theorem 1 (termination by dependency chains). An STTRS is terminating if and only if $\mathbf{DC} = \emptyset$.

Proof. By Lemma 8 and the definition of a dependency chain. $\square$

Corollary 1 (termination by $\mathbf{DC}_{\min}$). Let $\mathcal{R}$ be an STTRS such that DP is finite. Then $\mathcal{R}$ is terminating if and only if $\mathbf{DC}_{\min} = \emptyset$.

Proof. ($\Rightarrow$) By Theorem 1 and the inclusion $\mathbf{DC_{min}} \subseteq \mathbf{DC}$. ($\Leftarrow$) By Theorem 1 and the well-foundedness of the inclusion relation on the family of finite sets. $\square$

The following property of $\mathbf{DC_{min}}$ is often used in the subsequent sections.

Lemma 9 (characterization of $\mathbf{DC_{min}}$). Let D be a set of dependency pairs. $D \in \mathbf{DC_{min}}$ if and only if $D \in \mathbf{DC}$ and for any dependency chain $t_0, t_1, \ldots$ of D and any $l \rightarrowtail r \in D$ there exist infinitely many i such that $t_i \overset{\text{nh}}{\rightarrow}^* \cdot \rightarrowtail_{\{l \rightarrowtail r\}} t_{i+1}$.

Proof. ($\Rightarrow$) Suppose $D \in \mathbf{DC_{min}}$. If some dependency pair $l \rightarrowtail r \in D$ were used only finitely many times, then we would have $D \setminus \{l \rightarrowtail r\} \in \mathbf{DC}$, which contradicts the minimality of D. ($\Leftarrow$) For a proof by contradiction, suppose there exists $D' \subsetneq D$ such that $D' \in \mathbf{DC}$. Then there exists a dependency chain of D', which is also a dependency chain of D since $D' \subseteq D$. No dependency pairs in $D \setminus D'$ are used in this chain. Hence any $l \rightarrowtail r \in D \setminus D'$ can be used to obtain the desired contradiction. $\square$

5   Dependency Graph

The dependency graph is useful to find sets of dependency pairs from which dependency chains may arise. Any dependency chain corresponds to a path of the dependency graph. One can prove absence of dependency chains by showing that for every cycle in the graph there is no dependency chain corresponding to that cycle.

Definition 8 (dependency graph). Let $\mathcal{R}$ be an STTRS and D be a set of dependency pairs of $\mathcal{R}$. The *dependency graph* $\mathrm{DG}_{\mathcal{R}}(D)$ is the directed graph whose set of vertices is D in which $\langle l \rightarrowtail r, l' \rightarrowtail r' \rangle$ is an edge if and only if there exist substitutions σ, σ' such that $r\sigma \overset{\text{nh}}{\rightarrow}^* l'\sigma'$.

Since the edges of a dependency graph are not computable in general, its computable estimation needs to be considered. The most basic estimation in the first-order setting is the one using REN $\circ$ CAP and unifiability [3]. We are now going to extend this estimation for the simply typed framework.

Recall that in simply typed term rewriting, there may be a term consisting of a head symbol and partially applied arguments which cannot be rewritten until the remaining arguments are applied. To estimate such partially applied terms, we introduce a notion of pattern.

Definition 9 (defined and undefined pattern). The *pattern* $\mathrm{pat}(t)$ of a term t is the term obtained from t by replacing its arguments with fresh variables, recursively defined as follows: (1) $\mathrm{pat}(t) = t$ if $t \in \Sigma \cup V$, and (2) $\mathrm{pat}(t) = (\mathrm{pat}(t_0) \; x_1 \cdots x_n)$ if $t = (t_0 \; t_1 \cdots t_n)$ where $x_1, \ldots, x_n$ are distinct and fresh variables of appropriate types. A *subpattern* of t is defined as follows: (1) $\mathrm{pat}(t)$ is a subpattern of t, (2) if $t = (t_0 \; t_1 \cdots t_n)$ then any subpattern of t_0 is a subpattern of t. A term t has a *defined* pattern if either (1) $\mathrm{head}(t) \in V$ or (2) $\mathrm{head}(t) \in \Sigma_{\mathrm{d}}$ and there exists a subpattern p of t such that $p \precsim l$ for some $l \rightarrow r \in R$; otherwise, it has an *undefined* pattern.

Example 6 (defined and undefined patterns). Let $\mathcal{R}$ be the STTRS in Example 1. Let $s = ((\text{fold} + ((+ x)\ 0))\ [\ ])$. Then s has a pattern $((\text{fold}\ F\ y)\ z)$; its subpatterns are $((\text{fold}\ F\ y)\ z)$, $(\text{fold}\ F\ y)$, and fold. The term t has a defined pattern, while $(\text{fold} + ((+ x)\ 0))$ has an undefined pattern.

Lemma 10 (properties of patterns). Let s and t be terms. (1) $\text{pat}(s) \precsim s$. (2) If $s = \kappa[t]$ for some argument context κ and p is a subpattern of t then p is a subpattern of s. (3) If $\text{head}(s) \in \Sigma$, then, for any substitution σ, $\text{pat}(s) = \text{pat}(s\sigma)$ (up to the renaming of fresh variables).

Proof. (1) By induction on s. (2) By induction on κ. (3) By induction on s. □

Lemma 11 (properties of defined and undefined patterns). (1) If $s = (s_0\ s_1 \cdots s_n)$ has an undefined pattern then s_0 has an undefined pattern. (2) If s has an undefined pattern, then for any substitution σ, $s\sigma$ has an undefined pattern. (3) For any rewrite rule $l \to r$ and any substitution σ, $l\sigma$ has a defined pattern. (4) If s has a defined pattern and $C[s]$ has an undefined pattern then $C[x]$ has an undefined pattern where $x \in V$.

An approximation of a term is now defined by using the notion of pattern.

Definition 10 (approximation approx**).** Define $\text{approx}(t) = \text{REN}(\text{CAP}(t))$. Here, CAP replaces every outermost subterm in any argument which have a defined pattern by a fresh variable of appropriate type. REN replaces all occurrences of variables but the head occurrence[2] with distinct fresh variables.

Example 7 (approximation approx*).* Let $s = ((\text{fold} + ((+ x)\ y))\ (s\ 0))$. Then $\text{approx}(s) = ((\text{fold} + z)\ (s\ 0))$. Let $t = (F\ (G\ ((+ x)\ y)))$. Then $\text{approx}(t) = (F\ z)$.

Note that our CAP keeps a subterm with a defined head symbol if none of its instance can be rewriten by a head step because of the lack of arguments. This is contrasted with the CAP for first-order terms, which replaces every subterm with a defined root symbol by a variable. Our CAP is the same as the original first-order definition when it is applied to a first-order term.

Lemma 12 (properties of CAP **and** REN**).** (1) $\text{CAP}(s) \precsim s$ and $\text{REN}(s) \precsim s$ for any term s. (2) $\text{CAP}(s) \precsim \text{CAP}(s\sigma)$ and $\text{REN}(s) \precsim \text{REN}(s\sigma)$ for any term s and substitution σ. (3) For any $l \to r \in R$, $x \in V$, a context C, and a substitution σ, $\text{CAP}(C[l\sigma]) \precsim \text{CAP}(C[x])$.

Definition 11 (estimated dependency graph). Let $\mathcal{R}$ be an STTRS and D a set of dependency pairs of $\mathcal{R}$. The *estimated dependency graph* $\text{EDG}_{\mathcal{R}}(D)$ is the directed graph whose set of vertices is D in which $\langle l \rightarrowtail r, l' \rightarrowtail r' \rangle$ is an edge if and only if $\text{approx}(r)$ and l' are unifiable. The estimated dependency graph $\text{EDG}_{\mathcal{R}}(\text{DP}(\mathcal{R}))$ is abbreviated as $\text{EDG}(\mathcal{R})$.

---

[2] This is not at all essential, but it turns out to be useful to define *head instantiation* later.

Example 8 (estimated dependency graph). Let $\mathcal{R}$ be the STTRS of Example 1. The dependency pairs of $\mathcal{R}$ are presented in Example 5. The estimated dependency graph $\mathrm{EDG}(\mathcal{R})$ is shown below.

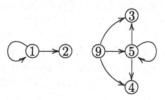

Lemma 13 (soundness of estimation). For any terms s, t and substitutions σ, τ, if $s\sigma \overset{\mathrm{nh}}{\to}^* t\tau$ then approx(s) and t are unifiable.

Proof. Let $u \to_\mathcal{R} v$ with $s, t \in \mathrm{T}(\Sigma, V)$. Then $u = C[l\sigma]$ and $C[r\sigma] = v$ for some rewrite rule $l \to r$, substitution σ and a context C such that head(C) $\in \Sigma_\mathrm{d} \cup V$. Then by (2) and (3) of Lemma 12, approx(u) $\precsim$ approx(v). Thus, by induction on the length of the reduction, it easily follows $s\sigma \overset{\mathrm{nh}}{\to}^* t\tau$ implies approx($s\sigma$) $\precsim$ approx($t\tau$). Then by (1) and (2) of Lemma 12, approx(s) $\precsim$ approx($s\sigma$) $\precsim$ approx($t\tau$) $\precsim t\tau \succsim t$. Since variables in approx(s) and t are disjoint, approx(s) and t are unifiable. □

Theorem 2 (approximation by estimated dependency graph). Let D be a finite set of dependency pairs of an STTRS $\mathcal{R}$. If $D' \subseteq D$ and $D' \in \mathbf{DC}_{\min}(\mathcal{R})$, then for any vertices $u, v \in D'$ there exists a nonempty path in D' from u to v.

Proof. By Lemma 9, Lemma 13, and the definition of an estimated dependency graph. □

6   Subterm Criterion and Head Instantiation

Simple projections and the subterm criterion have been introduced by Hirokawa and Middeldorp [10] to show the absence of dependency chains. We first extend these notions to the case of simply typed term rewriting. A simple projection maps a term to one of its arguments according to its head symbol. In the simply typed framework, a simple projection may not be well-defined because of partially applied arguments and head variables. For this reason, we have to speak of *feasible* simple projections. The head instantiation technique is introduced in order to handle dependency pairs with head variables.

Definition 12 (feasible simple projection). A *simple projection* is a mapping which maps each defined symbol to a positive integer. A simple projection π is *feasible* for a term t if (1) head(t) $\in \Sigma_\mathrm{d}$ and (2) $\pi(\mathrm{head}(t)) \leq |\mathrm{Arg}(t)|$. In that case, we define $\pi(t) = \mathrm{Arg}(t)_{|\pi(\mathrm{head}(t))}$. A simple projection π is feasible for a set D of dependency pairs if it is feasible for l and r for any $l \rightarrowtail r \in D$. In that case, we define $\pi(D) = \{\pi(l) \rightarrowtail \pi(r) \mid l \rightarrowtail r \in D\}$.

Example 9 (feasible simple projection). Let π_1 be a simple projection satisfying $\pi_1(+) = 1$. Then π_1 is feasible for $((+ \ (s \ 0)) \ 0)$ and for $(+ \ (s \ 0))$: we have $\pi_1((+ \ (s \ 0)) \ 0) = \pi_1(+ \ (s \ 0)) = (s \ 0)$. Let π_2 be a simple projection satisfying $\pi_2(+) = 2$. Then π_2 is feasible for $((+ \ (s \ 0)) \ 0)$ but not for $(+ \ (s \ 0))$: the latter lacks the second argument.

Definition 13 (subterm criterion). Let D be a set of dependency pairs. We say that D satisfies the *subterm criterion* if there exists a simple projection π such that π is feasible for D, $\pi(D) \subseteq \unrhd$, and $\pi(D) \cap \rhd \neq \emptyset$.

Theorem 3 (absence of dependency chains by subterm criterion). If a set of dependency pairs D satisfies the subterm criterion, then $D \notin \mathbf{DC}_{\min}$.

Proof. The proof proceeds in the same way as for first-order TRSs (see [10]). $\square$

Example 10 (termination by subterm criterion). Let $\mathcal{R}$ be the STTRS of Example 1. $\mathrm{DP}(\mathcal{R})$ and $\mathrm{EDG}(\mathcal{R})$ are presented in Examples 5 and 8, respectively. Let π be a simple projection satisfying $\pi(+) = 1$ and $\pi(\mathsf{fold}) = 3$. Then $\pi((1)), \pi((5)) \in \rhd$ and hence $\{(1)\}, \{(5)\} \notin \mathbf{DC}_{\min}$. Therefore $\mathcal{R}$ is terminating.

A simple projection is infeasible for a set of dependency pairs if some right-hand side has a variable as its head symbol. To deal with such a case, we introduce a head instantiation technique. Some instantiation techniques for first-order dependency pairs are discussed in [4, 6, 7].

Definition 14 (head instantiation). Let $l \rightarrowtail r$ and $l' \rightarrowtail r'$ be dependency pairs such that $\mathrm{head}(r) \in V$. Moreover, we assume that $l \rightarrowtail r$ and $l' \rightarrowtail r'$ do not share any variables in common (after renaming). Suppose that $\mathrm{approx}(r)$ and l' are unifiable and let $\sigma = \mathrm{mgu}(\mathrm{approx}(r), l')$. The *head substitution* of $l \rightarrowtail r$ for $l' \rightarrowtail r'$ is defined as $\{F \mapsto \mathrm{pat}(\sigma(F))\}$ where $F = \mathrm{head}(r) = \mathrm{head}(\mathrm{approx}(r))$. The set $\mathrm{HS}_D(l \rightarrowtail r)$ of head substitutions of $l \rightarrowtail r$ for a set D of dependency pairs is the singleton set of the identity substitution if $\mathrm{head}(r) \in \Sigma_\mathrm{d}$ and is the set $\{\sigma \mid \sigma$ is the head substitution of $l \rightarrowtail r$ for some $l' \rightarrowtail r' \in D\}$ if $\mathrm{head}(r) \in V$. The *head instantiation* $\mathrm{HI}(D)$ of a set D of dependency pairs is defined by:

$$\mathrm{HI}(D) = \bigcup_{l \rightarrowtail r \in D} \{l\sigma \rightarrowtail r\sigma \mid \sigma \in \mathrm{HS}_D(l \rightarrowtail r)\}.$$

Theorem 4 (soundness and completeness of head instantiation). Let D be a set of dependency pairs. Then $\mathrm{HI}(D) \in \mathbf{DC}$ if and only if $D \in \mathbf{DC}$.

Proof. ($\Rightarrow$) By the inclusion $\rightarrowtail_{\mathrm{HI}(D)} \subseteq \rightarrowtail_D$. ($\Leftarrow$) By the fact that $s \rightarrowtail_D t \xrightarrow{\mathrm{nh}}{}^* \cdot \rightarrowtail_D u$ implies $s \rightarrowtail_{\mathrm{HI}(D)} t$.

Example 11 (Termination by head instantiation). Let $\mathcal{R} = \langle \Sigma, R \rangle$ be an STTRS where $\Sigma = \{ \ []^\mathrm{o}, \ :^{\mathrm{o} \times \mathrm{o} \rightarrow \mathrm{o}}, \ \mathsf{map}^{(\mathrm{o} \rightarrow \mathrm{o}) \times \mathrm{o} \rightarrow \mathrm{o}}, \ \mathsf{pair}^{\mathrm{o} \times \mathrm{o} \rightarrow \mathrm{o}}, \ \mathsf{append}^{\mathrm{o} \times \mathrm{o} \rightarrow \mathrm{o}}, \ \mathsf{curry}^{(\mathrm{o} \times \mathrm{o} \rightarrow \mathrm{o}) \rightarrow \mathrm{o} \rightarrow \mathrm{o}}, \ \mathsf{cartprod}^{\mathrm{o} \times \mathrm{o} \rightarrow \mathrm{o}} \}$, and

$$R = \begin{cases} \text{append } [] \ xs & \to \quad xs \\ \text{append } (: x \ xs) \ ys & \to \quad : x \ (\text{append } xs \ ys) \\ \text{map } F \ [] & \to \quad [] \\ \text{map } F \ (: x \ xs) & \to \quad : (F \ x) \ (\text{map } F \ xs) \\ ((\text{curry } G) \ x) \ y & \to \quad G \ x \ y \\ \text{cartprod } [] \ ys & \to \quad [] \\ \text{cartprod } (: x \ xs) \ ys & \\ \qquad \to \quad \text{append } (\text{map } ((\text{curry pair}) \ x) \ ys) \ (\text{cartprod } xs \ ys) \end{cases}.$$

The function cartprod computes the cartesian product of two lists. For example, cartprod $[1,2]$ $[3,4] = [(1,3),(1,4),(2,3),(2,4)]$. We have $\mathrm{DP}(\mathcal{R}) =$

$$\begin{cases} (1) & \text{append } (: x \ xs) \ ys & \rightarrowtail & \text{append } xs \ ys \\ (2) & \text{map } F \ (: x \ xs) & \rightarrowtail & F \ x \\ (3) & \text{map } F \ (: x \ xs) & \rightarrowtail & \text{map } F \ xs \\ (4) & ((\text{curry } G) \ x) \ y & \rightarrowtail & G \ x \ y \\ (5) & \text{cartprod } (: x \ xs) \ ys & & \\ & & \rightarrowtail & \text{append } (\text{map } ((\text{curry pair}) \ x) \ ys) \ (\text{cartprod } xs \ ys) \\ (6) & \text{cartprod } (: x \ xs) \ ys & \rightarrowtail & \text{append} \\ (7) & \text{cartprod } (: x \ xs) \ ys & \rightarrowtail & \text{map } ((\text{curry pair}) \ x) \ ys \\ (8) & \text{cartprod } (: x \ xs) \ ys & \rightarrowtail & \text{map} \\ (9) & \text{cartprod } (: x \ xs) \ ys & \rightarrowtail & (\text{curry pair}) \ x \\ (10) & \text{cartprod } (: x \ xs) \ ys & \rightarrowtail & \text{curry pair} \\ (11) & \text{cartprod } (: x \ xs) \ ys & \rightarrowtail & \text{curry} \\ (12) & \text{cartprod } (: x \ xs) \ ys & \rightarrowtail & \text{cartprod } xs \ ys \end{cases}.$$

The strongly connected components of $\mathrm{EDG}(\mathcal{R})$ are $\{(1)\}$ and $\{ (2), (3), (4), (7), (12) \}$. It is easy to prove $\{(1)\} \notin \mathbf{DC}_{\min}$ by using a simple projection. Let $D = \{(2), (3), (4), (7), (12)\}$. Then $\mathrm{HI}(D) =$

$$\begin{cases} (2)^{\{F \mapsto ((\text{curry } G) \ y)\}} & \text{map } ((\text{curry } G) \ y) \ (: x \ xs) & \rightarrowtail & ((\text{curry } G) \ y) \ x \\ (3) & \text{map } F \ (: x \ xs) & & \rightarrowtail & \text{map } F \ xs \\ (4)^{\{G \mapsto \text{cartprod}\}} & ((\text{curry cartprod}) \ x) \ y & & \rightarrowtail & \text{cartprod } x \ y \\ (7) & \text{cartprod } (: x \ xs) \ ys & & \rightarrowtail & \text{map } ((\text{curry pair}) \ x) \ ys \\ (12) & \text{cartprod } (: x \ xs) \ ys & & \rightarrowtail & \text{cartprod } xs \ ys \end{cases}.$$

We now show $\mathrm{HI}(D) \notin \mathbf{DC}$ by using the recursive SCC algorithm of [9]. Take a simple projection π as $\pi(\text{curry}) = 3$, $\pi(\text{map}) = \pi(\text{cartprod}) = 2$. It is easy to check $\{\pi((2)^{\{F \mapsto ((\text{curry } G) \ y)\}}), \pi((3))\} \subseteq \rhd$ and $\{\pi((4)^{\{G \mapsto \text{cartprod}\}}), \pi((7)), \pi((12))\} \subseteq \unrhd$. Thus it suffices to show $D' = \{\pi((4)^{\{G \mapsto \text{cartprod}\}}), (7), (12)\} \notin \mathbf{DC}$. The only cycle in $\mathrm{EDG}(D')$ is $\{(12)\}$. By taking a simple projection π as $\pi(\text{cartprod}) = 1$, we know $D' \notin \mathbf{DC}$. Hence $\mathrm{HI}(D) \notin \mathbf{DC}$. Therefore $\mathcal{R}$ is terminating.

7 Experiments

Based on the techniques presented in this paper, we implemented a termination prover for STTRSs written in the functional programming language Haskell. We have tested 125 examples of STTRSs (of which 122 examples are terminating).

They include typical higher-order functions such as fold, map, rec, filter, etc. of various simple types. For example, fold of types $(o\times o\to o)\times o\times o\to o$, $(o\times o\to o)\times o\to o\to o$, $(o\to o\to o)\times o\to o\to o$, and $(o\to o\to o)\to o\to o\to o$. The table below is the number of STTRSs categorized by the number of functions, i.e. defined symbols. (Mutually recursive functions are counted as one function.)

number of functions	1	2	3	4	5	6	7	8
number of programs	40	21	28	15	7	5	6	3

In our first experiment, we count the number of STTRSs $\mathcal{R}$ whose termination can be inferred by checking that $EDG(\mathcal{R})$ is acyclic. In the second experiment, termination proof using subterm criterion is tried additionally; and in the third, both subterm criterion and head instantiation are used.

We have also tested the direct application of the first-order dependency pair method to a first-order encoding of STTRSs. In the experiment, we used the termination prover AProVE [8] of ver.1.1c-β where the non-overlappingness check is disabled (so that it does not attempt innermost termination proof, but attempts termination proof). Our encoding is the most naive one; the encoding Θ from a simply typed term to the corresponding first-order term is defined as $\Theta((t_0\ t_1\cdots t_n)) = a_n(\Theta(t_0),\Theta(t_1),\ldots,\Theta(t_n))$, where a_n $(n>0)$ are new function symbols of arity n.

The result is summarized in the table below.

	cycle check	+ subterm criterion	+ head instantiation	first-order encoding
success	6	56	98	67
failure/timeout(30 sec.)	119/0	69/0	27/0	55/3

This table shows that both the subterm criterion and the head instantiation are effective in simply typed dependency pair approach. Our approach proves more examples than the termination proof via the naive first-order encoding does. There are, however, 8 examples that succeed in the approach via first-order encoding but fail in our approach.

8   Concluding Remarks

We presented an extension of the dependency pair method of first-order term rewriting which enlarges the scope of automatic termination proofs for simply typed term rewriting systems. We clarified basic concepts of the dependency pair method in the simply typed term rewriting framework. We incorporated the subterm criterion into the extension and introduced the head instantiation. As a result, we obtained a dependency pair method which is effectively applicable even in the presence of function variables. We compared our method with the direct application of the first-order dependency pair method to a naive first-order encoding of simply typed term rewriting systems. Comparison with a more elaborate encoding such as [1] and [2] remains as a future work.

In contrast to the subterm criterion, the argument filtering is not likely to be incorporated in a straightforward way; for, in the presence of variables of function types, argument filtering functions are not closed under substitutions. Because of this difficulty, the reduction pair and the argument filtering techniques in the simply typed term rewriting framework still need to be investigated. Extensions of other techniques of the first-order dependency pair method such as the innermost termination proof have not yet been tried.

In [17, 18], dependency pair techniques for Nipkow's HRSs [15] have been proposed. In [18] the notions of the dependency pair and the dependency graph are incorporated into HRSs, while the difficulty of using the argument filtering are reported. In [17], the argument filtering has been incorporated at the cost of limiting HRSs to be non-duplicating and non-nested.

In [13, 14], Kusakari proposed path orderings and a dependency pair method for his higher-order framework TRS_{hv}. His method is based on the reduction pair and the argument filtering techniques. Our characterization of minimal non-terminating terms is similar to those used in his paper. However, we believe that our in-depth analysis is indispensable to make our development robust. In his method, in contrast to ours, the definition of the dependency chain is modified, while the definition of the dependency pair is kept same as that of the first-order dependency pair. By this difference, it seems not obvious whether our version of the dependency graph and the subterm criterion are incorporated to his dependency pair method or not.

Finally, we remark that experiments for a numerous collection of examples have not been likely intended by these existing works on higher-order dependency pair methods. The head instantiation and estimation of dependency graphs using patterns are proposed for the first time in this paper.

Acknowledgments

The authors thank the anonymous referees for their valuable comments. This work was partially supported by grants from JSPS, Nos. 14780187 and 16700030, and from Kayamori Foundation of Informational Science Advancement.

References

1. T. Aoto and T. Yamada. Termination of simply typed term rewriting systems by translation and labelling. In *Proceedings of the 14th International Conference on Rewriting Techniques and Applications*, volume 2706 of *LNCS*, pages 380–394. Springer-Verlag, 2003.
2. T. Aoto and T. Yamada. Termination of simply-typed applicative term rewriting systems. In *Proceedings of the 2nd International Workshop on Higher-Order Rewriting*, 2004.
3. T. Arts and J. Giesl. Termination of term rewriting using dependency pairs. *Theoretical Computer Science*, 236(1–2):133–178, 2000.
4. J. Giesl and T. Arts. Verification of Erlang processes by dependency pairs. *Applicable Algebra in Engineering, Communication and Computing*, 12(1/2):39–72, 2001.

5. J. Giesl, T. Arts, and E. Ohlebusch. Modular termination proofs for rewriting using dependency pairs. *Journal of Symbolic Computation*, 34(1):21–58, 2002.
6. J. Giesl, R. Thiemann, P. Schneider-Kamp, and S. Falke. Improving dependency pairs. In *Proceedings of the 10th International Conference on Logic for Programming, Artificial Intelligence and Reasoning*, volume 2850 of *LNAI*, pages 165–179. Springer-Verlag, 2003.
7. J. Giesl, R. Thiemann, P. Schneider-Kamp, and S. Falke. Mechanizing dependency pairs. Research Report 2003–08, Aachener Informatik-Bericht, 2003.
8. J. Giesl, R. Thiemann, P. Schneider-Kamp, and S. Falke. Automated termination proofs with AProVE. In *Proceedings of the 15th International Conference on Rewriting Techniques and Applications*, volume 3091 of *LNCS*, pages 210–220. Springer-Verlag, 2004.
9. N. Hirokawa and A. Middeldorp. Automating the dependency pair method. In *Proceedings of the 19th International Conference on Automated Deduction*, volume 2741 of *LNAI*. Springer-Verlag, 2003.
10. N. Hirokawa and A. Middeldorp. Dependency pairs revisited. In *Proceedings of the 15th International Conference on Rewriting Techniques and Applications*, volume 3091 of *LNCS*, pages 249–268. Springer-Verlag, 2004.
11. J.-P. Jouannaud and M. Okada. Executable higher-order algebraic specification languages. In *Proceedings of the 6th IEEE Symposium on Logic in Computer Science*, pages 350–361. IEEE Press, 1991.
12. J. W. Klop. *Combinatory Reduction Systems*. PhD thesis, Rijksuniversiteit, Utrecht, 1980.
13. K. Kusakari. On proving termination of term rewriting systems with higher-order variables. *IPSJ Transactions on Programming*, 42(SIG 7 PRO 11):35–45, 2001.
14. K. Kusakari. Higher-order path orders based on computability. *IEICE Transactions on Information and Systems*, E87–D(2):352–359, 2003.
15. R. Mayr and T. Nipkow. Higher-order rewrite systems and their confluence. *Theoretical Computer Science*, 192(1):3–29, 1998.
16. A. Middeldorp. Approximations for strategies and termination. *Electronic Notes in Theoretical Computer Science*, 70(6):1–20, 2002.
17. M. Sakai and K. Kusakari. On new dependency pair method for proving termination of higher-order rewrite systems. In *Proceedings of the International Workshop on Rewriting in Proof and Computation*, pages 176–187, 2001.
18. M. Sakai, Y. Watanabe, and T. Sakabe. An extension of dependency pair method for proving termination of higher-order rewrite systems. *IEICE Transactions on Information and Systems*, E84–D(8):1025–1032, 2001.
19. R. Thiemann, J. Giesl, and P. Schneider-Kamp. Improved modular termination proofs using dependency pairs. In *Proceedings of the 2nd International Joint Conference on Automated Reasoning*, volume 3097 of *LNAI*, pages 75–90. Springer-Verlag, 2004.
20. T. Yamada. Confluence and termination of simply typed term rewriting systems. In *Proceedings of the 12th International Conference on Rewriting Techniques and Applications*, volume 2051 of *LNCS*, pages 338–352. Springer-Verlag, 2001.

Universal Algebra
for Termination of Higher-Order Rewriting

Makoto Hamana

Department of Computer Science, Gunma University, Japan
hamana@cs.gunma-u.ac.jp

Abstract. We show that the structures of binding algebras and Σ-monoids by Fiore, Plotkin and Turi are sound and complete models of Klop's Combinatory Reduction Systems (CRSs). These algebraic structures play the same role of universal algebra for term rewriting systems. Restricting the algebraic structures to the ones equipped with well-founded relations, we obtain a complete characterisation of terminating CRSs. We can also naturally extend the characterisation to rewriting on meta-terms by using the notion of Σ-monoids.

1 Introduction

At RTA'98, Plotkin presented the theory of *binding algebras* [Plo98], which aimed to apply ideas in universal algebra to type theory. It is interesting that this was given as an invited talk at RTA. That is to say, in the context of rewriting, it can be read as a possibility of a new direction of foundation of *higher-order rewriting* as a type theoretic system. Plotkin's idea of binding algebras was inspired by Aczel's work [Acz78]. In the field of rewriting, also inspired by Aczel's same work, Klop invented a system of higher-order rewriting called *Combinatory Reduction System* (CRS) [Klo80]. It is natural to think that these two works, having a common origin, have some relationship. However, such a relationship is not obvious, especially about how the seemingly complex syntax of CRSs can be understood in the theory of binding algebras.

Plotkin's program of binding algebras later produced the notion of Σ-monoids [FPT99]. Interestingly, the *free Σ-monoids* constructed in [Ham04] is the same as the syntax of "meta-terms" of CRSs (cf. Theorem 5). This similarity suggests that the universal algebra for CRSs may be Σ-monoids. Based on this idea, the present paper provides a complete algebraic characterisation of CRSs.

Contribution. Complete characterisation of terminating CRSs obtained in this paper provides a method of proving the termination of CRSs by algebraic interpretation. The following CRS $\mathcal{R}$ for conversion into prenex normal form, i.e. pushing quantifiers outside, is a typical example of higher-order rewrite rules that require the feature of variable binding [Pol96, Raa]:

$$P \wedge \forall(x.Q[x]) \rightarrow \forall(x.P \wedge Q[x]) \qquad \neg\forall(x.Q[x]) \rightarrow \exists(x.\neg(Q[x]))$$
$$\forall(x.Q[x]) \wedge P \rightarrow \forall(x.P \wedge Q[x]) \qquad \neg\exists(x.Q[x]) \rightarrow \forall(x.\neg(Q[x]))$$

J. Giesl (Ed.): RTA 2005, LNCS 3467, pp. 135–149, 2005.

with the similar rewrite rules for $\vee$ and $\exists$ at the left column. Intuitively, rewriting using $\mathcal{R}$ and its termination are clear; notwithstanding, the application of existing proof methods in the theory of higher-order rewriting to the CRS $\mathcal{R}$ is not so straightforward [JR01], or it requires consideration of an involved function space to interpret binders [Pol96, Pol94]. The present paper provides a simpler method of showing termination of CRS such as $\mathcal{R}$ (cf. Example 24).

Organisation. This paper is organised as follows. We first review the definition of CRSs in Section 2. We then introduce the notion of "structural CRSs". define a class of structurally well-formed CRSs in Section 3. Section 4 gives algebraic semantics of CRSs syntax and valuations. Section 5 gives algebraic semantics of CRSs rewriting. Section 6 gives algebraic semantics of CRSs meta-rewriting. Finally, in Section 7, we show examples of termination proofs using a result of this paper.

Future Work. This work opens a new direction of model theoretic study of higher-order rewriting. An immediate application will be semantic labelling method [Zan94] for CRSs using the algebraic structure developed in this paper. Recursive path ordering on free structures in more general setting is also hopeful.

2   Combinatory Reduction Systems

We review the definition of CRSs. We use the definition of the standard reference [KOR93] of CRSs with a slight modification of syntax used in [DR98]: $-.-$ and $-[-]$ instead of ordinary ones $[-]-$ and $-(-)$ in [KOR93].

CRS. Assume a signature Σ of function symbols F^l with arity, metavariables z^l with arity (in both cases the superscript $l \in \mathbb{N}$ is the arity).

(i) CRS *terms* have the form

$$t ::= x \mid x.t \mid F^l(t_1, \ldots, t_l).$$

The three forms are respectively called *variables*, *abstractions*, and *function terms*.

(ii) CRS *meta-terms* extend CRS terms to

$$t ::= x \mid x.t \mid F^l(t_1, \ldots, t_l) \mid z^l[t_1, \ldots, t_l].$$

The last form is called a *meta-application*.

(iii) A *valuation* θ is a mapping that assigns to n-ary metavariable z an n-ary *substitute* (a meta-level lambda notation, cf. [KOR93]):

$$\theta : z \longmapsto \underline{\lambda}(x_1, \ldots, x_n).t \tag{1}$$

Valuations are extended to a function on meta-terms:

$$\theta(x) = x \qquad \theta(F(t_1, \ldots, t_l)) = F(\theta(t_1), \ldots, \theta(t_l))$$
$$\theta(x.t) = x.\theta(t) \qquad \theta(z[t_1, \ldots, t_l]) = \theta(z)(\theta(t_1), \ldots, \theta(t_l)) \tag{2}$$

Note that the rhs of the equation (2) uses an application at the meta-level to the substitute. The valuation is *safe* if there are no two substitutes $\theta(z)$ and $\theta(z')$ such that $\theta(z)$ contains a free variable x which appears also bound in $\theta(z')$.

(iv) CRS *rules*, written $l \to r$, consist of two meta-terms l and r with the following additional restrictions:

(iv-a) l and r are closed (w.r.t. variables) meta-terms,

(iv-b) l must be a "pattern", i.e. a function term where all meta-applications have the form $z[x_1, \ldots, x_n]$ with distinct x_i,

(iv-c) r can only contain meta-applications with meta-variables occurring in the left-hand side.

The rewrite rule $l \to r$ is *safe for θ*, if for no z in l and r, the substitute $\theta(z)$ has a free variable x occurring in an abstraction $x.-$ of l and r. A set of rewrite rules is called a CRS.

(v) The CRS *rewrite relation* $\to_{\mathcal{R}}$ is generated by context and safe valuation closure of a given CRS $\mathcal{R}$:

$$\frac{l \to r \in \mathcal{R}}{\theta(l) \to_{\mathcal{R}} \theta(l)} \text{ safe } \theta \qquad \frac{s \to_{\mathcal{R}} t}{x.s \to_{\mathcal{R}} x.t} \qquad \frac{s \to_{\mathcal{R}} t}{F(\ldots, s, \ldots) \to_{\mathcal{R}} F(\ldots, t, \ldots)}$$

where $l \to r$ must be safe for the safe valuation θ. The third rule means a rewriting at the i-th argument of F.

3   Structural CRSs

In this section, we introduce the notion of *structural* CRS as a class of well-formed CRSs. This idea of structural CRS is to treat only CRS (meta-)terms built from binding signature (cf. Aczel's contraction schemes [Acz78]). A binding signature specifies how many binders are taken in arguments of each function symbol.

Formally, a *binding signature* Σ is consisting of a set Σ of function symbols with an arity function $a : \Sigma \to \mathbb{N}^*$. A function symbol of *binding arity* $\langle n_1, \ldots, n_l \rangle$, denoted by $f : \langle n_1, \ldots, n_l \rangle$, has l arguments and binds n_i variables in the i-th argument ($1 \le i \le l$).

For a formal treatment of named variables modulo α-equivalence in CRSs, we assume the method of de Bruijn levels [dB72, LRD95, FPT99] for the naming convention of variables (N.B. not for metavariables) in CRSs. We also use the convention that $n \in \mathbb{N}$ denotes the set $\{1, \ldots, n\}$ (n is possibly 0). Under the method of de Bruijn levels, this n means the set of variables from 1 to n.

Definition 1. A (meta-)term t is called *structural* if t is built from a binding signature Σ and consistent with the binding arities of function symbols in Σ.

Schematically, structural meta-terms have the form:

$$t ::= x \mid F(x_1 \cdots x_{i_1}.t_1, \ldots, x_1 \cdots x_{i_l}.t_l) \mid z^l[t_1, \ldots, t_l]$$

where F has the binding arity $\langle i_1, \ldots, i_l \rangle$.

More precisely, structural meta-terms are defined as follows. Fix an $\mathbb{N}$-indexed set Z of metavariables defined by $Z(l) \triangleq \{z \mid z \text{ has arity } l\}$. A meta-term t is structural if $n \vdash t$ is derived from the following rules.

$$\frac{x \in n}{n \vdash x} \qquad \frac{F : \langle i_1, \ldots, i_l \rangle \in \Sigma \quad n+i_1 \vdash t_1 \ \cdots \ n+i_l \vdash t_l}{n \vdash F(\, n+1 \ldots n+i_1.t_1, \ \ldots, \ n+1 \ldots n+i_l.t_l \,)}$$

$$\frac{z \in Z(l) \quad n \vdash t_1 \ \cdots \ n \vdash t_l}{n \vdash z[t_1, \ldots, t_l]}$$

By these rules, a meta-term always follows the method of de Bruijn levels. Using only the first two rules (or equivalently, assuming $Z(l) = \varnothing$ for all l), we obtain *structural terms* under n.

The notion of structural is obviously extended to rewrite rules, CRS, and valuations. A rewrite rule is called structural if all meta-terms in the rule are structural. A CRS is structural if all rules are structural.

Definition 2. A valuation θ is *structural* if for any mapping by $\theta : z \mapsto \lambda(x_1, \ldots, x_n).t$, t is a structural term and all variables in t are included in $x_1, \ldots, x_n$.

Structual CRS is a fairly good assumption because we can easily find that almost all concrete examples of CRSs considered in the literature are structural; namely, we can easily find a suitable binding signature of a given "plain" CRS. Actually, in Raamsdonk's collection [Raa] of examples of higher-order rewrite systems *all CRSs are structural*.

Example 3 (CPS translation). The format of structural CRS is very similar to an "everyday" meta-language for expressing formal systems in computer science and logic. An example is the structural CRS $\mathcal{R}$ for prenex normal form in the introduction. Another example related to theory of programming languages is the following CRS $\mathcal{S}$ of a call-by-value CPS translation [DR98].

Assume the metavariables $Z = \{v^0, E^1, (E_0)^0, (E_1)^0\}$ and the binding signature Σ consisting of the function symbols $\lambda, \bar{\lambda} : \langle 1 \rangle$, $(-\ -), (-\bar{\ }-) : \langle 0, 0 \rangle$, CPS, $(\![-]\!) : \langle 0 \rangle$. We write the structural CRS $\mathcal{S}$ of CPS translation in two ways: the left column is written in the usual named notation, and the right column is written in de Bruijn level notation, which is the format we use in this paper.

CPS(E)	$\to \lambda k.(\![E]\!)^{-}(\bar\lambda m.km)$		CPS(E)	$\to \lambda 1.(\![E]\!)^{-}(\bar\lambda 2.12)$
$(\![v]\!)$	$\to \bar\lambda k.k^{-} v$		$(\![v]\!)$	$\to \bar\lambda 1.1^{-} v$
$(\![\lambda x.E[x]]\!)$	$\to \bar\lambda k.k^{-}(\lambda x.\lambda k.(\![E[x]]\!)^{-}(\bar\lambda m.km))$		$(\![\lambda 1.E[1]]\!)$	$\to \bar\lambda 1.1^{-}(\lambda 2.\lambda 3.(\![E[2]]\!)^{-}(\bar\lambda 4.34))$
$(\![E_0 E_1]\!) \to \bar\lambda k.$	$(\![E_0]\!)^{-}(\bar\lambda m.(\![E_1]\!)^{-}(\bar\lambda n.mn(\lambda a.k^{-} a)))$		$(\![E_0 E_1]\!) \to \bar\lambda 1.$	$(\![E_0]\!)^{-}(\bar\lambda 2.(\![E_1]\!)^{-}(\bar\lambda 3.23(\lambda 4.1^{-} 4)))$

A point is that de Bruijn level version is obtained by just renaming variable names with numbers according to their (de Bruijn's) levels. Notice that this completely differs from the more well-known method of de Bruijn *indexes*. Meta-terms in de Bruijn levels are just "normal forms" of α-equivalent meta-terms (e.g. $\bar\lambda k.k^{-} v =_\alpha \bar\lambda 1.1^{-} v$).

Is the structural CRS $\mathcal{S}$ terminating[1]? Intuitively, termination is clear because $(\![-]\!)$ recursively decomposes a λ-term. In this paper, we derive a formal

---

[1] This does not contain β-reduction rules, i.e. only for translation.

way of showing termination from an algebraic characterisation of rewriting of CRS. How this S is shown to be terminating will be given in Example 25 at the end of the paper.

4 Algebraic Semantics of Syntax

In this section and in the next section, we consider algebraic semantics of CRSs. As far as the author knows, this is the first algebraic consideration of CRSs. The basic idea is similar to the algebraic semantics of TRSs by monotone Σ-algebras popularized by Zantema [Zan94]. But the framework of usual first-order universal algebra is insufficient. We consider CRS's syntax in the framework of *binding algebras* by Fiore, Plotkin and Turi [FPT99].

4.1 Binding Algebras

We review the notion of binding algebras. For detail, see [FPT99]. Let $\mathbb{F}$ be the category which has finite cardinals $n = \{1, \ldots, n\}$ (n is possibly 0) as objects, and all functions between them as arrows $m \to n$. This is the category of object variables by the method of de Bruijn levels (i.e. natural numbers) and their renamings. We use the functor category $\mathbf{Set}^{\mathbb{F}}$. We define the functor $\delta : \mathbf{Set}^{\mathbb{F}} \to \mathbf{Set}^{\mathbb{F}}$ as follows: for $L \in \mathbf{Set}^{\mathbb{F}}, n \in \mathbb{F}, \rho \in \mathrm{arr}\ \mathbb{F}, (\delta L)(n) = L(n+1), \quad (\delta L)(\rho) = L(\rho + \mathrm{id}_1)$. To a binding signature Σ, we associate the *signature functor* $\Sigma :$ $\mathbf{Set}^{\mathbb{F}} \to \mathbf{Set}^{\mathbb{F}}$ given by $\Sigma A \triangleq \coprod_{f:\langle n_1,\ldots,n_l \rangle \in \Sigma} \prod_{1 \le i \le l} \delta^{n_i} A$. A Σ-*binding algebra* (or simply Σ-*algebra*) is a pair (A, α) consisting of a presheaf $A \in \mathbf{Set}^{\mathbb{F}}$ and a map ($[\ ]$ denotes a copair of coproducts) $\alpha = [f_A]_{f \in \Sigma} : \Sigma A \longrightarrow A$ called *algebra structure*, where f_A is an *operation* $f_A : \delta^{n_1} A \times \ldots \times \delta^{n_l} A \longrightarrow A$ defined for each function symbol $f : \langle n_1, \ldots, n_l \rangle \in \Sigma$.

The "the presheaf of variables" $\mathrm{V} \in \mathbf{Set}^{\mathbb{F}}$ is defined by $\mathrm{V}(n) = n, \mathrm{V}(\rho) = \rho$ ($\rho : m \to n \in \mathbb{F}$). Then, $(\mathbf{Set}^{\mathbb{F}}, \bullet, \mathrm{V})$ forms a monoidal category [Mac71], where the "substitution" monoidal product is defined as follows. For presheaves A and B, $(A \bullet B)(n) \triangleq (\coprod_{m \in \mathbb{N}} A(m) \times B(n)^m)/\sim$ where $\sim$ is the equivalence relation generated by $(t; u_{\rho 1}, \ldots, u_{\rho m}) \sim (A(\rho)(t); u_1, \ldots, u_l)$ for $\rho : m \to l \in \mathbb{F}$. Throughout the paper, we use the following notation: an element of $A(m) \times B(n)^m$ is denoted by $(t; u_1, \ldots, u_m)$ where $t \in A(m)$ and $u_1, \ldots, u_m \in B(m)$. A representative of an equivalence class in $A \bullet B(n)$ is also denoted by this notation.

Let Σ be a signature functor with strength st defined by a binding signature. A Σ-*monoid* $M = (\alpha, \eta, \mu)$ consists of a *monoid object* [Mac71] ($M, \eta : \mathrm{V} \to M, \mu : M \bullet M \to M$) in the monoidal category $(\mathbf{Set}^{\mathbb{F}}, \bullet, \mathrm{V})$ with a Σ-binding algebra $\alpha : \Sigma M \to M$ satisfying $\mu \circ (\alpha \bullet \mathrm{id}_M) = \alpha \circ \Sigma\mu \circ st$. A Σ-*monoid morphism* $(M, \alpha) \longrightarrow (M', \alpha')$ is a morphism in $\mathbf{Set}^{\mathbb{F}}$ which is both Σ-algebra homomorphism and monoid morphism.

4.2 Algebra of Structural CRS Terms

Structural terms and meta-terms have a good algebraic structure. We define the presheaf $T_{\Sigma}\mathrm{V} \in \mathbf{Set}^{\mathbb{F}}$ of all structural terms under n by $T_{\Sigma}\mathrm{V}(n) = \{t \mid n \vdash t,\ t$ is a term$\}$ with obvious arrow part [Ham04]. We also define the map

$\nu : V \longrightarrow T_\Sigma V$ in $\mathbf{Set}^\mathbb{F}$ by $\nu(n) : V(n) \longrightarrow T_\Sigma V(n)$, $x \longmapsto x$. We abbreviate $n+1, \ldots, n+k.t$ to $n+\vec{k}.t$. For every $f \in \Sigma$ with the arity $\langle i_1, \ldots, i_l \rangle$, we define the map $F_T : \delta^{i_1} T_\Sigma V \times \cdots \times \delta^{i_l} T_\Sigma V \longrightarrow T_\Sigma V$ in $\mathbf{Set}^\mathbb{F}$ by $(t_1, \ldots, t_l) \longmapsto F(n+\vec{i_1}.t_1, \ldots, n+\vec{i_l}.t_l)$.

Theorem 4. *Structural CRS terms $T_\Sigma V$ forms an initial $V+\Sigma$-binding algebra.*

Proof. Due to [FPT99]. The "syntactic algebra" in ([FPT99] Theorem 2.1) is nothing but the $V + \Sigma$-algebra $(T_\Sigma V, [\nu, [F_{T_\Sigma}]_{F \in \Sigma}])$. □

Moreover, let Z be an arbitrary $\mathbb{N}$-indexed set of metavariables (cf. Sect. 3). The *presheaf $M_\Sigma Z$ of meta-terms* is defined by

$$M_\Sigma Z(n) = \{t \mid n \vdash t\}.$$

There is the map $\beta : M_\Sigma Z \bullet M_\Sigma Z \longrightarrow M_\Sigma Z$ in $\mathbf{Set}^\mathbb{F}$, called *multiplication*, that performs a substitution for variables [Ham04].

Theorem 5. *Structural CRS meta-terms $M_\Sigma Z$ forms a free Σ-monoid over $\hat{Z}$, where $\hat{Z}(n) = \coprod_{k \in \mathbb{N}} \mathbb{F}(k, n) \times Z(k)$.*

Proof. Due to [Ham04]. For $\hat{Z} \in \mathbf{Set}^\mathbb{F}$, the free Σ-monoid constructed in [Ham04] is nothing but $(M_\Sigma Z, [F_{M_\Sigma}]_{F \in \Sigma}, \nu, \beta)$ by just identifying minor notational difference of terms: regard $\mathsf{ovar}(x), [n]t, \ulcorner z \urcorner \langle t_1, \ldots, t_l \rangle$ in [Ham04] as $x, n.t, z[t_1, \ldots, t_l]$ respectively in the present paper. Here, operations F_{M_Σ} are defined by the same as F_{T_Σ}. □

4.3   Algebraic Characterisation of Valuations

Definition 6. *An assignment $\phi : Z \longrightarrow A$ is a morphism of $\mathbf{Set}^\mathbb{F}$ whose target A has a Σ-monoid structure (A, ν, β).*

Notice that Z in the above definitions is a presheaf in $\mathbf{Set}^\mathbb{F}$. So just an $\mathbb{N}$-indexed set X of metavariables cannot be the source of this presentation of valuation. Fortunately, we can always construct a presheaf from an $\mathbb{N}$-indexed set X by defining $\hat{X}(n) \triangleq \coprod_{k \in \mathbb{N}} \mathbb{F}(k, n) \times X(k)$ (see [Ham04] Sect. 5.2). Hence, hereafter we abuse the notation to use X to denote its presheaf version $\hat{X} \in \mathbf{Set}^\mathbb{F}$ in an assignment.

An assignment ϕ is extended to a Σ-monoid morphism $\phi^* : M_\Sigma Z \longrightarrow A$:

$$M_\Sigma Z(n) \longrightarrow A(n)$$
$$x \longmapsto \nu(n)(x) \qquad (x \in n)$$
$$F(n+\vec{i_1}.t_1, \ldots, n+\vec{i_l}.t_l) \longmapsto F_A(n+\vec{i_1}.\phi^*(n+i_1)(t_1), \ldots, n+\vec{i_l}.\phi^*(n+i_l)(t_l))$$
$$z[t_1, \ldots, t_l] \longmapsto \beta(n)(\phi(l)(z); \phi^*(n)(t_1), \ldots \phi^*(n)(t_l))$$

where $f : \langle i_1, \ldots, i_l \rangle \in \Sigma$. In the special case $A = T_\Sigma V$, we have

Proposition 7. *An assignment* $\theta : Z \longrightarrow T_\Sigma V$ *gives a structural valuation, and* $\theta^* : M_\Sigma Z \longrightarrow T_\Sigma V$ *gives its "homomorphic" extension on meta-terms.*

To see why, first we note that the assignment θ is a family of maps $\theta(n) :$ $Z(n) \longrightarrow T_\Sigma V(n)$ such that

$$\theta(n) : z \longmapsto t \in T_\Sigma V(n).$$

Namely, it maps an n-ary metavariable z to some structural term t under n. Comparing the definition of structural valuation with this, and regarding the substitute $\underline{\lambda}(x_1, \ldots, x_n).t$ as $t \in T_\Sigma V(n)$ (because θ is structural), both definitions coincide. Hence, hereafter we use the word "valuation" in this sense:

Definition 8. A *valuation* is an assignment $\theta : Z \longrightarrow T_\Sigma V$ into the Σ-monoid of terms. Also, we use the following: a *meta-valuation* is an assignment $\theta :$ $Z \longrightarrow M_\Sigma X$ into the Σ-monoid of meta-terms.

Now we know in what sense θ^* is a "homomorphic" extension of a valuation θ (which is not explained formally in the ordinary definitions [KOR93, OR94, DR98, Oos94]). Namely θ^* is a Σ-*monoid morphism*, which preserves Σ-algebra structure (i.e. Σ-homomorphism) and monoid structure.

4.4   Structural Valuations are Sufficient

A valuation in the original sense (Sect. 2) was a map $\theta : z \mapsto \underline{\lambda}(x_1, \ldots, x_n).t$ where t is an *arbitrary* term, which means that $\underline{\lambda}(x_1, \ldots, x_n).t$ may have variables other than $x_1, \ldots, x_n$. But in the case of a structural valuation, variables in t are taken only from $x_1, \ldots, x_n$. We show that structural valuations are sufficient to generate CRS rewrite relation on terms if we make some weakening of rules.

For $m \leq m'$, let $\rho : \mathbb{N} \to \mathbb{N}$ be the function defined by $\rho(m + i) \triangleq m' + i$ for each $i \in \mathbb{Z}$. Suppose $\mathbb{N}$-indexed metavariable sets $Z' = Y \cup \{z^m\}$, $Z = Y \cup \{z^{m'}\}$, $z \notin Y$. The *weakening of the arity of the metavariable* z *by* ρ *from* m *to* m' is a function ρ_z on (unstructural) meta-terms defined as follows.

$$\rho_z(z[t_1, \ldots, t_m]) = z[1, \ldots, m' - m, \rho_z(t_1), \ldots, \rho_z(t_m)]$$
$$\rho_z(n.t) = \rho(n).\rho_z(t) \quad \rho_z(F(\vec{t})) = F(\rho_z(\vec{t})) \quad \rho_z(x) = \rho(x) \quad (x \in \mathbb{N}).$$

Notation 9. We may use the notation $Z|n \vdash s \to t$ for a rule or a rewrite step if metavariables and variables in s and t are included in Z and n respectively. We may also simply write $Z \vdash s \to t$ or $n \vdash s \to t$ if another part is not important.

Let $\mathcal{R}$ be a structural CRS that follows the method of de Bruijn levels. Then *weakening closure* of $\mathcal{R}$, denoted by $\mathcal{R}^\circ$, is defined by the following inference rules (i.e. the least set satisfying the rules):

$$\frac{l \to r \in \mathcal{R}}{l \to r \in \mathcal{R}^\circ} \qquad \frac{Y \cup \{z^m\} \vdash l \to r \in \mathcal{R}^\circ}{Y \cup \{z^{m+j}\} \vdash \vec{j}.\rho_z l \to \vec{j}.\rho_z r \in \mathcal{R}^\circ}$$

where ρ_z is weakening of the arity of the metavariable z from m to $m+j$ ($j \in \mathbb{N}$ is arbitrary). This means that although originally a metavariable z^m can be replaced with a term exactly containing m-variables, it will be weakened to z^{m+j}, which can be replaced with a term containing $m+j$-variables.

Then, we reformulate the generation of rewrite relation as follows:

$$\frac{Z \vdash \vec{n}.l \to \vec{n}.r \in \mathcal{R}}{n \vdash \theta^*(n)(l) \Rightarrow_{\mathcal{R}} \theta^*(n)(r)} \qquad \frac{n+i \vdash s \Rightarrow_{\mathcal{R}} t}{n \vdash F(\ldots, n+\vec{i}.s, \ldots) \Rightarrow_{\mathcal{R}} F(\ldots, n+\vec{i}.t, \ldots)}$$

where $\theta : Z : \longrightarrow T_\Sigma V$ is a valuation.

Proposition 10. *For a structural CRS $\mathcal{R}$ that follows the method of de Bruijn levels, the ordinary definition (cf. Sect. 2) and the above definition with $\mathcal{R}^\circ$ generate the same rewrite relation on structural terms, i.e. $s \Rightarrow_{\mathcal{R}^\circ} t$ iff $s \to_{\mathcal{R}} t$ for structural terms s, t.*

5   Algebraic Semantics of Rewriting

In this section, we interpret rewrite rules of structural CRSs by Σ-binding algebras, and give a complete characterisation of termination in this framework. *Hereafter, in this paper we only consider structural CRSs. So we just say "a CRS" for a structural CRS.*

For a presheaf A, we write $>_A$ for a family of transitive relations $\{>_{A(n)}\}_{n \in \mathbb{N}}$, where $>_{A(n)}$ is a transitive relation on the set $A(n)$ for each $n \in \mathbb{N}$. In this paper, we use the following notion of monotonicity [Zan94].

Definition 11. Let $(A_1, >_{A_1}), \ldots, (A_l, >_{A_l}), (B, >_B)$ be presheaves equipped with transitive relations. A map $f : A_1 \times \cdots \times A_l \longrightarrow B$ in $\mathbf{Set}^{\mathbb{F}}$ is *monotone* if all $a_1, b_1 \in A_1(n), \ldots, a_l, b_l \in A_l(n)$ with $a_k >_{A(n)} b_k$ for some k and $a_j = b_j$ for all $j \neq k$, then $f(n)(a_1, \ldots, a_l) >_{B(n)} f(n)(b_1, \ldots, b_l)$.

We interpret rewrite rules in a V+Σ-algebra.

Definition 12. Let A be a V + Σ-algebra. A *term-generated assignment* $\phi :$ $Z \longrightarrow A$ is a morphism of $\mathbf{Set}^{\mathbb{F}}$ that is expressed as the composite

$$Z \xrightarrow{\ \theta\ } T_\Sigma V \xrightarrow{\ !_A\ } A$$

for some valuation θ, where $!_A$ is the unique V + Σ-algebra homomorphism from the initial V+Σ-algebra $T_\Sigma V$. Throughout the paper, we denote by $!_A$ this unique V + Σ-homomorphism.

This means that an interpretation of a metavariable z by a term-generated assignment θ is performed by firstly assigning to z some term t and then interpreting the term in a V+Σ-algebra A. Why this is needed is that CRS rewrite relation is generated on terms (not on meta-terms). So, to interpret CRS rewrite rules, not all assignments are needed; only term-generated assignments are sufficient.

Definition 13. A *monotone* $V+\Sigma$-*algebra* $(A, >_A)$ is a $V+\Sigma$-algebra $A = (A, [\nu, [F_A]_{F\in\Sigma}])$, (where $\nu : V \longrightarrow A$), equipped with a transitive relation $>_{A(n)}$ on $A(n)$ for each $n \in \mathbb{N}$ such that every operation f_A is monotone. Moreover, if $>_{A(n)}$ is a well-founded strict partial order for each $n \in \mathbb{N}$, A is called *well-founded*.

Definition 14. Let $\mathcal{R}$ be a CRS. A monotone $V+\Sigma$-algebra $(A, >_A)$ *satisfies* a CRS rewrite rule $Z \vdash \vec{n}.l \to \vec{n}.r$ if

$$\phi^*(n)(l) >_{A(n)} \phi^*(n)(r)$$

for all term-generated assignments $\phi : Z \longrightarrow A$. A $(V+\Sigma, \mathcal{R})$-*algebra* A is a monotone $V+\Sigma$-algebra A that satisfies all rules in the weakening closure $\mathcal{R}^\circ$.

Define the $\mathbb{N}$-indexed transitive relation $\to^+_{\mathcal{R}(n)} \triangleq \{(s,t) \mid n \vdash s \Rightarrow^+_{\mathcal{R}\circ} t\}$, where the latter $(-)^+$ denotes the transitive closure.

Theorem 15. *For a CRS $\mathcal{R}$, $(T_\Sigma V, \to^+_{\mathcal{R}})$ is an initial $(V+\Sigma, \mathcal{R})$-algebra, i.e. for any $(V+\Sigma, \mathcal{R})$-algebra A, there exists a unique monotone homomorphism $T_\Sigma V \longrightarrow A$.*

Proof. Let $(A, >_A)$ be a $(V+\Sigma, \mathcal{R})$-algebra. Since $T_\Sigma V$ is an initial $V+\Sigma$-algebra (Theorem 4), $!_A : T_\Sigma V \longrightarrow A$ is a unique $V+\Sigma$-algebra homomorphism. So, the remaining task is to show $!_A$ is monotone. This is proved by induction on the structure of inference of $\Rightarrow_{\mathcal{R}}$ and induction on the length of $\Rightarrow^+$. Note that $!_A \circ \theta$ is term-generated and all operations F_A on A are monotone.   □

The following states that $(V+\Sigma, \mathcal{R})$-algebras are sound and complete for many-step rewrite relation (where Notation 9 is used).

Corollary 16. *Let $\mathcal{R}$ be a CRS. The followings are equivalent:*

(i) $n \vdash s \to^+_{\mathcal{R}} t$ holds,
(ii) $!_A(n)(s) >_{A(n)} !_A(n)(t)$ for all $(V+\Sigma, \mathcal{R})$-algebras $(A, >_A)$.

Proof. (i)$\Rightarrow$(ii): By Theorem 15.
(ii)$\Rightarrow$(i): Take $(A, >_A) = (T_\Sigma V, \to^+_{\mathcal{R}})$.   □

Restricting the above corollary to the case of well-founded monotone algebras, we obtain a complete characterisation of terminating CRSs.

Theorem 17. *A CRS $\mathcal{R}$ is terminating if and only if there is a well-founded $(V+\Sigma, \mathcal{R})$-algebra.*

Proof. ($\Leftarrow$): Let A be a well-founded $(V+\Sigma, \mathcal{R})$-algebra. Assume $\mathcal{R}$ is non-terminating, i.e. there exists an infinite reduction sequence $n \vdash t_1 \to_{\mathcal{R}} t_2 \to_{\mathcal{R}} \cdots$. By Corollary 16, we have $!_A(n)(t_1) >_{A(n)} !_A(n)(t_2) >_{A(n)} \cdots$. This contradicts well-foundedness of $>_A$.

($\Rightarrow$): When a CRS $\mathcal{R}$ is terminating, the initial $(V+\Sigma, \mathcal{R})$-algebra $(T_\Sigma V, \to^+_{\mathcal{R}})$ is a desired well-founded algebra, because the strict partial order $\to^+_{\mathcal{R}}$ is well-founded.   □

Example 18 (Incompleteness of functional interpretation [Pol96]). Assume the metavariables $Z = \{F^1, X^1\}$ and the binding signature $\Sigma = \{c : \langle 0 \rangle\}$. Consider the CRS $\mathcal{R}$ consisting of the following only:

$$c(F[F[X[1]]]) \rightarrow F[X[1]].$$

We want to show termination of $\mathcal{R}$. Intuitively, this termination seems easy to be proved because with any rewrite step the number of c-symbols decreases. Nevertheless the existing interpretation method of higher-order rewriting based on the model of hereditary monotone functionals *cannot show termination of $\mathcal{R}$* due to incompleteness of the model [Pol94, Pol96]. In contrast to it, we *can show* termination of $\mathcal{R}$ by using Theorem 17 as follows. Take the monotone $V + \Sigma$-algebra $(T_\Sigma V, \succ_{T_\Sigma V})$ where $s \succ_{T_\Sigma V(n)} t$ iff the number of c-symbols in s and t decreases. Notice that now all terms in $T_\Sigma V(n)$ are consisting of c and variables in n only. Hence, all assignments into $T_\Sigma V$ are of the forms $F \mapsto c^k(1)$, $X \mapsto c^m(1)$ (k-times and m-times c's). This gives a well-founded $(V + \Sigma, \mathcal{R})$-algebra $(T_\Sigma V, \succ_{T_\Sigma V})$, which implies termination of $\mathcal{R}$ by Theorem 17.

6   Algebraic Semantics of Meta-rewriting

We go beyond the standard definition of rewriting of CRS, and consider rewriting on *meta-terms*, which we call *meta-rewriting*. In the literature, although meta-rewriting has not been formally defined, Oostrom considered the notions of meta-CR and meta-SN of CRS and pointed out each of them is not derived from CR and SN of CRS respectively ([Oos94] Sect. 3.4).

We consider *meta-termination*, i.e. termination of meta-rewriting. In this section, we give algebraic semantics of meta-rewriting. Basically we repeat the semantics in Sect. 5, but we use Σ-monoids instead of Σ-binding algebras for the semantics structure.

Rewriting on Meta-terms. First we formally define meta-rewriting. Let Z be an $\mathbb{N}$-indexed set of metavariables. For a CRS $\mathcal{R}$ in which any two rules have disjoint metavariables taken from Z (if not, rename rules suitably), we denote the CRS by $(\mathcal{R}, Z)$. We define the meta-rewriting relation $\rightsquigarrow_\mathcal{R}$ as follows:

$$\frac{\vec{n}.l \rightarrow \vec{n}.r \in \mathcal{R}}{n \vdash \theta^*(n)(l) \rightsquigarrow_\mathcal{R} \theta^*(n)(l)} \qquad \frac{n+i \vdash s \rightsquigarrow_\mathcal{R} t}{n \vdash F(\ldots, n+\vec{i}.s, \ldots) \rightsquigarrow_\mathcal{R} F(\ldots, n+\vec{i}.t, \ldots)}$$

$$\frac{z \in Z(l) \quad n \vdash s \rightsquigarrow_\mathcal{R} t}{n \vdash z[\ldots, s, \ldots] \rightsquigarrow_\mathcal{R} z[\ldots, t, \ldots]}$$

where θ is a *meta-valuation* $Z \longrightarrow M_\Sigma X$ (Definition 8). We say that $\mathcal{R}$ is meta-terminating if $\rightsquigarrow_\mathcal{R}$ is well-founded.

Definition 19. A *monotone Σ-monoid* $(A, >_A)$ is a Σ-monoid A equipped with a transitive relation $>_{A(n)}$ on $A(n)$ for each $n \in \mathbb{N}$ such that every operation is monotone. Moreover, if $>_{A(n)}$ is a well-founded strict partial order for each $n \in \mathbb{N}$, A is called *well-founded*.

Let $\mathcal{R}$ be a CRS. A monotone Σ-monoid $A = (A, >_A)$ *satisfies* a rewrite rule $Z \vdash \vec{n}.l \to \vec{n}.r \in \mathcal{R}$ if

$$\phi^*(n)(l) >_{A(n)} \phi^*(n)(r)$$

for all assignments[2] $\phi : Z \longrightarrow A$. If A satisfies all rules in the weakening closure $\mathcal{R}^\circ$, it is called $(\Sigma, \mathcal{R})$-*monoid*.

An important example of $(\Sigma, \mathcal{R})$-monoid is $(M_\Sigma Z, \leadsto^+_\mathcal{R})$.

Definition 20. Let (A, ν, β) be a monotone Σ-monoid, and $\phi : Z \longrightarrow A$ an assignment. Define the map $\sigma : Z \bullet A \longrightarrow A$ by the composite

$$Z \bullet A \xrightarrow{\phi \bullet \mathrm{id}_A} A \bullet A \xrightarrow{\beta} A.$$

The assignment ϕ is called *admissible* if σ is monotone[3].

Notice that the multiplication β need not to be monotone. Actually, it is rather difficult to find a Σ-monoid whose multiplication is monotone. The unit $\nu : V \longrightarrow A$ is automatically monotone because V has no transitive relation.

The notion of admissible assignments is an important ingredient of interpretation of meta-rewriting. Arbitrary assignments are not suitable to interpret meta-rewriting because it may cause non-order preservation. For example, assume the constants $\Sigma = \{a, b, c\}$, the metavariable $Z = \{z^1\}$ and the CRS $\mathcal{R} = \{a \to b\}$. Then we have a meta-rewriting $z[a] \leadsto_\mathcal{R} z[b]$. We interpret this rewrite step in the $(\Sigma, \mathcal{R})$-monoid $(M_\Sigma Z, \leadsto^+_\mathcal{R})$. Take the assignment $\phi : z \longmapsto c$. Then, this does not preserve the order:

$$\phi^*(z[a]) = c \not\leadsto_\mathcal{R} c = \phi^*(z[b]).$$

We need "monotonic" interpretation of meta-rewriting to establish algebraic termination method. The idea of admissible assignment is motivated by to prohibit this kind of "non-monotonic" interpretation of a rewrite step.

This problem is already recognised by van de Pol [Pol94]. The notion of admissible assignments is analogue to his notion of strict functionals. Actually, we can show that hereditary monotone functionals in his model forms a monotone Σ-monoid and our admissible assignments into this monotone Σ-monoid is the same as the strict valuations at the second-order types. Hence, we can apply the method of termination proof using hereditary monotone functionals to CRSs. For instance, termination of the examples of higher-order rewrite systems given in [Pol94, Pol96] (and their CRS versions are in [Raa]) can be shown by using Σ-monoids of hereditary monotone functionals given in [Pol94, Pol96].

Now we show a theorem analogue to Theorem 15 stating $(M_\Sigma Z, \leadsto^+_\mathcal{R})$ is an "initial model". More precisely,

---

[2] Compare this definition with Definition 14 for rewriting.

[3] More precisely, $\sigma(n) : \coprod_{m \in \mathbb{N}} Z(m) \times A(n)^m / \sim \longrightarrow A(n)$ is monotone, i.e. if $z \in Z(m)$ and all $a_1, b_1 \in A(n), \dots, a_m, b_m \in A(n)$ with $a_k >_{A(n)} b_k$ for some k and $a_j = b_j$ for all $j \neq k$, we have $\sigma(n)(z; a_1, \dots, a_m) >_{A(n)} \sigma(n)(z; b_1, \dots, b_m)$.

Theorem 21
For a CRS $(\mathcal{R}, Z)$, $(M_\Sigma Z, \leadsto^+_\mathcal{R})$ is a free $(\Sigma, \mathcal{R})$-monoid over Z, i.e. for any admissible assignment ϕ from Z into a $(\Sigma, \mathcal{R})$-monoid $(A, >_A)$, there exists a unique Σ-monoid map ϕ^* that is monotone and makes the right diagram commute in $\mathbf{Set}^\mathbb{F}$, where $\eta_Z : z^l \longmapsto z[1, \ldots, l]$.

Proof. Let $\phi : Z \longrightarrow A$ be an admissible assignment into a $(\Sigma, \mathcal{R})$-monoid $(A, >_A)$. Since $M_\Sigma Z$ is a free Σ-monoid [Ham04], ϕ^* is a unique Σ-monoid morphism that makes the above diagram commute in $\mathbf{Set}^\mathbb{F}$. So, the remaining task is to show ϕ^* is monotone. This is proved by induction on the structure of inference of $\leadsto_\mathcal{R}$ and the length of $\leadsto^+_\mathcal{R}$. The case for instantiation of a rewrite rule, we use $(\phi^* \circ \theta)^* = \phi^* \circ \theta^*$, which is proved by induction on meta-terms. The crucial case is to show ϕ^* preserves the relation of $z[\ldots, s, \ldots] \leadsto_\mathcal{R} z[\ldots, t, \ldots]$. This holds because we have assumed that ϕ is admissible. □

Theorem 22. *A CRS $(\mathcal{R}, Z)$ is meta-terminating if and only if there is a well-founded $(\Sigma, \mathcal{R})$-monoid.*

Proof. ($\Leftarrow$): Let A be a well-founded $(\Sigma, \mathcal{R})$-monoid. Assume $\mathcal{R}$ is not meta-terminating, i.e. there exists an infinite meta-rewriting sequence

$$Z|n \vdash t_1 \leadsto_\mathcal{R} t_2 \leadsto_\mathcal{R} t_3 \leadsto_\mathcal{R} \cdots.$$

By Theorem 21, for any admissible assignment $\phi : Z \longrightarrow A$, we have

$$\phi^*(n)(t_1) >_{A(n)} \phi^*(n)(t_2) >_{A(n)} \phi^*(n)(t_3) >_{A(n)} \cdots.$$

This contradicts well-foundedness of $>_A$.
($\Rightarrow$): When a CRS $\mathcal{R}$ is meta-terminating, the free $(\Sigma, \mathcal{R})$-monoid $(M_\Sigma Z, \leadsto^+_\mathcal{R})$ over Z is a desired well-founded one, because the strict partial order $\leadsto^+_\mathcal{R}$ is well-founded. □

7   Termination of Binding CRSs

Let $(\mathcal{R}, X)$ be a CRS such that every meta-application in rules of $\mathcal{R}$ is always of the form $z^l[1, \ldots, l]$. We call such a CRS a *binding CRS* because it is essentially meta-application-free (cf. binding TRS [Ham03]). To interpret a rule and meta-rewriting in a binding CRS $\mathcal{R}$, we do not need the monoid structure of Σ-monoids, i.e. the multiplication β is not used. Because, for example, interpret the meta-term $z[1, 2]$ (for z^2) in a rule by an assignment $\phi : X \longrightarrow A$ into a Σ-monoid (A, ν, β):

$$\phi^*(z[1, 2]) = \beta(\phi(z); \nu(1), \nu(2)) = \phi(z).$$

This is due to $A \bullet V \cong A$, i.e. V is the unit of the monoidal category $\mathbf{Set}^\mathbb{F}$. So, to interpret a meta-term like $z[1, 2]$, we just need an assignment ϕ. Hence, we assume A to be a $X+V+\Sigma$-algebra for interpretation of binding CRSs. Then we can replete the discussion of interpretation of meta-rewriting: *A satisfies* a rule $\bar{n}.l \to \bar{n}.r \in \mathcal{R}$ if $\phi^*(n)(l) >_A \phi^*(n)(r)$ for all assignments $\phi : X \longrightarrow A$ into $X+V+\Sigma$-algebras. We denote by $B_\Sigma X$ an initial $X+V+\Sigma$-algebra and call an element of it a *binding meta-term*. Notice that a binding CRS is a CRS built only from binding meta-terms. We define the meta-rewriting on binding meta-terms by $\to_\mathcal{R} \triangleq \leadsto_\mathcal{R} \cap \bigcup_{n\in\mathbb{N}}(B_\Sigma X \times B_\Sigma X)(n)$. Then, $(B_\Sigma X, \to_\mathcal{R}^+)$ is an initial $(X+V+\Sigma, \mathcal{R})$-algebra.

Proposition 23. *A binding CRS $(\mathcal{R}, X)$ is meta-terminating on binding meta-terms if and only if there is a well-founded $(X+V+\Sigma, \mathcal{R})$-algebra.*

For a binding CRS $\mathcal{R}$, it is clear that meta-termination of $\mathcal{R}$ on binding meta-terms implies termination of $\mathcal{R}$ on terms because all terms are binding meta-terms (meta-application-free). Hence, in the case of binding CRSs this becomes an interesting termination proof method by interpretation because we do not need a monoid structure.

Example 24. We show termination of the CRS $\mathcal{R}$ for conversion into prenex normal form in the introduction. Formally, $\mathcal{R}$ is built from the binding signature $\Sigma = \{\forall, \exists : \langle 1 \rangle, \wedge, \vee : \langle 0, 0 \rangle, \neg : \langle 0 \rangle\}$ and the metavariables $X = \{\mathsf{P}^0, \mathsf{Q}^1\}$. The structural CRS $\mathcal{R}$ in de Bruijn levels is obtained by just replacing the variable x with 1.

$$\mathsf{P} \wedge \forall(1.\mathsf{Q}[1]) \to \forall(1.\mathsf{P} \wedge \mathsf{Q}[1]) \qquad \neg\forall(1.\mathsf{Q}[1]) \to \exists(1.\neg(\mathsf{Q}[1]))$$
$$\forall(1.\mathsf{Q}[1]) \wedge \mathsf{P} \to \forall(1.\mathsf{P} \wedge \mathsf{Q}[1]) \qquad \neg\exists(1.\mathsf{Q}[1]) \to \forall(1.\neg(\mathsf{Q}[1])).$$

We use Proposition 23 to show termination. Take the $X+V+\Sigma$-algebra K by $K(n) = \mathbb{N}$ with $>_{K(n)}$ by the usual order $>$ on $\mathbb{N}$ for all $n \in \mathbb{N}$. The operations are given by

$$\wedge_{K(n)}(x,y) = \vee_{K(n)}(x,y) = 2x + 2y$$
$$\neg_{K(n)}(x) = 2x \qquad \forall_{K(n)}(x) = \exists_{K(n)}(x) = x + 1.$$

All operations are monotone. We can show that K satisfies the rules: take an assignment $\phi : X \longrightarrow K$ by $\mathsf{P} \mapsto x \in \mathbb{N}$ and $\mathsf{Q} \mapsto y \in \mathbb{N}$, then

$$\phi^*(0)(\mathsf{P} \wedge \forall(1.\mathsf{Q}[1])) = 2x + 2(y+1) >_{K(0)} (2x + 2y) + 1 = \phi^*(0)(\forall(1.\mathsf{P} \wedge \mathsf{Q}[1]))$$
$$\phi^*(0)(\neg\exists(1.\mathsf{Q}[1])) = 2(y+1) >_{K(0)} 2y + 1 = \phi^*(0)(\forall(1.\neg(\mathsf{Q}[1]))).$$

Similar for other rules. Since $>_{K(0)}$ is well-founded, this shows K with ϕ is a well-founded $(X+V+\Sigma, \mathcal{R})$-algebra. Thus, the binding CRS $\mathcal{R}$ is meta-terminating on binding meta-terms by Proposition 23. Hence $\mathcal{R}$ is terminating on all CRS terms. This interpretation is simpler than the hereditary monotone functional model given in [Pol96].

Example 25. The CRS $\mathcal{S}$ for a CPS translation in Example 3 is also shown to be terminating by the following polynomial interpretation: take the $X + V + \Sigma$-algebra K by $K(n) = \mathbb{N}$ with the unit $\nu : V \to K$, $i \mapsto 0$ and the operations:

$$\mathsf{CPS}_{K(n)}(e) = 5e + 5 \quad (\![e]\!)_{K(n)} = 5e + 1 \quad \overline{\lambda}_{K(n)}(e) = e \quad \lambda_{K(n)}(e) = e + 1$$
$$(e_0 \, \overline{} e_1)_{K(n)} = e_0 + e_1 \quad (e_0 \, e_1)_{K(n)} = e_0 + e_1 + 1.$$

Checking this satisfies $\mathcal{S}$ is just by calculation. Hence $\mathcal{S}$ is terminating.

Namely, if a CRS is a binding CRS, we do not need functionals to interpret higher-order function symbols such as $\forall, \exists, \overline{\lambda}, \lambda$.

Acknowledgments

I am grateful to Gordon Plotkin, John Power, Neil Ghani, Aart Middeldorp, and the anonymous referees. This work is supported by the JSPS Grant-in-Aid for Scientific Research (16700005).

References

[Acz78]   P. Aczel. A general Church-Rosser theorem. Technical report, University of Manchester, 1978.

[dB72]   N. de Bruijn. Lambda calculus notation with nameless dummies, a tool for automatic formula manipulation, with application to the church-rosser theorem. *Indagationes Mathematicae*, 34:381–391, 1972.

[DR98]   O. Danvy and K.H. Rose. Higher-order rewriting and partial evaluation. In *Rewriting Techniques and Applications, 9th International Conference, (RTA'98)*, LNCS 1379, 1998.

[FPT99]   M. Fiore, G. Plotkin, and D. Turi. Abstract syntax and variable binding. In *Proc. 14th Annual Symposium on Logic in Computer Science*, pages 193–202, 1999.

[Ham03]   M. Hamana. Term rewriting with variable binding: An initial algebra approach. In *Fifth ACM-SIGPLAN International Conference on Principles and Practice of Declarative Programming (PPDP'03)*, pages 148–159, 2003.

[Ham04]   M. Hamana. Free Σ-monoids: A higher-order syntax with metavariables. In *Asian Symposium on Programming Languages and Systems (APLAS 2004)*, LNCS 3302, pages 348–363, 2004.

[JR01]   J.-P. Jouannaud and A. Rubio. Higher-order recursive path orderings à la carte. In *International Workshop on Rewriting in Proof and Computation (RPC'01)*, pages 161–175, 2001.

[Klo80]   J.W. Klop. *Combinatory Reduction Systems*. PhD thesis, CWI, Amsterdam, 1980. volume 127 of Mathematical Centre Tracts.

[KOR93]   J.W. Klop, V. van Oostrom, and F. van Raamsdonk. Combinatory reduction systems: Introduction and survey. *Theor. Comput. Sci.*, 121(1&2):279–308, 1993.

[LRD95]   P. Lescanne and J. Rouyer-Degli. Explicit substitutions with de Bruijn's levels. In *Rewriting Techniques and Applications, 6th International Conference (RTA-95)*, LNCS 914, pages 294–308. Springer, 1995.

[Mac71]   S. Mac Lane. *Categories for the Working Mathematician*, volume 5 of *Graduate Texts in Mathematics*. Springer-Verlag, New York, 1971.

[Oos94] V. van Oostrom. *Confluence for Abstract and Higher-Order Rewriting*. PhD thesis, Vrije Universiteit, Amsterdam, 1994.

[OR94] V. van Oostrom and F. van Raamsdonk. Comparing combinatory reduction systems and higher-order rewrite systems. In *the First International Workshop on Higher-Order Algebra, Logic and Term Rewriting (HOA'93)*, LNCS 816, 1994.

[Plo98] G. Plotkin. Binding algebras: A step between universal algebra and type theory (invited talk). In *Rewriting Techniques and Applications, 9th International Conference, RTA'98, Tsukuba, Japan*, 1998.

[Pol94] J. van de Pol. Termination proofs for higher-order rewrite systems. In *the First International Workshop on Higher-Order Algebra, Logic and Term Rewriting (HOA'93)*, LNCS 816, pages 305–325, 1994.

[Pol96] J. van de Pol. *Termination of Higher-order Rewrite Systems*. PhD thesis, Universiteit Utrecht, 1996.

[Raa] F. van Raamsdonk. Examples of higher-order rewriting systems. at http://www.cs.vu.nl/~femke/ps/.

[Zan94] H. Zantema. Termination of term rewriting: interpretation and type elimination. *Journal of Symbolic Computation*, 17:23–50, 1994.

Appendix: Elementary Description of The Category $\mathbf{Set}^{\mathbb{F}}$

For those who are not familiar with category theory, we devote this section to an elementary description of the central categorical structure used in this paper: the category $\mathbf{Set}^{\mathbb{F}}$ and related morphisms. The functor category $\mathbf{Set}^{\mathbb{F}}$ plays an central role in this paper. The objects of it are functors $\mathbb{F} \to \mathbf{Set}$ and the arrows are natural transformations between them. In more elementary term, an objects A of $\mathbf{Set}^{\mathbb{F}}$ (often written as $A \in \mathbf{Set}^{\mathbb{F}}$) is given by a $\mathbb{N}$-indexed set $\{A(n)\}_{n \in \mathbb{N}}$ with "the arrow part" i.e. for each function $\rho : m \to n \in \mathbb{F}$, we also need to give a function $A(\rho) : A(m) \longrightarrow A(n)$.

An arrow (or called a map, morphism) between objects $A, B \in \mathbf{Set}^{\mathbb{F}}$ is a natural transformation $f : A \longrightarrow B$; more elementary, it is given by a family of functions of the form $f(n) : A(n) \longrightarrow B(n)$ parameterised by all $n \in \mathbb{N}$ that satisfies the condition $\forall a \in A(m) . B(\rho)(f(m)(a)) = f(n)(A(\rho)(a))$ for all functions $\rho : m \to n$. This condition ("naturality") diagrammatically means the commutativity of the diagram

$$
\begin{array}{ccc}
m & A(m) \xrightarrow{\quad f(m) \quad} & B(m) \\
{\scriptstyle \rho}\downarrow & {\scriptstyle A(\rho)}\downarrow & \downarrow{\scriptstyle B(\rho)} \\
n & A(n) \xrightarrow{\quad f(n) \quad} & B(n)
\end{array}
$$

Very roughly, we can think of $A \in \mathbf{Set}^{\mathbb{F}}$ as an $\mathbb{N}$-indexed set equipped with "something", and a map $f : A \longrightarrow B$ of $\mathbf{Set}^{\mathbb{F}}$ as an $\mathbb{N}$-indexed function $f(n) : A(n) \longrightarrow B(n)$ with "some coherence law". These "something" precisely mean the above descriptions. We may ignore them to get a rough understanding (with keeping in mind that these have officially such conditions). An object $A \in \mathbf{Set}^{\mathbb{F}}$ is often called a *presheaf*.

Quasi-interpretations and Small Space Bounds

Guillaume Bonfante[1,2], Jean-Yves Marion[1,2], and Jean-Yves Moyen[3,4]

[1] Loria, Calligramme project, B.P. 239, 54506 Vandœuvre-lès-Nancy Cédex, France
[2] École Nationale Supérieure des Mines de Nancy, INPL, France
[3] Loria, Calligramme project, B.P. 239, 54506 Vandœuvre-lès-Nancy Cédex, France
[4] Université Henri Poincaré Nancy I, France

Abstract. Quasi-interpretations are an useful tool to control resources usage of term rewriting systems, either time or space. They not only combine well with path orderings and provide characterizations of usual complexity classes but also give hints in order to optimize the program. Moreover, the existence of a quasi-interpretation is decidable.

In this paper, we present some more characterizations of complexity classes using quasi-interpretations. We mainly focus on small space-bounded complexity classes. On one hand, by restricting quasi-interpretations to sums (that is allowing only affine quasi-interpretations), we obtain a characterization of LINSPACE. On the other hand, a strong tiering discipline on programs together with quasi-interpretations yield a characterization of LOGSPACE.

Lastly, we give two new characterizations of PSPACE: in the first, the quasi-interpretation has to be strictly decreasing on each rule and in the second, some linearity constraints are added to the system but no assumption concerning the termination proof is made.

1 Introduction

The quasi-interpretation over reals method is a decidable procedure to certify time or space complexity of a program based on rewriting semantics. This paper focuses on several small space complexity classes.

From a practical point of view, the bottom line is to perform a complexity analysis in order to measure space resources by static analysis. The knowledge of the resource consumption is an important information to compile more efficiently a program by certifying the space usage, this may also reinforce security. Dynamic approaches, like direct evaluation of the space or time usage by numerical recipes, as done by Benziger [1] for Nuprl, are the other way for controlling resources. The advantage of static analysis is that it can provide a certificate which guarantees the resource usage.

A program is a term rewriting system in which there are constructor terms defining the domain and the range of the computation. Quasi-interpretations are related to interpretation termination proofs. An interpretation is a quasi-interpretation. Conversely, a quasi-interpretation does not prove termination *a priori*. There are some interests to study programs, which have a polynomially

J. Giesl (Ed.): RTA 2005, LNCS 3467, pp. 150–164, 2005.

bounded quasi-interpretation because it delineates a not too big class of functions. Surprisingly, a function, which admits a polynomial interpretation, can run in doubly exponential time. This observation was made by Hofbauer and Lautemann [2] and Cichon and Lescanne [3]. The whole picture about polynomial interpretation has been clarified in [4]. One of the key ideas is to separate constructor terms from defined function symbols. In fact, the program complexity depends mainly on the quasi-interpretation of constructor terms.

From the point of view of complexity, quasi-interpretations are very exciting. Indeed quasi-interpretations give an upper bound on the size of the normal form of a term if it exists. We obtained different resource upper bounds by combining quasi-interpretations and termination proofs.

- If the program admits a quasi-interpretation and terminates by Product Path Ordering, (PPO) then it is computable in polynomial time. See [5].
- If the program admits a quasi-interpretation and terminates by Lexicographic Path Ordering (LPO), then the computation consumes a polynomial space. See [6].

The results mentioned above should be compared to the expressive power of both recursive path ordering PPO and LPO. Hofbauer [7] showed that Multiset Path Ordering (MPO) gives rise to a characterization of primitive recursive functions. It is not difficult to see that the same proof goes for PPO instead of MPO. Then, Weiermann [8] has shown that LPO characterizes the multiple recursive functions. So, quasi-interpretations allow us to tame the complexity of treated algorithms.

In this paper, we present new characterizations of small space complexity classes. The first result is a characterization of the class LINSPACE of functions, which are computable in linear space in the size of the inputs. For this, we restrict quasi-interpretations to purely affine functions, that is the quasi-interpretation of each symbol is a function like $aX + b$. The class of functions, which are computed by programs which terminate by LPO and admit a purely affine quasi-interpretations, is exactly LINSPACE. Let us mention here the first intrinsic characterization of LINSPACE. It is due to Ritchie [9].

The second result is a characterization of the class LOGSPACE which contains all functions which run in logarithmic space in the size of the inputs. Intuitively, this means that we cannot copy an input on some registers. However, we can use pointers to read and mark inputs. For this, we introduce the notion of tiers inside a rewrite system which has yet be suggested by Beckmann-Weiermann [10] and Marion [11]. The tiering discipline is due to independently Bellantoni-Cook [12] and to Leivant-Marion [13, 14] in the context of simply typed lambda-calculus. Our characterization of LOGSPACE is based on Jones [15] and is also related to the work of Gurevich [16] on primitive recursive function algebra over first order finite structures.

The last result is two characterizations of PSPACE, which is the class of functions computable in polynomial space. Such characterizations of PSPACE by functional programming have been already considered by Oitavem [17].

In this paper, we demonstrate the use of quasi-interpretations to bound memory during a program execution. These results do not give a meaningful upper bound on the runtime. However, it is interesting to have an upper bound on the memory in several practical cases related to embedded systems. In this direction, Amadio and al [18] applied those ideas successfully to resource bytecode verifier and to resource verifier for synchronous cooperative threads [19]. We consider quasi-interpretation over the reals as it gives us a procedure to synthesis quasi-interpretations. See [20, 21] for full details.

2   First Order Functional Programming

Throughout the following discussion, we consider three disjoint sets $\mathcal{X}, \mathcal{F}, \mathcal{C}$ of variables, function symbols and constructors.

2.1   Syntax of Programs

Definition 1. *The sets of* terms *and the* rules *are defined in the following way:*

$$
\begin{array}{llll}
(Constructor\ terms) & T(\mathcal{C}) \ni u & ::= & \mathbf{c} \mid \mathbf{c}(u_1, \cdots, u_n) \\
(terms) & T(\mathcal{C}, \mathcal{F}, \mathcal{X}) \ni t & ::= & \mathbf{c} \mid x \mid \mathbf{c}(t_1, \cdots, t_n) \mid f(t_1, \cdots, t_n) \\
(patterns) & \mathcal{P} \ni p & ::= & \mathbf{c} \mid x \mid \mathbf{c}(p_1, \cdots, p_n) \\
(rules) & \mathcal{D} \ni d & ::= & f(p_1, \cdots, p_n) \to t
\end{array}
$$

where $x \in \mathcal{X}$, $f \in \mathcal{F}$, and $\mathbf{c} \in \mathcal{C}$. We shall use a type writer font for function symbols and a bold face font for constructors.

Remark 2. Notice that function symbols (sometimes called defined symbols) do not appear in any pattern matching and may not have arity 0.

In the following, we will write $\overrightarrow{t}$ to denote the vector of terms $t_1, \cdots, t_n$ if those terms are not relevant for the current discussion.

Definition 3. *A program is a quadruple $f = \langle \mathcal{X}, \mathcal{C}, \mathcal{F}, \mathcal{E} \rangle$ such that $\mathcal{E}$ is a finite set of $\mathcal{D}$-rules. Each variable in the right-hand side of a rule also appears in the left hand side of the same rule. We distinguish among $\mathcal{F}$ a main function symbol whose name is given by the program name f.*

The size $|t|$ of a term t is defined by $|b| = 0$ and $|b(t_1, \ldots, t_n)| = 1 + \sum_i |t_i|$ where $b \in \mathcal{C} \cup \mathcal{F}$. The size of $\overrightarrow{t} = (t_1, \ldots, t_n)$ is $|\overrightarrow{t}| = \sum_i |t_i|$.

2.2   Semantics

The domain of the computed functions is the constructor algebra $T(\mathcal{C})$. The set of rules induces a rewriting relation $\to$. We use $\xrightarrow{+}$ for the transitive closure of $\to$ and $\xrightarrow{*}$ for the reflexive transitive closure of $\to$. $t \xrightarrow{!} s$ iff $t \xrightarrow{*} s$ and s is in normal form.

Definition 4. *If the system is confluent, then we say that the program is deterministic.*

In this case, the program $\mathbf{f}$ computes a partial function $[\![\mathbf{f}]\!] : \mathcal{T}(\mathcal{C})^n \to \mathcal{T}(\mathcal{C})$ defined as follows. For all $u_i \in \mathcal{T}(\mathcal{C}), [\![\mathbf{f}]\!](\overrightarrow{u}) = v$ iff $\mathbf{f}(\overrightarrow{u}) \xrightarrow{!} v$ and v is a constructor term. Otherwise $[\![\mathbf{f}]\!](\overrightarrow{u})$ is undefined. See, among others, Huet [22] for conditions ensuring the confluence of the system.

Definition 5. *If the system is not confluent, then we say that the program is* non-deterministic.

In this case, following Grädel and Gurevich [23] or Bonfante, Cichon, Marion and Touzet [4], the value of any term is the maximum value of the normal forms of the terms (for a given order on terms, usually lexicographic ordering). Notice that this includes the usual definition for decision problems by choosing **true** > **false**.

3   Orderings

3.1   Precedences and Ranks

Definition 6. *Let $f = \langle \mathcal{X}, \mathcal{C}, \mathcal{F}, \mathcal{E} \rangle$ be a program. A precedence over f is a pre-order $\preceq_{\mathcal{F}}$ over $\mathcal{F}$ compatible with the rules. By compatible, we mean that if $g(\overrightarrow{p}) \to r$ is a rule, then for any symbol h in r, we have $h \preceq_{\mathcal{F}} g$. Let us note $\approx_{\mathcal{F}}$ the induced equivalence relation, that is $g \approx_{\mathcal{F}} h$ iff $g \preceq_{\mathcal{F}} h \wedge h \preceq_{\mathcal{F}} g$. The strict induced ordering is noted $\prec_{\mathcal{F}}$, $g \prec_{\mathcal{F}} h$ iff $g \preceq_{\mathcal{F}} h \wedge \neg(h \preceq_{\mathcal{F}} g)$.*

We extend canonically the ordering on $\mathcal{F} \bigcup \mathcal{C}$ by choosing $c \prec_{\mathcal{F}} f$ for each constructor c and each function f and two constructors are mutually incomparable.

Lemma 7. *Let f be a program and $\preceq_{\mathcal{F}}$ be a precedence for that program. $\preceq_{\mathcal{F}}$ induces a rank function rk assigning an integer to each symbol such that $b \preceq_{\mathcal{F}} b'$ implies $rk(b) \leq rk(b')$ and $b \prec_{\mathcal{F}} b'$ implies $rk(b) < rk(b')$.*

Proof. By induction on the (finite) ordering.   □

In the following, we will often speak indifferently of either precedence or rank function, depending whether we want to focus on the ordering or on the "slicing" aspect.

Proposition 8. *Let $f = \langle \mathcal{X}, \mathcal{C}, \mathcal{F}, \mathcal{E} \rangle$ be a program. There exists a precedence $\preceq_{\mathcal{F}}$ for that program.*

Proof. For each rule $g(\overrightarrow{p}) \to r$ and each function symbol $h \in r$, put $h \preceq_{\mathcal{F}} g$. Build the reflexive transitive closure of this relation.   □

Thus, mutually recursive functions will have the same rank and subfunctions will have smaller rank than the caller.

Proposition 9. *Let f be a program and $\preceq_{\mathcal{F}}$ a precedence of f. Let $g(\overrightarrow{v}) \xrightarrow{*} t$ be such that the $v_i s$ are constructor terms. Then no symbol in t has a rank strictly higher than g, that is for all symbols $b \in t$, $b \preceq_{\mathcal{F}} g$.*

It is an immediate consequence of the compatibility and the transitivity of the precedence. This proposition is really important in the following because it allows some kind of computation "by rank": functions of high ranks will never interfere when computing the value of functions of low ranks. Thus, we will be able to design recursive procedure to compute functions depending on their ranks.

3.2 Termination Orderings

We now focus on termination orderings which we see as a kind of mold that capture some algorithmic patterns. Here, we consider Lexicographic Path Ordering which is a simplification ordering and so well-founded. See Dershowitz [24] for complete details about termination orderings. Krishnamoorthy and Narendran in [25] have proved that deciding whether a program terminates by Lexicographic Path Ordering is a NP-complete problem.

Let $\preceq$ be a partial ordering and $\prec$ be its strict part.

Definition 10. *The* lexicographic extension *of* $\prec$*, noted* $\prec^l$*, is defined as follows.*

We have $(m_1, \cdots, m_k) \prec^l (n_1, \cdots, n_{k'})$ *if and only if there exists an index j such that (i) $\forall i < j, m_i \preceq n_i$ and (ii) $m_j \prec n_j$.*

Lexicographic Path Ordering

Definition 11. *Given a precedence* $\preceq_\mathcal{F}$*, the* Lexicographic Path Ordering *(LPO)* $\prec_{lpo}$ *is defined in Figure 1.*

A rule $l \to r$ is decreasing if we have $r \prec_{lpo} l$. A program is ordered by $\prec_{lpo}$ if there is a precedence on $\mathcal{F}$ such that each rule is decreasing.

$$\frac{\forall i, s_i \preceq_{lpo} t_i \quad \exists j/s_j \prec_{lpo} t_j}{c(s_1, \cdots, s_n) \prec_{lpo} c(t_1, \cdots, t_n)} c \in \mathcal{C} \qquad \frac{s = t_i \ or \ s \prec_{lpo} t_i}{s \prec_{lpo} \mathbf{f}(\ldots, t_i, \ldots)} \mathbf{f} \in \mathcal{F} \bigcup \mathcal{C}$$

$$\frac{\forall i \ s_i \prec_{lpo} \mathbf{f}(t_1, \cdots, t_n)}{c(s_1, \cdots, s_m) \prec_{lpo} \mathbf{f}(t_1, \cdots, t_n)} \mathbf{f} \in \mathcal{F}, c \in \mathcal{C}$$

$$\frac{\forall i \ s_i \prec_{lpo} \mathbf{f}(t_1, \cdots, t_n) \quad \mathbf{g} \prec_\mathcal{F} \mathbf{f}}{g(s_1, \cdots, s_m) \prec_{lpo} \mathbf{f}(t_1, \cdots, t_n)} \mathbf{f}, \mathbf{g} \in \mathcal{F}$$

$$\frac{(s_1, \cdots, s_m) \prec^l_{rpo} (t_1, \cdots, t_n) \quad \mathbf{f} \approx_\mathcal{F} \mathbf{g} \quad \forall i \ s_i \prec_{lpo} \mathbf{f}(t_1, \cdots, t_n)}{g(s_1, \cdots, s_m) \prec_{lpo} \mathbf{f}(t_1, \cdots, t_n)} \mathbf{f}, \mathbf{g} \in \mathcal{F}$$

Fig. 1. Definition of $\prec_{lpo}$.

3.3   Quasi-interpretations

In order to control resources more precisely that termination orderings do, we suggest the use of quasi-interpretations. Quasi-interpretations have been introduced by Bonfante, Marion [11] and Marion-Moyen [5] and have proved themselves useful to obtain complexity bounds on programs.

The set of non-negative real numbers is noted $\mathbf{R}^+$.

Definition 12 (Assignment). *An assignment of a symbol $b \in \mathcal{F} \bigcup \mathcal{C}$ whose arity is n is a function $(\!|b|\!) : (\mathbf{R}^+)^n \to \mathbf{R}^+$ such that:*

(Subterm) $(\!|b|\!)(X_1, \cdots, X_n) \geq X_i$ *for all* $1 \leq i \leq n$.
(Weak Monotonicity) $(\!|b|\!)$ *is increasing (not necessarily strictly) with respect to each variable.*
(Additive) *For each constructor* **c**, $(\!|c|\!)(X_1, \cdots, X_n) = \sum_{i=1}^{n} X_i + \alpha_\mathbf{c}$, $\alpha_\mathbf{c} \geq 1$

We extend assignments $(\!|-|\!)$ to terms canonically. Given a term t with n variables, the assignment $(\!|t|\!)$ is a function $(\mathbf{R}^+)^n \to \mathbf{R}^+$ defined by the rules:

$$(\!|b(t_1, \cdots, t_n)|\!) = (\!|b|\!)((\!|t_1|\!), \cdots, (\!|t_n|\!))$$
$$(\!|x|\!) = X$$

Given two functions $f : (\mathbf{R}^+)^n \to \mathbf{R}^+$ and $g : (\mathbf{R}^+)^m \to \mathbf{R}^+$ such that $n \geq m$, we say that $f \geq g$ iff $\forall X_1, \ldots, X_n : f(X_1, \ldots, X_n) \geq g(X_1, \ldots, X_m)$.

We have $(\!|s|\!) \geq (\!|t|\!)$ if t is a subterm of s. We have also, for every substitution σ, $(\!|s|\!) \geq (\!|t|\!)$ implies that $(\!|s\sigma|\!) \geq (\!|t\sigma|\!)$.

Definition 13 (Quasi-interpretation). *A program assignment $(\!|-|\!)$ is an assignment of each program symbol. An assignment $(\!|-|\!)$ of a program is a quasi-interpretation if for each rule $l \to r$, $(\!|l|\!) \geq (\!|r|\!)$*

It is worth noticing that the above inequality is not strict which differs from the notion of interpretation used to prove termination. See, among others, [4].

Note that we restrict ourselves to additive quasi-interpretations, that is the form of the assignment of constructors is fixed. Allowing other (polynomial) assignment to constructors brings higher upper bound on complexity as shown in [21].

Definition 14. *The quasi-interpretation of a function symbol f is said to be: (i) affine if $(\!|f|\!)(X_1, \cdots, X_n) = \sum_{i=1}^{n} \alpha_{f,i} X_i + \alpha_f$ with $\alpha_f > 0$, (ii) multiplicative if $(\!|f|\!)$ is a polynomial of the form $P + \alpha_f$ with $\alpha_f > 0$ and $P(x_1, \ldots, x_n) \geq x_i$ for all $i \leq n$.*

The quasi-interpretation of a program is said to be purely affine (resp. purely multiplicative) if the quasi-interpretation of each function symbol is affine (resp. multiplicative).

Notice that for function symbols, even if $\alpha_\mathbf{f} < 1$, the quasi-interpretation may be affine. Whereas for constructors, we force $\alpha_\mathbf{c} \geq 1$.

Proposition 15 ([4]). *Suppose that f is a program with a quasi-interpretation. Assume t is a constructor term in $\mathcal{T}(\mathcal{C})$, we have $|t| \leq (\!|t|\!) \leq O(|t|)$.*

4   Call-Trees

We present now call-trees which are a tool that we shall use all along. Let $f = \langle \mathcal{X}, \mathcal{C}, \mathcal{F}, \mathcal{E} \rangle$ be a program. A call-tree gives a static view of an execution and captures all function calls. Hence, we can study dependencies between function calls without taking care of the extra details provided by the underlying rewriting relation. Call-trees are related to dependency graphs proposed by Aarts and Giesl [26].

Definition 16. *A* state *is a tuple* $\langle h, u_1, \cdots, u_p \rangle$, *sometimes noted* $\langle h, \overrightarrow{u} \rangle$, *where* h *is a function symbol of arity* p *and* $u_1, \ldots, u_p$ *are constructor terms. Assume that* $\eta_1 = \langle h, u_1, \cdots, u_p \rangle$ *and* $\eta_2 = \langle g, s_1, \cdots, s_m \rangle$ *are two states. A* transition *is a triplet* $\eta_1 \overset{e}{\rightsquigarrow} \eta_2$ *such that:*

  (i) e *is a rule* $h(q_1, \cdots, q_p) \to t$ *of* $\mathcal{E}$,
  (ii) *there is a substitution* σ *such that* $q_i\sigma = u_i$ *for all* $1 \leq i \leq p$,
  (iii) *there is a subterm* $g(v_1, \cdots, v_m)$ *of* t *such that* $v_i\sigma \overset{!}{\to} s_i$ *for all* $1 \leq i \leq m$.

Transition(f) *is the set of all transitions between states.* $\overset{*}{\rightsquigarrow}$ *is the reflexive transitive closure of* $\cup_{e \in \mathcal{E}} \overset{e}{\rightsquigarrow}$.

Definition 17. *The* $\langle f, t_1, \cdots, t_n \rangle$-*call tree is a tree defined as follows: (i) nodes are states, (ii) the root is the state* $\langle f, t_1, \cdots, t_n \rangle$, *(iii) for each state* η_1, *the children of* η_1 *are the states* $\{\eta_2 \mid \eta_1 \overset{e}{\rightsquigarrow} \eta_2 \in Transition(f)\}$.

Proposition 18. *Assume that* $\eta_1 = \langle h, \overrightarrow{u} \rangle \overset{*}{\rightsquigarrow} \eta_2 = \langle g, s_1, \cdots, s_m \rangle$ *and* $(\!|-|\!)$ *is a quasi-interpretation for that program. Then we have* $(\!|g(s_1, \cdots, s_m)|\!) \leq (\!|h(\overrightarrow{u})|\!)$ *and thus* $|s_i| \leq (\!|s_i|\!) \leq (\!|h(\overrightarrow{u})|\!)$ *for all* $1 \leq i \leq m$.

Proof. By virtue of the quasi-interpretation definition. Both the subterm property, the weak monotonicity property and Proposition 15 are necessary. □

The size of a state $\langle g, s_1, \cdots, s_m \rangle$ is $\sum_{i=1}^{m} |s_i|$.

Lemma 19. *The size of each state of the* $\langle f, t_1, \cdots, t_n \rangle$-*call tree is bounded by* $d \times (\!|f(\overrightarrow{t})|\!)$ *where* d *is the maximal arity of a function symbol.*

Proof. $|\langle g, s_1, \cdots, s_m \rangle| = \sum_i |s_i| \leq \sum_i (\!|s_i|\!) \leq m \times (\!|f(\overrightarrow{t})|\!) \leq d \times (\!|f(\overrightarrow{t})|\!)$ □

5   Linear Space

Let a LPO^{QI+}-program be a program that terminates by LPO and admits a purely affine quasi-interpretation.

Theorem 20. *The set of functions computed by deterministic LPO^{QI+}-programs is exactly* LinSpace.

Proof. The upper-bound on complexity is established by Theorem 27. The completeness of the characterization is established by Theorem 25. □

Theorem 21. *The set of functions computed by non-deterministic LPO^{QI+}-programs is exactly* NLinSpace.

5.1   Unary k-Stack Turing Machines

In this Section, we follow Gurevich [16], Leivant [13] and Marion et al. [4].

Lemma 22 ([4]). *A function is in* LINSPACE *if it is computed by a multi-stack Turing Machine (STM) over the unary alphabet in polynomial time.*

Actually, we will use another characterization.

Lemma 23. *A function is in* LINSPACE *if it is computed by a polynomial time multi-stack Turing Machine over the unary alphabet and each stack is bounded by the size of the input.*

Proof. Following Lemma 22, we take M a STM that computes a function $f \in$ LINSPACE. Let K be such that M works in time less than n^K. As a consequence, each stack is bounded by n^K. We can encode such numbers by K stacks, each representing a digit of the number in base n. Suppose that M has p stacks, let N have $2Kp$ stacks $\Pi_1^1, \Pi'_1^1, \ldots, \Pi_K^1, \Pi'_K^1, \ldots, \Pi_1^p, \Pi'_1^p, \ldots, \Pi_K^p, \Pi'_K^p$. The stacks $\Pi_1^1, \Pi_2^1, \ldots, \Pi_K^1$ of N represent the stack Π_1 of M. The primed stacks are the complement of their corresponding stack to n, the length of the input. To simulate the incrementing of a stack of M, say Π_1, find the first non-empty stack Π'_j^1, decrement it by one, increment Π_j^1 by one, then reset all Π_i^1 to 0 and all Π'_i^1 to n where $i < j$. Decrementing a stack is done in a similar way. So, the simulation is done within the space bound. $\square$

5.2   Simulation of Linspace by LPO^{QI+}-Programs

Given a function $f : \mathbb{W}^n \to \mathbb{W}$ in LINSPACE, we take a p-stack Turing Machine M which computes f, works on the unary alphabet $\{s\}$ and such that each stack is bounded by the length of the input. Say that M works in time n^k for some k. As we do the simulation of the machine step for step, we need to design a program whose derivation length is at least n^k on input n.

Lemma 24. *Given $k \in \mathbf{N}$, there is a (deterministic) LPO^{QI+}-programs that runs in n^k steps on input of size n.*

Proof. Let us first show how to have a program whose derivation length is the square of its argument.

$$C_2(\mathbf{0}, \mathbf{0}, y') \to \mathbf{0}$$
$$C_2(\mathbf{S}(x), \mathbf{0}, y') \to C_2(x, y', \mathbf{0})$$
$$C_2(x, \mathbf{S}(y), y') \to C_2(x, y, \mathbf{S}(y'))$$
$$P_2(x) \to C_2(x, x, \mathbf{0})$$

Let us define $(\!(\mathbf{0})\!) = 1$ and $(\!(\mathbf{S})\!)(X) = X + 1$. By choosing $(\!(C_2)\!)(X, Y, Y') = X + Y + Y' + 1$ and $(\!(P_2)\!)(X) = 2X + 2$, the program admits a purely affine quasi-interpretation. It is routine to check that it terminates by LPO.

Now, to have a program whose derivation length is n^k, we use the same trick: have a program C_k with $2k$ (actually $2k - 1$ is sufficient) variables x_i, x'_i

and adapts the rules of $\mathtt{C_2}$ to allow $\mathtt{C}_k$ to count in base n. x_i' is always the complementary of x_i to n, thus allowing to reinitialize the values. The generic rule is:

$$\mathtt{C}_k(\mathtt{S}(x_1), x_1', \ldots, \mathtt{S}(x_i), x_i', \mathbf{0}, x_{i+1}', \ldots) \to \mathtt{C}_k(\mathtt{S}(x_1), x_1', \ldots, x_i, \mathtt{S}(x_i'), x_{i+1}', \mathbf{0}, \ldots)$$

Then, $\mathtt{P}_k(x)$ simply calls $\mathtt{C}_k$ with each $x_i = x$ and $x_i' = \mathbf{0}$. The resulting program has a derivation length equal to n^k, $(\!|\mathtt{C}_k|\!)(X_1, \ldots, X_{2k}) = \sum_{i=1}^{2k} X_i + 1$ and $(\!|\mathtt{P}_k|\!)(X) = kX + k + 1$, both are purely affine for a fixed k. □

Theorem 25. *Suppose that f is in* LINSPACE, *then it is computable by a* LPO^{QI+}-*program.*

Proof. Let M be a p-Stack Turing machine computing f. As Lemma 23 shows it, we can suppose that the stacks are bounded by the size of the input along the computation. Let us consider the following system. To the template rules above, we add some new variables, one variable for the state of the machine and two for each stack of M. The first one (Π_i) represents the content of the stack (the ith), the second one (Π_i') its complement to n, the size of the input.

Take now for instance a transition that increments the stack Π_1 and goes to state q' if Π_2 is empty in state q, we transform each rule $\mathtt{C}_k(\overrightarrow{x}) \to \mathtt{C}_k(\overrightarrow{y})$ as follows:

$$\mathtt{C}_k(\overrightarrow{x}, \mathtt{q}, \Pi_1, \mathtt{S}(\Pi_1'), \mathbf{0}, \Pi_2', \ldots) \to \mathtt{C}_k(\overrightarrow{y}, \mathtt{q}', \mathtt{S}(\Pi_1), \Pi_1', \mathbf{0}, \Pi_2', \ldots)$$

The program terminates by LPO because there is a decrease between ($\overrightarrow{x}$) and ($\overrightarrow{y}$). $(\!|\mathtt{C}_k|\!)$ is the sum of its arguments. □

5.3   Computation of LPO^{QI+}-Programs in Linear Space

Let $\mathtt{f}$ be a program admitting a purely affine quasi-interpretation. If each function symbol $\mathtt{f}$ has a quasi-interpretation $(\!|\mathtt{f}|\!)(X_1, \cdots, X_n) = \sum \alpha_{\mathtt{f},i} X_i + \alpha_{\mathtt{f}}$ with $\alpha_{\mathtt{f}} > 0$, then the following proposition holds:

Proposition 26. *The size of any term t is bounded by $K(\!|t|\!)$ for some constant K.*

Proof. Take $\alpha = \min_{a \in \mathcal{F} \cup \mathcal{C}}(\alpha_a)$. As $\mathcal{F} \cup \mathcal{C}$ is finite, $\alpha > 0$. By induction, we have: $|t| \leq (\!|t|\!)/\alpha$.

Theorem 27. LPO^{QI+}-*Programs are computable by* LINSPACE *machines.*

Proof. First of all notice that if $\mathtt{f}(\overrightarrow{t}) \overset{*}{\to} s$, then $(\!|s|\!) \leq (\!|\mathtt{f}(\overrightarrow{t})|\!)$. At the same time, we have $|s| \leq K(\!|s|\!)$ with respect to Proposition 26. And finally, Proposition 15 shows that $(\!|\mathtt{f}(\overrightarrow{t})|\!) \leq (\!|\mathtt{f}|\!)(B(|\overrightarrow{t}|) + 1)$ for some constant B. But, as $(\!|\mathtt{f}|\!)$ is affine, we have $(\!|\mathtt{f}(\overrightarrow{t})|\!) \leq (\!|\mathtt{f}|\!)(B(|\overrightarrow{t}|) + 1) \leq A(|\overrightarrow{t}| + 1)$ for some constant A.

Putting things together, we conclude that the size of the terms obtained along the computation is bounded by: $|s| \leq KA(|t| + 1)$.

A reduction step may be performed within linear space and this ensure that the whole computation can. Indeed, finding a redex needs finite memory (plus a pointer in the term) and replacing the redex may take finitely many times the size of the input if one has to recopy some variables, like for instance in the rule: $D(x) \to \mathtt{add}(x,x)$. So, the linear constant of the cost of replacing a term can be bounded by: $M = \max_{l \to r \in \mathcal{E}}(|r|)$. In other words, the computation can be done in space $AMKn$ on inputs of size n.   □

6   Logarithmic Space

From now on, we do not make any supposition concerning the quasi-interpretation of functions. This means that the quasi-interpretation of a function must only respect the rules (Subterm) and (Weak Monotonicity) of Definition 12 and this allow, for example, $(\!|\mathtt{f}|\!)(X,Y) = \max(X,Y)$.

In this section, we follow the work of Jones [15] in order to get a characterization of LOGSPACE. The arguments of a function are separated into three distinct sets:

- An input set. This set is read-only and contains binary words that are subterms of the initial inputs.
- A working set. This set is read-write and contains unary integers. This condition, together with the quasi-interpretation, has the net effect of allowing these variables to be only pointers on a part of the input.
- An output set. This set is write-only and ensures that the result of a recurrence is not allowed to control another recurrence (at the same rank). This is done by preventing any pattern matching on this set. These conditions are forced by a tiering-rule similar to the one used by Bellantoni and Cook [12].

We consider programs whose inputs are binary words (that is built from $\{\epsilon, 0, 1\}$) and outputs are unary integers (that is built from $\{\mathbf{Z}, \mathbf{S}\}$). We now define a strong tiering discipline on those programs, separating arguments of functions into three sets: inputs, working and simple as described above.

Definition 28. *Formally, let $LinPO_{tier}^{QI}$ be the set of programs f such that:*

(Cons) $\mathcal{C} = \{\epsilon, 0, 1, \mathbf{Z}, \mathbf{S}\}$;
(LPO) f terminates by LPO.
  (QI) f admits a quasi-interpretation.
  (Lin) If $f(\overrightarrow{u}) \to t$, there is at most one function symbol with the same precedence as f in t. That is $\#\{g \in \mathcal{F}, g \approx_{\mathcal{F}} f$ and $g \in t\} \leq 1$.

Second, it is tiered, that is

(Rule) Each rule has the form $f(p_1, \cdots, p_n; q_1, \cdots, q_m; y_1, \cdots, y_l) \to t$ where p_i is a pattern over the binary words, q_i is a pattern over the unary integers and y_i is a variable.
(Input) If there is a subterm $g(p'_1, \cdots, p'_k; \overrightarrow{q'}; \overrightarrow{y'}) \trianglelefteq t$ then for all i, there exists j such that $p'_i \trianglelefteq p_j$.

(Output) *If there is a subterm* $g(\overrightarrow{p'}; q'_1, \cdots, q'_{m'}; y'_1, \cdots, y'_{l'}) \trianglelefteq t$, *then neither* **0, 1**, ϵ *nor a variable present in any* p'_i *appears in any of the* q'_i *or the* y'_j.

(Tier) *If there is a subterm* $g(\overrightarrow{p'}; q'_1, \cdots, q'_{m'}; \overrightarrow{y'}) \trianglelefteq t$ *then for all i and j,* $y_j \notin q'_i$.

(Simple) *If there are two subterms* $g(\overrightarrow{p'}; q'_1, \cdots, q'_{m'}; \overrightarrow{y'}) \trianglelefteq t$ *and* $h(\overrightarrow{p''}; \overrightarrow{q''}; \overrightarrow{y''}) \trianglelefteq q'_i$ *then* $h \prec_{\mathcal{F}} g$. *That is the result of a function can only be matched by functions of strictly higher rank.*

The condition (Rule) separates the three sets of arguments for each function. The condition (Input) states that the input set is read-only. The condition (Output) says that inputs have not the same type as other variables and thus may no be used anywhere. The conditions (Tier) and (Simple) ensure the tiering discipline with respect to the precedence.

Theorem 29. *The set of functions computed by* $LinPO_{tier}^{QI}$-*programs is exactly* LOGSPACE, *the set of functions computable in logarithmic space on a Turing machine.*

We only give here the sketch of proof. A more complete proof can be found in [27].

Proof (Sketch of proof). Jones has shown that LOGSPACE is the set of functions computed by polynomial time counter machines with counter no bigger than the input [15]. It is easy to simulate such a machine by a $LinPO_{tier}^{QI}$-program. The polynomial clock is done in a way similar to Lemma 24 and the counters can easily be stored into the read-write variables.

In order to show that $LinPO_{tier}^{QI}$-programs can be computed in logarithmic space, there are crucial points:

- the constructor terms in input position are always subterms of the initial term. As a consequence, they can be stored by a pointer. This is due to Hypothesis (Input);
- the constructor terms in working and simple position have a polynomial size (wrt the input data); it is a consequence of Hypothesis (QI). As they are unary (Hypothesis (Output)), they also can be stored in logarithmic space.
- consider the deriving strategy where you rewrite the symbol of minimal rank that is leftmost innermost. The tiering discipline shows that this strategy is complete (Hypotheses (Tier, Simple)); that is give the normal form of the term. Notice that the system terminates due to Hypothesis (LPO).
- due to the Hypothesis (Lin), the rewriting strategy above is such that there are only finitely many function symbols in terms along the computation.

In other words, each terms along the computation can be coded in logarithmic space and as each step can be performed in logarithmic space, the computation can be done within the space bound.   □

7   Polynomial Space

7.1   Strict Quasi-interpretations

We say that a program admits a strict-quasi-interpretation if it admits a quasi-interpretation and (i) each rule verifies: $(\!|l|\!) > (\!|r|\!)$, (ii) $(\!|\mathbf{f}|\!)(x_1, \ldots, x_n) \geq x_i + \alpha_{\mathbf{f}}$ for some constant $\alpha_{\mathbf{f}} > 0$ and all $i \leq n$.

Remark 30. If the interpretations are over rational numbers, condition (ii) can be reformulated: $(\!|\mathbf{f}|\!)(x_1, \ldots, x_n) > x_i$ for all i.

Recall that a polynomial interpretation verifies all the properties above. They have the further property: the interpretations are necessarily polynomials. This prevent in particular the use of max in assignments. In that case, we have:

Theorem 31 (Bonfante, Cichon, Marion, Touzet [4]). *Programs which admit polynomial interpretations compute exactly* PTIME.

Theorem 32 (Dershowitz [28]). *A program with a strict-quasi-interpretation terminates.*

Proposition 33. *The height of a term t of a program that admits a strict-quasi-interpretation is bounded by $(\!|t|\!)$.*

Theorem 34. *A function is computed by a program that admits a strict quasi-interpretation iff it is in* PSPACE.

Proof. The fact that a function in PSPACE can be computed by a strict-quasi-interpretation is shown by Lemma 60 of [21]. The quasi-interpretation given there need to be slightly modified in order to be strict.

Let $\mathbf{f}$ be a program which admits a strict quasi-interpretation and $\mathbf{f}(\overrightarrow{t}) \xrightarrow{*} s$. The hypothesis on the assignment induces a polynomial bound on the height of terms. So, by considering a leftmost innermost strategy, we have a polynomial bound on the size of the term. Actually, there are at most $\alpha \times P(n)$ (with $P(n)$ the bound on the height of the term and α the maximal size of a right hand side of a rule) defined symbols in terms along the computation. As a consequence, the size of the term is $O(P(n)^2)$.

Computing the next configuration can be done clearly in polynomial space. As a consequence, the computation can be done within PSPACE.   □

There is actually an other characterization of PSPACE due to Bonfante [29]. It shows the role of the two parameters that we are considering here: the choice of $l > r$ or $l \geq r$ and the choice of interpretations over $+, \times$ or $+, \times, \max$.

Theorem 35 (Bonfante [29]). *Functions computed by programs such that*

1. *the program admits a purely multiplicative quasi-interpretation,*
2. *the program terminates,*

are exactly PSPACE *functions.*

7.2   Linear Systems

Definition 36. *Let f be a program and $(\prec_{\mathcal{F}}, \approx_{\mathcal{F}})$ be a precedence. f is* linear *with respect to $(\prec_{\mathcal{F}}, \approx_{\mathcal{F}})$ if for each rule $f(\overrightarrow{p}) \to r$ there is at most one symbol in r with the same precedence as f.*

When the precedence is unspecified, we implicitly refer to a precedence compatible with the program such as the one used for termination orderings or the one build in Proposition 9. In those cases, we simply say that the program is linear.

Let a LinQI-program be a program that (i) terminates, (ii) is linear with respect to the implicit precedence, and (iii) admits a quasi interpretation.

Theorem 37. *The set of functions computed by LinQI-programs is exactly* PSPACE.

Proof. The proof will goes by induction on the rank of the head symbol of nodes in a call-tree.

If the rank is 0 (which can be achieved by a trivial extension of the definition of call-trees), then no computations are needed in order to have the value of $\langle f, t_1, \cdots, t_n \rangle$ (remember that in a node, t_is are constructors terms).

Otherwise, we introduce sequents of the form $\langle f, t_1, \cdots, t_n \rangle; t \vdash \langle g, s_1, \cdots, s_m \rangle; s$ meaning "$f(\overrightarrow{t}) \overset{!}{\to} t$ if $g(\overrightarrow{s}) \overset{!}{\to} s$", and we define the covering graph of f in the $\langle f, t_1, \cdots, t_n \rangle$-call tree as the sub-graph obtained by keeping only nodes labelled by symbols with the same rank as f.

As the program is linear, the covering graph is a chain. As the program admits a quasi-interpretation, it has at most $O(2^B)$ nodes, where $B = (\!|f(\overrightarrow{t})|\!)$. Moreover, the size of each node of the call-tree is bounded by B.

To compute the value of $\langle f, t_1, \cdots, t_n \rangle$, proceed as follows:

1. Guess a node $\langle g, s_1, \cdots, s_m \rangle$ with $g \approx_{\mathcal{F}} f$ and $\langle g, s_1, \cdots, s_m \rangle$ having no child with the same precedence, guess two values t and s and introduce the sequent $\langle g, s_1, \cdots, s_m \rangle; s \vdash \langle f, t_1, \cdots, t_n \rangle; t$.
2. Perform a reduction step on $\langle f, t_1, \cdots, t_n \rangle$. If $\langle g, s_1, \cdots, s_m \rangle$ is a child of $\langle f, t_1, \cdots, t_n \rangle$ in the call-tree, then one just need to compute the values of other children (using s when necessary) to verify that t was correctly guessed. This can be done in polynomial space by induction hypothesis.
3. If $\langle g, s_1, \cdots, s_m \rangle$ is not a child of $\langle f, t_1, \cdots, t_n \rangle$, then guess a new node $\langle h, u_1, \cdots, u_p \rangle$ and a new value u. Replace the current sequent by the two sequents $\langle h, u_1, \cdots, u_p \rangle; u \vdash \langle f, t_1, \cdots, t_n \rangle; t$ and $\langle g, s_1, \cdots, s_m \rangle; s \vdash \langle h, u_1, \cdots, u_p \rangle; u$.

The depth of the proof tree for any sequent will be logarithmic in the length of the covering graph, thus polynomial in the size of the inputs. Each sequent has a polynomial size. Thus a clever depth-first search will allow to compute the value of a term in polynomial space. □

The three characterizations of PSPACE may be obtained as a corollary of the following conjecture.

Claim. The set of functions computed by programs which terminate and admit a quasi-interpretation is exactly PSPACE.

References

1. Benzinger, R.: Automated complexity analysis of NUPRL extracts. PhD thesis, Cornell University (1999)
2. Hofbauer, D., Lautemann, C.: Termination proofs and the length of derivations. In: RTA. Number 355 in Lecture Notes in Computer Science (1988)
3. Cichon, E., Lescanne, P.: Polynomial interpretations and the complexity of algorithms. In: CADE'11. Number 607 in Lecture Notes in Artificial Intelligence (1992) 139–147
4. Bonfante, G., Cichon, A., Marion, J.Y., Touzet, H.: Algorithms with polynomial interpretation termination proof. Journal of Functional Programming **11** (2000)
5. Marion, J.Y., Moyen, J.Y.: Efficient first order functional program interpreter with time bound certifications. In: LPAR 2000. Volume 1955 of Lecture Notes in Computer Science., Springer (2000) 25–42
6. Bonfante, G., Marion, J.Y., Moyen, J.Y.: On lexicographic termination ordering with space bound certifications. In: PSI 2001, Ershov Memorial Conference. Volume 2244 of Lecture Notes in Computer Science., Springer (2001)
7. Hofbauer, D.: Termination proofs with multiset path orderings imply primitive recursive derivation lengths. Theoretical Computer Science **105** (1992) 129–140
8. Weiermann, A.: Termination proofs by lexicographic path orderings yield multiply recursive derivation lengths. Theoretical Computer Science **139** (1995) 335–362
9. Ritchie, R.: Classes of predictably computable functions. Transaction of the American Mathematical Society **106** (1963) 139–173
10. Beckmann, A., Weiermann, A.: A term rewriting characterization of the polytime functions and related complexity classes. Archive for Mathematical Logic **36** (1996) 11–30
11. Marion, J.Y.: Analysing the implicit complexity of programs. Information and Computation **183** (2003) 2–18
12. Bellantoni, S., Cook, S.: A new recursion-theoretic characterization of the polytime functions. Computational Complexity **2** (1992) 97–110
13. Leivant, D.: Predicative recurrence and computational complexity I: Word recurrence and poly-time. In Clote, P., Remmel, J., eds.: Feasible Mathematics II. Birkhäuser (1994) 320–343
14. Leivant, D., Marion, J.Y.: Lambda calculus characterizations of poly-time. Fundamenta Informaticae **19** (1993) 167,184
15. Jones, N.D.: LOGSPACE and PTIME characterized by programming languages. Theoretical Computer Science **228** (1999) 151–174
16. Gurevich, Y.: Algebras of feasible functions. In: Twenty Fourth Symposium on Foundations of Computer Science, IEEE Computer Society Press (1983) 210–214
17. Oitavem, I.: A term rewriting characterization of the functions computable in polynomial space. Archive for Mathematical Logic **41** (2002) 35–47
18. Amadio, R., Coupet-Grimal, S., Zilio, S.D., Jakubiec, L.: A functional scenario for bytecode verification of resource bounds. In: CSL. (2004) to appear.
19. Amadio, R.M., Dal-Zilio, S.: Resource control for synchronous cooperative threads. In: CONCUR. (2004) 68–82
20. Bonfante, G., Marion, J.Y., Moyen, J.Y., Péchoux, R.: Synthesis of quasi-interpretations. Technical report, Loria (2005) Submited to RULE '05. Available at http://www.loria.fr/~bonfante/publis/RULE05.pdf.
21. Bonfante, G., Marion, J.Y., Moyen, J.Y.: Quasi-interpretations. Technical report, Loria (2004) Submited to Theoretical Computer Science, accessible: http://www.loria.fr/~moyen/.

22. Huet, G.: Confluent reductions: Abstract properties and applications to term rewriting systems. Journal of the ACM **27** (1980) 797–821
23. Grädel, E., Gurevich, Y.: Tailoring recursion for complexity. Journal of Symbolic Logic **60** (1995) 952–969
24. Dershowitz, N.: Orderings for term-rewriting systems. Theoretical Computer Science **17** (1982) 279–301
25. Krishnamoorthy, M.S., Narendran, P.: On recursive path ordering. Theoretical Computer Science **40** (1985) 323–328
26. Arts, T., Giesl, J.: Termination of term rewriting using dependency pairs. Theoretical Computer Science **236** (2000) 133–178
27. Bonfante, G., Marion, J.Y., Moyen, J.Y.: Quasi-interpretations and small space bounds. Technical report, Loria (2005)
    `http://www.loria.fr/~bonfante/publis/RTA05f.pdf`.
28. Dershowitz, N.: A note on simplification ordering. Information Processing Letters **9** (1979) 212–215
29. Bonfante, G.: Constructions d'ordres, analyse de la complexité. Thèse, Institut National Polytechnique de Lorraine (2000)

A Sufficient Completeness Reasoning Tool
for Partial Specifications*

Joe Hendrix[1], Manuel Clavel[2], and José Meseguer[1]

[1] University of Illinois at Urbana-Champaign, USA
[2] Universidad Complutense de Madrid, Spain

Abstract. We present the Maude sufficient completeness tool, which explicitly supports sufficient completeness reasoning for partial conditional specifications having sorts and subsorts and with domains of functions defined by conditional memberships. Our tool consists of two main components: (i) a sufficient completeness analyzer that generates a set of proof obligations which if discharged, ensures sufficient completeness; and (ii) Maude's inductive theorem prover (ITP) that is used as a backend to try to automatically discharge those proof obligations.

1 Introduction

In computer science practice, equational specifications are often *partial*. That is, some of the relevant operations are only defined on an adequate subset of data. Simple examples of undefinedness include computing the top of an empty stack, division by zero, and many operations on data structures. This has led to the design of increasingly more expressive equational formalisms to deal with partiality (see [1] for a survey). In particular, the papers [1–3] proposed membership equational logic (MEL) as a framework logic for the equational specification of partial functions. The key idea is that the domain of definition of a partial function is axiomatized by conditional membership axioms stating when the function is defined. However, since conditional memberships may have arbitrarily complex conditions and equations may be conditional, in this setting the *sufficient completeness problem* is undecidable in general.

The Maude sufficient completeness tool (SCC), which analyzes MEL theories specified in Maude, is therefore not a decision procedure. Instead it is a reasoning tool consisting of two main components: (i) a *sufficient completeness analyzer* that generates a set of *proof obligations* which if discharged, ensures sufficient completeness of confluent, sort decreasing and reductive specifications; (ii) Maude's *inductive theorem prover* (ITP), that is used as a backend to try to automatically discharge those proof obligations.

Our tool has a number of useful applications. Two obvious ones are: (i) checking that the defined functions of a MEL specification will always evaluate to data

---

* Research supported by Grants ONR N00014-02-1-0715, NSF CCR-0234524, and Spanish MCYT Projects TIC2002-01167 and TIC2003-01000.

J. Giesl (Ed.): RTA 2005, LNCS 3467, pp. 165–174, 2005.

built with the constructors; and (ii) for inductive theorem proving purposes, ensuring the correctness of the chosen proof technique (e.g. structural induction, cover set induction, inductionless induction, etc.) which typically depends on sufficient completeness. There are two other applications for which our tool has proved useful: (iii) checking that a rewrite theory specifying a concurrent system is *deadlock-free*, which is needed for verifying temporal logic properties using abstraction techniques [4]; the point is that deadlock-freeness can be characterized as the sufficient completeness of an associated MEL specification; and (iv) supporting more powerful *cover set induction schemes* in the style of [5] that can prove general conjectures of the form $\varphi(f(t_1, \ldots, t_n))$, where φ is a formula containing the expression $f(t_1, \ldots, t_n)$ with f a defined function symbol and the $t_1, \ldots, t_n$ constructor terms; the point here is that the sufficient completeness checker can be used to generate base cases in the induction scheme which are needed because in general the $t_1, \ldots, t_n$ may be nonvariable terms. This last application is a "turning of the tables" in the interoperation between Maude's ITP and SCC: in the second tool, the ITP plays an auxiliary role in discharging proof obligations, whereas in the ITP itself (which supports cover set induction) the SCC plays an auxiliary role in generating induction schemes.

2   Preliminaries

A MEL *signature* Σ is a triple $\Sigma = (\mathcal{K}, \Sigma, \mathcal{S})$, where $\mathcal{K}$ is a set of *kinds*, $\mathcal{S}$ is a disjoint $\mathcal{K}$-kinded family $\mathcal{S} = \{\mathcal{S}_k\}_{k \in \mathcal{K}}$ of sets of *sorts*, and $\Sigma = \{\Sigma_{w,s}\}_{(w,s) \in \mathcal{K}^* \times \mathcal{K}}$ is a $\mathcal{K}$-kinded signature of function symbols. Given a $\mathcal{K}$-kinded disjoint family of finite sets of variables $\vec{x} = x_1 : k_1, \ldots, x_n : k_n$, where $x_1, \ldots x_n$ are disjoint from the constants in Σ and the kinds $k_1, \ldots k_n$ in the list can be repeated, a Σ-*equation* is a formula $t = t'$, with $t, t' \in T_\Sigma(\vec{x})$, $T_\Sigma(\vec{x})$ being the free Σ-algebra on the variables $\vec{x}$, and such that t, t' have the same kind, i.e. $t, t' \in T_\Sigma(\vec{x})_k$ for some $k \in \mathcal{K}$. A Σ-*membership* is a formula $t : s$ such that if $t \in T_\Sigma(\vec{x})_k$, then $s \in \mathcal{S}_k$. Σ-*sentences* are universally quantified Horn clauses of the form

$$(\forall \vec{x})\ A \text{ if } A_1 \wedge \cdots \wedge A_n$$

where A and the A_i are either Σ-equations or Σ-memberships. If A is a Σ-equation, we call the sentence a *conditional equation*; and if A is a Σ-membership, we call it a *conditional membership*. A MEL theory is a pair $\mathcal{E} = (\Sigma, \Gamma)$ with Σ a MEL signature and Γ a set of Σ-sentences. A *model* of a MEL signature $(\mathcal{K}, \Sigma, \mathcal{S})$ is a $(\mathcal{K}, \Sigma)$-algebra $\mathcal{A}$ together with a subset $\mathcal{A}_s \subseteq \mathcal{A}_k$, for each sort $s \in \mathcal{S}_k$. Then, models of a MEL theory $\mathcal{E} = (\Sigma, \Gamma)$ are models of Σ satisfying the axioms Γ. There is a sound and complete inference system to derive all theorems of a MEL theory (Σ, Γ) [1]. We denote the initial algebra of $\mathcal{E} = (\Sigma, \Gamma)$ by $T_\mathcal{E}$. There is a unique Σ-homomorphism $h : T_\mathcal{E} \to A$ for every model A of $\mathcal{E}$.

Under appropriate assumptions on the MEL theory $\mathcal{E}$ the conditional equations can be used from left to right as *rewrite rules* [2]. This is the way in which MEL is efficiently implemented in the Maude language [6]. An inference system for MEL reasoning is described in detail in Figure 7, page 57 of [2]. The notions of confluence and termination of term rewriting can be generalized to conditional MEL theories by corresponding notions of *confluence* and *reductiveness* [2]. Since sort computations are involved, a third important notion is *sort decreasingness*. Assuming that $\mathcal{E}$ is confluent and reductive, we can characterize sort decreasingness as the property that for each term t if we can infer $\mathcal{E} \vdash t : s$ with the rewrite inference system in Figure 7 of [2], then we can also infer $\mathcal{E} \vdash \text{can}_{\mathcal{E}}(t) : s$ where $\text{can}_{\mathcal{E}}(t)$ denotes the canonical form of t obtained by applying the confluent and reductive rewrite rules in $\mathcal{E}$. Intuitively, the more we simplify a term with the equations, the easier it becomes to compute its sort without having to remember any intermediate terms in the rewrite computation.

3   A Partial Specification Example

In Misra's data type of *powerlists* [7], a powerlist must be of length 2^n for some $n \in \mathbb{N}$, and the *zip* operator $\bowtie$ is only fully defined on powerlists of equal length. We can specify powerlists in MEL as a Maude functional module as follows:

```
fmod POWERLIST is protecting NAT . sort Pow .
    op [_] : Nat -> Pow [ctor] . ops _|_ _X_ : [Pow] [Pow] -> [Pow] .
    op len : Pow -> Nat .
    vars I J : Nat .              vars P Q R S : Pow .
    cmb (P | Q) : Pow if len(P) = len(Q) .
    cmb (P X Q) : Pow if len(P) = len(Q) [metadata "dfn"].
    eq [I] X [J] = [I] | [J] .    eq (P | Q) X (R | S) = (P X R) | (Q X S) .
    eq len([I]) = 1 .             eq len(P | Q) = len(P) + len(Q) .
endfm
```

The functional module POWERLISTincludes the predefined module NAT, which declares the natural numbers with the expected arithmetic operations and relations. In the sort declaration section we introduce the sort Pow, which we will reserve for those terms representing powerlists; Maude automatically introduces also the kind [Pow] to denote the kind of the sort Pow. In the operator declaration section we introduce four operators: [_] for representing the operation that forms powerlist elements; _|_ for representing the powerlist *tie* operation; _X_ for representing the powerlist *zip* operation; and len for representing the operation that computes the length of a powerlist. Since we know that not all terms constructed with the operators _|_ and _X_ will represent powerlists, we declare those operators at the kind level. For example, $[4] \bowtie ([2] \mid [3])$ is not a powerlist. This is represented in POWERLIST by the fact that the term [4] X ([2] | [3]) has kind [Pow], but it does not belong to the sort Pow. On the other hand, since we want to use the [_] operator to *construct* powerlists (in particular, powerlists with only one element), we declare this operator at the sort level and with the ctor attribute. Finally, since we expect that the len operator applied to a powerlist will always evaluate to a natural number, we declare this operator at the sort level, but without the ctor attribute.

In the variable declaration section, we associate to the variables I and J the sort Nat, and to the variables P, Q, R, and S the sort Pow. By doing this, we are in fact declaring: i) that I and J are variables of the kind [Nat], and P, Q, R, and S of the kind [Pow], and ii) that in all memberships and equations in which those variables appear, there is an extra condition stating that those variables only range over the set of terms belonging to their associated sort. Finally, in the membership declaration section, we declare that both the *tie* and the *zip* of two powerlists are powerlists if they have equal length; however, since we do not want to use the _X_ operator as a *constructor* operator for terms representing powerlists, but rather as a *defined* operator, we declare the membership for the _X_ operator with the dfn attribute. In fact, if we go back to the operator declarations section, we can realize that

```
op [_] : Nat -> Pow [ctor] .   op len : Pow -> Nat .
```

is just syntactic sugar for the following declarations:

```
op [_] : [Nat] -> [Pow] .      op len : [Pow] -> [Nat] .
mb [I]: Pow .                   mb len(P): Nat  [metadata "dfn"] .
```

As we will explain in the following section, the sufficient completeness problem for POWERLIST reduces to proving that all terms P X Q and $\text{len}(P)$, where P and Q are terms built with our *constructor* memberships, can be proved to be of sort Pow without using the *defined* memberships.

4   Sufficient Completeness for MEL Specifications

The definition of sufficient completeness for MEL specifications is somewhat subtle, in that in its most general form it cannot be given only in terms of a subsignature Ω of constructors. The point is that, when specifying the conditional memberships for constructor operators in Ω, other nonconstructor function symbols may appear in the condition. This is illustrated in the powerlist example by the conditional membership for the constructor _|_ of powerlists. The definition below strictly generalizes that in [2], which ruled out the use of nonconstructor symbols in conditions of constructor memberships.

Definition 1. *Let* $\mathcal{E} = ((\mathcal{K}, \Sigma, \mathcal{S}), E \cup M_< \cup M_\Sigma)$ *be a MEL specification where* E *contains the conditional equations,* $M_<$ *contains the memberships corresponding to subsort declarations explained below, and* M_Σ *contains the conditional memberships specifying the sorts of function symbols in* Σ. *Subsort declarations* $s < s'$ *with* $s \neq s'$ *and* $s, s' \in S_k$ *for some* k *are axiomatized by the conditional membership:*

$$(\forall x : k) \; x : s' \text{ if } x : s$$

Finally, we assume that any conditional membership in M_Σ *is of the form:*

$$(\forall \vec{x}) \; f(t_1, \ldots, t_n) : s \text{ if } t_1 : s_1 \wedge \cdots \wedge t_n : s_n \wedge \mathcal{C} \tag{1}$$

where $f \in \Sigma$, $\vec{x} = \text{var}(f(t_1, \ldots, f_n))$, *and* $\mathcal{C}$ *is a (possibly empty) conjunction of* Σ-*equations and* Σ-*memberships,* $\text{var}(\mathcal{C}) \subseteq \vec{x}$, *and if* f *is a constant in* $\Sigma_{\epsilon,k}$ *then* $\mathcal{C}$ *is empty.*

Given a subset of memberships $M_\Omega \subseteq M_\Sigma$, called constructor memberships, *we define a* constructor subtheory *to be $\mathcal{E}_\Omega = ((\mathcal{K}, \Sigma, \mathcal{S}), E \cup M_< \cup M_\Omega)$. Furthermore, we say that $\mathcal{E}$ is* sufficiently complete *relative to M_Ω iff $\mathcal{E}_\Omega$ is such that the unique Σ-homomorphism*

$$h : T_{\mathcal{E}_\Omega} \to T_\mathcal{E}$$

is an isomorphism. Finally, we define M_Δ to be $M_\Sigma - M_\Omega$.

To illustrate these notions, we can use (the desugared version of) POWERLIST. In this specification: M_Σ is the set containing the memberships

(1)   mb 0 : Nat .
(2)   cmb s N : Nat if N : Nat .
(3)   mb [I]: Pow .
(4)   cmb (P | Q) : Pow if len(P) = len(Q) .
(5)   mb len(P): Nat [metadata "dfn"] .
(6)   cmb (P X Q) : Pow if len(P) = len(Q) [metadata "dfn"] .;

$M_<$ is the empty set; M_Δ is the set containing (5) and (6), that is the memberships labeled with dfn; and M_Ω is the set containing (1)–(4).

The soundness of the Maude sufficient completeness tool is based on the following theorem, which we have proven in the technical report [8].

Theorem 1. *Let $\mathcal{E} = (\Sigma, E \cup M_< \cup M_\Omega \cup M_\Delta)$ be a MEL specification satisfying:*

(i) $\mathcal{E}$ and $\mathcal{E}_\Omega$ are reductive, ground confluent, and ground sort-decreasing.
(ii) Each membership in $M_\Omega \cup M_\Delta$ is of the restricted form (1).

Then the two statements below are equivalent:

(a) $\mathcal{E}$ is sufficiently complete relative to constructor memberships M_Ω
(b) For each membership $(\forall \vec{x})\ t : s$ if C in M_Δ and ground substitution $\theta : \vec{x} \to T_\Sigma$ such that $\mathcal{E}_\Omega \models C\theta$, either $t\theta$ is $\mathcal{E}_\Omega$-reducible or there is a membership in M_Ω of the form $(\forall \vec{y})\ u : s'$ if C' with $s' \leq s$ and a substitution $\tau : \vec{y} \to T_\Sigma$ such that $t\theta = u\tau$ and $\mathcal{E}_\Omega \models C'\tau$.

Due to space constraints, we do not reproduce the proof here: consult [8] for the detailed proof.

5   The Maude Sufficient Completeness Tool

The Maude Sufficient Completeness tool (SCC) is itself written in Maude using *reflection*. (More details on reflection in Maude in Sect. 5.2.) The soundness of the tool is based on Theorem 1. There are two major components to the tool: a *Sufficient Completeness Analyzer*, which generates proof obligations for sufficient completeness problems, and the Maude *Inductive Theorem Prover* (ITP), extended with additional commands to try to automatically prove those proof obligations. The tool has been run on a variety of different MEL specifications, and is available for download with source, documentation, and examples (including MEL specifications of ordered lists with sorting functions, stacks, and binary trees) from the tool's webpage: http://maude.cs.uiuc.edu/tools/scc/

5.1   The Maude Sufficient Completeness Analyzer

The Maude Sufficient Completeness Analyzer follows the incremental construct-or-based narrowing of patterns approach, but generalized to handle conditional specifications. Given a MEL theory $\mathcal{E} = (\Sigma, E \cup M_< \cup M_\Sigma)$ in Maude, conveniently annotated to indicate a constructor subtheory $\mathcal{E}_\Omega$, the Maude sufficient completeness analyzer generates, in a two phase process, a set of proof obligations which if discharged, ensures the sufficient completeness of $\mathcal{E}$ relative to M_Ω. The sufficient completeness analyzer assumes that $\mathcal{E}$ satisfies the requirements (i) and (ii) in Theorem 1.

The Narrowing Procedure. In its first phase, the analyzer returns a set $\Delta = \{(t, s, C)_i\}_{i \in \mathbb{N}}$ such that, if t' is a counterexample for sufficient completeness, then there exists a triple $(t, s, C) \in \Delta$ and a substitution $\theta : \mathrm{var}(t) \to T_\Sigma$ such that $t' = \theta t$ and $\mathcal{E}_\Omega \models \theta C$. The set Δ is generated from the initial set $\{(t, s, C) \mid (\forall \vec{x})\ t : s\ \textbf{if}\ C \in M_\Delta\}$ by applying rule (2) below until it cannot be applied anymore. The rule (2) uses the *expandability relation* ◀ and the *expand function* exp which are defined as follows:

Definition 2. *Let* t, t' *be terms in* $T_\Sigma(\vec{x})$ *such that* $\mathrm{var}(t) \cap \mathrm{var}(t') = \emptyset$, *and* $x \in \mathrm{var}(t)$. *Then,* $t \blacktriangleleft_x t'$ *iff there exists a substitution* θ *such that* θ *is a most general unifier of* t *and* t', *and* $\theta(x)$ *is not a variable.*

Definition 3. *Let* $t \in T_\Sigma(\vec{x})_k$, $s \in S_k$, C *a conjunction of atomic formulas,* $x \in \vec{x}$ *with* $x : s' \in C$, *and* M *a set of memberships whose variables have all been renamed to be disjoint from* $\vec{x}$. *Then,*

$$\exp(t, s, C, x, M) = \{(t\theta, s, C\theta \wedge C') \mid (\forall \vec{y})\ u : s'\ \textbf{if}\ C' \in M, \theta = (x \mapsto u)\}$$

Finally, we define the inference rule that generates the set Δ. Note that this rule will only be applied a finite number of times, because of the condition $t \blacktriangleleft_x t'$ on the rule.

Δ-rule For any $(\forall \vec{y})\ t' = t''\ \textbf{if}\ C'$ in E,

$$\frac{\Delta' \cup \{(t, s, C)\}}{\Delta' \cup \exp(t, s, C, x, M_< \cup M_\Omega)} \quad \text{if } x \in \mathrm{var}(t), t \blacktriangleleft_x t' \tag{2}$$

The Proof Obligation Generator. In its second phase, the SCC produces, from the set Δ, a set of proof obligations which if discharged, guarantees that $\mathcal{E}$ is sufficiently complete with respect to M_Ω. Since a triple $(t, s, C) \in \Delta$ represents a set of potential counterexamples, the proof obligation generator produces a sentence which if proven in $\mathcal{E}_\Omega$, implies that for every substitution $\theta : \mathrm{var}(t) \to T_\Sigma$ at least one of the following holds and, therefore by Theorem 1, that $\mathcal{E}$ is sufficiently complete with respect to $\mathcal{E}_\Omega$:

a) $\mathcal{E}_\Omega \not\models C\theta$
b) $t\theta$ is *reducible*
c) There exists a membership $(\forall \vec{y})\ u : s'\ \textbf{if}\ C'$ in M_Ω with $s' < s$ and a substitution $\tau : \vec{y} \to T_\Sigma$ such that $t\theta = u\tau$ and $\mathcal{E}_\Omega \models C'\tau$.

In particular, for each $(t, s, \mathcal{C}) \in \Delta$, the proof obligation generator constructs the sentence:

$$(\forall x \in \text{var}(t)) \left(\neg \mathcal{C} \vee \bigvee_{\substack{t'=t'' \text{ if } \mathcal{C}' \in E, \\ \theta \text{ s.t. } \mathbb{C}[t'\theta]=t}} \mathcal{C}'\theta \vee \bigvee_{\substack{u:s' \text{ if } \mathcal{C}' \in M_\Omega \text{ s.t. } s'<s, \\ \theta \text{ s.t. } u\theta=t}} \mathcal{C}'\theta \right) \quad (3)$$

where $\mathbb{C}$ denotes a context.

5.2   The Maude ITP

The ITP [9] tool is an experimental interactive tool for proving properties of MEL specifications in Maude. The ITP tool has been written entirely in Maude, and is in fact an *executable* specification in MEL of the formal inference system that it implements. The ITP inference system treats MEL specifications as *data* – for example, one inference may add to the specification an induction hypothesis as a new equational axiom. This makes a *reflective* design, in which Maude equational specifications become data at the metalevel, ideally suited for implementing the ITP. Using reflection to implement the ITP tool has one important additional advantage, namely, the ease to rapidly extend it by integrating other tools implemented in Maude using reflection, as it is the case of the SCC.

In the ITP, the user introduces commands which are interpreted as actions that may change the state of the proof, specifically the set of goals that remain to be proved, with each goal consisting of a formula to be proved and the MEL specification in which the formula must be proved. After executing the action requested by the user, the tool reports the resulting state of the proof. The main module implementing the ITP is the ITP-TOOL module. In this module, states of proofs, sets of goals, goals and formulas are represented by terms of different sorts, and the actions interpreting the ITP commands are represented as different, equationally defined functions over those terms.

To integrate the SCC in the ITP we have added two new commands, scc and scc*, to the ITP; the scc* command is an extension of scc that takes into account the information obtained by this command *at run-time*. We begin with the scc command. This command is implemented by extending the module ITP-TOOL with a new, equationally defined function that, given an equational specification $\mathcal{E}$, does the following:

- first, it calls on $\mathcal{E}$ the function checkCompleteness, which implements the sufficient completeness analyzer described in Sect. 5.1;
- then, it converts the resulting proof obligations into a set of ITP goals, which are all associated with $\mathcal{E}_\Omega$;
- finally, it eliminates from the state of the proof those goals that can be proved automatically using the ITP auto* command[1].

---

[1] The current implementation of the auto* command integrates its rewriting-based simplification strategy with a decision procedure for linear arithmetic with uninterpreted function symbols; this theory includes many of the formulas that one tends to encounter in proof obligations generated by the SCC tool.

As an example, we can use the scc command to check the sufficient completeness of POWERLIST. After introducing in Maude the command line (scc POWERLIST .), the ITP tool reports the resulting state of the proof:

```
====================================
CTOR-POWERLIST$1.0
====================================
|- A{P:Pow ; Q':Pow}
    ((~(len(P:Pow)= len(Q':Pow)))V(~(len(P:Pow)+ len(Q':Pow)= 1)))
====================================
CTOR-POWERLIST$2.0
====================================
|- A{P:Pow ; Q:Pow}
    ((~(len(P:Pow)= len(Q:Pow)))V(~(len(Q:Pow)+ len(P:Pow)= 1)))
```

In this case, the auto* command failed to discharge the above goals corresponding to proof obligations generated by SCC, despite the fact that the formulas associated to those goals are obviously true in $\mathcal{E}_\Omega$ The reason is the following. In $\mathcal{E}_\Omega$, the len operator is declared at the kind level: it takes a term of the kind [Pow] and returns a term of the kind [Nat]. In this situation, the decision procedure cannot recognize the formulas associated to the above goals as belonging to the class of formulas that it can solve. Therefore, to discharge the proof obligations it is necessary to prove that the operator len always returns a term of sort Nat when it is called on terms of sort Pow. This is, however, implied by the fact that SCC has generated no proof obligations for the len operator.

Since the situation described above is a rather common one, we have also implemented the command scc* that associates all the goals corresponding to the proof obligations generated by SCC with $\mathcal{E}_\Omega$, but extended this time with all the operator declarations in $\mathcal{E}$ that SCC has found unproblematic. In the case of powerlists, scc* discharges all the proof obligations automatically.

6   Related Work

We cannot survey here the extensive literature on sufficient completeness: we just mention some related work to place things in context. Sufficient completeness of MEL specifications was first studied in [2]; the definition and methods on which the present tool is based are strictly general than those in [2], allowing a much wider class of MEL specifications to be checked. Sufficient completeness itself goes back to Guttag's thesis [10] (but see [11] for a more formal, direct treatment of this notion). An early algorithm for handling unconditional linear specifications is due to Nipkow and Weikum [12]. For a good review of the literature up to the late 80s, as well as some important decidability/undecidability and complexity results, see [13, 14]. A more recent development is the casting of the decidable cases of sufficient completeness as tree automata decision problems: see Chapter 4 of [15] and references there. Two sufficient completeness tools having a similar approach to ours, namely the incremental constructor-based narrowing of patterns, are the sufficient completeness checkers of the Spike [16] and RRL [17] theorem provers, both of which are based on many-sorted equational logic. By contrast, our approach is based on a more expressive partial equational logic (MEL). However, RRL [17], although based on a total many-sorted logic, can address some partiality issues in a different way: incompleteness can be due to *omissions*, yielding real counterexample patterns, or can be *intentional*, due to partiality, in which case the partial function's domain of definition can be specified by a quantifier-free formula, which can be used to ascertain if a counterexample pattern is relevant in that domain.

7   Conclusions and Future Work

At present, the SCC can handle specifications where some symbols have been declared commutative. Future work will extend the tool to handle equations modulo different combinations of associativity, commutativity, and identity. It is well-known that sufficient completeness is undecidable in the presence of associative axioms, even for left-linear confluent and terminating equations [14]. However, equational tree automata techniques in the style of [18] can still make the problem decidable for some subclasses, and the ITP can support reasoning to discharge proof obligations for the general case.

As already mentioned, the tool assumes MEL specifications $\mathcal{E}$ that are ground confluent, reductive, and sort-decreasing. Although Maude already has tools to check these properties in the special case where $\mathcal{E}$ is an *order-sorted* specifications [19], tools to discharge the corresponding obligations for general MEL specifications need to be developed. For termination of MEL specifications there is already a tool prototype [20] and supporting theory [20, 21]. For checking confluence and sort-decreasingness of *general* MEL specifications detailed supporting theory can be found in [2], but a tool needs to be developed.

References

1. Meseguer, J.: Membership algebra as a logical framework for equational specification. In: In 12th International Workshop on Recent Trends in Algebraic Development Techniques (WADT'97). Volume 1376 of Lecture Notes in Computer Science., Springer-Verlag (1998) 18–61
2. Bouhoula, A., Jouannaud, J.P., Meseguer, J.: Specification and proof in membership equational logic. Theoretical Computer Science **236** (2000) 35–132
3. Meseguer, J., Roşu, G.: A total approach to partial algebraic specification. In: Proceedings of ICALP. Volume 2380 of Lecture Notes in Computer Science., Springer (2002) 572–584
4. Meseguer, J., Palomino, M., Martí-Oliet, N.: Equational abstractions. In: Proceedings of CADE. Volume 2741 of Lecture Notes in Computer Science., Springer (2003) 2–16
5. Kapur, D., Subramaniam, M.: New uses of linear arithmetic in automated theorem proving by induction. Journal of Automated Reasoning **16** (1996) 39–78
6. Clavel, M., Durán, F., Eker, S., Lincoln, P., Martí-Oliet, N., Meseguer, J., Quesada, J.: Maude: Specification and programming in rewriting logic. Theoretical Computer Science **285** (2002) 187–243

7. Misra, J.: Powerlist: a structure for parallel recursion. ACM Transactions on Programming Languages and Systems **16** (1994) 1737–1767
8. Hendrix, J., Clavel, M., Meseguer, J.: A sufficient completeness reasoning tool for partial specifications (extended technical report). Available on tool website at http://maude.cs.uiuc.edu/tools/scc/ (2005)
9. Clavel, M.: The ITP tool's home page. http://maude.sip.ucm.es/itp (2005)
10. Guttag, J.: The Specification and Application to Programming of Abstract Data Types. PhD thesis, University of Toronto (1975) Computer Science Department, Report CSRG-59.
11. Guttag, J.V., Horning, J.J.: The algebraic specification of abstract data types. Acta Inf. **10** (1978) 27–52
12. Nipkow, T., Weikum, G.: A decidability result about sufficient-completeness of axiomatically specified abstract data types. In: Proc. 6th GI-Conf. Theoretical Computer Science. Volume 145 of Lecture Notes in Computer Science., Springer (1983) 257–268
13. Kapur, D., Narendran, P., Zhang, H.: On sufficient-completeness and related properties of term rewriting systems. Acta Informatica **24** (1987) 395–415
14. Kapur, D., Narendran, P., Rosenkrantz, D.J., Zhang, H.: Sufficient-completeness, ground-reducibility and their complexity. Acta Informatica **28** (1991) 311–350
15. Comon, H., Dauchet, M., Gilleron, R., Jacquemard, F., Lugiez, D., Tison, S., Tommasi, M.: Tree automata techniques and applications. Available on: http://www.grappa.univ-lille3.fr/tata (1997) release October, 1st 2002.
16. Bouhoula, A., Rusinowitch, M.: SPIKE: A system for automatic inductive proofs. In: Algebraic Methodology and Software Technology, AMAST '95, Proceedings. Volume 936 of Lecture Notes in Computer Science., Springer (1995) 576–577
17. Kapur, D.: An automated tool for analyzing completeness of equational specifications. In: Proceedings of the 1994 International Symposium on Software Testing and Analysis (ISSTA), August 17-19, 1994, Seattle, WA, USA. Software Engineering Notes, Special Issue, ACM Press (1994) 28–43
18. Ohsaki, H., Seki, H., Takai, T.: Recognizing boolean closed a-tree languages with membership conditional rewriting mechanism. In: Rewriting Techniques and Applications, 14th International Conference, RTA 2003, Valencia, Spain, June 9-11, 2003, Proceedings. Volume 2706 of Lecture Notes in Computer Science., Springer (2003) 483–498
19. Clavel, M., Durán, F., Eker, S., Meseguer, J.: Building equational proving tools by reflection in rewriting logic. In: Cafe: An Industrial-Strength Algebraic Formal Method. Elsevier (2000)
20. Durán, F., Lucas, S., Meseguer, J., Marché, C., Urbain, X.: Proving termination of membership equational programs. In: Proceedings of the 2004 ACM SIGPLAN Workshop on Partial Evaluation and Semantics-based Program Manipulation, 2004, Verona, Italy, August 24-25, 2004, ACM Press (2004) 147–158
21. Lucas, S., Meseguer, J., Marché, C.: Operational termination of generalized conditional term rewriting systems. Submitted. (2004)

Tyrolean Termination Tool*

Nao Hirokawa and Aart Middeldorp

Institute of Computer Science,
University of Innsbruck,
6020 Innsbruck, Austria
{nao.hirokawa,aart.middeldorp}@uibk.ac.at

1 Introduction

This paper describes the Tyrolean Termination Tool (T$_T$T in the sequel), the successor of the Tsukuba Termination Tool [12]. We describe the differences between the two and explain the new features, some of which are not (yet) available in any other termination tool, in some detail. T$_T$T is a tool for automatically proving termination of rewrite systems based on the dependency pair method of Arts and Giesl [3]. It produces high-quality output and has a convenient web interface. The tool is available at

http://cl2-informatik.uibk.ac.at/ttt

T$_T$T incorporates several new improvements to the dependency pair method. In addition, it is now possible to run the tool in *fully automatic mode* on a *collection* of rewrite systems. Moreover, besides ordinary (first-order) rewrite systems, the tool accepts simply-typed applicative rewrite systems which are transformed into ordinary rewrite systems by the recent method of Aoto and Yamada [2].

In the next section we describe the differences between the semi automatic mode and the Tsukuba Termination Tool. Section 3 describes the fully automatic mode. In Section 4 we show a termination proof of a simply-typed applicative system obtained by T$_T$T. In Section 5 we describe how to input a collection of rewrite systems and how to interpret the resulting output. Some implementation details are given in Section 6. The final section contains a short comparison with other tools for automatically proving termination.

2 Semi-automatic Mode

Figure 1 shows the web interface.

This menu corresponds to the options that were available in the Tsukuba Termination Tool. A first difference is that we now support the dependency pair method for innermost termination [3]. A second difference is that dependency pairs that are covered by the subterm criterion of Dershowitz [7] are excluded. The other differences are described in the following paragraphs.

---

* A preliminary description of the Tyrolean Termination Tool appeared in the proceedings of the 7th International Workshop on Termination, Technical Report AIB-2004-07, RWTH Aachen, pages 249–268, 2004.

J. Giesl (Ed.): RTA 2005, LNCS 3467, pp. 175–184, 2005.
© Springer-Verlag Berlin Heidelberg 2005

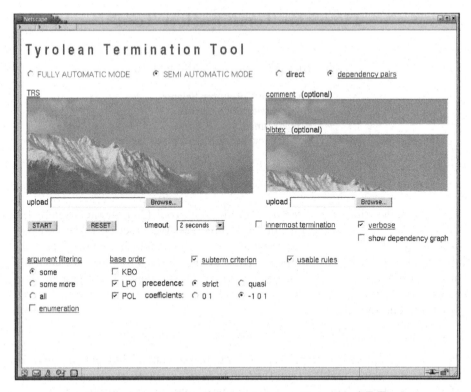

Fig. 1. A screen shot of the semi automatic mode of T$_T$T.

First of all, when approximating the (innermost) dependency graph the original estimations of [3] are no longer available since the approximations described in [15] generally produce smaller graphs while the computational overhead is negligible.

Secondly, the user can no longer select the cycle analysis method (all cycles separately, all strongly connected components separately, or the recursive SCC algorithm of [15]). Extensive experiments reveal that the latter method outperforms the other two, so this is now the only supported method in T$_T$T.

Finally, the default method to search for appropriate argument filterings has been changed from *enumeration* to the *divide and conquer* algorithm of [15]. By using dynamic programming techniques, the divide and conquer method has been improved (cf. [15]) to the extent that for most examples it is more efficient than the straightforward enumeration method. Still, there are TRSs where enumeration is more effective, so the user has the option to change the search strategy (by clicking the enumerate box).

New features include (1) a very useful criterion based on the subterm relation to discard SCCs of the dependency graph without considering any rewrite rules and (2) a very powerful modularity criterion for termination inspired by the *usable rules* of [3] for innermost termination. These features are described in detail in [13]. The first one is selected by clicking the *subterm criterion* box and the second by clicking the *usable rules* box. In addition, linear polynomial interpretations with coefficients from $\{-1, 0, 1\}$ can be used as base order. In [14] it is explained how polynomial interpretations with negative coefficients, like $x - 1$ for a unary function symbol or $x - y$ for a binary function symbol, can be effectively used in connection with the dependency pair method.

3 Fully Automatic Mode

In this mode TᴛT uses a simple strategy to (recursively) solve the ordering constraints for each SCC of the approximated dependency graph. The strategy is based on the new features described in the previous section and uses LPO (both with strict and quasi-precedence) with *some* argument filterings [15] and linear polynomial interpretations with coefficients from $\{-1, 0, 1\}$ as base orders.

After computing the SCCs of the approximated (innermost) dependency graph, the strategy subjects each SCC to the following algorithm:

1. First we check whether the new *subterm criterion* is applicable.
2. If the subterm criterion was unsuccessful, we compute the *usable rules*.
3. The resulting (usable rules and dependency pairs) constraints are subjected to the *natural* (see [14]) polynomial interpretation with coefficients from $\{0, 1\}$.
4. If the constraints could not be solved in step 3, we employ the *divide and conquer* algorithm for computing suitable argument filterings with respect to the *some* heuristic [15] and LPO with *strict* precedence.
5. If the previous step was unsuccessful, we repeat step 3 with *arbitrary* polynomial interpretations with coefficients from $\{0, 1\}$.
6. Next we repeat step 4 with the variant of LPO based on *quasi-precedences* and a small increase in the search space for argument filterings (see below).
7. If the constraints could still not be solved, we try linear polynomial interpretations with coefficients from $\{-1, 0, 1\}$.

If only part of an SCC could be handled, we subject the resulting new SCCs recursively to the same algorithm.

If the current set of constraints can be solved in step 3 or 4, then they can also be solved in step 5 or 6, respectively, but the reverse is not true. The sole reason for adopting LPO and polynomial interpretations in alternating layers is efficiency; the search space in steps 3 and 4 is significantly smaller than in steps 5 and 6. The reason for putting the subterm criterion first is that with this criterion many SCCs can be eliminated very quickly, cf. the third paragraph of Section 6. The extension of the search space for argument filterings mentioned in step 6 is obtained by also considering the full *reverse* argument filtering $[n, \ldots, 1]$ for every n-ary function symbol. The advantage of this extension is that there is no need for a specialized version of LPO with right-to-left status.

The effectiveness of the automatic strategy can be seen from the data presented in Figure 2, which were obtained by running TᴛT in fully automatic mode on the 89 terminating TRSs (66 in Section 3 and 23 in Section 4) of [4]. An explanation of the data is given in Section 5.

Individual Data

TRS	status	TRS	status	TRS	status	TRS	status	TRS	status	TRS	status	TRS	status	TRS	status
3.01	0.00	3.05b	0.04	3.07	0.01	3.10	0.49	3.15	0.01	3.19	0.03	3.24	0.01	3.29	0.00
3.02	0.02	3.05c	0.06	3.08a	0.02	3.11	0.58	3.16	0.01	3.20	0.01	3.25	0.00	3.30	0.02
3.03	0.03	3.06a	0.05	3.08b	0.02	3.12	0.02	3.17a	0.02	3.21	0.01	3.26	0.01	3.31	0.00
3.04	0.03	3.06b	0.04	3.08c	0.05	3.13	0.42	3.17b	0.09	3.22	0.01	3.27	0.00	3.32	0.00
3.05a	0.04	3.06c	0.08	3.09	0.05	3.14	0.02	3.18	0.02	3.23	0.00	3.28	0.02	3.33	0.01

TRS	status	TRS	status	TRS	status	TRS	status	TRS	status	TRS	status	TRS	status	TRS	status
3.34	0.00	3.39	0.35	3.44	0.00	3.49	0.01	3.53b	0.01	3.57	0.04	4.23	0.02	4.28	0.02
3.35	0.01	3.40	0.62	3.45	0.01	3.50	0.01	3.53c	0.01	4.20a	0.01	4.24	0.02	4.29	(1)
3.36	0.03	3.41	0.01	3.46	0.02	3.51	0.02	3.54	0.01	4.20b	0.01	4.25	0.02	4.30a	(1)
3.37	0.00	3.42	0.04	3.47	0.01	3.52	0.01	3.55	0.67	4.21	0.02	4.26	(1)	4.30b	0.05
3.38	0.15	3.43	(1)	3.48	0.05	3.53a	0.08	3.56	0.01	4.22	0.01	4.27	0.02	4.30c	(1)

TRS	status	TRS	status
4.30d	(1)	4.35	0.96
4.31	0.11	4.36	0.22
4.32	0.01	4.37a	0.01
4.33	0.03	4.37b	0.01
4.34	0.02		

Summary

success	failure	timeout	total
80 (0.06)	3 (0.36)	6 (1)	89 (12.10)

Fig. 2. Output produced by T$_T$T.

Our automatic strategy differs from the "Meta-Combination Algorithm" described in [11]; we avoid transforming SCC constraints using techniques like narrowing and instantiation because they tend to complicate the produced termination proofs. Instead, we rely on techniques (subterm criterion and polynomial interpretations with negative coefficients) that lead to termination proofs that are (relatively) easy to understand.

4   Simply-Typed Applicative Rewrite Systems

Besides ordinary first-order TRSs, T$_T$T accepts *simply-typed applicative rewrite systems* (STARSs) [1]. Applicative terms are built from variables, constants, and a single binary operator $\cdot$, called application. Constants and variables are equipped with a simple type such that the rewrite rules typecheck. A typical example is provided by the following rules for the map function

$$(\mathsf{map} \cdot f) \cdot \mathsf{nil} \to \mathsf{nil}$$
$$(\mathsf{map} \cdot f) \cdot ((\mathsf{cons} \cdot x) \cdot y) \to (\mathsf{cons} \cdot (f \cdot x)) \cdot ((\mathsf{map} \cdot f) \cdot y)$$

with the type declaration nil: α, cons: $\beta \to \alpha \to \alpha$, map: $(\beta \to \beta) \to \alpha \to \alpha$, $f: \beta \to \beta$, $x: \beta$, and $y: \alpha$. Here α is the list type and β the type of elements of lists. STARSs are useful to model higher-order functions in a first-order setting. As usual, the application operator $\cdot$ is suppressed in the notation and parentheses are removed under the "association to the left" rule. The above rules then become

$$\mathsf{map}\ f\ \mathsf{nil} \to \mathsf{nil}$$

$$\mathsf{map}\ f\ (\mathsf{cons}\ x\ y) \to \mathsf{cons}\ (f\ x)\ (\mathsf{map}\ f\ y)$$

This corresponds to the syntax of STARSs in T⊤T. The types of constants must be declared by the keyword **TYPES**. The types of variables is automatically inferred when typechecking the rules, which follow the **RULES** keyword. So the above STARS would be inputted to T⊤T as

```
TYPES
  nil : a                                    ;
cons : b => (a => a)                         ;
  map : (b => b) => a => a                   ;


RULES
map f nil           -> nil                   ;
map f (cons x y) -> cons (f x) (map f y) ;
```

   In order to prove termination of STARSs, T⊤T uses the two-phase transformation developed by Aoto and Yamada [2]. In the first phase all head variables (e.g. f in f x) are removed by the *head variable instantiation* technique. The soundness of this phase relies on the *ground term existence condition*, which basically states that all simple types are inhabited by at least one ground term. Users need not be concerned about this technicality as T⊤T automatically adds fresh constants of the appropriate types to the signature so that the ground term existence condition is satisfied. (Moreover, the termination status of the original STARS is not affected by adding fresh constants.) After the first phase an ordinary TRS is obtained in which the application symbol is the only non-constant symbol. Such TRSs are not easily proved terminating since the root symbol of every term that has at least two symbols is the application symbol and thus provides no information which could be put to good use. In the second phase applicative terms are transformed into ordinary terms by the *translation to functional form* technique. This technique removes all occurrences of the application symbol. We refer to [2] for a complete description of the transformation. We contend ourselves with showing the Postscript output (in Figure 3) produced by T⊤T on the following variation of combinatory logic (inspired by a recent question posted on the TYPES Forum by Jeremy Dawson):

```
TYPES
I : o => o                                   ;
W : (o => o => o) => o => o                  ;
S : (o => o => o) => (o => o) => o => o ;
```

```
RULES
I x      -> x       ;
W f x    -> f x x   ;
S x y z -> x z (y z) ;
```

Note that the types are crucial for termination; the untyped version admits the cyclic rewrite step W W W → W W W.

5   A Collection of Rewrite Systems as Input

Single TRSs (or STARSs) are inputted by typing (the type declarations and) the rules into the upper left text area or by uploading a file via the browse button. Besides the original TₜT syntax (which is obtained by clicking the <u>TRS</u> link), TₜT supports the official format[1] of the Termination Problems Data Base. The user can also upload a zip archive. All files ending in .trs are extracted from the archive and the termination prover runs on each of these files in turn. The results are collected and presented in two tables. The first table lists for each TRS the execution time in seconds together with the status: bold green indicates success, *red italics* indicates failure, and gray indicates timeout. By clicking green (*red*) entries the user can view the termination proof (attempt) in HTML or high-quality Postscript format. The second table gives the number of successes and failures, both with the average time spent on each TRS, the number of timeouts, and the total number of TRSs extracted from the zip archive together with the total execution time. Figure 2 shows the two tables for the 89 terminating TRSs in Sections 3 and 4 of [4]. Here we used TₜT's fully automatic mode with a timeout of 1 second (for each TRS). The experiment was performed on a PC equipped with a 2.20 GHz Mobile Intel Pentium 4 Processor - M and 512 MB of memory, using native-compiled code for Linux/Fedora.

6   Some Implementation Details

The web interface of TₜT is written in Ruby[2] and the termination prover underlying TₜT is written in Objective Caml (OCaml)[3], using the third-party libraries[4] findlib, extlib, and pcre-ocaml. We plan to make the OCaml source code available in the near future.

The termination prover consists of about 13,000 lines of OCaml code. About 20% is used for the manipulation of terms and rules. Another 15% is devoted to graph manipulations. This part of the code is not only used to compute dependency graph approximations, but also for precedences in KBO and LPO, and for the dependency relation which is used to compute the usable rules. The various termination methods that are provided by TₜT account for less than

---

[1] http://www.lri.fr/~marche/wst2004-competition/format.html
[2] http://www.ruby-lang.org/
[3] http://www.ocaml.org/
[4] http://caml.inria.fr/humps/

Termination Proof Script[a]

Consider the simply-typed applicative TRS

$$
\begin{aligned}
\mathsf{I}\, x &\to x \\
\mathsf{W}\, f\, x &\to f\, x\, x \\
\mathsf{S}\, x\, y\, z &\to x\, z\, (y\, z)
\end{aligned}
$$

over the signature $\mathsf{I}\colon \mathsf{o} \Rightarrow \mathsf{o}$, $\mathsf{W}\colon (\mathsf{o} \Rightarrow \mathsf{o} \Rightarrow \mathsf{o}) \Rightarrow \mathsf{o} \Rightarrow \mathsf{o}$, and $\mathsf{S}\colon (\mathsf{o} \Rightarrow \mathsf{o} \Rightarrow \mathsf{o}) \Rightarrow (\mathsf{o} \Rightarrow \mathsf{o}) \Rightarrow \mathsf{o} \Rightarrow \mathsf{o}$. In order to satisfy the ground term existence condition we extend the signature by $\mathsf{c}\colon \mathsf{o} \Rightarrow \mathsf{o} \Rightarrow \mathsf{o}$ and $\mathsf{c}'\colon \mathsf{o}$. Instantiating all head variables yields the following rules:

$$
\begin{aligned}
\mathsf{I}\, x &\to x \\
\mathsf{W}\, \mathsf{c}\, x &\to \mathsf{c}\, x\, x \\
\mathsf{S}\, \mathsf{c}\, \mathsf{I}\, z &\to \mathsf{c}\, z\, (\mathsf{I}\, z) \\
\mathsf{S}\, \mathsf{c}\, (\mathsf{W}\, w)\, z &\to \mathsf{c}\, z\, (\mathsf{W}\, w\, z) \\
\mathsf{S}\, \mathsf{c}\, (\mathsf{S}\, w\, v)\, z &\to \mathsf{c}\, z\, (\mathsf{S}\, w\, v\, z) \\
\mathsf{S}\, \mathsf{c}\, (\mathsf{c}\, w)\, z &\to \mathsf{c}\, z\, (\mathsf{c}\, w\, z)
\end{aligned}
$$

By transforming terms into functional form the TRS

$$
\begin{aligned}
1:&& \mathsf{I}_1(x) &\to x \\
2:&& \mathsf{W}_2(\mathsf{c}, x) &\to \mathsf{c}_2(x, x) \\
3:&& \mathsf{S}_3(\mathsf{c}, \mathsf{I}, z) &\to \mathsf{c}_2(z, \mathsf{I}_1(z)) \\
4:&& \mathsf{S}_3(\mathsf{c}, \mathsf{W}_1(w), z) &\to \mathsf{c}_2(z, \mathsf{W}_2(w, z)) \\
5:&& \mathsf{S}_3(\mathsf{c}, \mathsf{S}_2(w, v), z) &\to \mathsf{c}_2(z, \mathsf{S}_3(w, v, z)) \\
6:&& \mathsf{S}_3(\mathsf{c}, \mathsf{c}_1(w), z) &\to \mathsf{c}_2(z, \mathsf{c}_2(w, z))
\end{aligned}
$$

is obtained. There are 3 dependency pairs:

$$
\begin{aligned}
7:&& \mathsf{S}_3^\sharp(\mathsf{c}, \mathsf{I}, z) &\to \mathsf{I}_1^\sharp(z) \\
8:&& \mathsf{S}_3^\sharp(\mathsf{c}, \mathsf{W}_1(w), z) &\to \mathsf{W}_2^\sharp(w, z) \\
9:&& \mathsf{S}_3^\sharp(\mathsf{c}, \mathsf{S}_2(w, v), z) &\to \mathsf{S}_3^\sharp(w, v, z)
\end{aligned}
$$

The approximated dependency graph contains one SCC: $\{9\}$.

– Consider the SCC $\{9\}$. By taking the simple projection π with $\pi(\mathsf{S}_3^\sharp) = 2$, the dependency pair simplifies to

$$
9: \quad \mathsf{S}_2(w, v) \to v
$$

and is compatible with the proper subterm relation.

---

[a] Tyrolean Termination Tool (0.03 seconds) — November 18, 2004.

Fig. 3. Example output.

10% each. Most of the remaining code deals with I/O: parsing the input and producing HTML and Postscript output. For the official Termination Problems Data Base format we use parsers written in OCaml by Claude Marché. A rich OCaml library for the manipulation of terms (or rose trees) and graphs would have made our task much easier!

It is interesting to note that two of the original techniques that make TᴛT fast, the recursive SCC algorithm and the subterm criterion, account for just 13 and 20 lines, respectively. Especially the latter should be the method of first choice in any termination prover. To wit, of the 628 (full) termination problems for pure first-order term and string rewrite systems in the Termination Problems Data Base, 215 are proved terminating by the subterm criterion; the total time to check the whole collection is a mere 32 seconds (on the architecture mentioned in the previous section). Several of these 215 rewrite systems cannot be proved terminating by the latest release of C𝑖ME [5]. (See the next section for a comparison between TᴛT and other termination provers.)

Concerning the implementation of simply-typed applicative rewrite systems, we use the Damas-Milner type reconstruction algorithm (see e.g. [17]) to infer the types of variables.

We conclude this section with some remarks on the implementation of base orders in TᴛT. The implementation of LPO follows [12] but we first check whether the current pair of terms can be oriented by the embedding order in every recursive call to LPO. This improves the efficiency by about 20%. The implementation of KBO is based on [16]. We use the "method for complete description" [8] to compute a suitable weight function. The implementation of polynomial interpretations with coefficients from $\{0, 1\}$ is based on [6, Figure 1] together with the simplification rules described in Section 4.4.1 of the same paper. The current implementation of polynomial interpretations with coefficients from $\{-1, 0, 1\}$ in TᴛT is rather naive. We anticipate that the recent techniques of [6] can be extended to handle negative coefficients.

7 Comparison

Needless to say, TᴛT is not the only available tool for proving termination of rewrite systems. In this final section we compare our tool with the other systems that participated in the TRS category[5] of the termination competition that was organized as part of the 7th International Workshop on Termination[6].

- We start our discussion with C𝑖ME [5], the very first tool for automatically proving termination of rewrite systems that is still available. C𝑖ME is a tool with powerful techniques for finding termination proofs based on polynomial interpretations in connection with the dependency pair method. Since C𝑖ME does not support (yet) the most recent insights in the dependency pair method, it is less powerful than AProVE (described below) or TᴛT. In contrast to TᴛT, C𝑖ME can handle rewrite systems with AC operators. As a

---

[5] http://www.lri.fr/~marche/wst2004-competition/webform.cgi?command=trs
[6] http://www-i2.informatik.rwth-aachen.de/WST04/

matter of fact, termination is only a side-issue in CiME. Its main strength lies in completing equational theories modulo theories like AC and C.

- CARIBOO [9] is a tool specializing in termination proofs for a particular evaluation strategy, like innermost evaluation or the strategies used in OBJ-like languages. The underlying proof method is based on an inductive process akin to narrowing, but its termination proving power comes from CiME, which is used as an external solver. T$_T$T supports innermost termination, but no other strategies.
- Matchbox [19] is a tool that is entirely based on methods from formal language theory. These methods are especially useful for proving termination of string rewrite systems. Matchbox tries to establish termination or non-termination by using recent results on match-bounded rewriting [10]. Matchbox is not intended as a general-purpose termination prover (as its author writes in [19]).
- AProVE is the most powerful tool. Besides ordinary TRSs, it can handle logic programs, conditional rewrite systems, context-sensitive rewrite systems, and it supports rewriting modulo AC. Version 1.0 of AProVE is described in [11]. Of all existing tools, AProVE supports the most base orders and even offers several different algorithms implementing these. It incorporates virtually all recent refinements of the dependency pair method. AProVE has several methods that are not available in any other tool. We mention here the size-change principle [18], transformations for dependency pairs like narrowing and instantiation, and a modular refinement where the set of usable rules is determined after a suitable argument filtering has been computed. Despite all this, last year's termination competition version of AProVE, which further includes the methods derived from match-bounded rewriting, could handle only a few more systems than T$_T$T.

We conclude the paper by listing what we believe to be the main attractions of T$_T$T (in no particular order):

- T$_T$T comes equipped with a very user-friendly web interface,
- T$_T$T produces readable and beautifully typeset proofs,
- T$_T$T is a very fast termination tool,
- T$_T$T is a very powerful tool based on relatively few techniques.

References

1. T. Aoto and T. Yamada. Termination of simply typed term rewriting by translation and labelling. In *Proc. 14th RTA*, volume 2706 of *LNCS*, pages 380–394, 2003.
2. T. Aoto and T. Yamada. Termination of simply-typed applicative term rewriting systems. In *Proc. 2nd HOR*, Technical Report AIB-2004-03, RWTH Aachen, pages 61–65, 2004.
3. T. Arts and J. Giesl. Termination of term rewriting using dependency pairs. *Theoretical Computer Science*, 236:133–178, 2000.
4. T. Arts and J. Giesl. A collection of examples for termination of term rewriting using dependency pairs. Technical Report AIB-2001-09, RWTH Aachen, 2001.

5. E. Contejean, C. Marché, B. Monate, and X. Urbain. C*i*ME version 2, 2000. Available at `http://cime.lri.fr/`.
6. E. Contejean, C. Marché, A.-P. Tomás, and X. Urbain. Mechanically proving termination using polynomial interpretations. Research Report 1382, LRI, 2004.
7. N. Dershowitz. Termination by abstraction. In *Proc. 20th ICLP*, volume 3132 of *LNCS*, pages 1–18, 2004.
8. J. Dick, J. Kalmus, and U. Martin. Automating the Knuth-Bendix ordering. *Acta Infomatica*, 28:95–119, 1990.
9. O. Fissore, I. Gnaedig, and H. Kirchner. CARIBOO: An induction based proof tool for termination with strategies. In *Proc. 4th PPDP*, pages 62–73. ACM Press, 2002.
10. Alfons Geser, Dieter Hofbauer, and Johannes Waldmann. Match-bounded string rewriting. *Applicable Algebra in Engineering, Communication and Computing*, 15:149–171, 2004.
11. J. Giesl, R. Thiemann, P. Schneider-Kamp, and S. Falke. Automated termination proofs with AProVE. In *Proc. 15th RTA*, volume 3091 of *LNCS*, pages 210–220, 2004.
12. N. Hirokawa and A. Middeldorp. Tsukuba termination tool. In *Proc. 14th RTA*, volume 2706 of *LNCS*, pages 311–320, 2003.
13. N. Hirokawa and A. Middeldorp. Dependency pairs revisited. In *Proc. 15th RTA*, volume 3091 of *LNCS*, pages 249–268, 2004.
14. N. Hirokawa and A. Middeldorp. Polynomial interpretations with negative coefficients. In *Proc. 7th AISC*, volume 3249 of *LNAI*, pages 185–198, 2004.
15. N. Hirokawa and A. Middeldorp. Automating the dependency pair method. *Information and Computation*, 2005. To appear. A preliminary version appeared in *Proc. 19th CADE*, volume 2741 of *LNAI*, pages 32–46, 2003.
16. K. Korovin and A. Voronkov. Orienting rewrite rules with the Knuth-Bendix order. *Information and Computation*, 183:165–186, 2003.
17. B.C. Pierce. *Types and Programming Languages*. MIT Press, 2002.
18. R. Thiemann and J. Giesl. Size-change termination for term rewriting. In *Proc. 14th RTA*, volume 2706 of *LNCS*, pages 264–278, 2003.
19. J. Waldmann. Matchbox: A tool for match-bounded string rewriting. In *Proc. 15th RTA*, volume 3091 of *LNCS*, pages 85–94, 2004.

Call-by-Value Is Dual to Call-by-Name Reloaded

Philip Wadler

Edinburgh University

Abstract. We consider the relation of the dual calculus of Wadler (2003) to the $\lambda\mu$-calculus of Parigot (1992). We give translations from the $\lambda\mu$-calculus into the dual calculus and back again. The translations form an equational correspondence as defined by Sabry and Felleisen (1993). In particular, translating from $\lambda\mu$ to dual and then 'reloading' from dual back into $\lambda\mu$ yields a term equal to the original term. Composing the translations with duality on the dual calculus yields an involutive notion of duality on the $\lambda\mu$-calculus. A previous notion of duality on the $\lambda\mu$-calculus has been suggested by Selinger (2001), but it is not involutive.

Note: This paper uses color to clarify the relation of types and terms, and of source and target calculi. If the URL below is not in blue please download the color version from

http://homepages.inf.ed.ac.uk/wadler/

or google 'wadler dual reloaded'.

1 Introduction

Sometimes less is more. Implication is a key connective of logic, but for some purposes it is better to define it in terms of other connectives, taking $A \supset B \equiv \neg A \vee B$. This is helpful if one wishes to understand de Morgan duality. The dual of & is $\vee$, and $\neg$ is self dual, but the dual of an implication $A \supset B$ is the difference operator, $B - A \equiv B \& \neg A$, which is not particularly familiar.

Church (1932) introduced the call-by-name λ-calculus, and a few years later Bernays (1936) proposed the call-by-value variant. A line of work, including that of Filinski (1989), Griffin (1990), Parigot (1992), Danos, Joinet, and Schellinx (1995), Barbanera and Berardi (1996), Streicher and Reuss (1998), Selinger (1998,2001), and Curien and Herbelin (2000), has led to a startling conclusion: call-by-value is the de Morgan dual of call-by-name.

Wadler (2003) presents a dual calculus that corresponds to the classical sequent calculus of Gentzen (1935) in the same way that the lambda calculus of Church (1932,1940) corresponds to the intuitionistic natural deduction of Gentzen (1935). The calculus possesses an involutive duality, which takes call-by-value into call-by-name and vice-versa. A key to achieving this is to not take implication as primitive, but to define it by taking $A \supset B \equiv \neg A \vee B$ under call-by-name, or $A \supset B \equiv \neg(A \& \neg B)$ under call-by-value.

Wadler (2003) included a discussion of call-by-value and call-by-name CPS translations from the dual calculus into the λ-calculus. Here we complete the

J. Giesl (Ed.): RTA 2005, LNCS 3467, pp. 185–203, 2005.

story by discussing a translation from the $\lambda\mu$-calculus of Parigot (1992) into the dual calculus, together with an inverse translation. We will show that there is a translation from the $\lambda\mu$-calculus into the dual calculus which forms an *equational correspondence*, as defined by Sabry and Felleisen (1993).

Say we have a source and target calculus with equations defined on them, writing

$$M =_v N, \quad M =^v N$$

for equality in the source and target respectively, and

$$M^*, \quad M_*$$

for translations from source to target and target to source respectively. We have an *equational correspondence* if the following four conditions hold.

- The translation from source to target preserves equations,

$$M =_v N \text{ implies } M^* =^v N^*,$$

  with M, N source terms.
- The translation from target to source preserves equations,

$$M =^v N \text{ implies } M_* =_v N_*,$$

  with M, N target terms.
- Translating for source to target and then 'reloading' from target to source yields a term equal to the original term,

$$(M^*)_* =_v M,$$

  with M a source term.
- Translating for target to source and then 'reloading' from source to target yields a term equal to the original term,

$$(M_*)^* =^v M,$$

  with M a target term.

The existence of an equational correspondence shows in a strong sense that the translation is both *sound* and *complete* with respect to equations. In particular an equation holds in the source if and only if its translation holds in the target.

Wadler (2003) also presents a CPS translation from the dual calculus into λ-calculus, again in both call-by-value and call-by-name variants. Composing the translation from the $\lambda\mu$-calculus to the dual calculus with the CPS translation for the dual calculus yields the usual call-by-value and call-by-name CPS translations for $\lambda\mu$, as studied by Hoffman and Streicher (1997) and Selinger (2001).

Following the technique introduced in Sabry and Wadler (1997), it is shown that the CPS translation for the dual calculus is a *reflection*, that is it both preserves and reflects reductions. Every reflection is trivially an equational correspondence, where equality is the reflexive, symmetric, and transitive closure of reduction. Since equational correspondences compose, it follows immediately that the CPS translation for $\lambda\mu$-calculus is also an equational correspondence.

Fujita (2003) also shows that the call-by-value CPS translation for $\lambda\mu$-calculus is an equational correspondence; but says nothing about call-by-name. The advantage of the proof here is that the CPS translation for $\lambda\mu$ can be computed by composing other translations, and that its properties follow immediately from its construction by composition rather than requiring separate proof.

Duality is a translation that takes the call-by-value dual calculus into the call-by-name dual calculus, and conversely; that is, if two terms are equal in the call-by-value calculus then their duals are equal call-by-name. Duality is an involution; that is, the dual of the dual is the identity. It follows immediately that duality is an equational correspondence.

Our type system corresponds to minimal logic, with types $A \& B$, $A \vee B$, $\neg A$, and $A \supset B$ corresponding to 'and', 'or', 'not', and 'implies'. (We would have $\neg A = A \supset \bot$, if we had defined a type $\bot$ corresponding to 'false'.) Duality exchanges 'and' with 'or', and 'not' is self dual. The dual of implication $A \supset B = \neg A \vee B$ is difference $B - A = B \& \neg A$. (One can confirm this by checking $B - A = \neg(\neg A \supset \neg B)$.) We choose not to include difference in our type system, because its computational interpretation is not familiar. (For one exploration of what the computational interpretation of difference might be, see Crolard (2004).) It follows that before we consider duality, we first must translate away implications. We use the translation $A \supset B = \neg(A \& \neg B)$ for call-by-value and $A \supset B = \neg A \vee B$ for call-by-name.

We may derive a duality transform from $\lambda\mu$-calculus to itself by forming the threefold composition of (i) the translation from $\lambda\mu$-calculus to dual calculus with (ii) the duality translation from dual calculus to itself with (iii) the reloading translation from dual calculus back to $\lambda\mu$-calculus; and follows immediately that this is an equational correspondence. The same duality transform works for both call-by-value and call-by-name.

Selinger (2001) also presents a duality transformation for $\lambda\mu$-calculus. Selinger's duality required some cleverness to construct — it answered an open question of Streicher and Reuss (1998).

As one would hope, Selinger's duality is an involution for the types corresponding to 'and' and 'or'. However, Selinger has no type corresponding directly to 'not', so he is forced to consider what the dual of an implication might be. Since he has no type corresponding to difference, he is forced to require two distinct mappings, one from call-by-value into call-by-name and one from call-by-name into call-by-value. Further, the composition of these maps does not yield the identity but only the identity up to isomorphism of types. Here we avoid the problem by adding a negation type to the $\lambda\mu$-calculus, requiring that one translate implications before computing the dual. The result is that for us duality on $\lambda\mu$ becomes a proper involution.

The advantage of the proof here is that duality for $\lambda\mu$ can be computed by composing other translations, and that its properties follow immediately from its construction by composition rather than requiring separate proof. Also, the work here uses purely syntactic techniques, depending only on equations in the $\lambda\mu$ and dual calculi, with no reference to control categories or other semantic frameworks.

Wadler (2003) considers reductions, while this paper considers equations. One advantage of considering equations is that it is then easy to add (η) rules, which are problematic for reductions in the presence of sums (see Balat, di Cosmo, and Fiore (2004)). An interesting open question is whether one can replace the equations of this paper by reductions (possibly omitting the (η) rules), and refine the equational correspondence to a reflection.

This paper contains almost entirely new material as compared with Wadler (2003). The description of the dual calculus overlaps with that paper, but the relationship with $\lambda\mu$ is entirely new, as is the treatment of η laws.

2   The $\lambda\mu$-Calculus

The syntax and type rules of the $\lambda\mu$-calculus are shown in Figure 1. Following Parigot (1993), we distinguish two main constructs, *terms* and *statements* (Parigot called these *unnamed terms* and *named terms*.)

As usual, we require the body of a μ-abstraction to be a statement. We provide two variants of λ-abstraction, one where the body is a statement (corresponding to negation), and one where the body is an expressions (corresponding to implication). Informally, one can think of these as related by the equation $\neg A = A \supset \bot$.

Let A, B range over types. A type is atomic X; a conjunction $A \,\&\, B$; a disjunction $A \vee B$; a negation $\neg A$; or an implication $A \supset B$.

Let x, y, z range over variables, α, β, γ range over covariables, M, N, O range over terms, and S, T range over statements. A term is a variable x; a λ-abstraction $\lambda x.\, S$ or $\lambda x.\, N$; a negation application $O\, M$ (where $O : \neg A$); or a μ-abstraction $\mu\alpha.\, S$. A statement is a function application $O\, M$ (where $O : A \supset B$); or a covariable application $[\alpha]M$. The computational interpretation of a μ-abstraction $\mu\alpha.\, S$ is to bind the covariable α and then evaluate statment S; if during evaluation of S the covariable α is applied to a value, then that value is returned as the value of the μ-abstraction; this is similar to the behaviour of *callcc* in Scheme.

We also have products and sums. Products are constructed with pairing $\langle M, N \rangle$ and decontstructed with projections fst O and snd O. Following Selinger (2001), we construct sums with a variant of the mu abstraction $\mu[\alpha, \beta].\, S$, and deconstruct sums with a variant of covariable application $[\alpha, \beta]O$. The term $\mu[\alpha, \beta].\, S$ constructs a sum: if α is passed a value of type A then the μ-abstraction returns a left injection into the sum type $A \vee B$, and if β is passed a value of type B then the μ-abstraction returns a right injection into the sum type $A \vee B$. Conversely, the statement $[\alpha, \beta]O$ deconstructs a sum; the term O has a sum type $A \vee B$, and if it returns a left summand then covariable α is passed the value of type A, while if it returns a right summand then covariable β is passed the value of type B.

Substitution of a term for a variable is standard, but substitution for a covariable is slightly tricky. The notation used here is adapted from Selinger (2001).

Type	A, B	$::= X \mid A \& B \mid A \vee B \mid \neg A \mid A \supset B$
Term	$M, N, O ::=$	$x \mid \langle M, N \rangle \mid \operatorname{fst} O \mid \operatorname{snd} O \mid \mu[\alpha, \beta]. S \mid$
		$\lambda x. S \mid \lambda x. N \mid O M \mid \mu\alpha. S$
Statement	S, T	$::= [\alpha]M \mid [\alpha, \beta]O \mid O M$
Antecedent	Γ	$::= x_1 : A_1, \ldots, x_m : A_m$
Succedent	Θ	$::= \beta_1 : B_1, \ldots, \beta_n : B_n$

$$\text{Right sequent} \quad \Gamma \to \Theta \mid M : A$$
$$\text{Center sequent} \quad \Gamma \mid S \vdash\!\!\mid \Theta$$

$$\frac{}{\Gamma, x : A \to \Theta \mid x : A} \; \text{Id}$$

$$\frac{\Gamma \to \Theta \mid M : A \qquad \Gamma \to \Theta \mid N : B}{\Gamma \to \Theta \mid \langle M, N \rangle : A \& B} \; \&\text{I}$$

$$\frac{\Gamma \to \Theta \mid O : A \& B}{\Gamma \to \Theta \mid \operatorname{fst} O : A} \; \&\text{E} \qquad \frac{\Gamma \to \Theta \mid O : A \& B}{\Gamma \to \Theta \mid \operatorname{snd} O : B} \; \&\text{E}$$

$$\frac{\Gamma \mid S \vdash\!\!\mid \Theta, \alpha : A, \beta : B}{\Gamma \to \Theta \mid \mu[\alpha, \beta]. S : A \vee B} \; \vee\text{I}$$

$$\frac{\Gamma \to \Theta, \alpha : A, \beta : B \mid O : A \vee B}{\Gamma \mid [\alpha, \beta]O \vdash\!\!\mid \Theta, \alpha : A, \beta : B} \; \vee\text{E}$$

$$\frac{x : A, \Gamma \mid S \vdash\!\!\mid \Theta}{\Gamma \to \Theta \mid \lambda x. S : \neg A} \; \neg\text{I} \qquad \frac{\Gamma \to \Theta \mid O : \neg A \qquad \Gamma \to \Theta \mid M : A}{\Gamma \mid O M \vdash\!\!\mid \Theta} \; \neg\text{E}$$

$$\frac{x : A, \Gamma \to \Theta \mid N : B}{\Gamma \to \Theta \mid \lambda x. B : A \supset B} \; \supset\text{I} \qquad \frac{\Gamma \to \Theta \mid O : A \supset B \qquad \Gamma \to \Theta \mid M : A}{\Gamma \to \Theta \mid O M : B} \; \supset\text{E}$$

$$\frac{\Gamma \mid S \vdash\!\!\mid \Theta, \alpha : A}{\Gamma \to \Theta \mid \mu\alpha. S : A} \; \text{Activate} \qquad \frac{\Gamma \to \Theta, \alpha : A \mid M : A}{\Gamma \mid [\alpha]M \vdash\!\!\mid \Theta, \alpha : A} \; \text{Passivate}$$

Fig. 1. Syntax and types of the $\lambda\mu$-calculus.

Definition 1. *(Substitution for a covariable) Let S be a statement, α a covariable of type A, and $T\{-\}$ be a statement context with a hole accepting a term of type A. We write*

$$S\{T\{-\}/[\alpha]\{-\}\}$$

for the substitution that makes the recursive replacements

$$[\alpha]M \quad \mapsto \quad T\{M\},$$
$$[\alpha, \beta]O \quad \mapsto \quad T\{\mu\alpha.\,[\alpha, \beta]O\},$$
$$[\beta, \alpha]O \quad \mapsto \quad T\{\mu\alpha.\,[\beta, \alpha]O\}.$$

Call-by-value equalities, written $=_v$ are shown in Figure 2, and call-by-name equalities, written $=_n$ are shown in Figure 3.

For the call-by-value calculus we need a notion of value, and notions of evaluation and statement contexts. Let V, W range over values, E range over evaluation contexts, and D range over statement contexts. A value is a variable, a pair of values, an injection of a value, a function, or a projection from a value. An evaluation context is a term with a hole, and a statement context is a statement with a hole, such that any term substituted into the hole will be the next to be evaluated. We write $\{-\}$ for the hole; the result of placing term M into the hole in an evaluation context E is written $E\{M\}$, similarly for statement contexts.

The rules are grouped as (β) rules, which reduce a deconstructor applied to a contructor; (η) rules, which introduce a constructor applied to a deconstructor; and some additional rules. In the call-by-value calculus, three rules are stated with statement contexts. It is easy to prove, using $(\eta\mu)$, that the rules also hold when the statement context D is replaced with an evaluation context E. The (name) rule introduces a name for the next term to be evaluated; it is similar to the rules (let.1) and (let.2) in the λ_c-calculus of Moggi (1988) and the various (let) rules in Selinger (2001). The (comp) rule is similar to the associativity rule in the the λ_c-calculus of Moggi (1988), and the (let) rule in Selinger (2001).

Values	$V, W ::= x \mid \langle V, W \rangle \mid \mu[\alpha, \beta].\,[\alpha]V \mid \mu[\alpha, \beta].\,[\beta]W \mid$	
	$\quad \lambda x.\,S \mid \lambda x.\,N \mid \text{fst } V \mid \text{fst } W$	
Evaluation context E	$::= \{-\} \mid \langle E, N \rangle \mid \langle V, E \rangle \mid \text{fst } E \mid \text{snd } E \mid E\,M \mid V\,E$	
Statement context D	$::= [\alpha]E \mid [\alpha, \beta]E \mid E\,M \mid V\,E$	

$(\beta\&)$	$\text{fst } \langle V, W \rangle$	$=_v V$
$(\beta\&)$	$\text{snd } \langle V, W \rangle$	$=_v W$
$(\beta\vee)$	$[\alpha, \beta]\mu[\alpha', \beta'].\,S$	$=_v S\{\alpha/\alpha', \beta/\beta'\}$
$(\beta\neg)$	$(\lambda x.\,S)\,V$	$=_v S\{V/x\}$
$(\beta\supset)$	$(\lambda x.\,N)\,V$	$=_v N\{V/x\}$
$(\beta\mu)$	$[\alpha]\mu\alpha'.\,S$	$=_v S\{\alpha'/\alpha\}$
$(\eta\&)$	$V : A \,\&\, B$	$=_v \langle \text{fst } V, \text{snd } V \rangle$
$(\eta\vee)$	$M : A \vee B$	$=_v \mu[\alpha, \beta].\,[\alpha, \beta]M$
$(\eta\neg)$	$V : \neg A$	$=_v \lambda x.\,V\,x$
$(\eta\supset)$	$V : A \supset B$	$=_v \lambda x.\,V\,x$
$(\eta\mu)$	M	$=_v \mu\alpha.\,[\alpha]M$
(name)	$D\{M\}$	$=_v (\lambda x.\,D\{x\})\,M$
(comp)	$D\{(\lambda x.\,N)\,M\}$	$=_v (\lambda x.\,D\{N\})\,M$
(ς)	$D\{\mu\alpha.\,S\}$	$=_v S\{D\{-\}/[\alpha]\{-\}\}$

Fig. 2. Equations of the call-by-value $\lambda\mu$-calculus.

$(\beta\&)$	fst $\langle M, N \rangle$	$=_n M$
$(\beta\&)$	snd $\langle M, N \rangle$	$=_n N$
$(\beta\vee)$	$[\alpha, \beta]\mu[\alpha', \beta'].\, S$	$=_n S\{\alpha/\alpha', \beta/\beta'\}$
$(\beta\neg)$	$(\lambda x.\, S)\, M$	$=_n S\{M/x\}$
$(\beta\supset)$	$(\lambda x.\, N)\, M$	$=_n N\{M/x\}$
$(\beta\mu)$	$[\alpha]\mu\alpha'.\, S$	$=_n S\{\alpha'/\alpha\}$
$(\eta\&)$	$M : A \,\&\, B$	$=_n \langle \text{fst } M, \text{snd } M \rangle$
$(\eta\vee)$	$M : A \vee B$	$=_n \mu[\alpha, \beta].\, [\alpha, \beta]M$
$(\eta\neg)$	$M : \neg A$	$=_n \lambda x.\, M\,x$
$(\eta\supset)$	$M : A \supset B$	$=_n \lambda x.\, M\,x$
$(\eta\mu)$	M	$=_n \mu\alpha.\, [\alpha]M$
$(\varsigma\vee)$	$[\alpha, \beta](\mu\gamma.\, S)$	$=_n S\{[\alpha, \beta]\{-\}/[\gamma]\{-\}\}$
$(\varsigma\&)$	fst $(\mu\gamma.\, S)$	$=_n \mu\alpha.\, S\{[\alpha]\text{fst }\{-\}/[\gamma]\{-\}\}$
$(\varsigma\&)$	snd $(\mu\gamma.\, S)$	$=_n \mu\beta.\, S\{[\beta]\text{snd }\{-\}/[\gamma]\{-\}\}$
$(\varsigma\neg)$	$(\mu\gamma.\, S)\, M$	$=_n S\{\{-\}\, M/[\gamma]\{-\}\}$
$(\varsigma\supset)$	$(\mu\gamma.\, S)\, M$	$=_n \mu\beta.\, S\{[\beta]\{-\}\, M/[\gamma]\{-\}\}$

Fig. 3. Equations of the call-by-name $\lambda\mu$-calculus.

(let.1)	$O\,M$	$=_v$ let $z = O$ in $z\,N$
(let.2)	$V\,M$	$=_v$ let $x = M$ in $V\,x$
(comp)	let $y = ($let $x = M$ in $N)$ in O	$=_v$ let $x = M$ in let $y = N$ in O

The (ς) rules of the call-by-value and call-by-name calculi are similar to the (ς) rules of Selinger (2001).

As noted, implication can be defined in terms of the other connectives, but different definitions must be used for call-by-value or call-by-name.

Proposition 1. *Under call-by-value, implication may be defined by*

$$A \supset B \equiv \neg(A \,\&\, \neg B)$$
$$\lambda x.\, N \equiv \lambda z.\, (\lambda x.\, (\text{snd}\,z)\, N)\, (\text{fst}\,z)$$
$$O\,M \equiv \mu\beta.\, O\,\langle M, \lambda y.\, [\beta]y \rangle$$

validating $(\beta\supset)$, $(\eta\supset)$, *and the other equations for functions, and where the translation of a function abstraction is a value.*

Proposition 2. *Under call-by-name, implication may be defined by*

$$A \supset B \equiv \neg A \vee B$$
$$\lambda x.\, N \equiv \mu[\gamma, \beta].\, [\gamma]\lambda x.\, [\beta]N$$
$$O\,M \equiv \mu\beta.\, (\mu\gamma.\, [\gamma, \beta]O)\, M$$

validating $(\beta\supset)$, $(\eta\supset)$, *and* $(\varsigma\supset)$.

3   The Dual Calculus

Figure 4 presents the syntax and inference rules of the dual calculus. Types, variables, and covariables are the same as the $\lambda\mu$-calculus.

Let M, N range over terms, which yield values. A term is either a variable x; a pair $\langle M, N \rangle$; an injection on the left or right of a sum $\langle M \rangle$inl or $\langle N \rangle$inr; a complement of a coterm $[K]$not; a function abstraction $\lambda x.N$, with x bound in N; or a covariable abstraction $(S).\alpha$, with α bound in S.

Let K, L range over coterms, which consume values. A coterm is either a covariable α; a projection from the left or right of a product fst$[K]$ or snd$[L]$; a case $[K, L]$; a complement of a term not$\langle M \rangle$; a function application $M @ L$; or a variable abstraction $x.(S)$, with x bound in S.

Finally, let S, T range over statements. A statement is a cut of a term against a coterm, $M \bullet K$. Note that angle brackets always surround terms, square brackets always surround coterms, and round brackets always surround statements. Curly brackets are used for substitution and holes in contexts.

The type rules given here differ slightly from Wadler (2003), in that they are presented in syntax-directed form; so thinning, exchange, and contraction are built into the form of the rules rather than given as separate structural rules.

A cut of a term against a variable abstraction, or a cut of a covariable abstraction against a coterm, corresponds to substitution. This suggests the following reduction rules.

$$(\beta\text{L}) \qquad M \bullet x.(S) = S\{M/x\}$$
$$(\beta\text{R}) \qquad (S).\alpha \bullet K = S\{K/\alpha\}$$

Here substitution in a statement of a term for a variable is written $S\{M/x\}$, and substitution in a statement of a coterm for a covariable is written $S\{K/\alpha\}$.

A critical pair occurs when a covariable abstraction is cut against a variable abstraction.

$$(S).\alpha \bullet x.(T)$$

Sometimes such reductions are confluent.

$$(x \bullet \alpha).\alpha \bullet y.(y \bullet \beta)$$

$$x \bullet y.(y \bullet \beta) \qquad\qquad (x \bullet \alpha).\alpha \bullet \beta$$

$$x \bullet \beta$$

But sometimes they are not.

$$(x \bullet \alpha).\beta \bullet y.(z \bullet \gamma)$$

$$x \bullet \alpha \qquad\qquad z \bullet \gamma$$

To restore confluence we must limit reductions, and this is achieved by adopting call-by-value or call-by-name.

Type	$A, B ::= X \mid A \& B \mid A \vee B \mid \neg A \mid A \supset B$

Term	$M, N ::= x \mid \langle M, N \rangle \mid \langle M \rangle \mathrm{inl} \mid \langle N \rangle \mathrm{inr} \mid [K]\mathrm{not} \mid \lambda x.N \mid (S).\alpha$
Coterm	$K, L ::= \alpha \mid [K, L] \mid \mathrm{fst}[K] \mid \mathrm{snd}[L] \mid \mathrm{not}\langle M \rangle \mid M @ L \mid x.(S)$
Statement	$S, T ::= M \bullet K$

Antecedent	Γ	$::= x_1 : A_1, \ldots, x_m : A_m$
Succedent	Θ	$::= \beta_1 : B_1, \ldots, \beta_n : B_n$

Right sequent $\quad \Gamma \to \Theta \mid M : A$
Left sequent $\quad\;\; K : A \mid \Gamma \to \Theta$
Center sequent $\quad \Gamma \mid S \longmapsto \Theta$

$$\frac{}{x : A, \Gamma \to \Theta \mid x : A} \; \text{IdR} \qquad \frac{}{\alpha : A \mid \Gamma \to \Theta, \alpha : A} \; \text{IdL}$$

$$\frac{\Gamma \to \Theta \mid M : A \qquad \Gamma \to \Theta \mid N : B}{\Gamma \to \Theta \mid \langle M, N \rangle : A \& B} \; \text{\&R}$$

$$\frac{K : A \mid \Gamma \to \Theta}{\mathrm{fst}[K] : A \& B \mid \Gamma \to \Theta} \; \text{\&L} \qquad \frac{L : B \mid \Gamma \to \Theta}{\mathrm{snd}[L] : A \& B \mid \Gamma \to \Theta} \; \text{\&L}$$

$$\frac{\Gamma \to \Theta \mid M : A}{\Gamma \to \Theta \mid \langle M \rangle \mathrm{inl} : A \vee B} \; \text{∨R} \qquad \frac{\Gamma \to \Theta \mid N : B}{\Gamma \to \Theta \mid \langle N \rangle \mathrm{inr} : A \vee B} \; \text{∨R}$$

$$\frac{K : A \mid \Gamma \to \Theta \qquad L : B \mid \Gamma \to \Theta}{[K, L] : A \vee B \mid \Gamma \to \Theta} \; \text{∨L}$$

$$\frac{K : A \mid \Gamma \to \Theta}{\Gamma \to \Theta \mid [K]\mathrm{not} : \neg A} \; \text{¬R} \qquad \frac{\Gamma \to \Theta \mid M : A}{\mathrm{not}\langle M \rangle : \neg A \mid \Gamma \to \Theta} \; \text{¬L}$$

$$\frac{x : A, \Gamma \to \Theta \mid N : B}{\Gamma \to \Theta \mid \lambda x.N : A \supset B} \; \text{⊃R} \qquad \frac{\Gamma \to \Theta \mid M : A \qquad L : B \mid \Gamma \to \Theta}{M @ L : A \supset B \mid \Gamma \to \Theta} \; \text{⊃L}$$

$$\frac{\Gamma \mid S \longmapsto \Theta, \alpha : A}{\Gamma \to \Theta \mid (S).\alpha : A} \; \text{RI} \qquad \frac{x : A, \Gamma \mid S \longmapsto \Theta}{x.(S) : A \mid \Gamma \to \Theta} \; \text{LI}$$

$$\frac{\Gamma \to \Theta \mid M : A \qquad K : A \mid \Gamma \to \Theta}{\Gamma \mid M \bullet K \longmapsto \Theta} \; \text{Cut}$$

Fig. 4. Syntax and types of the dual calculus.

Value	$V, W ::= x \mid \langle V, W \rangle \mid \langle V \rangle\mathrm{inl} \mid \langle W \rangle\mathrm{inr} \mid [K]\mathrm{not} \mid$
	$\lambda x.N \mid (V \bullet \mathrm{fst}[\alpha]).\alpha \mid (V \bullet \mathrm{snd}[\beta]).\beta$
Evaluation context E	$::= \{-\} \mid \langle E, N \rangle \mid \langle V, E \rangle \mid \langle E \rangle\mathrm{inl} \mid \langle E \rangle\mathrm{inr}$

$(\beta\&)$	$\langle V, W \rangle \bullet \mathrm{fst}[K]$	$=^v V \bullet K$
$(\beta\&)$	$\langle V, W \rangle \bullet \mathrm{snd}[L]$	$=^v W \bullet L$
$(\beta\vee)$	$\langle V \rangle\mathrm{inl} \bullet [K, L]$	$=^v V \bullet K$
$(\beta\vee)$	$\langle W \rangle\mathrm{inr} \bullet [K, L]$	$=^v W \bullet L$
$(\beta\neg)$	$[K]\mathrm{not} \bullet \mathrm{not}\langle M \rangle$	$=^v M \bullet K$
$(\beta\supset)$	$\lambda x.N \bullet M @ L$	$=^v M \bullet x.(N \bullet L)$
$(\beta\mathrm{L})$	$V \bullet x.(S)$	$=^v S\{V/x\}$
$(\beta\mathrm{R})$	$(S).\alpha \bullet K$	$=^v S\{K/\alpha\}$
$(\eta\&)$	$V : A \& B$	$=^v \langle (V \bullet \mathrm{fst}[\alpha]).\alpha, (V \bullet \mathrm{snd}[\beta]).\beta \rangle$
$(\eta\vee)$	$K : A \vee B$	$=^v [x.(\langle x \rangle\mathrm{inl} \bullet K), y.(\langle y \rangle\mathrm{inr} \bullet K)]$
$(\eta\neg)$	$V : \neg A$	$=^v [x.(V \bullet \mathrm{not}\langle x \rangle)]\mathrm{not}$
$(\eta\supset)$	$V : A \supset B$	$=^v \lambda x.((V \bullet x @ \beta).\beta)$
$(\eta\mathrm{L})$	K	$=^v x.(x \bullet K)$
$(\eta\mathrm{R})$	M	$=^v (M \bullet \alpha).\alpha$
(name)	$E\{M\} \bullet K$	$=^v M \bullet x.(x \bullet K)$

Fig. 5. Equations of the call-by-value dual calculus.

Call-by-value only reduces a cut of a value against a variable abstraction, but reduces a cut of a covariable abstraction against any coterm.

$$(\beta\mathrm{L}) \quad V \bullet x.(S) =^v S\{V/x\}$$
$$(\beta\mathrm{R}) \quad (S).\alpha \bullet K =^v S\{K/\alpha\}$$

Value V replaces term M in rule $(\beta\mathrm{L})$. A value cannot be a covariable abstraction, so this avoids the critical pair.

Call-by-name only reduces a cut of a covariable abstraction against a covalue, but reduces a cut of any coterm against a variable abstraction.

$$(\beta\mathrm{L}) \quad V \bullet x.(S) =^v S\{V/x\}$$
$$(\beta\mathrm{R}) \quad (S).\alpha \bullet K =^v S\{K/\alpha\}$$

Covalue P replaces coterm K in rule $(\beta\mathrm{R})$. A covalue cannot be a variable abstraction, so this avoids the critical pair.

In λ-calculus, the move from call-by-value to call-by-name generalizes values to terms. In dual calculus, the move from call-by-value to call-by-name generalizes values to terms but restricts coterms to covalues, clarifying the duality.

Call-by-value equalities, written $=^v$, are shown in Figure 5 and call-by-name equalities, written $=^n$, are shown in Figure 6.

Let V, W range over values. A value is a variable, a pair of values, a left or right injection of a value, any complement, any function, or a projection from a value.

Covalue	$P, Q ::= \alpha \mid [Q, P] \mid \text{snd}[P] \mid \text{fst}[Q] \mid \text{not}\langle M \rangle \mid$
	$M @ Q \mid x.(\langle x \rangle \text{inr} \bullet P) \mid y.(\langle y \rangle \text{inl} \bullet P)$
Coevaluation context	$F \quad ::= \{-\} \mid [L, F] \mid [F, P] \mid \text{snd}[F] \mid \text{fst}[F]$

$(\beta\vee)$	$\langle M \rangle \text{inr} \bullet [Q, P]$	$=^n M \bullet P$
$(\beta\vee)$	$\langle N \rangle \text{inl} \bullet [Q, P]$	$=^n N \bullet Q$
$(\beta\&)$	$\langle N, M \rangle \bullet \text{snd}[P]$	$=^n M \bullet P$
$(\beta\&)$	$\langle N, M \rangle \bullet \text{fst}[Q]$	$=^n N \bullet Q$
$(\beta\neg)$	$[K]\text{not} \bullet \text{not}\langle M \rangle$	$=^n M \bullet K$
$(\beta\supset)$	$\lambda x.N \bullet M @ L$	$=^n M \bullet x.(N \bullet L)$
(βR)	$(S).\alpha \bullet P$	$=^n S\{P/\alpha\}$
(βL)	$M \bullet x.(S)$	$=^n S\{M/x\}$
$(\eta\vee)$	$P : A \vee B$	$=^n [y.(\langle y \rangle \text{inl} \bullet P), x.(\langle x \rangle \text{inr} \bullet P)]$
$(\eta\&)$	$M : A \& B$	$=^n \langle (M \bullet \text{fst}[\beta]).\beta, (M \bullet \text{snd}[\alpha]).\alpha \rangle$
$(\eta\neg)$	$P : \neg A$	$=^n \text{not}\langle ([\alpha]\text{not} \bullet P).\alpha \rangle$
$(\eta\supset)$	$M : A \supset B$	$=^n \lambda x.((M \bullet x @ \beta).\beta)$
(ηR)	M	$=^n (M \bullet \alpha).\alpha$
(ηL)	K	$=^n x.(x \bullet K)$
(name)	$M \bullet F\{K\}$	$=^n (M \bullet \alpha).\alpha \bullet K$

Fig. 6. Equations of the call-by-name dual calculus.

Let P, Q range over covalues. A covalue is a covariable, a first or second projection of a covalue, a case over a pair of covalues, any complement, an application context where the second component is a covalue, or a left or right injection into a covalue. Covalues correspond to a strict context, one that is guaranteed to demand the value passed to it.

As before, the reduction rules are grouped into (β), $(\eta,)$ (name) and (ς) rules. The (name) rules correspond to the (ς) rules of Wadler (2003).

As before, implication can be defined in terms of the other connectives, but different definitions must be used for call-by-value or call-by-name. Under call-by-value function abstractions must translate to values, while under call-by-name function applications must translate to covalues, and this is what forces different definitions for the two reduction disciplines.

Proposition 3. *Under call-by-value, implication may be defined by*

$$A \supset B \equiv \neg(A \& \neg B)$$
$$\lambda x.N \equiv [z.(z \bullet \text{fst}[x.(z \bullet \text{snd}[\text{not}\langle N \rangle])])]\text{not}$$
$$M @ L \equiv \text{not}\langle\langle M, [L]\text{not} \rangle\rangle.$$

validating $(\beta\supset)$, $(\eta\supset)$, and the other equations for functions, and where the translation of a function abstraction is a value.

Proposition 4. *Under call-by-name, implication can be defined by*

$$A \supset B \equiv \neg A \lor B$$
$$\lambda x.N \equiv (\langle [x.(\langle N \rangle inr \bullet \gamma)] not \rangle inl \bullet \gamma).\gamma$$
$$M @ L \equiv [not\langle M \rangle, L].$$

validating $(\beta \supset)$, $(\eta \supset)$, *and the other equations for functions, and where the translation of a function application is a covalue.*

4   Translations

We now consider the translation from the $\lambda\mu$-calculus to the dual calculus and its inverse translation. The results of this section apply to all types, including implication.

Definition 2. *The translation from the $\lambda\mu$-calculus into the dual calculus is given in Figure 7. It consists of two operations,*

$$M^*, \quad S^*.$$

- *If M is a $\lambda\mu$ term of type A, then M^* is a dual term of type A.*

$$\frac{\Gamma \to \Theta \mid M : A}{\Gamma \to \Theta \mid M^* : A}$$

- *If S is a $\lambda\mu$ statement, then S^* is a dual statement.*

$$\frac{\Gamma \mid S \vdash\!\!\!\to \Theta}{\Gamma \mid S^* \vdash\!\!\!\to \Theta}$$

Definition 3. *The translation from the dual calculus into the $\lambda\mu$-calculus is given in Figure 8. It consists of three operations,*

$$M_*, \quad K_*\{O\}, \quad S_*.$$

$$
\begin{aligned}
(x)^* &\equiv x \\
(\langle M, N \rangle)^* &\equiv \langle M^*, N^* \rangle \\
(\text{fst } O)^* &\equiv (O^* \bullet \text{fst}[\alpha]).\alpha &\qquad ([\alpha, \beta]O)^* &\equiv O^* \bullet [\alpha, \beta] \\
(\text{snd } O)^* &\equiv (O^* \bullet \text{snd}[\beta]).\beta &\qquad (O\,M)^* &\equiv O^* \bullet not\langle M^* \rangle \\
(\mu[\alpha, \beta].\,S)^* &\equiv (\langle\langle\langle(S^*).\beta\rangle inr \bullet \gamma).\alpha\rangle inl \bullet \gamma).\gamma &\qquad ([\alpha]M)^* &\equiv M^* \bullet \alpha \\
(\lambda x.\,S)^* &\equiv [x.(S^*)]not \\
(\lambda x.\,N)^* &\equiv \lambda x.N^* \\
(O\,M)^* &\equiv (O^* \bullet M^* @ \beta).\beta \\
(\mu\alpha.\,S)^* &\equiv (S^*).\alpha
\end{aligned}
$$

Fig. 7. Translation from $\lambda\mu$-calculus to dual calculus.

$$
\begin{aligned}
(x)_* &\equiv x \\
(\langle M, N \rangle)_* &\equiv \langle M_*, N_* \rangle \\
(\langle M \rangle \mathrm{inl})_* &\equiv \mu[\alpha, \beta].\,[\alpha]M_* \\
(\langle N \rangle \mathrm{inr})_* &\equiv \mu[\alpha, \beta].\,[\beta]N_* \\
([K]\mathrm{not})_* &\equiv \lambda x.\,K_*\{x\} \\
(\lambda x.N)_* &\equiv \lambda x.\,N_* \\
((S).\alpha)_* &\equiv \mu\alpha.\,S_*
\end{aligned}
\qquad
\begin{aligned}
(\alpha)_*\{O\} &\equiv [\alpha]O \\
([K,L])_*\{O\} &\equiv L_*\{\mu\beta.\,K_*\{\mu\alpha.\,[\alpha,\beta]O\}\} \\
(\mathrm{fst}[K])_*\{O\} &\equiv K_*\{\mathrm{fst}\,O\} \\
(\mathrm{snd}[L])_*\{O\} &\equiv L_*\{\mathrm{snd}\,O\} \\
(\mathrm{not}\langle M \rangle)_*\{O\} &\equiv O\,M_* \\
(M @ L)_*\{O\} &\equiv L_*\{O\,M_*\} \\
(x.(S))_*\{O\} &\equiv (\lambda x.\,S_*)\,O
\end{aligned}
$$

$$
(M \bullet K)_* \equiv K_*\{M_*\}
$$

Fig. 8. Translation from dual calculus to $\lambda\mu$-calculus.

- If M is a dual term of type A, then M_* is a $\lambda\mu$ term of type A.

$$
\frac{\Gamma \to \Theta \mid M : A}{\Gamma \to \Theta \mid M_* : A}
$$

- If K is a dual coterm of type A, and O is a $\lambda\mu$ term of type A, then $K_*\{O\}$ is a $\lambda\mu$ statement.

$$
\frac{K : A \mid \Gamma \to \Theta \qquad \Gamma \to \Theta \mid O : A}{\Gamma \mid K_*\{O\} \mapsto \Theta}
$$

- If S is a dual statement, then S_* is a $\lambda\mu$ statement.

$$
\frac{\Gamma \mid S \mapsto \Theta}{\Gamma \mid S_* \mapsto \Theta}
$$

In general, these translations do not preserve reductions, but they do preserve equalities. We now present the detailed results to show that the translations form an equational correspondence between the call-by-value $\lambda\mu$ calculus and the call-by-value dual calculus.

Proposition 5. *($\lambda\mu$ reloaded) Translating from the $\lambda\mu$-calculus into the dual calculus and then 'reloading' into the $\lambda\mu$-calculus gives a term equal to the original under call-by-value,*

$$
\begin{aligned}
(M^*)_* &=_v M \\
(S^*)_* &=_v S
\end{aligned}
$$

with M a term and S a statement in $\lambda\mu$.

The two lines are shown by case analysis on terms and statements of $\lambda\mu$.

Proposition 6. *(dual reloaded) Translating from the dual calculus into the $\lambda\mu$-calculus and then 'reloading' into the dual calculus gives a term equal to the original under call-by-value,*

$$
\begin{aligned}
(M_*)^* &=^v M \\
(K_*\{O\})^* &=^v O^* \bullet K \\
(S_*)^* &=^v S,
\end{aligned}
$$

with M a term, K a coterm, and S a statement in dual, and O a term in $\lambda\mu$.

The three lines are shown by case analysis on terms, coterms, and statements of dual.

Proposition 7. *($\lambda\mu$ to dual preserves equalities) Translating from the $\lambda\mu$-calculus into the dual calculus preserves call-by-value equalities,*

$$M =_v N \quad implies \quad M^* =^v N^*$$
$$S =_v T \quad implies \quad S^* =^v T^*,$$

with M, N terms and S, T statements in $\lambda\mu$.

The two lines are shown by case analysis on the equations of $\lambda\mu$ that apply to terms and statements respectively.

Proposition 8. *(dual to $\lambda\mu$ preserves equalities) Translating from the dual calculus into the $\lambda\mu$-calculus preserves call-by-value equalities,*

$$M =^v N \quad implies \quad M_* =_v N_*$$
$$K =^v L \quad implies \quad K_*\{O\} =_v L_*\{O\}$$
$$S =^v T \quad implies \quad S_* =_v T_*,$$

with M, N terms, K, L coterms, and S, T statements in dual, and O a term in $\lambda\mu$.

The three lines are shown by case analysis on the equations of dual that apply to terms, coterms, and statements respectively.

The four propositions above also hold for call-by-name. The restatement is easy, simply replace $=_v$ and $=^v$ everywhere by $=_n$ and $=^n$. However, while the structure of the proofs is essentially the same, the new sets of reductions require that one repeat the proofs entirely, since there is no simple, systematic relation between the call-by-value and call-by-name reductions of $\lambda\mu$.

However, there is a systematic relation between the call-by-value and call-by-name reductions of dual. We next consider how to characterize and exploit this regularity.

5   Duality

We now review the results about duality for the dual from Wadler (2003), and use these to derive similar results concering duality for the $\lambda\mu$-calculus. Since duality is not defined for implication, before applying the results of this section any occurrences of implication must be translated away, using the results given previously.

The dual calculus is designed to exploit duality. Variables are dual to covariables, pairs are duals to sums, complement is self dual, term abstraction is dual to coterm abstraction, and cut is self dual. This can be captured in a translation from the dual calculus into itself. The translation is involutive – that is, it is its own inverse – and it carries call-by-value equations into call-by-name equations, and vice versa. So it is an equational correspondence.

We assume a one-to-one correspondence between variables and covariables. Each variable x corresponds to a covariable $\bar{x}$, and each covariable α corresponds to a variable $\bar{\alpha}$, such that $\bar{\bar{x}} \equiv x$ and $\bar{\bar{\alpha}} \equiv \alpha$. For instance, we might take $\bar{x} \equiv \alpha, \bar{y} \equiv \beta, \bar{z} \equiv \gamma$, and hence $\bar{\alpha} = x, \bar{\beta} = y, \bar{\gamma} = z$.

$$
\begin{aligned}
(X)^\circ &\equiv X \\
(A \mathbin{\&} B)^\circ &\equiv B^\circ \vee A^\circ \\
(A \vee B)^\circ &\equiv B^\circ \mathbin{\&} A^\circ \\
(\neg A)^\circ &\equiv \neg A^\circ
\end{aligned}
$$

$$
\begin{aligned}
(x)^\circ &\equiv \bar{x} & (\alpha)^\circ &\equiv \bar{\alpha} \\
(\langle M, N \rangle)^\circ &\equiv [N^\circ, M^\circ] & ([K, L])^\circ &\equiv \langle L^\circ, K^\circ \rangle \\
(\langle M \rangle \mathrm{inl})^\circ &\equiv \mathrm{snd}[M^\circ] & (\mathrm{fst}[K])^\circ &\equiv \langle K^\circ \rangle \mathrm{inr} \\
(\langle N \rangle \mathrm{inr})^\circ &\equiv \mathrm{fst}[N^\circ] & (\mathrm{snd}[L])^\circ &\equiv \langle L^\circ \rangle \mathrm{inl} \\
([K]\mathrm{not})^\circ &\equiv \mathrm{not}\langle K^\circ \rangle & (\mathrm{not}\langle M \rangle)^\circ &\equiv [M^\circ]\mathrm{not} \\
((S).\alpha)^\circ &\equiv \bar{\alpha}.(S^\circ) & (x.(S))^\circ &\equiv (S^\circ).\bar{x}
\end{aligned}
$$

$$
(M \bullet K)^\circ \equiv K^\circ \bullet M^\circ
$$

Fig. 9. Duality for the dual calculus.

$$
\begin{aligned}
(x)_\circ \{O'\} &\equiv [\bar{x}]O' \\
(\langle M, N \rangle)_\circ \{O'\} &\equiv N_\circ \{\mu\beta.\, M_\circ \{\mu\alpha.\, [\beta, \alpha]O'\}\} \\
(\mathrm{fst}\, O)_\circ \{O'\} &\equiv (\lambda x.\, O_\circ \{\mu[\beta, \alpha].\, [\alpha]x\})\, O' & ([\alpha]M)_\circ &\equiv M_\circ \{\bar{\alpha}\} \\
(\mathrm{snd}\, O)_\circ \{O'\} &\equiv (\lambda y.\, O_\circ \{\mu[\beta, \alpha].\, [\beta]y\})\, O' & ([\alpha, \beta]O)_\circ &\equiv O_\circ \{\langle \bar{\alpha}, \bar{\beta} \rangle\} \\
(\mu[\alpha, \beta].\, S)_\circ \{O'\} &\equiv (\lambda z.\, (\lambda\bar{\beta}.\, (\lambda\bar{\alpha}.\, S_\circ)\, (\mathrm{fst}\, z))\, (\mathrm{snd}\, z))\, O' & (O\, M)_\circ &\equiv O_\circ \{\lambda x.\, M_\circ \{x\}\} \\
(\lambda x.\, S)_\circ \{O'\} &\equiv O'\, (\mu\bar{x}.\, S_\circ) \\
(\mu\alpha.\, S)_\circ \{O'\} &\equiv (\lambda\bar{\alpha}.\, S_\circ)\, O'
\end{aligned}
$$

Fig. 10. Duality for the $\lambda\mu$-calculus.

Definition 4. *The duality translation from the dual calculus to itself is given in Figure 9. It consists of operations on types, terms, coterms, and statements,*

$$
A^\circ, \quad M^\circ, \quad K^\circ, \quad S^\circ.
$$

- *If A is a type, then A° is the dual type. This extends to environments and coenvironments. If $\Gamma \equiv x_1 : A_1, \ldots, x_m : A_m$, its dual is $\Gamma^\circ \equiv \bar{x}_m : A_m^\circ, \ldots, \bar{x}_1 : A_1^\circ$, and similarly for coenvironments.*
- *If M is a dual term of type A, then M° is a dual coterm of type A.*

$$
\frac{\Gamma \to \Theta \mid M : A}{M^\circ : A \mid \Theta^\circ \to \Gamma^\circ}
$$

- *If K is a dual coterm of type A, and K° is a dual term of type A.*

$$
\frac{K : A \mid \Gamma \to \Theta}{\Theta^\circ \to \Gamma^\circ \mid K^\circ : A}
$$

- *If S is a dual statement, then S° is a dual statement.*

$$
\frac{\Gamma \mid S \vdash\mathrel{\mkern-5mu}\to \Theta}{\Theta^\circ \mid S^\circ \vdash\mathrel{\mkern-5mu}\to \Gamma^\circ}
$$

It is immediate from the definition that duality is its own inverse.

Proposition 9. *(Involution) Duality is an involution up to identity,*

$$(A^\circ)^\circ \equiv A$$
$$(M^\circ)^\circ \equiv M$$
$$(K^\circ)^\circ \equiv K$$
$$(S^\circ)^\circ \equiv S,$$

with A a type of dual, M a term of dual, K a coterm of dual, and S a statement of dual.

For the dual calculus, call-by-value is dual to call-by-name. This is easily confirmed by inspection of the reduction rules; indeed, it was the principle guiding their design.

Proposition 10. *(Call-by-value is dual to call-by-name) Duality takes call-by-value equalities into call-by-name equalities, and vice versa.*

$$M =^v N \quad \text{iff} \quad M^\circ =^n N^\circ$$
$$K =^v L \quad \text{iff} \quad K^\circ =^n L^\circ$$
$$S =^v T \quad \text{iff} \quad S^\circ =^n T^\circ,$$

with M, N terms, K, L coterms, and S, T statements of dual.

An immediate consequence is that duality is an equational correspondence between the call-by-value dual calculus and the call-by-name dual calculus.

We now extend the above results from the dual calculus to the $\lambda\mu$-calculus.

Using the translations of the previous section, we can compute duals for the $\lambda\mu$-calculus by translating from $\lambda\mu$ to dual, taking the dual, and then 'reloading' back into $\lambda\mu$.

Definition 5. *The duality transformation from the $\lambda\mu$ calculus to itself is given in Figure 10. It consists of operations on types, terms, and statements, defined as follows,*

$$A_\circ \quad \equiv A^\circ$$
$$M_\circ\{O\} \equiv ((M^*)^\circ)_*\{O\}$$
$$S_\circ \quad \equiv ((S^*)^\circ)_*$$

 - *If A is a type, then $A_\circ \equiv A^\circ$ is the dual type.*
 - *If M is a $\lambda\mu$ term of type A and O is a $\lambda\mu$ term of type $A_\circ$, then $M_\circ\{O\}$ is a $\lambda\mu$ statement.*

$$\frac{\Gamma \to \Theta \mid M : A \qquad \Theta_\circ \to \Gamma_\circ \mid O : A_\circ}{\Theta_\circ \mid M_\circ\{O\} \vdash\!\!\!\vdash \Gamma_\circ}$$

 - *If S is a $\lambda\mu$ statement, then $S_\circ$ is a $\lambda\mu$ statement.*

$$\frac{\Gamma \mid S \vdash\!\!\!\vdash \Theta}{\Theta_\circ \mid S_\circ \vdash\!\!\!\vdash \Gamma_\circ}$$

In effect, we compose three equational correspondences (from $\lambda\mu$ to dual, from dual to itself, and from dual to $\lambda\mu$) to yield a new equational correspondence (from $\lambda\mu$ to itself).

It follows immediately that duality on $\lambda\mu$ takes call-by-value into call-by-name.

Proposition 11. *(Call-by-value is dual to call-by-name, reloaded) Duality takes call-by-value equalities on $\lambda\mu$ into call-by-name equalities, and vice versa.*

$$M =_v N \quad iff \quad M_\circ\{O\} =_n N_\circ\{O\}$$
$$S =_v T \quad iff \quad S_\circ =_n T_\circ$$

Here M, N are terms and S, T are statements of $\lambda\mu$.

The proof is easy. For the first line, we have

$$
\begin{aligned}
&\quad M =_v N \\
iff \quad &M^* =^v N^* \\
iff \quad &(M^*)^\circ =^n (N^*)^\circ \\
iff \quad &((M^*)^\circ)_*\{O\} =_n ((N^*)^\circ)_*\{O\} \\
iff \quad &M_\circ\{O\} =_n N_\circ\{O\}
\end{aligned}
$$

The second line is similar.

Proposition 12. *(Involution, reloaded) Duality on $\lambda\mu$ is an involution up to equality,*

$$
\begin{aligned}
\mu\alpha.\,(M_\circ\{\bar{\alpha}\})_\circ &=_v M \\
(M_\circ\{O\})_\circ &=_v O_\circ\{M\} \\
(S_\circ)_\circ &=_v S,
\end{aligned}
$$

with M, O terms and S a statement of $\lambda\mu$.

This follows from Propositions 5, 6, and 9 We will prove the lines in inverse order. The third line is easy,

$$
\begin{aligned}
&\quad (S_\circ)_\circ \\
\equiv\ &(((((S^*)^\circ)_*)^*)^\circ)_* \\
=_v\ &(((S^*)^\circ)^\circ)_* \\
\equiv\ &(S^*)_* \\
=_v\ &S.
\end{aligned}
$$

The second line is only slightly harder,

$$
\begin{aligned}
&\quad (M_\circ\{O\})_\circ \\
\equiv\ &(((((M^*)^\circ)_*\{O\})^*)^\circ)_* \\
=_v\ &((O^* \bullet (M^*)^\circ)^\circ)_* \\
\equiv\ &(M^* \bullet (O^*)^\circ)_* \\
\equiv\ &((O^*)^\circ)_*\{(M^*)_*\} \\
=_v\ &O_\circ\{M\}.
\end{aligned}
$$

The first line follows from the second,

$$
\begin{aligned}
&\quad \mu\alpha.\,(M_\circ\{\bar{\alpha}\})_\circ \\
=_v\ &\mu\alpha.\,\bar{\alpha}_\circ\{M\} \\
\equiv\ &\mu\alpha.\,[\alpha]M \\
=_v\ &M
\end{aligned}
$$

Since all of the results of the preceding section hold with $=_v$ replaced by $=_n$, the same holds for the above. However, unlike the preceding section, we don't need to redo any complex case analyses; the additional results follow immediately from the work done previously.

Selinger (2001) gives a duality for $\lambda\mu$ that takes call-by-value into call-by-name, but it is *not* involutive. There are two distinct translations to take call-by-value into call-by-name and call-by-name into call-by-value. Futher, one translation followed by the other does not preserve types up to identity, only up to isomorphism. However, closer inspection shows that the two translations are identical on all components except function types, and agree with the duality translation on $\lambda\mu$ given here. The key difference is that here we have replaced implication by negation in $\lambda\mu$, yielding a cleaner version of duality. Sometimes less is more!

Acknowledgements

The work on dual calculus owes a great deal to Pierre-Louis Curien, Hugo Herbelin, and Peter Selinger, for their original work and for subsequent discussions. For other comments related to this work, I thank Sebastian Carlier, Olivier Danvy, Roberto Di Cosmo, Marcelo Fiore, Ken-etsu Fujita, Vladimir Gapeyev, Tim Griffin, Daisuke Kimura, Paul Levy, Sam Lindley, William Lovas, Robert McGrail, Rex Page, Bernard Reuss, Ken Shan, Thomas Streicher, Josef Svenningsson, and Steve Zdancewic

References

1. Zena Ariola and Hugo Herbelin (2003). Minimal classical logic and control operators. In *30'th International Colloquium on Automata, Languages and Programming*, Eindhoven, The Netherlands.
2. Vincent Balat, Roberto Di Cosmo and Marcelo Fiore (2004). Extensional normalisation and type-directed partial evaluation for typed lambda-calculus with sums. *Principles of Programming Languages*, 2004.
3. F. Barbanera and S. Berardi (1996). A symmetric lambda calculus for classical program extraction. *Information and Computation*, 125(2):103–117.
4. P. Bernays (1936). Review of "Some Properties of Conversion" by Alonzo Church and J. B. Rosser. *Journal of Symbolic Logic*, 1:74–75.
5. Alonzo Church (1932). A set of postulates for the foundation of logic. *Annals of Mathematics*, II.33:346–366.
6. Alonzo Church (1940). A formulation of the simple theory of types. *Journal of Symbolic Logic*, 5:56–68.
7. Tristan Crolard (2004). A formulae-as-types interpretation of Subtractive Logic. *Journal of Logic and Computation*, 14(4):529–570.
8. P.-L. Curien and H. Herbelin (2000). The duality of computation. In *5'th International Conference on Functional Programming*, pages 233–243, ACM, September 2000.

9. V. Danos, J-B. Joinet and H. Schellinx (1995). LKQ and LKT: Sequent calculi for second order logic based upon linear decomposition of classical implication. In *Advances in Linear Logic*, J-Y. Girard, Y Lafont and L. Regnier editors, London Mathematical Society Lecture Note Series 222, Cambridge University Press, pp. 211-224.

10. Andrzej Filinski (1989). Declarative continuations and categorical duality. Master's thesis, University of Copenhagen, Copenhagen, Denmark, August 1989. (DIKU Report 89/11.)

11. Ken-etsu Fujita (2003). A Sound and Complete CPS-Translation for $\lambda\mu$-Calculus. In Martin Hofmann (editor), *Typed Lambda Calculi and Applications*, Valencia, Spain, 10-12 June 2003, Springer-Verlag, LNCS 2701.

12. Gerhard Gentzen (1935). Investigations into Logical Deduction. *Mathematische Zeitschrift* 39:176–210,405–431. Reprinted in M. E. Szabo, editor, *The Collected Papers of Gerhard Gentzen*, North-Holland, 1969.

13. Timothy Griffin (1990). A formulae-as-types notion of control. In *17'th Symposium on Principles of Programming Languages*, San Francisco, CA, ACM, January 1990.

14. M. Hofmann and T. Streicher (1997). Continuation models are universal for $\lambda\mu$-calculus. In *Proceedings of the Twelfth Annual IEEE Symposium on Logic in Computer Science*, pages 387–397.

15. Austin Melton, Bernd S. W. Schröder, and George E. Strecker (1994). Lagois connections, a counterpart to Galois connections. *Theoretical Computer Science*, 136(1):79–107, 1994.

16. Eugenio Moggi (1988). Computational lambda-calculus and monads. Technical Report ECS-LFCS-88-66, Edinburgh University, Department of Computer Science.

17. C.-H. L. Ong (1996). A semantic view of classical proofs: Type-theoretic, categorical, and denotational characterizations. In *Proceedings of the Eleventh Annual IEEE Symposium on Logic in Computer Science*, pages 230–241.

18. C.-H. L. Ong and C. A. Stewart (1997). A Curry-Howard foundation for functional computation with control. In *Proceedings of the Symposium on Principles of Programming Languages*, pages 215–227.

19. M. Parigot (1992). $\lambda\mu$-calculus: an algorithmic interpretation of classical natural deduction. In LPAR 1992, pages 190–201, Springer-Verlag, LNCS 624.

20. G. D. Plotkin (1975). Call-by-name, call-by-value and the λ-calculus. In *Theoretical Computer Science*, 1:125–159.

21. David Pym and Eike Ritter (2001). On the Semantics of Classical Disjunction. *Journal of Pure and Applied Algebra* 159:315–338.

22. Amr Sabry and Matthias Felleisen (1993). Reasoning about programs in continuation-passing style. *Lisp and Symbolic Computation*, 6(3/4):289–360.

23. Amr Sabry and Philip Wadler (1997). A reflection on call-by-value. In *ACM Transactions on Programming Languages and Systems*, 19(6):916-941.

24. Peter Selinger (1998). Control categories and duality: an axiomatic approach to the semantics of functional control. Talk presented at *Mathematical Foundations of Programming Semantics*, London, May 1998.

25. Peter Selinger (2001). Control categories and duality: on the categorical semantics of the lambda-mu calculus. In *Mathematical Structures in Computer Science*, 11:207–260.

26. Thomas Streicher and Bernard Reuss (1998). Classical logic, continuation semantics and abstract machines. *Journal of Functional Programming*, 8(6):543–572, November 1998.

27. Philip Wadler (2003). Call-by-name is dual to call-by-value. In *International Conference on Functional Programming*, Uppsala, Sweden, 25–29 August 2003.

λμ-Calculus and Duality:
Call-by-Name and Call-by-Value

Jérôme Rocheteau

[1] ESTAS – INRETS, 20 rue Élisée Reclus – BP 317,
F-59666 Villeneuve d'Ascq Cedex
jerome.rocheteau@inrets.fr
[2] Preuves, Programmes et Systèmes, CNRS – Université de Paris VII,
UMR 7126 – Case 7014, 175 rue du Chevaleret – 75013 Paris – France

Abstract. Under the extension of Curry-Howard's correspondence to classical logic, Gentzen's NK and LK systems can be seen as syntax-directed systems of simple types respectively for Parigot's $\lambda\mu$-calculus and Curien-Herbelin's $\bar{\lambda}\mu\tilde{\mu}$-calculus. We aim at showing their computational equivalence. We define translations between these calculi. We prove simulation theorems for an undirected evaluation as well as for call-by-name and call-by-value evaluations.

1 Introduction

Key systems for classical logic in proof theory are Gentzen's NK and LK. The logical equivalence between the latter was proved in [Gentzen, 1934]. We deal with the extension of Curry-Howard's correspondence between proofs and programs through the systems of simple types for the $\lambda\mu$ and $\bar{\lambda}\mu\tilde{\mu}$-calculi. This extension concerns some other calculi. It is initially Felleisen's λc-calculus. Its type system is the intuitionistic natural deduction with the double negation axiom. Griffin proposed this axiom as the type for the c-operator in [Griffin, 1990]. However, we focus on calculi that correspond closer to Gentzen's systems. The $\lambda\mu$-calculus was defined for NK in [Parigot, 1992]. The $\bar{\lambda}\mu\tilde{\mu}$-calculus was designed for LK in [Curien and Herbelin, 2000]. In the general case, these two calculi are not deterministic. There exists critical pairs. The $\bar{\lambda}\mu\tilde{\mu}$-calculus admits two deterministic projections depending on choosing one of the two possible symmetric orientations of a critical pair. They correspond to the call-by-name/call-by-value duality.

We aim at proving the computational equivalence between $\lambda\mu$ and $\bar{\lambda}\mu\tilde{\mu}$-calculi. A major step was reached with the proof of the simulation of the $\lambda\mu$-calculus by the $\bar{\lambda}\mu\tilde{\mu}$-calculus in [Curien and Herbelin, 2000]. It holds both for call-by-name and call-by-value evaluations. We present the call-by-name/call-by-value projections of the $\lambda\mu$-calculus in the same way as for the $\bar{\lambda}\mu\tilde{\mu}$ in [Curien and Herbelin, 2000] . It consists of choosing one of the two possible orientations of a critical pair. We prove that the $\lambda\mu$-calculus simulates backwards the $\bar{\lambda}\mu\tilde{\mu}$-calculus in such a way that we obtain easily the same result for the call-by-name,

J. Giesl (Ed.): RTA 2005, LNCS 3467, pp. 204–218, 2005.

for the call-by-value and for the simple type case. The $\bar{\lambda}\mu\tilde{\mu}$-calculus is composed of three syntactic categories: terms, contexts (or environments) and commands. The $\lambda\mu$-calculus is basically composed of terms and commands. We add contexts to the $\lambda\mu$-calculus. It eases mappings between the $\lambda\mu$ and $\bar{\lambda}\mu\tilde{\mu}$-calculi. We extend the translation from the $\lambda\mu$-calculus to the $\bar{\lambda}\mu\tilde{\mu}$-calculus defined in [Curien and Herbelin, 2000] over the $\lambda\mu$-contexts. We define backwards a translation from the $\bar{\lambda}\mu\tilde{\mu}$-calculus to the $\lambda\mu$-calculus.

In section 2 we present the $\lambda\mu$-calculus. In section 3 we present the $\bar{\lambda}\mu\tilde{\mu}$-calculus. In section 4 we define translations between these two calculi. In section 5 we prove simulation theorems that hold for call-by-name and call-by-value.

2 $\lambda\mu$-Calculus

We follow the definition given in [Parigot, 1992]. Firstly, we present the grammar of terms and commands. Secondly, we present the system of simple types. Thirdly, we present generic reductions and their call-by-name and call-by-value projections. Fourthly, we extend both the grammar and the type system to the contexts.

Basically, the $\lambda\mu$-calculus is composed of terms and commands. They are defined by mutual induction:

$$t \ ::= \ x \mid \lambda x.t \mid (t)\,t \mid \mu\alpha.c \qquad c \ ::= \ [\alpha]\,t$$

Symbols x range over λ-variables, symbols α range over μ-variables. We note $x \in t$ or $\alpha \in t$ the fact that x or α has a free occurrence in t. Symbols λ and μ are binders. Two terms are equal modulo α-equivalence.

The system of simple types for the $\lambda\mu$-calculus is based on two kinds of sequents. The first $\Gamma \vdash t : T \mid \Delta$ concerns the terms and the second $c : (\Gamma \vdash \Delta)$ concerns the commands in which T is a simple type obtained by the grammar $T \ ::= \ X \mid T \to T$, Γ is a finite domain application from λ-variables to simple types and Δ is a finite domain application from μ-variables to simple types. Γ, Γ' denotes the union of the applications Γ and Γ'. System rules are:

$$\frac{}{x{:}A \vdash x{:}A \mid} \qquad \frac{\Gamma \vdash t{:}B \mid \Delta}{\Gamma\backslash\{x{:}A\} \vdash \lambda x.t{:}A{\to}B \mid \Delta} \qquad \frac{\Gamma \vdash u{:}A{\to}B \mid \Delta \qquad \Gamma' \vdash v{:}A \mid \Delta'}{\Gamma,\Gamma' \vdash (u)\,v{:}B \mid \Delta,\Delta'} (*)$$

$$\frac{\Gamma \vdash t{:}A \mid \Delta}{[\alpha]\,t{:}(\Gamma \vdash \Delta,\alpha{:}A)} (*) \qquad \frac{c{:}(\Gamma \vdash \Delta)}{\Gamma \vdash \mu\alpha.c{:}A \mid \Delta\backslash\{\alpha{:}A\}}$$

The restriction $(*)$ requires that Γ and Γ' match each other on the intersection of their domains. This holds for Δ and Δ' too.

The category of contexts is introduced in order to ease comparisons with the homonymous category of the $\bar{\lambda}\mu\tilde{\mu}$-calculus. $\lambda\mu$-contexts are defined by mutual induction with the terms:

$$e \ ::= \ \alpha \mid \beta(t) \mid t \cdot e$$

We can see contexts as commands with a hole to fill. The first construction α expects a term t in order to provide the command $[\alpha]\,t$. The second $\beta(t)$ expects a term u in order to provide the command $[\beta]\,(t)\,u$. The last $t \cdot h$ puts the term t on a stack and expects another term to fill the hole.

Definition 1. *Let t a term and e a context. The command $e\{t\}$ is defined by induction on e:*

$$e\{t\} = \begin{cases} [\alpha]\,t & \text{if } e = \alpha \\ [\beta]\,(u)\,t & \text{if } e = \beta(u) \\ h\{(t)\,u\} & \text{if } e = u \cdot h \end{cases}$$

The type system is extended to another kind of sequents $\Gamma \mid e : T \vdash \Delta$. The typing rules give the context e the type of the term t that fills the hole of e:

$$\frac{}{\mid \alpha{:}A \vdash \alpha{:}A} \qquad \frac{\Gamma \vdash t{:}(A \to B) \mid \Delta}{\Gamma \mid \beta(t){:}A \vdash \Delta, \beta{:}B} \qquad \frac{\Gamma \vdash t{:}A \mid \Delta \qquad \Gamma' \mid e{:}B \vdash \Delta'}{\Gamma,\Gamma' \mid t \cdot e{:}(A \to B) \vdash \Delta, \Delta'}$$

A sequent calculus like cut-rule can then be derived in this system as a term against context application.

Lemma 1. *The rule* $\dfrac{\Gamma \vdash t{:}A \mid \Delta \qquad \Gamma' \mid e{:}A \vdash \Delta'}{e\{t\}{:}(\Gamma,\Gamma' \vdash \Delta, \Delta')}$ *holds in $\lambda\mu$.*

Proof. By induction on e.

- if $e = \alpha$ then $e\{t\} = [\alpha]\,t$ and $\dfrac{\Gamma \vdash t{:}A \mid \Delta}{[\alpha]\,t{:}(\Gamma \vdash \Delta, \alpha{:}A)}$
- if $e = \beta(u)$ then $e\{t\} = [\beta]\,(u)\,t$ and

$$\frac{\dfrac{\Gamma \vdash u{:}(A \to B) \mid \Delta \qquad \Gamma' \vdash t{:}A \mid \Delta'}{\Gamma, \Gamma' \vdash (u)\,t{:}B \mid \Delta, \Delta'}}{[\beta]\,(u)\,t{:}(\Gamma,\Gamma' \vdash \Delta, \Delta', \beta{:}B)}$$

- if $e = u \cdot h$ then $e\{t\} = h\{(t)\,u\}$ and

$$\frac{\dfrac{\Gamma \vdash t{:}(A \to B) \mid \Delta \qquad \Gamma' \vdash u{:}A \mid \Delta'}{\Gamma, \Gamma' \vdash (t)\,u{:}B \mid \Delta, \Delta'} \qquad \Gamma'' \mid h{:}B \vdash \Delta''}{h\{(t)\,u\}{:}(\Gamma, \Gamma', \Gamma'' \vdash \Delta, \Delta', \Delta'')} \text{ ind. hyp.}$$

Definition 2. *Let t a term, e a context and α a μ-variable, The term $t[\alpha \leftarrow e]$ – the substitution of α by e in t – is defined by induction on t:*

$$t[\alpha \leftarrow e] = \begin{cases} x & \text{if } t = x \\ \lambda x.u[\alpha \leftarrow e] & \text{if } t = \lambda x.u \\ (u[\alpha \leftarrow e])\,v[\alpha \leftarrow e] & \text{if } t = (u)\,v \\ \mu\beta.c[\alpha \leftarrow e] & \text{if } t = \mu\beta.c \end{cases}$$

$$c[\alpha \leftarrow e] = \begin{cases} e\{t[\alpha \leftarrow e]\} & \text{if } c = [\alpha]\,t \\ [\beta]\,t[\alpha \leftarrow e] & \text{if } c = [\beta]\,t \end{cases}$$

The computation notion is based on reductions. We remind one-step reduction rules:

$$(\lambda x.u)\, t \to_\beta u[x \leftarrow t] \qquad (\mu\alpha.c)\, t \to_\mu \mu\alpha.c[\alpha \leftarrow t \cdot \alpha]$$
$$\mu\delta.[\delta]\, t \to_\theta t \text{ (if } \delta \notin t) \qquad (t)\, \mu\alpha.c \to_{\mu'} \mu\alpha.c[\alpha \leftarrow \alpha(t)]$$
$$[\beta]\, \mu\alpha.c \to_\rho c[\alpha \leftarrow \beta]$$

The reduction $\overset{*}{\to}_\gamma$ stands for the reflexive and transitive closure of $\to_\gamma$ and the reduction $\overset{*}{\to}$ stands for the union of $\overset{*}{\to}_\gamma$ for $\gamma \in \{\beta, \mu, \mu', \rho, \theta\}$.

Some of these reductions are linear. Both of the ρ and θ-reductions are linear because they correspond to the identity in NK. The β-reduction from the term $(\lambda x.t)\, y$ is linear because it consists of replacing a variable by another variable inside a term. It corresponds to a normalisation against an axiom rule in NK. The β-reduction from the term $(\lambda x.t)\, u$ where x has a single free occurrence in t is linear too because it consists either of substituting a single variable occurrence by any term. It corresponds either to a normalisation without a proof-tree branch duplication.

Reductions $\leadsto_\gamma$, $\overset{*}{\leadsto}_\gamma$ and $\overset{*}{\leadsto}$ have the same meanings as in the general case. The relation $\approx$ is defined as the reflexive, transitive and symmetric closure of $\overset{*}{\leadsto}$.

There exists a critical pair for computation determinism. Applicative terms $(\lambda x.t)\, \mu\beta.d$ and $(\mu\alpha.c)\, \mu\beta.d$ can be β or μ'-rewritten in the first case and μ or μ'-rewritten in the second case. We can see the call-by-name and call-by-value disciplines as restrictions of the generic reductions.

The call-by-name evaluation consists of allowing every reduction but the μ'-rule. The β-reduction holds in the first case and the μ-reduction in the second. Formally the call-by-name reduction is $\overset{*}{\to}_n = \overset{*}{\to} \setminus \overset{*}{\to}_{\mu'}$.

The call-by-value evaluation consists of prohibiting β and μ-reductions in which the argument is a μ-abstraction. Formally we define a subset of terms called values by this grammar: $v ::= x \mid \lambda x.t$. β_v and μ_v-reductions are defined instead of generic β and μ ones:

$$(\lambda x.u)\, v \to_{\beta_v} u[x \leftarrow v] \qquad (\mu\alpha.c)\, v \to_{\mu_v} \mu\alpha.c[\alpha \leftarrow v \cdot \alpha]$$

The call-by-value reduction $\overset{*}{\to}_v$ is the union of $\overset{*}{\to}_\gamma$ for $\gamma \in \{\beta_v, \mu_v, \mu', \rho, \theta\}$. Critical pairs are then μ'-rewritten.

There is another way to define call-by-value into the λμ-calculus. The solution is detailed in [Ong and Stewart, 1997]. It consists of restricting the μ'-rule to values instead of the μ:

$$(v)\, \mu\alpha.c \to_{\mu'_v} \mu\alpha.c[\alpha \leftarrow \alpha(v)]$$

Formally $\overset{*}{\to}_v$ becomes the union of $\overset{*}{\to}_\gamma$ for $\gamma \in \{\beta_v, \mu, \mu'_v, \rho, \theta\}$. In fact terms $(\lambda x.t)\, \mu\alpha.c$ and $(\mu\alpha.c)\, \mu\alpha'.c'$ are respectively μ' and μ-reduced because $\mu\alpha.c$ is not a value in these cases. However, we follow Curien-Herbelin's call-by-value definition.

We finish this section by a lemma. It is useful for the section 5 simulation theorems. Any command of the form $e\{\mu\alpha.c\}$ is a redex. However, some can not be reduced in call-by-name nor in call-by-value.

Lemma 2. $e\{\mu\alpha.c\} \stackrel{*}{\rightarrow} c[\alpha \leftarrow e]$

Proof. By induction on e.

- if $e = \beta$ then $e\{\mu\alpha.c\} = [\beta]\,\mu\alpha.c \rightsquigarrow_\rho c[\alpha \leftarrow \beta]$
- if $e = \beta(t)$ then

$$
\begin{aligned}
e\{\mu\alpha.c\} &= [\beta]\,(t)\,\mu\alpha.c \\
&\rightarrow_{\mu'} [\beta]\,\mu\alpha.c[\alpha \leftarrow \alpha(t)] \\
&\rightsquigarrow_\rho c[\alpha \leftarrow \alpha(t)][\alpha \leftarrow \beta] \\
&= c[\alpha \leftarrow \beta(t)]
\end{aligned}
$$

- if $e = t \cdot h$ then

$$
\begin{aligned}
e\{\mu\alpha.c\} &= h\{(\mu\alpha.c)\,t\} \\
&\rightarrow_\mu h\{\mu\alpha.c[\alpha \leftarrow t \cdot \alpha]\} \\
&\stackrel{*}{\rightarrow} c[\alpha \leftarrow t \cdot \alpha][\alpha \leftarrow h] \\
&= c[\alpha \leftarrow t \cdot h]
\end{aligned}
$$

This lemma does not hold in call-by-name for the $\beta(t)$ induction case because no μ'-rule is allowed. It holds in call-by-value if t is a value for the $h \cdot t$ induction case.

3 $\bar{\lambda}\mu\tilde{\mu}$-Calculus

The $\bar{\lambda}\mu\tilde{\mu}$-calculus has the same relation against LK as the $\lambda\mu$-calculus against NK. Reductions of $\bar{\lambda}\mu\tilde{\mu}$-calculus correspond to the cut elimination steps in LK as well as the $\lambda\mu$-reductions correspond to the NK-normalisation. We follow the definition given in [Curien and Herbelin, 2000]. Firstly, we present the grammar of the $\bar{\lambda}\mu\tilde{\mu}$-calculus. Secondly, we present the simple type system. Thirdly, we present generic reductions and their call-by-name and call-by-value projections.

The $\bar{\lambda}\mu\tilde{\mu}$-calculus is basically composed of terms, commands and contexts[1]. They are defined by mutual induction:

$$
t ::= x \mid \lambda x.t \mid \mu\alpha.c \qquad c ::= \langle t \mid e \rangle \qquad e ::= \alpha \mid t \cdot e \mid \tilde{\mu}x.c
$$

As in the $\lambda\mu$, symbols x range over λ-variables, symbols α range over μ-variables and symbols λ, μ and $\tilde{\mu}$ are binders. Terms are equal modulo α-equivalence.

This calculus symmetry looks like LK's left/right symmetry. It is confirmed by its system of simple types. This system shares types with the $\lambda\mu$-calculus. It shares the same kinds of sequents too. Its rules are:

$$
\frac{}{x{:}A \vdash x{:}A \mid} \qquad \frac{}{\mid \alpha{:}A \vdash \alpha{:}A} \qquad \frac{\Gamma \vdash t{:}B \mid \Delta}{\Gamma\backslash\{x{:}A\} \vdash \lambda x.t{:}A{\rightarrow}B \mid \Delta} \qquad \frac{\Gamma \vdash t{:}A \mid \Delta \qquad \Gamma' \mid e{:}B \vdash \Delta'}{\Gamma,\Gamma' \mid t{\cdot}e{:}A{\rightarrow}B \vdash \Delta,\Delta'} \; (*)
$$

$$
\frac{c{:}(\Gamma \vdash \Delta)}{\Gamma \vdash \mu\alpha.c{:}A \mid \Delta\backslash\{\alpha{:}A\}} \qquad \frac{c{:}(\Gamma \vdash \Delta)}{\Gamma\backslash\{x{:}A\} \mid \tilde{\mu}x.c{:}A \vdash \Delta} \qquad \frac{\Gamma \vdash t{:}A \mid \Delta \qquad \Gamma' \mid e{:}A \vdash \Delta'}{\langle t \mid e \rangle{:}(\Gamma,\Gamma' \vdash \Delta,\Delta')} \; (*)
$$

The restriction $(*)$ is the same as that of $\lambda\mu$.

---

[1] In [Dougherty et al., 2004] these are referred to respectively callers, callees and capsules. We kept the terminology in [Curien and Herbelin, 2000] that sounds closer to its meaning: terms are programs, contexts are environments and commands represent "a closed system containing both the program and its environment".

We present one-step reduction rules. Substitutions inside the $\bar{\lambda}\mu\tilde{\mu}$-calculus are supposed to be known. Each rule concerns a command but the θ-rule:

$$\langle \lambda x.u \mid t \cdot e \rangle \rightarrow_\beta \langle t \mid \tilde{\mu}x.\langle u \mid e \rangle \rangle \qquad \langle \mu\alpha.c \mid e \rangle \rightarrow_\mu c[\alpha \leftarrow e]$$
$$\mu\delta.\langle t \mid \delta \rangle \rightarrow_\theta t \ (\delta \notin t) \qquad \langle t \mid \tilde{\mu}x.c \rangle \rightarrow_{\tilde{\mu}} c[x \leftarrow t]$$

μ and $\tilde{\mu}$-reductions are duals of each other. They correspond to the structural rules in LK. Reductions $\xrightarrow{*}_\gamma$ and $\rightsquigarrow_\gamma$ have the same meanings as in the $\lambda\mu$-calculus. The β-rule is a mere term modification without term duplication. It is therefore a linear reduction. The θ-reduction is linear too. There is no ρ-reduction. It is a μ-rule particular case in which $e = \beta$.

This system is not deterministic. There is a single critical pair $\langle \mu\alpha.c \mid \tilde{\mu}x.d \rangle$. It can be both μ or $\tilde{\mu}$-rewritten so that Church-Rosser's property does not hold. In fact $\langle \mu\alpha.\langle x \mid y \cdot \alpha \rangle \mid \tilde{\mu}x.\langle z \mid x \cdot \beta \rangle \rangle$ is μ-rewritten as $\langle x \mid y \cdot \tilde{\mu}x.\langle z \mid x \cdot \beta \rangle \rangle$ and is $\tilde{\mu}$-rewritten as $\langle z \mid \mu\alpha.\langle x \mid y \cdot \alpha \rangle \cdot \beta \rangle$. These are two different normal forms.

Call-by-name and call-by-value disciplines still deal with this problem. They both consist of restricting the context construction. The first new grammar is called $\bar{\lambda}\mu\tilde{\mu}_T$ and the second is called $\bar{\lambda}\mu\tilde{\mu}_Q$.

The call-by-name evaluation consists of restricting the μ-rule to a subset of contexts that are called stacks. $\bar{\lambda}\mu\tilde{\mu}_T$-grammar is:

$$t ::= x \mid \lambda x.t \mid \mu\alpha.c \quad c ::= \langle t \mid e \rangle \quad s ::= \alpha \mid t \cdot s \quad e ::= s \mid \tilde{\mu}x.c$$

The μ_n-rule is restricted to the stacks:

$$\langle \mu\alpha.c \mid s \rangle \rightarrow_{\mu_n} c[\alpha \leftarrow s]$$

Call-by-name reduction $\xrightarrow{*}_n$ is the union of $\xrightarrow{*}_\gamma$ for $\gamma \in \{\beta, \mu_n, \tilde{\mu}, \theta\}$. The critical pair can then only be $\tilde{\mu}$-rewritten. This reduction was proved confluent and stable in the $\bar{\lambda}\mu\tilde{\mu}_T$-calculus in [Curien and Herbelin, 2000].

The call-by-value oriented grammar consists of allowing the $t \cdot e$ context construction only for values. $\bar{\lambda}\mu\tilde{\mu}_Q$-grammar is:

$$t ::= x \mid \lambda x.t \mid \mu\alpha.c \quad v ::= x \mid \lambda x.t \quad c ::= \langle t \mid e \rangle \quad e ::= \alpha \mid v \cdot e \mid \tilde{\mu}x.c$$

The $\tilde{\mu}_v$-rule is restricted to values:

$$\langle v \mid \tilde{\mu}x.c \rangle \rightarrow_{\tilde{\mu}_v} c[x \leftarrow v]$$

Call-by-value reduction $\xrightarrow{*}_v$ is the union of $\xrightarrow{*}_\gamma$ for $\gamma \in \{\beta, \mu, \tilde{\mu}_v, \theta\}$. The command $\langle \mu\alpha.c \mid \mu\alpha'.c' \rangle$ can then only be μ-rewritten. This reduction was proved confluent and stable in the $\bar{\lambda}\mu\tilde{\mu}_Q$-calculus in [Curien and Herbelin, 2000].

The β'-rule contracts as shortcut for both a linear β-rule and a $\tilde{\mu}$-rule:

$$\langle \lambda x.u \mid t \cdot e \rangle \rightarrow_{\beta'} \langle u[x \leftarrow t] \mid e \rangle$$

This β'-rule is obviously compatible with the call-by-name evaluation. It is also compatible with the call-by-value because t is a value by definition of $\bar{\lambda}\mu\tilde{\mu}_Q$.

4 Translations Between $\lambda\mu$ and $\bar{\lambda}\mu\tilde{\mu}$-Calculi

We define a translation $(\ )^{\dagger}$ from $\lambda\mu$ to $\bar{\lambda}\mu\tilde{\mu}$. It extends that of Curien-Herbelin to the $\lambda\mu$-contexts. We define backwards a translation $(\ )^{\circ}$ from $\bar{\lambda}\mu\tilde{\mu}$ to $\lambda\mu$. We prove properties about their compatibilities with the simple type system and about their compositions.

Definition 3. *Application* $(\ )^{\dagger}$ *maps any $\lambda\mu$-term t, command c and context e respectively to a $\bar{\lambda}\mu\tilde{\mu}$-term, command and context.* $(\ )^{\dagger}$ *is defined by induction on t, c and e:*

$$
t^{\dagger} = \begin{cases}
x & \text{if } t = x \\
\lambda x.u^{\dagger} & \text{if } t = \lambda x.u \\
\mu\beta.\langle v^{\dagger} \mid \tilde{\mu}y.\langle u^{\dagger} \mid y \cdot \beta \rangle \rangle & \text{if } t = (u)\,v \quad (\star) \\
\mu\alpha.c^{\dagger} & \text{if } t = \mu\alpha.c
\end{cases}
$$

$$
c^{\dagger} = [\alpha]\,t^{\dagger} = \langle t^{\dagger} \mid \alpha \rangle
$$

$$
e^{\dagger} = \begin{cases}
\alpha & \text{if } e = \alpha \\
\tilde{\mu}y.\langle t^{\dagger} \mid y \cdot \beta \rangle & \text{if } e = \beta(t) \quad (\star\star) \\
t^{\dagger} \cdot h^{\dagger} & \text{if } e = t \cdot h
\end{cases}
$$

Condition $(\star)$ requires that variables y and β have no free occurrence in u neither in v. Condition $(\star\star)$ requires that $y \notin t$. A straightforward induction leads us to state that t and $t^{\dagger}$ have the same free variables set.

It seems more natural to translate $(u)\,v$ by $\mu\beta.\langle u^{\dagger} \mid v^{\dagger} \cdot \beta \rangle$. This shorter term corresponds in LK to the arrow elimination rule in NK too. But it would not be compatible with the call-by-value evaluation. For example, $(x)\,\mu\alpha.c$ would be translated as $\mu\beta.\langle x \mid \mu\alpha.c^{\dagger} \cdot \beta \rangle$ in this case. It can not be reduced by any rule in the $\bar{\lambda}\mu\tilde{\mu}$-calculus. However, $(x)\,\mu\alpha.c$ can be μ'-reduced in the $\lambda\mu$-calculus.

$(u)\,v$ should be translated as $\mu\beta.\langle u^{\dagger} \mid \tilde{\mu}y.\langle v^{\dagger} \mid \tilde{\mu}x.\langle y \mid x \cdot \beta \rangle \rangle \rangle$ with Ong and Stewart's call-by-value definition in [Ong and Stewart, 1997].

We show that translation $(\ )^{\dagger}$ is compatible with the type system. If a typing environment for a term t exists, it holds for $t^{\dagger}$.

Lemma 3. $\Gamma \vdash t : A \mid \Delta \implies \Gamma \vdash t^{\dagger} : A \mid \Delta$

Proof. By a straightforward induction on t. We show the less than obvious cases.

- if $t = (u)\,v$ then $t^{\dagger} = \mu\beta.\langle v^{\dagger} \mid \tilde{\mu}y.\langle u^{\dagger} \mid y \cdot \beta \rangle \rangle$ and

$$
\cfrac{\Gamma' \vdash v^{\dagger} : A \mid \Delta' \qquad \cfrac{\Gamma \vdash u^{\dagger} : A \to B \mid \Delta \qquad \cfrac{\cfrac{y : A \vdash y : A \mid \qquad \mid \beta : B \vdash \beta : B}{y : A \mid y \cdot \beta : (A \to B) \vdash \beta : B}}{\langle u^{\dagger} \mid y \cdot \beta \rangle : (\Gamma, y : A \vdash \Delta, \beta : B)}}{\Gamma \mid \tilde{\mu}y.\langle u^{\dagger} \mid y \cdot \beta \rangle \vdash \Delta, \beta : B}}{\cfrac{\langle v^{\dagger} \mid \tilde{\mu}y.\langle u^{\dagger} \mid y \cdot \beta \rangle \rangle : (\Gamma, \Gamma' \vdash \Delta, \Delta', \beta : B)}{\Gamma, \Gamma' \vdash \mu\beta.\langle v^{\dagger} \mid \tilde{\mu}y.\langle u^{\dagger} \mid y \cdot \beta \rangle \rangle : B \mid \Delta, \Delta'}}
$$

– if $e = \beta(t)$ then $e^{\dagger} = \tilde{\mu}y.\langle t^{\dagger} \,|\, y \cdot \beta\rangle$ and

$$
\frac{\dfrac{\dfrac{}{y:A\vdash y:A\ |}\quad \dfrac{}{|\ \beta:B\vdash\beta:B}}{\Gamma\vdash t^{\dagger}:(A{\rightarrow}B)\ |\ \Delta \qquad y:A\ |\ y\cdot\beta:(A{\rightarrow}B)\vdash\beta:B}}{\dfrac{\langle t^{\dagger}\,|\,y\cdot\beta\rangle:(\Gamma,y:A\vdash\Delta,\beta:B)}{\Gamma\ |\ \tilde{\mu}y.\langle t^{\dagger}\,|\,y\cdot\beta\rangle:A\vdash\Delta,\beta:B}}
$$

Definition 4. *Application $(\ )^{\circ}$ maps backwards any $\bar{\lambda}\mu\tilde{\mu}$-term t to a $\lambda\mu$-term. Definition 1 is used to translate any $\bar{\lambda}\mu\tilde{\mu}$-command c. Definition of the $\lambda\mu$-contexts is used to map the $\bar{\lambda}\mu\tilde{\mu}$-contexts e as well. $(\ )^{\circ}$ is built by induction on t, c and e:*

$$
t^{\circ} = \begin{cases} x & \text{if } t = x \\ \lambda x.u^{\circ} & \text{if } t = \lambda x.u \\ \mu\alpha.c^{\circ} & \text{if } t = \mu\alpha.c \end{cases}
$$

$$
c^{\circ} = \langle t\,|\,e\rangle^{\circ} = e^{\circ}\{t^{\circ}\}
$$

$$
e^{\circ} = \begin{cases} \alpha & \text{if } e = \alpha \\ t^{\circ} \cdot h^{\circ} & \text{if } e = t \cdot h \\ \beta(\lambda x.\mu\delta.c^{\circ}) & \text{if } e = \tilde{\mu}x.c \quad (*) \end{cases}
$$

Condition $(*)$ requires that $\delta \notin c$. t and t° have the same free variables set. Application $(\ )^{\circ}$ is compatible with the type system too.

Lemma 4. $\Gamma \vdash t : A \,|\, \Delta \implies \Gamma \vdash t^{\circ} : A \,|\, \Delta$

Proof. By a straightforward induction on t. We give two cases.

– if $c = \langle t\,|\,e\rangle$ then $c^{\circ} = e^{\circ}\{t^{\circ}\}$ and

$$
\frac{\Gamma\vdash t:A\ |\ \Delta \qquad \Gamma'\ |\ e:A\vdash\Delta'}{\langle t\,|\,e\rangle:(\Gamma,\Gamma'\vdash\Delta,\Delta')}\ ^{\circ} = \frac{\Gamma\vdash t^{\circ}:A\ |\ \Delta \qquad \Gamma'\ |\ e^{\circ}:A\vdash\Delta'}{e^{\circ}\{t^{\circ}\}:(\Gamma,\Gamma'\vdash\Delta,\Delta')}\ \text{lem. 1}
$$

– if $e = \tilde{\mu}x.c$ then $e^{\circ} = \beta(\lambda x.\mu\delta.c^{\circ})$ and

$$
\frac{c:(\Gamma\vdash\Delta)}{\Gamma\backslash\{x:A\}\ |\ \tilde{\mu}x.c:A\vdash\Delta}\ ^{\circ} = \frac{\dfrac{\dfrac{c^{\circ}:(\Gamma\vdash\Delta)}{\Gamma\vdash\mu\delta.c^{\circ}:B\ |\ \Delta}}{\Gamma\backslash\{x:A\}\vdash\lambda x.\mu\delta.c^{\circ}:(A{\rightarrow}B)}}{\Gamma\backslash\{x:A\}\ |\ \beta(\lambda x.\mu\delta.c^{\circ}):A\vdash\Delta,\beta:B}
$$

We focus on properties about the composition of $(\ )^{\dagger}$ and $(\ )^{\circ}$. We want to state that $t^{\dagger\circ} = t$ and that $t^{\circ\dagger} = t$ for any term. But it is not the case, these results hold modulo linear reductions.

Theorem 1. $t^{\dagger\circ} \overset{*}{\leadsto} t$

Proof. By a straightforward induction on t. Every cases is obtained successively by expanding definitions 3, 1, 4 and by applying the induction hypothesis. We give the case which uses linear reductions additionally.

– if $t = (u)\,v$ then

$$
\begin{aligned}
(u)\,v^{\dagger\circ} &= \mu\beta.\langle v^\dagger \mid \tilde{\mu}y.\langle u^\dagger \mid y\cdot\beta\rangle\rangle^\circ \\
&= \mu\beta.[\gamma]\,(\lambda y.\mu\delta.[\beta]\,(u^{\dagger\circ})\,y)\,v^{\dagger\circ} \\
&\overset{*}{\rightsquiggle} \mu\beta.[\gamma]\,(\lambda y.\mu\delta.[\beta]\,(u)\,y)\,v \\
&\rightsquiggle_\beta \mu\beta.[\gamma]\,\mu\delta.[\beta]\,(u)\,v \\
&\rightsquiggle_\rho \mu\beta.[\beta]\,(u)\,v \\
&\rightsquiggle_\theta (u)\,v
\end{aligned}
$$

We prove two lemmas before stating backwards that $(\ )^{\circ\dagger}$ is the identity modulo linear reductions. The first lemma is useful to prove the second.

Lemma 5. $\langle t_0 t_1 \ldots t_n{}^\dagger \mid e\rangle \overset{*}{\rightsquiggle} \langle t_0{}^\dagger \mid t_1{}^\dagger \cdot \ldots \cdot t_n{}^\dagger \cdot e\rangle$

Proof. By induction on n.

– if $n = 0$ then it is obvious
– if $n = m + 1$ then

$$
\begin{aligned}
\langle t_0 t_1 \ldots t_m t_{m+1}{}^\dagger \mid e\rangle &= \langle \mu\beta.\langle t_{m+1}{}^\dagger \mid \tilde{\mu}y.\langle t_0 t_1 \ldots t_m{}^\dagger \mid y\cdot\beta\rangle\rangle \mid e\rangle \\
&\rightsquiggle_\mu \langle t_{m+1}{}^\dagger \mid \tilde{\mu}y.\langle t_0 t_1 \ldots t_m{}^\dagger \mid y\cdot e\rangle\rangle \\
&\rightsquiggle_{\tilde{\mu}} \langle t_0 t_1 \ldots t_m{}^\dagger \mid t_{m+1}{}^\dagger \cdot e\rangle \\
&\overset{*}{\rightsquiggle} \langle t_0{}^\dagger \mid t_1{}^\dagger \cdot \ldots \cdot t_m{}^\dagger \cdot t_{m+1}{}^\dagger \cdot e\rangle
\end{aligned}
$$

The second lemma shows how to map a definition 1 command.

Lemma 6. $e\{t\}^\dagger \overset{*}{\rightsquiggle} \langle t^\dagger \mid e^\dagger\rangle$

Proof. By induction on e.

– if $e = \alpha$ then it is obvious by definitions 1 and 3
– if $e = \beta(u)$ then

$$
\begin{aligned}
\beta(u)\{t\}^\dagger &= [\beta]\,(u)\,t^\dagger \\
&= \langle \mu\gamma.\langle t^\dagger \mid \tilde{\mu}y.\langle u^\dagger \mid y\cdot\gamma\rangle\rangle \mid \beta\rangle \\
&\rightsquiggle_\mu \langle t^\dagger \mid \tilde{\mu}y.\langle u^\dagger \mid y\cdot\beta\rangle\rangle \\
&= \langle t^\dagger \mid \beta(u)^\dagger\rangle
\end{aligned}
$$

– if $e = u \cdot h$ then

$$
\begin{aligned}
u \cdot h\{t\}^\dagger &= h\{(t)\,u\}^\dagger \\
&\overset{*}{\rightsquiggle} \langle (t)\,u^\dagger \mid h^\dagger\rangle \\
&\overset{*}{\rightsquiggle} \langle t^\dagger \mid u^\dagger \cdot h^\dagger\rangle \\
&= \langle t^\dagger \mid u \cdot h^\dagger\rangle
\end{aligned}
$$

Theorem 2. $t^{\circ\dagger} \overset{*}{\rightsquiggle} t$

Proof. By induction on t. We apply definitions 3, 4 successively and the induction hypothesis. We give a typical case and another which needs either the previous lemma or linear reductions.

– if $c = \langle t \,|\, e\rangle$ then

$$\begin{aligned}
\langle t \,|\, e\rangle^{\circ\dagger} &= e^{\circ}\{t^{\circ}\}^{\dagger} \\
&\overset{*}{\rightsquigarrow} \langle t^{\circ\dagger} \,|\, e^{\circ\dagger}\rangle \\
&\overset{*}{\rightsquigarrow} \langle t \,|\, e\rangle
\end{aligned}$$

– if $e = \tilde{\mu}x.c$ then

$$\begin{aligned}
\tilde{\mu}x.c^{\circ\dagger} &= \beta(\lambda x.\mu\beta.c^{\circ})^{\dagger} \\
&= \tilde{\mu}y.\langle \lambda x.\mu\beta.c^{\circ\dagger} \,|\, y \cdot \beta\rangle \\
&\overset{*}{\rightsquigarrow} \tilde{\mu}y.\langle \lambda x.\mu\beta.c \,|\, y \cdot \beta\rangle \\
&\rightsquigarrow_{\beta} \tilde{\mu}y.\langle y \,|\, \tilde{\mu}x.\langle \mu\beta.c \,|\, \beta\rangle\rangle \\
&\rightsquigarrow_{\tilde{\mu}} \tilde{\mu}x.\langle \mu\beta.c \,|\, \beta\rangle \\
&\rightsquigarrow_{\mu} \tilde{\mu}x.c
\end{aligned}$$

5   Simulations Between $\lambda\mu$ and $\bar{\lambda}\mu\tilde{\mu}$-Calculi

We want to prove that the $\lambda\mu$-calculus simulates and is simulated backwards by the $\bar{\lambda}\mu\tilde{\mu}$-calculus. We focus on the undirected evaluation. Call-by-name and call-by-value are drawn from this.

We begin with the simulation of the $\lambda\mu$ by the $\bar{\lambda}\mu\tilde{\mu}$. The next four lemmas show results of a $\lambda\mu$-substitution after a β, μ, μ' and ρ-reduction. Each proof consists successively of

– expanding the $\lambda\mu$-substitution
– expanding the definition of $(\ )^{\dagger}$
– applying the induction hypothesis if necessary
– factorising the $\bar{\lambda}\mu\tilde{\mu}$-substitution
– factorising the definition of $(\ )^{\dagger}$

We give basic cases and those which use lemmas additionally for any proof.

Lemma 7. $t[x \leftarrow u]^{\dagger} = t^{\dagger}[x \leftarrow u^{\dagger}]$

Proof. By induction on t.

– if $t = x$ then $x[x \leftarrow u]^{\dagger} = u^{\dagger} = x^{\dagger}[x \leftarrow u^{\dagger}]$
– if $t = y$ then $y[x \leftarrow u]^{\dagger} = y = y^{\dagger}[x \leftarrow u^{\dagger}]$
– if $t = (v)\,w$ then

$$\begin{aligned}
(v)\,w[x \leftarrow u]^{\dagger} &= (v[x \leftarrow u])\,w[x \leftarrow u]^{\dagger} \\
&= \mu\beta.\langle w[x \leftarrow u]^{\dagger} \,|\, \tilde{\mu}y.\langle v[x \leftarrow u]^{\dagger} \,|\, y \cdot \beta\rangle\rangle \\
&= \mu\beta.\langle w^{\dagger}[x \leftarrow u^{\dagger}] \,|\, \tilde{\mu}y.\langle v^{\dagger}[x \leftarrow u^{\dagger}] \,|\, y \cdot \beta\rangle\rangle \\
&= \mu\beta.\langle w^{\dagger} \,|\, \tilde{\mu}y.\langle v^{\dagger} \,|\, y \cdot \beta\rangle\rangle[x \leftarrow u^{\dagger}] \\
&= (v)\,w^{\dagger}[x \leftarrow u^{\dagger}]
\end{aligned}$$

Lemma 8. $t[\alpha \leftarrow u \cdot \alpha]^{\dagger} \overset{*}{\rightsquigarrow} t^{\dagger}[\alpha \leftarrow u^{\dagger} \cdot \alpha]$

Proof. By induction on t.

– if $t = (a) b$ then

$$
\begin{aligned}
(a) b[\alpha \leftarrow u \cdot \alpha]^{\dagger} &= (a[\alpha \leftarrow u \cdot \alpha]) b[\alpha \leftarrow u \cdot \alpha]^{\dagger} \\
&= \mu\beta.\langle b[\alpha \leftarrow u \cdot \alpha]^{\dagger} \mid \tilde{\mu}y.\langle a[\alpha \leftarrow u \cdot \alpha]^{\dagger} \mid y \cdot \beta\rangle\rangle \\
&\overset{*}{\leadsto} \mu\beta.\langle b^{\dagger}[\alpha \leftarrow u^{\dagger} \cdot \alpha] \mid \tilde{\mu}y.\langle a^{\dagger}[\alpha \leftarrow u^{\dagger} \cdot \alpha] \mid y \cdot \beta\rangle\rangle \\
&= \mu\beta.\langle b^{\dagger} \mid \tilde{\mu}y.\langle a^{\dagger} \mid y \cdot \beta\rangle\rangle[\alpha \leftarrow u^{\dagger} \cdot \alpha] \\
&= (a) b^{\dagger}[\alpha \leftarrow u^{\dagger} \cdot \alpha]
\end{aligned}
$$

– if $c = [\alpha] w$ then

$$
\begin{aligned}
[\alpha] w[\alpha \leftarrow u \cdot \alpha]^{\dagger} &= [\alpha] w[\alpha \leftarrow u \cdot \alpha]^{\dagger} \\
&= u \cdot \alpha\{w[\alpha \leftarrow u \cdot \alpha]\}^{\dagger} \\
&= \langle w[\alpha \leftarrow u \cdot \alpha]^{\dagger} \mid u^{\dagger} \cdot \alpha\rangle \\
&\overset{*}{\leadsto} \langle w^{\dagger}[\alpha \leftarrow u^{\dagger} \cdot \alpha] \mid u^{\dagger} \cdot \alpha\rangle \\
&= \langle w^{\dagger} \mid \alpha\rangle[\alpha \leftarrow u^{\dagger} \cdot \alpha] \\
&= [\alpha] w^{\dagger}[\alpha \leftarrow u^{\dagger} \cdot \alpha]
\end{aligned}
$$

Lemma 9. $t[\alpha \leftarrow \alpha(u)]^{\dagger} \overset{*}{\leadsto} t^{\dagger}[\alpha \leftarrow \tilde{\mu}y.\langle y \mid y \cdot \alpha\rangle]u^{\dagger}$

Proof. By induction on t.

– if $t = (a) b$ then

$$
\begin{aligned}
(a) b[\alpha &\leftarrow \alpha(u)]^{\dagger} \\
&= (a[\alpha \leftarrow \alpha(u)]) b[\alpha \leftarrow \alpha(u)]^{\dagger} \\
&= \mu\beta.\langle b[\alpha \leftarrow \alpha(u)]^{\dagger} \mid \tilde{\mu}y.\langle a[\alpha \leftarrow \alpha(u)]^{\dagger} \mid y \cdot \beta\rangle\rangle \\
&\overset{*}{\leadsto} \mu\beta.\langle b^{\dagger}[\alpha \leftarrow \tilde{\mu}y.\langle u^{\dagger} \mid y \cdot \alpha\rangle] \mid \tilde{\mu}y.\langle a^{\dagger}[\alpha \leftarrow \tilde{\mu}y.\langle u^{\dagger} \mid y \cdot \alpha\rangle] \mid y \cdot \beta\rangle\rangle \\
&= \mu\beta.\langle b^{\dagger} \mid \tilde{\mu}y.\langle a^{\dagger} \mid y \cdot \beta\rangle\rangle[\alpha \leftarrow \tilde{\mu}y.\langle u^{\dagger} \mid y \cdot \alpha\rangle] \\
&= (a) b^{\dagger}[\alpha \leftarrow \tilde{\mu}y.\langle u^{\dagger} \mid y \cdot \alpha\rangle]
\end{aligned}
$$

– if $c = [\alpha] w$ then

$$
\begin{aligned}
[\alpha] w[\alpha \leftarrow \alpha(u)]^{\dagger} &= [\alpha] (w[\alpha \leftarrow \alpha(u)]) u^{\dagger} \\
&= \langle (w[\alpha \leftarrow \alpha(u)]) u^{\dagger} \mid \alpha\rangle \\
&\overset{*}{\leadsto} \langle w[\alpha \leftarrow \alpha(u)]^{\dagger} \mid u^{\dagger} \cdot \alpha\rangle \\
&\overset{*}{\leadsto} \langle w^{\dagger}[\alpha \leftarrow \tilde{\mu}y.\langle u^{\dagger} \mid y \cdot \alpha\rangle] \mid u^{\dagger} \cdot \alpha\rangle \\
&= \langle w^{\dagger} \mid \alpha\rangle[\alpha \leftarrow \tilde{\mu}y.\langle u^{\dagger} \mid y \cdot \alpha\rangle] \\
&= [\alpha] w^{\dagger}[\alpha \leftarrow \tilde{\mu}y.\langle u^{\dagger} \mid y \cdot \alpha\rangle]
\end{aligned}
$$

Lemma 10. $t[\alpha \leftarrow \beta]^{\dagger} = t^{\dagger}[\alpha \leftarrow \beta]$

Proof. By induction on t.

– if $c = [\alpha] u$ then

$$
\begin{aligned}
[\alpha] u[\alpha \leftarrow \beta]^{\dagger} &= [\beta] u[\alpha \leftarrow \beta]^{\dagger} \\
&= \langle u[\alpha \leftarrow \beta]^{\dagger} \mid \beta\rangle \\
&= \langle u^{\dagger}[\alpha \leftarrow \beta] \mid \beta\rangle \\
&= \langle u^{\dagger} \mid \alpha\rangle[\alpha \leftarrow \beta] \\
&= [\alpha] u^{\dagger}[\alpha \leftarrow \beta]
\end{aligned}
$$

Theorem 3 (simulation of the $\lambda\mu$-calculus by the $\bar{\lambda}\mu\tilde{\mu}$-calculus).

$$t \rightarrow_\gamma v \implies \exists u \ t^\dagger \xrightarrow{*} u \wedge v^\dagger \overset{*}{\leadsto} u$$

Proof. By cases on γ.

- if $\gamma = \beta$ then

$$
\begin{aligned}
(\lambda x.u)\, v^\dagger &= \mu\beta.\langle v^\dagger \mid \tilde{\mu}y.\langle \lambda x.u^\dagger \mid y \cdot \beta \rangle \rangle \\
&\leadsto_\beta \mu\beta.\langle v^\dagger \mid \tilde{\mu}y.\langle y \mid \tilde{\mu}x.\langle u^\dagger \mid \beta \rangle \rangle \rangle \\
&\leadsto_{\tilde{\mu}} \mu\beta.\langle v^\dagger \mid \tilde{\mu}x.\langle u^\dagger \mid \beta \rangle \rangle \\
&\rightarrow_{\tilde{\mu}} \mu\beta.\langle u^\dagger[x \leftarrow v^\dagger] \mid \beta \rangle \\
&\leadsto_\theta u^\dagger[x \leftarrow v^\dagger] \\
&= u[x \leftarrow v]^\dagger
\end{aligned}
$$

- if $\gamma = \mu$ then

$$
\begin{aligned}
(\mu\alpha.c)\, v^\dagger &= \mu\beta.\langle v^\dagger \mid \tilde{\mu}y.\langle \mu\alpha.c^\dagger \mid y \cdot \beta \rangle \rangle \\
&\leadsto_{\tilde{\mu}} \mu\alpha.\langle \mu\alpha.c^\dagger \mid v^\dagger \cdot \alpha \rangle \\
&\rightarrow_\mu \mu\alpha.c^\dagger[\alpha \leftarrow v^\dagger \cdot \alpha] \\
&\approx \mu\alpha.c[\alpha \leftarrow v \cdot \alpha]^\dagger
\end{aligned}
$$

- if $\gamma = \mu'$ then

$$
\begin{aligned}
(v)\,\mu\alpha.c^\dagger &= \mu\beta.\langle \mu\alpha.c^\dagger \mid \tilde{\mu}y.\langle v^\dagger \mid y \cdot \beta \rangle \rangle \\
&\rightarrow_\mu \mu\alpha.c^\dagger[\alpha \leftarrow \tilde{\mu}y.\langle v^\dagger \mid y \cdot \alpha \rangle] \\
&\approx \mu\alpha.c[\alpha \leftarrow \alpha(v)]^\dagger
\end{aligned}
$$

- if $\gamma = \rho$ then

$$
\begin{aligned}
[\beta]\,\mu\alpha.c^\dagger &= \langle \mu\alpha.c^\dagger \mid \beta \rangle \\
&\leadsto_\mu c^\dagger[\alpha \leftarrow \beta] \\
&= c[\alpha \leftarrow \beta]^\dagger
\end{aligned}
$$

- if $\gamma = \theta$ then $\mu\delta.[\delta]\, t^\dagger = \mu\delta.\langle t^\dagger \mid \delta \rangle \leadsto_\theta t^\dagger$

Corollary 1 (call-by-name case). $t \rightarrow_n v \implies \exists u \ t^\dagger \xrightarrow{*}_n u \wedge v^\dagger \overset{*}{\leadsto}_n u$

Proof. By cases on β and μ-rules.

$(\lambda x.u)\, v$ is β-reduced in call-by-name without any restriction. It is simulated in the $\bar{\lambda}\mu\tilde{\mu}$-calculus by a $\tilde{\mu}$-reduction. The latter is in call-by-name without any restriction too.

$(\mu\alpha.c)\, v$ is μ-reduced in call-by-name without any restriction. It is simulated in the $\bar{\lambda}\mu\tilde{\mu}$-calculus by a μ-reduction. The latter is in call-by-name if $v^\dagger \cdot \alpha$ is a stack. It is the case by definition 3.

Corollary 2 (call-by-value case). $t \rightarrow_v v \implies \exists u \ t^\dagger \xrightarrow{*}_v u \wedge v^\dagger \overset{*}{\leadsto}_v u$

Proof. By cases on β, μ and μ'-rules.

$(\lambda x.u)\,v$ is β-reduced in call-by-value if v is a value. It is simulated in the $\bar{\lambda}\mu\tilde{\mu}$-calculus by a $\tilde{\mu}$-reduction. The latter is in call-by-value if $v^\dagger$ is a value. It is the case by the definition of $\bar{\lambda}\mu\tilde{\mu}_Q$.

$(\mu\alpha.c)\,v$ is μ-reduced in call-by-value if v is a value. It is simulated in the $\bar{\lambda}\mu\tilde{\mu}$-calculus by a μ-reduction. The latter is in call-by-value without any restriction.

$(v)\,\mu\alpha.c$ is μ'-reduced in call-by-value without any restriction. It is simulated in the $\bar{\lambda}\mu\tilde{\mu}$-calculus by a μ-reduction. The latter is in call-by-value without any restriction as well.

The $\bar{\lambda}\mu\tilde{\mu}$-simulation by the $\lambda\mu$-calculus requires preliminary lemmas showing that $(\ )^\circ$ commutes over $\lambda\mu$ and $\bar{\lambda}\mu\tilde{\mu}$-substitutions. Each proof consists of

- expanding the $\bar{\lambda}\mu\tilde{\mu}$-substitution
- expanding the definition of $(\ )^\circ$
- applying the induction hypothesis if necessary
- factorising the $\lambda\mu$-substitution
- factorising the definition of $(\ )^\circ$

Lemma 11. $t[x \leftarrow u]^\circ = t^\circ[x \leftarrow u^\circ]$

Proof. By induction on t.

- if $t = x$ then $x[x \leftarrow u]^\circ = u^\circ = x^\circ[x \leftarrow u^\circ]$
- if $t = y$ then $y[x \leftarrow u]^\circ = y = y^\circ[x \leftarrow u^\circ]$
- if $t = \langle t \,|\, e \rangle$ then

$$\begin{aligned}
\langle t\,|\,e\rangle[x \leftarrow u]^\circ &= \langle t[x \leftarrow u]\,|\,e[x \leftarrow u]\rangle^\circ \\
&= e[x \leftarrow u]^\circ\{t[x \leftarrow u]^\circ\} \\
&= e^\circ[x \leftarrow u^\circ]\{t^\circ[x \leftarrow u^\circ]\} \\
&= e^\circ\{t^\circ\}[x \leftarrow u^\circ] \\
&= \langle t\,|\,e\rangle^\circ[x \leftarrow u^\circ]
\end{aligned}$$

Lemma 12. $t[\alpha \leftarrow h]^\circ = t^\circ[\alpha \leftarrow h^\circ]$

Proof. By induction on t.

- if $c = \langle t\,|\,e\rangle$ then

$$\begin{aligned}
\langle t\,|\,e\rangle[\alpha \leftarrow h]^\circ &= \langle t[\alpha \leftarrow h]\,|\,e[\alpha \leftarrow h]\rangle^\circ \\
&= e[\alpha \leftarrow h]^\circ\{t[\alpha \leftarrow h]^\circ\} \\
&= e^\circ[\alpha \leftarrow h^\circ]\{t^\circ[\alpha \leftarrow h^\circ]\} \\
&= e^\circ\{t^\circ\}[\alpha \leftarrow h^\circ] \\
&= \langle t\,|\,e\rangle^\circ[\alpha \leftarrow h^\circ]
\end{aligned}$$

- if $e = \alpha$ then $\alpha[\alpha \leftarrow h]^\circ = h^\circ = \alpha^\circ[\alpha \leftarrow h^\circ]$
- if $e = \beta$ then $\beta[\alpha \leftarrow h]^\circ = \beta^\circ = \beta^\circ[\alpha \leftarrow h^\circ]$

Theorem 4 (simulation of the $\bar{\lambda}\mu\tilde{\mu}$-calculus by the $\lambda\mu$-calculus).

$$t \to_\gamma v \Longrightarrow \exists u\ t^\circ \xrightarrow{*} u \overset{*}{\rightsquigarrow} v^\circ$$

Proof. By cases on γ.

- if $\gamma = \beta'$ then
$$
\begin{aligned}
\langle \lambda x.u \,|\, v \cdot e \rangle^\circ &= v^\circ \cdot e^\circ \{\lambda x.u^\circ\} \\
&= e^\circ \{(\lambda x.u^\circ)\, v^\circ\} \\
&\to_\beta e^\circ \{u^\circ[x \leftarrow v^\circ]\} \\
&= e^\circ \{u[x \leftarrow v]^\circ\} \\
&= \langle u[x \leftarrow v] \,|\, e \rangle^\circ
\end{aligned}
$$

- if $\gamma = \mu$ then
$$
\begin{aligned}
\langle \mu\alpha.c \,|\, e \rangle^\circ &= e^\circ \{\mu\alpha.c^\circ\} \\
&\xrightarrow{*} c^\circ[\alpha \leftarrow e^\circ] \\
&= c[\alpha \leftarrow e]^\circ
\end{aligned}
$$

- if $\gamma = \tilde{\mu}$ then
$$
\begin{aligned}
\langle t \,|\, \tilde{\mu}x.c \rangle^\circ &= [\beta]\,(\lambda x.\mu\delta.c^\delta)\, t^\circ \\
&\to_\beta [\beta]\,\mu\delta.c^\circ[x \leftarrow t^\circ] \\
&\rightsquigarrow_\rho c^\circ[x \leftarrow t^\circ] \\
&= c[x \leftarrow t]^\circ
\end{aligned}
$$

- if $\gamma = \theta$ then $\mu\delta.\langle t \,|\, \delta \rangle^\circ = \mu\delta.[\delta]\,t^\circ \rightsquigarrow_\theta t^\circ$

Corollary 3 (call-by-name case). $t \to_n v \Longrightarrow \exists u\ t^\circ \xrightarrow{*}_n u \overset{*}{\rightsquigarrow}_n v^\circ$

Proof. By cases on β', μ and $\tilde{\mu}$-rules.

$\langle \lambda x.u \,|\, v \cdot e \rangle$ is β'-reduced in call-by-name without any restriction. It is simulated in the $\lambda\mu$-calculus by a β-reduction. The latter is in call-by-name without any restriction too.

$\langle \mu\alpha.c \,|\, e \rangle$ is μ-reduced in call-by-name if $e \neq \tilde{\mu}x.c'$ else it were $\tilde{\mu}$-reduced. It is simulated in the $\lambda\mu$-calculus with the help of lemma 2. The latter is in call-by-name if $e^\circ \neq \beta(t)$ i.e. if $e \neq \tilde{\mu}x.c'$. It is the case by definition 4.

$\langle t \,|\, \tilde{\mu}x.c \rangle$ is $\tilde{\mu}$-reduced in call-by-name without any restriction. It is simulated in the $\lambda\mu$-calculus by a β-reduction. The latter is in call-by-name without any restriction as well.

Corollary 4 (call-by-value case). $t \to_v v \Longrightarrow \exists u\ t^\circ \xrightarrow{*}_v u \overset{*}{\rightsquigarrow}_v v^\circ$

Proof. By cases on β', μ and $\tilde{\mu}$-rules.

$\langle \lambda x.u \,|\, v \cdot e \rangle$ is β'-reduced in call-by-value if v is a value. It is simulated in the $\lambda\mu$-calculus by a β-reduction. The latter is in call-by-value if v° is a value. It is the case by definition 4.

$\langle \mu\alpha.c \,|\, e \rangle$ is μ-reduced in call-by-value if e is either a μ-variable or a context of the form $v \cdot h$ where v is a value or a μ-abstraction by the definition of $\bar{\lambda}\mu\tilde{\mu}_Q$. It is simulated in the $\lambda\mu$-calculus with the help of lemma 2. The latter is in call-by-value if v° is a value in a context of the form $h^\circ \cdot v^\circ$ i.e. if v is a value in a $v \cdot h$ context. It is the case by definition 4.

$\langle t \mid \tilde{\mu}x.c \rangle$ is $\tilde{\mu}$-reduced in call-by-value if t is a value. It is simulated in the $\lambda\mu$-calculus by a β-reduction. The latter is in call-by-value if t° is a value. It is the case by definition 4.

6   Conclusion

Analysis of the $\lambda\mu$ and $\bar{\lambda}\mu\tilde{\mu}$-calculi has shown their computational equivalence. It holds for undirected evaluations of pure calculi (see theorems 3 and 4). This result is then easily obtained for call-by-name and call-by-value evaluations (see corollaries 1, 2, 3 and 4). It concerns the simple type system too (see lemmas 3 and 4).

The simulation of the $\bar{\lambda}\mu\tilde{\mu}$-calculus by the $\lambda\mu$-calculus is smoother than the simulation of the $\lambda\mu$-calculus by the $\bar{\lambda}\mu\tilde{\mu}$-calculus. The first is obtained with the help of linear *reductions* whereas the second is obtained with the help of linear *expansions*.

This work can be extended in three ways. The first consists of proving the same results for the call-by-value evaluation of the $\lambda\mu$-calculus defined in [Ong and Stewart, 1997]. The second consists of defining CPS translations to λ-calculus in order to complete [Curien and Herbelin, 2000]. The third consists of extending the type system to the other logical constants.

References

[Curien and Herbelin, 2000]   Curien, P.-L. and Herbelin, H. (2000). The Duality of Computation. In *Proceedings of the International Conference on Functional Programming*.

[Dougherty et al., 2004]   Dougherty, D. J., Ghilezan, S., and Lescanne, P. (2004). Characterizing strong normalization in a language with control operators. In *Proceedings of the 6th International ACM SIGPLAN Conference on Principles and Practice of Declarative Programming*, pages 155–166.

[Gentzen, 1934]   Gentzen, G. (1934). Investigations into Logical Deduction. In Szabo, M., editor, *Collected Papers of Gerhard Gentzen*. North Holland.

[Griffin, 1990]   Griffin, T. G. (1990). The Formulae-as-Types Notion of Control. In *Proceedings of the 17th ACM Symposium on Principles of Programming Languages*, pages 47–57. ACM Press.

[Ong and Stewart, 1997]   Ong, L. and Stewart, C. (1997). A Curry-Howard foundation for functional computation with control. In *Proceedings of the 24th Annual ACM Symposium on Principles of Programming Languages*, pages 215–227. ACM Press.

[Parigot, 1992]   Parigot, M. (1992). $\lambda\mu$-calculus: an Algorithmic Interpretation of Classical Natural Deduction. In *Proceedings of Internationnal Conference on Logic Programming and Automated Deduction*, volume 624 of *Lectures Notes in Computer Science*, pages 190–201. Springer.

Reduction in a Linear Lambda-Calculus with Applications to Operational Semantics

Alex Simpson

LFCS, School of Informatics, University of Edinburgh, UK

Abstract. We study beta-reduction in a linear lambda-calculus derived from Abramsky's linear combinatory algebras. Reductions are classified depending on whether the redex is in the computationally active part of a term ("surface" reductions) or whether it is suspended within the body of a thunk ("internal" reductions). If surface reduction is considered on its own then any normalizing term is strongly normalizing. More generally, if a term can be reduced to surface normal form by a combined sequence of surface and internal reductions then every combined reduction sequence from the term contains only finitely many surface reductions. We apply these results to the operational semantics of Lily, a second-order linear lambda-calculus with recursion, introduced by Bierman, Pitts and Russo, for which we give simple proofs that call-by-value, call-by-name and call-by-need contextual equivalences coincide.

1 Introduction

The language Lily was introduced by Bierman, Pitts and Russo in [3]. It is a typed lambda-calculus based on a second-order intuitionistic linear type theory with recursion. What makes it interesting from a programming language perspective is that, following ideas of Plotkin [10], the language is able to encode a remarkably rich range of datatype constructs (eager products, lazy products, coproducts, polymorphism, abstract types, recursive types, etc.). Furthermore, its linearity makes it potentially useful for modelling single-threadedness and other state and resource-related concepts, cf. [7].

The main achievement of [3] was to establish direct operational techniques for reasoning about Lily up to contextual equivalence. Such techniques include useful extensionality properties, and a powerful framework for establishing program equalities using an adaptation (based on [8]) of Reynolds' relational parametricity (first introduced in [11]). In order to get this machinery to work, the authors of [3] need to first establish one key result about Lily, a result which pervades all further developments in their paper. This result, the so-called Strictness Theorem, asserts the (surprising at first sight) fact that call-by-name and call-by-value operational semantics for Lily both give rise to the same notion of contextual equivalence.

The outline proof of the Strictness Theorem in [3] makes rather heavy use of the well-stocked armoury of known operational techniques. In particular it

J. Giesl (Ed.): RTA 2005, LNCS 3467, pp. 219–234, 2005.

uses Howe's method [4] to obtain a version of Mason and Talcott's *ciu theorem* [6]. The starting point for the research in this paper was the realisation that basic techniques from rewriting could be applied to obtain an alternative, self-contained and essentially simple proof of the Strictness Theorem.

In Sec. 2, we review Lily and its operational semantics. Then, in Secs. 3 and 4, we present our alternative proof of the Strictness Theorem. We translate Lily into a simple untyped linear lambda-calculus containing: *linear* lambda abstractions, $\lambda x.M$; *non-linear* lambda abstractions, $\lambda!x.M$, which require their arguments to be suspended as "thunks"; and *thunks* themselves, $!M$. We study beta-reduction in this untyped calculus, making the restriction that, as thunks are considered suspended, reductions should not take place within a thunk. This restricted relation, which we call *surface reduction*, turns out to be extremely well behaved: as well as the expected confluence property, it holds that every normalizing term is strongly normalizing. The Strictness Theorem for Lily follows easily from this latter fact, using straightforward simulations under surface reduction of call-by-name and call-by-value evaluation for Lily.

In Sec. 5, we (temporarily) turn attention away from Lily and take a deeper look at our untyped linear lambda-calculus and its reduction properties. In order to obtain a conversion relation between terms that is a congruence, it is necessary to consider also reductions inside thunks. We call such reductions *internal reductions*, and we call arbitrary reductions (either surface or internal) *combined reductions*. As well as the expected confluence properties (for both internal and combined reductions), we show that internal reductions can always be postponed until after surface reductions. Further, we show that if a term reduces (under combined reduction) to a surface normal form then any sequence of combined reductions contains only finitely many surface reductions.

Next, in Sec. 6, we return to Lily and show that the results of Sec. 5 again have applications to operational semantics. We use them to establish the equivalence of the call-by-name operational semantics of Lily with an implementation-oriented call-by-need semantics. Once again, the equivalence of these two semantics had previously been established by the authors of [3], but with an intricate and lengthy proof (private communication). Our proof turns out to be relatively straightforward.

Finally, in Sec. 7, we observe that our untyped linear lambda-calculus is exactly the lambda-calculus counterpart of Abramsky's *linear combinatory algebras*, presented in [1]. This connection makes us believe that the linear lambda-calculus introduced in this paper is rather natural. Accordingly, it is plausible that the properties of reduction established in Secs. 3 and 5 may turn out to have other applications, perhaps again in the area of operational semantics, but possibly more widely.

2   Lily and Its Operational Semantics

In this section, we review the language Lily, a typed λ-calculus, based on second-order intuitionistic linear type theory with recursion, introduced in [3].

The language of types for `Lily` contains just three type constructors: linear function space $\sigma \multimap \tau$; linear "exponentials" $!\sigma$, used to type thunks; and universally quantified types $\forall \alpha. \sigma$, used for polymorphism. Types $\sigma, \tau, \ldots$ are thus built up from type variables $\alpha, \beta, \ldots$, according to the grammar:

$$\sigma ::= \alpha \mid \sigma \multimap \tau \mid !\tau \mid \forall \alpha. \tau .$$

As usual, α is bound in $\forall \alpha. \tau$. We write $\mathrm{ftv}(\sigma)$ for the set of free type variables in σ (and below apply the same notation to terms and contexts in the evident way). If $\mathrm{ftv}(\sigma) = \emptyset$ then σ is said to be *closed*.

Although simple, the above language of types is remarkably rich. For example, the other type constructors of intuitionistic linear logic can all be encoded: non-linear (intuitionistic) function space, $\sigma \to \tau$, using Girard's $!\sigma \multimap \tau$; tensor, $\otimes$, product, $\&$ and sum, $\oplus$. One can also encode basic ground types (booleans, natural numbers, etc.), and existentially quantified types $\exists \alpha. \sigma$, and, due to the recursion operator in `Lily`, arbitrary recursive types. These encodings are due to Plotkin [10], see [3] for details.

The term language of `Lily` is the expected typed λ-calculus associated with the above types, together with a recursion operator[1]. Raw terms $s, t, \ldots$ are built from term variables $x, y, \ldots$ according to the grammar:

$$t ::= x \mid \lambda x{:}\sigma.\, t \mid s(t) \mid !t \mid \text{let } !x = s \text{ in } t \mid \Lambda \alpha.\, t \mid t(\sigma) \mid \text{rec } x{:}\sigma.\, t .$$

Here, x is bound in $\lambda x{:}\sigma.\, t$, in let $!x = s$ in t [2] and in rec $x{:}\sigma.\, t$, and α is bound in $\Lambda \alpha.\, t$. We write $\mathrm{fv}(t)$ for the set of free variables in a term t. We identify terms up to α-equivalence.

$$\Gamma; x{:}\sigma \vdash x : \sigma \qquad \Gamma, x{:}\sigma; - \vdash x : \sigma$$

$$\frac{\Gamma; \Delta, x{:}\sigma \vdash t : \tau}{\Gamma; \Delta \vdash \lambda x{:}\sigma.\, t : \sigma \multimap \tau} \qquad \frac{\Gamma; \Delta \vdash s : \sigma \multimap \tau \quad \Gamma; \Delta' \vdash t : \sigma}{\Gamma; \Delta, \Delta' \vdash s(t) : \tau}$$

$$\frac{\Gamma; - \vdash t : \tau}{\Gamma; - \vdash !t : !\tau} \qquad \frac{\Gamma; \Delta \vdash s : !\sigma \quad \Gamma, x{:}\sigma; \Delta' \vdash t : \tau}{\Gamma; \Delta, \Delta' \vdash \text{let } !x = s \text{ in } t : \tau} \; \Delta \# x{:}\sigma$$

$$\frac{\Gamma; \Delta \vdash t : \tau}{\Gamma; \Delta \vdash \Lambda \alpha.\, t : \forall \alpha. \tau} \; \alpha \notin \mathrm{ftv}(\Gamma, \Delta) \qquad \frac{\Gamma; \Delta \vdash t : \forall \alpha. \tau}{\Gamma; \Delta \vdash t(\sigma) : \tau[\sigma/\alpha]} \qquad \frac{\Gamma, x{:}\sigma; - \vdash t : \sigma}{\Gamma; - \vdash \text{rec } x{:}\sigma.\, t : \sigma}$$

Fig. 1. Typing rules for `Lily`.

---

[1] We depart from [3] by building an explicit recursion operator into `Lily`, instead of incorporating recursion within thunks. This is an inessential difference.

[2] For simplicity, we place an inessential restriction in the typing rules ensuring that the term let $!x = s$ in t is well typed only when x does not occur free in s.

$$\frac{s \rightarrow s'}{s(t) \rightarrow s'(t)} \qquad \frac{t \rightarrow t'}{t(\sigma) \rightarrow t'(\sigma)} \qquad \frac{}{(\Lambda\alpha.\, t)(\sigma) \rightarrow t[\sigma/\alpha]}$$

$$\frac{s \rightarrow s'}{\text{let } !x = s \text{ in } t \rightarrow \text{let } !x = s' \text{ in } t} \qquad \frac{}{\text{let } !x = !s \text{ in } t \rightarrow t[s/x]} \qquad \frac{}{\text{rec } x\!:\!\sigma.\, t \rightarrow t[\text{rec } x\!:\!\sigma.\, t/x]}$$

$$\frac{t \rightarrow_{\text{vl}} t'}{(\lambda x\!:\!\sigma.\, s)(t) \rightarrow_{\text{vl}} (\lambda x\!:\!\sigma.\, s)(t')} \qquad \frac{}{(\lambda x\!:\!\sigma.\, s)(v) \rightarrow_{\text{vl}} s[v/x]} \qquad \frac{}{(\lambda x\!:\!\sigma.\, s)(t) \rightarrow_{\text{nm}} s[t/x]}$$

Fig. 2. Call-by-value and Call-by-name Evaluation for `Lily`.

The typing rules for `Lily` are based on Barber and Plotkin's DILL [2]. We use $\Gamma, \Delta, \ldots$ to range over "contexts", which are finite functions from term variables to types. We write $\Gamma \# \Delta$ to say that the domains of Γ and Δ are disjoint. The typing rules manipulate sequents $\Gamma; \Delta \vdash t\!:\!\sigma$ where $\Gamma \# \Delta$. Here, Γ types the "intuitionistic" variables appearing in the term t, which have no restriction on how they occur, and Δ types the "linear" variables, each of which occurs exactly once in t, not within the scope of a ! or rec operator. The typing rules are presented in Fig. 1. In them, a comma always denotes a disjoint union of contexts and a dash denotes the empty context. We write $t : \tau$ to mean that the sequent $\vdash t\!:\!\tau$ is derivable, where τ is a closed type (t is necessarily a closed term).

Following [3], we define two operational semantics for `Lily`, one using a call-by-value evaluation of function application, and one using call-by-name. In both cases, the operational semantics reduces terms to *values* $v, \ldots$, which are terms of the form:

$$v ::= \lambda x\!:\!\sigma.\, t \mid !t \mid \Lambda\alpha.\, t \ .$$

In contrast to [3], we give the operational semantics in a small-step style. This facilitates our proofs, but only in an inessential way, the equivalence of big-step and small-step definitions being anyway easy to establish.

Figure 2 defines two small-step evaluation relations $t \rightarrow_{\text{vl}} t'$ and $t \rightarrow_{\text{nm}} t'$ between `Lily` terms. The call-by-value (or strict) relation $t \rightarrow_{\text{vl}} t'$ is inductively defined by the two specific $\rightarrow_{\text{vl}}$ rules for application together with all rules written using the neutral $\rightarrow$ notation. Similarly, the call-by-name (or non-strict) relation $t \rightarrow_{\text{nm}} t'$ is defined by the specific $\rightarrow_{\text{nm}}$ rule for application together with the neutral rules. Note that both operational semantics are deterministic.

Our interest lies in the operational semantics of `Lily` *programs*, i.e. of closed terms of closed type. It is easily seen that if $t : \sigma$ and $t \rightarrow_{\text{vl}} t'$ then $t' : \sigma$ (and similar if $t \rightarrow_{\text{nm}} t'$). Also, by induction on the structure of t, one sees that if $t : \sigma$ then t does not reduce under $\rightarrow_{\text{vl}}$ if and only if t is a value (and similar for $\rightarrow_{\text{nm}}$). We write $t \downarrow_{\text{vl}}$ (resp. $t \downarrow_{\text{nm}}$) for the "termination" property: there exists a value v such that $t \rightarrow^*_{\text{vl}} v$ (resp. $t \rightarrow^*_{\text{nm}} v$), where, as usual, R^* (resp. R^+) denotes the reflexive-transitive (resp. transitive) closure of the relation R.

The program below shows that sometimes call-by-name evaluation terminates when call-by-value does not (cf. [3, Example 2.2]).

$$(\lambda f : \forall \alpha.\alpha \multimap \forall \alpha.\alpha. \; \lambda x : \forall \alpha.\alpha. \; f(x))(\text{rec } g : \forall \alpha.\alpha \multimap \forall \alpha.\alpha. \; g) \qquad (1)$$

This program has type $\forall \alpha.\alpha \multimap \forall \alpha.\alpha$. An important insight of [3], is that the most useful notion of contextual equivalence for Lily is obtained by only observing termination for programs of exponential type $!\tau$. The restriction to such observations corresponds to observing termination at ground types (such as booleans, naturals, etc.), it yields desirable extensionality properties for contextual equivalence, and it is crucial to the correctness of Plotkin's [10] encodings of datatype constructions in Lily.

The key result of [3] that underpins its entire study of contextual equivalence for Lily is the "Strictness Theorem".

Theorem 2.1 (Strictness Theorem [3][3]). *If $t : !\tau$ then $t \downarrow_{\text{vl}}$ if and only if $t \downarrow_{\text{nm}}$.*

When termination observations are restricted to exponential types, it follows immediately from the theorem that both call-by-value and call-by-name operational semantics induce the same contextual equivalence.

We remark that the Strictness Theorem is stated in the most general form possible: the result holds for no types other than exponential types, as simple adaptations of (1) readily show. This suggests that any proof of Theorem 2.1 has to uncover some crucial property of exponential types. The machinery used in [3] to this end has already been mentioned in Section 1. In this paper, we shall instead prove Theorem 2.1 using surprisingly elementary techniques from rewriting, translating Lily into a very simple untyped linear λ-calculus in which (the appropriate notion of) β-reduction simulates both call-by-value and call-by-name operational semantics. This untyped linear λ-calculus includes explicit thunks, and it is the treatment of these thunks that will reflect the all-important behaviour of Lily at exponential type.

3   A Linear Lambda-Calculus and Surface Reduction

In this section, we intruduce our untyped linear λ-calculus. Its main ingredients are: applications MN; linear lambda abstractions, $\lambda x.M$; non-linear lambda abstractions, $\lambda!x.M$, which require their arguments to be suspended as thunks; and thunks themselves, $!M$. Formally, raw terms $M, N, \ldots$ are built up from variables $x, y, \ldots$ according to the grammar:

$$M ::= x \mid MN \mid \lambda x.M \mid \lambda!x.M \mid !M \; .$$

The variable x is bound in both $\lambda x.M$ and $\lambda!x.M$. We write $\equiv$ for syntactic equality of terms modulo α-equivalence.

---

[3] The theorem as stated here is easily shown to be equivalent to the original [3, Theorem 2.3].

$$(\lambda x.M)(N) \rightarrow M[N/x] \qquad (\lambda !x.M)(!N) \rightarrow M[N/x]$$

$$\frac{M \rightarrow M'}{MN \rightarrow M'N} \qquad \frac{N \rightarrow N'}{MN \rightarrow MN'} \qquad \frac{M \rightarrow M'}{\lambda x.M \rightarrow \lambda x.M'} \qquad \frac{M \rightarrow M'}{\lambda !x.M \rightarrow \lambda !x.M'}$$

Fig. 3. Surface Reduction.

We say that x is *linear in* M if x occurs free exactly once in M and, moreover, this free occurence of x does not lie within the scope of a ! operator in M. A term M is said to be *linear* if, in every subterm of M the form $\lambda x.M'$, it holds that x is linear in M'. *Henceforth, we consider linear terms only.*

In Fig. 3, we define a version of β-reduction for our calculus. The important points are the two types of redex, and that no reduction occurs under the scope of a ! operator. The latter restriction reflects the idea that thunks are suspended computations. We call the reduction defined in Fig. 3 *surface reduction*. It is easily shown that when M is linear and $M \rightarrow N$ then N is linear. From now on, all similar observations about linearity will be omitted. All operations we consider will respect the linearity of terms.

A term is said to be in *surface normal form* if there is no surface reduction from the term. Trivially, any term $!M$ is in surface normal form. A *reduction sequence* from M is a finite or infinite sequence $M \equiv M_0 \rightarrow M_1 \rightarrow M_2 \rightarrow \dots$. A *completed* reduction sequence is a reduction sequence that is either infinite or is finite with the last term in the sequence in surface normal form.

The linearity restriction on terms combines with the disallowance of reduction within thunks to ensure that the basic well-behavedness properties of surface reduction are almost trivial to establish. The main, though very simple, results of this section are Corollaries 3.3 and 3.4 below. (Only the latter is used in the proof of Theorem 2.1.)

Lemma 3.1

1. If $M \rightarrow M'$ then $M[N/x] \rightarrow M'[N/x]$.
2. If $N \rightarrow N'$ and x is linear in M then $M[N/x] \rightarrow M[N'/x]$.

Proposition 3.2. *If $M \rightarrow L$ and $M \rightarrow L'$ then either $L \equiv L'$ or there exists N such that $L \rightarrow N$ and $L' \rightarrow N$.*

Proof. By induction on the structure of M, considering all possible cases for $M \rightarrow L$ and $M \rightarrow L'$. We consider only the two redex cases.

If $M \equiv (\lambda x.M_1)(M_2) \rightarrow M_1[M_2/x] \equiv L$ and $L \not\equiv L'$ then either $L' \equiv (\lambda x.L_1')(M_2)$ where $M_1 \rightarrow L_1'$ or $L' \equiv (\lambda x.M_1)(L_2')$ where $M_2 \rightarrow L_2'$. In the first case, we have $L \rightarrow L_1'[M_2/x]$, by Lemma 3.1.1, and also $L' \rightarrow L_1'[M_2/x]$. In the second, we have $L \rightarrow M_1[L_2'/x]$, by Lemma 3.1.2, and also $L' \rightarrow M_1[L_2'/x]$.

If $M \equiv (\lambda !x.M_1)(!M_2) \rightarrow M_1[M_2/x] \equiv L$ and $L \not\equiv L'$ then $L' \equiv (\lambda !x.L_1')(!M_2)$ where $M_1 \rightarrow L_1'$. Thus $L \rightarrow L_1'[M_2/x]$, by Lemma 3.1.1, and also $L' \rightarrow L_1'[M_2/x]$. $\qquad\square$

Corollary 3.3 (Confluence). *If $M \to^* M_1$ and $M \to^* M_2$ then there exists N such that $M_1 \to^* N$ and $M_2 \to^* N$.*

Corollary 3.4 (Uniform normalization). *If $M \to^* V$ is a k-step reduction sequence, where V is in surface normal form, then every reduction sequence from M has at most k steps, and every completed reduction sequence has exactly k steps and terminates with V. In particular, if a term is normalizing under surface reduction then it is strongly normalizing.*

4   Proof of the Strictness Theorem

The proof is based on a simple translation of Lily into the untyped linear λ-calculus of Sec. 3. The translation uses an untyped recursion construct, defined by:

$$\mu x.M =_{\text{def}} (\lambda!x.M[x(!x)/x])(!\,\lambda!x.M[x(!x)/x]) \ .$$

Observe that $\mu x.M \to M[(\mu x.M)/x]$.

To every raw term t of Lily, we define a raw term t^*, in the grammar from Sec. 3, by induction on the structure of t. In the definition, we make use of a distinguished variable u, used as a dummy translation for types.

$$x^* =_{\text{def}} x \qquad (\text{let }!x = s \text{ in } t)^* =_{\text{def}} (\lambda!x.t^*)(s^*) \cdot$$
$$(\lambda x{:}\sigma.\ t)^* =_{\text{def}} \lambda x.t^* \qquad (\Lambda\alpha.\ t)^* =_{\text{def}} \lambda!w.t^* \qquad w \notin \text{fv}(t^*)$$
$$(s(t))^* =_{\text{def}} s^*\,t^* \qquad (t(\sigma))^* =_{\text{def}} t^*(!u)$$
$$(!t)^* =_{\text{def}} !\,t^* \qquad (\text{rec }x{:}\sigma.\ t)^* =_{\text{def}} \mu x.t^* \ .$$

The four lemmas below are straightforward.

Lemma 4.1. $(s[t/x])^* \equiv s^*[t^*/x]$.

Lemma 4.2. *If $\Gamma;\Delta \vdash t{:}\ \sigma$ then the raw term t^* is linear.*

Lemma 4.3. *If $t_1 \to_{\text{vl}} t_2$ then $t_1^* \to t_2^*$.*

Lemma 4.4. *If $t_1 \to_{\text{nm}} t_2$ then $t_1^* \to t_2^*$.*

Corollary 4.5. *If $t : !\tau$ then the following are equivalent:*

1. $t \downarrow_{\text{vl}}$,
2. $t \downarrow_{\text{nm}}$,
3. t^* is surface normalizing.

Proof. To show that 1 implies 3, suppose that $t \downarrow_{\text{vl}}$. Then there exists v with $v : !\tau$ such that $t \to_{\text{vl}}^* v$. As $v : !\tau$, it holds that $v \equiv !t'$. By Lemma 4.3, $t^* \to^* (!t')^* \equiv !(t'^*)$. But $!(t'^*)$ is in surface normal form, hence t^* is surface normalizing.

For the converse, suppose $t \not\downarrow_{\text{vl}}$. Then there exists an infinite sequence of call-by-value evaluation steps $t \equiv t_0 \to_{\text{vl}} t_1 \to_{\text{vl}} t_2 \to_{\text{vl}} \dots$. Whence, by Lemma 4.3, t^* has an infinite surface reduction sequence. Thus, by Corollary 3.4, t^* is not normalizing under surface reduction.

The equivalence of 2 and 3 is shown in the same way, using Lemma 4.4. $\square$

$$\frac{M \to M'}{!M \dashrightarrow !M'} \qquad \frac{M \dashrightarrow M'}{MN \dashrightarrow M'N} \qquad \frac{N \dashrightarrow N'}{MN \dashrightarrow MN'}$$

$$\frac{M \dashrightarrow M'}{\lambda x.M \dashrightarrow \lambda x.M'} \qquad \frac{M \dashrightarrow M'}{\lambda !x.M \dashrightarrow \lambda !x.M'} \qquad \frac{M \dashrightarrow M'}{!M \dashrightarrow !M'}$$

Fig. 4. Internal Reduction.

Theorem 2.1 is immediate from the corollary. Note that the point that fails for Lily programs $t : \sigma$ of arbitrary type is that it is not in general the case that $t \downarrow_{\text{vl}}$ (or $t \downarrow_{\text{nm}}$) implies that t^* is surface normalizing, because, apart from at exponential type, Lily values do not necessarily translate to surface normal forms, indeed not even to surface normalizing terms (for example, $\lambda x{:}\sigma.\ \text{rec}\ y{:}\tau.\ y$).

It is worth remarking that the techniques of this section can similarly be used to show that variant operational semantics for Lily, in which evaluation takes place under Λ- and/or λ-abstractions, also give rise to the same contextual equivalence.

5   Internal and Combined Reduction

In this section, we undertake a deeper study of reduction in our untyped linear λ-calculus. While surface reduction is computationally motivated, the disallowance of reduction inside thunks means that the conversion relation induced by surface reduction is not a congruence. To obtain a conversion relation that is a congruence, it is necessary to consider reduction inside thunks.

We implement reduction inside thunks using *internal reduction*, $M \dashrightarrow M'$, defined in Figure 4. *Combined reduction* $M \Rightarrow M'$ is defined by: $M \Rightarrow M'$ if $M \to M'$ or $M \dashrightarrow M'$. Note that it is possible that both $M \to M'$ and $M \dashrightarrow M'$ (for example, $\Omega(!\Omega) \to \Omega(!\Omega)$ and $\Omega(!\Omega) \dashrightarrow \Omega(!\Omega)$, where $\Omega =_{\text{def}} \mu x.x$, using the notation of Section 4). Accordingly, when we consider mixed reduction sequences containing both surface and internal reductions, we shall assume that each step comes with a distinguished status (as surface or internal).

The main technical effort of this section will go into the proof of Propositions 5.1 and 5.2 below.

Proposition 5.1 (Confluence)

1. *If $M \dashrightarrow^* M_1$ and $M \dashrightarrow^* M_2$ then there exists N such that $M_1 \dashrightarrow^* N$ and $M_2 \dashrightarrow^* N$.*
2. *If $M \Rightarrow^* M_1$ and $M \Rightarrow^* M_2$ then there exists N such that $M_1 \Rightarrow^* N$ and $M_2 \Rightarrow^* N$.*

By the proposition, the conversion relation defined by $M =_\beta M'$ if there exists N such that $M \Rightarrow^* N$ and $M' \Rightarrow^* N$ is an equivalence relation. It is, moreover, a congruence. Thus surface and internal reduction together provide an oriented decomposition of the natural β-conversion between terms of the untyped linear calculus. The next result exhibits natural structure within this decomposition.

Proposition 5.2 (Internal Postponement). *If $M \Rightarrow^* N$, by a reduction sequence containing k surface reductions, then there exists L such that $M \rightarrow^* L \dashrightarrow^* N$, where the surface reduction sequence $M \rightarrow^* L$ contains at least k reductions.*

The proofs of the two propositions above make use of the (standard) technology of parallel reduction relations. Before giving these, we apply Proposition 5.2 to derive further properties of and interactions between surface, internal and combined reduction. The main result of the section is Theorem 5.5 below.

Lemma 5.3. *If $M \rightarrow N$ and $M \dashrightarrow M'$ then there exists N' such that $M' \rightarrow N'$.*

Proof. By induction on the derivation of $M \rightarrow N$. We consider one case.
    Suppose $M \equiv (\lambda!x.M_1)(!M_2) \rightarrow M_1[M_2/x] \equiv N$. Either $M' \equiv (\lambda!x.M_1')(!M_2)$ where $M_1 \dashrightarrow M_1'$, or $M' \equiv (\lambda!x.M_1)(!M_2')$ where $M_2 \Rightarrow M_2'$. In the first case, $M' \rightarrow M_1'[M_2/x]$. In the second, $M' \rightarrow M_1[M_2'/x]$. $\qquad\square$

Corollary 5.4. *If V is in suface normal form then:*

1. *$V \dashrightarrow N$ implies N is in surface normal form.*
2. *$M \dashrightarrow V$ implies M is in surface normal form.*

Proof. Statement 1 follows from Proposition 5.2, and statement 2 from Lemma 5.3. $\qquad\square$

Theorem 5.5. *If $M \Rightarrow^* V$, where V is in surface normal form, then each infinite $\Rightarrow$ reduction sequence from M contains only finitely many $\rightarrow$ reductions.*

Proof. By Proposition 5.2, there exists U such that $M \rightarrow^* U \dashrightarrow^* V$. By Corollary 5.4.2, U is in surface normal form. Let k be the number of reductions in the sequence $M \rightarrow^* U$. We show that every $\Rightarrow$ reduction sequence from M contains at most k surface reductions. Consider any reduction sequence $M \Rightarrow^* N$ with l surface reductions. By Proposition 5.2, there exists L such that $M \rightarrow^* L$ with at least l reductions. But, by Corollary 3.4, any $\rightarrow$ reduction sequence from M has at most k reductions. Thus indeed $l \le k$. $\qquad\square$

We now turn to the proofs of Propositions 5.1 and 5.2, which use the parallel versions of combined and internal reduction defined in Figs. 5 and 6 respectively.

Lemma 5.6

1. *$M \not\Rightarrow M$ and $M \dashrightarrow\!\!\!\!/\; M$.*
2. *If $M \Rightarrow M'$ then $M \not\Rightarrow M'$. Conversely, if $M \not\Rightarrow M'$ then $M \Rightarrow^* M'$.*
3. *If $M \dashrightarrow M'$ then $M \dashrightarrow\!\!\!\!/\; M'$. Conversely, if $M \dashrightarrow\!\!\!\!/\; M'$ then $M \dashrightarrow^* M'$.*
4. *If $M \dashrightarrow\!\!\!\!/\; M'$ then $M \not\Rightarrow M'$.*

Lemma 5.7

1. *If $M \not\Rightarrow M'$ and $N \not\Rightarrow N'$ then $M[N/x] \not\Rightarrow M'[N'/x]$.*
2. *If $M \dashrightarrow\!\!\!\!/\; M'$ and $N \dashrightarrow\!\!\!\!/\; N'$ then $M[N/x] \dashrightarrow\!\!\!\!/\; M'[N'/x]$.*

$$\frac{}{x \not\Rightarrow x} \qquad \frac{M \not\Rightarrow M' \quad N \not\Rightarrow N'}{(\lambda x.M)(N) \not\Rightarrow M'[N'/x]} \qquad \frac{M \not\Rightarrow M' \quad N \not\Rightarrow N'}{(\lambda!x.M)(!N) \not\Rightarrow M'[N'/x]}$$

$$\frac{M \not\Rightarrow M' \quad N \not\Rightarrow N'}{MN \not\Rightarrow M'N'} \qquad \frac{M \not\Rightarrow M'}{\lambda x.M \not\Rightarrow \lambda x.M'} \qquad \frac{M \not\Rightarrow M'}{\lambda!x.M \not\Rightarrow \lambda!x.M'} \qquad \frac{M \not\Rightarrow M'}{!M \not\Rightarrow !M'}$$

Fig. 5. Parallel Combined Reduction.

$$\frac{}{x -\not\to x} \qquad \frac{M -\not\to M' \quad N -\not\to N'}{MN -\not\to M'N'}$$

$$\frac{M -\not\to M'}{\lambda x.M -\not\to \lambda x.M'} \qquad \frac{M -\not\to M'}{\lambda!x.M -\not\to \lambda!x.M'} \qquad \frac{M \not\Rightarrow M'}{!M -\not\to !M'}$$

Fig. 6. Parallel Internal Reduction.

Lemma 5.8

1. If $M \not\Rightarrow M_1$ and $M \not\Rightarrow M_2$ then there exists N such that $M_1 \not\Rightarrow N$ and $M_2 \not\Rightarrow N$.
2. If $M -\not\to M_1$ and $M -\not\to M_2$ then there exists N such that $M_1 -\not\to N$ and $M_2 -\not\to N$.

Proof. The proof, which is by induction on the structure of M, is a routine analysis of all possible cases, cf. [9]. □

Proposition 5.1 is a straightforward consequence the last lemma.

The remaining lemmas are directed towards the proof of Proposition 5.2.

Sub-lemma 5.9. *If* $M -\not\to M'$, $N \not\Rightarrow N'$ *and* $N \to^* N'' -\not\to N'$ *then there exists* L *such that* $M[N/x] \to^* L -\not\to M'[N'/x]$.

Proof. By a straightforward induction on the derivation of $M -\not\to M'$. □

Lemma 5.10. *If* $M \not\Rightarrow M'$ *then there exists* L *such that* $M \to^* L -\not\to M'$.

Proof. By induction on the derivation of $M \not\Rightarrow M'$. The most interesting case is when $M \equiv (\lambda!x.M_1)(!M_2) \not\Rightarrow M_1'[M_2'/x] \equiv M'$, where $M_1 \not\Rightarrow M_1'$ and $M_2 \not\Rightarrow M_2'$. Then, by induction hypothesis, there exist L_1, L_2 such that $M_1 \to^* L_1 -\not\to M_1'$ and $M_2 \to^* L_2 -\not\to M_2'$. By Sub-lemma 5.9, there exists L such that $L_1[M_2/x] \to^* L -\not\to M_1'[M_2'/x]$. Thus $M \equiv (\lambda!x.M_1)(!M_2) \to M_1[M_2/x] \to^* L_1[M_2/x] \to^* L -\not\to M_1'[M_2'/x] \equiv M'$, as required. □

Lemma 5.11. *If* $M -\not\to L \to N$ *then there exists* L' *such that* $M \to L' \not\Rightarrow N$.

Proof. By induction on the derivation of $L \to N$. We consider two cases.

If $L \equiv (\lambda !x.L_1)(!L_2) \to L_1[L_2/x] \equiv N$, then $M \equiv (\lambda !x.M_1)(!M_2)$ where $M_1 \text{-}\!\!\not\to L_1$ and $M_2 \not\Rightarrow L_2$. Thus $M \to M_1[M_2/x]$ and, by Lemmas 5.6 and 5.7, we have that $M_1[M_2/x] \not\Rightarrow L_1[L_2/x] \equiv N$. Hence the result holds with $L' =_{\text{def}} M_1[M_2/x]$.

If $L \equiv L_1L_2 \to N_1L_2 \equiv N$, where $L_1 \to N_1$, then $M \equiv M_1M_2$ where $M_1 \text{-}\!\!\not\to L_1$ and $M_2 \text{-}\!\!\not\to L_2$. By induction hypothesis, there exists L_1' such that $M_1 \to L_1' \not\Rightarrow N_1$. Thus $M \to L_1'M_2 \not\Rightarrow N_1L_2$, hence the result holds with $L' =_{\text{def}} L_1'M_2$. □

Proof (of Proposition 5.2). We have a reduction sequence $M \Rightarrow^* N$, possibly consisting of both $\to$ and $\dashrightarrow$ rewrites. This can equally well be viewed as a sequence of $\to$ and $\text{-}\!\!\not\to$ rewrites. We begin by associating a complexity measure to any such reduction sequence of $\to$ and $\text{-}\!\!\not\to$ rewrites. To do this, first assign to to each $\text{-}\!\!\not\to$ rewrite in the sequence the number of $\to$ rewrites that occur to the right of it. We thus obtain a sequence of numbers, one for each $\text{-}\!\!\not\to$ rewrite, which we write in ascending order (equivalently, we write in sequence starting with the rightmost $\text{-}\!\!\not\to$ rewrite and working leftwards). For example, the rewrite sequence

$$M \equiv M_0 \text{-}\!\!\not\to M_1 \text{-}\!\!\not\to M_2 \to M_3 \to M_4 \text{-}\!\!\not\to M_5 \to M_6 \text{-}\!\!\not\to M_7 \equiv N$$

gets assigned the sequence $0, 1, 3, 3$. This sequence is our complexity measure.

Now take the sequence of $\to$ and $\text{-}\!\!\not\to$ rewrites reducing M to N. If this sequence does not contain a subsequence $M_i \text{-}\!\!\not\to M_{i+1} \to M_{i+2}$, then we have $M \to^* M' \text{-}\!\!\not\to^* N$, and hence $M \to^* M' \dashrightarrow^* N$ as required.

Otherwise, select a two-step subsequence $M_i \text{-}\!\!\not\to M_{i+1} \to M_{i+2}$. Using Lemma 5.11 followed by 5.10, replace this with a sequence $M_i \to M' \to^* M'' \text{-}\!\!\not\to M_{i+2}$. One thus obtains a new reduction sequence from M to N containing the same number of $\text{-}\!\!\not\to$ rewrites and at least as many $\to$ rewrites (possibly more). However, because the identified $\text{-}\!\!\not\to$ rewrite is shifted to the right, the complexity measure of the new sequence is below that of the original in the lexicographic ordering. Thus by repeatedly selecting two-step subsequences, we repeatedly reduce the complexity measure until we obtain a reduction sequence $M \to^* M' \text{-}\!\!\not\to^* N$ containing at least as many surface rewrites as the original sequence. Therefore $M \to^* M' \dashrightarrow^* N$, as required. □

6   Call-by-Need Operational Semantics for Lily

In the Lily expressions let $!x = s$ in t and $\text{rec}\, x : \sigma.\, t$, the variable x may occur zero, one or several times in t. Because of this, the natural implementation mechanism is call-by-need, whereby the evaluation of the terms substituted for such variables is shared. (In contrast, in an application $(\lambda x : \sigma.\, t)(s)$, the variable x occurs exactly once in t, and there is no call for sharing.) An operational semantics implementing such a call-by-need evaluation strategy is presented in [3], and the authors have proved that the call-by-need semantics does not affect the notion of contextual equivalence (private communication). In this section, we outline a straightforward proof of this result.

1. $(S,\, s(t),\, H) \to_{nd} (S \langle (-)(t) \rangle,\, s,\, H)$

2. $(S,\, \text{let }!x = s \text{ in } t,\, H) \to_{nd} (S \langle \text{let }!x = (-) \text{ in } t \rangle,\, s,\, H)$

3. $(S,\, t(\sigma),\, H) \to_{nd} (S \langle (-)(\sigma) \rangle,\, t,\, H)$

4.* $(S,\, \text{rec }x\!:\!\sigma.\, t,\, H) \to_{nd} (S \langle x \rangle,\, t,\, [x \mapsto t]H)$

5.* $(S \langle (-)(t) \rangle,\, \lambda x\!:\!\sigma.\, s,\, H) \to_{nd} (S,\, s[t/x],\, H)$

6.* $(S \langle \text{let }!x = (-) \text{ in } t \rangle,\, !s,\, H) \to_{nd} (S,\, t,\, [x \mapsto s]H)$

7.* $(S \langle (-)(\sigma) \rangle,\, \Lambda \alpha.\, t, H) \to_{nd} (S,\, t[\sigma/\alpha],\, H)$

8.* $(S,\, x,\, H) \to_{nd} (S \langle x \rangle,\, H(x),\, H)$

9. $(S \langle x \rangle,\, v,\, H) \to_{nd} (S,\, v,\, H[v/x])$

* *active reductions*, see Appendix A.

Fig. 7. Call-by-need Evaluation for Lily.

Again, rather than using the big-step operational semantics of [3], which is based on [5, 12], it is convenient for our purposes to use a small-step version, following [13]. We use $S, \ldots$ to range over *variable/frame stacks*, which are sequences of items of two forms: (i) $\langle F \rangle$, where F is an "evaluation frame",

$$F ::= (-)(t) \mid \text{let }!x = (-) \text{ in } t \mid (-)(\sigma) \; ;$$

(ii) or $\langle x \rangle$, for a variable x. We use H to range over *heaps*, which are finite sequences of assignments of the form $[x \mapsto t]$, with all variables x distinct.

The call-by-need evaluation relation is defined in Fig. 7. It implements a single-step relation of the form $(S, t, H) \to_{nd} (S', t', H')$. Roughly, this is interpreted as saying that the Lily term built up from t using the nested evaluation frames in S evaluates in a single step to the term built from t' using the frames in S'. In Fig. 7, when we write $[x \mapsto t]H$, we assume that x is not in the domain of H. We treat heaps H as functions, writing $H(x)$ for the value assigned to x, and writing $H[v/x]$ for the heap obtained from H by replacing the existing term assigned to x (which is assumed to be in the domain of H) with v.

The call-by-need evaluation of a Lily program $t : \sigma$ starts off with the configuration $(\varepsilon, t, \varepsilon)$ (where ε is the empty sequence) and then proceeds deterministically according to the rules in Fig. 7. Either an infinite sequence of $\to_{nd}$ reductions results, or the evaluation terminates in a configuration of the form (ε, v, H) for some (possibly open) value v. If the latter case holds, we write $t \downarrow_{nd}$. The main result of this section states that, for programs of arbitrary type, the call-by-need semantics terminates if and only if the call-by-name semantics does.

Theorem 6.1. *If $t : \sigma$ then $t \downarrow_{nd}$ if and only if $t \downarrow_{nm}$.*

The sharing of recursion implemented in Fig. 7, introduces cycles into the heap, and this makes it hard to give a direct operational proof of the equivalence of call-by-name and call-by-need, see [12] for discussion. This difficulty has, in fact, been overcome by the authors of [3], but their proof is highly involved (private communication).

We give a significantly simpler proof that call-by-name and call-by-need co-incide. First, we define an almost trivial translation of Lily into itself, which serves the purpose of "padding out" the call-by-name semantics (sic) for the purpose of facilitating its comparison with the call-by-need semantics. The remaining step is to prove that the "almost trivial" translation really is trivial. For this last step, we again translate into the untyped linear λ-calculus of Sec. 3, this time applying Theorem 5.5.

The almost trivial translation from Lily to itself, is the identity everywhere, except for the translation of thunks, which are padded with a dummy recursion, acting as delay.

$$(!s)^\dagger =_{\text{def}} !(\text{rec } z : \tau.\ s^\dagger) \qquad\qquad z \notin \text{fv}(s)\ .$$

Here, we are translating well-typed terms $\Gamma; \Delta \vdash t : \sigma$, to well-typed terms $\Gamma; \Delta \vdash t^\dagger : \sigma$, and the type τ introduced above is determined by this requirement.

Lemma 6.2. *If $t : \sigma$ then $t \downarrow_{\text{nd}}$ if and only if $t^\dagger \downarrow_{\text{nm}}$.*

To prove Lemma 6.2, one shows that the call-by-name evaluation of $t^\dagger$ simulates the call-by-need evaluation of t. Crucially, the padding of thunks ensures that rule 8 of Fig. 7 always corresponds to a $\rightarrow_{\text{nm}}$ reduction for the term generated from $t^\dagger$ by inserting it in the context determined by F and substituting, for each variable x with associated heap assignment $[x \mapsto s]$, a term $\text{rec } x : \sigma.\ s_0$, where s_0 is the term originally assigned to x when it was first added to the heap. More details are given in Appendix A.

Theorem 6.1 now follows from the lemma below, which is an easy application of Theorem 5.5.

Lemma 6.3. *If $t : \sigma$ then $t \downarrow_{\text{nm}}$ if and only if $t^\dagger \downarrow_{\text{nm}}$.*

Proof. We give another translation from Lily into our untyped linear λ-calculus.

$$
\begin{aligned}
x^\ddagger &=_{\text{def}} x & (\text{let } !x = s \text{ in } t)^\ddagger &=_{\text{def}} (\lambda!x.t^\ddagger)(s^\ddagger) \\
(\lambda x : \sigma.\ t)^\ddagger &=_{\text{def}} !(\lambda!x.t^\ddagger) & (\Lambda\alpha.\ t)^\ddagger &=_{\text{def}} !t^\ddagger \\
(s(t))^\ddagger &=_{\text{def}} (\lambda!w.w(!t^\ddagger))(s^\ddagger) & (t(\sigma))^\ddagger &=_{\text{def}} (\lambda!z.z)(t^\ddagger) \\
(!t)^\ddagger &=_{\text{def}} !t^\ddagger & (\text{rec } x : \sigma.\ t)^\ddagger &=_{\text{def}} \mu x.\ t^\ddagger
\end{aligned}
$$

It is easily established that, for $t : \sigma$ we have that $t \downarrow_{\text{nm}}$ if and only if $t^\ddagger$ is surface normalizing. However, we have $(t^\dagger)^\ddagger \dashrightarrow^* t^\ddagger$. Therefore, by Theorem 5.5, $t^\ddagger$ is surface normalizing if and only if $(t^\dagger)^\ddagger$ is. Thus indeed $t \downarrow_{\text{nm}}$ if and only if $t^\dagger \downarrow_{\text{nm}}$. $\qquad\square$

More generally, a similar application of Theorem 5.5 shows that call-by-name termination is preserved by the congruence relation on Lily terms generated by the call-by-name reductions. In other words, the natural "conversion relation" on Lily terms is correct with respect to contextual equivalence. Of course, the use of rewriting methods for establishing such simple results goes back to [9].

7 Linear Combinatory Algebras

The aim of this short final section is to demonstrate that our untyped linear λ-calculus is the λ-calculus counterpart of Abramsky's *linear combinatory algebras*, see [1]. This gives some evidence that our calculus arises reasonably naturally, independently of is applications to operational semantics.

Definition 7.1. A *!-applicative structure* is an algebra $(A, \cdot, !)$ where $\cdot$ is a binary operation on the set A and $!$ is a unary operation.

As is standard, we usually omit the "application" operation '$\cdot$', using a simple juxtaposition xy for $x \cdot y$. Application associates to the left (i.e. $xyz = (xy)z$).

Definition 7.2 ([1]). A *linear combinatory algebra* is a !-applicative structure $(A, \cdot, !)$ in which there exist elements $\mathsf{I}, \mathsf{B}, \mathsf{C}, \mathsf{K}, \mathsf{W}, \mathsf{D}, \delta, \mathsf{F} \in A$ satisfying:

$$
\begin{aligned}
\mathsf{I}x &= x & \mathsf{W}x(!y) &= x(!y)(!y) \\
\mathsf{B}xyz &= x(yz) & \mathsf{D}(!x) &= x \\
\mathsf{C}xyz &= xzy & \delta(!x) &= !!x \\
\mathsf{K}x(!y) &= x & \mathsf{F}(!x)(!y) &= !(xy) \ .
\end{aligned}
$$

The main result of this section asserts that linear combinatory algebras are characterized by a form of combinatory completeness in which the forms of implicit λ-abstraction available correspond to the two forms $\lambda x.M$ and $\lambda!x.M$ of our untyped linear λ-calculus. Moreover, the equalities associated with the implicit abstractions agree with the two redex forms in Fig. 3.

A *!-applicative polynomial* over a set A is a syntactic expression built up using elements of A as constants, variables $x, y, \ldots$, and operator symbols '$\cdot$', and '$!$'. Any !-applicative structure $(A, \cdot, !)$ induces an evident equality relation between polynomials.

We say that a variable x is *linear* in a !-applicative polynomial e, if it occurs exactly once, and not within the scope of a '$!$'-operator symbol. We write vars(e) for the set of variables occurring in e, and linvars(e) for the set of variables that are linear in e.

Theorem 7.3 (Linear combinatory completeness). *For any !-applicative structure $(A, \cdot, !)$, the following are equivalent.*

1. *$(A, \cdot, !)$ is a linear combinatory algebra.*
2. *For any !-applicative polynomial e over A,*
   (a) *if $x \in$ linvars(e) then there exists a polynomial $\lambda^* x. e$ with vars$(\lambda^* x. e) =$ vars$(e) - \{x\}$ and linvars$(\lambda^* x. e) =$ linvars$(e) - \{x\}$ such that the equality $(\lambda^* x. e)(x) = e$ holds;*
   (b) *there exists a polynomial $\lambda!^* x. e$ with vars$(\lambda!^* x. e) =$ vars$(e) - \{x\}$ and linvars$(\lambda!^* x. e) =$ linvars$(e) - \{x\}$ such that $(\lambda!^* x. e)(!x) = e$.*

It follows easily from the theorem that the closed linear terms of our untyped λ-calculus, considered modulo $=_\beta$ (see Sec. 5), themselves form a linear combinatory algebra.

References

1. S. Abramsky, E. Haghverdi, and P. Scott. Geometry of interaction and linear combinatory algebras. *Math. Struct. in Comp. Sci.*, 12:625–665, 2002.
2. A. Barber. *Linear Type Theories, Semantics and Action Calculi*. PhD thesis, Department of Computer Science, University of Edinburgh, 1997.
3. G.M. Bierman, A.M. Pitts, and C.V. Russo. Operational properties of Lily, a polymorphic linear lambda calculus with recursion. *Elect. Notes in Theor. Comp. Sci.*, 41, 2000.
4. D.J. Howe. Proving congruence of bisimulation in functional programming languages. *Inf. and Comp.*, 124:103–112, 1996.
5. J. Launchbury. A natural semantics for lazy evaluation. In *Proc. 20th POPL*, pages 144–154, 1993.
6. I.A. Mason and C.L. Talcott. Equivalence in functional languages with effects. *J. Functional Programming*, 1:287–327, 1991.
7. P.W. O'Hearn and J.C. Reynolds. From Algol to polymorphic linear lambda-calculus. *Journal of the ACM*, 47:167–223, 2000.
8. A.M. Pitts. Parametric polymorphism and operational equivalence. *Math. Struct. in Comp. Sci.*, 10:321–359, 2000.
9. G.D. Plotkin. Call-by-name, call-by-value and the λ-calculus. *Theor. Comp. Sci.*, 1:125–159, 1975.
10. G.D. Plotkin. Type theory and recursion. Invited talk at *8th Symposium on Logic in Comp. Sci.*, 1993.
11. J.C. Reynolds. Types, abstraction and parametric polymorphism. In *Information Processing '83*, pages 513–523. North Holland, 1983.
12. J. Seaman and S.P. Iyer. An operational semantics of sharing in lazy evaluation. *Science of Computer Programming*, 27:289–322, 1996.
13. P. Setstoft. Deriving a lazy abstract machine. *J. Functional Programming*, 7:231–248, 1999.

A   Outline Proof of Lemma 6.2

The main technical lemma we need is Lemma A.1 below. This concerns configurations (S, t, H) arrived at by a sequence $(\varepsilon, s, \varepsilon) \rightarrow_{\mathrm{nd}}^* (S, t, H)$ for some program $s : \sigma$. Given such a sequence, and any term t' with $\mathrm{fv}(t')$ contained in the domain of H, (the term t is one such), we define $t'[H]$ as follows. If $H = \varepsilon$ then $t'[H] =_{\mathrm{def}} t$. If $H = [x \mapsto u]H'$ then $t'[H] =_{\mathrm{def}} (t'[(\mathrm{rec}\, x : \sigma'.\ u_0)/x])[H']$, where u_0 is the first value assigned to x in a heap occuring along the sequence $(\varepsilon, s, \varepsilon) \rightarrow_{\mathrm{nd}}^* (S, t, H)$, and σ' is the appropriate type. Here $t'[H]$ is an abuse of notation since the value does not solely depend on H. In fact, for any two heaps H_1, H_2 occurring in the sequence $(\varepsilon, s, \varepsilon) \rightarrow_{\mathrm{nd}}^* (S, t, H)$ and containing $\mathrm{fv}(t')$, it holds that $t'[H_1] \equiv t'[H_2]$. Also, for any term t' we define $[S]t'$ as follows. If $S = \varepsilon$ then $[S]t' =_{\mathrm{def}} t'$. If $S = S' \langle (-)(s') \rangle$ then $[S]t' =_{\mathrm{def}} [S'](t'(s))$. If $S = S' \langle \mathrm{let}\ !x = - \ \mathrm{in}\ s' \rangle$ then $[S]t' =_{\mathrm{def}} [S'](\mathrm{let}\ !x = t'\ \mathrm{in}\ s')$. If $S = S' \langle (-)(\sigma') \rangle$ then $[S]t' =_{\mathrm{def}} [S'](t'(\sigma'))$. If $S = S' \langle x \rangle$ then $[S]t' =_{\mathrm{def}} [S']t'$. Finally, we call reductions number 4–8, in Fig. 7, *active*, and the others *passive*.

Lemma A.1. *Suppose $s : \sigma$ and $(\varepsilon, s, \varepsilon) \to_{nd}^{*} (S, t, H)$.*

1. *If x is declared in H then $(x[H])^{\dagger} \to_{nm}^{+} ((H(x))[H])^{\dagger}$.*
2. *If $S = S_0 \langle x \rangle S_1$ then $(x[H])^{\dagger} \to_{nm}^{+} (([S_1]t)[H])^{\dagger}$.*
3. *If $(S, t, H) \to_{nd} (S', t', H')$, where $S = S_0 S_1$ and $S' = S_0 S_1'$, then it holds that $(([S_1]t)[H])^{\dagger} \to_{nm}^{*} (([S_1']t')[H'])^{\dagger}$. Moreover, if the call-by-need reduction step is active then the call-by-name sequence contains at least one reduction.*

All three statements are proved simultaneously, by induction on the length of the reduction sequence $(\varepsilon, s, \varepsilon) \to_{nd}^{*} (S, t, H)$. For space reasons, we omit the details.

Proof (of Lemma 6.2). If $t \downarrow_{nd}$ then it follows easily from Lemma A.1.3 that $t^{\dagger} \downarrow_{nm}$. If $t \not\downarrow_{nd}$ then there exists an infinite $\to_{nd}$ reduction sequence from $(\varepsilon, t, \varepsilon)$. Because the four passive reductions either strictly reduce the size of the term component in a configuration, or retain the same term and reduce the size of the stack, the infinite sequence cannot contain infinitely many consecutive passive reductions. Therefore, it must contain infinitely many active reductions. Thus, again by Lemma A.1.3, $t^{\dagger}$ has an infinite $\to_{nm}$ reduction sequence. So indeed $t^{\dagger} \not\downarrow_{nm}$. □

Higher-Order Matching
in the Linear Lambda Calculus
in the Absence of Constants Is NP-Complete

Ryo Yoshinaka

[1] Graduate School of Interdisciplinary Information Studies, University of Tokyo,
7–3–1, Hongo, Bunkyo-ku, Tokyo, 113–0033, Japan
[2] National Institute of Informatics,
2–1–2 Hitotsubashi, Chiyoda-ku, Tokyo 101–8430, Japan
ry@nii.ac.jp

Abstract. A lambda term is linear if every bound variable occurs exactly once. The same constant may occur more than once in a linear term. It is known that higher-order matching in the linear lambda calculus is NP-complete (de Groote 2000), even if each unknown occurs exactly once (Salvati and de Groote 2003). Salvati and de Groote (2003) also claim that the interpolation problem, a more restricted kind of matching problem which has just one occurrence of just one unknown, is NP-complete in the linear lambda calculus. In this paper, we correct a flaw in Salvati and de Groote's (2003) proof of this claim, and prove that NP-hardness still holds if we exclude constants from problem instances. Thus, multiple occurrences of constants do not play an essential role for NP-hardness of higher-order matching in the linear lambda calculus.

1 Introduction

While the second-order unification problem modulo β and $\beta\eta$ [5] and the sixth-order matching problem modulo β are undecidable [9] (matching modulo $\beta\eta$ is still open), some subclasses of the matching problem are known to be decidable. In the second-order case [6], third-order case [4], and fourth-order case [10], the matching problem is decidable. These results as well as others we will mention below hold for both β and $\beta\eta$-matching.

Another kind of restriction on matching concerns the number of occurrences of variables in solutions (and in problem instances). A λ-term is k-*duplicating* if every variable (other than unknowns) occurs free at most k times in each of its subterms[1]. Dougherty and Wierzbicki [3] show that the matching problem of determining whether there is a k-duplicating solution for a given problem instance is decidable. Moreover, they show matching in the 1-duplicating (i.e., affine) lambda calculus is NP-complete by extending the result by de Groote [1] that matching in the *linear* lambda calculus is NP-complete.

---

[1] This definition slightly differs from the definition given by Dougherty and Wierzbicki [3]. But the difference is not essential.

J. Giesl (Ed.): RTA 2005, LNCS 3467, pp. 235–249, 2005.

An affine λ-term is linear if every lambda abstractor binds exactly one occurrence of the bound variable. Salvati and de Groote [12] prove that NP-hardness of the second-order matching problem in the linear lambda calculus still holds if the linearity is imposed on occurrences of unknowns, i.e., each unknown occurs exactly once in the problem.

There is another result on matching involving the linearity given by Levy [8]. He gives some conditions under which second-order unification with the restriction that a solution must substitute linear λ-terms for unknowns is decidable.

In Salvati and de Groote's paper [12], they also discuss a more restricted kind of matching problem, called the *interpolation problem*, which has just one occurrence of just one unknown. Interpolation in the linear lambda calculus plays an important role for parsing and generation with a certain grammar formalism [2, 11], but we will not present this formalism here. Regrettably, Salvati and de Groote's proof of NP-completeness of third-order interpolation in the linear lambda calculus contains an error. In this paper we correct the flaw and prove NP-completeness of third-order interpolation in the linear lambda calculus. While no free variable occurs twice or more in a linear λ-term, the number of occurrences of constants is not constrained. Since constants behave like free variables, the natural question arises which asks whether NP-hardness still holds when we exclude constants from problem instances. This paper shows that fourth-order interpolation in the linear lambda calculus is NP-complete even in the absence of constants. Therefore, multiple occurrences of constants do not play an essential role for NP-hardness of higher-order matching in the linear lambda calculus.

2   Basic Definitions

Definition 1. Let $\mathcal{A}$ be a finite set of *atomic types*. The set $\mathbb{T}(\mathcal{A})$ of *types* built on $\mathcal{A}$ is defined as follows:

- An atomic type $o \in \mathcal{A}$ is a type in $\mathbb{T}(\mathcal{A})$.
- If $\gamma, \delta \in \mathbb{T}(\mathcal{A})$, then $\gamma \to \delta \in \mathbb{T}(\mathcal{A})$.

The *order* Od of a type is defined as follows:

- The order $\mathrm{Od}(o)$ of an atomic type $o \in \mathcal{A}$ is 1.
- The order of $\gamma \to \delta$ is defined by $\mathrm{Od}(\gamma \to \delta) = \max\{\mathrm{Od}(\gamma) + 1, \mathrm{Od}(\delta)\}$.

$\gamma_1 \to \gamma_2 \to \cdots \to \gamma_n \to \delta$ abbreviates $\gamma_1 \to (\gamma_2 \to (\cdots \to (\gamma_n \to \delta)\ldots)))$ and $\gamma^n \to \delta$ abbreviates $\underbrace{\gamma \to \cdots \to \gamma}_{n \text{ times}} \to \delta$.

Definition 2. A *higher-order signature* Σ is a triple $\langle \mathcal{A}, \mathsf{C}, \mathrm{Tp} \rangle$ where $\mathcal{A}$ is a finite set of *atomic types*, C is a finite set of *constants*, and Tp is a function from C to $\mathbb{T}(\mathcal{A})$.

Definition 3. For a higher-order signature Σ, we obtain a countably infinite set $\mathcal{X}$ of *variables* and a countably infinite set $\mathcal{U}$ of *unknowns* such that for every type $\gamma \in \mathbb{T}(\mathcal{A})$, there are infinitely many variables of type γ and infinitely many unknowns of type γ. The set of λ-*terms* constructed on $\Sigma = \langle \mathcal{A}, \mathsf{C}, \mathbb{T}\mathfrak{p} \rangle$ and the extension of $\mathbb{T}\mathfrak{p}$ to λ-terms are recursively defined as follows:

- A constant $\mathsf{a} \in \mathsf{C}$ is a λ-term.
- A variable $x \in \mathcal{X}$ of type $\gamma \in \mathbb{T}(\mathcal{A})$ is a λ-term. We write $\mathbb{T}\mathfrak{p}(x) = \gamma$.
- An unknown $X \in \mathcal{U}$ of type $\gamma \in \mathbb{T}(\mathcal{A})$ is a λ-term. We write $\mathbb{T}\mathfrak{p}(X) = \gamma$.
- For two λ-terms M and N, if $\mathbb{T}\mathfrak{p}(M) = \gamma \to \delta$ and $\mathbb{T}\mathfrak{p}(N) = \gamma$ then MN is a λ-term of type $\mathbb{T}\mathfrak{p}(MN) = \delta$.
- For a variable $x \in \mathcal{X}$ of a type γ and a λ-term M, $\lambda x.M$ is a λ-term of type $\mathbb{T}\mathfrak{p}(\lambda x.M) = \gamma \to \mathbb{T}\mathfrak{p}(M)$.

The set of *free variables* of a λ-term denoted by FV is defined as follows:

- $\mathrm{FV}(\mathsf{a}) = \varnothing$ for $\mathsf{a} \in \mathsf{C}$. $\mathrm{FV}(x) = \{x\}$ for $x \in \mathcal{X}$. $\mathrm{FV}(X) = \varnothing$ for $X \in \mathcal{U}$.
- $\mathrm{FV}(MN) = \mathrm{FV}(M) \cup \mathrm{FV}(N)$ for λ-terms M and N.
- $\mathrm{FV}(\lambda x.M) = \mathrm{FV}(M) - \{x\}$ for a variable $x \in \mathcal{X}$ and a λ-term M.

Definition 4. A λ-term M is *closed* iff (if and only if) $\mathrm{FV}(M) = \varnothing$. M is *affine* iff for every subterm N of M, any variable x occurs free in N at most once. M is *relevant* iff for every subterm $\lambda x.N$ of M, x occurs free in N at least once. M is *linear* iff M is affine and relevant.

Note that a closed λ-term can contain unknowns and that constants and unknowns may occur any number of times in a linear λ-term.

To represent types of linear λ-terms, we use $\multimap$ instead of $\to$ according to the custom. We use upper case italic letters $A, B, C, \ldots$ for λ-terms, but X and Y for unknowns, lower case italic letters $x, y, z, \ldots$ for variables, and sanserif $\mathsf{a}, \mathsf{b}, \ldots$ for constants. We abbreviate $(\ldots (M_1 M_2) \ldots) M_n$ to $M_1 M_2 \ldots M_n$, and $\lambda x_1.(\ldots (\lambda x_n.M) \ldots)$ to $\lambda x_1 \ldots x_n.M$ or $\lambda \vec{x}.M$ if $\vec{x}$ means the sequence $x_1, \ldots, x_n$. Also for a finite set $\mathcal{Y}$ of variables, we write $\lambda \mathcal{Y}.M$ under the assumption that the variables in $\mathcal{Y}$ are ordered in an appropriate manner. We define $MN^0 = M$ and $MN^{n+1} = MN^n N$. We write $M[x := N]$ for the usual capture-avoiding substitution of a λ-term N for a variable $x \in \mathcal{X}$, and similarly, $M[X := N]$ for capture-avoiding substitution for an unknown $X \in \mathcal{U}$. We abbreviate a simultaneous substitution $[x_1 := N_1, \ldots, x_n := N_n]$ to $[\vec{x} := \vec{N}]$ when $\vec{x}$ means $x_1, \ldots, x_n$ and $\vec{N}$ means $N_1, \ldots, N_n$. For a substitution σ, $M_1 \ldots M_n \sigma$ means $(M_1 \ldots M_n)\sigma$. If a λ-term M is of the form $M = M_0 M_1 \ldots M_n$ where $M_0 \in \mathsf{C} \cup \mathcal{X} \cup \mathcal{U}$, we say that M_0 is the *head* of M and that M_i is an *argument* of M_0 in M for $i \geq 1$. We define β-normal form, η-long form, β-reduction $\to_\beta$, $\beta\eta$-equivalence $=_{\beta\eta}$ etc., in the usual way. We simply say "normal form" for β-normal form and "long form" for η-long form.

Definition 5. A *unification equation* is a pair of closed λ-terms $\langle L, R \rangle$ of the same type. Let $\vec{X}$ be the unknowns in $\langle L, R \rangle$. A substitution $\sigma = [\vec{X} := \vec{N}]$ where $\vec{N}$ contain no unknowns is a *solution* for $\langle L, R \rangle$ modulo β iff $L\sigma =_\beta R\sigma$. σ is a *solution* for $\langle L, R \rangle$ modulo $\beta\eta$ iff $L\sigma =_{\beta\eta} R\sigma$. A *matching equation* is a unification equation $\langle L, R \rangle$ such that R has no unknowns. An *interpolation* *equation* is a matching equation $\langle L, R \rangle$ such that L has just one occurrence of just one unknown as its head, i.e., $L = X L_1 \ldots L_m$, where each L_i contains no unknowns. The *order* of a unification equation is defined to be the maximum of the orders of the types of the unknowns appearing in the equation. The *unification, matching,* and *interpolation problem modulo β ($\beta\eta$)* are decision problems which ask whether or not there is a solution modulo β ($\beta\eta$) for given unification, matching, interpolation equations respectively. The unification, matching, and interpolation problem *in the linear lambda calculus* allow only linear λ-terms as problem instances and solutions.

3 NP-Complete Varieties of the Satisfiability Problem

As a preparation for showing NP-completeness of interpolation in the linear lambda calculus, we introduce some NP-complete problems.

Definition 6. Let $\mathcal{V}$ be a finite set of *Boolean variables*. Let $\mathcal{V}$ be the set of *positive literals* and $\neg\mathcal{V} = \{\neg v \mid v \in \mathcal{V}\}$ the set of *negative literals*. We then define the set of *literals* as $\mathcal{V} \cup \neg\mathcal{V}$. An *instance of the satisfiability problem (or Sat in short)* $\mathcal{S}$ on $\mathcal{V}$ is a collection of *clauses* which are non-empty subsets of $\mathcal{V} \cup \neg\mathcal{V}$. A *valuation* ψ on $\mathcal{V}$ is a mapping from $\mathcal{V} \cup \neg\mathcal{V}$ to $\{0, 1\}$ such that $\psi(v) + \psi(\neg v) = 1$ for all $v \in \mathcal{V}$. A clause $\mathcal{C} \in \mathcal{S}$ is *satisfied by a valuation* ψ *via a literal* $w \in \mathcal{V} \cup \neg\mathcal{V}$ iff $w \in \mathcal{C}$ and $\psi(w) = 1$. A Sat $\mathcal{S}$ is *satisfied by a valuation* ψ iff every $\mathcal{C} \in \mathcal{S}$ is satisfied by ψ. A Sat $\mathcal{S}$ is *satisfiable* iff there is ψ that satisfies $\mathcal{S}$.

It is well known that the question of whether a given $\mathcal{S}$ is satisfiable is NP-complete.

Definition 7. An *mP-Sat* $\mathcal{S}$ is a Sat such that each positive literal $v \in \mathcal{V}$ occurs exactly m times in $\mathcal{S}$. An *nN-Sat* $\mathcal{S}$ is a Sat such that each negative literal $\neg v \in \neg\mathcal{V}$ occurs exactly n times in $\mathcal{S}$.

A Sat $\mathcal{S}$ is *polarized* iff each clause $\mathcal{C} \in \mathcal{S}$ contains only positive literals or only negative literals. We say $\mathcal{C}$ is *positive* if $\mathcal{C} \subseteq \mathcal{V}$, and $\mathcal{C}$ is *negative* if $\mathcal{C} \subseteq \neg\mathcal{V}$.

By combining the above definitions we define an *mPnN-Sat* to be a Sat which is at the same time an *mP-Sat* and an *nN-Sat*.

Theorem 1. *The question of whether a given 2P1N-Sat $\mathcal{S}$ is satisfiable or not is NP-complete.*

Proof. For a given Sat $\mathcal{S}$ on $\mathcal{V}$, we construct a 2P1N-Sat $\mathcal{S}'$ on $\mathcal{V}'$ such that $\mathcal{S}$ is satisfiable iff $\mathcal{S}'$ is satisfiable.

(i) For each $v_i \in \mathcal{V}$, let m_i be the number of occurrences of positive literal v_i in $\mathcal{S}$ and n_i be the number of occurrences of negative literal $\neg v_i$ in $\mathcal{S}$.
(ii) Introduce new Boolean variables $v_{i,j}$ for $1 \leq j \leq m_i$ and $u_{i,k}$ for $1 \leq k \leq n_i$.
(iii) Replace the j-th occurrence of the positive literal v_i with $v_{i,j}$ for $1 \leq j \leq m_i$.
(iv) Replace the k-th occurrence of the negative literal $\neg v_i$ with $u_{i,k}$ for $1 \leq k \leq n_i$.

(v) Add clauses $\{v_{i,j}, \neg v_{i,j+1}\}$ for $1 \leq j < m_i$, $\{v_{i,m_i}, u_{i,1}\}$, $\{\neg u_{i,k}, u_{i,k+1}\}$ for $1 \leq k < n_i$, and $\{\neg u_{i,n_i}, \neg v_{i,1}\}$. (If $m_i = 0$, then add the clause $\{\neg u_{i,n_i}, u_{i,1}\}$. If $n_i = 0$, then add the clause $\{v_{i,m_i}, \neg v_{i,1}\}$.)

By the construction, $\mathcal{S}'$ is a 2P1N-Sat. It is clear that if $\mathcal{S}$ is satisfied by ψ, then $\mathcal{S}'$ is also satisfied by ψ' where $\psi'(v_{i,j}) = \psi(v_i)$ and $\psi'(u_{i,k}) = \psi(\neg v_i)$. Conversely, suppose that $\mathcal{S}'$ is satisfied by a valuation ψ'. The condition (v) ensures that $\psi'(v_{i,1}) = 0$ implies $\psi'(v_{i,2}) = \cdots = \psi'(v_{i,m_i}) = \psi'(\neg u_{i,1}) = \cdots = \psi'(\neg u_{i,n_i}) = 0$ and that $\psi'(v_{i,1}) = 1$ implies $\psi'(\neg u_{i,n_i}) = \cdots = \psi'(\neg u_{i,1}) = \psi'(v_{i,m_i}) = \cdots = \psi'(v_{i,2}) = 1$. Therefore, $\mathcal{S}$ is also satisfied by ψ for $\psi(v_i) = \psi'(v_{i,1})$. □

Theorem 2 (Kilpeläinen and Mannila [7]). *The question of whether a given polarized 1N-Sat $\mathcal{S}$ is satisfiable or not is NP-complete[2].*

Proof. For a given Sat $\mathcal{S}$ on $\mathcal{V}$, we construct a polarized 1N-Sat $\mathcal{S}'$ on $\mathcal{V}'$ such that $\mathcal{S}$ is satisfiable iff $\mathcal{S}'$ is satisfiable.

$$\mathcal{V}' = \{v_i, u_i \mid v_i \in \mathcal{V}\}$$
$$\mathcal{S}' = \{\mathcal{C}' \mid \mathcal{C} \in \mathcal{S}\} \cup \{\{v_i, u_i\}, \{\neg v_i, \neg u_i\} \mid v_i \in \mathcal{V}\}$$
$$\text{where } \mathcal{C}' = \{v_i \mid v_i \in \mathcal{C}\} \cup \{u_j \mid \neg v_j \in \mathcal{C}\}$$
□

Definition 8. For a polarized 1N-Sat $\mathcal{S} = \{\mathcal{C}_1, \ldots, \mathcal{C}_m, \mathcal{D}_1, \ldots, \mathcal{D}_n\}$ on $\mathcal{V} = \{v_1, \ldots, v_l\}$ where each $\mathcal{C}_j$ is a positive clause and each $\mathcal{D}_k$ is a negative clause, we define the following two functions $\mu_{\mathcal{S}}$ and $\nu_{\mathcal{S}}$ which represent the positive and negative occurrences of each Boolean variable respectively:

$$\mu_{\mathcal{S}}(i,j) = \begin{cases} j & \text{if } v_i \in \mathcal{C}_j \\ 0 & \text{otherwise} \end{cases}, \quad \nu_{\mathcal{S}}(i) = k \text{ for } \neg v_i \in \mathcal{D}_k$$

for $1 \leq i \leq l$ and $1 \leq j \leq m$.

Since each negative literal occurs exactly once in a 1N-Sat, the unary function $\nu_{\mathcal{S}}$ is well-defined. We omit the subscript $\mathcal{S}$ of $\mu_{\mathcal{S}}$ and $\nu_{\mathcal{S}}$ if no confusion occurs.

4   Interpolation in the Linear Lambda Calculus

While every interpolation equation $\langle X L_1 \ldots L_m, R \rangle$ has a trivial solution $[X := \lambda x_1 \ldots x_m.R]$ in the (general) lambda calculus, the interpolation problem in the *linear* lambda calculus is not trivial. That the interpolation problem in the linear lambda calculus is in NP is an immediate corollary of the following theorem.

Theorem 3 (de Groote [1]). *The matching problem in the linear lambda calculus is NP-complete.*

---

[2] In the original paper, they do not mention the restriction on polarity, though their proof entails it.

Note that the definition of matching equations by de Groote [1] allows λ-terms which are not closed and he proves that the problem is in NP under the assumption that R is closed. He claims that if R has a free variable x, $\langle L, R \rangle$ admits a solution iff $\langle \lambda x.L, \lambda x.R \rangle$ admits a solution. That is not the case for $L = X$ and $R = x$, since a substitution for an unknown avoids capturing free variables. But this is not a serious error at all. Clearly $\langle L, R \rangle$ admits a solution iff $\langle L', R' \rangle$ admits a solution where L' and R' are obtained by replacing each free variable x in L and R with a fresh constant a_x.

Theorem 4 (Salvati and de Groote [12]). *Second-order matching in the linear lambda calculus is NP-complete even if every unknown appears exactly once.*

As already mentioned in the introduction, Salvati and de Groote [12, Proposition 3] claim that *third-order interpolation in the linear lambda calculus is NP-complete* by a reduction from a 1N-Sat. For a given 1N-Sat $\mathcal{S} = \{\mathcal{C}_1, \ldots, \mathcal{C}_m\}$ on $\mathcal{V} = \{v_1, \ldots, v_l\}$, they define a third-order interpolation equation $\langle L, R \rangle$ on $\Sigma = \langle \mathcal{A}, \mathsf{C}, \mathrm{Tp} \rangle$ as follows[3]:

$$\mathcal{A} = \{o\}, \ \mathsf{C} = \{\mathsf{a}, \mathsf{c}_j, \mathsf{g} \mid 1 \leq j \leq m\},$$
$$\mathrm{Tp}(x_j) = \mathrm{Tp}(\mathsf{a}) = o, \ \mathrm{Tp}(\mathsf{c}_j) = o^m \multimap o, \ \mathrm{Tp}(\mathsf{g}) = o^l \multimap o,$$
$$L = X(\lambda x_1 \ldots x_m.\mathsf{c}_1 x_1 \ldots x_m) \ldots (\lambda x_1 \ldots x_m.\mathsf{c}_m x_1 \ldots x_m),$$
$$R = \mathsf{g} V_1 \ldots V_l \text{ for } V_i = N_i P_{i,1} \ldots P_{i,m}$$
$$\text{where } N_i = \mathsf{c}_j \text{ for } \neg v_i \in \mathcal{C}_j \text{ and } P_{i,j} = \begin{cases} \mathsf{c}_j \mathsf{a}^m & \text{if } v_i \in \mathcal{C}_j \\ \mathsf{a} & \text{otherwise} \end{cases}.$$

We present a counterexample to their claim that $\mathcal{S}$ is satisfiable iff $\langle L, R \rangle$ has a solution. Consider their reduction from the following 1N-Sat $\mathcal{S}$:

INSTANCE $\mathcal{S}$: $\mathcal{C}_1 = \{v_1\}$, $\mathcal{C}_2 = \{\neg v_1\}$
REDUCTION $\langle L, R \rangle$: $L = X(\lambda x_1 x_2.\mathsf{c}_1 x_1 x_2)(\lambda x_1 x_2.\mathsf{c}_2 x_1 x_2)$
$$R = \mathsf{g}(\mathsf{c}_2(\mathsf{c}_1 \mathsf{a}\mathsf{a})\mathsf{a})$$

They state that $\langle L, R \rangle$ has no solution, since $\mathcal{S}$ is not satisfiable. In fact, however, $[X := \lambda y_1 y_2.\mathsf{g}(y_2(y_1 \mathsf{a}\mathsf{a})\mathsf{a})]$ is a solution for $\langle L, R \rangle$.

In this section, we present a correct proof for the proposition that *third-order interpolation in the linear lambda calculus is NP-complete* by a reduction from a *polarized* 1N-Sat.

Definition 9. Suppose that a polarized 1N-Sat $\mathcal{S} = \{\mathcal{C}_1, \ldots, \mathcal{C}_m, \mathcal{D}_1, \ldots, \mathcal{D}_n\}$ on $\mathcal{V} = \{v_1, \ldots, v_l\}$ is given where each $\mathcal{C}_j$ is a positive clause and each $\mathcal{D}_k$ is a negative clause. We define an interpolation equation $\langle L_\mathcal{S}, R_\mathcal{S} \rangle$ on a higher-order signature $\Sigma = \langle \mathcal{A}, \mathsf{C}, \mathrm{Tp} \rangle$, where $\mathcal{A} = \{o\}$, $\mathsf{C} = \{\mathsf{c}_j, \mathsf{d}_k, \mathsf{f}, \mathsf{g} \mid 0 \leq j \leq m$

---

[3] An inessential change is made to the original reduction to facilitate comparison with our reduction.

and $1 \leq k \leq n\}$, and $\mathrm{Tp}(x_j) = \mathrm{Tp}(c_j) = o$, $\mathrm{Tp}(d_k) = o^m \multimap o$, $\mathrm{Tp}(f) = o \multimap$ o, $\mathrm{Tp}(g) = o^l \multimap o$. Let

$$L_S = XC_1 \ldots C_m D_1 \ldots D_n$$

$$\text{where} \quad \begin{cases} C_j = \mathsf{f}\mathsf{c}_j \\ D_k = \lambda x_1 \ldots x_m.\mathsf{d}_k(\mathsf{f} x_1) \ldots (\mathsf{f} x_m) \\ \mathrm{Tp}(X) = o^m \multimap (o^m \multimap o)^n \multimap o \text{ and } \mathrm{Od}(X) = 3 \end{cases} ,$$

$$R_S = \mathsf{g} V_1 \ldots V_l$$

$$\text{where} \quad V_i = \mathsf{d}_{\nu(i)}(\mathsf{f}\mathsf{c}_{\mu(i,1)}) \ldots (\mathsf{f}\mathsf{c}_{\mu(i,m)}).$$

The intuition behind the reduction is the following. Each C_j in L_S represents the positive clause $\mathcal{C}_j$ and each D_k in L_S represents the negative clause $\mathcal{D}_k$. Each V_i in R_S represents the occurrences of the Boolean variable $v_i \in \mathcal{V}$ in the clauses of $\mathcal{S}$. V_i contains c_j for $j \neq 0$ (respectively d_k) iff v_i appears in $\mathcal{C}_j$ (resp. $\neg v_i$ appears in $\mathcal{D}_k$). If $\mathcal{S}$ is satisfied by a valuation ψ, then for each $\mathcal{C}_j$ (resp. $\mathcal{D}_k$), there is $v_i \in \mathcal{C}_j$ (resp. $\neg v_i \in \mathcal{C}_k$) such that $\psi(v_i) = 1$ (resp. $\psi(\neg v_i) = 1$). In this case, we can construct a solution $[X := \lambda y_1 \ldots y_m z_1 \ldots z_n.gU_1 \ldots U_l]$ which puts the argument C_j (resp. D_k) of X into U_i via y_j (resp. z_k) and makes U_i equivalent to V_i by β-reduction (see Example 1). Conversely if $\langle L_S, R_S \rangle$ has a solution $[X := S]$, then S must be of the form $\lambda y_1 \ldots y_m z_1 \ldots z_n.gU_1 \ldots U_l$ and must put each argument C_j (resp. D_k) of X into U_i for some i via y_j (resp. z_k) by the linearity. Then, one can find a valuation ψ such that if S puts the argument C_j (resp. D_k) into U_i, then ψ satisfies $\mathcal{C}_j$ via v_i (resp. $\mathcal{D}_k$ via $\neg v_i$). The presence of the constant f is the essential difference between Salvati and de Groote's reduction [12] and ours. Due to the number of occurrences of f in V_i for each i, C_j and D_k cannot simultaneously be put into the same U_i for any j and k. This corresponds to the fact that any valuation ψ on $\mathcal{V}$ cannot simultaneously satisfy $\mathcal{C}_j$ via v_i and $\mathcal{D}_k$ via $\neg v_i$ (see Example 2).

Example 1. INSTANCE $\mathcal{S}$: $\mathcal{C}_1 = \{v_1\}$, $\mathcal{C}_2 = \{v_1, v_2\}$, $\mathcal{D}_1 = \{\neg v_1, \neg v_2\}$

REDUCTION $\langle L_S, R_S \rangle$: $L_S = X(\mathsf{f}\mathsf{c}_1)(\mathsf{f}\mathsf{c}_2)(\lambda x_1 x_2.\mathsf{d}_1(\mathsf{f} x_1)(\mathsf{f} x_2))$

$$R_S = \mathsf{g}(\mathsf{d}_1(\mathsf{f}\mathsf{c}_1)(\mathsf{f}\mathsf{c}_2))(\mathsf{d}_1(\mathsf{f}\mathsf{c}_0)(\mathsf{f}\mathsf{c}_2))$$

Let ψ be defined as $\psi(v_1) = 1, \psi(v_2) = 0$. Corresponding to the fact that ψ satisfies $\mathcal{C}_1$ via v_1, $\mathcal{C}_2$ via v_1, and $\mathcal{D}_1$ via $\neg v_2$, we give a solution $[X := \lambda y_1 y_2 z_1.gU_1 U_2]$ which puts the argument C_1 of X into U_1, C_2 into U_1, and D_1 into U_2. That is, we give a solution $[X := S]$ where

$$S = \lambda y_1 y_2 z_1.\mathsf{g}(\mathsf{d}_1 y_1 y_2)(z_1 \mathsf{c}_0 \mathsf{c}_2).$$

Indeed, we obtain

$$L_S[X := S] = SC_1 C_2 D_1 \twoheadrightarrow_\beta \mathsf{g}(\mathsf{d}_1 C_1 C_2)(D_1 \mathsf{c}_0 \mathsf{c}_2) \twoheadrightarrow_\beta R_S.$$

Example 2. INSTANCE $\mathcal{S}$: $\mathcal{C}_1 = \{v_1\}$, $\mathcal{C}_2 = \{v_2\}$, $\mathcal{D}_1 = \{\neg v_1, \neg v_2\}$

REDUCTION $\langle L_{\mathcal{S}}, R_{\mathcal{S}} \rangle$: $L_{\mathcal{S}} = X(\mathsf{fc}_1)(\mathsf{fc}_2)\big(\lambda x_1 x_2.\mathsf{d}_1(\mathsf{f} x_1)(\mathsf{f} x_2)\big)$

$$R_{\mathcal{S}} = \mathsf{g}\big(\mathsf{d}_1(\mathsf{fc}_1)(\mathsf{fc}_0)\big)\big(\mathsf{d}_1(\mathsf{fc}_0)(\mathsf{fc}_2)\big)$$

$\mathcal{S}$ is not satisfiable and $\langle L_{\mathcal{S}}, R_{\mathcal{S}} \rangle$ has no solution. e.g.,

$$L_{\mathcal{S}}[X := \lambda y_1 y_2 z_1.\mathsf{g}\big(\mathsf{d}_1 y_1(\mathsf{fc}_0)\big)\big(z_1 \mathsf{c}_0 y_2\big)]$$
$$\twoheadrightarrow_\beta \mathsf{g}\big(\mathsf{d}_1(\mathsf{fc}_1)(\mathsf{fc}_0)\big)\big(\mathsf{d}_1(\mathsf{fc}_0)(\mathsf{f}(\mathsf{fc}_2))\big)$$
$$\neq_\beta R_{\mathcal{S}}.$$

Lemma 1. $\langle L_{\mathcal{S}}, R_{\mathcal{S}} \rangle$ *admits a solution whenever* $\mathcal{S}$ *is satisfiable.*

Proof. Suppose that $\mathcal{S}$ is satisfied by a valuation ψ. Then, for each clause of $\mathcal{S}$, one can choose a literal via which ψ satisfies the clause. Let a function ϕ from $\mathcal{S}$ to $\mathcal{V} \cup \neg\mathcal{V}$ be such a choice. That is, $\phi(\mathcal{C}_j) \in \mathcal{C}_j$, $\psi(\phi(\mathcal{C}_j)) = 1$, $\phi(\mathcal{D}_k) \in \mathcal{D}_k$, and $\psi(\phi(\mathcal{D}_k)) = 1$. Define S by

$$S = \lambda y_1 \ldots y_m z_1 \ldots z_n.\mathsf{g}U_1 \ldots U_l$$

$$U_i = \begin{cases} z_{\nu(i)}\mathsf{c}_{\mu(i,1)} \cdots \mathsf{c}_{\mu(i,m)} & \text{if } \phi(\mathcal{D}_{\nu(i)}) = \neg v_i \\ \mathsf{d}_{\nu(i)}U_{i,1} \ldots U_{i,m} & \text{otherwise} \end{cases}$$

$$U_{i,j} = \begin{cases} y_j & \text{if } \phi(\mathcal{C}_j) = v_i \\ \mathsf{fc}_{\mu(i,j)} & \text{otherwise} \end{cases}.$$

First we confirm the linearity of S. Each z_k indeed occurs exactly once in S, since z_k appears in U_i iff $\phi(\mathcal{D}_k) = \neg v_i$. For y_j, let i be such that $\phi(\mathcal{C}_j) = v_i$. Then, $\psi(v_i) = 1$ and thus $\phi(\mathcal{D}_{\nu(i)}) \neq \neg v_i$ because $\psi(\phi(\mathcal{D}_{\nu(i)})) = 1$. Hence, $U_i = \mathsf{d}_{\nu(i)}U_{i,1} \ldots U_{i,m}$ and the only occurrence of y_j in S is in $U_{i,j}$. Therefore, S is a linear λ-term.

In order to see that $L_{\mathcal{S}}[X := S] \twoheadrightarrow_\beta R_{\mathcal{S}}$, it is enough to check that $U_i\sigma \twoheadrightarrow_\beta V_i$ for the substitution $\sigma = [y_1 := C_1, \ldots, y_m := C_m, z_1 := D_1, \ldots, z_n := D_n]$. If $\phi(\mathcal{D}_{\nu(i)}) = \neg v_i$, then

$$U_i\sigma = D_{\nu(i)}\mathsf{c}_{\mu(i,1)} \cdots \mathsf{c}_{\mu(i,m)}$$
$$= (\lambda x_1 \ldots x_m.\mathsf{d}_{\nu(i)}(\mathsf{f} x_1) \ldots (\mathsf{f} x_m))\mathsf{c}_{\mu(i,1)} \cdots \mathsf{c}_{\mu(i,m)}$$
$$\twoheadrightarrow_\beta \mathsf{d}_{\nu(i)}(\mathsf{fc}_{\mu(i,1)}) \cdots (\mathsf{fc}_{\mu(i,m)})$$
$$= V_i.$$

If $\phi(\mathcal{D}_{\nu(i)}) \neq \neg v_i$, then it is easy to see that $U_{i,j}\sigma = \mathsf{fc}_{\mu(i,j)}$. If $\phi(\mathcal{C}_j) \neq v_i$, $U_{i,j}\sigma = \mathsf{fc}_{\mu(i,j)}\sigma = \mathsf{fc}_{\mu(i,j)}$. Otherwise, $\phi(\mathcal{C}_j) = v_i$ implies $\mu(i,j) = j$ and $U_{i,j}\sigma = y_j\sigma = C_j = \mathsf{fc}_j = \mathsf{fc}_{\mu(i,j)}$. Thus,

$$U_i\sigma = \mathsf{d}_{\nu(i)}U_{i,1} \ldots U_{i,m}\sigma \twoheadrightarrow_\beta \mathsf{d}_{\nu(i)}(\mathsf{fc}_{\mu(i,1)}) \cdots (\mathsf{fc}_{\mu(i,m)}) = V_i. \qquad \square$$

Lemma 2. $\mathcal{S}$ *is satisfiable whenever* $\langle L_{\mathcal{S}}, R_{\mathcal{S}} \rangle$ *admits a solution.*

Proof. Suppose that $[X := S]$ is a solution for $\langle L_S, R_S \rangle$. We can assume that S is in long normal form and $S = \lambda y_1 \ldots y_m z_1 \ldots z_n . S'$. Let σ denote the substitution $[\vec{y} := \vec{C}, \vec{z} := \vec{D}]$. Since $S'\sigma \twoheadrightarrow_\beta R_S$, S' must be equal to $g U_1 \ldots U_l$ for some λ-terms U_i such that $U_i \sigma \twoheadrightarrow_\beta V_i$. It is obvious that the head of each U_i is either $z_{\nu(i)}$ or $d_{\nu(i)}$. We show that S is satisfied by the valuation ψ defined as follows:

$$
\psi(v_i) = \begin{cases} 0 & \text{if the head of } U_i \text{ is } z_{\nu(i)} \\ 1 & \text{otherwise} \end{cases}
$$

We show that each positive clause $C_j \in S$ is satisfied by ψ. Suppose that y_j appears in U_i. Then, the head of U_i cannot be z_k for any k, because if both y_j and z_k are in U_i, $U_i \sigma$ contains at least $(m+1)$ occurrences of f, so $U_i \sigma$ never β-reduces to V_i, which contains exactly m occurrences of f. Thus, the head of U_i is $d_{\nu(i)}$ and $\psi(v_i) = 1$. Since $y_j \sigma = f c_j$, $U_i \sigma$ and V_i must contain c_j and this implies $\mu(i,j) = j$, i.e., $v_i \in C_j$. So ψ satisfies C_j via v_i.

We show that each negative clause $D_k \in S$ is satisfied by ψ. Suppose that z_k appears in U_i. Since $z_k \sigma$ contains d_k, $k = \nu(i)$ and the head of U_i is z_k. Therefore, $\psi(v_i) = 0$ and $\neg v_i \in D_k$. So ψ satisfies D_k via $\neg v_i$. $\square$

Proposition 1. *Third-order interpolation modulo β ($\beta\eta$) in the linear lambda calculus is NP-complete.*

Proof. By Theorem 3 and Lemmas 1 and 2. $\square$

The above proof also entails NP-hardness of third-order interpolation in the relevant lambda calculus.

It is clear that the second-order interpolation problem in the linear lambda calculus is in P. The following proposition demonstrates the essential difference between the second-order case and the third-order case.

Proposition 2. *Third-order interpolation modulo β ($\beta\eta$) in the linear lambda calculus is NP-complete even if all the constants, variables (other than the unknown) and arguments of the unknown in the problem instances have types o or $o \multimap o$.*

Proof. By a reduction from a 2P1N-Sat $S = \{C_1, \ldots, C_m\}$ on $V = \{v_1, \ldots, v_l\}$. We define an interpolation equation $\langle L, R \rangle$ on $\Sigma = \{A, C, \text{Tp}\}$ as follows:

$A = \{o\}$, $C = \{c_j, f \mid 1 \leq j \leq m\}$, $\text{Tp}(c_j) = \text{Tp}(f) = o \multimap o$,

$L = X C_1 \ldots C_m$

where $C_j = \lambda x. f(c_j (f x))$ for $\text{Tp}(x) = o$ and $\text{Tp}(X) = (o \multimap o)^m \multimap o \multimap o$,

R is the normal form of $\lambda x. V_1(\ldots(V_l x)\ldots)$

where $V_i = \lambda x.(f(c_{\pi(i,1)}(f(c_{\nu(i)}(f(c_{\pi(i,2)}(f x)))))))$,

$\pi(i,k) = j$ if the k-th occurrence of the positive literal v_i is in C_j, and

$\nu(i) = j$ if the occurrence of the negative literal $\neg v_i$ is in C_j.

The intuition behind the reduction is similar to the previous reduction. To facilitate discussion, assume that $\pi(i,1) \neq \nu(i) \neq \pi(i,2)$ for every i. If there is a solution σ for $\langle L, R \rangle$, then one can find a valuation ψ which satisfies S such that whenever σ assign c_j in C_j to $c_{\pi(i,k)}$ in V_i, ψ satisfies C_j via v_i, and whenever σ assign c_j in C_j to $c_{\nu(i)}$ in V_i, ψ satisfies C_j via $\neg v_i$. The occurrences of the constant f makes it impossible for both $C_{\nu(i)}$ and $C_{\pi(i,k)}$ to be simultaneously used to construct V_i. This constraint corresponds to the fact that no valuation simultaneously satisfies $C_{\pi(i,k)}$ via v_i and $C_{\nu(i)}$ via $\neg v_i$. Conversely, we easily see that if S is satisfiable, then $\langle L, R \rangle$ has a solution. □

The problem in Proposition 2 is almost the same problem of determining whether there is a permutation θ of a sequence of strings $\langle w_1, \ldots, w_m \rangle$ over an alphabet C such that the concatenation of $w_{\theta(1)}, \ldots, w_{\theta(m)}$ coincides with a string w_R over C. Though the problem seems fundamental, the author could not find a reference on the complexity of it and thus has invented his own proof of its NP-completeness. The idea of the proof is used in the proof of Proposition 2.

The reader may wonder why we have presented a reduction in Definition 9, though the new reduction in the proof of Proposition 2 is equally simple and Proposition 2 is stronger than Proposition 1. The reason is that in the next section, we will eliminate constants from $\langle L_S, R_S \rangle$ defined in Definition 9 with a certain technique, which does not work for the new reduction.

5 Elimination of Constants

In this section, we show that the interpolation problem in the linear lambda calculus is NP-hard even if there are no constants by eliminating constants in $\langle L_S, R_S \rangle$ in Definition 9.

For general unification problem, it is easy to construct a *constant-free* $\langle P^*, Q^* \rangle$ from a unification equation $\langle P, Q \rangle$, such that $\langle P, Q \rangle$ has a solution iff $\langle P^*, Q^* \rangle$ has a solution. We obtain P^* and Q^* by successive transformations performed on P and Q: first we replace the constants $\vec{a}$ by fresh variables $\vec{a}$, then we replace each unknown X_i with $Y_i \vec{a}$ and finally we abstract the free variables. If $[X_i := S_i]$ is a solution for $\langle P, Q \rangle$, then $[Y_i := \lambda \vec{a}.S_i^*]$ is a solution for $\langle P^*, Q^* \rangle$, where S^* is obtained by replacing each constant a by the variable a. If $[Y_i := S_i^*]$ is a solution for $\langle P^*, Q^* \rangle$, then $[X_i := S_i^* \vec{a}]$ is a solution for $\langle P, Q \rangle$.

However, such a transformation does not work for the unification problem in the *linear* lambda calculus, because free variables can occur at most once in a linear λ-term, while constants can occur any number of times. To construct $\langle L_S^* = Y C_1^* \ldots C_m^* D_1^* \ldots D_n^*, R_S^* \rangle$ by eliminating constants from $\langle L_S, R_S \rangle$ defined in Definition 9, we adopt the following strategy:

- Let T be among $C_1, \ldots, C_m, D_1, \ldots, D_n, R_S$.
- Let T' be the result of replacing occurrences of each constant in T with *suitable* free variables or λ-terms constructed from free variables.
- Let $T^* = \lambda \vec{x}.T'$ for a sequence $\vec{x}$ of the elements of $\text{FV}(T')$.

The main issue is what variable or λ-term each occurrence of a constant should be replaced with. The formal definition and an example are given in Definition 10 and Example 3. We give the basic strategy of our transformation here. First, for each constant a, we provide an atomic type p_a. Second we replace the i-th occurrence of a in T with the application $x_{a,i} y_{a,i}$ of variables $x_{a,i}$ of type $p_a \multimap \mathrm{Tp}(a)$ and $y_{a,i}$ of type p_a. Finally, we close T by $\lambda \vec{x}\vec{y}$. This way, we can avoid identifying λ-terms which are surrogates for distinct constants of the same type, while preserving the well-typedness and the linearity of the interpolation equation. If $\langle L_S, R_S \rangle$ has a solution $[X := S]$, then $\langle L_S^*, R_S^* \rangle$ also has a solution $[X^* := S^*]$ which does not essentially differ from $[X := S]$. Each occurrence of a constant a in S is replaced with the application $x_{a,i} y_{a,i}$ of two bound variables of R_S^* for the appropriate i. Moreover, $[X^* := S^*]$ lets the arguments T^* of the unknown Y be applied to bound variables of R_S^* which constitute the surrogates for constants which appear in T.

A major problem is that one may construct a solution which causes a substitution of a complex λ-term for the bound variable $x_{a,i}$ of an argument T^* of the unknown Y, since a λ-term which has the type $p_a \multimap \mathrm{Tp}(a)$ is not necessarily a bound variable $x_{a,j}$ in R_S^* for some j. For instance, consider the interpolation equation $\langle X(fa), f(ga) \rangle$ where the types of constants are as follows: $\mathrm{Tp}(a) = o$, $\mathrm{Tp}(f) = \mathrm{Tp}(g) = o \multimap o$. Clearly it has no solution. By applying the above conversion to $\langle X(fa), f(ga) \rangle$, we obtain the following equation:

$$\langle Y\left(\lambda x_f y_f x_a y_a . x_f y_f (x_a y_a) \right), \ \lambda x_f y_f x_g y_g x_a y_a . x_f y_f \left(x_g y_g (x_a y_a) \right) \rangle$$

where the types of variables are defined by $\mathrm{Tp}(x_a) = p_a \multimap o$, $\mathrm{Tp}(y_a) = p_a$, $\mathrm{Tp}(x_f) = p_f \multimap o \multimap o$, $\mathrm{Tp}(y_f) = p_f$, $\mathrm{Tp}(x_g) = p_g \multimap o \multimap o$, and $\mathrm{Tp}(y_g) = p_g$. But, this has a solution

$$[Y := \lambda z.\lambda x_f y_f x_g y_g x_a y_a . z x_f y_f \left(\lambda a . x_g y_g (x_a a) \right) y_a]$$

where $\mathrm{Tp}(a) = p_a$. This problem may ruin our naive transformation from constants to applications of bound variables, or at least, makes our discussion very complicated. This is the reason why we have not employed the reduction in the proof of Proposition 2 as a basis for a constant-free reduction. Fortunately, we can tightly restrict λ-terms that can be substituted for bound variables in the arguments T^* of Y, when we employ the reduction in Definition 9 as a basis for the new reduction. The constants in $\langle L_S, R_S \rangle$ in Definition 9 are *stratified* in the sense that c_j is always an argument of f, fc_j is of d_k for some k, and a λ-term whose head is d_k is of g. So, we can let λ-terms in $\langle L_S^*, R_S^* \rangle$ which are surrogates for c_j have type q, surrogates for f have type $q \multimap r$, surrogates for d_k have type $r^m \multimap t$, and surrogates for g have type $t^l \multimap u$.

Now, we give a formal definition of the linear interpolation equation $\langle L_S^*, R_S^* \rangle$ for a given polarized 1N-Sat $S = \{C_1, \ldots, C_m, D_1, \ldots, D_n\}$ on $V = \{v_1, \ldots, v_l\}$.

Definition 10. Let the set of atomic types be $\mathcal{A} = \{p_j, q, r, s_k, t, u \,|\, 0 \le j \le m,$ $1 \le k \le n\}$. The types of variables in $\langle L_S^*, R_S^* \rangle$ are given as follows for $1 \le i \le l$, $1 \le j \le m$ and $1 \le k \le n$:

$$\mathrm{Tp}(c_j) = p_j, \quad \mathrm{Tp}(c_{i,j}) = p_{\mu(i,j)}, \quad \mathrm{Tp}(\bar{c}_j) = p_j \multimap q, \quad \mathrm{Tp}(\bar{c}_{i,j}) = p_{\mu(i,j)} \multimap q,$$
$$\mathrm{Tp}(d_k) = \mathrm{Tp}(d_{i,k}) = s_k, \quad \mathrm{Tp}(\bar{d}_k) = \mathrm{Tp}(\bar{d}_{i,k}) = s_k \multimap r^m \multimap t$$
$$\mathrm{Tp}(f) = \mathrm{Tp}(f_j) = \mathrm{Tp}(f_{i,j}) = q \multimap r, \quad \mathrm{Tp}(g) = t^l \multimap u, \quad \mathrm{Tp}(x_j) = q.$$

We denote the set of variables in $R_{\mathcal{S}}^*$ by $\mathcal{X}_R$.

$$\mathcal{X}_R = \{c_{i,j}, \bar{c}_{i,j}, f_{i,j}, d_{i,\nu(i)}, \bar{d}_{i,\nu(i)}, g \mid 1 \le i \le l \text{ and } 1 \le j \le m\}$$

$\langle L_{\mathcal{S}}^*, R_{\mathcal{S}}^* \rangle$ is defined by

$$L_{\mathcal{S}}^* = Y C_1^* \dots C_m^* D_1^* \dots D_n^* \quad \text{where}$$

$$\begin{cases} C_j^* = \lambda f \bar{c}_j c_j . f(\bar{c}_j c_j) \\ D_k^* = \lambda \bar{d}_k d_k f_1 \dots f_m x_1 \dots x_m . \bar{d}_k d_k (f_1 x_1) \dots (f_m x_m) \\ \mathrm{Tp}(Y) = \mathrm{Tp}(C_1^*) \multimap \dots \multimap \mathrm{Tp}(C_m^*) \multimap \mathrm{Tp}(D_1^*) \multimap \dots \multimap \mathrm{Tp}(D_n^*) \multimap \mathrm{Tp}(R_{\mathcal{S}}^*) \end{cases}$$

$$R_{\mathcal{S}}^* = \lambda \mathcal{X}_R . g V_1^* \dots V_l^* \quad \text{where}$$

$$V_i^* = \bar{d}_{i,\nu(i)} d_{i,\nu(i)} \big(f_{i,1}(\bar{c}_{i,1} c_{i,1}) \big) \dots \big(f_{i,m}(\bar{c}_{i,m} c_{i,m}) \big).$$

It is not difficult to check that the above λ-terms are well-typed and that

$$\mathrm{Tp}(C_j^*) = (q \multimap r) \multimap (p_j \multimap q) \multimap p_j \multimap r,$$
$$\mathrm{Tp}(D_k^*) = (s_k \multimap r^m \multimap t) \multimap s_k \multimap (q \multimap r)^m \multimap q^m \multimap t,$$
$$\mathrm{Od}(Y) = 4.$$

We use y_j for variables of type $\mathrm{Tp}(C_j^*)$ and z_k of type $\mathrm{Tp}(D_k^*)$ in this section.

Example 3. INSTANCE $\mathcal{S} : \mathcal{C}_1 = \{v_1\}, \; \mathcal{C}_2 = \{v_2\}, \; \mathcal{D}_1 = \{\neg v_1, \neg v_2\}$

REDUCTION WITH CONSTANTS $\langle L_{\mathcal{S}}, R_{\mathcal{S}} \rangle$:

$$L_{\mathcal{S}} = X(\mathsf{fc}_1)(\mathsf{fc}_2)\big(\lambda x_1 x_2 . \mathsf{d}_1(\mathsf{f}x_1)(\mathsf{f}x_2)\big)$$
$$R_{\mathcal{S}} = \mathsf{g}\big(\mathsf{d}_1(\mathsf{fc}_1)(\mathsf{fc}_0)\big)\big(\mathsf{d}_1(\mathsf{fc}_0)(\mathsf{fc}_2)\big)$$

REDUCTION WITHOUT CONSTANTS $\langle L_{\mathcal{S}}^*, R_{\mathcal{S}}^* \rangle$:

$$L_{\mathcal{S}}^* = Y\big(\lambda f \bar{c}_1 c_1 . f(\bar{c}_1 c_1)\big)\big(\lambda f \bar{c}_2 c_2 . f(\bar{c}_2 c_2)\big)$$
$$\big(\lambda \bar{d}_1 d_1 f_1 f_2 x_1 x_2 . \bar{d}_1 d_1 (f_1 x_1)(f_2 x_2)\big)$$
$$R_{\mathcal{S}}^* = \lambda \mathcal{X}_R . g \Big(\bar{d}_{1,1} d_{1,1} \big(f_{1,1}(\bar{c}_{1,1} c_{1,1}) \big)\big(f_{1,2}(\bar{c}_{1,2} c_{1,2}) \big) \Big)$$
$$\Big(\bar{d}_{2,1} d_{2,1} \big(f_{2,1}(\bar{c}_{2,1} c_{2,1}) \big)\big(f_{2,2}(\bar{c}_{2,2} c_{2,2}) \big) \Big)$$

$\mathcal{S}$ is not satisfiable and $\langle L_{\mathcal{S}}, R_{\mathcal{S}} \rangle$ and $\langle L_{\mathcal{S}}^*, R_{\mathcal{S}}^* \rangle$ have no solution. Note that

$$Y := \lambda y_1 y_2 z_1 \mathcal{X}_R . g\big(z_1 \bar{d}_{1,1} d_{1,1} f_{1,1} f_{1,2}(\bar{c}_{1,1} c_{1,1})(\bar{c}_{1,2} c_{1,2})\big)$$
$$\big(\bar{d}_{2,1} d_{2,1} (y_1 f_{2,1} \bar{c}_{2,1} c_{2,1})(y_2 f_{2,2} \bar{c}_{2,2} c_{2,2})\big)$$

is not a solution for $\langle L_{\mathcal{S}}^*, R_{\mathcal{S}}^* \rangle$. $(y_1 f_{2,1} \bar{c}_{2,1} c_{2,1})$ is not a λ-term of the simply typed lambda calculus, for $\mathrm{Tp}(y_1) = (q \multimap r) \multimap (p_1 \multimap q) \multimap p_1 \multimap r$ but $\mathrm{Tp}(\bar{c}_{2,1}) = p_{\mu(2,1)} \multimap q = p_0 \multimap q$ and $\mathrm{Tp}(c_{2,1}) = p_{\mu(2,1)} = p_0$.

Lemma 3. $\langle L_S^*, R_S^* \rangle$ *admits a solution whenever S is satisfiable.*

Proof. Suppose that ψ is a valuation which satisfies S. As in the proof of Lemma 1, we can find $\phi : S \rightarrow (\mathcal{V} \cup \neg\mathcal{V})$ which indicates a literal via which each clause is satisfied. We show that a solution is given by $[Y := S]$ where

$$S = \lambda y_1 \dots y_m z_1 \dots z_n \mathcal{X}_R . g U_1 \dots U_l \quad \text{where}$$

$$U_i = \begin{cases} z_{\nu(i)} \bar{d}_{i,\nu(i)} d_{i,\nu(i)} f_{i,1} \dots f_{i,m} (\bar{c}_{i,1} c_{i,1}) \dots (\bar{c}_{i,m} c_{i,m}) & \text{if } \phi(\mathcal{D}_{\nu(i)}) = \neg v_i \\ \bar{d}_{i,\nu(i)} d_{i,\nu(i)} U_{i,1} \dots U_{i,m} & \text{otherwise} \end{cases}$$

$$U_{i,j} = \begin{cases} y_j f_{i,j} \bar{c}_{i,j} c_{i,j} & \text{if } \phi(\mathcal{C}_j) = v_i \\ f_{i,j} (\bar{c}_{i,j} c_{i,j}) & \text{otherwise} \end{cases} .$$

Indeed S is a well-typed closed linear λ-term. The linearity of S can be checked as in the proof of Lemma 1. We check the well-typedness of S. Recall that $\text{Tp}(z_{\nu(i)}) = (s_{\nu(i)} \multimap r^m \multimap t) \multimap s_{\nu(i)} \multimap (q \multimap r)^m \multimap q^m \multimap t$, $\text{Tp}(\bar{d}_{i,\nu(i)}) = s_{\nu(i)} \multimap r^m \multimap t$, $\text{Tp}(d_{i,\nu(i)}) = s_{\nu(i)}$, $\text{Tp}(f_{i,j}) = q \multimap r$ and $\text{Tp}(\bar{c}_{i,j} c_{i,j}) = q$. Hence, U_i is well-typed and has the type t if $\phi(\mathcal{D}_{\nu(i)}) = \neg v_i$. If $\phi(\mathcal{D}_{\nu(i)}) \neq \neg v_i$, it is enough to check that each $U_{i,j}$ has type r. If $\phi(\mathcal{C}_j) = v_i$ and $U_{i,j} = y_j f_{i,j} \bar{c}_{i,j} c_{i,j}$, then $v_i \in \mathcal{C}_j$, $\text{Tp}(\bar{c}_{i,j}) = p_{\mu(i,j)} \multimap q = p_j \multimap q$ and $\text{Tp}(c_{i,j}) = p_{\mu(i,j)} = p_j$. Recall that $\text{Tp}(y_j) = (q \multimap r) \multimap (p_j \multimap q) \multimap p_j \multimap r$ and $\text{Tp}(f_{i,j}) = q \multimap r$. Hence, $U_{i,j}$ is well-typed and has the type r. If $\phi(\mathcal{C}_j) \neq v_i$, it is clear that $U_{i,j}$ is well-typed and has the type r. Therefore, S is well-typed.

Second we show that $L_S^*[Y := S] \twoheadrightarrow_\beta R_S^*$. By the definition, $L_S^*[Y := S] \twoheadrightarrow_\beta \lambda \mathcal{X}_R . g U_1 \dots U_l \sigma$ for $\sigma = [\vec{y} := \vec{C}^*, \vec{z} := \vec{D}^*]$. It is enough to show that $U_i \sigma \twoheadrightarrow_\beta V_i^*$. If $\phi(\mathcal{D}_{\nu(i)}) = \neg v_i$, then

$$U_i \sigma = D_{\nu(i)}^* \bar{d}_{i,\nu(i)} d_{i,\nu(i)} f_{i,1} \dots f_{i,m} (\bar{c}_{i,1} c_{i,1}) \dots (\bar{c}_{i,m} c_{i,m})$$
$$\twoheadrightarrow_\beta \bar{d}_{i,\nu(i)} d_{i,\nu(i)} \big(f_{i,1}(\bar{c}_{i,1} c_{i,1})\big) \dots \big(f_{i,m}(\bar{c}_{i,m} c_{i,m})\big)$$
$$= V_i^*.$$

Otherwise, $U_i = \bar{d}_{i,\nu(i)} d_{i,\nu(i)} U_{i,1} \dots U_{i,m}$. It is obvious that $U_{i,j} \sigma \twoheadrightarrow_\beta f_{i,j}(\bar{c}_{i,j} c_{i,j})$ for all j, since $C_j^* f_{i,j} \bar{c}_{i,j} c_{i,j} \twoheadrightarrow_\beta f_{i,j}(\bar{c}_{i,j} c_{i,j})$. □

Lemma 4. *Suppose that a linear λ-term M contains no free variables other than the elements of $\mathcal{X}_R \cup \{y_j, z_k \mid 1 \leq j \leq m, 1 \leq k \leq n\}$.*

If M has an atomic type δ and contains a subterm N of type $\gamma_1 \multimap \cdots \multimap \gamma_i \multimap \gamma$ with $\gamma \in \mathcal{A}$, then $\gamma \leq \delta$ where $\leq$ is the partial order on $\mathcal{A}$ such that $p_j \lesssim q \lesssim r \lesssim t \lesssim u$ for $0 \leq j \leq m$ and $s_k \lesssim t \lesssim u$ for $1 \leq k \leq n$.

If $\text{Tp}(M) = q \multimap r$, then $M =_\eta f_{i,j}$ for some i and j.

Lemma 5. *Suppose that a subterm of a linear λ-term M has an atomic type p. Then, for every N such that $M \twoheadrightarrow_\beta N$, N has a subterm of type p.*

Lemma 6. *S is satisfiable whenever $\langle L_S^*, R_S^* \rangle$ admits a solution.*

Proof. Suppose that $[Y := S]$ is a solution. We can assume that S is normal and $S = \lambda y_1 \ldots y_m z_1 \ldots z_n \mathcal{X}_R . S'$. Because of the type of the variable g in R^*_S, the head of S' is neither y_j nor z_k. S' must be $g U_1 \ldots U_l$ for some $U_1, \ldots, U_l$ of type t. Let $\sigma = [\vec{y} := \vec{C}^*, \vec{z} := \vec{D}^*]$. Since $S C^*_1 \ldots C^*_m D^*_1 \ldots D^*_n \twoheadrightarrow_\beta \lambda \mathcal{X}_R . g U_1 \ldots U_l \sigma \twoheadrightarrow_\beta \lambda \mathcal{X}_R . g V^*_1 \ldots V^*_l$, we have $U_i \sigma \twoheadrightarrow_\beta V^*_i$. Note that

(*) U_i contains no free variables other than the elements of

$$\mathcal{X}_i = \mathrm{FV}(V^*_i) \cup \{ y_j, z_k \mid 1 \le j \le m,\, 1 \le k \le n \}$$
$$= \{ \bar{d}_{i,\nu(i)}, d_{i,\nu(i)}, f_{i,j}, \bar{c}_{i,j}, c_{i,j}, y_j, z_k \mid 1 \le j \le m,\, 1 \le k \le n \}.$$

Since U_i has type t, the head of U_i must be $\bar{d}_{i,\nu(i)}$ or z_k for some k.

First, we show that if U_i contains z_k, then z_k is the head of U_i and U_i contains neither y_j nor $z_{k'}$ for any j and $k' \ne k$. Since $z_k \sigma = D^*_k$ contains a variable of type s_k, $U_i \sigma$ and its normal form V^*_i contain a subterm of type s_k by Lemma 5. This implies that $k = \nu(i)$ and V^*_i contains no subterm of type $s_{k'}$ for any $k' \ne k$ and thus $U_i \sigma$ does so by Lemma 5 again. Therefore, U_i does not contain $z_{k'}$ for any $k' \ne k$. If $U_i = \bar{d}_{i,\nu(i)} M_{\nu(i)} R_1 \ldots R_m$ for $\mathrm{Tp}(M_{\nu(i)}) = s_{\nu(i)}$ and $\mathrm{Tp}(R_h) = r$ for $1 \le h \le m$, then z_k cannot occur in U_i by Lemma 4. So U_i is of the form $z_k N_k M_k F_1 \ldots F_m Q_1 \ldots Q_m$ where $\mathrm{Tp}(N_k) = s_k \multimap r^m \multimap t$, $\mathrm{Tp}(M_k) = s_k$, $\mathrm{Tp}(F_h) = q \multimap r$, and $\mathrm{Tp}(Q_h) = q$ for $1 \le h \le m$. We check that y_j cannot occur in N_k, M_k, Q_h or F_h for any h. It is obvious that y_j is not in M_k, Q_h or F_h by (*) and Lemma 4. Since N_k does not contain $z_{k'}$ for any k', if N_k is in long normal form, then N_k is equal to $\lambda d_k w_1 \ldots w_m . \bar{d}_{i,\nu(i)} d_k R_1 \ldots R_m$ where $\mathrm{Tp}(d_k) = s_k$ and $\mathrm{Tp}(w_h) = \mathrm{Tp}(R_h) = r$ for all h (provided that $\nu(i) = k$). One can check that for all h there is h' such that $R_h = w_{h'}$. So y_j does not appear in N_k. Therefore, y_j cannot occur in U_i unless the head of U_i is $\bar{d}_{i,\nu(i)}$.

Now, we show that S is satisfied by the valuation ψ defined as follows:

$$\psi(v_i) = \begin{cases} 0 & \text{if the head of } U_i \text{ is a variable } z_k \text{ for some } k \\ 1 & \text{otherwise} \end{cases}$$

By the linearity of S, each variable $y_1, \ldots, y_m, z_1, \ldots, z_n$ appears in U_i for some i. To show that each positive clause $\mathcal{C}_j$ is satisfied by ψ, suppose that y_j occurs in U_i. The above discussion claims that the head of U_i is $\bar{d}_{i,\nu(i)}$ and thus $\psi(v_i) = 1$. Since $y_j \sigma = C^*_j$ contains a variable of type p_j, $U_i \sigma$ and its normal form V^*_i contain a subterm of type p_j by Lemma 5. Therefore, $\mathrm{Tp}(c_{i,j}) = p_{\mu(i,j)} = p_j$ and $v_i \in \mathcal{C}_j$. So ψ satisfies $\mathcal{C}_j$ via v_i.

To show that each negative clause $\mathcal{D}_k$ is satisfied by ψ, suppose that z_k occurs in U_i. The above discussion claims that z_k is the head of U_i and $k = \nu(i)$. Therefore, ψ satisfies $\mathcal{D}_k$ via $\neg v_i$. $\quad\square$

Theorem 5. *Fourth-order interpolation modulo β ($\beta\eta$) in the linear lambda calculus in the absence of constants is NP-complete.*

Proof. By Theorem 3 and Lemmas 3 and 6. $\quad\square$

The above proof also entails NP-hardness of fourth-order interpolation in the relevant lambda calculus in the absence of constants.

It is easy to see that Theorem 5 does not hold for the third-order case unless $P = NP$. For an interpolation equation $\langle X L_1 \ldots L_m, R \rangle$, if L_i is a constant-free closed linear λ-term whose type is at most second-order, then $L_i = \lambda x_i.x_i$ and $\mathrm{Tp}(L_i) = p_i \multimap p_i$ for some atomic type $p_i \in \mathcal{A}$. Hence $\langle X L_1 \ldots L_m, R \rangle$ has a linear solution modulo β (*resp.* $\beta\eta$) iff (*resp.* the long form of) R has a subterm of type p_i for every i by Lemma 5.

Acknowledgement

The author would like to thank anonymous referees and Makoto Kanazawa for their valuable comments and advice.

References

[1] Philippe de Groote. Linear higher-order matching is NP-complete. In *11th International Conference on Rewriting Techniques and Applications*, volume 1833 of *Lecture Notes in Computer Science*, pages 127–140. Springer-Verlag, 2000.

[2] Philippe de Groote. Towards abstract categorial grammars. In *Association for Computational Linguistics, 30th Annual Meeting and 10th Conference of the European Chapter, Proceedings of the Conference*, pages 148–155, 2001.

[3] Daniel Dougherty and ToMasz Wierzbicki. A decidable variant of higher order matching. In *13th International Conference on Rewriting Techniques and Applications*, volume 2378 of *Lecture Notes in Computer Science*, pages 340–351. Springer-Verlag, 2002.

[4] Gilles Dowek. Third order matching is decidable. *Annals of Pure and Applied Logic*, 69:135–155, 1994.

[5] Warren D. Goldfarb. The undecidability of the second-order unification problem. *Theoretical Computer Science*, 13:225–230, 1981.

[6] Gepard Huet and Bernard Lang. Proving and applying program transformations expressed with second order patterns. *Acta Informatica*, 11:31–55, 1978.

[7] Pekka Kilpeläinen and Heikki Mannila. Ordered and unordered tree inclusion. *SIAM Journal of Computing*, 24(2):340–356, 1995.

[8] Jordi Levy. Linear second-order unification. In *7th International Conference on Rewriting Techniques and Applications*, volume 1103 of *Lecture Notes in Computer Science*, pages 332–346. Springer-Verlag, 1996.

[9] Ralph Loader. Higher order β matching is undecidable. *Logic Journal of the IGPL*, 11(1):51–68, 2003.

[10] Vincent Padovani. Decidability of fourth order matching. *Mathematical Structures in Computer Science*, 3(1):361–372, 2000.

[11] Sylvain Pogodalla. Using and extending ACG technology: Endowing categorial grammars with an underspecified semantic representation. In *Proceedings of Categorial Grammars 2004, Montpellier*, pages 197–209, June 2004.

[12] Sylvain Salvati and Philippe de Groote. On the complexity of higher-order matching in the linear λ-calculus. In *14th International Conference on Rewriting Techniques and Applications*, volume 2706 of *Lecture Notes in Computer Science*, pages 234–245. Springer-Verlag, 2003.

Localized Fairness: A Rewriting Semantics

José Meseguer

CS Department, University of Illinois at Urbana-Champaign, USA

Abstract. Fairness is a rich phenomenon: we have weak and strong fairness, and many different *variants* of those concepts: transition fairness, object/process fairness, actor fairness, position fairness, and so on, associated with specific *models* or *languages*, but lacking a common theoretical framework. This work uses *rewriting semantics* as a common theoretical framework for fairness. A common thread tying together the different fairness variants is the notion of *localization*: fairness must often be localized to specific entities in a system. For systems specified as rewrite theories localization can be formalized by making explicit the *subset of variables* in a rule corresponding to the items that must be localized. In this way, localized fairness becomes a *parametric* notion, that can be easily specialized to model a very wide range of fairness phenomena. After formalizing these concepts and proving basic results, the paper studies in detail both a relative and an absolute LTL semantics for rewrite theories with localized fairness requirements, and shows that it is always possible to pass from the relative to the absolute semantics by means of a theory transformation. This allows using a standard LTL model checker to check properties under fairness assumptions.

1 Introduction

A key motivation of rewriting logic [13] is to provide a general semantic framework for specifying concurrent systems. Indeed, most known models of concurrency have been specified quite naturally as rewrite theories [11]. In this way, rewriting techniques – with rewrite rules no longer understood as oriented equalities, but as *concurrent transitions* – play a crucial role in concurrency theory. *Fairness* is an area ripe for the application of rewriting techniques. There has been some initial work in this direction, by Francez and Porat [18, 19] – who focused on what I call rule fairness and gave partial results on the decidability of fair termination for ground TRSs – and by Tison [20], who proved the decidability of fair termination for general ground TRSs using tree automata.

However, fairness is a rich phenomenon with many different variants. First of all, there is weak and strong fairness (called, respectively, *justice* and *fairness* by Manna and Pnueli [10]). Justice and fairness are *properties of infinite computations*. Justice means that, *if a certain kind of transition is continuously enabled beyond a certain point, then it is taken infinitely often.* Assuming predicates, *enabled.τ* and *taken.τ* that specify when a given transition τ is enabled, resp. has been taken, then justice for τ can be expressed, using the $\Diamond$ (eventually) and $\square$ (always) operators, by the linear time temporal logic (LTL) formula,

$$\Diamond\square enabled.\tau \rightarrow \square\Diamond taken.\tau$$

J. Giesl (Ed.): RTA 2005, LNCS 3467, pp. 250–263, 2005.

Similarly, fairness means that *if a certain kind of transition is infinitely often enabled, then it is taken infinitely often.* In temporal logic terms, this is expressed by the formula,

$$\Box\Diamond enabled.\tau \rightarrow \Box\Diamond taken.\tau$$

There are also many different *variants* of the justice and fairness concepts: there is transition fairness, object/process fairness, actor fairness, position fairness, and so on. At present these different variants have a *pre-theoretic* status, in the sense that they are associated with specific *models* or *languages*, but they lack a common theoretical framework. For example, process fairness is discussed by Francez [6] in terms of the CSP language. Similarly, actor fairness is discussed by Agha in terms of the Actor model of computation [1]. The opportunity now at hand, and the main goal of this work, is to use *rewriting semantics as a common theoretical framework for fairness*, that can explain and do justice to the different fairness variants and phenomena. This is an essential task in order to provide a rewriting semantics framework for concurrency.

A key intuition is that the common thread tying together the different variants of fairness is the notion of *localization*: that fairness must often be localized to specific entities in a system. This idea can be illustrated with a distributed object-oriented system example, exhibiting the need for object/process fairness. Objects can be represented as record-like terms of the general form $\langle o : C \mid a_1 : v_1, \ldots a_n : v_n \rangle$, where o is an *object-identifier*, C its class name, and the $a_i : v_i$ the object's attribute-value pairs. The objects and messages collectively form a *multiset*, called a *configuration*, which is associative, commutative, and has *none* as an identity, and where the multiset union operator can be denoted with empty syntax (juxtaposition) [14]. An interesting example is a *dining philosophers* specification with n philosophers, where object identifiers can be modeled as the integers modulo n. Each philosopher can be in one of three *modes*: t (thinking), h (hungry), and e (eating), and can *hold* none, one, or two chopsticks of the form chop([I],[I+1]), with chop commutative, which are placed between the i-th and the $i+1$-th philosopher on the table. To avoid deadlocks, philosophers can only think in a library (a subconfiguration encapsulated by a library operator), and can go from the library to the dining room table only when they get hungry; however, there must always be at least one philosopher in the library. The relevant rewrite rules (see [12] for the full specification) to pass from thinking to hungry [t2h], to pick the left [pickl] or right [pickr] fork, or go back to the library [e2t], are the following (the Maude [4] keyword rl (resp. crl) declares a rule (resp. a conditional rule)):

```
crl [t2h] : library(< [I] : Phil | mode : t , holds :  none > C) =>
  < [I] : Phil | mode : h , holds :  none > library(C) if C =/= none .

crl [pickl] :  < [I] : Phil | mode : h , holds : none > chop([I],[J])
  => < [I] : Phil | mode : h , holds :  chop([I],[J]) > if [J] = [s(I)]
     [fair(I)] .

rl [pickr] :  < [I] : Phil | mode : h , holds : chop([I],[J]) >
  chop([I],[K]) => < [I] : Phil | mode : e , holds :  chop([I],[J])
  chop([I],[K]) > [fair(I)] .
```

```
rl [e2t]  :  < [I] : Phil | mode : e , holds :   chop([I],[J])
     chop([I],[K]) > library(C) =>   chop([I],[J]) chop([I],[K])
     library(< [I] : Phil | mode : t , holds :   none > C) [just(I)]  .
```

The justice and fairness requirements are specified as attributes of each rule using a Maude-like syntax. There is one justice requirement (for [e2t]) and fairness requirements for [pickl] and [pickr]. Notice that they are all *localized* to the variable I of the object identifier expression (here an equivalence class expression [I]). This means that each such requirement is *local to each philosopher*. For example, the fair(I) requirement for [pickl] and [pickr] means that *each philosopher* should have a fair chance of picking the left and right forks. It is not enough to require that *some* philosopher should have that chance: this is what would have been demanded if one had specified the attribute fair, instead of fair(I), for those two rules. Under the above requirements, the rules ensure *nonstarvation*. But notice again that nonstarvation must be *localized to each philosopher*. That is, it is not enough to require that *some* hungry philosopher eventually eats (because of the deadlock free nature of the specification, this always happens without any fairness assumptions). Instead, what should be required is that *all* hungry philosophers eventually get to eat; and for this each of the above requirements' localization to I is essential.

The general idea is then that, given a rewrite rule $l : t \longrightarrow t'$, say with variables $\bar{x}$, one can *localize* a justice or fairness requirement to a *subset* $\bar{y} \subseteq \bar{x}$ of those variables. In the above example the subset $\bar{y}$ just happened to be a single variable I, but in general the subset could also be empty (the rule justice/fairness case) or could contain more than one variable. Furthermore, since several rules *can share the same label*, justice and fairness requirements can be specified not just for one rule, but for a group of rules with the same label. In this way, localized fairness emerges as a *parametric* notion, that can be easily specialized to model a very wide range of fairness phenomena. There are essentially *two parameters* supporting this flexibility: (i) how fine-grained one chooses to be in the choice of *labels* for the rewrite rules (the coarsest choice is to use the *same* label for all rules, and the finest to use different labels for different rules); and (ii) what *subset of variables* $\bar{x}$ in a rule one chooses to localize the fairness requirements to (the more variables used, the more localized).

For a rewrite theory $\mathcal{R}$ having labeled conditional rewrite rules of the form $l : t \longrightarrow t'$ *if cond*, I now give a precise notation for localized fairness specifications. For convenience I assume a common set X of variables, each with its type information (a sort or a kind), so that the variables $\bar{x}$ of each rule in $\mathcal{R}$ are a subset $\bar{x} \subseteq X$. This allows one to choose a set $\bar{y}$ of variables to localize the fairness requirements of one or more rules sharing the same label l, provided that for each such rule, say having variables $\bar{x}$, one always has $\bar{y} \subseteq \bar{x}$.

Definition 1. *A* basic fairness specification *for a rewrite theory $\mathcal{R}$ is an expression of the form $j(l(\bar{y}))$ (called a basic* justice specification*) or $f(l(\bar{y}))$ (called a basic* fairness specification*) such that l is a label used in $\mathcal{R}$, and $\bar{y}$ is a set of variables contained in the variables of all rules with label l in $\mathcal{R}$. A fairness specification for $\mathcal{R}$ is a finite set of basic fairness specifications. One can always* *decompose such a specification as* $\mathcal{J} \cup \mathcal{F}$, *where* $\mathcal{J}$ *is the set of basic justice specifications, and* $\mathcal{F}$ *is the set of basic fairness specifications. In general,* $\mathcal{J}$ *or* $\mathcal{F}$ *(or both) can be empty. The pair* $(\mathcal{R}, \mathcal{J} \cup \mathcal{F})$ *is called a* rewrite theory with fairness specifications $\mathcal{J} \cup \mathcal{F}$. □

The best way to get a feeling for the different fairness phenomena that can be formalized by the above definition is to discuss some concrete instances.

Label Justice and Fairness. This is the least localized notion. One only cares about whether eventually there is always *some* rule among those having a given label l enabled at each step (resp. at each step in an infinite set $J \subseteq \mathbb{N}$) and then one wants rules with that label to be infinitely often taken.

Rule Justice and Fairness. This is the special case of label justice, resp. fairness, in which *different rules have different labels*; therefore here the justice and fairness requirements are localized on a rule-by-rule basis. Maude's `rewrite` command [4] ensures rule justice in this sense.

Object/Process Justice and Fairness. Here one is dealing with distributed objects (sometimes called processes) that interact with each other either through messages or synchronizing directly. The above dining philosophers example is a typical example of the general pattern: (i) each rule has a different label; and (ii) each rule is localized by the variable(s) of its object identifier expression. I call this *object/process justice*, resp. *object/process fairness*.

Actor Justice. Actors [1] are the special class of distributed object systems where all object interactions are achieved by asynchronous message passing. In their rewriting semantics one object and one message rewrites to a new state for the object, with possibly new messages and new objects being created [14]. The Actor model [1] requires that any message sent to an object should be eventually received. This can be expressed as a localized justice requirement by: (i) making all rule labels different; and (ii) giving a justice requirement for each rule localized to the variables of the object identifier expression and of the message it receives. For object-oriented systems, Maude's `frewrite` command [4] ensures actor justice in addition to rule and position justice (see below).

Ground Justice and Fairness. In a sense this is the most localized possible variant (but see the discussion on position fairness below!) in which: (i) all rule labels are different; and (ii) all rule variables are localized.

Position Justice and Fairness. Consider for example a rewrite system with Σ having constants a, b and a binary symbol f, and with rules $l_1 : a \longrightarrow f(a,a)$ and $l_2 : f(x,y) \longrightarrow b$. *Position* justice, resp. fairness, means that if a given term position is eventually always enabled, resp. enabled infinitely often, then that position is rewritten infinitely often. Notice that the above system has infinite rule fair computations, but is *terminating* under the position justice assumption. Since *positions* are a metalevel concept, not explicitly present in terms or rules, one should make them explicit at the object level by decorating terms with position strings. For example, the term $f(a, b)$ can be decorated as $f(a@1, b@2)@nil$.

Rules may likewise be decorated as: $\tilde{l}_1 : a@p \longrightarrow f(a@p.1, b@p.2)@p$ and $\tilde{l}_2 :$ $f(x@p.1, y@p.2)@p \longrightarrow b@p$, where p is a variable of sort *Position*, and where the string concatenation operator $\_\_$ is declared associative and with identity *nil*. In this way, a position-aware variant of our system is obtained. Position justice or fairness then corresponds to: (i) using the same label for all rules (in our example, identifying $\tilde{l}_1$ and $\tilde{l}_2$); and (ii) localizing all rules to the explicit position variable p. This can be done in general for any left-linear rewrite system, as shown in detail in [12]. One can of course have a more fine-grained combination of *rule and position* justice or fairness; in this case different rules must have different labels, and one again localizes only the position variable p. Maude's frewrite command ensures both rule and position justice in all computations. The most localized notions possible are *ground and position* justice or fairness, where all rule labels are different, and all rule variables, including the position variable p, are localized.

The paper makes the following contributions:

– Gives a precise notation for specifying localized justice and fairness for concurrent systems modeled as rewrite theories; defines the semantics of such specifications; and proves basic results about this semantics. To the best of my knowledge *the general concept of localized fairness* presented in this paper *is new*, although specific instances of that concept have been previously known in an ad-hoc way.
– Shows that a rich variety of fairness phenomena, such as all those listed above, are all special cases of localized fairness.
– Studies in detail both a relative and an absolute LTL semantics for systems specified as rewrite theories with localized justice/fairness specifications. In the relative semantics, such specifications are built into the model and one proves an LTL formula φ relatively to such a model; instead, in the absolute semantics justice/fairness specifications are turned into an explicit LTL formula θ and one proves that an LTL formula φ holds under those requirements by proving that the implication $\theta \rightarrow \varphi$ holds in a *standard* model (Kripke structure) not having any fairness requirements built in. I show that it is always possible to reduce the relative to the absolute semantics by means of a rewrite theory transformation. One can then use a standard LTL model checker to model check properties under justice/fairness requirements.

Section 2 recalls basic concepts on Kripke structures, LTL, and rewrite theories; Section 3 defines the semantics of localized fairness and studies basic properties; Section 4 studies the relative and absolute LTL semantics under localized fairness assumptions; and Section 5 ends with some concluding remarks.

2 Kripke Structures, LTL, and Rewrite Theories

I recall basic notions on Kripke structures, linear time temporal logic (LTL) and rewrite theories needed in the rest of the paper. *AP* denotes a set of atomic propositions used in LTL formulas and in Kripke structures.

A *Kripke structure* is a triple $\mathcal{A} = (A, \rightarrow_A, L_A)$, where A is a set of states, $\rightarrow_A \subseteq A \times A$ is a *total* transition relation, and $L_A : A \rightarrow \mathcal{P}(AP)$ is a function mapping each state to the set of atomic propositions holding in it. The notation $a \rightarrow_A b$ abbreviates $(a, b) \in \rightarrow_A$. Note that the transition relation must be *total*, that is, for each $a \in A$ there is a $b \in A$ such that $a \rightarrow_A b$. For $\rightarrow$ an arbitrary relation, $\rightarrow^\bullet$ denotes the total relation that extends $\rightarrow$ by adding a pair $a \rightarrow^\bullet a$ for each a such that there is no b with $a \rightarrow b$. A *path* in a Kripke structure $\mathcal{A}$ is a function $\pi : \mathbb{N} \longrightarrow A$ such that, for each $i \in \mathbb{N}$, $\pi(i) \rightarrow_A \pi(i+1)$.

The syntax of LTL(AP) is given by the following grammar:

$$\varphi = p \in AP \mid \varphi \vee \varphi \mid \neg\varphi \mid \bigcirc \varphi \mid \varphi \mathcal{U} \varphi.$$

The semantics of the logic, specifying the satisfaction relation $\mathcal{A}, a \models \varphi$ between a Kripke structure $\mathcal{A}$, an initial state $a \in A$, and $\varphi \in$ LTL(AP), is defined as usual (see for example [3, Sect. 3.1], where $\varphi \mathcal{U} \psi$ and $\bigcirc \varphi$ are expressed in CTL* notation as $\mathbf{A}(\varphi\mathbf{U}\psi)$ and $\mathbf{AX}\varphi$). Other Boolean and temporal operators (e.g., $\top$, $\bot$, $\wedge$, $\rightarrow$, $\square$, $\diamond$, $\mathcal{R}$, and $\rightsquigarrow$) can be defined as syntactic sugar.

Given Kripke structures $\mathcal{A} = (A, \rightarrow_A, L_A)$ and $\mathcal{B} = (B, \rightarrow_B, L_B)$, on the same set AP of atomic propositions, an *AP-simulation* $H : \mathcal{A} \longrightarrow \mathcal{B}$ of $\mathcal{A}$ by $\mathcal{B}$ is a binary relation $H \subseteq A \times B$ such that:

- if $a \rightarrow_A a'$ and aHb, then there is $b' \in B$ such that $b \rightarrow_B b'$ and $a'Hb'$, and
- $(\forall a \in A)(\forall b \in B)$ aHb $\Rightarrow$ $L_B(b) = L_A(a)$.

If both H and H^{-1} are *AP*-simulations, then H is called an *AP-bisimulation*. It is well-known (see [3]) that if $H : \mathcal{A} \longrightarrow \mathcal{B}$ is a simulation, aHb, and for an LTL formula φ one has $\mathcal{B}, b \models \varphi$, this then implies $\mathcal{A}, a \models \varphi$. The implication becomes an equivalence if H is a bisimulation.

A *rewrite theory* is a four-tuple $\mathcal{R} = (\Sigma, \phi, E \cup A, R)$, where $(\Sigma, E \cup A)$ is an equational theory, that is assumed decomposed into a set E of (possibly conditional) equations, and a set A of equational axioms such as associativity, commutativity, etc., so that the equations E are applied *modulo A*. $(\Sigma, E \cup A)$ specifies a set of *states* as the algebraic data type $T_{\Sigma/E\cup A,k}$ associated to the initial algebra $T_{\Sigma/E\cup A}$ of $(\Sigma, E \cup A)$ by the choice of a type k of states in Σ [1]. The system's *transitions* are axiomatized by the *conditional rewrite rules* R which are of the form $l : t \longrightarrow t'$ *if cond*, with l a label, t and t' Σ-terms, possibly with variables, and *cond* a condition[2]. Finally, the map ϕ associates

---

[1] $(\Sigma, E \cup A)$ can be assumed to be an equational theory in *membership equational logic* [15], that can have types, subtypes defined by semantic conditions, and operator overloading; order-sorted, many-sorted, or unsorted specifications are all then special cases. The desired set of states is then described by the carrier $T_{\Sigma/E\cup A,k}$ of the initial algebra $T_{\Sigma/E\cup A}$ for one of those types k, technically called either *sorts* or *kinds* in [15]. The elements of $T_{\Sigma/E\cup A}$ are $E \cup A$-equivalence classes of terms $[t]_{E\cup A}$; that is, two terms are equal iff they can be proved equal using $E \cup A$.

[2] In this paper I assume that the condition *cond* can involve a conjunction of equations $u = v$ and *memberships* of the form $w : s$, stating that the term w has sort s. The conjunction must hold for a substitution instance θ before one is allowed to rewrite $\theta(t)$ to $\theta(t')$. I also assume that $vars(t') \cup vars(cond) \subseteq vars(t)$.

to each n-ary operator f in Σ a subset $\phi(f) \subseteq \{1, \ldots, n\}$ of *frozen* argument positions, so that it is forbidden to apply rules in subterms under those positions. I will assume throughout that if f is a defined function (not a constructor), then $\phi(f) = \{1, \ldots, n\}$.

Rewriting logic has inference rules to infer all the possible concurrent computations in a system [2, 13], in the sense that, given two states $[u], [v] \in T_{\Sigma/E \cup A, k}$, one can *reach* $[v]$ from $[u]$ by some possibly complex concurrent computation iff one can prove $\mathcal{R} \vdash u \longrightarrow v$ in the logic. In particular one can easily define the *one-step $\mathcal{R}$-rewriting relation*, which is a binary relation $\to^1_{\mathcal{R},k}$ on $T_{\Sigma,k}$ that holds between terms $u, v \in T_{\Sigma,k}$ iff there is a one-step proof of $\mathcal{R} \vdash u \longrightarrow v$, that is, a proof in which only one rewrite rule in R is applied to a single subterm. One can get a binary relation (with the same name) $\to^1_{\mathcal{R},k}$ on $T_{\Sigma/E \cup A,k}$ by defining $[u] \to^1_{\mathcal{R},k} [v]$ iff $u' \to^1_{\mathcal{R},k} v'$ for some $u' \in [u], v' \in [v]$.

$\mathcal{R} = (\Sigma, \phi, E \cup A, R)$ is *computable* if: (1) there exists a *matching algorithm modulo A* producing a finite number of A-matching substitutions, or failing otherwise, that can implement rewriting in A-equivalence classes; (2) the equational theory $(\Sigma, E \cup A)$ is *ground confluent and terminating modulo A*; and (3) the rules R are *ground coherent* relative to the equations E modulo A in a somewhat stronger sense than in [21], namely, if one has a one-step rewrite proof $\mathcal{R} \vdash u \longrightarrow^1 v$ using a rule $l : t \longrightarrow t'$ *if* $cond$ at some position in u with a substitution θ, then there are terms $u' \in can_{E/A}(u)$, and v', and a one-step rewrite proof $\mathcal{R} \vdash u' \longrightarrow^1 v'$ using the same rule at some position in u' with substitution θ' such that $\theta'(x) \in can_{E/A}(\theta(x))$ for each variable x, and such that $can_{E/A}(v) = can_{E/A}(v')$, where $can_{E/A}(t)$ denotes the canonical form of t modulo A as an A-equivalence class. Conditions (1–2) ensure that $T_{\Sigma/E \cup A,k}$ is a computable set, since each ground term t can be simplified by applying the equations E from left to right modulo A to reach the *canonical form* $can_{E/A}(t)$. Condition (3) then implies that $(\to^1_{\mathcal{R},k})^\bullet$ is a *computable binary relation* on $T_{\Sigma/E \cup A,k}$: one can decide $[t]_{E \cup A} \to^1_{\mathcal{R}} [u]_{E \cup A}$ generating the finite set of all one-step $\mathcal{R}$-rewrites modulo A of $can_{E/A}(t)$ and testing for equal canonical forms.

To associate a Kripke structure to a rewrite theory one must specify the state predicates Π. Their syntax can be specified in a subsignature $\Pi \subseteq \Sigma$ of function symbols p of the general form $p : s_1 \ldots s_n \longrightarrow Prop$, allowing state predicates to be *parametric*. Their semantics can be defined by means of equations $D \subseteq E \cup A$ with the help of an auxiliary operator $\_ \models \_ : k \ Prop \longrightarrow Result$ in Σ, with *Result* a supersort (supertype) of *Bool*. By definition, given ground terms $u_1, \ldots, u_n$, one says that the state predicate $p(u_1, \ldots, u_n)$ *holds* in the state $[t]$ iff $E \cup A \vdash t \models p(u_1, \ldots, u_n) = true$. One can then associate to $\mathcal{R}$ a Kripke structure $\mathcal{K}(\mathcal{R}, k)_\Pi = (T_{\Sigma/E,k}, (\to^1_{\mathcal{R},k})^\bullet, L_\Pi)$, whose atomic predicates are specified by the set $AP_\Pi = \{\theta(p) \mid p \in \Pi, \theta \text{ ground substitution}\}$ [3], and where $L_\Pi([t]) = \{\theta(p) \in AP_\Pi \mid \theta(p) \text{ holds in } [t]\}$. If the rewrite theory $\mathcal{R}$ is computable, one then obtains a *computable* Kripke structure $\mathcal{K}(\mathcal{R}, k)_\Pi$ which, if it has finite reachability sets, can be used for model checking LTL formulas. The Maude 2.0 system has an on-the-fly, explicit-state LTL model checker [5] which supports the methodology just described.

---

[3] By convention, if p has n parameters, $\theta(p)$ denotes the term $\theta(p(x_1, \ldots, x_n))$.

3   The Semantics of Localized Fairness

I now define what it means for a fairness specification to be satisfied in the Kripke structure $\mathcal{K}(\mathcal{R}, k)_\Pi$ associated to a rewrite theory $\mathcal{R}$. The definition makes precise the exact sense in which the justice and fairness notions become *localized* by the choices of the parameters l and $\bar{y}$.

Definition 2. *(Satisfaction of Fairness Specifications). A path*[4] *π in $\mathcal{K}(\mathcal{R}, k)_\Pi$ satisfies $j(l(\bar{y}))$, written $\pi \models j(l(\bar{y}))$, if whenever there is a natural number m and a ground substitution θ of the variables $\bar{y}$ such that, for all natural numbers i, there is a rule with label l in $\mathcal{R}$ and a substitution ρ_i extending θ to the variables $\bar{x}_i$ of that rule such that the rule is enabled (can be applied modulo A) at some position in u_{m+i}, then for an infinite set $J \subseteq \mathbb{N}$ of indices j there is indeed a rewrite $u_{m+j} \rightarrow^1_{\mathcal{R},k} v_{m+j+1}$ with the enabled rule and a substitution ρ'_j such that $\rho'_j|_{\bar{y}} = \theta$, and $[v_{m+j+1}] = [u_{m+j+1}]$.*

Similarly, π satisfies $f(l(\bar{y}))$, written $\pi \models f(l(\bar{y}))$, if whenever there is a ground substitution θ of the variables $\bar{y}$ such that, for an infinite set $J \subseteq \mathbb{N}$ of indices j, there is a rule with label l in $\mathcal{R}$ and a substitution ρ_j extending θ to the variables $\bar{y}_j$ of that rule such that the rule is enabled at some position in u_j, then for an infinite set $J' \subseteq J$ of indices j' there is indeed a rewrite $u_{j'} \rightarrow^1_{\mathcal{R},k} v_{j'+1}$ with the enabled rule and a substitution $\rho'_{j'}$ such that $\rho'_{j'}|_{\bar{y}} = \theta$, and $[v_{j'+1}] = [u_{j'+1}]$.

Finally, if $\mathcal{J} \cup \mathcal{F}$ is a fairness specification for $\mathcal{R}$, one writes $\pi \models \mathcal{J} \cup \mathcal{F}$ if π satisfies each of the basic fairness specifications in $\mathcal{J} \cup \mathcal{F}$; and $\mathcal{K}(\mathcal{R}, k)_\Pi, [t] \models \mathcal{J} \cup \mathcal{F}$ iff $\pi \models \mathcal{J} \cup \mathcal{F}$ for each path π in $\mathcal{K}(\mathcal{R}, k)_\Pi$ such that $\pi(0) = [t]$. One then also writes: $Path(\mathcal{R})^{\mathcal{J}}_{\mathcal{F}} = \{\pi \text{ in } \mathcal{K}(\mathcal{R}, k)_\Pi \text{ s.t. } \pi \models \mathcal{J} \cup \mathcal{F}\}$. $Path(\mathcal{R})^{\mathcal{J}}$, resp. $Path(\mathcal{R})_{\mathcal{F}}$, abbreviate $Path(\mathcal{R})^{\mathcal{J}}_{\varnothing}$, resp. $Path(\mathcal{R})^{\varnothing}_{\mathcal{F}}$.   □

Due to the very definition of their semantics, adding more localized justice/fairness requirements to a specification has the effect of further reducing the set of paths satisfying all such requirements. This can be expressed by the formula,

$$Path(\mathcal{R})^{\mathcal{J}}_{\mathcal{F}} = (\bigcap_{j(l(\bar{y})) \in \mathcal{J}} Path(\mathcal{R})^{j(l(\bar{y}))}) \cap (\bigcap_{f(l'(\bar{y}')) \in \mathcal{F}} Path(\mathcal{R})_{f(l'(\bar{y}'))}).$$

Since fairness implies justice, one always has, $Path(\mathcal{R})_{f(l(\bar{y}))} \subseteq Path(\mathcal{R})^{j(l(\bar{y}))}$. At first sight it would seem that the more localized a justice or fairness requirement is, the *stronger* it is logically; that is, one might conjecture that if $\bar{y} \subseteq \bar{z}$, then $Path(\mathcal{R})^{j(l(\bar{z}))} \subseteq Path(\mathcal{R})^{j(l(\bar{y}))}$, and $Path(\mathcal{R})_{f(l(\bar{z}))} \subseteq Path(\mathcal{R})_{f(l(\bar{y}))}$. However, there are rewrite theories $\mathcal{R}$ with rules such that $\bar{y} \subset \bar{z}$ are subsets of variables for rules labeled l, and neither $Path(\mathcal{R})^{j(l(\bar{z}))} \subseteq Path(\mathcal{R})^{j(l(\bar{y}))}$ nor

---

[4] Notice that, by construction, this is either an infinite sequence $\{[u_i]\}_{i \in \mathbb{N}}$ such that for each i one has $[u_i] \rightarrow^1_{\mathcal{R},k} [u_{i+1}]$, or a finite such sequence ending in an equivalence class $[u_m]$ that cannot be further rewritten, so that then $[u_m] = [u_{m+i}]$ for all i; this second case is due to the totalization $(\rightarrow^1_{\mathcal{R},k})^\bullet$ of the relation $\rightarrow^1_{\mathcal{R},k}$ in $\mathcal{K}(\mathcal{R}, k)_\Pi$. I will assume throughout that all representatives $u_i \in [u_i]$ are in E/A-canonical form.

$Path(\mathcal{R})^{j(l(\bar{y}))} \subseteq Path(\mathcal{R})^{j(l(\bar{z}))}$; likewise, neither $Path(\mathcal{R})_{f(l(\bar{z}))} \subseteq Path(\mathcal{R})_{f(l(\bar{y}))}$, nor $Path(\mathcal{R})_{f(l(\bar{y}))} \subseteq Path(\mathcal{R})_{f(l(\bar{z}))}$. This is demonstrated by the following counterexample. Let $\mathcal{R}$ be an object-oriented rewrite theory in which object identifiers are terms of the form $o(n)$, with n a natural number, and holding just a natural number value in a single attribute val. There are two rules $inc : o(n) \longrightarrow o(s(n))$, and $double : \langle o(n) \mid val : m \rangle \longrightarrow \langle o(n) \mid val : 2 * m \rangle$. The path π for the computation,

$$\langle o(0) \mid val : 1 \rangle \, \langle o(1) \mid val : 1 \rangle \to \langle o(0) \mid val : 2 \rangle \, \langle o(1) \mid val : 1 \rangle \to$$
$$\langle o(0) \mid val : 2 \rangle \, \langle o(2) \mid val : 1 \rangle \ldots \to \langle o(0) \mid val : 2^{n-1} \rangle \, \langle o(n) \mid val : 1 \rangle \to \ldots$$

is such that $\pi \in Path(\mathcal{R})_{f(double)} = Path(\mathcal{R})^{j(double)}$ (the set equality holds because both rules are always enabled for nonempty object configurations), but $\pi \notin Path(\mathcal{R})_{f(double(n))} = Path(\mathcal{R})^{j(double(n))}$. Similarly, the path π' for the computation,

$$\langle o(1) \mid val : 1 \rangle \to \langle o(2) \mid val : 1 \rangle \to \langle o(3) \mid val : 1 \rangle \ldots \to \langle o(n) \mid val : 1 \rangle \to \ldots$$

is such that $\pi' \in Path(\mathcal{R})_{f(double(n))} = Path(\mathcal{R})^{j(double(n))}$, but $\pi' \notin Path(\mathcal{R})_{f(double)} = Path(\mathcal{R})^{j(double)}$.

To be able to model check LTL properties of a rewrite theory $\mathcal{R}$ one should assume that $\mathcal{R}$ has *finite reachability sets*, that is, that for each initial state $[t]$, the set of states reachable by rewriting with $\mathcal{R}$ is finite. The following result is useful in those circumstances.

Proposition 1. *If a computable rewrite theory $\mathcal{R}$ has finite reachability sets, then, if $\bar{y} \subset \bar{z}$ are subsets of the variables of all rules labeled l in $\mathcal{R}$, one has $Path(\mathcal{R})_{f(l(\bar{z}))} \subseteq Path(\mathcal{R})_{f(l(\bar{y}))}$.*

Proof. Let $\pi \in Path(\mathcal{R})_{f(l(\bar{z}))}$. By the finite reachability assumption, any path involves only a finite number of states $[u_1], \ldots, [u_k]$. Given the variables $\bar{z}$ and the label l, by the computability of $\mathcal{R}$, there is then only a finite number of A-matching substitutions, say $\theta_1, \ldots, \theta_n$, on the variables $\bar{z}$ for which some rule with label l is enabled in some state in π. Suppose now that, for some substitution ρ on the variables $\bar{y}$ some rules with label l are enabled infinitely often in π. There can then only be a finite number of possible extensions of ρ to a subset $\{\theta_{j_1}, \ldots, \theta_{j_m}\} \subseteq \{\theta_1, \ldots, \theta_n\}$ of substitutions of the variables $\bar{z}$ for which some rule with label l is enabled in some state in π. Therefore, there is some θ_{j_i} in that subset for which some rules with label l are enabled infinitely often in π. Since $\pi \in Path(\mathcal{R})_{f(l(\bar{z}))}$ some rules with label l are infinitely often taken in π with that substitution θ_{j_i} on $\bar{z}$, and therefore with substitution ρ on $\bar{y}$. Therefore, $\pi \in Path(\mathcal{R})_{f(l(\bar{y}))}$, as desired. $\square$

Under finite reachability, a similar containment doesn't hold for justice. That is, there are computable rewrite theories $\mathcal{R}$ with finite reachability sets and subsets $\bar{y} \subset \bar{z}$ of the variables of all rules labeled by some l in $\mathcal{R}$, such that $Path(\mathcal{R})^{j(l(\bar{z}))} \not\subseteq Path(\mathcal{R})^{j(l(\bar{y}))}$. This is demonstrated by the following counterexample. Let $\mathcal{R}$ be a rewrite theory with a sort $Conf$ of states having a binary associative and commutative multiset union operator with empty syntax, a constant $[]$, and unary operators $[\_], \{\_\} : Conf \longrightarrow Conf$, and $[\_] : Nat \longrightarrow Conf$. There are four rules: $in : [n] \; [] \longrightarrow [[n]], \; in' : [n] \; [] \longrightarrow \{[n]\}, \; out : [[n]] \longrightarrow [n] \; [],$ and $out' : \{[n]\} \longrightarrow [n] \; []$. Let π be the path associated to the cyclic computation

$$[1] \, [2] \, [] \, [] \to \{[1]\} \, [2] \, [] \to [1] \, [2] \, [] \, [] \to \{[2]\} \, [1] \, [] \to [1] \, [2] \, [] \, [] \to \cdots$$

Note that $\pi \in Path(\mathcal{R})^{j(in(n))}$, but $\pi \notin Path(\mathcal{R})^{j(in)}$.

4 Relative vs. Absolute LTL Semantics

Given a computable rewrite theory $\mathcal{R}$ with fairness specifications $\mathcal{J} \cup \mathcal{F}$ and with associated Kripke structure $\mathcal{K}(\mathcal{R}, k)_\Pi$, one can *relativize* to the fairness specifications $\mathcal{J} \cup \mathcal{F}$ the semantics of LTL formulas. In this way, one can build the specifications $\mathcal{J} \cup \mathcal{F}$ into the LTL model. As it is well-known [3, 10], the semantics of an LTL formula φ is first defined on paths by specifying recursively the relation $\pi \models \varphi$. Then, the satisfaction relation $\mathcal{K}(\mathcal{R}, k)_\Pi, [t] \models \varphi$ is defined by a universal quantification over paths: $\mathcal{K}(\mathcal{R}, k)_\Pi, [t] \models \varphi$ iff for all paths $\pi \in Path(\mathcal{R})$ with $\pi(0) = [t]$, $\pi \models \varphi$. To build in the fairness specifications $\mathcal{J} \cup \mathcal{F}$, thus relativizing to them the LTL semantics, one can define a new satisfaction relation $\models_\mathcal{F}^\mathcal{J}$ as follows: (i) $\pi \models_\mathcal{F}^\mathcal{J} \varphi$ iff $\pi \models \varphi$; and (ii) $\mathcal{K}(\mathcal{R}, k)_\Pi, [t] \models_\mathcal{F}^\mathcal{J} \varphi$ iff for all paths $\pi \in Path(\mathcal{R})_\mathcal{F}^\mathcal{J}$ with $\pi(0) = [t]$, $\pi \models \varphi$. Of course, one has a more expressive model, but one then needs deductive and model checking methods to reason in this richer, relativized semantics. There are two ways to address this need: (1) to develop deductive (see, for example, [16, 17]) or model checking (see, for example, [7, 9]) techniques that build in reasoning under the fairness specifications; or (2) to develop techniques transforming the deductive or model checking reasoning under fairness into standard deductive or model checking temporal logic reasoning; at the deductive level this is advocated, for example, in [8]. The first are obviously relative semantics approaches, whereas the second seek a reduction to absolute semantics. Each approach has its advantages and usefulness for different purposes. Generally speaking, building in such information is useful and efficient, but not all tools support this, and, furthermore, at present none to my knowledge supports reasoning about localized fairness in the general sense presented here. Also, some deductive approaches are based on quite specific models or languages not trivially extensible to our general framework.

There is therefore sufficient interest in investigating reductions to absolute semantics, both to support standard deductive reasoning, and to be able to use an LTL model checker for rewrite theories like Maude's [5] to model check properties under localized fairness assumptions. The essence of all absolute approaches, whether deductive or model checking, is to *represent the fairness specifications as an LTL formula* $\theta_\mathcal{F}^\mathcal{J}$, so that one can reduce the relative satisfaction problem $\mathcal{K}(\mathcal{R}, k)_\Pi, [t] \models_\mathcal{F}^\mathcal{J} \varphi$ to the absolute satisfaction of the formula $(\theta_\mathcal{F}^\mathcal{J} \to \varphi)$. Recall that justice of a transition τ means: $\Diamond\Box enabled.\tau \to \Box\Diamond taken.\tau$; and likewise fairness means: $\Box\Diamond enabled.\tau \to \Box\Diamond taken.\tau$. Therefore, if one can properly define the *enabled* and *taken* predicates, it should be possible to obtain the desired formula $\theta_\mathcal{F}^\mathcal{J}$. Defining the appropriate localized forms of the *enabled* predicate for any computable rewrite theory $\mathcal{R}$ is relatively easy. Suppose a justice requirement $j(l(\bar{y}))$, where the sorts of the variables $\bar{y}$ are, say, $s_1, \ldots, s_n$. One can then define a parameterized predicate *enabled.l* : $s_1, \ldots, s_n \longrightarrow$ *Prop* and define its semantics in terms of a family of auxiliary predicates *enabled.l.aux* : $ks_1, \ldots, s_n \longrightarrow$ *Result* for each kind (type) k in Σ (recall that *Result* is a supersort of *Bool*) by giving the following equations: (i) $x \models$ *enabled.l*$(\bar{y}) =$ *enabled.l.aux*$(x, \bar{y})$; (ii) for each rewrite rule with label l, say, $l : t \longrightarrow t'$ *if cond*, whose variables must contain those in $\bar{y}$, a conditional equation *enabled.l.aux*$(t, \bar{y}) = true$ *if cond*; and (iii) for each constructor symbol $f : k_1 \ldots k_n \longrightarrow k$ in Σ and each nonfrozen position $i \notin \phi(f)$ an equation *enabled.l.aux*$(f(x_1, \ldots, x_n), \bar{y}) = true$ *if enabled.l.aux*$(x_i, \bar{y}) = true$. One can check that these are ground confluent and terminating equations, and that for each term t of the state sort or kind one has an equivalence,

$$(\exists t') \, t \xrightarrow{l(\theta)}_{\mathcal{R},k}^{1} t' \quad \Leftrightarrow \quad t \models \text{enabled.l}(\theta) = true.$$

The serious challenge is defining the *taken.l* predicate. The point is that this must be a predicate on a given state $[t]$; but it is in general *undecidable* whether, for a given ground substitution θ, there is a ground term t' such that $t' \xrightarrow{l(\theta)}_{\mathcal{R},k}^{1} t$. Even in cases in which this question could be decided, one would only know that $[t]$ *could* have been reached in one step from another state $[t']$ with a rule with label l applied at some position with partial substitution θ. But many other previous states, labels, and partial substitutions could likewise have been used: in general one would have no way to know whether the step $l(\theta)$ was the one *actually taken*. The upshot is that defining *taken.l* on the original set of states is entirely hopeless: something else must be done. The solution I propose is to transform $\mathcal{R}$ into another theory $\mathcal{R}_\mathcal{F}^\mathcal{J}$ having a Kripke structure bisimilar to $\mathcal{K}(\mathcal{R}, k)_\Pi$. [12] shows that this can always be done for any computable $\mathcal{R}$. Here I explain the main intuition – in a simpler setting than that required to define the transformation for a general theory $\mathcal{R}$ – using our running dining philosophers example. In this case, the transformed theory $\mathcal{R}_\mathcal{F}^\mathcal{J}$ is obtained as follows. The original signature is first extended with the following new sorts and operations:

```
sorts  Label LState .
op mt : -> Label .
ops pickl pickr e2t : Nat/(N) -> Label .
op {_,_} : Configuration Label -> LState .
```

That is, states in the new state sort, LState will be pairs consisting of a configuration of philosophers and forks, and a *label* indicating the last rewrite and which object performed it, i.e., label expressions are unary operators with the same name as the corresponding rule label having object identifiers (in the parametric sort Nat/(N) of naturals modulo n) as arguments. If no rule has yet been taken, or if t2h is taken, one uses the empty label mt. The equations are unchanged, and rewrite rules are a modified version of the original rules in which the label expression for the rule application is recorded in the resulting state:

```
crl [t2h] : {C' library(< [N] : Phil | mode : t , holds :  none > C), L}
=> {C' < [N] : Phil | mode : h , holds :  none > library(C), mt}
                                      if C =/= none .
```

```
crl [pickl] : {C < [N] : Phil | mode : h , holds : none > chop([N],[M]),
L} => {C < [N] : Phil | mode : h , holds :  chop([N],[M]) >, pickl([N])}
    if [M] = [s(N)] .
rl [pickr] : {C < [N] : Phil | mode : h , holds : chop([N],[M]) >
    chop([N],[K]), L} =>
    {C < [N] : Phil | mode : e , holds :  chop([N],[M]) chop([N],[K]) >,
    pickr([N])} .
rl [e2t] : {C' < [N] : Phil | mode : e , holds :  chop([N],[M])
    chop([N],[K]) > library(C), L} => {C' chop([N],[M]) chop([N],[K])
    library(< [N] : Phil | mode : t , holds :  none > C), e2t([N])} .
```

The fact that the above rules always rewrite at the top of the state expression make the expression of the *enabled.l* and *taken.l* predicates particularly simple. For example, for $l = $ pickr they are defined as follows:

```
op enabled.pickr : Nat/(N) -> Prop .
op taken : Label -> Prop .
eq {C < [N] : Phil | mode : h , holds : chop([N],[M]) >
                    chop([N],[K]), L} |= enabled.pickr([N]) = true .
eq {C,L} |= taken(L) = true .
```

where the operator |= is used to define equationally the labeling function as a binary predicate between states and propositions (see [12] for the full specification). A similar pattern can be followed to define the *enabled.l* predicates for the remaining labels. Furthermore, one can lift the definition of all predicates $p \in \Pi$ on the states of the original system to $\mathcal{R}_{\mathcal{F}}^{\mathcal{J}}$ by adding to the equations defining p in $\mathcal{R}$ the equation (note the ad-hoc overloading of $\models$) $\{x, l\} \models y = x \models y$, where x, y, and l are variables of, respectively: the old state sort (or kind), $\texttt{Prop}^5$, and Label. In this way, one obtains a Kripke structure $\mathcal{K}(\mathcal{R}_{\mathcal{F}}^{\mathcal{J}}, k)_{\Pi}$, which can be further extended to a Kripke structure $\mathcal{K}(\mathcal{R}_{\mathcal{F}}^{\mathcal{J}}, k)_{\Pi'}$ by adding to the predicates in Π the new localized *taken* predicates. In this example, since we there are three basic fairness specifications in $\mathcal{J} \cup \mathcal{F}$, the desired LTL formula $\theta_{\mathcal{F}}^{\mathcal{J}}$ characterizing the fairness specifications is a conjunction $\theta_{pickl} \wedge \theta_{pickr} \wedge \theta_{e2t}$. For example, for an instance with n philosophers, the formula θ_{pickr} becomes:

$$\bigwedge_{0 \leq i \leq n-1} \Box \Diamond enabled.pickr([i]) \rightarrow \Box \Diamond taken(pickr([i]))$$

The formulas θ_{pickl} and θ_{e2t} are defined in an entirely analogous way. What one can now do is to use the rewrite theory $\mathcal{R}_{\mathcal{F}}^{\mathcal{J}}$ to reason in an *absolute semantics* way about the relative (to $\mathcal{J} \cup \mathcal{F}$) semantics of LTL properties in $\mathcal{R}$. The key observations are: (1) $\mathcal{K}(\mathcal{R}_{\mathcal{F}}^{\mathcal{J}}, [\texttt{LState}])'_{\Pi}$ is also finitely reachable; and (2) there is a surjective map $H : [\{t, u\}] \mapsto [t]$, sending each state in $\mathcal{R}_{\mathcal{F}}^{\mathcal{J}}$ to a corresponding state in $\mathcal{R}$ and satisfying the following properties: (i) H is a *bisimulation* $\mathcal{K}(\mathcal{R}_{\mathcal{F}}^{\mathcal{J}}, [\texttt{LState}])_{\Pi} \cong \mathcal{K}(\mathcal{R}, [\texttt{Configuration}])_{\Pi}$, and (ii) for each LTL formula φ on the state predicates of $\mathcal{R}$ one has an *equivalence*:

$$\mathcal{K}(\mathcal{R}_{\mathcal{F}}^{\mathcal{J}}, [\texttt{LState}])_{\Pi'}, [\{t, u\}] \models (\theta_{\mathcal{F}}^{\mathcal{J}} \rightarrow \varphi) \Leftrightarrow \mathcal{K}(\mathcal{R}, [\texttt{Configuration}])_{\Pi}, [t] \models_{\mathcal{F}}^{\mathcal{J}} \varphi.$$

---

[5] One may assume that the *enabled.l* predicates are in Π and have been extended from the old states to the new ones this way; in the above example I defined `enabled.pickr` directly on the new states for the sake of brevity.

For example, one can now use the transformed theory to verify a formula φ asserting the non-starvation of our original dining philosophers system under the specified fairness assumptions for some chosen value of n, by model checking instead the formula $\theta_{\mathcal{F}}^{\mathcal{J}} \to \varphi$ in our transformed system. The pattern followed for the above theory transformation remains valid *for any object-oriented rewrite theory* whose configurations are a flat multiset [14]. The above equivalence holds in fact true for *any* computable rewrite theory $\mathcal{R}$, but the general construction is somewhat more involved, due to the fact that in general rewriting can happen not just at the top, but anywhere in a term. That is, as explained in detail in [12], there is a general theory transformation $\mathcal{R} \mapsto \mathcal{R}_{\mathcal{F}}^{\mathcal{J}}$ such that: (1) it preserves finite reachability of the corresponding Kripke structures; and (2) there is a bisimulation $H : \mathcal{K}(\mathcal{R}_{\mathcal{F}}^{\mathcal{J}}, \hat{k})_{\Pi} \cong \mathcal{K}(\mathcal{R}, k)_{\Pi}$ (with $\hat{k}$ a kind of state-label pairs, see [12]), and one has:

Theorem 1. *For any computable $\mathcal{R}$ specifying a Kripke structure and having fairness specifications $\mathcal{J} \cup \mathcal{F}$, and for any LTL formula φ in its state predicates, one can define a formula $\theta_{\mathcal{F}}^{\mathcal{J}}$ characterizing the fairness requirements so that there is an equivalence,*

$$\mathcal{K}(\mathcal{R}_{\mathcal{F}}^{\mathcal{J}}, \hat{k})_{\Pi'}, [t] \models (\theta_{\mathcal{F}}^{\mathcal{J}} \to \varphi) \quad \Leftrightarrow \quad \mathcal{K}(\mathcal{R}, k)_{\Pi}, H([t]) \models_{\mathcal{F}}^{\mathcal{J}} \varphi. \qquad \square$$

In general, the formula $\theta_{\mathcal{F}}^{\mathcal{J}}$ may have universally quantified *parameters*, so that a deductive, instead of a model checking, verification may be needed. In our example this could be avoided because of the finite number n of philosophers; but the universal quantification was implicit in the conjunction over all philosophers used to define, e.g., θ_{pickr}. Rule justice/fairness is of course *parameterless*.

5   Concluding Remarks

I have proposed the notion of localized fairness, shown its usefulness, given a simple rewriting semantics for it, and proved basic results about it and about LTL reasoning under such fairness assumptions. Several research directions seem ready to be explored. First, deductive methods to reason about localized fairness, including techniques for proving termination under such assumptions – extending for example the termination proof techniques under rule fairness in [18, 19] – should be developed. Second, model checking methods that build in localized fairness – extending model checking algorithms for fairness such as [7, 9] – would also be quite helpful for efficiency reasons, because for sophisticated fairness requirements the formula $\theta_{\mathcal{F}}^{\mathcal{J}}$ can be big, and there is an exponential blowup when building the Büchi automaton for an LTL formula [3].

Acknowledgments

I thank Steven Eker and Miguel Palomino for very helpful discussions and suggestions on these ideas, which led to significant additions and improvements in the exposition. I also thank Jayadev Misra for very helpful comments, and the referees for an excellent list of improvements. Research supported by Grants ONR N00014-02-1-0715 and NSF CCR-0234524.

References

1. G. Agha. *Actors*. MIT Press, 1986.
2. R. Bruni and J. Meseguer. Generalized rewrite theories. In J. Baeten, J. Lenstra, J. Parrow, and G. Woeginger, editors, *Proceedings of ICALP 2003, 30th International Colloquium on Automata, Languages and Programming*, volume 2719 of *Springer LNCS*, pages 252–266, 2003.
3. E. Clarke, O. Grumberg, and D. Peled. *Model Checking*. MIT Press, 2001.
4. M. Clavel, F. Durán, S. Eker, P. Lincoln, N. Martí-Oliet, J. Meseguer, and C. Talcott. Maude 2.0 Manual. June 2003, http://maude.cs.uiuc.edu.
5. S. Eker, J. Meseguer, and A. Sridharanarayanan. The Maude LTL model checker. In F. Gadducci and U. Montanari, editors, *Proc. 4th. Intl. Workshop on Rewriting Logic and its Applications*. ENTCS, Elsevier, 2002.
6. N. Francez. *Fairness*. Springer-Verlag, 1986.
7. V. Gyuris and A. Sistla. On-the-fly model-checking under fairness that exploits symmetry. *Formal Methods in System Design*. To appear.
8. L. Lamport. A temporal logic of actions. *ACM Trans. on Prog. Lang. and Systems*, 16(3):872–923, 1994.
9. T. Latvala and K. Heljanko. Coping with strong fairness. *Fundamenta Informaticae*, 34:1–19, 2000.
10. Z. Manna and A. Pnueli. *The Temporal Logic of Reactive and Concurrent Systems – Specification*. Springer-Verlag, 1992.
11. N. Martí-Oliet and J. Meseguer. Rewriting logic: roadmap and bibliography. *Theoretical Computer Science*, 285:121–154, 2002.
12. J. Meseguer. Localized fairness. Manuscript, UIUC, February 2005, http://maude.cs.uiuc.edu.
13. J. Meseguer. Conditional rewriting logic as a unified model of concurrency. *Theoretical Computer Science*, 96(1):73–155, 1992.
14. J. Meseguer. A logical theory of concurrent objects and its realization in the Maude language. In G. Agha, P. Wegner, and A. Yonezawa, editors, *Research Directions in Concurrent Object-Oriented Programming*, pages 314–390. MIT Press, 1993.
15. J. Meseguer. Membership algebra as a logical framework for equational specification. In F. Parisi-Presicce, editor, *Proc. WADT'97*, pages 18–61. Springer LNCS 1376, 1998.
16. J. Misra. *A Discipline of Multiprogramming*. Springer-Verlag, 2001.
17. A. Pnueli and Y. Kesten. Algorithmic and deductive verification methods for CTL*. In M. Broy and M. Pizka, editors, *Models, Algebras, and Logic of Engineering Software, NATO Advanced Study Institute, Marktoberdorf, Germany, July 30 – August 11, 2002*, pages 109–131. IOS Press, 2003.
18. S. Porat and N. Francez. Fairness in term rewriting systems. In *Proc. Intl. Conf. on Rewriting Techniques and Applications (RTA 1985)*, pages 287–300. Springer LNCS 202, 1985.
19. S. Porat and N. Francez. Full communication and fair termination in equational (and combined) term-rewriting systems. In *Proc. 8th Intl. Conf. on Automated Deduction (CADE-8)*, pages 21–41. Springer LNCS 230, 1986.
20. S. Tison. Fair termination is decidable for ground systems. In N. Dershowitz, editor, *Rewriting Techniques and Applications, Chappel Hill, North Carolina*, pages 462–476. Springer LNCS 355, 1989.
21. P. Viry. Equational rules for rewriting logic. *Theoretical Computer Science*, 285:487–517, 2002.

Partial Inversion
of Constructor Term Rewriting Systems*

Naoki Nishida, Masahiko Sakai, and Toshiki Sakabe

Graduate School of Information Science, Nagoya University
Furo-cho, Chikusa-ku, Nagoya 464-8603, Japan
{nishida,sakai,sakabe}@is.nagoya-u.ac.jp

Abstract. Partial-inversion compilers generate programs which compute some unknown inputs of given programs from a given output and the rest of inputs whose values are already given. In this paper, we propose a partial-inversion compiler of constructor term rewriting systems. The compiler automatically generates a conditional term rewriting system, and then unravels it to an unconditional system. To improve the efficiency of inverse computation, we show that innermost strategy is usable to obtain all solutions if the generated system is right-linear.

1   Introduction

Roughly speaking, an *inverse* of a program P with one argument is a program P' such that $P(a) = b$ coincides $P'(b) = a$ for any data a and b. In case of a program P with two arguments, its *full inverse* program P' satisfies that $P(a, b) = c$ coincides $P'(c) = (a, b)$ for any data a, b and c. On the other hand, a *partial inverse* of P with respect to the first argument is a program P'' such that $P(a, b) = c$ coincides $P''(c, a) = b$.

We can find inverse programs in practical cases. Data compression and extraction commands (for example, `gzip` and `gunzip`) are examples of a program with one argument and its inverse. For a cryptographic encoder $E(x, k)$ with a symmetric key k, the decoder $D(y, k)$ can be seen as a partial inverse of $E(x, k)$ with respect to the second argument. Other typical examples of partial-inverses are subtraction for addition and division for multiplication.

The inversion sometimes helps us to generate a program from a specification given by equations. For example, the function gcd that computes the greatest common divisor is specified by the equations $gcd(x+y, y) = gcd(x, y)$, $gcd(x, 0) = x$ and $gcd(x, y) = gcd(y, x)$. From these equations, we can construct the following program:

$$\left\{ \begin{array}{ll} gcd(z, y) \to gcd(z - y, y), & x - 0 \to x, \\ gcd(x, 0) \to x, & s(x) - s(y) \to x - y, \\ gcd(x, y) \to gcd(y, x). & \end{array} \right.$$

---

* This work is partly supported by MEXT. KAKENHI #15500007 and #16300005.

J. Giesl (Ed.): RTA 2005, LNCS 3467, pp. 264–278, 2005.

Subtraction used in the above program is the partial inverse of addition in the specification. In this case, we fortunately know the definition of the subtraction. However, we do not always know the definition of the needed inverse. For example, it is not easy to code division as a recursive function by hand, while the function of division is well-known.

Can we automatically construct an inverse from a given program? This naive question motivates the study of *inversion compilers*. A (partial-)inversion compiler is an algorithm which automatically generates a (partial) inverse for a given program. In this paper, we present a partial-inversion compiler in the framework of term rewriting systems (TRSs), prove its correctness, and discuss the efficiency of the computation done by inverses generated by the compiler.

The partial-inversion compiler which we propose generates a partial-inverse EV-TRS (a TRS with *extra variables* that appear only in right-hand sides of rewrite rules) from a given constructor TRS. The compiler consists of two stages. In the first stage, the compiler generates a conditional TRS (a CTRS) as a partial-inverse program. In the second stage, the compiler transforms it to an equivalent EV-TRS, being based on the method of *unraveling* [14, 17, 19]. We prove the correctness of the compiler, that is, the generated EV-TRS is really a partial-inverse of the given TRS.

In inverse computation, it is not easy to handle values which are erased in the forward computation. In order to represent a guess of such a value, we exploit extra variables. Although the reduction of an EV-TRS is essentially infinitely-branching and non-terminating, the reduction can be simulated by *narrowing* sequences starting from ground terms [16]. The termination problem of such narrowing is closer to that of the TRS reduction than that of the ordinary narrowing. Therefore, the existence of extra variables in the generated systems is not disadvantageous for the compiler in this paper. The compiler sometimes generates a TRS (which has no extra variable), and it is terminating if it is lucky.

In logic programs like Prolog, inverse computation is realized by narrowing-based computation. However, the execution of inverse computation in Prolog does not terminate in some practical cases, when solving erased values in the forward computation. By contrast, the corresponding generated EV-TRS of our compiler terminates and bring all solutions by the depth-first search.

Unraveling brings a problem into the compiler. The reductions of CTRSs cannot be always simulated by the unraveled CTRSs completely. Unravelings [14, 19] which are developed in order to analyze properties of CTRSs, are not suitable for the simulation of the CTRS reduction. In this paper, we show that the combination of *membership conditional* [25] and *context-sensitive* [12] reductions brings the completeness. We also introduce another syntactic condition of CTRSs [17] for the completeness.

Another important issue for inverse computation is the efficiency. It is often necessary to find all normal forms of a given term. However, the exhaustive search is inefficient. In this paper, we show that the generated EV-TRSs always satisfy *ILRJ* property which is a part of conditions in [21] for the completeness of the innermost strategy on TRSs.

This paper is organized as follows. Section 2 prepares notations. In Section 3, we define a partial-inversion compiler of constructor TRSs, and prove its correctness. In Section 4, we discuss the computation of the generated EV-TRSs, and in Section 5 the improvement of the efficiency of partial-inverse computation. In Section 6, we state some related works briefly and give concluding remarks. The missing proofs of theorems will be shown in the full version of this paper [18].

2   Preliminaries

This paper follows general notations of term rewriting [2, 20].

Let $\mathcal{V}$ be a countably infinite set of *variables*. The set of all *terms* over a *signature* $\mathcal{F}$ and $\mathcal{V}$ is denoted by $\mathcal{T}(\mathcal{F}, \mathcal{V})$. Especially, we abbreviate the set $\mathcal{T}(\mathcal{F}, \emptyset)$ of all *ground terms* to $\mathcal{T}(\mathcal{F})$. The set of variables occurring in one of terms $t_1, \ldots, t_n$ is represented by $Var(t_1, \ldots, t_n)$. *Identity* of terms is denoted by $\equiv$. The *root* symbol of a term t is represented by $\mathbf{root}(t)$. For a *context* $C[, \ldots,]$ with n holes $\square$ at positions $p_1, \ldots, p_n$ and for terms $t_1, \ldots, t_n$, the notation $C[t_1, \ldots, t_n]_{p_1, \ldots, p_n}$ (or simply $C[t_1, \ldots, t_n]$) denotes the term obtained by replacing $\square$ at p_i with t_i for $i = 1, \ldots, n$. We denote the (proper) *subterm relation* by $\trianglelefteq$ ($\lhd$). The term $\sigma(t)$ obtained by applying a substitution σ to a term t is abbreviated to $t\sigma$. The *composition* $\sigma\sigma'$ of substitutions σ and σ' is defined as $t\sigma\sigma' \equiv \sigma'(\sigma(t))$.

An *(oriented) conditional rewrite rule* $l \to r \Leftarrow c$ over a signature $\mathcal{F}$ consists of the left-hand side l ($\in \mathcal{T}(\mathcal{F}, \mathcal{V}) \setminus \mathcal{V}$) (lhs), the right-hand side r ($\in \mathcal{T}(\mathcal{F}, \mathcal{V})$) (rhs) and the conditional part c which is a sequence $s_1 \to t_1, \ldots, s_k \to t_k$ of *oriented conditions* with $\{s_1, t_1, \ldots, s_k, t_k\} \subseteq \mathcal{T}(\mathcal{F}, \mathcal{V})$. We write $l \to r$ instead of $l \to r \Leftarrow c$ if it is unconditional (that is, if $k = 0$). We denote by $\rho : l \to r \Leftarrow c$ the rule $l \to r \Leftarrow c$ with a unique label ρ. The set $\mathcal{E}Var(\rho)$ of all *extra variables* of $\rho : l \to r \Leftarrow c$ is defined as $\mathcal{E}Var(\rho) = Var(r, c) \setminus Var(l)$.

An *(oriented) conditional rewriting system* (CTRS) over a signature $\mathcal{F}$ is a finite set of oriented conditional rewrite rules over $\mathcal{F}$. A CTRS is a *term rewriting system with extra variables* (EV-TRS) if its every rule is unconditional, and then it is a *term rewriting system* (TRS) if it is an EV-TRS without extra variables. We use $R^{\Leftarrow}$ and $S^{\Leftarrow}$ for CTRSs (possibly EV-TRSs), and R and S for EV-TRSs. For a CTRS $R^{\Leftarrow}$, the *rewrite relation* of $R^{\Leftarrow}$ is denoted by $\to_{R^{\Leftarrow}}$. To specify the position p for $s \to_{R^{\Leftarrow}} t$, we write $s \to_{R^{\Leftarrow}}^{p} t$. The set of all *normal forms* of $R^{\Leftarrow}$ is denoted by $NF_{R^{\Leftarrow}}(\mathcal{F}, \mathcal{V})$. Conditional rewrite rules $\rho : l \to r \Leftarrow c$ are classified according to the distribution of variables among l, r and c, as follows: ρ is in *type 3* if $Var(r) \subseteq Var(l, c)$, and in *type 4* if no restriction is imposed. An *i-CTRS* contains only conditional rewrite rules of type i. A CTRS $R^{\Leftarrow}$ is said to be *deterministic* if every rule $\rho : l \to r \Leftarrow s_1 \to t_1, \ldots, s_k \to t_k \in R^{\Leftarrow}$ is deterministic, that is, $Var(s_i) \subseteq Var(l, t_1, \ldots, t_{i-1})$ for $1 \leq i \leq k$.

The set of *defined symbols* for a CTRS $R^{\Leftarrow}$ over a signature $\mathcal{F}$ is $\mathcal{D}_{R^{\Leftarrow}} = \{ \mathbf{root}(l) \mid l \to r \Leftarrow c \in R^{\Leftarrow} \}$. The signature is partitioned as $\mathcal{F} = \mathcal{D}_{R^{\Leftarrow}} \uplus \mathcal{C}_{R^{\Leftarrow}}$ where $\uplus$ is the disjoint union of sets. Function symbols in $\mathcal{C}_{R^{\Leftarrow}}$ are called *constructors* of $R^{\Leftarrow}$. A term t in $\mathcal{T}(\mathcal{C}_{R^{\Leftarrow}}, \mathcal{V})$ is called a *constructor term* of $R^{\Leftarrow}$. A *constructor system* is a CTRS $R^{\Leftarrow}$ such that every rule $f(t_1, \ldots, t_n) \to r \Leftarrow c \in R^{\Leftarrow}$ satisfies $\{t_1, \ldots, t_n\} \subseteq \mathcal{T}(\mathcal{C}_{R^{\Leftarrow}}, \mathcal{V})$. A CTRS $R^{\Leftarrow}$ is *convergent* if it is confluent and strongly normalizing with respect to $\to_{R^{\Leftarrow}}$. We use a, b, c as constructors, f, g, h as defined symbols, and x, y, z as variables.

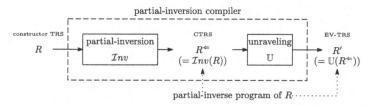

Fig. 1. The structure of the partial-inversion compiler proposed in this paper.

3   Partial-Inversion Compiler

In this section, we present a partial-inversion compiler of constructor TRSs which consists of two stages (Fig. 1). The first stage is an actual partial-inversion which generates a partial-inverse CTRS from a given constructor TRS. The second is the unraveling which transforms the CTRS to an equivalent EV-TRS [20].

We prepare special constructors $\{\mathsf{tp}_0, \mathsf{tp}_1, \ldots\}$ to denote the tuple $(t_1, \ldots, t_i)$ of terms $t_1, \ldots, t_i$ as $\mathsf{tp}_i(t_1, \ldots, t_i)$. The tuple $\mathsf{tp}_1(t)$ of a term t may be abbreviated to the term t. The reason why introducing such symbols is that inverses of n-ary functions return tuples of some terms.

An *index* for an n-ary defined symbol f is a natural number i such that $1 \leq i \leq n$, which intuitively stands for an argument position of f. We use sets of indexes for f to represent which arguments of f are given. For a set I of indexes for f, we denote by $|I|$ the cardinality of I. The j-th index of I in the increasing order is denoted by I_j. That is, if $I = \{i_1, \ldots, i_m\}$ and $i_j < i_{j+1}$ for all $1 \leq j < m$, then I_j represents i_j. The notation $\bar{I}$ denotes the complement of I, that is, $\bar{I} = \{1, \ldots, n\} \setminus I$. For a set $\mathcal{D}$ of defined symbols, the set of all pairs of a defined symbol and a set of indexes for the symbol is denoted by $\mathbb{I}_{\mathcal{D}}$: $\mathbb{I}_{\mathcal{D}} = \{(g, I) \mid g \in \mathcal{D}, I \subseteq \{1, \ldots, |g|\}\}$ where $|g|$ is the arity of g.

3.1   Definition of Partial-Inverses

Here we give a concrete definition of *partial-inverses*.

Definition 1. *Let $R^{\Leftarrow}$ be a CTRS over a signature $\mathcal{F}$, and $S^{\Leftarrow}$ be a CTRS over a signature $\mathcal{F}'$ satisfying $\mathcal{C}_{R^{\Leftarrow}} \subseteq \mathcal{C}_{S^{\Leftarrow}}$. Let f and g be defined symbols of $R^{\Leftarrow}$ and $S^{\Leftarrow}$, respectively, and I be a set of indexes for f. Then, g is a* partial inverse *of f with respect to a set I if the following holds:*

> *for all ground constructor terms $t, t_1, \ldots, t_n$ of $R^{\Leftarrow}$, $f(t_1, \ldots, t_n) \xrightarrow{*}_{R^{\Leftarrow}} t$ if and only if $g(t, t_{I_1}, \ldots, t_{I_{|I|}}) \xrightarrow{*}_{S^{\Leftarrow}} \mathsf{tp}_{|\bar{I}|}(t_{\bar{I}_1}, \ldots, t_{\bar{I}_{|\bar{I}|}})$,*

where $\mathsf{tp}_{|\bar{I}|} \in \mathcal{C}_{S^{\Leftarrow}}$. In particular, g is a full inverse of f if $I = \emptyset$. The CTRS $S^{\Leftarrow}$ is called a partial-inverse system *of $R^{\Leftarrow}$ with respect to a set $\mathcal{D}_I \subseteq \mathbb{I}_{\mathcal{D}_{R^{\Leftarrow}}}$ (simply called a* partial-inverse system *of $R^{\Leftarrow}$ when $\mathcal{D}_I = \mathbb{I}_{\mathcal{D}_{R^{\Leftarrow}}}$) if for every $(f, I) \in \mathcal{D}_I$, there exists $g \in \mathcal{D}_{S^{\Leftarrow}}$ such that g is a partial-inverse of f with respect to I. $R^{\Leftarrow}$ is sometimes called the* forward(-computation) system *for $S^{\Leftarrow}$.*

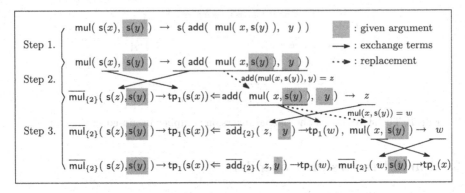

Fig. 2. Sketch of partial-inversion of the third rule of R_1.

Example 2. Consider the following convergent constructor TRSs over the signature $\{0, \mathsf{s}, \mathsf{add}, \mathsf{mul}, \mathsf{minus}\}$:

$$R_1 = \begin{cases} \mathsf{add}(0, y) \to y, \\ \mathsf{add}(\mathsf{s}(x), y) \to \mathsf{s}(\mathsf{add}(x, y)), \\ \mathsf{mul}(0, y) \to 0, \quad \mathsf{mul}(x, 0) \to 0, \\ \mathsf{mul}(\mathsf{s}(x), \mathsf{s}(y)) \to \mathsf{s}(\mathsf{add}(\mathsf{mul}(x, \mathsf{s}(y)), y)) \end{cases} \quad R_2 = \begin{cases} \mathsf{minus}(x, 0) \to x, \\ \mathsf{minus}(\mathsf{s}(x), \mathsf{s}(y)) \to \mathsf{minus}(x, y). \end{cases}$$

R_2 is a partial-inverse TRS of R_1 with respect to $\{(\mathsf{add}, \{1\})\}$.

3.2 Idea of Partial Inversion

This subsection intuitively explains how to generate a partial-inverse CTRS from a given constructor TRS, by using R_1 in Example 2 and the pair $(\mathsf{mul}, \{2\})$.

Roughly speaking, we generate a conditional rewrite rule for a rewrite rule in a given system R, and a set of indexes for the root symbol of the lhs. The idea of the generation is based on the essential property of inverses: $f(v_1, \dots, v_n) = v$ if and only if $f^{-1}(v, v_1, \dots, v_i) = (v_{i+1}, \dots, v_n)$. For a given pair (f, I) of a defined symbol f and a set I of indexes for f, we use $\overline{f}_I$ as a symbol of the partial-inverse of f with respect to I. We add a special rewrite rule for each (f, I) (as we will show the detail later). The partial-inverse CTRS of R_1 with respect to $\{(\mathsf{mul}, \{2\})\}$ is generated as follows:

$$R_3^{\Leftarrow} = \begin{cases} \overline{\mathsf{add}}_{\{2\}}(y, y) \to \mathsf{tp}_1(0), \\ \overline{\mathsf{add}}_{\{2\}}(\mathsf{s}(z), y) \to \mathsf{tp}_1(\mathsf{s}(x)) \Leftarrow \overline{\mathsf{add}}_{\{2\}}(z, y) \to \mathsf{tp}_1(x), \\ \overline{\mathsf{mul}}_{\{2\}}(0, y) \to \mathsf{tp}_1(0), \qquad \overline{\mathsf{mul}}_{\{2\}}(0, 0) \to \mathsf{tp}_1(x), \\ \overline{\mathsf{mul}}_{\{2\}}(\mathsf{s}(z), \mathsf{s}(y)) \to \mathsf{tp}_1(\mathsf{s}(x)) \\ \qquad\qquad \Leftarrow \overline{\mathsf{add}}_{\{2\}}(z, y) \to \mathsf{tp}_1(w), \overline{\mathsf{mul}}_{\{2\}}(w, \mathsf{s}(y)) \to \mathsf{tp}_1(x), \\ \overline{\mathsf{add}}_{\{2\}}(\mathsf{add}(x, y), y) \to \mathsf{tp}_1(x), \qquad \overline{\mathsf{mul}}_{\{2\}}(\mathsf{mul}(x, y), y) \to \mathsf{tp}_1(x). \end{cases}$$

We give an intuitive explanation by using the third rule of mul. To generate the fifth rule in R_3, we apply the following three steps to it (see Fig. 2):

(Step 1). *This step analyzes the rule and classifies variables in the rule into the given and the unknown, depending on whether value is assigned in future execution.* The result of this step is illustrated in the second line of Fig. 2. The pair $(\mathsf{mul}, \{2\})$ means that the second argument $\mathsf{s}(y)$ of mul in the lhs is given. Hence, the value of y is assigned in the execution. On the contrary, the value of x is unknown. Therefore, the second arguments of add and mul in the rhs are given, and their first arguments are unknown[1].

(Step 2). *This step creates a rule for inverses by exchanging parts of both the left- and right-hand sides of the rule, except the given arguments in the lhs.* When exchanging, we replace the term $\mathsf{add}(\mathsf{mul}(x, \mathsf{s}(y)), y)$ in the rhs with a fresh variable z (not used in the rule), and add the condition $\mathsf{add}(\mathsf{mul}(x, \mathsf{s}(y)), y) \to z$ caused by the replacement to the conditional part. This transformation makes the original rhs a constructor term. We also replace the root symbol mul in the lhs with the symbol $\overline{\mathsf{mul}}_{\{2\}}$.

(Step 3). *This step applies Step 2 to the condition part until it becomes deterministic.* By applying Step 2 twice to the condition part, we obtain the conditional rewrite rule of $\overline{\mathsf{mul}}_{\{2\}}$ illustrated in the bottom line of Fig. 2.

By applying the above three steps to all rules in R_1 and all needed pairs $((\mathsf{add}, \{2\})$ and $(\mathsf{mul}, \{2\}))$, we obtain the first five rules in $R_3^{\Leftarrow}$.

In addition to the above three steps, we generate for every pair $(f, I) \in \mathbb{I}_{\mathcal{D}}$ the special rewrite rule $\overline{f}_I(f(x_1, \ldots, x_n), x_{I_1}, \ldots, x_{I_{|I|}}) \to \mathsf{tp}_{|\bar{I}|}(x_{\bar{I}_1}, \ldots, x_{\bar{I}_{|\bar{I}|}})$, called the *inverse-property* rule of f with respect to I. These rules are necessary for inverse computation of functions that invoke partial functions. Consider the TRS $R_4 = R_1 \cup \{\ \mathsf{half}(0) \to 0,\ \mathsf{half}(\mathsf{s}^2(x)) \to \mathsf{s}(\mathsf{half}(x)),\ \mathsf{g}(x) \to \mathsf{mul}(x, \mathsf{half}(\mathsf{s}(x)))\ \}$. Since we have a derivation $\mathsf{g}(0) \xrightarrow{*}_{R_4} 0$, the inverse computation $\overline{\mathsf{g}}_{\emptyset}(0) \xrightarrow{*} \mathsf{tp}_1(0)$ should hold. If $\overline{\mathsf{half}}_{\emptyset}(\mathsf{half}(x)) \to \mathsf{tp}_1(x)$ were missing, $\overline{\mathsf{g}}_{\emptyset}(0)$ could not be reachable to $\mathsf{tp}_1(0)$ because the only applicable rule to $\overline{\mathsf{g}}_{\emptyset}(0)$ is $\overline{\mathsf{g}}_{\emptyset}(y) \to \mathsf{tp}_1(x) \Leftarrow \overline{\mathsf{mul}}_{\emptyset}(y) \to \mathsf{tp}_2(x, z), \overline{\mathsf{half}}(z) \to \mathsf{tp}_1(\mathsf{s}(x))$ and we must find z satisfying $\mathsf{half}_{\emptyset}(z) \to \mathsf{tp}_1(\mathsf{s}(0))$.

3.3   Generation of Partial-Inverse CTRSs

In this subsection, we formalize the idea described in Subsection 3.2. To simplify the presentation, we focus on generating partial-inverse CTRSs with respect to $\mathbb{I}_{\mathcal{D}}$. We can easily get the CTRS with respect to a subset of $\mathbb{I}_{\mathcal{D}}$ by collecting usable rules after the generation.

We first provide definitions (Definition 3–5) necessary to show the first stage of our compiler. We write $C[\![t_1, \ldots, t_n]\!]$ instead of a term $C[t_1, \ldots, t_n]$ if C is a constructor context (that is, $C[, \ldots,] \in \mathcal{T}(\mathcal{C} \cup \{\square\}, \mathcal{V})$) and $\mathsf{root}(t_i)$ is a defined symbol for every $i \in \{1, \ldots, n\}$. It is clear that every term s can be represented in the form of $C'[\![s_1, \ldots, s_k]\!]$.

Definition 3. *Let* $\rho : f(w_1, \ldots, w_n) \to C[\![r_1, \ldots, r_m]\!]$ *be a rewrite rule over a signature* $\mathcal{D} \uplus \mathcal{C}$ *and* I *be a set of indexes for* f. *The set* $\mathcal{UV}ar(\rho, I)$ *of variables in* ρ, *whose values are* unknown, *is defined as follows:*

---

[1] The value of $\mathsf{mul}(x, \mathsf{s}(y))$ is not given because this term has the unknown variable x.

$$\mathcal{U}Var\,(\,f(w_1,\dots,w_n) \to C[\![r_1,\dots,r_m]\!], I\,)$$
$$= Var(w_{\bar{I}_1},\dots,w_{\bar{I}_{|\bar{I}|}}) \setminus Var(w_{I_1},\dots,w_{I_{|I|}}, C[\,,\dots,]).$$

The above definition corresponds to the preparation of Step 1 in Subsection 3.2. An example of the above definition will be found after Definition 5.

For a term t and the set X of unknown variables, we define a label attachment to each defined symbol in t. The label of the defined symbol g for a subterm $g(t_1,\dots,t_n)$ is a set I of indexes for g such that $i \in I$ if all variables in t_i are known (no variable in t is not in X).

Definition 4. *Let $\mathcal{F}\,(=\mathcal{C}\uplus\mathcal{D})$ be a signature, $f(t_1,\dots,t_n)$ be a term with $f \in \mathcal{D}$, and X be a set of variables. We define the set of indexes which specifies positions of arguments t_is not containing any variable in X, as $\mathcal{I}\,(f(t_1,\dots,t_n),X)$ $= \{\,i \mid Var(t_i) \cap X = \emptyset\,\}$.*

For a term t, the labeled term $\mathcal{L}ab(t,X)$ in which defined symbols are labeled with a set of indexes, is recursively defined as follows:

- *$\mathcal{L}ab(x,X) = x$ where x is a variable,*
- *$\mathcal{L}ab(c(t_1,\dots,t_n),X) = c\,(\mathcal{L}ab(t_1,X),\dots,\mathcal{L}ab(t_n,X))$ where $c \in \mathcal{C}$, and*
- *$\mathcal{L}ab(f(t_1,\dots,t_n),X) = f_I\,(\mathcal{L}ab(t_1,X),\dots,\mathcal{L}ab(t_n,X))$ where $f \in \mathcal{D}$ and $I = \mathcal{I}(f(t_1,\dots,t_n),X)$.*

The transformation from a rewrite rule $f(w_1,\dots,w_n) \to r$ to the labeled rule $f_I(w_1,\dots,w_n) \to \mathcal{L}ab(r,\mathcal{U}Var(\rho,I))$ corresponds to Step 1 in Subsection 3.2.

We here define the procedure that produces a term and a sequence of conditions, which are parts of constructing a conditional rule from a given rule.

Definition 5. *Let $\mathcal{F}\,(=\mathcal{C}\uplus\mathcal{D})$ be a signature. The procedure $\mathcal{T}$, which outputs a pair of a term and a condition part from an input labeled term, is inductively defined as follows:*

(a) *$\mathcal{T}(x) = \langle\, x\,;\,\epsilon\,\rangle$ where x is a variable,*

(b) *$\mathcal{T}(c(t_1,\dots,t_n)) = \langle c(u_1,\dots,u_n);Cond_1,\dots,Cond_n\rangle$ where $c \in \mathcal{C}$ and $\mathcal{T}(t_i)$ $= \langle\, u_i\,;\,Cond_i\,\rangle$ for $1 \le i \le n$,*

(c) *$\mathcal{T}(f_I(t_1,\dots,t_n)) =$*
    $\langle\, y\,;\,\overline{f}_I(y,u_{I_1},\dots,u_{I_{|I|}}) \to tp_{|\bar{I}|}(s_{\bar{I}_1},\dots,s_{\bar{I}_{|\bar{I}|}}),Cond'_{\bar{I}_1},\dots,Cond'_{\bar{I}_{|\bar{I}|}}\,\rangle$
    where $f \in \mathcal{D}$, $|I| < n$, y is a 'fresh'[2] variable, $\mathcal{T}(t_i) = \langle\, u_i\,;\,Cond_i\,\rangle$ for $1 \le i \le n$,
    - *$u_{\bar{I}_j} = C_{\bar{I}_j}[u_{\bar{I}_j,1},\dots,u_{\bar{I}_j,m_{\bar{I}_j}}]$, $s_{\bar{I}_j} = C_{\bar{I}_j}[z_{\bar{I}_j,1},\dots,z_{\bar{I}_j,m_{\bar{I}_j}}]$,*
    - *$Cond'_{\bar{I}_j} = Cond_{\bar{I}_j}, u_{\bar{I}_j,1} \to z_{\bar{I}_j,1},\dots,u_{\bar{I}_j,m_{\bar{I}_j}} \to z_{\bar{I}_j,m_{\bar{I}_j}}$, and $z_{\bar{I}_j,k}$ is a 'fresh'[2] variable, and,*

(d) *$\mathcal{T}(f_{\{1,\dots,n\}}(t_1,\dots,t_n)) = \langle\, f(u_1,\dots,u_n)\,;\,Cond_1,\dots,Cond_n\,\rangle$ where $f \in \mathcal{D}$ and $\mathcal{T}(t_i) = \langle\, u_i\,;\,Cond_i\,\rangle$ for $1 \le i \le n$,*

where we write ϵ to represent the empty sequence of conditions.

---

[2] This means that
   - in the case (c), $y \notin \bigcup_{i=1}^{n} Var(t_i, u_i, Cond_i)$, $z_{\bar{I}_j,k} \notin \{y\} \cup \bigcup_{i=1}^{n} Var(t_i, u_i, Cond_i)$, $z_{\bar{I}_j,k'} \ne z_{\bar{I}_j,k''}$ ($k' \ne k''$), and
   - in the both (b) and (c), variables introduced in each $\mathcal{T}(t_i) = \langle\, u_i\,;\,Cond_i\,\rangle$ are disjoint, that is, $(Var(u_i, Cond_i) \setminus Var(t_i)) \cap Var(t_j, u_j, Cond_j) = \emptyset$ for $i \ne j$.

It is clear that the above procedure $\mathscr{T}$ always terminates and returns a pair of a term and a conditional part. Note that for $\mathscr{T}(t) = \langle\, u \,;\, Cond\,\rangle$, u can be represented as $C[\![u_1, \ldots, u_n]\!]$ $(\in T(\mathcal{F}, \mathcal{V}))$.

Example 6. $\mathcal{U}Var$, $\mathcal{I}$, $\mathcal{L}ab$ and $\mathscr{T}$ perform for the fifth rule of R_1 as follows:

$$\mathcal{U}Var(\,\mathsf{mul}(\mathsf{s}(x), \mathsf{s}(y)) \to \mathsf{s}(\mathsf{add}(\mathsf{mul}(x, \mathsf{s}(y)), y)),\ \{2\}\,) = \{x\},$$
$$\mathcal{I}(\,\mathsf{add}(\mathsf{mul}(x, \mathsf{s}(y)), y),\ \{x\}\,) = \{2\}, \quad \mathcal{I}(\,\mathsf{mul}(x, \mathsf{s}(y)),\ \{x\}\,) = \{2\},$$
$$\mathcal{L}ab(\,\mathsf{s}(\mathsf{add}(\mathsf{mul}(x, \mathsf{s}(y)), y)),\ \{x\}\,) = \mathsf{s}(\mathsf{add}_{\{2\}}(\mathsf{mul}_{\{2\}}(x, \mathsf{s}(y)), y)),$$
$$\mathscr{T}(\,\mathsf{s}(\mathsf{add}_{\{2\}}(\mathsf{mul}_{\{2\}}(x, \mathsf{s}(y)), y))\,)$$
$$= \langle\, \mathsf{s}(z)\,;\, \overline{\mathsf{add}}_{\{2\}}(z, y) \to w, \overline{\mathsf{mul}}_{\{2\}}(w, \mathsf{s}(y)) \to x \,\rangle.$$

We finally define the partial-inversion from constructor TRSs into CTRSs.

Definition 7. *Let R be a constructor TRS over a signature $\mathcal{F}$. For a rewrite rule $\rho : f(w_1, \ldots, w_n) \to r \in R$ and a set I of indexes for f, its corresponding conditional rewrite rule $InvRule(\rho, I)$ of $\overline{f}_I$ is defined as follows:*

$$InvRule(\,f(w_1, \ldots, w_n) \to r,\ I\,) =$$
$$\overline{f}_I(C[y_1, \ldots, y_m], w_{I_1}, \ldots, w_{I_{|I|}}) \to \mathsf{tp}_{|\bar{I}|}(w_{\bar{I}_1}, \ldots, w_{\bar{I}_{|\bar{I}|}})$$
$$\Leftarrow u_1 \to y_1, \ldots, u_m \to y_m, Cond$$

where $\mathscr{T}(\,\mathcal{L}ab(r, \mathcal{U}Var(\rho, I))\,) = \langle\, C[\![u_1, \ldots, u_m]\!] \,;\, Cond\,\rangle$, each y_i is a variable with $y_i \notin Var(w_1, \ldots, w_n)$, and $(Var(C[\![u_1, \ldots, u_m]\!], Cond) \setminus Var(r)) \cap Var(y_1, \ldots, y_m, w_1, \ldots, w_n) = \emptyset$.
The partial-inverse CTRS $Inv(R)$ is defined as follows:

$$Inv(R) = \{\, InvRule(\rho, I) \mid \rho : f(w_1, \ldots, w_n) \to r \in R, I \subseteq \{1, \ldots, n\}\,\}$$
$$\cup \{\, \overline{f}_I(f(x_1, \ldots, x_n), x_{I_1}, \ldots, x_{I_{|I|}}) \to \mathsf{tp}_{|\bar{I}|}(x_{\bar{I}_1}, \ldots, x_{\bar{I}_{|\bar{I}|}}) \mid (f, I) \in \mathbb{I}_{\mathcal{D}_R}\,\} \cup R.$$

The extended signature $\overline{\mathcal{F}}$ is defined as $\overline{\mathcal{F}} = \mathcal{F} \uplus \{\, \overline{f}_I \mid (f, I) \in \mathbb{I}_{\mathcal{D}_R}\,\}$.

The generation of conditional rules by $InvRule$ corresponds to Step 2 and Step 3 in Subsection 3.2. Note that our approach does not require the left-linearity of input systems. It is clear that $InvRule(\rho, I)$ is exactly a deterministic conditional rewrite rule and then $Inv(R)$ is exactly a deterministic CTRS over the signature $\overline{\mathcal{F}}$. For a CTRS $S^{\Leftarrow}$ and a set D of defined symbols of $S^{\Leftarrow}$, we use $(S^{\Leftarrow})|_D$ for all rules in $S^{\Leftarrow}$ that are necessary to calculate terms in $T(D \cup \mathcal{C}_{S^{\Leftarrow}}, \mathcal{V})$.

Example 8. Consider R_1 again. For $(\mathsf{mul}, \{2\})$ and $(\mathsf{mul}, \{1, 2\})$, we obtain by Inv the following CTRS:

$$R_5^{\Leftarrow} = Inv(R_1)|_{\{\overline{\mathsf{mul}}_{\{2\}}, \overline{\mathsf{mul}}_{\{1,2\}}\}}$$
$$= R_3^{\Leftarrow} \cup \left\{ \begin{array}{l} \overline{\mathsf{mul}}_{\{1,2\}}(0, 0, y) \to \mathsf{tp}_0, \quad \overline{\mathsf{mul}}_{\{1,2\}}(0, x, 0) \to \mathsf{tp}_0, \\ \overline{\mathsf{mul}}_{\{1,2\}}(\mathsf{s}(z), \mathsf{s}(x), \mathsf{s}(y)) \to \mathsf{tp}_0 \Leftarrow \mathsf{add}(\mathsf{mul}(x, \mathsf{s}(y)), y) \to z, \\ \overline{\mathsf{mul}}_{\{1,2\}}(\mathsf{mul}(x, y), x, y) \to \mathsf{tp}_0 \end{array} \right\} \cup R_1.$$

The following theorem shows that the CTRS obtained by Inv from a given constructor TRS can perform the inverse computation of innermost derivations of the TRS. For a TRS R, we write $\xrightarrow[\mathrm{in}]{} R$ as the *innermost reduction* of R.

Theorem 9. *Let R be a constructor TRS over a signature $\mathcal{F}$. Let $(f, I) \in \mathbb{I}_{\mathcal{D}_R}$ and $t, t_1, \ldots, t_n$ be normal forms of R. Then, $f(t_1, \ldots, t_n) \xrightarrow[\text{in}]{*}_R t$ if and only if $\bar{f}_I(t, t_{I_1}, \ldots, t_{I_{|I|}}) \rightarrow_{\mathcal{I}nv(R)} \mathsf{tp}_{|\bar{I}|}(t_{\bar{I}_1}, \ldots, t_{\bar{I}_{|\bar{I}|}})$.*

It is clear that every normalizing reduction sequence of convergent TRSs can be simulated by an innermost sequence. Hence, the above theorem implies the following corollary:

Corollary 10. *Let R be a convergent constructor TRS over a signature $\mathcal{F}$. Let $(f, I) \in \mathbb{I}_{\mathcal{D}_R}$ and $t, t_1, \ldots, t_n$ be normal forms of R. Then, $f(t_1, \ldots, t_n) \xrightarrow{*}_R t$ if and only if $\bar{f}_I(t, t_{I_1}, \ldots, t_{I_{|I|}}) \rightarrow_{\mathcal{I}nv(R)} \mathsf{tp}_{|\bar{I}|}(t_{\bar{I}_1}, \ldots, t_{\bar{I}_{|\bar{I}|}})$.*

From the above corollary and the fact $\mathcal{T}(\mathcal{C}_R) \subseteq NF_R(\mathcal{F}, \mathcal{V})$, the CTRS $\mathcal{I}nv(R)$ is exactly a partial-inverse system of the convergent constructor TRS R with respect to $\mathbb{I}_{\mathcal{D}_R}$ in the sense of Definition 1. In addition, it is clear that $(\mathcal{I}nv(R))|_{D_{\mathcal{I}}}$ for some subset $D_{\mathcal{I}} \subseteq \mathbb{I}_{\mathcal{D}_R}$ is a partial-inverse CTRS of R with respect to $D_{\mathcal{I}}$.

3.4   Unraveling to Unconditional Systems

As the second stage of the compiler proposed in this paper, we here give the unraveling for deterministic CTRSs [20]. For a set $A = \{a_1, \ldots, a_n\}$, $\vec{A}$ denotes a list $a_1, \ldots, a_n$ for a unique representation of A.

Definition 11 ([20]). *Let $R^{\Leftarrow}$ be a deterministic CTRS over a signature $\mathcal{F}$. For every conditional rewrite rule $\rho : l \rightarrow r \Leftarrow s_1 \rightarrow t_1, \ldots, s_k \rightarrow t_k$ in $R^{\Leftarrow}$, we prepare k fresh function symbols $\mathsf{u}_1^\rho, \ldots, \mathsf{u}_k^\rho$, called $\mathbb{U}$ symbols, neither of which appears in $\mathcal{F}$. Then, the set of rewrite rules determined by ρ is defined as follows:*

$$\mathbb{U}(\rho) = \{\ l \rightarrow \mathsf{u}_1^\rho(s_1, \vec{X}_1),\ \mathsf{u}_1^\rho(t_1, \vec{X}_1) \rightarrow \mathsf{u}_2^\rho(s_2, \vec{X}_2),\ \cdots,\ \mathsf{u}_k^\rho(t_k, \vec{X}_k) \rightarrow r\ \}$$

where $X_i = Var(l, t_1, \ldots, t_{i-1}) \cap Var(r, t_i, s_{i+1}, t_{i+1}, \ldots, s_k, t_k)$. The system $\mathbb{U}(R^{\Leftarrow}) = \bigcup_{\rho \in R^{\Leftarrow}} \mathbb{U}(\rho)$ is an EV-TRS over the extended signature $\mathcal{F}_{\mathbb{U}(R)} = \mathcal{F} \uplus \bigcup_{\rho \in R^{\Leftarrow}} \{\ \mathsf{u}_i^\rho \mid 1 \le i \le |\rho|\ \}$ where $|\rho|$ denotes the number of the conditions of ρ.

It is clear that $R^{\Leftarrow}$ is a 3-CTRS if and only if $\mathbb{U}(R^{\Leftarrow})$ is a TRS.

Example 12. The CTRS $R_3^{\Leftarrow}$ in Subsection 3.2 is unraveled by $\mathbb{U}$ as follows:

$$\mathbb{U}(R_3^{\Leftarrow}) = \begin{cases} \overline{\mathsf{add}}_{\{2\}}(y, y) \rightarrow \mathsf{tp}_1(0), & \\ \overline{\mathsf{add}}_{\{2\}}(\mathsf{s}(z), y) \rightarrow \mathsf{u}_1(\overline{\mathsf{add}}_{\{2\}}(z, y)), & \mathsf{u}_1(\mathsf{tp}_1(x)) \rightarrow \mathsf{tp}_1(\mathsf{s}(x)), \\ \overline{\mathsf{mul}}_{\{2\}}(0, y) \rightarrow \mathsf{tp}_1(0), & \mathsf{mul}_{\{2\}}(0, 0) \rightarrow \mathsf{tp}_1(x), \\ \overline{\mathsf{mul}}_{\{2\}}(\mathsf{s}(z), \mathsf{s}(y)) \rightarrow \mathsf{u}_2(\overline{\mathsf{add}}_{\{2\}}(z, y), y), & \\ \mathsf{u}_2(\mathsf{tp}_1(w), y) \rightarrow \mathsf{u}_3(\overline{\mathsf{mul}}_{\{2\}}(w, \mathsf{s}(y))), & \mathsf{u}_3(\mathsf{tp}_1(x)) \rightarrow \mathsf{tp}_1(\mathsf{s}(x)), \\ \overline{\mathsf{add}}_{\{2\}}(\mathsf{add}(x, y), y) \rightarrow \mathsf{tp}_1(x), & \mathsf{mul}_{\{2\}}(\mathsf{mul}(x, y), y) \rightarrow \mathsf{tp}_1(x). \end{cases}$$

The unraveled CTRS cannot always simulate any rewrite sequence of an original CTRS completely. It holds for every deterministic CTRS $R^{\Leftarrow}$ over a signature $\mathcal{F}$ that, for every terms s and t in $\mathcal{T}(\mathcal{F}, \mathcal{V})$, $s \xrightarrow{*}_{R^{\Leftarrow}} t$ implies $s \xrightarrow{*}_{\mathbb{U}(R^{\Leftarrow})} t$.

However, the converse does not hold in general (see Example 7.2.14 in [20]). The converse is called the *simulation-completeness* of $\mathbb{U}$ for $R^{\Leftarrow}$. That is, for every terms s and t in $\mathcal{T}(\mathcal{F}, \mathcal{V})$, $s \xrightarrow{*}_{\mathbb{U}(R^{\Leftarrow})} t$ implies $s \xrightarrow{*}_{R^{\Leftarrow}} t$. In Section 4, we will show two solutions of this problem.

The unraveled CTRSs may produce some garbage normal forms, which contain $\mathbb{U}$ symbols. When $\mathbb{U}$ is simulation-complete, we can easily recognize whether a obtained normal form is garbage or not.

4 Computation of Partial-Inverse EV-TRSs

The rewrite relation of the generated systems has two problems to be solved. One is the existence of extra variables, and the other is the simulation-*incompleteness* of the unraveling $\mathbb{U}$. In this section, we state briefly that the first problem does not matter, and then we deal with the second problem from two approaches: a restriction against the rewrite relation, and a syntactic constraint on CTRSs.

4.1 On Extra Variables

Extra variables cause infinitely-branching and non-termination of the rewrite relation. However, these troubles are solved by *narrowing* [10] *starting from ground terms* – every *EV-normalized* rewrite sequence with the ground initial term has a narrowing sequence which starts from the same initial term, and of which the last term is more general than the last of the rewrite sequence, and vice versa [16]. Here a rewrite sequence is said to be *EV-normalized*, denoted by $\xrightarrow[\text{ev}]{}$, if a normal form is substituted for each extra variable in every reduction step of the rewrite sequence. In the subproof of Theorem 9 (see [18]), the inverse computation $\overline{f}_I(t, t_{I_1}, \ldots, t_{I_{|I|}}) \rightarrow_{\mathcal{I}nv(R)} \mathsf{tp}_{|\bar{I}|}(t_{\bar{I}_1}, \ldots, t_{\bar{I}_{|\bar{I}|}})$ is constructed from the innermost derivation $f(t_1, \ldots, t_n) \xrightarrow[\text{in}]{*}_R t$, substituting normal forms for extra variables. In addition, such a reduction can be easily simulated by the unraveled CTRS. These facts mean that every partial-inverse computation for innermost normalizing derivation can be represented by a (ground) EV-normalized rewrite sequence of the generated EV-TRSs.

Proposition 13. *Let R be a constructor TRS over a signature $\mathcal{F}$. Let $(f, I) \in \mathbb{I}_{\mathcal{D}_R}$ and $t, t_1, \ldots, t_n$ be normal forms of R. If $f(t_1, \ldots, t_n) \xrightarrow[\text{in}]{*}_R t$, then there exists an EV-normalized rewrite sequence such that $\overline{f}_I(t, t_{I_1}, \ldots, t_{I_{|I|}}) \xrightarrow[\text{ev}]{*}_{\mathbb{U}(\mathcal{I}nv(R))} \mathsf{tp}_{|\bar{I}|}(t_{\bar{I}_1}, \ldots, t_{\bar{I}_{|\bar{I}|}})$.*

Thus, the necessary rewrite sequences for partial-inverse computation can be simulated by narrowing starting from ground terms. Every step of narrowing is finitely branching up to renaming, and the non-termination is less serious in the narrowing starting from ground terms than in the ordinary narrowing. In fact, some EV-TRSs terminate with respect to narrowing starting from ground terms, while they do not terminates with respect to the ordinary narrowing.

4.2   On Simulation-Incompleteness of Unraveling

The second problem stated in the beginning of this section is caused from the
disordered evaluation of rules that originate in conditional parts (as shown in
Example 7.2.14 of [20]). It is solved by combining the *membership conditional*
and *context-sensitive* reductions.

We first give definitions of three reductions for the unraveled CTRSs: the
membership conditional reduction [25], the *context-sensitive reduction* [12] and
their combined reduction.

Definition 14. *Let $R^{\Leftarrow}$ be a deterministic CTRS over a signature $\mathcal{F}$. The* mem-
bership conditional rewrite relation $\xrightarrow[\mathrm{M}]{}_{\mathrm{U}(R^{\Leftarrow})}$ on $\mathcal{T}(\mathcal{F}_{\mathrm{U}(R^{\Leftarrow})}, \mathcal{V})$ is defined as
$$\xrightarrow[\mathrm{M}]{}_{\mathrm{U}(R^{\Leftarrow})} = \{\, (s,t) \mid s \xrightarrow{p}_{\mathrm{U}(R^{\Leftarrow})} t, (\,\forall u, u \lhd s|_p \text{ implies } u \in \mathcal{T}(\mathcal{F}, \mathcal{V})\,)\,\}.$$
*Let μ be a mapping from $\mathcal{F}_{\mathrm{U}(R^{\Leftarrow})}$ to a set of argument positions for f such
that $\mu(\mathsf{u}) = \{1\}$ for all $\mathsf{u} \in \mathcal{F}_{\mathrm{U}(R^{\Leftarrow})} \setminus \mathcal{F}$, and $\mu(f) = \{1, \dots, n\}$ for all n-ary
function symbol $f \in \mathcal{F}$. The set $\mathcal{O}_\mu(t)$ for a term $t \in \mathcal{T}(\mathcal{F}_{\mathrm{U}(R^{\Leftarrow})}, \mathcal{V})$ is defined
recursively as follows:*

- $\mathcal{O}_\mu(x) = \emptyset$ *where x is a variable, and*
- $\mathcal{O}_\mu(f(t_1, \dots, t_n)) = \{\, i_j q \mid 1 \le j \le m, q \in \mathcal{O}_\mu(t_{i_j})\,\}$ *where $f \in \mathcal{F}_{\mathrm{U}(R^{\Leftarrow})}$ and
  $\mu(f) = \{i_1, \dots, i_m\}$.*

The context-sensitive rewrite relation $\xrightarrow[\mathrm{CS}]{}_{\mathrm{U}(R^{\Leftarrow})(\mu)}$ on $\mathcal{T}(\mathcal{F}_{\mathrm{U}(R^{\Leftarrow})}, \mathcal{V})$ is defined
as $\xrightarrow[\mathrm{CS}]{}_{\mathrm{U}(R^{\Leftarrow})(\mu)} = \{\, (s,t) \mid s \xrightarrow{p}_{\mathrm{U}(R^{\Leftarrow})} t, p \in \mathcal{O}_\mu(s)\,\}.$
The membership context-sensitive (MCS) rewrite relation $\xrightarrow[\mathrm{MCS}]{}_{\mathrm{U}(R^{\Leftarrow})(\mu)}$ *is*
defined as $\xrightarrow[\mathrm{MCS}]{}_{\mathrm{U}(R^{\Leftarrow})(\mu)} = (\xrightarrow[\mathrm{M}]{}_{\mathrm{U}(R^{\Leftarrow})}) \cap (\xrightarrow[\mathrm{CS}]{}_{\mathrm{U}(R^{\Leftarrow})(\mu)}).$

Remark that these three reductions on narrowing are defined similarly. It is clear
that $\xrightarrow[\mathrm{MCS}]{}_{\mathrm{U}(R^{\Leftarrow})(\mu)} \subseteq \xrightarrow[\mathrm{M}]{}_{\mathrm{U}(R^{\Leftarrow})} \subseteq \to_{\mathrm{U}(R^{\Leftarrow})}$, and $\xrightarrow[\mathrm{MCS}]{}_{\mathrm{U}(R^{\Leftarrow})(\mu)} \subseteq \xrightarrow[\mathrm{CS}]{}_{\mathrm{U}(R^{\Leftarrow})(\mu)}$
$\subseteq \to_{\mathrm{U}(R^{\Leftarrow})}$. From these facts, both the termination of $\to_{\mathrm{U}(R^{\Leftarrow})}$ and the μ-
termination of $\xrightarrow[\mathrm{CS}]{}_{\mathrm{U}(R^{\Leftarrow})(\mu)}$ guarantee the termination of $\xrightarrow[\mathrm{M}]{}_{\mathrm{U}(R^{\Leftarrow})}$, $\xrightarrow[\mathrm{CS}]{}_{\mathrm{U}(R^{\Leftarrow})(\mu)}$
and $\xrightarrow[\mathrm{MCS}]{}_{\mathrm{U}(R^{\Leftarrow})(\mu)}$. To prove the termination of them, we can use existing tools,
such as APROVE [5], TTT [8], MU-TERM [13]. The MCS reduction above is im-
plemented as the particular case of MEP [4], and then we can use the technique
in [4] to prove the termination of $\xrightarrow[\mathrm{MCS}]{}_{\mathrm{U}(R^{\Leftarrow})(\mu)}$.

We here discuss the simulation-completeness with respect to the MCS reduc-
tion. Consider the rule $\rho : \mathsf{f}(x, y) \to x \Leftarrow \mathsf{g}(x) \to z, \mathsf{g}(y) \to z$ and the correspond-
ing unraveled rules $\mathbb{U}(\rho) = \{\, \mathsf{f}(x, y) \to \mathsf{u}_1^\rho(\mathsf{g}(x), x, y), \mathsf{u}_1^\rho(z, x, y) \to \mathsf{u}_2^\rho(\mathsf{g}(y), x, z),$
$\mathsf{u}_2^\rho(z, x, z) \to x\,\}$. The $\mathbb{U}$ symbol u_1^ρ is used to evaluate in the first argument the
first condition $\mathsf{g}(x) \to z$, and to deliver the value of variables x and y via the
rest of its arguments (that is, the second and third). From this observation, no
redex in either of k-th argument ($2 \le k \le n$) of u should be reduced until u
is reduced, and u should not be evaluated until the evaluation of the first argument is finished. These evaluations are kept by the context-sensitive and the
membership reductions, respectively. Hence, conditional parts of rules in $R^{\Leftarrow}$ are
evaluated in proper order on $\xrightarrow[\mathrm{MCS}]{}_{\mathrm{U}(R^{\Leftarrow})(\mu)}$.

Theorem 15. *For every deterministic CTRS $R^{\Leftarrow}$, $\mathbb{U}$ is simulation-complete
with respect to $\xrightarrow[\mathrm{MCS}]{}_{\mathrm{U}(R^{\Leftarrow})(\mu)}$.*

Proof (Sketch). This theorem can be proved straightforward by induction on the length of the rewrite sequences of $\to_{\mathbb{U}(R^{\Leftarrow})}$.   □

As another solution of the second problem stated in the beginning of this section, we have shown some results on the simulation-completeness of $\mathbb{U}$ with respect to the ordinary rewrite relation [17].

Theorem 16 ([17]). *Let $R^{\Leftarrow}$ be a deterministic CTRS.*

- *If either $\mathbb{U}(R^{\Leftarrow})$ is a left-linear TRS or $\mathbb{U}(R^{\Leftarrow})$ is right-linear and non-erasing, then $\mathbb{U}$ is simulation-complete for $R^{\Leftarrow}$ with respect to $\to_{\mathbb{U}(R^{\Leftarrow})}$.*
- *If $\mathbb{U}(R^{\Leftarrow})$ is left-linear, then $\mathbb{U}$ is simulation-complete for $R^{\Leftarrow}$ with respect to $\xrightarrow{\text{ev}}_{\mathbb{U}(R^{\Leftarrow})}$.*

Note that syntactic conditions of CTRSs that the unraveled CTRSs are left-linear, right-linear and non-erasing, respectively, are shown in [17]. Theorem 16 may not be applicable to other unravelings, while $\xrightarrow{\text{MCS}}_{\mathbb{U}(R^{\Leftarrow})}(\mu)$ may provide the simulation-completeness to them. According to Proposition 13, the restriction of the rewrite sequences to $\xrightarrow{\text{ev}}_{\mathbb{U}(R^{\Leftarrow})}$ does not affect the computation of the generated EV-TRSs.

From Theorem 9 and the discussion in this section, we can conclude that our method is correct.

5   Improving Efficiency of Partial-Inverse Computation

In this section, we show that the efficiency of partial-inverse computation can be improved by the innermost strategy without loss of completeness if the systems are right-linear.

It has been shown in [23] that for every right-linear overlay TRS, all normal forms of terminating terms can be obtained by innermost strategy. As shown in the following theorem, this result has been extended.

Theorem 17 ([21]). *Let R be an ILRJ and right-linear TRS, and s be a terminating term. For all normal forms t of s, $s \xrightarrow{*}_{\text{in}}_{R} t$.*

Here the *inside critical pairs* of a CTRS $R^{\Leftarrow}$ are the critical pairs obtained from rules overlap at non-root positions. The CTRS $R^{\Leftarrow}$ is said to be *inside left-to-right joinable* (ILRJ, for short) if every inside critical pair $\langle s, t \rangle$ ($s \leftarrow^{\varepsilon <} \cdot \to^{\varepsilon} t$) satisfies $s \xrightarrow{*}_{R^{\Leftarrow}} t$. Note that overlay systems are ILRJ.

Our partial-inverse EV-TRSs are sometimes not overlay. For example, $R_5^{\Leftarrow}$ in Example 8 is not overlay because they have the defined symbols add and mul in the first arguments of the lhs's of the inverse-property rules. From the same reason, the combination of forward and inverse programs such as $R_1 \cup \mathbb{U}(R_3^{\Leftarrow})$ are not always overlay. However, $R_3^{\Leftarrow}$ in Subsection 3.2 and $\mathbb{U}(R_3^{\Leftarrow})$ in Example 12 are ILRJ. In fact, the generated partial-inverse systems are always ILRJ.

Theorem 18. *Let R be a constructor TRS over a signature $\mathcal{F}$. Assume that for every rule $l \to r \in R$, the rhs r is weakly normalizing for the innermost reduction. Then, $\mathcal{I}nv(R) \cup R$ and $\mathbb{U}(\mathcal{I}nv(R)) \cup R$ are ILRJ.*

Proof. By the construction of $\mathcal{I}nv(R)$, inside overlaps in $\mathcal{I}nv(R) \cup R$ happens between rules $f(w_1, \ldots, w_n) \to r \in R$ and $\overline{f}_I(f(x_1, \ldots, x_n), x_{I_1}, \ldots, x_{I_{|I|}}) \to \mathsf{tp}_{|\bar{I}|}(x_{\bar{I}_1}, \ldots, x_{\bar{I}_{|\bar{I}|}}) \in \mathcal{I}nv(R)$. Hence, we have inside critical pairs only in the form $\langle \overline{f}_I(r, w_{I_1}, \ldots, w_{I_{|I|}}), \mathsf{tp}_{|\bar{I}|}(w_{\bar{I}_1}, \ldots, w_{\bar{I}_{|\bar{I}|}}) \rangle$. By the assumption, there exists a normal form t of R such that $r \xrightarrow{*}_R t$. Then, we have $f(w_1, \ldots, w_n) \xrightarrow{}_{\text{in}} {}_R r \xrightarrow{*}_{\text{in}} {}_R t$. It follows Theorem 9 that $\overline{f}_I(t, w_{I_1}, \ldots, w_{I_{|I|}}) \to_{\mathcal{I}nv(R)} \mathsf{tp}_{|\bar{I}|}(w_{\bar{I}_1}, \ldots, w_{\bar{I}_{|\bar{I}|}})$, and hence $\overline{f}_I(r, w_{I_1}, \ldots, w_{I_{|I|}}) \xrightarrow{*}_{\mathcal{I}nv(R) \cup R} \mathsf{tp}_{|\bar{I}|}(w_{\bar{I}_1}, \ldots, w_{\bar{I}_{|\bar{I}|}})$. Therefore, $\mathcal{I}nv(R) \cup R$ is ILRJ. The case of $\mathbb{U}(\mathcal{I}nv(R)) \cup R$ is similar to the above case.   □

In this case, it is clear that $(\mathcal{I}nv(R))|_{D_\mathcal{I}} \cup R$ and $\mathbb{U}((\mathcal{I}nv(R))|_{D_\mathcal{I}}) \cup R$ for $D_\mathcal{I} \subseteq \mathbb{I}_{\mathcal{D}_R}$ are also ILRJ. Now we suppose that forward computation is convergent, and then the assumption holds. Hence, the assumption is not really a restriction for the generated systems. It also holds that the CTRS $S^{\Leftarrow} \cup (R \cup (\mathcal{I}nv(R))|_{D_\mathcal{I}})$ (or $S^{\Leftarrow} \cup (R \cup \mathbb{U}((\mathcal{I}nv(R))|_{D_\mathcal{I}}))$) is ILRJ if $S^{\Leftarrow}$ is ILRJ, because the assumption that they are *constructor-sharing systems* is adequate.

From Theorem 17 and 18, innermost strategy is effective to improve the efficiency of reductions by the right-linear partial-inverse systems, without loss of the reachability to solutions.

The MCS reduction $\xrightarrow[\text{MCS}]{} \mathbb{U}(R^{\Leftarrow})(\mu)$ does not eliminate any necessary reduction sequence starting from a given term. Innermost strategy does not also eliminate such a sequence, when all of the conditions in Theorem 17 are satisfied. Therefore, in such cases, the MCS reduction with innermost strategy (that is, $\xrightarrow[\text{MCS,in}]{} \mathbb{U}(R^{\Leftarrow})(\mu)$) is not less efficient than either $\xrightarrow[\text{in}]{} \mathbb{U}(R^{\Leftarrow})$ or $\xrightarrow[\text{MCS}]{} \mathbb{U}(R^{\Leftarrow})(\mu)$.

6   Related Works and Conclusion

Full-inversion compilers have been studied in [6, 7, 11, 22] which are applicable to several functional languages, and of which the correctness was not discussed (not proved). By contrast, we have shown the correctness of our method. Moreover, the discussion in Section 5 seems the first work on improving the efficiency of inverse computation. A partial-inversion compiler is considered in [22], which is for the programming language *Refal* (as like constructor normal CTRSs), and in which the non-determinism of inverses is solved by representing output of functions as a set. *Bidirectional transformation* [9] based on the *bidirectional updating* in the field of database, uses *bidirectional languages* which is a similar notion to the partial inversion.

There is another approach to inverse computation. *Inverse interpreters* are procedures that compute unknown inputs from the program and a given output. Several kinds of interpreters have been studied in [1, 3, 24]. Inverse interpreters seem to deal easily with partial-inverse problem. The algorithm in [3] (which is consists of inference rules) essentially resembles our method in the sense that the input class is (ground-)convergent constructor TRSs. The algorithm terminates if the input TRS is *constructing*. We believe that the EV-TRS generated by our compiler from a constructing TRS is terminating with respect to narrowing starting from ground terms. Moreover, we have the example TRS R_4 which is not constructing but whose inverse is terminating.

The inverse computation in this paper can handle general solutions by variables that represent arbitrary terms. The compiler proposed in this paper is of course applicable to functions which returns multiple values, because the compiler can handle them by rules in the form of $f(\cdots) \rightarrow tuple(v_1, \ldots, v_m)$. The resulted TRS of the unraveling may be optimized. For example, u_1 and u_3 in $\mathbb{U}(R_3^{\Leftarrow})$ are nonsense.

We have encountered some examples that the efficiency of $\overrightarrow{\mathrm{MCS,in}}\,\mathbb{U}(\mathcal{I}nv(R))(\mu)$ is equal to that of $\overrightarrow{\mathrm{in}}\,\mathbb{U}(\mathcal{I}nv(R))$. It is a future work to analyze the detail of the efficiency. We are also interested in relationships between syntactic properties of an input TRS and the generated EV-TRS, for example, a condition of R inducing the right-linearity of $\mathbb{U}(\mathcal{I}nv(R))$.

Which is more effective, full inverses or partial inverses? Full inverses are included in partial inverses, and partial inverses can compute by the corresponding full-inverses because the results of the full inverses contain all solutions of partial inverses. Full inverses seems to be less efficient than partial inverses, but we know some desirable properties of syntactic relationships between forward and full-inverse programs. Analysis for this problem is another future work.

References

1. Abramov, S., Glück, R.: The universal resolving algorithm: Inverse computation in a functional language. In Proceedings of the 5th International Conference on Mathematics of Program Construction. Volume 1837 of LNCS, Springer (2000) 187–212.
2. Baader, F., Nipkow, T.: *Term Rewriting and All That*. Cambridge University Press (1998).
3. Dershowitz, N., Mitra, S.: Jeopardy. In *Proceedings of the 10th International Conference on Rewriting Techniques and Applications*. Volume 1631 of LNCS, Springer (1999) 16–29.
4. Durán, F., Lucas, S., Meseguer, J., Marché, C., Urbain, X.: Proving termination of membership equational programs. In *Proceedings of the ACM SIGPLAN 2004 Symposium on Partial Evaluation and Program Manipulation*. ACM Press (2004) 147–158.
5. Giesl, J., Thiemann, R., Schneider-Kamp, P., Falke, S.: Automated termination proofs with AProVE. In *Proceedings of the 15th International Conference on Rewriting Techniques and Applications*. Volume 3091 of LNCS, Springer (2004) 210–220.
6. Glück, R., Kawabe, M.: A program inverter for a functional language with equality and constructors. In *Proceedings of the first Asian Symposium on Programming Languages and Systems*. Volume 2895 of LNCS, Springer (2003) 246–264.
7. Harrison, P.G.: Function inversion. In *Proceedings of the IFIP TC2 Workshop on Partial Evaluation and Mixed Computation*. North-Holland (1988) 153–166.
8. Hirokawa, N., Middeldorp, A.: Tsukuba termination tool. In *Proceedings of the 14th International Conference on Rewriting Techniques and Applications*. Volume 2706 of LNCS, Springer (2003) 311–320.
9. Hu, Z., Mu, S.-C., Takeichi, M.: A programmable editor for developing structured documents based on bidirectional transformations. In *Proceedings of the 2004 ACM SIGPLAN Workshop on Partial Evaluation and Semantics-based Program Manipulation*. ACM Press (2004) 178–189.

10. Hullot, J.-M.: Canonical forms and unification. In *Proceedings of the 5th International Conference on Automated Deduction*. Volume 87 of LNCS, Springer (1980) 318–334.
11. Khoshnevisan, H., Sephton, K.M.: InvX: An automatic function inverter. In *Proceedings of the 3rd International Conference on Rewriting Techniques and Applications*. Volume 355 of LNCS, Springer (1989) 564–568.
12. Lucas, S.: Context-sensitive computations in functional and functional logic programs. Journal of Functional and Logic Programming **1998** (1). The MIT Press (1998) 1–61.
13. Lucas, S.: MU-TERM: A tool for proving termination of context-sensitive rewriting. In *Proceedings of the 15th International Conference on Rewriting Techniques and Applications*. Volume 3091 of LNCS, Springer (2004) 200–209.
14. Marchiori, M.: Unravelings and ultra-properties. In *Proceedings of the 5th International Conference on Algebraic and Logic Programming*. Volume 1139 of LNCS, Springer (1996) 107–121.
15. Nishida, N., Sakai, M., Sakabe, T.: Generation of inverse term rewriting systems for pure treeless functions. In *Proceedings of the International Workshop on Rewriting in Proof and Computation*. Sendai, Japan (2001) 188–198.
16. Nishida, N., Sakai, M., Sakabe, T.: Narrowing-based simulation of term rewriting systems with extra variables and its termination proof. In *Functional and Constraint Logic Programming*. Volume 86 of ENTCS, Issue 3, Elsevier (2003) 18 pages.
17. Nishida, N., Sakai, M., Sakabe, T.: On simulation-completeness of unraveling for conditional term rewriting systems. Technical Report SS 2004-18 of IEICE (2004) 25–30.
18. Nishida, N., Sakai, M., Sakabe, T.: Partial inversion of constructor term rewriting systems. Full version of this paper including the proofs of theorems (2005). Available from http://www.sakabe.i.is.nagoya-u.ac.jp/~nishida/papers/.
19. Ohlebusch, E.: Termination of logic programs: Transformational methods revisited. Applicable Algebra in Engineering, Communication and Computing **12** (2001) 73–116.
20. Ohlebusch, E.: *Advanced Topics in Term Rewriting*. Springer (2002).
21. Okamoto, K., Sakai, M., Sakabe, T.: Completeness of innermost strategy for inside-LR-joinable right-linear terminating TRSs. In *Record of 2003 Tokai-section Joint Conference of the Eight Institutes of Electrical and Related Engineers*. (2003) 564 (in Japanese). This is an extended result of [23], and the English version of this paper is available from http://www.sakabe.i.is.nagoya-u.ac.jp/~sakai/papers/.
22. Romanenko, A.: Inversion and metacomputation. In *Proceedings of the Symposium on Partial Evaluation and Semantics-Based Program Manipulation*. Volume 26 of SIGPLAN Notices, ACM Press (1991) 12–22.
23. Sakai, M., Okamoto, K., Sakabe, T.: Innermost reductions find all normal forms on right-linear terminating overlay TRSs. In *Proceedings of the 3rd International Workshop on Reduction Strategies in Rewriting and Programming*. Valencia, Spain (2003) 79–88.
24. Secher, J.P., Sørensen, M.H.: From checking to inference via driving and dag grammars. In *Proceedings of the 2002 ACM SIGPLAN Workshop on Partial Evaluation and Semantics-Based Program Manipulation*. Volume 37 of SIGPLAN Notices, ACM (2002) 41–51.
25. Toyama, Y.: Confluent term rewriting systems with membership conditions. In *Proceedings of the 1st International Workshop on Conditional Term Rewriting Systems*. Volume 308 of LNCS, Springer (1987) 228–241.

Natural Narrowing
for General Term Rewriting Systems

Santiago Escobar[1], José Meseguer[2], and Prasanna Thati[3]

[1] Universidad Politécnica de Valencia, Spain
sescobar@dsic.upv.es
[2] University of Illinois at Urbana-Champaign, USA
meseguer@cs.uiuc.edu
[3] Carnegie Mellon University, USA
thati@cs.cmu.edu

Abstract. For narrowing to be an efficient evaluation mechanism, several *lazy* narrowing strategies have been proposed, although typically for the restricted case of left-linear constructor systems. These assumptions, while reasonable for functional programming applications, are too restrictive for a much broader range of applications to which narrowing can be fruitfully applied, including applications where rules have a non-equational meaning either as *transitions* in a concurrent system or as *inferences* in a logical system. In this paper, we propose an efficient lazy narrowing strategy called *natural narrowing* which can be applied to general term rewriting systems with no restrictions whatsoever. An important consequence of this generalization is the *wide range of applications* that can now be efficiently supported by narrowing, such as symbolic model checking and theorem proving.

1 Introduction

Rewriting is currently recognized as a very general declarative formalism to specify, program, and reason about computational systems. The more traditional applications have been in the context of equational reasoning and of equational/functional programming, where the rewriting relation is understood as *oriented equality*. But there is an increasing awareness of the usefulness of rewriting in non-equational contexts, where a rewrite is understood as a *transition* or an *inference*: for example to specify and program concurrent systems or logical inference systems. This has led, for example, to theoretical developments such as rewriting logic [28], and to the development of language implementations supporting non-equational rewriting such as ELAN [6] and Maude [7].

A similar widening of the scope is needed for *narrowing*, which generalizes rewriting by performing unification in nonvariable positions instead of the usual matching. Narrowing can in this way endow rewriting languages with new programming and reasoning capabilities in a much wider setting and for a much wider range of applications than those based on equational logic. The traditional understanding of narrowing [15, 23] has been as a mechanism for equational *unification*, that is, for solving equational goals $E \vdash (\exists \overrightarrow{x})\ t = t'$; as

J. Giesl (Ed.): RTA 2005, LNCS 3467, pp. 279–293, 2005.

a consequence, properties such as confluence and termination have often been assumed, and many equational reasoning applications, as well as a number of functional/logic programming languages supporting narrowing, have been developed. As proposed for example in [29], in a non-equational setting narrowing can instead be understood as a powerful mechanism for solving *reachability goals* $\mathcal{R} \vdash (\exists \overrightarrow{x})\ t \rightarrow^* t'$ in a rewrite theory $\mathcal{R}$. The traditional equational interpretation can still be kept as a special case since, as explained in Section 4.2, solving equations becomes a special case of solving more general reachability goals. However, in this wider setting traditional assumptions such as confluence and termination are in general no longer reasonable (see Section 5 and [29, 34] for a discussion of completeness issues for narrowing in this setting).

A key challenge for narrowing is the danger of *combinatorial explosion* in exploring the narrowing tree. It becomes in fact essential in practice to use adequate *narrowing strategies* that are as greedy as possible, yet remain complete. One important breakthrough in this direction was the realization that one could extend the work on optimal lazy *reduction* strategies originating with Huet and Levy [22], and extended in different ways by other researchers (see, e.g., [1, 3, 11, 32]) to obtain efficient lazy *narrowing* strategies that only instantiate those positions that are really needed. This was first achieved by Antoy, Echahed and Hanus, who extended their (weakly) outermost-needed rewriting strategy to a *(weakly) needed narrowing* strategy [3, 4]. Recently, both (weakly) outermost-needed rewriting and (weakly) outermost-needed narrowing have been further improved by Escobar by means of the *natural rewriting* and *natural narrowing* strategies [11, 12]. We postpone a more detailed discussion of related work in this area until Section 5. For the moment, the main point to bear in mind is that most of the work on lazy rewriting and lazy narrowing strategies [1–4, 11, 12, 17, 19, 25, 32] has taken place within the context of functional (logic) programming languages, and depends on assumptions such as having *left-linear* and *constructor* rules. These assumptions are reasonable for some functional (logic) programming languages, but they substantially limit the expressive power of equational languages such as OBJ [18], CafeOBJ [16], ASF+SDF [10], and the equational subset of Maude [7], where non-linear left-hand sides which need not be constructor-based are perfectly acceptable. Such assumptions become even more restrictive and unreasonable for non-equational rewriting languages such as ELAN [6] and Maude [7], where a rewrite $t \rightarrow t'$ is no longer understood as a step of equational simplification but as a transition, and where the rules need not be confluent nor terminating, need not be left-linear, and the constructor assumption is utterly unreasonable and almost never holds.

The goal of this paper is to propose an efficient lazy narrowing strategy called *generalized natural narrowing* that greatly extends the natural narrowing strategy of [11] and keeps all the good properties while overcoming all the above limitations. In fact our strategy can be applied to *completely general* rewrite systems with no restrictions whatsoever: even rewrite rules with extra variables in their righthand sides are allowed; furthermore, we allow rewritings to be *context-sensitive* [26] according to a function ϕ specifying which argument positions in each function symbol are *frozen*, so that rewriting in the subterms at those positions is forbidden. In this way, we obtain a general lazy narrowing strategy applicable in the broader setting of solving reachability goals for rewrite systems whose rules can have a non-equational semantics. Furthermore, this generalization is obtained together with *actual gains in efficiency*, in the sense that our natural narrowing strategy, besides being efficiently implementable, performs strictly better than previously proposed lazy strategies when specialized to their setting (see Section 5). A further efficiency advantage, generalizing that of [3, 4, 11, 12], is that, as explained in Section 3, natural narrowing computes substitutions in an *incremental* way, without explicit use of a unification algorithm. Perhaps the most important consequence of the generality of our natural narrowing strategy is the *wide range of applications* that can now be supported. To give the reader a better feeling for some of these application areas, including symbolic model checking and theorem proving, we discuss those areas, and the respective benefits of using natural narrowing for each of them, in Section 4. What emerges, in summary, is a general and efficient *unified mechanism*, seamlessly integrating rewriting and narrowing, and making it available for a very wide range of programming and proving applications.

To give the reader a first intuitive feeling for how our generalized natural narrowing works, and for the difficulties that it resolves, we illustrate some of the key issues by means of a simple example.

Example 1. Consider the following rewrite system for proving equality ($\approx$) of arithmetic expressions built using modulus or remainder (%), subtraction ($-$), and minimum (min) operations on natural numbers.

(1) M % s(N) $\rightarrow$ (M$-$s(N)) % s(N)   (5) min(0, N) $\rightarrow$ 0
(2) (0 $-$ s(M)) % s(N) $\rightarrow$ N $-$ M   (6) min(s(N),0) $\rightarrow$ 0
(3) M $-$ 0 $\rightarrow$ M   (7) min(s(N),s(M)) $\rightarrow$ s(min(M,N))
(4) s(M) $-$ s(N) $\rightarrow$ M$-$N   (8) X $\approx$ X $\rightarrow$ True

Note that this rewrite system is not left-linear because of rule (8) and it is not constructor-based because of rule (2). Furthermore, note that it is neither terminating nor confluent due to rule (1).

The aim of natural rewriting and narrowing strategies [11] is to lazily compute head-normal forms of a given term t. Specifically, given a term that is not a head-normal form, the strategy reduces (narrows) to the extent possible, only those redexes (narroxes) that are necessary for a rule to be applied at the root. We would like to generalize natural rewriting and narrowing to a version that enjoys the good optimality properties of natural rewriting and natural narrowing (see [11]) and that can also handle non-left-linear and non-constructor rules such as (2) and (8). This is accomplished for rewriting in the *generalized natural rewriting strategy* of [14]. For example, consider the term[1] $t_1 = 10! \% \min(X, X-0) \approx 10! \% 0$ and the following two narrowing sequences we are interested in amongst all possible. First, the following sequence leading to True, that starts by unifying subterm $t_1|_{1.2}$ with left-hand side (lhs) l_5:

---

[1] The subterm 10! represents factorial of $s^{10}(0)$ but we do not include the rules for ! because we are only interested in the fact that it has a remarkable computational cost, and therefore we would like to avoid its reduction in the examples whenever possible.

$$10! \% \underline{\min(\text{X},\text{X}-0)} \approx 10! \% 0 \rightsquigarrow_{[\text{X} \mapsto 0]} \underline{10! \% 0 \approx 10! \% 0} \rightsquigarrow_{id} \text{True}$$

Second, the following sequence not leading to **True**, that starts by reducing subterm $t_1|_{1.2.2}$ with lhs l_3 and that early instantiates variable X:

$$10! \% \min(\text{X},\underline{\text{X}-0}) \approx 10! \% 0 \rightsquigarrow_{[\text{X} \mapsto \text{s}(\text{X'})]} 10! \% \underline{\min(\text{s}(\text{X'}),\text{s}(\text{X'}))} \approx 10! \% 0$$
$$\rightsquigarrow_{id} \quad\quad 10! \% \text{s}(\min(\text{X'},\text{X'})) \approx 10! \% 0$$

Note that although it is possible to further narrow the last term, we are not interested in doing so, since such term is already a head-normal form. In the following, we informally introduce the key points of our strategy:

1. (*Demanded positions*). This notion is relative to a lhs l and determines which positions in a term t should be narrowed in order to be able to apply lhs l at root position. For the term $t_1 = 10! \% \min(\text{X},\text{X}-0) \approx 10! \% 0$ and lhs $l_8 = \text{X} \approx \text{X}$, only subterm $\min(\text{X},\text{X}-0)$ is demanded, since it is the only disagreeing part in $t_1|_1$ w.r.t. $t_1|_2$.

2. (*Failing term*). This notion is relative to a lhs l and stops further wasteful narrowing steps. Specifically, the last term $10! \% \text{s}(\min(\text{X'},\text{X'})) \approx 10! \% 0$ of the second former sequence fails w.r.t. l_8, since the subterm $\text{s}(\min(\text{X'},\text{X'}))$ is demanded by l_8 but there is no possible narrowing step above it that would convert it into term 0.

3. (*Most frequently demanded positions*). This notion determines those demanded positions w.r.t. non-failing lhs's that are demanded by the maximum number of rules and that cover all such non-failing lhs's. It provides the optimality properties of our natural rewriting and narrowing strategies, since it substantially reduces the set of positions to be considered. If we look closely at lhs's l_5, l_6, and l_7 defining min, we can see that position 1 in the term $\min(\text{X},\text{X}-0)$ is more demanded than position 2, i.e., position 1 is disagreeing w.r.t. l_5, l_6, and l_7, whereas position 2 is disagreeing only w.r.t. l_6 and l_7. Thus, position 1 is the most frequently demanded position for all rules defining min that also covers such rules. Note that position 1 is rooted by a variable and this motivates the following point.

4. (*Lazy instantiation*). This notion relates to an incremental construction of unifiers without the explicit use of a unification algorithm. This is necessary in the previous example, since subterm $\min(\text{X},\text{X}-0)$ does not unify with lhs l_6 and l_7. However, we can deduce that narrowing at subterm $\text{X}-0$ is only necessary when substitution $[\text{X} \mapsto \text{s}(\text{X'})]$, inferred from l_6 and l_7, has been applied. Thus, we early construct the appropriate substitutions $[\text{X} \mapsto 0]$ and $[\text{X} \mapsto \text{s}(\text{X'})]$ in order to reduce the search space.

In Section 2, we present the preliminary background. In Section 3 we define our generalized natural narrowing strategy. In Section 4, we motivate our work by illustrating various applications of the generalized narrowing strategy. In Section 5, we compare our work with related approaches, and we conclude in Section 6. Proofs of all results can be found in [13].

2   Preliminaries

We assume some familiarity with term rewriting and narrowing (see [33] for missing definitions). We assume a finite alphabet (function symbols) Σ and a countable set of variables X. We denote the set of terms built from Σ and X by $T_\Sigma(X)$ and write T_Σ for ground terms. A term is said to be linear if it has no multiple occurrences of a single variable. We use finite sequences of integers to denote a position in a term. Given a set $S \subseteq \Sigma \cup X$, $\mathcal{P}os_S(t)$ denotes positions in t where symbols or variables in S occur. We write $\mathcal{P}os_f(t)$ and $\mathcal{P}os(t)$ as a shorthand for $\mathcal{P}os_{\{f\}}(t)$ and $\mathcal{P}os_{\Sigma \cup X}(t)$, respectively. We denote the *root position* by Λ. Given positions p, q, we denote its concatenation as $p.q$. For sets of positions P, Q we define $P.Q = \{p.q \mid p \in P \land q \in Q\}$. We write $P.q$ as a shorthand for $P.\{q\}$ and similarly for $p.Q$. The subterm of t at position p is denoted as $t|_p$, and $t[s]_p$ is the term t with the subterm at position p replaced by s. We define $t|_P = \{t|_p \mid p \in P\}$. The symbol labeling the root of t is denoted as $root(t)$.

A *substitution* is a function $\sigma : X \to T_\Sigma(X)$ which maps variables to terms, and which is different from the identity only for a finite subset $\mathcal{D}om(\sigma)$ of X. We homomorphically extend substitutions to terms. We denote by id the identity substitution: $id(x) = x$ for all $x \in X$. Terms are ordered by the preorder $\leq$ of "relative generality", i.e., $s \leq t$ if there exists σ s.t. $\sigma(s) = t$. We write $\sigma^{-1}(x) = \{y \in \mathcal{D}om(\sigma) \mid \sigma(y) = x\}$.

A rewrite rule is an ordered pair (l, r) of terms, also written $l \to r$, with $l \notin X$. A *rewrite system*[2] is a triple $\mathcal{R} = (\Sigma, \phi, R)$ with Σ a signature, R a set of rewrite rules, and $\phi : \Sigma \to \mathcal{P}(\mathbb{N})$ specifies the *frozen* arguments $\phi(f) \subseteq \{1, \ldots, k\}$ for the arity k of f. We say position p in t is *frozen* if $\exists q < p$ such that $p = q.i.q'$ and $i \in \phi(root(t|_q))$. $L(\mathcal{R})$ denotes the set of *lhs*'s of $\mathcal{R}$. A rewrite system $\mathcal{R}$ is left-linear if for all $l \in L(\mathcal{R})$, l is a linear term. Given $\mathcal{R} = (\Sigma, \phi, R)$, we assume that Σ is defined as the disjoint union $\Sigma = \mathcal{C} \uplus \mathcal{D}$ of symbols $c \in \mathcal{C}$, called *constructors*, and symbols $f \in \mathcal{D}$, called *defined symbols*, where $\mathcal{D} = \{root(l) \mid l \to r \in R\}$ and $\mathcal{C} = \Sigma - \mathcal{D}$. A pattern is a term $f(l_1, \ldots, l_k)$ where $f \in \mathcal{D}$ and $l_i \in T_\mathcal{C}(X)$, for $1 \leq i \leq k$. A rewrite system $\mathcal{R} = (\mathcal{C} \uplus \mathcal{D}, \phi, R)$ is a constructor system (CS) if every $l \in L(\mathcal{R})$ is a pattern.

A term t rewrites to s at a non-frozen position $p \in \mathcal{P}os_\mathcal{D}(t)$ using the rule $l \to r \in R$, called a *rewrite step* and written $t \to_{\langle p, l \to r \rangle} s$ ($t \xrightarrow{p} s$ or simply $t \to s$), if $t|_p = \sigma(l)$ and $s = t[\sigma(r)]_p$. The pair $\langle p, l \to r \rangle$ is called a *redex*. A term t is a *normal form* if it contains no redex. A term t is a *head-normal form* if it cannot be reduced to a redex. We denote by $\xrightarrow{>\Lambda}$ a rewrite step at a position $p > \Lambda$. A substitution σ is called *normalized* if $\sigma(x)$ is a normal form for all variables x. Similarly, a *narrowing step* is defined as $t \rightsquigarrow_{\langle p, l \to r, \sigma \rangle} s$ (or simply $t \rightsquigarrow_\sigma s$) if $\sigma(t) \to_{\langle p, l \to r \rangle} s$. Note that we do not require $p \in \mathcal{P}os_\mathcal{D}(t)$ for a narrowing step, which is a usual condition for left-linear constructor systems. Instead, we require substitutions computed by narrowing to be normalized.

3   Generalizing Natural Narrowing

In [14], we have generalized the natural rewriting strategy of [11] to the larger class of rewrite systems that need not be left-linear and constructor-based. In

---

[2] What we call here a rewrite system is a special case of a *rewrite theory* [28], which is a 4-tuple (Σ, ϕ, E, R) with E a set of equations.

this paper, we define, again for this larger class of systems, a generalized natural narrowing strategy that, to the extent possible, performs only those narrowing steps that are essential for applying some rule at root position. That is, if a term t is not a head-normal form (or some substitution making the term not a head-normal form exists), then we know that after a (possibly empty) sequence of narrowing steps at positions other than the root, a narrowing step at the root position with a rule $l \rightarrow r$ is possible. We use the notions of demanded positions, failing terms, and most frequently demanded positions, as in [14]. First, we recall the notion of demanded positions.

Definition 1. *For a term s and a set of terms $T = \{t_1, \ldots, t_n\}$ we say that s is a* context *of the terms in T if $s \leq t_i$ for all $1 \leq i \leq n$. There is always a* least general context *s of T, i.e., one such that for any other context s' we have $s' \leq s$; furthermore s is unique up to renaming of variables. For $1 \leq i \leq n$, let the substitution σ_i be such that $\sigma_i(s) = t_i$ and $Dom(\sigma_i) \subseteq Var(s)$. We define the set $Pos_{\neq}(T)$ of* disagreeing positions *between the terms in T as those $p \in Pos_X(s)$ such that there is an i with $\sigma_i(s|_p) \neq s|_p$.*

Definition 2 (Demanded positions). *For terms l and t, let s be the least general context of l and t, and let σ be the substitution such that $\sigma(s) = l$. We define the set of* demanded positions *in t w.r.t. l as*

$$DP_l(t) = \bigcup_{x \in Var(s)} \begin{array}{l} \text{if } \sigma(x) \notin X \text{ then } Pos_x(s) \text{ else } Q.Pos_{\neq}(t|_Q) \\ \text{where } Q = Pos_{\sigma^{-1}(\sigma(x))}(s) \end{array}$$

Intuitively, the set $DP_l(t)$ returns a set of positions in t at which t necessarily has to be "changed" (either by a narrowing step if it is a non-variable position, or by an instantiation if it is a variable position) before the rule $l \rightarrow r$ can be applied at the root position, i.e., before l can match the term under consideration.

Example 2. Consider the left-hand side $l_8 = X \approx X$ and the term $t_1 = 10! \% \min(X, X-0) \approx 10! \% 0$ of Example 1. The least general context of l_8 and t_1 is $s = W \approx Z$. Now, for $\sigma = \{W \mapsto X, Z \mapsto X\}$, we have $\sigma(s) = l_8$. Then, while computing $DP_{l_8}(t_1)$, we compute the set of disagreeing positions between the subterms in t_1 corresponding to the non-linear variable X in l_8, i.e., the set $Pos_{\neq}(T)$ for $T = t_1|_{\{1,2\}} = \{10! \% \min(X, X-0), 10! \% 0\}$. According to Definition 1, the least general context of T is the term $10! \% Y$ and the set of disagreeing positions between terms in T is then $Pos_{\neq}(T) = \{2\}$. Thus, we obtain that $DP_{l_8}(t_1) = \{1, 2\}.\{2\} = \{1.2, 2.2\}$.

Now, four points have to be addressed. First, note that the symbol at a position $p \in DP_l(t)$ in t can be changed not only by a narrowing step or by instantiation at p, but also by a narrowing step or instantiation at a position $q < p$. Thus, besides considering the positions in $DP_l(t)$ as candidates for evaluation, we also need to consider the positions q in t that are above some position in $DP_l(t)$. Thus, for a position q in a term t, we define $D_t^\uparrow(q) = \{p \mid p \leq q \wedge p \in Pos_D(t) \wedge p \text{ is not frozen}\}$. We lift this to sets of positions as

$D_t^\uparrow(Q) = \cup_{q \in Q} D_t^\uparrow(q)$. This gives us the following useful result that shows how demandedness captures the neededness of positions in rewrite sequences for the application of a rule at root position.

Lemma 1. [14] *Consider a rewrite sequence* $t \to_{\langle p_1, l_1 \to r_1 \rangle} t_1 \cdots \to_{\langle p_n, l_n \to r_n \rangle} t_n$ *such that* $p_n = \Lambda$. *Then, either* $l_n \leq t$ *or there is a* k, $1 \leq k < n$, *such that* $p_k \in D_t^\uparrow(DP_{l_n}(t))$.

Example 3. Continuing Example 2, we have $D_{t_1}^\uparrow(DP_{l_8}(t_1)) = D_{t_1}^\uparrow(\{1.2, 2.2\}) = \{\Lambda, 1, 1.2, 2\}$, since position 2.2 is rooted by a constructor symbol, and thus removed.

Second, we only compute $DP_l(t)$ for those left-hand sides l such that t does not *fail* w.r.t. l. Roughly, we say t fails w.r.t. l if no sequence of narrowing steps or instantiations in t will help to produce a term to which the rule $l \to r$ can be applied at the root; this notion is undecidable but we provide a safe and computable approximation.

For a position p and a term t, we define the set $R_t(p)$ of *reflections* of p w.r.t. t as follows: if p is under a variable position in t, i.e., $p = q.q'$ for some q such that $t|_q = x$, then $R_t(p) = \mathcal{P}os_x(t).q'$, else $R_t(p) = \{p\}$. We say that the path to p in t is *stable* (or simply p is stable) if $root(t|_p) \notin X$ and $D_t^\uparrow(p) \setminus \{\Lambda\} = \varnothing$.

Definition 3 (Failing term). *Given terms* l, t, *we say* t *fails w.r.t.* l, *denoted by* $l \blacktriangleleft t$, *if there is* $p \in DP_l(t)$ *such that* p *is stable, and one of the following holds: (i)* $R_l(p) \cap DP_l(t) = \{p\}$; *or (ii) there is* $q \in R_l(p) \cap DP_l(t)$ *with* $root(t|_p) \neq root(t|_q)$, *and* q *is also stable. We denote by* $l \not\blacktriangleleft t$ *that* t *is not failing w.r.t.* l.

Example 4. Consider the subterm $t_1|_2 = 10! \% 0$ and the lhs's $l_1 = M \% s(N)$ and $l_2 = (0 - s(M)) \% s(N)$ in Example 1. We have that $l_1 \blacktriangleleft t_1|_2$ and $l_2 \blacktriangleleft t_1|_2$, because position 2 in $t_1|_2$ belongs to $DP_{l_1}(t_1|_2)$ and $DP_{l_2}(t_1|_2)$, is stable, and $R_{l_1}(2) = R_{l_2}(2) = \{2\}$. Similarly, the term $t' = 10! \% s(\min(X', X')) \approx 10! \% 0$ fails w.r.t. l_8 since $1.2, 2.2 \in DP_{l_8}(t')$, $1.2, 2.2$ are stable, and $R_{l_8}(1.2) = R_{l_8}(2.2) = \{1.2, 2.2\}$.

Third, we do not consider all positions in each $DP_l(t)$ but a subset called the *most frequently demanded positions*. The idea behind this is that narrowing or instantiating at those positions is enough for being able to reduce at root position. In this way, we can substantially reduce (or optimize) the set of positions to be considered.

Definition 4 (Set cover). *For a set of positions* P, *a sequence of lhs's* $l_1, ..., l_n$, *and a sequence of sets of positions* $Q_1, ..., Q_n$, *we say that* P *covers* $l_1, ..., l_n$ *and* $Q_1, ..., Q_n$ *if for all* $1 \leq i \leq n$, *there is a position* $p \in P \cap Q_i$ *such that* $R_{l_i}(p) \subseteq P$.

Definition 5 (Most frequently demanded positions). *We define the filtered set of demanded positions of a term* t *by the set* $FP(t)$ *returning one of the minimal sets of positions that cover* $l_1, ..., l_n$ *and* $DP_{l_1}(t), ..., DP_{l_n}(t)$, *where* $\{l_1, ..., l_n\} = \{l \in L(\mathcal{R}) \mid l \not\blacktriangleleft t\}$.

Example 5. Consider the subterm $t_1|_1 = 10! \% \min(\text{X},\text{X}-0)$ and the lhs's l_1 and l_2 in Example 1. The reader can check that $DP_{l_1}(t_1|_1) = \{2\}$, $DP_{l_2}(t_1|_1) = \{1,2\}$, and $DP_l(t_1|_1) = \{\Lambda\}$ for any other lhs l. Then, the set $P = \{2,\Lambda\}$ covers all lhs's and we obtain that $FP(t_1|_1) = \{\Lambda, 2\}$. On the other hand, consider the term t_1 and the lhs l_8 in Example 1. From Example 2, we have $DP_{l_8}(t_1) = \{1.2, 2.2\}$ and $DP_l(t_1) = \{\Lambda\}$ for any other lhs l. Thus, $FP(t_1) = \{\Lambda, 1.2, 2.2\}$, since this set is closed by reflection.

Fourth, whenever t matches l for a rule $l \to r$, in addition to reducing t with $l \to r$ we have to consider as candidates for evaluation those positions q in t that have a defined symbol and that are above a variable position in l; see [14] for further explanations. Now, we are able to define the set of *sufficient demanded positions* that collects the four previous ideas.

Definition 6 (Sufficient demanded positions). *We define the* sufficient set of demanded positions *of a term t as* $SP(t) = \cup_{l \in L(\mathcal{R}) \wedge l \leq t} D_l^\uparrow(\mathcal{P}os_X(l))$ $\cup$ $D_t^\uparrow(FP(t))$.

Example 6. Consider term t_1 in Example 1. Since t_1 is not a redex, we have $SP(t_1) = D_{t_1}^\uparrow(FP(t_1))$. By Example 5, we have $SP(t_1) = D_{t_1}^\uparrow(\{\Lambda, 1.2, 2.2\})$. And by Example 3, we have $SP(t_1) = \{\Lambda, 1, 1.2, 2\}$.

Before defining generalized natural narrowing, we have to define a set of demanded substitutions for narrowing, to address the question of what substitutions the variables at demanded positions are to be instantiated with.

Definition 7 (Demanded substitutions). *We define the set $DSub(t)$ of demanded substitutions for narrowing t at the root position as follows. For each pair of $p \in FP(t) \cap \mathcal{P}os_X(t)$ and $l \in L(\mathcal{R})$ such that $p \in DP_l(t)$ and $l \not\blacktriangleleft t$, we construct the substitution σ as explained below, and stipulate that $\sigma \in DSub(t)$.*
  If $p \in \mathcal{P}os_\Sigma(l)$, then $\sigma = [t|_p \mapsto root(l|_p)(\overrightarrow{w})]$ for distinct fresh variables $\overrightarrow{w}$. On the other hand, if $p \notin \mathcal{P}os_\Sigma(l)$, then we know that p is under a non-linear variable position in l, i.e., $|R_l(p) \cap DP_l(t)| > 1$, and let $Q = R_l(p) \cap DP_l(t)$. There are two cases: (i) if all the terms in $t|_Q$ are variables, then we define σ to be such that for every $q \in Q$ we have $\sigma(t|_q) = w$ for a fresh variable w, (ii) if every non-variable term in $t|_Q$ is rooted by the same symbol f, then we define $\sigma = [t|_p \mapsto f(\overrightarrow{w})]$.

Note that, in the previous definition, if there are two non-variable positions $p_1, p_2 \in Q$ such that $t|_{p_1}$ and $t|_{p_2}$ are rooted with different symbols, then no substitution can resolve the conflict between the disagreeing positions p_1 and p_2 and they are demanded for evaluation.
  We can deduce the following useful result that ensures that appropriate substitutions are inferred from left-hand sides for demanded variable positions.

Lemma 2. *Let l, t, σ, and $p \in FP(t) \cap \mathcal{P}os_X(t)$. If $l \not\blacktriangleleft t$, $p \in DP_l(t)$, and $p \notin DP_l(\sigma(t))$, then there exists $\theta \in DSub(t)$ s.t. $\theta(t|_p) \neq t|_p$ and $\theta|_{Var(t)} \leq \sigma|_{Var(t)}$.*

Example 7. Consider the subterm $t_1|_{1.2} = \min(\text{X},\text{X}-0)$ and the lhs's $l_5 = \min(0,\text{N})$, $l_6 = \min(\text{s}(\text{N}),0)$, and $l_7 = \min(\text{s}(\text{N}),\text{s}(\text{M}))$ of Example 1. The reader can check that $DP_{l_5}(t_1|_{1.2}) = \{1\}$, $DP_{l_6}(t_1|_{1.2}) = \{1,2\}$, $DP_{l_7}(t_1|_{1.2}) = \{1,2\}$, and $DP_l(t_1|_{1.2}) = \{\Lambda\}$ for any other lhs l. Thus $FP(t_1|_{1.2}) = \{\Lambda, 1\}$ according to Definition 5. Then, position 1 is a variable position and we have that the demanded substitutions for its variable X are $DSub(t_1|_{1.2}) = \{[\text{X} \mapsto 0, \text{X} \mapsto \text{s(X')}]\}$, since $1 \in \mathcal{P}os_\Sigma(l_5)$, $1 \in \mathcal{P}os_\Sigma(l_6)$, $root(l_5|_1) = 0$, and $root(l_6|_1) = \text{s}$.

Thus, substitutions are computed in a lazy and *incremental* fashion *without resorting to an explicit unification algorithm*. Now, we formally define our generalized natural narrowing strategy.

Definition 8 (Generalized Natural Narrowing). *We define the set of demanded narroxes of a term t as*

$$DN(t) = \{\langle \Lambda, l \to r, id \rangle \mid l \in L(\mathcal{R}) \wedge l \leq t\} \ \cup \ \bigcup_{q \in SP(t) \setminus \{\Lambda\}} q.DN(t|_q) \ \cup$$
$$\bigcup_{\sigma \in DSub(t)} DN(\sigma(t))@\sigma$$

where for a set of narroxes S we define $q.S = \{\langle q.p, l \to r, \theta \rangle \mid \langle p, l \to r, \theta \rangle \in S\}$ and $S@\sigma = \{\langle p, l \to r, \theta \circ \sigma \rangle \mid \langle p, l \to r, \theta \rangle \in S\}$. We say that term t reduces by natural narrowing to term s, denoted by $t \overset{m}{\leadsto}_{\langle p, l \to r, \sigma \rangle} s$ (or simply $t \overset{m}{\leadsto} s$) if $t \leadsto_{\langle p, l \to r, \sigma \rangle} s$, $\langle p, l \to r, \sigma \rangle \in DN(t)$, and $p \in \mathcal{P}os_D(t)$.

In the following, we omit the rule $l \to r$ in a narrowing step $\langle p, l \to r, \sigma \rangle$, whenever there is no scope for ambiguity about the rule.

Example 8. Consider again the term $t_1 = \text{10! \% min(X,X-0)} \approx \text{10! \% 0}$ from Example 1 and the computation of $DN(t_1)$. Since t_1 is not a redex, we have that $DN(t_1) = \bigcup_{q \in SP(t_1) \setminus \{\Lambda\}} q.DN(t_1|_q) \ \cup \ \bigcup_{\sigma \in DSub(t_1)} DN(\sigma(t_1))@\sigma$. By Example 6, $SP(t_1) = \{\Lambda, 1, 1.2, 2\}$. We also have $DSub(t_1) = \varnothing$, since no position in $FP(t_1)$ is a variable. Thus, $DN(t_1) = 1.DN(t_1|_1) \cup 2.DN(t_1|_2) \cup 1.2.DN(t_1|_{1.2})$. This implies that we recursively compute $DN(t_1|_1)$, $DN(t_1|_{1.2})$ and $DN(t_1|_2)$.

Now consider $DN(t_1|_{1.2}) = DN(\text{min(X,X-0)})$. Since it is not a redex, we have $DN(t_1|_{1.2}) = \bigcup_{q \in SP(t_1|_{1.2}) \setminus \{\Lambda\}} q.DN(t_1|_{1.2.q}) \ \cup \ \bigcup_{\sigma \in DSub(t_1|_{1.2})} DN(\sigma(t_1|_{1.2}))@\sigma$. By Example 7, we have $SP(t_1|_{1.2}) = D^\uparrow_{t_1|_{1.2}}(FP(t_1|_{1.2})) = D^\uparrow_{t_1|_{1.2}}(\{\Lambda, 1\}) = \{\Lambda\}$ and $DSub(t_1|_{1.2}) = \{\sigma, \sigma'\}$ for $\sigma = [\text{X} \mapsto 0]$ and $\sigma' = [\text{X} \mapsto \text{s(X')}]$. Then, $DN(t_1|_{1.2}) = DN(\sigma(t_1|_{1.2}))@\sigma \cup DN(\sigma'(t_1|_{1.2}))@\sigma'$, and we recursively call to $DN(\text{min(0,0-0)})$ and $DN(\text{min(s(X'),s(X')-0)})$. Now, the reader can check that $DN(\text{min(0,0-0)}) = \{ \langle \Lambda, id \rangle \}$, since term min(0,0-0) matches lhs l_5, and $DN(\text{min(s(X'),s(X')-0)}) = 2.DN(\text{s(X')-0}) = \{ \langle 2, id \rangle \}$, since term s(X')-0 matches lhs l_3. Hence, we can conclude $DN(t_1|_{1.2}) = \{\langle \Lambda, id \rangle\}@\sigma \cup \{\langle 2, id \rangle\}@\sigma' = \{ \langle \Lambda, [\text{X} \mapsto 0] \rangle, \langle 2, [\text{X} \mapsto \text{s(X')}] \rangle \}$.

Now consider $DN(t_1|_1) = DN(\text{10! \% min(X,X-0)})$. Since it is not a redex, we have $DN(t_1|_1) = \bigcup_{q \in SP(t_1|_1) \setminus \{\Lambda\}} q.DN(t_1|_{1.q}) \ \cup \ \bigcup_{\sigma \in DSub(t_1|_1)} DN(\sigma(t_1|_1))@\sigma$ and $SP(t_1|_1) = D^\uparrow_{t_1|_1}(FP(t_1|_1))$. By Example 5, $FP(t_1|_1) = \{\Lambda, 2\}$, and then $SP(t_1|_1) = D^\uparrow_{t_1|_1}(\{\Lambda, 2\}) = \{\Lambda, 2\}$. Moreover, we have $DSub(t_1|_1) = \varnothing$, since no position in $FP(t_1|_1)$ is rooted by a variable. Thus, we have $DN(t_1|_1) = 2.DN(t_1|_{1.2})$. However, $DN(t_1|_{1.2})$ was already computed before, and we obtain $DN(t_1|_1) = \{ \langle 2, [\text{X} \mapsto 0] \rangle, \langle 2.2, [\text{X} \mapsto \text{s(X')}] \rangle \}$.

Finally, consider $DN(t_1|_2) = DN(10! \% 0)$. Since it is not a redex, we have $DN(t_1|_2) = \cup_{q \in SP(t_1|_2) \setminus \{\Lambda\}} q.DN(t_1|_{2.q}) \cup \cup_{\sigma \in DSub(t_1|_2)} DN(\sigma(t_1|_2))@\sigma$. But by Example 4, we have $l_1 \blacktriangleleft t_1|_2$, $l_2 \blacktriangleleft t_1|_2$, and $DP_l(t_1|_2) = \{\Lambda\}$ for any other lhs l. Thus, we can conclude $DN(t_1|_2) = \varnothing$. Finally, putting everything together we have $DN(t_1) = \{ \langle 1.2, [X \mapsto 0] \rangle, \langle 1.2.2, [X \mapsto s(X')] \rangle \}$. Note these narroxes correspond to the optimal narrowing sequences in Example 1.

We are now ready to state the correctness and completeness properties of our generalized natural narrowing strategy. The key fact used in establishing these properties is the correspondence between generalized natural rewriting and generalized natural narrowing, as stated by the following lemma.

Lemma 3 (Completeness w.r.t. rewriting). *For a normalized substitution σ, if $\sigma(t) \xrightarrow{m}_{\langle p,l \to r \rangle} s$, then there are η, θ, s' such that $t \overset{m}{\leadsto}_{\langle p,l \to r, \theta \rangle} s'$, $\sigma|_{Var(t)} = (\eta \circ \theta)|_{Var(t)}$, $s = \eta(s')$, and η is normalized.*

Using Lemmas 2 and 3 we get the following correctness and completeness results for generalized natural narrowing.

Theorem 1 (Correctness). *If t is not a variable and is a $\overset{m}{\leadsto}$ -normal form, then for every normalized σ we have $\sigma(t)$ is a head-normal form.*

Theorem 2 (Completeness). *If $\sigma(t) \to^* s$ and σ is a normalized substitution, then there are s', θ, θ' s.t. $t \overset{m}{\leadsto}_{\theta}^* s'$, $\theta'(s') \xrightarrow{\geq \Lambda}^* s$, and $\sigma|_{Var(t)} = (\theta' \circ \theta)|_{Var(t)}$.*

The following example shows that our completeness result needs not hold for non-normalized substitutions.

Example 9. Consider the rewrite system from [29] with rules: (i) f(b,c)→d, (ii) a→b, and (iii) a→c. For the term $t = $ f(X,X) and substitution $\sigma = [X \mapsto a]$ we have $\sigma(t) \to^*$ d. But the generalized natural narrowing strategy cannot compute the normal form d. Specifically, positions 1 and 2 in t are demanded by rule (i), and the variable X is instantiated with substitutions $[X \mapsto b]$ and $[X \mapsto c]$, which are inferred from rule (i). However, both f(b,b) and f(c,c) are failing w.r.t. the left-hand side of rule (i). Thus $DN(t) = \varnothing$, and the strategy does not narrow the term f(X,X) any further.

4   Application Areas

In this section, we show how narrowing can be used as a unified mechanism for programming and proving, and explain informally how the generalized natural narrowing strategy makes the specific applications more efficient. Our purpose in this section is motivational. More applications, details and examples are given in [13].

4.1   Symbolic Model Checking

Narrowing can be used as a technique for symbolic reachability analysis of concurrent systems with an infinite state space. Specifically, a concurrent system can naturally be expressed as a rewrite system $\mathcal{R} = (\Sigma, \phi, R)$, where terms represent states, and a rewrite rule $t \rightarrow t'$ is understood as a (parametric) local transition [27, 28]. We can then formalize a reachability problem for a concurrent system thus specified as solving an existential formula $(\exists \overrightarrow{x}) \; t(\overrightarrow{x}) \rightarrow^* t'(\overrightarrow{x})$ where the *source* $t(\overrightarrow{x})$ is a term with variables representing a possibly infinite set of *initial* states (namely all its instances by *ground substitutions*) and the *target* $t'(\overrightarrow{x})$ represents a likewise possibly infinite set of *final* states that we want to reach by a sequence of transitions. *Solutions* to this reachability problem can then be described by substitutions σ for which indeed we have, $\sigma(t(\overrightarrow{x})) \rightarrow^* \sigma(t'(\overrightarrow{x}))$. More generally, we may consider conjunctive reachability goals of the form $G = (\exists \overrightarrow{x}) \; t_1(\overrightarrow{x}) \rightarrow^* t'_1(\overrightarrow{x}) \wedge \ldots \wedge t_n(\overrightarrow{x}) \rightarrow^* t'_n(\overrightarrow{x})$.

We can reduce solving a conjunctive reachability goal such as G, to the case of solving a *single* reachability goal by means of a theory transformation associating to a rewrite system $\mathcal{R} = (\Sigma, \phi, R)$, a corresponding rewrite system $\hat{\mathcal{R}} = (\hat{\Sigma}, \hat{\phi}, \hat{R})$, where $\hat{\Sigma} = \Sigma \cup \{\wedge, \twoheadrightarrow, \mathtt{True}\}$, $\hat{\phi}$ extends ϕ with $\hat{\phi}(\twoheadrightarrow) = \{2\}$, and $\hat{R} = R \cup \{x \twoheadrightarrow x \rightarrow \mathtt{True}, \; \mathtt{True} \wedge \mathtt{True} \rightarrow \mathtt{True}\}$. Note that the second argument of $\twoheadrightarrow$ is frozen, since only the sources of a goal are to be rewritten. Then, σ is a solution of the conjunctive goal G in the rewrite system R if and only if σ is a solution of the *single* reachability goal $(\exists \overrightarrow{x}) \hat{G} \rightarrow^* \mathtt{True}$ in the rewrite system $\hat{R}$, where $\hat{G}$ is the $\hat{\Sigma}$-term $\hat{G} = t_1 \twoheadrightarrow t'_1 \wedge \ldots \wedge t_n \twoheadrightarrow t'_n$.

Now, since the term $\mathtt{True}$ in the transformed theory $\hat{\mathcal{R}}$ is a head normal form, any head normalizing strategy $\mathcal{S}$ [30], including our efficient generalized natural narrowing strategy, gives us a semi-decision procedure to find all the normalized solutions of a goal G. Specifically, the algorithm incrementally builds the narrowing tree starting from the term $\hat{G}$, and searches, using $\mathcal{S}$, for narrowing derivations that result in $\mathtt{True}$. When one such derivation is found, the composition of substitutions accumulated during the narrowing derivation in the reverse order gives us a solution of the goal G. Of course, the narrowing tree generated by $\mathcal{S}$ has to be explored in a *fair* manner, since the narrowing derivations can be infinitely long. The reader is referred to [29] for further details and an example where the above technique is applied for analysis of safety properties of security protocols.

Further, note that since the set of initial and final states specified in a goal can be infinite, and likewise there is no restriction on the number of reachable states, one can view the above narrowing procedure as a new form of "symbolic model checking" for infinite state systems.

4.2   Theorem Proving

Equational Unification: Narrowing was originally introduced as a complete method for generating all solutions of an equational unification problem, i.e., for goals F of the form $(\exists \overrightarrow{x}) \; t_1(\overrightarrow{x}) = t'_1(\overrightarrow{x}) \wedge \ldots \wedge t_n(\overrightarrow{x}) = t'_n(\overrightarrow{x})$ in free algebras modulo a set of equations that can be described by a set of confluent and termi- nating rewrite rules [15, 23, 24]. We note that the problem of solving reachability goals in rewrite systems generalizes the problem of equational unification. Specifically, suppose we are to solve the equational goal F above in the equational theory $\mathcal{E} = (\Sigma, E)$ where the equations E are confluent and terminating. Note that σ is a solution of F if and only if both $\sigma(t_i)$ and $\sigma(t_i')$ can be reduced by the (oriented) equations E to a common term. We can thus consider the rewrite system $\mathcal{R}_{\mathcal{E}} = (\tilde{\Sigma}, \phi, R_E)$, where $\tilde{\Sigma} = \Sigma \cup \{\approx, \texttt{True}\}$, $\phi(f) = \varnothing$ for all $f \in \tilde{\Sigma}$, and $R_E = E \cup \{x \approx x \to \texttt{True}\}$. Then σ is a solution of the system of equations F in the equational theory $\mathcal{E}$ if and only if it is a solution of the reachability goal $G = (\exists \overrightarrow{x})\, t_1 \approx t_1' \to^* \texttt{True} \wedge \ldots \wedge t_n \approx t_n' \to^* \texttt{True}$ in the rewrite system $\mathcal{R}_{\mathcal{E}}$.

Inductive Theorem Proving: The just-described reduction of existential equality goals to reachability goals, when combined with the reduction described in Section 4.1 of conjunctive reachability goals to a single goal has important applications to *inductive theorem proving.* Specifically, it is useful in proving existentially quantified inductive theorems like $E \vdash_{ind} (\exists \overrightarrow{x})\, t = t'$ in the initial model defined by the equations E. Natural narrowing can, using the two reductions just mentioned, provide a very efficient semi-decision procedure (and even a decision procedure for some restricted theories [31]) for proving such inductive goals because it will *detect failures* to unify, stopping with a counterexample instead of blindly expanding the narrowing tree. Furthermore, natural narrowing can make *inductionless induction* provers, particularly in the most recent formulations in [8, 9], more effective and efficient, and can be used in such provers to prove also universal inductive theorems like $E \vdash_{ind} (\forall \overrightarrow{x})\, t = t'$ (or, more generally, clauses). The point is that natural narrowing's complete narrowing strategy can be used instead of the unrestricted narrowing carried out by *superposition* to compute a smaller set of deductions, which can increase the chances of termination of the inductionless induction procedure without loss of soundness. The extended version of this work [13] illustrates the previous point with an example where inductionless induction is able to prove an inductive theorem with natural narrowing, but not with unrestricted narrowing.

5   Related Work

Lazy rewriting strategies are based on the original *strongly needed reduction* strategy of Huet and Levy [22]. This strategy is optimal (computes only needed steps), correct, and complete for *strongly sequential rewrite systems* (SS), a subclass of orthogonal rewrite systems. Note that this strategy does not apply to Example 1, since orthogonality implies left-linearity. Sekar and Ramakrishnan proposed the *parallel needed reduction* strategy [32] as an extension of Huet and Levy's strongly needed reduction to make it correct and complete for a larger rewrite system class, though still optimal for the former class. This larger rewrite system class is *constructor weakly orthogonal systems* (CB-WO), and thus it is not applicable to Example 1. Antoy proposed the *outermost-needed rewriting* [1] as an efficient implementation of strongly needed reduction for *inductively sequential rewrite systems* (IS), which are equivalent to SS's when instantiated to left-linear constructor systems [21]. In [1], Antoy also provides the *weakly outermost-needed rewriting* to make outermost-needed rewriting correct and complete for CB-WO's. Thus, both are not applicable to Example 1.

The continuous interest in the unification of functional and logic programming in a seamless way attracted much attention (see [20] for a survey) and Antoy, Echahed and Hanus extended Antoy's outermost-needed rewriting strategy to the *needed narrowing* strategy [4], becoming the best narrowing strategy for functional logic programming languages over other narrowing strategies such as [17, 19, 25] (see [4] for a detailed comparison). Antoy, Echahed, and Hanus extended their needed narrowing strategy to the *weakly needed narrowing strategy* [3] in order to cope with CB-WO's. Note that these strategies are not applicable to Example 1.

In recent work [11], we have proposed refinements of (weakly) outermost-needed rewriting and (weakly) needed narrowing, called *natural rewriting* and *natural narrowing*. These strategies compute less (or exactly the same) steps than outermost-needed rewriting and needed narrowing for IS's. However, they are not applicable to Example 1 because of left-linearity and constructor conditions.

This work is part of a broader joint effort to generalize narrowing from equational logic to rewriting logic, so as to make possible a much wider range of applications. It builds on our previous work extending natural rewriting to general term rewriting systems [14], and also on work by Meseguer and Thati on narrowing for rewrite theories [29, 34]. As shown in [29, 34], for general rewrite theories which need not be confluent and need not be terminating, the issue of *completeness*, that is, of narrowing being a complete semi-decision procedure for solving reachability goals, is nontrivial and does not always hold, essentially because rewrites can take place in the substitutions. The paper [29] characterizes several classes of rewrite theories for which narrowing is complete. This paper offers a narrowing strategy, which could be the basis of [29, 34], to make narrowing efficient, and proves that this strategy is sound and complete in a different sense of "completeness," namely that any solution found by unrestricted narrowing (outside substitutions) will also be reachable using the strategy.

6   Conclusions and Future Work

We have generalized the narrowing strategy of [11] so that it can be applied to a much broader class of term rewrite systems. Specifically, the generalization drops the requirement that the rewrite system under consideration is left-linear and constructor, which is a typical assumption in functional (logic) programming. As a result of this generality, a much broader range of applications such as, symbolic reachability analysis of concurrent systems, and theorem proving, can be efficiently supported by our strategy. Since our generalization is conservative, we inherit all the optimality results presented in [11] for the class of left-linear constructor systems; note that the strategies in [11] are the best known for the class of left-linear constructor systems. An important problem for future research is to investigate optimality results of the generalized strategies for a larger class of rewrite systems.

Another interesting issue is to further generalize the strategies to the case where rewriting and narrowing are performed *modulo* a set of axioms (such as associativity, commutativity, and identity), as in languages such as ELAN [6] and Maude [7]. Specifically we are interested in strategies for rewrite theories of the form $\mathcal{R} = (\Sigma, \phi, E, R)$ where E is a set of axioms. A generalized narrowing strategy for such rewrite theories, that computes substitutions in an *incremental* manner, would have a very important efficiency advantage, since it will not explicitly use unification algorithms for the axioms E.

Acknowledgements

S. Escobar has been partially supported by projects MEC TIN 2004-07943-C04-02, EU ALA/95/23/2003/077-054, GV Grupos03/025 and grant 2667 of *Universidad Politécnica de Valencia* during a stay at Urbana-Champaign, USA.

References

1. S. Antoy. Definitional trees. In *Proc. of ALP'92*, LNCS 632:143–157. Springer, 1992.
2. S. Antoy. Constructor-based conditional narrowing. In *Proc. of PPDP'01*, pages 199–206. ACM, 2001.
3. S. Antoy, R. Echahed, and M. Hanus. Parallel evaluation strategies for functional logic languages. In *Proc. of ICLP'97*, pages 138–152. MIT Press, 1997.
4. S. Antoy, R. Echahed, and M. Hanus. A needed narrowing strategy. In *Journal of the ACM*, 47(4):776–822, 2000.
5. S. Antoy and S. Lucas. Demandness in rewriting and narrowing. In *Proc. of WFLP'02*, ENTCS 76. Elsevier, 2002.
6. P. Borovanský, C. Kirchner, H. Kirchner, and P.-E. Moreau. ELAN from a rewriting logic point of view. *Theoretical Computer Science*, 285:155–185, 2002.
7. M. Clavel, F. Durán, S. Eker, P. Lincoln, N. Martí-Oliet, J. Meseguer, and J. Quesada. Maude: specification and programming in rewriting logic. *Theoretical Computer Science*, 285:187–243, 2002.
8. H. Comon. Inductionless induction. In A. Robinson and A. Voronkov, editors, *Handbook of Automated Reasoning*, 1:913–962. Elsevier, 2001.
9. H. Comon and R. Nieuwenhuis. Induction = I-Axiomatization + First-Order Consistency. *Information and Computation*, 159(1/2):151–186, 2000.
10. A. Deursen, J. Heering, and P. Klint. *Language Prototyping: An Algebraic Specification Approach*. World Scientific, 1996.
11. S. Escobar. Refining weakly outermost-needed rewriting and narrowing. In *Proc. of PPDP'03*, pages 113–123. ACM, 2003.
12. S. Escobar. Implementing natural rewriting and narrowing efficiently. In *Proc. of FLOPS 2004*, LNCS 2998:147–162. Springer, 2004.
13. S. Escobar, J. Meseguer, and P. Thati. Natural narrowing as a general unified mechanism for programming and proving. Technical Report DSIC-II/16/04, DSIC, Universidad Politécnica de Valencia, 2004. Available at http://www.dsic.upv.es/users/elp/papers.html.
14. S. Escobar, J. Meseguer, and P. Thati. Natural rewriting for general term rewriting systems. In *Proc. of LOPSTR'04*, 2004. To appear.
15. M. Fay. First order unification in equational theories. In *Proc. of CADE-04*, LNCS 87:161–167. Springer, 1979.

16. K. Futatsugi and R. Diaconescu. *CafeOBJ Report.* World Scientific, AMAST Series, 1998.
17. E. Giovannetti, G. Levi, C. Moiso, and C. Palamidessi. Kernel Leaf: A Logic plus Functional Language. *Journal of Computer and System Sciences,* 42(2):139–185, 1991.
18. J. Goguen, T. Winkler, J. Meseguer, K. Futatsugi, and J.-P. Jouannaud. Introducing OBJ. In *Software Engineering with OBJ: Algebraic Specification in Action,* pages 3–167. Kluwer, 2000.
19. J. C. González-Moreno, M. T. Hortalá-González, F. J. López-Fraguas, and M. Rodríguez-Artalejo. An approach to declarative programming based on a rewriting logic. *Journal of Logic Programming,* 40(1):47–87, 1999.
20. M. Hanus. The integration of functions into logic programming: From theory to practice. *Journal of Logic Programming,* 19&20:583–628, 1994.
21. M. Hanus, S. Lucas, and A. Middeldorp. Strongly sequential and inductively sequential term rewriting systems. *Information Processing Letters,* 67(1):1–8, 1998.
22. G. Huet and J.-J. Lévy. Computations in Orthogonal Term Rewriting Systems, Part I + II. In *Computational logic: Essays in honour of J. Alan Robinson,* pages 395–414 and 415–443. MIT Press, 1992.
23. J. Hullot. Canonical forms and unification. In *Proc. of CADE-05,* LNCS 87:318–334. Springer, 1980.
24. J.-P. Jouannaud, C. Kirchner, and H. Kirchner. Incremental construction of unification algorithms in equational theories. In *Proc. of ICALP'83,* LNCS 154:361–373. Springer, 1983.
25. R. Loogen, F. López-Fraguas, and M. Rodríguez-Artalejo. A Demand Driven Computation Strategy for Lazy Narrowing. In *Proc. of PLILP'93,* LNCS 714:184–200. Springer, 1993.
26. S. Lucas. Context-sensitive rewriting strategies. *Information and Computation,* 178(1):294–343, 2002.
27. N. Martí-Oliet and J. Meseguer. Rewriting logic as a logical and semantic framework. In *Handbook of Philosophical Logic,* 9:1–88. Kluwer, 2001.
28. J. Meseguer. Conditional rewriting logic as a unified model of concurrency. *Theoretical Computer Science,* 96(1):73–155, 1992.
29. J. Meseguer and P. Thati. Symbolic reachability analysis using narrowing and its application to analysis of cryptographic protocols. In *Proc. of WRLA'04,* ENTCS 117:153–182. Elsevier, 2004.
30. A. Middeldorp. Call by Need Computations to Root-Stable Form. In *Proc. of POPL'97,* pages 94–105. ACM, 1997.
31. R. Nieuwenhuis. Basic paramodulation and decidable theories. In *Proc. of LICS'96,* pages 473–483. IEEE Computer Society, 1996.
32. R. Sekar and I. Ramakrishnan. Programming in equational logic: Beyond strong sequentiality. *Information and Computation,* 104(1):78–109, 1993.
33. TeReSe, editor. *Term Rewriting Systems.* Cambridge University Press, Cambridge, 2003.
34. P. Thati and J. Meseguer. Complete symbolic reachability analysis using back-and-forth narrowing. 2005. Submitted for publication.

The Finite Variant Property:
How to Get Rid of Some Algebraic Properties[*]

Hubert Comon-Lundh[2] and Stéphanie Delaune[1,2]

[1] France Télécom R&D
[2] Laboratoire Spécification & Vérification,
ENS de Cachan & CNRS UMR 8643,
61, avenue du Président Wilson,
94235 Cachan Cedex, France
{comon,delaune}@lsv.ens-cachan.fr

Abstract. We consider the following problem: Given a term t, a rewrite system $\mathcal{R}$, a finite set of equations E' such that $\mathcal{R}$ is E'-convergent, compute finitely many instances of t: $t_1, \ldots, t_n$ such that, for every substitution σ, there is an index i and a substitution θ such that $t\sigma\!\downarrow =_{E'} t_i\theta$ (where $t\sigma\!\downarrow$ is the normal form of $t\sigma$ w.r.t. $\rightarrow_{E'\backslash\mathcal{R}}$).

The goal of this paper is to give equivalent (resp. sufficient) conditions for the finite variant property and to systematically investigate this property for equational theories, which are relevant to security protocols verification. For instance, we prove that the finite variant property holds for Abelian Groups, and a theory of modular exponentiation and does not hold for the theory $ACUNh$ (Associativity, Commutativity, Unit, Nilpotence, homomorphism).

1 Introduction

In our recent work on the verification of cryptographic protocols [3, 5] we came twice across the following problem:

Given an AC-convergent rewrite system $\mathcal{R}$, is it possible (and how) to compute from any term t a finite set of instances $t\sigma_1, \ldots, t\sigma_n$ such that

$$\{t\sigma\!\downarrow_{\mathcal{R}} \mid \sigma \in \Sigma\} = \bigcup_{i=1}^{n}\{t\sigma_i\!\downarrow_{\mathcal{R}}\theta \mid \theta \in \Sigma\}$$

where Σ is the set of normalized substitutions and $u\!\downarrow_{\mathcal{R}}$ is the AC-normal form of u w.r.t. $\mathcal{R}$.

In other words, the reductions in $t\sigma$ only depend on reductions in finitely many (fixed) instances of t. This is typically what we will call the *finite variant property*: compute in advance all possible normal forms of an instance of t,

---

[*] This work has been partly supported by the RNTL project PROUVÉ 03V360 and the ACI-SI Rossignol.

independently of that instance. In [3], this problem is solved in an ad hoc way when $\mathcal{R}$ is the theory of exclusive or (also called the $ACUN$ theory), given by the rewrite rules:

$$x + x \to 0$$
$$x + x + y \to y$$
$$x + 0 \to x$$

and the associativity and commutativity axioms for $+$. Such a property, together with the finiteness of equivalence classes modulo E' is claimed to be the key property for decidability results in cryptographic protocols verification, in presence of algebraic properties [2]. That is why we are especially interested in studying the finite variant property for equational theories which are relevant to cryptography and which define infinite equivalence classes.

When $E' = \emptyset$, it is not difficult to see that the finite variant property is implied by the termination of basic narrowing. This is not so easy in general. Assume for instance that E' consists in the axioms of associativity and commutativity and E is defined by an AC-convergent rewrite system $\mathcal{R}$. On one hand, general AC-narrowing does not terminate, even for a single rule $y + x + x \to y$ and, on the other hand, basic narrowing is incomplete for E-unification. We didn't find any reference for the incompleteness of basic AC-narrowing, hence we show it in Section 3.2. E. Viola already noticed in [19] that the standard completeness proof of basic narrowing does not extend to the AC-case and proposes another narrowing strategy, introducing extensions of rules. This notion of narrowing restores completeness. However, termination is lost, even in simple cases. Even for equational theories presented by E'-convergent rewrite systems, basic narrowing might not terminate, while E has the finite variant property. This is the case for Abelian Groups, as we will see in Section 6.2.

The first contribution of this paper is to state a property (called *boundedness*) equivalent to the finite variant property in case of theories defined by convergent rewrite systems (Section 5.2). This is very similar to the existence of "narrowing bounds" in [19]. We differ in two respects: first we consider only terms (not unification problems) and second, there is a quantifier switch. Roughly, in [19], the "narrowing bound" is equivalent to "there exists a normalized θ such that $(t\theta\downarrow =_{AC} u$ and) all (inner) derivations starting from $t\theta$ are bounded". In our case, boundedness is equivalent to "for every normalized θ, there is a derivation from $t\theta$ to its normal form whose length is bounded".

Second, we give sufficient conditions for the boundedness property, which do not necessarily imply the termination of narrowing (Section 6.2) and prove that these conditions are met for several equational theories, which are relevant to cryptographic protocols. Our sufficient criteria is related to the notion of *optimally reducing* (AC)-term rewriting system introduced in [14]. Indeed being an optimally reducing rewrite system is a sufficient condition to satisfy our criteria, and therefore the boundedness property. We provide however with strictly weaker sufficient conditions and therefore new applications. For instance, we show that the theory of Abelian Groups has the boundedness property, relying on the unusual orientation of the inverse rule (Section 6.2). We use proof techniques which are similar to those of [12]. We also show in Section 7 that there are equational theories for which unifiability is in PTIME, while there is no convergent AC-rewrite system for the theory yielding the finite variant property.

Finally, we give some side-applications of the finite variant property: for instance the existential fragment of the theory of $\mathcal{T}(\mathcal{F})/=_E$ is decidable for the theories E under study.

We start with recalling some definitions in Section 2. We state in Section 3 some results on basic and equational narrowing (for instance the incompleteness of basic AC-narrowing). Next, we list in Section 4, some examples of equational theories, which are relevant to cryptographic protocols, explaining briefly where they come from. In Section 5, we state formally a definition of the finite variant property and give a characterization (the *boundedness property*) when the equational theory is presented by a finite E'-convergent rewrite system. Then, we briefly consider the case of $E' = \emptyset$ in Section 6.1. In Section 6.2 we give sufficient conditions for the boundedness property and then apply them to the relevant theories listed in Section 4. In Section 7, we prove that the theory $ACUNh$ (Associativity, Commutativity, Unit, Nilpotence, homomorphism), for which unifiability is in PTIME [13], does not have the finite variant property. In Section 8, we show other applications of the finite variant property, and we conclude in Section 9.

Missing proofs can be found in [4].

2   Preliminaries

2.1   Terms, Substitutions, Unification

We use classical notations and terminology from [7] on terms, unification, rewrite systems. $\mathcal{T}(\mathcal{F}, \mathcal{X})$ is the set of terms built over the finite (ranked) alphabet $\mathcal{F}$ of function symbols and the set of variable symbols $\mathcal{X}$. $\mathcal{T}(\mathcal{F}, \emptyset)$ is also written $\mathcal{T}(\mathcal{F})$. The set of positions of a term t is written $O(t)$, and $\bar{O}(t)$ is the set of non-variable positions of t. The empty sequence Λ denotes the top-most position. The subterm of $t \in \mathcal{T}(\mathcal{F}, \mathcal{X})$ at position $p \in O(t)$ is written $t|_p$. The term obtained by replacing $t|_p$ with u is denoted $t[u]_p$. The set of variables occurring in t is denoted $vars(t)$.

A substitution σ is a mapping from a finite subset of $\mathcal{X}$ called its domain and written $dom(\sigma)$ to $\mathcal{T}(\mathcal{F}, \mathcal{X})$. Substitutions are extended to endomorphisms of $\mathcal{T}(\mathcal{F}, \mathcal{X})$ as usual. We use a postfix notation for their application.

If E is a set of equations (unordered pair of terms), $=_E$ is the least congruence on $\mathcal{T}(\mathcal{F}, \mathcal{X})$ such that $u\sigma =_E v\sigma$ for all pairs $u = v \in E$ and substitutions σ. E is *regular* if, for every equation $t_1 = t_2 \in E$, $vars(t_1) = vars(t_2)$. Two terms s, t are E-unifiable if there is a substitution σ such that $s\sigma =_E t\sigma$. Such a substitution is called an E-unifier of s, t. We say that there is an E-unification algorithm if it is possible, for any two terms s, t, to compute a finite set $\sigma_1, \ldots, \sigma_n$ of E-unifiers of s, t, such that, for every E-unifier σ of s, t, there is an index i and a substitution θ such that, for every variable $x \in vars(s) \cup vars(t)$, $x\sigma =_E x\sigma_i\theta$.

2.2   Equational Rewriting

A *term rewriting system* (TRS) is a finite set of *rewrite rules* $l \rightarrow r$ where $l \in \mathcal{T}(\mathcal{F}, \mathcal{X})$ and $r \in \mathcal{T}(\mathcal{F}, vars(l))$. A term $s \in \mathcal{T}(\mathcal{F}, \mathcal{X})$ rewrites to t by a TRS $\mathcal{R}$, denoted $s \rightarrow_{\mathcal{R}} t$, if there is $l \rightarrow r$ in $\mathcal{R}$, $p \in O(s)$ and a substitution σ such that $s|_p = l\sigma$ and $t = s[r\sigma]_p$. The term $l\sigma$ is called a redex and we say that t rewrites to s by contracting the redex $l\sigma$. An innermost redex does not contain other redexes and in an innermost reduction sequence only innermost redexes are contracted. $\mathcal{R}^=$ is the symmetric closure of $\mathcal{R}$. $\xrightarrow{*}_{\mathcal{R}}$ is the reflexive and transitive closure of $\rightarrow_{\mathcal{R}}$. We write $t \xrightarrow{\leq n}_{\mathcal{R}} u$ if there is a reduction sequence of at most n steps from t to u. A TRS $\mathcal{R}$ is *terminating* if there are no infinite chains $t_1 \rightarrow_{\mathcal{R}} t_2 \rightarrow_{\mathcal{R}} \ldots$.

As in [7], given a set of rewrite rules $\mathcal{R}$ and a set of equations E, *rewriting modulo* E, is the relation $\rightarrow_{E\backslash\mathcal{R}}$ (others have used $\rightarrow_{\mathcal{R},E}$) defined as follows: $s \rightarrow_{E\backslash\mathcal{R}} t$ iff there exists a position $p \in O(s)$ such that $s|_p =_E l\sigma$ and $t = s[r\sigma]_p$ for some substitution σ and rule $l \rightarrow r \in \mathcal{R}$.

A rewrite system $\mathcal{R}$ is *E-confluent* if and only if for every s, t such that $s =_{\mathcal{R}=\cup E} t$, there exists s', t' such that $s \xrightarrow{*}_{E\backslash\mathcal{R}} s'$, $t \xrightarrow{*}_{E\backslash\mathcal{R}} t'$, and $s' =_E t'$. It said to be *E-convergent* if, in addition, $=_E \circ \rightarrow_{\mathcal{R}} \circ =_E$ is well founded.

A term t is in *normal form* (w.r.t. $\rightarrow_{E\backslash\mathcal{R}}$) if there is no term s such that $t \rightarrow_{E\backslash\mathcal{R}} s$. If $t \xrightarrow{*}_{E\backslash\mathcal{R}} s$ and s is in normal form then we say that s is a normal form of t. When this normal form is unique, we write $s = t\downarrow_{E\backslash\mathcal{R}}$ or shortly $s = t\downarrow$ when $E\backslash\mathcal{R}$ is clear from the context. A substitution σ is called *normalized* if for every $x \in dom(\sigma)$, $x\sigma$ is in normal form. We write $\sigma =_E \theta$ if $\forall x \in dom(\sigma) \cup dom(\theta)$ $x\sigma =_E x\theta$. For an E-convergent rewrite system $\mathcal{R}$ and a substitution σ, we write $\sigma\downarrow_{E\backslash\mathcal{R}}$ the substitution whose domain is $dom(\sigma)$ and such that $x(\sigma\downarrow_{E\backslash\mathcal{R}}) = (x\sigma)\downarrow_{E\backslash\mathcal{R}}$ for all $x \in dom(\sigma)$.

3   Narrowing

Given a TRS $\mathcal{R}$, we say that a term t *narrows to* t' *with the substitution* σ, at $p \in \bar{O}(t)$, by $l \rightarrow r \in \mathcal{R}$ if there exists a renaming $l' \rightarrow r'$ of $l \rightarrow r \in \mathcal{R}$ such that σ is a unifier of $t|_p$ and l' and $t' = (t[r]_p)\sigma$. In this case, we write $t \rightsquigarrow_\sigma t'$. We write $t \rightsquigarrow^*_\sigma t'$ if there exists a narrowing derivation $t = t_1 \rightsquigarrow_{\sigma_1} t_2 \ldots \rightsquigarrow_{\sigma_{n-1}} t_n = t'$ such that $\sigma = \sigma_1 \ldots \sigma_{n-1}$.

3.1   Equational Narrowing

If E is a set of equations such that an E-unification algorithm exists, we define E-narrowing as expected (σ is an E-unifier of $t|_p$ and l).

The following lemma states that every rewrite derivation ($\xrightarrow{*}_{E\backslash\mathcal{R}}$) can be lifted to a narrowing derivation.

Lemma 1 (lifting lemma). *Let E be a regular presentation for which an E-unification algorithm exists. Let t be a term, θ be a normalized substitution and $t\theta \xrightarrow{*}_{E\backslash\mathcal{R}} s'$. Then there exists a term t', a substitution σ and a normalized substitution θ' such that:*

1. $t \stackrel{*}{\leadsto}_\sigma t'$,
2. $t'\theta' =_E s'$,

Furthermore, the narrowing derivation $t \stackrel{*}{\leadsto}_\sigma t'$ and the rewrite sequence from $t\theta$ to s' use the same rewrite rules at the same positions.

We didn't find this lemma in the litterature. A similar lemma, but only for a one step derivation, and without the regularity assumption, is proved in [11] for instance. The proof does not extend to an arbitrary derivation length. Actually, we do not know whether or not the lemma would still hold without the regularity assumption (which we indeed use in the proof).

3.2   Basic Narrowing

Definition 1 (basic positions). *Let $t_1 \leadsto_{\sigma_1} t_2 \leadsto_{\sigma_2} \ldots \leadsto_{\sigma_{n-1}} t_n$ be a narrowing derivation. We assume that the ith step has been done at position p_i with the rule $l_i \rightarrow r_i$. We inductively define sets of positions $B_1, \ldots, B_n$ as follows:*

$$B_1 = \bar{O}(t) \qquad B_{i+1} = \mathcal{B}(B_i, p_i, r_i) \qquad for \ 1 \le i < n.$$

Here $\mathcal{B}(B_i, p_i, r_i)$ abbreviates $(B_i - \{q \in B_i | p_i \le q\}) \cup \{p_i.q | q \in \bar{O}(r_i)\}$. Positions in B_i are referred to as basic positions. We say that the above narrowing derivation is basic if $p_i \in B_i$ for $1 \le i < n$.

In the same way, a rewrite sequence (w.r.t. $E\backslash \mathcal{R}$) $t_1 \rightarrow t_2 \rightarrow \ldots \rightarrow t_n$ is based on a set of positions $B_1 \subseteq \bar{O}(t_1)$ if $p_i \in B_i$ for $1 \le i < n$ with $B_2, \ldots, B_{n-1}$ defined as above.

Note that the latter is well-defined since $\rightarrow_{E\backslash\mathcal{R}}$ preserves the positions which are not in the redex.

In case of non-equational narrowing, there are several well-known results, for instance:

Lemma 2 ([8]). *Let t be a term and σ a normalized substitution. Every innermost derivation sequence (w.r.t $\mathcal{R}$) starting from $t\sigma$ is based on $\bar{O}(t)$.*

It follows that basic narrowing is a complete unification procedure when $\mathcal{R}$ is a convergent rewrite system. The situation is quite different for equational narrowing. For instance in the case of AC-narrowing, Lemma 2 fails (contrary to what is suggested in [11]), as shown by the following example (this has also been noticed in [19]).

Example 1. Let $\mathcal{R}_+ = \{x + 0 \rightarrow x, x + x \rightarrow 0, x + x + y \rightarrow y\}$, which is known to be AC-convergent. Let $t = x_1 + x_2$ and $\sigma = \{x_1 \mapsto a + b, x_2 \mapsto a + b\}$. Consider the following innermost derivation (w.r.t. $AC\backslash\mathcal{R}_+$) starting from $t\sigma$.

$$(a + b) + (a + b) \xrightarrow[x+(x+y)\rightarrow y]{\Lambda} b + b \xrightarrow[x+x\rightarrow 0]{\Lambda} 0$$

The first rewriting step takes place at position $\Lambda \in B_1 = \bar{O}(t)$ with the rewriting rule $x + (x + y) \rightarrow y$. Hence the set B_2 is empty. So the above rewrite sequence is not based on $\bar{O}(t)$ although it is an innermost derivation.

This example can be generalized in such a way that there is a derivation from $t\sigma$ whose length is arbitrarily long. However, there is also another derivation whose length is short (1 in the above example).

Not only Lemma 2 fails, but actually basic AC-narrowing is not complete, as shown by the following example.

Example 2. We consider the following rewrite system $\mathcal{R}$, in which $+$ is an AC-symbol and a, b are constants:

$$a + a \rightarrow 0 \qquad b + b \rightarrow 0 \qquad a + a + x \rightarrow x$$
$$b + b + x \rightarrow x \qquad 0 + x \rightarrow x$$

$\mathcal{R}$ is AC-convergent. $\sigma = \{x_1 \mapsto a + b; \ x_2 \mapsto a + b\}$ is a solution of the equation $x_1 + x_2 = 0$, whereas there is no narrowing derivation yielding a more general solution. Indeed, narrowing with one of the first two rules yields $x_1 = x_2 = a$ or $x_1 = x_2 = b$, narrowing with the last rule yields $x = 0 \wedge x + 0 = x_1 + x_2$, which do not subsume σ. Narrowing with one of the two other rules, for example $a + a + x \rightarrow x$, yields $x = 0 \wedge a + a + x = x_1 + x_2$, again not wanted.

4   Some Relevant Equational Theories

We list here some algebraic theories which are relevant to cryptographic protocols and which we investigate in Section 6. We only consider theories for which equivalence classes are infinite. We use the notations which are customary in cryptographic protocol descriptions. In particular, the pairing symbol $\langle \_, \_ \rangle$ is used in infix notation and encrypting m with k is written $\{m\}_k$.

4.1   Explicit Destructors

The *Axiomatized Dolev-Yao Theory* (DYT) is the classical Dolev-Yao model with explicit destructors such as decryption and projections. Here is a presentation of this theory:

$$\pi_i(\langle x_1, x_2 \rangle) = x_i \ \text{ for } i = 1, 2 \qquad d(\{x\}_y, y^{-1}) = x \qquad x^{-1^{-1}} = x$$

In words, projections are inverses of pairing, and decrypting with k^{-1} a message encrypted with a key k gives back the plain text message. Alternatively, projections and decryption symbols are not part of the alphabet and such properties are part of the intruder deduction rules. Putting such rules in the equational theory or in the intruder deduction rules seems to be a matter of taste. However, there are subtle differences between the two approaches; some protocols can be attacked if we consider explicit destructors, while they cannot otherwise (see for instance [6]). This relies on the ability to apply the decryption algorithm $d(\_, \_)$ on a message x with a key y, even when x is not a cyphertext.

Proposition 1. *Orienting equations of* DYT *from left to right and adding* $d(\{x\}_{y^{-1}}, y) \rightarrow x$, *we get a convergent rewrite system* $\mathcal{R}_{\mathrm{DYT}}$. *Furthermore (basic) narrowing w.r.t.* $\mathcal{R}_{\mathrm{DYT}}$ *terminates.*

The *Key Inverse Theory* (KIT) is obtained by extending DYT with the equation $\{d(x,y)\}_{y^{-1}} = x$. It expresses that decryption and encryption with the inverse key are inverse of each other. This property holds when decryption is just an encryption with the inverse key, as for the cryptosystem RSA.

Proposition 2. *Orienting equations of* KIT *from left to right and adding the rules* $d(\{x\}_{y^{-1}}, y) \rightarrow x$ *and* $\{d(x, y^{-1})\}_y \rightarrow x$, *we get a convergent rewrite system* $\mathcal{R}_{\mathrm{KIT}}$. *Furthermore (basic) narrowing w.r.t.* $\mathcal{R}_{\mathrm{KIT}}$ *terminates.*

4.2   Exclusive Or Theory ($ACUN$)

This theory has been given in introduction. It is mandatory when protocols rely on exclusive or ([15] vs [17]). As recalled in Example 1, the rewrite system $\mathcal{R}_+$ for this theory is AC-convergent.

4.3   Abelian Groups Theory ($\mathcal{AG}$)

The Abelian Groups theory is defined by the following set of equations:

$$x * (y * z) = (x * y) * z \qquad x * x^{-1} = 1$$
$$x * y = y * x \qquad x * 1 = x$$

Proposition 3. *Adding the consequences:* $1^{-1} = 1$, $x^{-1^{-1}} = x$, $(x * y)^{-1} = x^{-1} * y^{-1}$, $x * (y * x^{-1}) = y$ *and orienting the rules from left to right, we get* $\mathcal{R}_*$, *an AC-convergent rewrite system for* $\mathcal{AG}$.

Note that, AC-narrowing (even basic) is not terminating w.r.t. $\mathcal{R}_*$, as we have an infinite derivation starting from x^{-1} by using repeatedly $(x * y)^{-1} \rightarrow x^{-1} * y^{-1}$.

4.4   Diffie-Hellman Theory ($\mathcal{DH}$)

This theory contains the axioms of the Abelian Groups theory for the symbol $*$ and two others equations concerning the modular exponentiation's symbol:

$$exp(x, 1) = x \qquad exp(exp(x, y), z) = exp(x, y * z)$$

This theory takes into account simple properties of product and exponentiation, which are widely used in protocol constructions. Exponentiation has more properties, which should be considered to capture to whole power of an attacker. However, we only consider the two above axioms since, as shown in [10], many extensions yield undecidable unification problems, hence undecidability of confidentiality, even for a bounded number of sessions.

4.5   Combinations

The theory $ACUNh$ consists of the axioms of $ACUN$ for $+$ and the equation $h(x + y) = h(x) + h(y)$. This theory is used in protocols such as the TMN protocol (h is used to model an encryption with the public-key of the server S).

The equation $h(x + y) = h(x) + h(y)$ can be oriented in both directions, yielding two AC-convergent rewrite systems, which are displayed in Figure 1: depending on the orientation, we get either 5 rules ($\mathcal{R}_1$) or 6 rules ($\mathcal{R}_2$).

Proposition 4. $\mathcal{R}_1$ *and* $\mathcal{R}_2$ *are AC-convergent.*

$$
\begin{array}{ll}
x + 0 \to x & \mathcal{R}_1 : h(x+y) \to h(x) + h(y) \\
x + x \to 0 & \\
x + x + y \to y & \mathcal{R}_2 : h(x) + h(y) \to h(x+y) \\
h(0) \to 0 & \qquad\; h(x) + h(y) + z \to h(x+y) + z
\end{array}
$$

Fig. 1. The Rewrite Systems $\mathcal{R}_1$ and $\mathcal{R}_2$ for the *ACUNh* Theory.

5   The Finite Variant Property

We come to the central notion of our paper: a property, which allows to reduce equational theories to some (supposedly simpler) other theory. Let us first recall the definitions given in introduction.

5.1   Definition and a First Characterization

We assume given a well founded ordering $\geq$ on terms, which is total on ground terms. Given a theory E and a ground term t, we write $t\!\downarrow_E$ the smallest term in the equivalence class of t. It will serve as a representative of the class.

Definition 2 (*E*-variants). *Given two sets of equations E, E', t' is an E-variant of a term t if there is a substitution θ such that $t\theta =_E t'$. A complete set of E-variants modulo E' of t (w.r.t. $\geq$) is a set S of E-variants of t such that, for every substitution σ, there is a term $t' \in S$ and a substitution θ such that $t\sigma\!\downarrow_E =_{E'} t'\theta$.*

Example 3. Assume $E = ACUN$ and $E' = AC$. Consider the term $x + f(x+y)$. A complete set of E-variants modulo AC is given by the single variable z. Indeed,

$$
(x+f(x+y))\{x \mapsto f(z)+z; y \mapsto f(z)\} =_{AC} f(z)+z+f(f(z)+z+f(z)) =_{ACUN} z
$$

hence z is a variant of $x + f(x + y)$. This is a complete set since, for every normalized substitution σ, $(x + f(x+y))\sigma\!\downarrow =_{AC} z\theta$ for some θ.

Definition 3 (finite variant property). *The pair (E, E') has the* finite variant property *(w.r.t. $\geq$) if for every term t, we can effectively compute a finite complete set of E-variants modulo E'.*

Sometimes, we will simply say variants and complete set of variants when E and E' are clear from the context.

Now, we need a (uniform) way to compute the E-variants of a term. That is why we will restrict our attention to theories E for which there exists $\mathcal{R}$ and E' such that $\mathcal{R}$ is an E'-convergent system for E. Then the ordering $\geq$ will be chosen in such a way that $\to_{E'\backslash\mathcal{R}} \subseteq \geq$. To summarize now, our aim is, given a theory E, to find a splitting of E in $(\mathcal{R}, E')$ and an ordering $\geq$ such that:

1. $\mathcal{R}$ is an E'-convergent system for E and $\to_{E'\backslash\mathcal{R}} \subseteq \geq$ is a decidable relation,
2. for every term t, there is a finite set of variants $t_1, \ldots, t_n$, effectively computable, such that, for every substitution σ, there is an index i and a substitution θ such that $t\sigma\!\downarrow_{E'\backslash\mathcal{R}} =_{E'} t_i\theta$.

We will simply say that $(\mathcal{R}, E')$ is a *decomposition* of E satisfying the *finite variant property* if the two above properties are satisfied. There are several well-known techniques to obtain presentations satisfying the first condition. Hence, we focus on the second condition.

The following lemma shows that, if $(\mathcal{R}, E')$ has the finite variant property, we may not only compute in advance some instances t_i of t such that $t\sigma\!\downarrow$ is always an instance of some t_i, but actually compute in advance substitutions θ_i such that $t_i = t\theta_i\!\downarrow$ is a complete set of variants and every normalized substitution can be factorized through θ_i.

Lemma 3. *A decomposition $(\mathcal{R}, E')$ has the finite variant property iff*

For every term t, there is a finite set of substitutions $\Sigma(t)$ such that

$$\forall\sigma\ \exists\theta\in\Sigma(t),\ \exists\tau.\ \sigma\!\downarrow\ =_{E'}\theta\tau\ \wedge\ (t\sigma)\!\downarrow\ =_{E'}(t\theta)\!\downarrow\tau$$

Proof Sketch: The if part is straightforward. Conversely, let T be the term $\langle t, \langle x_0, \langle \ldots, x_n\rangle\rangle\rangle$ where $\{x_0, \ldots, x_n\} = vars(t)$ and $\langle \_, \_\rangle$ is a free binary function symbol. We apply the hypothesis to T. This yields a definition of $\Sigma(t)$.   □

5.2   The Boundedness Condition

In what follows we assume we are given a theory E for which there exists $\mathcal{R}$ and E' such that $\mathcal{R}$ is an E'-convergent system for E.

Definition 4 (boundedness property). $(\mathcal{R}, E')$ *satisfies the* boundedness property *if for every term t, there exists an integer n such that for every normalized substitution σ, the normal form of $t\sigma$ is reachable by a derivation whose length can be bounded by n (thus independently of σ):*

$$\forall t, \exists n, \forall\sigma.\ t(\sigma\!\downarrow) \xrightarrow{\ \leq n\ }_{E'\backslash\mathcal{R}} (t\sigma)\!\downarrow$$

The following theorem shows the relationships between the boundedness condition and the finite variant property.

Theorem 1. *Let E' be a regular presentation for which an E'-unification algorithm exists. If moreover $(\mathcal{R}, E')$ satisfies the boundedness property then $(\mathcal{R}, E')$ is a decomposition of E satisfying the finite variant property.*

Conversely, if $(\mathcal{R}, E')$ satisfies the finite variant property, then it satisfies the boundedness property.

Proof Sketch: The first implication is actually similar to a result in [19]: we use narrowing, however bounding the length of derivation. For the converse, let t be any term. We first apply Lemma 3. Then we let n be such that $t\theta \xrightarrow{\ \leq n\ }_{E'\backslash\mathcal{R}} (t\theta)\!\downarrow$ for every $\theta \in \Sigma(t)$. Then we prove that, for every normalized substitution σ, $t\sigma \xrightarrow{\ \leq n\ }_{E'\backslash\mathcal{R}} (t\sigma)\!\downarrow$.   □

It must be emphasized that the proof of this theorem provides us with an effective way of computing the variants: simply narrow t at most n times, where n is given by the boundedness property.

6   Proving Boundedness

6.1   The Case $E' = \emptyset$

Thanks to Lemma 2, the narrowing derivation associated by Lemma 1 to an innermost derivation is basic. Moreover, since $\mathcal{R}$ is a convergent system, we can always choose an innermost derivation. Hence we have the following proposition:

Proposition 5. *If basic narrowing terminates for $\mathcal{R}$ then $(\mathcal{R}, \emptyset)$ is a decomposition of E satisfying the boundedness property.*

This proposition allows us to conclude that the decomposition $(\mathcal{R}_{\text{DYT}}, \emptyset)$ (resp. $(\mathcal{R}_{\text{KIT}}, \emptyset)$) of DYT (resp. KIT) presented in Section 4.1 satisfies the boundedness property and, by Theorem 1 we conclude that these decompositions satisfy the finite variant property.

6.2   Non-orientable Axioms

Because of non-orientable equations (typically AC), we need to consider equational rewriting. Unfortunately, we cannot extend directly the results of the previous section, as shown by Example 1. Anyway, for Abelian Groups and Diffie-Hellman theories, independently of the orientation of $x^{-1} * y^{-1} = (x * y)^{-1}$, AC-narrowing (even basic) does not terminate. That is why we need to develop refined criteria, which will be satisfied by these two theories, yielding a finite variant property.

Let us first give a simple decidable sufficient condition for boundedness.

Lemma 4. *If $(\mathcal{R}, E')$ is a decomposition of E which satisfies:*

$$\forall f \in \mathcal{F} \ \exists c \ \forall t_1, \ldots t_n \in \mathcal{T}(\mathcal{F}, \mathcal{X}). \ f(t_1{\downarrow}, \ldots, t_n{\downarrow}) \xrightarrow{\leq c}_{E' \backslash \mathcal{R}} f(t_1, \ldots, t_n){\downarrow}.$$

Then $(\mathcal{R}, E')$ satisfies the boundedness property.

Note that being an optimally reducing rewrite systems (see [14]) is a sufficient condition for the boundedness property. Indeed such systems actually satisfy the conditions of Lemma 4, with a constant $c = 1$. However, we are going to need (for instance for Abelian Groups) to apply Lemma 4 with constants larger than 1. Furthermore, even if we can apply Lemma 4, with $c = 1$, the rewrite system might not be optimally reducing, simply because there are extra rules not satisfying the required condition. Finally, in [14], the authors assume that the root symbol of any left hand side is not associative-commutative, which we do not. So, our condition, which is strictly weaker, provides us with new applications.

We show successively that Lemma 4 can be applied to the theories of exclusive or, Abelian Groups and Diffie-Hellman.

Lemma 5. *Let t_1 and t_2 be irreducible terms (w.r.t. $AC \backslash \mathcal{R}_+$), $t_1 + t_2$ can be reduced to its normal form, using at most 1 reduction step.*

A similar lemma does not hold for the Abelian Groups decomposition $(\mathcal{R}_*, AC)$ of Section 4.3. Even worse, this decomposition does not satisfy the boundedness property: consider the term $t = x^{-1}$ and the substitution $\sigma = \{x \mapsto a_0 * \ldots * a_n\}$, $t\sigma$ requires at least n reduction steps before we reach its normal form.

However, an unusual orientation of some rules yields a presentation for which the finite variant property holds. This orientation has first been proposed by Lankford (see [9]). We get the following rewrite system:

$$\mathcal{R}'_* = \left\{ \begin{array}{ll} x * 1 \rightarrow x & x^{-1^{-1}} \rightarrow x \\ 1^{-1} \rightarrow 1 & (x^{-1} * y)^{-1} \rightarrow x * y^{-1} \\ x * x^{-1} \rightarrow 1 & x * (x^{-1} * y) \rightarrow y \\ x^{-1} * y^{-1} \rightarrow (x * y)^{-1} & x^{-1} * (y^{-1} * z) \rightarrow (x * y)^{-1} * z \\ (x * y)^{-1} * y \rightarrow x^{-1} & (x * y)^{-1} * (y * z) \rightarrow x^{-1} * z \end{array} \right.$$

This rewrite system is an AC-convergent system for $A\mathcal{G}$ [9] and even though basic narrowing does not terminate, we can show that:

Lemma 6. Let t_1 and t_2 be irreducible terms (w.r.t. $AC \backslash \mathcal{R}'_*$), t_1^{-1} and $t_1 * t_2$ can be reduced to their normal form, using at most 1 (resp. 2) reduction step.

Example 4. Let $t_1 = a * (b * c)^{-1}$ and $t_2 = a^{-1} * b$. We have the following derivation from $t_1 * t_2$ to its normal form c^{-1}.

$$(a * (b * c)^{-1}) * (a^{-1} * b) \rightarrow_{AC \backslash \mathcal{R}'_*} ((b * c) * a)^{-1} * (a * b) \rightarrow_{AC \backslash \mathcal{R}'_*} c^{-1}$$

Now consider the Diffie-Hellman theory. We orient the two additional equations and get the following rewrite system:

$$\mathcal{R}_{\mathcal{DH}} = \mathcal{R}'_* \cup \left\{ \begin{array}{l} exp(x, 1) \rightarrow x \\ exp(exp(x, y), z) \rightarrow exp(x, y * z) \end{array} \right.$$

Proposition 6. $\mathcal{R}_{\mathcal{DH}}$ is an AC-convergent rewrite system for $\mathcal{DH}$.

Lemma 7. Let t_1 and t_2 be irreducible terms (w.r.t. $AC \backslash \mathcal{R}_{\mathcal{DH}}$), t_1^{-1}, $t_1 * t_2$ and $exp(t_1, t_2)$ can be reduced to their normal form, using at most 1 (resp. 2 and 4) reduction step.

We illustrate the worst case for which we need the 4 reduction steps to obtain the normal form.

Example 5. Let $t_1 = exp(e, a^{-1} * b)$ and $t_2 = b^{-1} * a$, $t = exp(t_1, t_2)$ can be reduced to its normal form (w.r.t. $AC \backslash \mathcal{R}_{\mathcal{DH}}$) by a derivation using 4 reduction steps. Indeed, we have:

$$\begin{aligned} exp(exp(e, a^{-1} * b), b^{-1} * a) &\rightarrow exp(e, (a^{-1} * b) * (b^{-1} * a)) \\ &\rightarrow exp(e, (a * b)^{-1} * (a * b)) \\ &\rightarrow exp(e, 1) \\ &\rightarrow e \end{aligned}$$

To sum up, as consequences of Theorem 1, Lemmas 4, 5, 6 and 7:

Corollary 1. The decompositions $(\mathcal{R}_+, AC)$, $(\mathcal{R}'_*, AC)$ and $(\mathcal{R}_{\mathcal{DH}}, AC)$ have the finite variant property.

7   *ACUNh* Does Not Satisfy the Finite Variant Property

We prove here that the theory *ACUNh*, introduced in Section 4.5 does not have the finite variant property.

Let us recall that, depending on the orientation of $h(x + y) = h(x) + h(y)$, we get two *AC*-convergent rewrite systems displayed in Figure 1. However, none of them yields an appropriate decomposition:

Lemma 8. *The decompositions $(\mathcal{R}_1, AC)$ and $(\mathcal{R}_2, AC)$ of the theory ACUNh do not satisfy the boundedness property.*

Proof: First, we consider the case of $(\mathcal{R}_1, AC)$, and we show the result by contradiction. Let $t = h(x)$ and n be such that $\forall \sigma. \ h(x)(\sigma\downarrow) \xrightarrow{\leq n}_{E'\backslash\mathcal{R}} (h(x)\sigma)\downarrow$. We consider the substitution $\sigma = \{x \mapsto a + h(a) + \ldots + h^{n+1}(a)\}$. It is easy to see that we need $n + 1$ rewriting steps (with the rule $h(x + y) \to h(x) + h(y)$) to rewrite $h(x)\sigma$ to its normal form $h(a) + \ldots + h^{n+2}(n)$. Hence contradiction.

The result for $(\mathcal{R}_2, AC)$ can be obtained in a similar way with the term $t = x + y$ and the substitution $\sigma = \{x \mapsto h^n(a); y \mapsto h^n(b)\}$.   □

There are not many other choices than $\mathcal{R}_1$ and $\mathcal{R}_2$ and we get the following:

Theorem 2. *There is no decomposition $(\mathcal{R}, AC)$ of ACUNh which satisfies the boundedness property and such that the right members of the rules in $\mathcal{R}$ are irreducible (w.r.t. $AC\backslash\mathcal{R}$).*

The idea is to prove first that, for any *AC*-convergent rewrite system $\mathcal{R}$, either $\to_{AC\backslash\mathcal{R}_1} \subseteq \xrightarrow{*}_{AC\backslash\mathcal{R}}$ or $\to_{AC\backslash\mathcal{R}_2} \subseteq \xrightarrow{*}_{AC\backslash\mathcal{R}}$. Next, we prove that there is an integer n such that $\to_{AC\backslash\mathcal{R}} \subseteq \xrightarrow{\leq n}_{AC\backslash\mathcal{R}_1}$ or $\to_{AC\backslash\mathcal{R}} \subseteq \xrightarrow{\leq n}_{AC\backslash\mathcal{R}_2}$ and we conclude by Lemma 8.

Corollary 2. *There is no decomposition $(\mathcal{R}, AC)$ of ACUNh which satisfies the finite variant property and such that the right members of the rules in $\mathcal{R}$ are irreducible (w.r.t. $AC\backslash\mathcal{R}$).*

The property required on the right members of the rules of $\mathcal{R}$ seems to be unnecessary. This assumption has been taken to make easier the proof.

8   Other Applications of the Finite Variant Property

Assume that (E, E') has the finite variant property. This can be used to reduce disunification problems modulo E to disunification problems modulo E':

Theorem 3. *The Σ_1 fragment of the first-order theory of $\mathcal{T}(\mathcal{F})/\!=_E$ is decidable whenever the Σ_1 fragment of the first-order theory of $\mathcal{T}(\mathcal{F})/\!=_{E'}$ is decidable.*

To prove this, simply compute the variants $\phi_1, \ldots, \phi_n$ of the formula ϕ. (In such a computation, logical connectives are seen as free symbols). For every substitution σ, there is an index i and a substitution θ such that $\phi\sigma\downarrow_E =_{E'} \phi_i\theta$. In particular, ϕ is solvable modulo E iff one of the ϕ_i is solvable modulo E'.

Then, since the Σ_1 fragment of the theory of $\mathcal{T}(\mathcal{F})/\!=_{AC}$ is decidable [1], we get the following new results:

Corollary 3. *The Σ_1 fragments of the first-order theories of quotient term algebras $\mathcal{T}(\mathcal{F})/=_{ACUN}$, $\mathcal{T}(\mathcal{F})/=_{AG}$, $\mathcal{T}(\mathcal{F})/=_{DH}$ are decidable.*

Such results cannot be derived from the decidability of unification. Even in the dismatching case this is not so trivial to get a decision procedure. Consider for instance $x + f(x + y) \neq a$ in the theory $ACUN$. A most general solution of the matching problem is $x = f(z) + a \wedge y = a + z + f(z)$. Complementing the solutions of the matching equation involves quantifier elimination : $\forall z.x \neq a + f(z) \vee y \neq a + z + f(z)$.

In the case of Abelian Groups, it is actually known that the first-order theory of finitely generated Abelian Groups is decidable [16]. However, adding a binary free function symbol, it might become undecidable. Actually, the status of the first order theories of above-mentioned quotient algebras is unknown. On the undecidability side, the method described in [18] can not be applied in a straightforward way. On the decidability side, the finite variant property does not help since the first-order theory of $\mathcal{T}(\mathcal{F})/=_{AC}$ is undecidable [18].

9   Conclusion

We believe that the finite variant property is important in many applications. It allows us to reduce problems modulo an equational theory E to problems modulo an equational theory $E' \subseteq E$. It is often useless for solving equations; for instance, unification modulo $ACUN$ is simpler than unification modulo AC. However, for other constraint solving problems such as intruder derivability constraints [5] or disunification problems mentioned in the previous section, this property can be crucial.

We have proposed some criteria for the finite variant property, which have been applied to several equational theories. The techniques are inspired by narrowing, though, as in [19], we do not rely directly on narrowing sequences, but rather on innermost reductions of instances.

An open question is to design other criteria (both for the finite variant property or its negation), which would not assume an E'-convergent rewrite system for E. For instance, does $(AC, \emptyset)$ have the finite variant property ? We are tempted to answer no, but the proof is challenging.

Acknowledgement

We would like to acknowledge P. Narendran and the anonymous referees who gave relevant comments which helped in improving the paper.

References

1. H. Comon. Complete axiomatizations of some quotient term algebras. *Theoretical Computer Science*, 118(2):167–191, 1993.
2. H. Comon-Lundh. Intruder theories (ongoing work). In *7th International Conference on Foundations of Software Science and Computation Structures (FoSSaCS'04)*, Barcelona, Spain, 2004. Invited talk, slides available at
http://www.lsv.ens-cachan.fr/~comon/biblio.html.

3. H. Comon-Lundh and V. Cortier. New decidability results for fragments of first-order logic and application to cryptographic protocols. In *Proc. 14th International Conference on Rewriting Techniques and Applications (RTA'03)*, volume 2706 of *LNCS*, pages 148–164, Valencia, Spain, 2003. Springer-Verlag.
4. H. Comon-Lundh and S. Delaune. The finite variant property: How to get rid of some algebraic properties. Research Report LSV-04-17, Laboratoire Spécification et Vérification, ENS Cachan, France, 2004. 21 pages.
5. H. Comon-Lundh and V. Shmatikov. Intruder deductions, constraint solving and insecurity decision in presence of exclusive or. In *Proc. of 18th Annual IEEE Symposium on Logic in Computer Science (LICS'03)*, pages 271–280, Ottawa, Canada, 2003. IEEE Comp. Soc. Press.
6. S. Delaune and F. Jacquemard. A decision procedure for the verification of security protocols with explicit destructors. In *Proc. 11th ACM Conference on Computer and Communications Security (CCS'04)*, pages 278–287, Washington, USA, 2004. ACM.
7. N. Dershowitz and J.-P. Jouannaud. Rewrite systems. In J. van Leeuwen, editor, *Handbook of Theoretical Computer Science*, volume B, chapter 6. Elsevier and MIT Press, 1990.
8. J.-M. Hullot. Canonical forms and unification. In *Proc. 5th Conference on Automated Deduction, (CADE'80)*, volume 87 of *LNCS*, pages 318–324, Les Arcs, France, 1980. Springer.
9. J.-M. Hullot. A catalogue of canonical term rewriting systems. Technical Report CSL-114, Computer Science Laboratory, SRI, CA, USA, 1980.
10. D. Kapur, P. Narendran, and L. Wang. An E-unification algorithm for analyzing protocols that use modular exponentiation. In *Proc. 14th International Conference on Rewriting Techniques and Applications (RTA'03)*, volume 2706 of *LNCS*, pages 165–179, Valencia, Spain, 2003. Springer-Verlag.
11. C. Kirchner. *Méthodes et Outils de Conception Systématique d'Algorithmes d'Unification dans les Théories Équationnelles*. PhD thesis, Université de Nancy I, 1985.
12. C. Meadows and P. Narendran. A unification algorithm for the group Diffie-Hellman protocol. In *Proc. of the Workshop on Issues in the Theory of Security (WITS'02)*, Portland, USA, 2002.
13. P. Narendran, Q. Guo, and D. Wolfram. Unification and matching modulo nilpotence. In *Proc. of the 13th International Conference on Automated Deduction, (CADE'96)*, volume 1104 of *LNCS*, pages 261–274, New Brunswick, USA, 1996. Springer-Verlag.
14. P. Narendran, F. Pfenning, and R. Statman. On the unification problem for cartesian closed categories. *Journal of Symbolic Logic*, 62(2):636–647, 1997.
15. L. Paulson. Mechanized proofs for a recursive authentication protocol. In *Proc. 10th Computer Security Foundations Workshop (CSFW'97)*, pages 84–95, Rockport, USA), 1997. IEEE Comp. Soc. Press.
16. C. Rackoff. On the complexity of the theories of weak direct products (preliminary report). In *Proc. of the 6th Annual ACM Symposium on Theory of Computing*, pages 149–160. ACM Press, 1974.
17. P. Y. A. Ryan and S. A. Schneider. An attack on a recursive authentication protocol: A cautionary tale. *Information Processing Letters*, 65(1):7–10, 1998.
18. R. Treinen. A new method for undecidability proofs of first order theories. *Journal of Symbolic Computation*, 14(5):437–457, 1992.
19. E. Viola. E-unifiability via narrowing. In *Proc. of the 7th Italian Conference on Theoretical Computer Science, (ICTCS'01)*, volume 2202 of *LNCS*, pages 426–438, Torino, Italy, 2001. Springer.

Intruder Deduction for AC-Like Equational Theories with Homomorphisms

Pascal Lafourcade[1,2], Denis Lugiez[2], and Ralf Treinen[1,*]

[1] LSV, ENS de Cachan & CNRS UMR 8643 & INRIA Futurs project SECSI,
94235 Cachan, France
http://www.lsv.ens-cachan.fr/~{lafourca,treinen}
[2] LIF, Université Aix-Marseille 1 & CNRS UMR 6166,
13453 Marseille Cedex 13, France
http://www.cmi.univ-mrs.fr/~lugiez

Abstract. Cryptographic protocols are small programs which involve a high level of concurrency and which are difficult to analyze by hand. The most successful methods to verify such protocols rely on rewriting techniques and automated deduction in order to implement or mimic the process calculus describing the protocol execution.

We focus on the intruder deduction problem, that is the vulnerability to passive attacks, in presence of several variants of AC-like axioms (from AC to Abelian groups, including the theory of *exclusive or*) and homomorphism which are the most frequent axioms arising in cryptographic protocols. Solutions are known for the cases of *exclusive or*, of Abelian groups, and of homomorphism alone. In this paper we address the combination of these AC-like theories with the law of homomorphism which leads to much more complex decision problems.

We prove decidability of the intruder deduction problem in all cases considered. Our decision procedure is in EXPTIME, except for a restricted case in which we have been able to get a PTIME decision procedure using a property of one-counter and pushdown automata.

1 Introduction

Cryptographic protocols are ubiquitous in distributed computing applications. They are employed for instance in internet banking, video on demand services, wireless communication, or secure UNIX services like ssh or scp. Cryptographic protocols can be described as relatively simple programs which are executed in an untrusted environment. These protocols use cryptographic primitives in order to implement symmetric (shared-key) encryption, and asymmetric (public-key) encryption and signatures.

Verifying protocols is notoriously difficult, and even very simple protocols which look completely harmless may have serious security flaws, as it was dramatically demonstrated by the bug of the Needham-Schroeder protocol found

---

* This work was partially supported by the research programs ACI-SI Rossignol, and RNTL PROUVÉ (n° 03 V 360).

J. Giesl (Ed.): RTA 2005, LNCS 3467, pp. 308–322, 2005.

by Lowe [14] using a model-checking tool. It took 17 years since the protocol was published to find the flaw, a so-called *man in the middle attack*. An overview of *authentication protocols* known a decade ago can be found in [5], a more recent data base of protocols and known flaws is [11]. These protocols are often implemented in small variants which differ from the originally proposed protocol, or are used in combination with other protocols. As a consequence, there is a multitude of verification problems, which raises the need for *automatic* tools.

There are different approaches to modeling cryptographic protocols and analyzing their security properties: process calculi like the *spi-calculus* [1], so-called cryptographic proofs (see, for instance, [2]), and the approach of Dolev and Yao [10] which consists in modeling an attacker by a deduction system. This deduction system specifies how the attacker can obtain new information from previous knowledge, which he has either obtained by silently eavesdropping the communication between honest protocol participants (in case of a *passive* attacker), or by eavesdropping and fraudulently emitting messages, thus provoking honest protocol participants to reply according to the protocol rules (this is the case of a so-called *active* attacker). We call *intruder deduction problem* the question whether a passive eavesdropper can obtain a certain information from knowledge that he observes on the network. The Dolev-Yao approach lends itself to automation since the question whether the intruder can obtain a certain information now reduces to the question whether this information can be deduced using a certain deduction system.

Classically, the verification of cryptographic protocols was based on the so-called *perfect cryptography assumption* which states that it is impossible to obtain any information about an encrypted message without knowing the exact key necessary to decrypt this message. This assumption allowed a separation of verification tasks into proving lower bounds for the cryptanalysis of the cryptographic primitives on the one hand, and verification of a distributed program on the other hand. Unfortunately, this perfect cryptography assumption has proven too idealistic: there are protocols which can be proven secure under the perfect cryptography assumption, but which are in reality insecure since an attacker can use properties of the cryptographic primitives in combination with the protocol rules in order to obtain knowledge of a secret. These properties are typically expressed as equational axioms (so-called algebraic properties), like for instance associativity and commutativity of certain operators. Algebraic properties may be essential for the executability of the protocol, or may just come into play because the cryptographic primitives employed by the protocol happen to satisfy these properties. A recent overview of algebraic properties of cryptographic primitives, their use to mount attacks on protocols, and existing results on verification of cryptographic protocols in presence of equational axioms can be found in [8].

A number of results have been obtained, both for the intruder deduction problem and for the preservation of secrecy under active attacks. We here only mention some results which are of particular relevance to the problems studied in this work: the intruder deduction problem in case of the equational axioms of *exclusive or* is decidable [6] in polynomial time [4], and in case of the equational

axioms of Abelian groups is decidable [6][1] in polynomial time [19]. Likewise, the
intruder deduction problem is decidable in polynomial time [7] in the case of
the equational theory of an homomorphism. Note that the two equational theories of *exclusive or* and of homomorphism model basic properties of important
cryptographic primitives:

- *Exclusive or* is a basic building block in many symmetric encryption methods
  (for instance DES or the more recent AES) or even used directly as an
  encryption method;
- Homomorphisms are ubiquitous in cryptography, by example the ElGamal
  encryption method has this property. Note that many protocols combine
  symmetric and asymmetric encryption.
- Symmetric encryption methods which often work on data blocks of fixed
  size are in the simplest of cases (the so-called *electronic codebook mode*)
  homomorphically extended to data streams of arbitrary size.

Some examples of attacks against protocols using the equational theories considered in this paper can be found in [8].

In this paper we investigate the intruder deduction problem in presence of
several variants of the equational theory of associativity and commutativity
(short AC) of a binary operator $\otimes$, plus the homomorphism property of a unary
function symbol over the AC operator. The variants of AC which we consider
are: pure AC, the theory of *exclusive or* (also called $ACUN$), and the theory
of Abelian groups. We are furthermore interested in the combination of these
AC-like theories with a generalization of one homomorphic function to some
form of distributivity of the encryption operator over the binary operator $\otimes$.
The homomorphism law is then replaced by a law stating that the encryption of
the $\otimes$ of two messages is equal to the $\otimes$ of the encryptions of the two messages
using the same encryption key. We do not assume that the set of encryption keys
is finite. Rather, any term can be used as an encryption key. This can be seen
as the extension to an infinite family of homomorphisms, one for each possible
encoding key. Our results can be summarized as follows:

1. The intruder deduction problem is decidable. It is NP-complete in case of
   the theory AC plus homomorphism, and we have an exponential-time upper
   bound for the equational theory $ACUN$ plus homomorphism and Abelian
   groups plus homomorphism.
2. The intruder deduction problem is in all three cases decidable in polynomial time if we restrict the class of problems to the so-called *binary* case,
   that is the case where the set of assumptions and the goal do not contain
   applications of $\otimes$ to more than two terms.
3. The first two sets of results carry over to the generalization which consists
   in replacing the homomorphic function by an encryption operation which
   distributes over $\otimes$.

---

[1] In fact, the NP-decision procedure in the case of Abelian groups given by [6] can
also be improved to *deterministic* polynomial time using the techniques explained
in this report.

We follow the approach of [6] and [7] which consists in a generalization of McAllester's *locality* method explained in Section 3.

Plan of the paper: We present in Section 2 the Dolev-Yao model of intruder capacities extended by a rewrite system modulo AC and list the rewrite systems investigated in this paper. In Section 3 we explain the generalization of McAllester's proof technique. We apply this technique in Sections 4, 5 and 6 to obtain decidability and complexity results for the case of *exclusive or* plus homomorphism. We discuss in Section 7 how these results can be transfered to some other related rewrite systems. Finally, we conclude in Section 8.

The full version of this paper with all proofs can be found at [13]. We use standard notation from rewriting. The reader may consult [3, 9] if necessary.

2   A Dolev-Yao Model for Rewriting Modulo AC

We consider the classic model of deduction rules [10] introduced by Dolev and Yao in order to model the deductive capacities of a passive intruder. We present here an extension of this model where we assume an associative and commutative operator $\otimes$, and an equational theory E which can be exploited by the intruder to mount an attack. Knowledge of the intruder is represented by terms built over a finite signature Σ of the form

$$\Sigma = \{\langle \cdot, \cdot \rangle, \{\cdot\}., \otimes, f\} \uplus \Sigma_0$$

where Σ_0 is a set of constant symbols. The term $\langle u, v \rangle$ represents the pair of the two terms u and v, and $\{u\}_v$ represents the encryption of the term u by the term v. For the sake of simplicity we here only consider symmetric encryption; the results and techniques can be easily transferred to the case of asymmetric encryption.

The equational theory E is represented by a convergent rewrite system R modulo AC, that is R is terminating and confluent modulo associativity and commutativity of $\otimes$, and for all terms $t, s \in T(\Sigma)$ we have that $t =_E s$ iff $t\downarrow_{R/AC} =_{AC} s\downarrow_{R/AC}$.

The deduction system describing the deductive capacities of an intruder is given in Figure 1. This deduction system is composed of the following rules: (A) the intruder may use any term which is in his initial knowledge, (P) the intruder can build a pair of two messages, (UL, UR) he can extract each member of a pair, (C) he can encrypt a message u with a key v, (D) if he knows a key v he can decrypt a message encrypted by the same key, (F) he can construct a new term using the function symbol f. Since we distinguish a special binary operator $\otimes$ we here furthermore add a family of rules (GX) which allows the intruder to build a new term from an arbitrary number of already known terms by using the (associative) $\otimes$ operator. The need for such a variadic rule (instead of just a binary rule) will become apparent in Section 3.

In fact, this deductive system is equivalent in deductive power to a variant of the system in which terms are not automatically normalized, but in which arbitrary equational proofs are allowed at any moment of the deduction. The

$$(A) \quad \frac{u \in T}{T \vdash u \downarrow_{R/AC}} \qquad\qquad (UL) \quad \frac{T \vdash r}{T \vdash u \downarrow_{R/AC}} \quad if \langle u, v \rangle = r \downarrow_{R/AC}$$

$$(P) \quad \frac{T \vdash u \quad T \vdash v}{T \vdash \langle u, v \rangle \downarrow_{R/AC}} \qquad\qquad (UR) \quad \frac{T \vdash r}{T \vdash v \downarrow_{R/AC}} \quad if \langle u, v \rangle = r \downarrow_{R/AC}$$

$$(C) \quad \frac{T \vdash u \quad T \vdash v}{T \vdash \{u\}_v \downarrow_{R/AC}} \qquad\qquad (D) \quad \frac{T \vdash r \quad T \vdash v}{T \vdash u \downarrow_{R/AC}} \quad if \{u\}_v = r \downarrow_{R/AC}$$

$$(F) \quad \frac{T \vdash u}{T \vdash f(u) \downarrow_{R/AC}} \qquad\qquad (GX) \quad \frac{T \vdash u_1 \quad \cdots \quad T \vdash u_n}{T \vdash (u_1 \otimes \ldots \otimes u_n) \downarrow_{R/AC}}$$

Fig. 1. A Dolev-Yao proof system working on normal forms by a rewrite system R modulo AC.

$$
\begin{array}{ccc}
 & & 0 \otimes x \to x \\
 & & x \otimes x \to 0 \\
 & & I(0) \to 0 \\
 & & I(x \otimes y) \to I(x) \otimes I(y) \\
 & 0 \otimes x \to x & I(I(x)) \to x \\
 & x \otimes x \to 0 & f(I(x)) \to I(f(x)) \\
 & f(0) \to 0 & f(0) \to 0 \\
f(x \otimes y) \to f(x) \otimes f(y) & f(x \otimes y) \to f(x) \otimes f(y) & f(x \otimes y) \to f(x) \otimes f(y) \\
\text{(a) } ACh & \text{(b) } ACUNh & \text{(c) } AGh
\end{array}
$$

Fig. 2. The three rewrite systems modulo AC

equivalence of the two proof systems has been shown in [7] without AC axioms; in [13] this has been extended to the case of a rewrite system modulo AC.

In the rest of the paper, we will investigate the Dolev-Yao deduction system modulo the rewrite systems presented in Figure 2, which correspond respectively to AC plus homomorphism of f over $\otimes$, the theory of *exclusive or* plus homomorphism of f over $\otimes$, and the theory of Abelian groups plus homomorphism of f over $\otimes$. We will omit the index R/AC and write $\to$ instead of $\to_{R/AC}$.

3   Locality and Complexity of Deduction Problems

Our starting point is the locality technique introduced by David McAllester [15]. He considers deduction systems which are represented by finite sets of Horn clauses. He shows that there exists a polynomial-time algorithm to decide the deducibility of a term w from a finite set of terms T_0 if the deduction system has the so-called locality property. A deduction system has the *locality property* if any proof of $T_0 \vdash w$ can be transformed into a local proof where a *local proof* is a proof where all the nodes are syntactic subterms of T_0 and w.

The idea of his proof is as follows: Checking existence of a proof amounts to checking existence of a local proof. Let us call for the moment a *relevant instance* of a deduction rule an instance of a rule where all terms are syntactic subterms of T_0 or w. Only these relevant instances are needed to construct a local proof.

We say that w is *one-step deducible* from some set T, if we can obtain w from T with only one application of a rule of the proof system. To check the existence of a local proof of $T_0 \vdash w$ it is now sufficient to saturate T_0 by the one-step deduction relation, where in addition it is sufficient to just consider the relevant instances of the deduction rules.

This approach suffers from two main restrictions:

- The deduction system must be finite.
- The notion of locality is restricted to syntactic subterms.

These restrictions raise a serious problem when we want to work modulo *AC*. If we used only a binary rule (GX) we would have to consider all possible subterms modulo *AC*. Unfortunately, there is in general an exponential number of subterms modulo *AC* of a given term. The solution proposed in [6], and which we also adopt here, is to use the rule (GX) with an arbitrary number of hypotheses. In this way, we can avoid the exponential number of subterms. However, we are now stuck with an infinite number of rules. Fortunately, we can still obtain an polynomial algorithm by implementing in a clever way the test whether a term w is one-step deducible from a set T.

Definition 1. *Let S be a function which maps a set of terms to a set of terms. A proof P of $T \vdash w$ is S-local if all nodes are labeled by some $T \vdash v$, with $v \in S(T \cup \{w\})$. A proof system is S-local if whenever there is a proof of $T \vdash w$ then there also is some S-local proof of $T \vdash w$.*

Theorem 1. *Let S be a function mapping a set of terms to a set of terms, and P a proof system. If*

- *the set $S(T)$ can be constructed in time $\mathcal{K}_1$,*
- *P is S-local,*
- *one-step deducibility in P is decidable in time $\mathcal{K}_2$,*

then provability in the proof system P is decidable in time $\max(\mathcal{K}_1, \mathcal{K}_2)$.

This theorem generalizes McAllester's result because in his case the size of the set of syntactic subterms of the set T is polynomial in the size of T, and since one-step deducibility is decidable in polynomial time for a finite proof system. Hence, in McAllester's case, it remained only the S-locality to show.

4   Proof Transformations

The following definitions and transformations can be applied to the cases *ACh* and *ACUNh*. The case of *AGh* requires an extension briefly discussed in Subsection 7.2.

Definition 2. *The size of a proof P is the number of nodes in P, denoted by $|P|$. A proof P of $T \vdash u$ is minimal if there is no proof P' of $T \vdash u$ such that $|P'| < |P|$.*

Definition 3. *Let P be a proof of $T \vdash w$, P is a*

- *simple proof if each node $T \vdash v$ occurs at most once on each branch.*
- *flat proof if there is no (GX) rule immediately above another (GX) rule,*
- *$\otimes$-lazy proof if P is flat and there is no (GX) rule immediately above an (F) rule in P,*
- *$\otimes$-eager proof if P is flat and if there is at most one (F) rule immediately above a (GX) rule in P.*

Since two successive (GX) rules can be merged into a single (GX) rule a minimal proof is a flat proof. Obviously any minimal proof is simple. Intuitively, in a $\otimes$-lazy proof the (GX) rule is applied as late as possible, and in a $\otimes$-eager proof the (GX) rule is applied as early as possible.

Lemma 1. *If there is a proof of $T \vdash w$ then there is also a $\otimes$-lazy proof and a $\otimes$-eager proof of $T \vdash w$.*

Proof. Successive (GX) rules can obviously be merged. We can obtain a $\otimes$-lazy proof by applying the following proof transformation rule:

$$
\text{(F)} \cfrac{\text{(GX)} \cfrac{T \vdash x_1 \ldots T \vdash x_n}{T \vdash x_1 \otimes \ldots \otimes x_n}}{T \vdash f(x_1) \otimes \ldots \otimes f(x_n)} \implies \text{(GX)} \cfrac{\text{(F)} \cfrac{T \vdash x_1}{T \vdash f(x_1)} \cdots \text{(F)} \cfrac{T \vdash x_n}{T \vdash f(x_n)}}{T \vdash f(x_1) \otimes \ldots \otimes f(x_n)}
$$

We obtain a $\otimes$-eager proof by applying the following proof transformation, where the rules (G_i) are all different from (F):

$$
\text{(GX)} \cfrac{\text{(F)} \cfrac{T \vdash x_1}{T \vdash f(x_1)} \cdots \text{(F)} \cfrac{T \vdash x_n}{T \vdash f(x_n)} \quad (G_1) \cfrac{T \vdash y_1}{T \vdash z_1} \cdots (G_m) \cfrac{T \vdash y_m}{T \vdash z_m}}{T \vdash f(x_1) \otimes \ldots \otimes f(x_n) \otimes z_1 \otimes \ldots \otimes z_m}
$$

$$\Downarrow$$

$$
\text{(GX)} \cfrac{\text{(F)} \cfrac{\text{(GX)} \cfrac{T \vdash x_1 \ldots T \vdash x_n}{T \vdash x_1 \otimes \ldots \otimes x_n}}{T \vdash f(x_1) \otimes \ldots \otimes f(x_n)} \quad (G_1) \cfrac{T \vdash y_1}{T \vdash z_1} \cdots (G_m) \cfrac{T \vdash y_m}{T \vdash z_m}}{T \vdash f(x_1) \otimes \ldots \otimes f(x_n) \otimes z_1 \otimes \ldots \otimes z_m}
$$

5   Locality for the Rewrite System $\mathcal{ACUNh}$

Definition 4. *Let u be a term in normal form, u is* headed with $\otimes$ *if u is of the form $u_1 \otimes \ldots \otimes u_n$ with $n > 1$. Otherwise u is* not headed with $\otimes$.

We define the function atoms(u) *as following:*

- *If* $u = u_1 \otimes \ldots \otimes u_n$, *where each of the* u_i *is not headed with* $\otimes$, *then* $atoms(u) = \{u_1, \ldots, u_n\}$. *The terms* u_i *are called the* atoms *of* u.
- *If* u *is not headed with* $\otimes$, *then* $atoms(u) = \{u\}$.

The definition of atoms(T) generalizes in a natural way to sets of terms T in normal form by atoms(T) := $\bigcup_{t \in T}$ atoms(t).

Definition 5. *We define for any* $T \subseteq T(\Sigma)$ *the set* $S_T(T)$ *as the smallest set which contains* T, *is closed under syntactic subterms, and such that if* $f(u_1) \otimes \ldots \otimes f(u_n) \in S_T(T)$ *then* $u_1 \otimes \ldots \otimes u_n \in S_T(T)$.

Lemma 2. *Let* P *be a proof which is minimal among all* $\otimes$-*lazy proofs of* $T \vdash w$, *and such that the last rule applied in* P *is of the form* $(X) \frac{T \vdash N_1 \ldots T \vdash N_n}{T \vdash w}$, *where* (X) *is one of* (UL), (UR), *or* (D). *Then* $N_i \in S_T(T)$ *for all* i.

This has been shown [7] in the setting of *exclusive or* without an homomorphism. The proof is very easily extended (see [13]) to our setting of *ACUNh*.

Lemma 3. *Let* P *be a proof which is minimal among all* $\otimes$-*lazy proofs of* $T \vdash w$, *and let* P' *be a subproof of* P *with root label* $T \vdash N$. *If the last rule applied in* P' *is* (P), (C), *or* (GX) *then* $N \in S_T(T \cup \{w\})$.

This is a central technical lemma. The proof is given in [13].

Lemma 4. *Let* P *be a proof which is minimal among all* $\otimes$-*lazy proofs of* $T \vdash w$, *and let* P' *be a subproof of* P *with root label* $T \vdash N$ *such that the last rule applied in* P' *is* (F). *If either*

1. *all nodes from the root of* P' *to the root of* P *are* (F),
2. *or if the first successor not labeled by* (F) *of the root of* P' *in* P *is labeled by a rule different from* (GX),

then $N \in S_T(T \cup \{w\})$.

The two cases of the lemma can be illustrated like this:

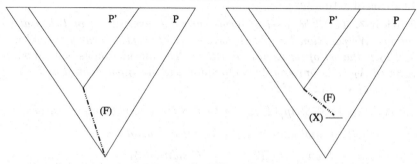

In the right picture, (X) denotes a rule different from (F) and from (GX). The lemma states that (F) nodes are in $S_T(T \cup \{w\})$ as long as they do not produce an hypothesis of a (GX) rule via a succeeding sequence of (F) nodes. This follows easily from Lemma 2 and Lemma 3 (see [13]).

Example 1 *The following proof of $T = \{u \otimes v, f(v)\} \vdash f(u)$ is minimal:*

$$
(GX) \; \cfrac{(F) \; \cfrac{(A) \; \cfrac{u \otimes v \in T}{T \vdash u \otimes v}}{T \vdash f(u) \otimes f(v)} \qquad (A) \; \cfrac{f(v) \in T}{T \vdash f(v)}}{T \vdash f(u)}
$$

We obtain $S_T(T \cup \{w\}) = \{u, v, u \otimes v, f(u), f(v)\}$. This proof is not S_T-local since $f(u) \otimes f(v) \notin S_T(T \cup \{w\})$.

As can be seen in the above example, the problem in defining S-locality for a polynomial-size S is to bound the number of applications of the (F) proof rule when constructing hypotheses to a (GX) rule.

5.1 Locality in the Binary Case

In the *binary* case, that is when all terms in $S_T(T\cup\{w\})$ have at most two atoms, we can actually find an upper bound for the number of applications of (F).

Definition 6. *A term t is binary if every $s \in S_T(t)$ either is not headed with $\otimes$, or is of the form $s_1 \otimes s_2$ where s_1, s_2 are not headed with $\otimes$. A set of terms is binary if each of its elements is binary. A proof is binary if each of its nodes is labeled by a sequent $T \vdash w$ where T and w are binary.*

Proposition 1. *If T and w are binary then every proof which is minimal among the $\otimes$-lazy proofs of $T \vdash w$ is binary.*

We define for any term t the term $Strip_f(f(t)) = Strip_f(t)$, and $Strip_f(t) = t$ if t does not have root symbol f. Furthermore, $\#_f(f(t)) = 1 + \#_f(t)$, and $\#_f(t) = 0$ when t is not headed by f. In the binary case we associate a one-counter automaton to the set $S_T(T \cup \{w\})$. The idea is that states of the automaton are terms in $Strip_f(atoms(S_T(T\cup\{w\})))$, and the counter represents the number of applications of f to a term.

Definition 7. *Let T be a set of terms such that every term in T has at most two atoms. We partition $T = T_1 \uplus T_2$ where T_1 is the set of terms not headed with $\otimes$, and T_2 is the set of terms headed with $\otimes$. The automaton associated with T, abbreviated A_T, is a one-counter automaton without input defined as follows: The set of states Q_T of A_T is*

$$\{\text{INIT}\} \cup \{p' \mid p \in Strip_f(T_1)\} \cup \{r \mid r \in Strip_f(T_1) \cup Strip_f(atoms(T_2))\}$$

where INIT *is the initial state of A_T. The set of transitions is:*

	From	To	Condition	Action
$\forall t \in T_1:$	INIT	$(Strip_f(t))'$	$c \geq 0$	$c := c$
$\forall t \in T_1:$	$(Strip_f(t))'$	$(Strip_f(t))'$	$c \geq 0$	$c := c+1$
$\forall t \in T_1:$	$(Strip_f(t))'$	$Strip_f(t)$	$c \geq \#_f(t)$	$c := c$
$\forall t \otimes s \in T_2:$	$Strip_f(t)$	$Strip_f(s)$	$c \geq \#_f(t)$	$c := c - \#_f(t) + \#_f(s)$

Note that in the last line of the above transition table the statement "$t \otimes s \in T_2$" is to be understood modulo *AC*, such that we obtain from a binary clause a back and a forth transition.

Example 2 *The automaton A_T for $T = \{a \otimes f^2(b), a\}$ is as follows, where I denotes the initial state:*

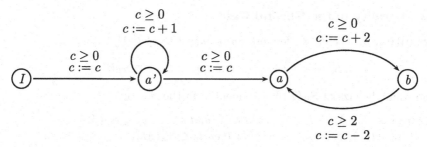

One of the two lemmata relating the proof system with our automata construction is:

Lemma 5. *Let T be a set of binary terms. For all $t_0, \ldots, t_n \in Strip_f(atoms(T))$ and all natural numbers $c_0, \ldots, c_n$ we have that*

$$A_T \models (t_0, c_0) \rightarrow (t_1, c_1) \rightarrow \ldots \rightarrow (t_n, c_n)$$

iff there are terms $s_1, \ldots, s_n \in T$ and natural numbers $d_1, \ldots, d_n$ such that:

1. *for $1 \leq i \leq n$ the term s_i is headed with $\otimes$ and has exactly two atoms, that is $s_i = s_i^1 \otimes s_i^2$*
2. $\forall 1 \leq i \leq n : f^{d_i}(s_i^1) = f^{c_{i-1}}(t_{i-1})$
3. $\forall 1 \leq i \leq n : f^{d_i}(s_i^2) = f^{c_i}(t_i)$

As a consequence and using the axiom $x \otimes x = 0$, we obtain that

$$\bigoplus_{i=1}^{n} f^{d_i}(s_i) \downarrow = f^{d_1}(s_1^1) \otimes f^{d_n}(s_n^2) = f^{c_0}(t_0) \otimes f^{c_n}(t_n)$$

Lemma 6. *Let A be a one-counter automaton and $\pi : (q, c_q) \rightarrow^* (r, c_r)$ a path between the state q with the counter $c_q \geq 0$ and the state r with the counter $c_r \geq 0$. Then there exists a path from (q, c_q) to (r, c_r) such that everywhere along the path the value of the counter is bounded by $p(|A|)$, where p is a polynomial function.*

We believe this lemma to be folklore but were unable to find a proof in the literature. A proof, along with a definition of the polynomial function p, is included in the complete version [13]. We can now define:

Definition 8. *We define for any finite subset U of $T(\Sigma)$:*

$$S_f(U) = \{f^i(u) \downarrow \mid u \in S_T(U), 0 \leq i \leq p(|A_T|)\}$$

where the function p is as in Lemma 6.

Note that the size of $S_f(U)$ is polynomial in the size of U. Combining Lemmata 2 through 6 we obtain:

Lemma 7. *Let $T \subseteq T(\Sigma)$ and $w \in T(\Sigma)$ be binary, and let P be a proof which is minimal among all $\otimes$-lazy proofs of $T \vdash w$. All nodes of P are in $S_f(T \cup \{w\})$.*

5.2 Locality in the General Case

Definition 9. *We define for any finite subset U of $T(\Sigma)$:*

$$S_\otimes(U) = \{u_1 \otimes \ldots \otimes u_n | u_1, \ldots, u_n \in S_T(U)\}$$

Note that the size of $S_\otimes(T)$ is exponential in the size of T.

Lemma 8. *Let $M \subseteq T(\Sigma)$, $t_0 \in T(\Sigma)$, and $t_1, \ldots, t_n \in S_T(M)$. If $(t_0 \otimes t_1 \otimes \ldots \otimes t_n) \downarrow \in S_\otimes(M)$ then $t_0 \in S_\otimes(M)$.*

The easy proof can be found in [13]. This lemma, together with the previous lemmata, is the key for proving the following lemma which states that any proof which is minimal among the $\otimes$-eager proofs of $T \vdash w$ contains only nodes in $S_\otimes(T \cup \{w\})$.

Lemma 9. *The Dolev-Yao proof system in case of ACUNh is $S_\otimes$-local.*

6   One-Step Deducibility in Case of *ACUNh*

We follow the well-known method for solving unification problems modulo *AC*-like theories [16]. We only show how to decide one-step deducibility for the family of rules (GX), since checking one-step deducibility for the remaining deduction rules is straightforward. We transform the problem of testing one-step deducibility into the satisfiability of a system of linear Diophantine equations.

Let $t \in T(\Sigma)$ and $u \in T(\Sigma)$ not headed with $\otimes$. We denote by $\delta(u, t)$ the number of occurrences of u in atoms(t) (which is, in the case *ACUNh*, either 0 or 1).

Definition 10. *Let $s \in T(\Sigma)$ and $T = \{t_1, \ldots, t_n\}$ be a finite subset of $T(\Sigma)$. Let atoms($T \cup \{s\}$) = $\{a_1, \ldots, a_m\}$. The equation system $D(T, s)$ over the variables $x_1, \ldots, x_n$ is*

$$D(T, s) := \bigwedge_{i=1}^{m} \sum_{j=1}^{n} \delta(a_i, t_j) * x_j = \delta(a_i, s)$$

Example 3 *Let $T = \{a_1 \otimes a_2 \otimes a_3, a_1 \otimes a_4, a_2 \otimes a_4\}$ and $s = a_1 \otimes a_2$, where all the a_i are not headed with $\otimes$. We introduce numerical variables x_1, x_2, x_3, that is one numerical variable for each element of T :*

$$
\begin{array}{ll}
x_1 & \text{for } a_1 \otimes a_2 \otimes a_3 \\
x_2 & \text{for } a_1 \otimes a_4 \\
x_3 & \text{for } a_2 \otimes a_4
\end{array}
$$

For every atom a_i we create an equation. This yields the following equation system:

$$\begin{cases} a_1 : x_1 + x_2 = 1 \\ a_2 : x_1 + x_3 = 1 \\ a_3 : x_1 = 0 \\ a_4 : x_2 + x_3 = 0 \end{cases}$$

Lemma 10. *Let $s \in T(\Sigma)$ and T a finite subset of $T(\Sigma)$. Then s is deducible with one application of a rule (GX) from T if and only if $D(T, s)$ is solvable over $\mathbb{Z}/2\mathbb{Z}$.*

Since satisfiability of a system of linear Diophantine equations over $\mathbb{Z}/2\mathbb{Z}$ is in PTIME [12], we obtain from Lemma 10, Theorem 1, and Lemma 9 that:

Theorem 2. *The question whether $T \vdash w$ is deducible form T in case of the rewrite system ACUNh is decidable in EXPTIME.*

In the binary case we obtain from Lemma 10, Theorem 1, Lemma 7, and Proposition 1 that:

Theorem 3. *The question whether $T \vdash w$ is deducible form T in case of the rewrite system ACUNh, where T and w are binary, is decidable in PTIME.*

7 Variants and Extensions

7.1 The Rewrite System *ACh*

The case of the rewrite system *ACh* is much simpler than the case *ACUNh* since with *ACh* it is not possible that terms are canceled out when applying the constructor $\otimes$. Hence we do not get the difficulty seen in Example 1.

Lemma 11. *The extended Dolev-Yao proof system in case of ACh is S_T-local.*

The downside is that, in order to decide one-step deducibility, we now have to solve linear Diophantine equation systems over $\mathbb{N}$. This problem is in general NP-complete [17]. Furthermore, it is quite easy to reduce satisfiability of linear Diophantine equations over $\mathbb{N}$ to the intruder deduction problem modulo *ACh*.

An exception is again the binary case, where one-step deducibility is decidable in polynomial time (which is trivial to prove in this case). We hence obtain:

Theorem 4. *The problem whether $T \vdash w$ in case of the rewrite system ACh is NP-complete, and decidable in PTIME if we restrict the problem to the binary case.*

7.2 The Rewrite System *AGh*

The case of the rewrite system *AGh* is very similar to the case of *ACUNh*. The lemmata and techniques can be adapted easily when we change the definitions of S_T, S_f, and $S_\otimes$ and require now in addition that they are closed under application of the inversion function and subsequent normalization of the term.

We can test one-step deducibility essentially as in Section 6. The major difference is that we now have to check our equation system $D(T, s)$ for satisfiability in $\mathbb{Z}$, which again is in PTIME [18].

Theorem 5. *The problem whether $T \vdash w$ in case of the rewrite system AGh is decidable in EXPTIME, and decidable in PTIME if we restrict the problem to the binary case.*

7.3   Extension to an Encryption Operation Which is Homomorphic over $\otimes$

This extension consists of replacing, in the three rewrite systems given at the end of Section 2, the rewrite rule

$$f(x \otimes y) \rightarrow f(x) \otimes f(y)$$

by the new rule

$$\{x \otimes y\}_z \rightarrow \{x\}_z \otimes \{y\}_z$$

On a technical level, this introduces the additional difficulty that we can now decompose in certain cases a sum built by $\otimes$, as for instance

$$\text{(D)} \quad \frac{T \vdash \{a\}_k \otimes \{b\}_k \otimes \{c\}_k \qquad T \vdash k}{T \vdash a \otimes b \otimes c}$$

However, we obtain for this extension lemmata and results which are analogous to the ones in the previous sections. The construction of the automaton for the binary case explained in Section 5 has now to be generalized since we now have an a priori infinite family of homomorphisms. In the case of Section 5 one counter was enough to count the number of applications of the homomorphic function f. In the extended case, we have to represent the sequence of encryption keys used in a stack of encryption operations, which can now be done with a pushdown automaton. We can find a lemma analogous to Lemma 6 also for the class of pushdown automata. The only remaining difficulty is to show that the stack alphabet, which consists of the encryption keys used in a minimal and $\otimes$-lazy proof, is finite. This is not obvious since we may use any term as an encryption key. However, we obtain easily by the Lemmata which correspond to Lemmata 2, 3, and 4 that:

Lemma 12. *Let P be a proof which is minimal among the $\otimes$-lazy proofs of $T \vdash w$. All the encryption keys used in the proof P are in $S_T(T \cup \{w\})$.*

As a consequence, the Theorems 2, 3, 4, and 5 still hold for this extension.

8   Conclusion

A summary of the results obtained on the complexity of the intruder deduction problem modulo AC-like equational theories with homomorphism is given in the following table. The results for homomorphism only (without AC axioms) have been shown in a different paper [7] and are here cited only for completeness.

Complexity of the intruder deduction problem

	Binary case	General case
h		*PTIME* [7]
ACh	*PTIME*	*NP*-Complete
ACUNh	*PTIME*	*EXPTIME*
AGh	*PTIME*	*EXPTIME*

The reason for the high complexity in the general case is a different one for the different equational theories considered, as shown in the following table:

Complexity in the general case

	Computation of subterms	One step deducibility	General deducibility
h	*PTIME* [7]	*PTIME* [7]	*PTIME* [7]
ACh	*PTIME*	*NP-Complete*	*NP-Complete*
ACUNh	*EXPTIME*	*PTIME*	*EXPTIME*
AGh	*EXPTIME*	*PTIME*	*EXPTIME*

As future work, we plan to investigate the case of an active intruder. We can yet observe that it has been shown in [8] that decidability of unification modulo an equational theory E is a necessary condition for the decidability of the security of a protocol for a bounded number of sessions and in presence of this equational theory E. Since unification modulo *AC* plus homomorphism is known undecidable [16], the security against active attackers is undecidable at least for this equational theory as well.

Acknowledgments

We are grateful for the numerous useful remarks and hints we obtained from our colleagues at LSV, in particular (in alphabetical order) Mathieu Baudet, Hubert Comon-Lundh, Stéphanie Delaune, Jean GoubaultLarrecq, Florent Jacquemard, Claudine Picaronny, and Philippe Schnoebelen. Philippe provided us with a first proof of Lemma 6.

References

[1] M. Abadi and A. D. Gordon. A calculus for cryptographic protocols: The spi calculus. *Information and Computation*, 148(1):1–70, Jan. 1999.

[2] M. Abadi and P. Rogaway. Reconciling two views of cryptography (the computational soundness of formal encryption). In *Proc. 1st IFIP International Conference on Theoretical Computer Science (IFIP–TCS)*, volume 1872 of *LNCS*, pages 3–22. Springer-Verlag, 2000.

[3] F. Baader and T. Nipkow. *Term Rewriting and All That*. Cambridge University Press, 1998.

[4] Y. Chevalier, R. Küsters, M. Rusinowitch, and M. Turuani. An NP decision procedure for protocol insecurity with XOR. In *Proc. of 18th Annual IEEE Symposium on Logic in Computer Science (LICS'03)*, pages 261–270, Ottawa (Canada), 2003. IEEE Comp. Soc. Press.

[5] J. Clark and J. Jacob. A survey of authentication protocol literature.
http://www.cs.york.ac.uk/~jac/papers/drareviewps.ps, 1997.

[6] H. Comon-Lundh and V. Shmatikov. Intruder deductions, constraint solving and insecurity decision in presence of exclusive or. In *Proc. of 18th Annual IEEE Symposium on Logic in Computer Science (LICS'03)*, pages 271–280, Ottawa (Canada), 2003. IEEE Comp. Soc. Press.

[7] H. Comon-Lundh and R. Treinen. Easy intruder deductions. In N. Dershowitz, editor, *Verification: Theory & Practice, Essays Dedicated to Zohar Manna on the Occasion of His 64th Birthday*, volume 2772 of *LNCS*, pages 225–242. Springer-Verlag, 2003.

[8] V. Cortier, S. Delaune, and P. Lafourcade. A survey of algebraic properties used in cryptographic protocols. Research Report LSV-04-15, LSV, ENS de Cachan, Sept. 2004. Available at http://www.lsv.ens-cachan.fr/Publis/RAPPORTS_LSV/rapports-year-2004-list.php.

[9] N. Dershowitz and J.-P. Jouannaud. Rewrite systems. In J. van Leeuwen, editor, *Handbook of Theoretical Computer Science*, volume B - Formal Models and Semantics, chapter 6, pages 243–320. Elsevier Science Publishers and The MIT Press, 1990.

[10] D. Dolev and A. Yao. On the security of public-key protocols. In *Transactions on Information Theory*, volume 29, pages 198–208. IEEE Comp. Soc. Press, 1983.

[11] F. Jacquemard. Security protocols open repository. Available at http://www.lsv.ens-cachan.fr/spore/index.html.

[12] E. Kaltofen, M. S. Krishnamoorthy, and B. D. Saunders. Fast parallel computation of hermite and smith forms of polynomial matrices. *SIAM J. Algebraic Discrete Methods*, 8(4):683–690, 1987.

[13] P. Lafourcade, D. Lugiez, and R. Treinen. Intruder deduction for ac-like equational theories with homomorphisms. Research Report LSV-04-16, LSV, ENS de Cachan, Nov. 2004. Available at http://www.lsv.ens-cachan.fr/Publis/RAPPORTS_LSV/rapports-year-2004-list.php.

[14] G. Lowe. An attack on the Needham-Schroeder public key authentication protocol. *Information Processing Letters*, 56(3):131–133, 1995.

[15] D. A. McAllester. Automatic recognition of tractability in inference relations. *JACM*, 40(2):284–303, April 1993.

[16] P. Narendran. Solving linear equations over polynomial semirings. In *Proc. of 11th Annual Symposium on Logic in Computer Science (LICS)*, pages 466–472, July 1996.

[17] C. H. Papadimitriou. *Computational Complexity*. Addison-Wesley, 1994.

[18] A. Schrijver. *Theory of Linear and Integer Programming*. Wiley, 1986.

[19] M. Turuani. Personal communication, 2003.

Proving Positive Almost-Sure Termination

Olivier Bournez and Florent Garnier

LORIA/INRIA, 615 Rue du Jardin Botanique,
54602 Villers lès Nancy Cedex, France

Abstract. In order to extend the modeling capabilities of rewriting systems, it is rather natural to consider that the firing of rules can be subject to some probabilistic laws. Considering rewrite rules subject to probabilities leads to numerous questions about the underlying notions and results.
We focus here on the problem of termination of a set of probabilistic rewrite rules. A probabilistic rewrite system is said almost surely terminating if the probability that a derivation leads to a normal form is one. Such a system is said positively almost surely terminating if furthermore the mean length of a derivation is finite. We provide several results and techniques in order to prove positive almost sure termination of a given set of probabilistic rewrite rules. All these techniques subsume classical ones for non-probabilistic systems.

1 Introduction

Since 30 years, term rewriting has shown to be a very powerful tool in several contexts where efficient methods for reasoning with equations are required [1, 13]. In the last decade, term rewriting has also shown to provide a very elegant framework for specifying concurrency models and deduction systems [16, 17].

When specifying probabilistic systems, it is rather natural to consider that the firing of a rewrite rule can be subject to some probabilistic rules. For that purpose, we proposed in [4] to add basic probabilistic strategies to rule based languages. The idea of adding probabilities to rewrite rules has also been explored in [9] in the context of probabilistic constraint handling rules, or in [18]. The idea of adding probabilities to high level models of reactive systems has also been explored for models like Petri Nets [2, 22], automata based models [6, 26], or process algebra [11].

Considering rewrite rules subject to probabilities leads to numerous questions about the underlying notions and results. In [4], we introduced probabilistic abstract reduction systems, and we introduced notions like almost-sure termination or probabilistic confluence, with relations between all these notions. In [3], we proved that, unlike what happens for classical rewriting logic, there is no hope to build a sound and complete proof system with probabilities in the general case. We however proposed a rather natural notion of rewriting logic which is sound and complete when proof terms are explicit [3].

This paper is a contribution devoted to a next step: understand and provide proof techniques for proving termination of a set of probabilistic rewrite rules.

J. Giesl (Ed.): RTA 2005, LNCS 3467, pp. 323–337, 2005.

As in [4], we propose to call a deterministic probabilistic rewrite system almost surely terminating if the probability that a term leads to a normal form is one. However, unlike in [4], we also allow non-deterministic systems. A non-deterministic probabilistic rewrite system is said almost surely terminating if the probability that a term leads to a normal form is one whatever the reduction strategy is.

The idea of mixing probabilities with non-determinism in several other high level models for reactive systems has quite extensively been discussed in literature. To solve semantical problems, discussed for example in [19] or [12], several approaches exist. One of them, called the generative approach [25], consists in ruling out non-determinism by means of a probability distribution that assigns a probability to each possible action. The reactive approach [25] consists in allowing both non-deterministic and probabilistic transitions. The present non-determinism is solved using the notion of schedulers [26]. Note that there exist intermediate approaches such that stratified approaches [25] or simple or fully probabilistic transition systems [24] that are variants or combinations of these two approaches. Our approach is close to the reactive approach, and what we call a probabilistic abstract reduction system is also called a Markov decision process in other contexts [20].

Termination is a desirable interesting notion. However, in the probabilistic context, we think we should distinguish "reasonable" termination from general termination.

Indeed, consider a system like a symmetric random walk on the set $\mathbb{Z}^k$ of integers. For $k = 1$ or 2, it visits almost surely all the points [5, 7]. Hence, whatever the current position is, if one wants to go to the origin, a strategy is to evolve like a symmetric random walk and stop at the origin. However, even if one is almost sure to reach the origin, the expected time before reaching the origin is infinite [5, 7].

Coming back to termination, the point is that in an almost surely terminating system, with probability one a term leads to a normal form, but if the mean number of a derivation is infinite then this information is rather useless.

Hence, we believe that the following notion is more interesting: a system will be said positively almost surely terminating if the mean length of a derivation is finite. After formally introducing all these notions, we will see that positive almost sure termination implies almost sure termination. The rest of this paper is then devoted to proof techniques that can be used to prove positive almost sure termination.

In particular, in the classical non-probabilistic case, a simple and often used criteria for proving termination consists in embedding the underlying abstract reduction system into the set of natural integers, in such a way that each transition corresponds to a decreasing transition. This technique is sound in the general case, and is complete for finitely branching systems [1].

We show that this technique has an equivalent for probabilistic abstract reduction systems: we prove that a probabilistic abstract reduction system is positively almost sure terminating if it can be embedded into the set of non-negative reals in such a way that each transition corresponds to a decreasing transition in mean. The technique is proved sound in the general case, and complete for finitely branching systems.

Benefiting from the possibility of considering non-deterministic probabilistic abstract reduction systems, we then define probabilistic rewrite systems. The idea is to allow in right hand sides of probabilistic rules a distribution on classical right hand sides of classical rewrite rules. The proposed notions are intended to subsume classical rewrite systems. In that spirit, they seem rather natural (at least for the rewrite community) and probabilistic rewrite systems provide an alternative to the numerous probabilistic high level formalisms for specifying reactive systems.

We then discuss the equivalent of the classical result that says that a rewrite system is terminating iff there is a reduction order monotone on each rewrite rule.

The paper is organized as follows: in Section 2, we recall classical non-probabilistic theory. Sections 3 and 4 recall basic probability and Markov chain theory, and Foster's theorem respectively. Section 5 introduces probabilistic abstract reduction systems. Section 6 defines positive almost sure termination. Section 7 provides techniques for proving positive almost sure termination of a probabilistic abstract reduction system. Probabilistic rewrite systems are introduced in Section 8. Techniques for proving their positive almost sure termination are discussed in Section 9.

2   Termination and Abstract Reduction Systems

We first come back to the classical setting: see for example [1, 13]. An *abstract reduction system (ARS)* is $\mathcal{A} = (A, \rightarrow)$ consisting of a set A and a binary relation $\rightarrow \subset A \times A$ on A. A *derivation* is a finite, or infinite sequence $\pi = \pi_0 \rightarrow \pi_1 \cdots \rightarrow \pi_n$ with $(\pi_i, \pi_{i+1}) \in \rightarrow$ for all i. An abstract reduction system is said *terminating* iff there is no infinite chain $a_0 \rightarrow a_1 \rightarrow \cdots$.

As said in [1], the most basic method for proving termination of some $\mathcal{A} = (A, \rightarrow)$ is to embed it into another abstract reduction system $\mathcal{B} = (B, >)$ which is known to terminate. This require a monotone mapping $V : A \rightarrow B$, where monotone means that $x \rightarrow x'$ implies $V(x) > V(x')$. Now $\rightarrow$ terminates because an infinite chain

$$a_0 \rightarrow a_1 \rightarrow \cdots$$

would induce an infinite chain

$$V(x_0) > V(x_1) > \ldots$$

The most popular choice for termination proofs is an embedding into $(\mathbb{N}, >)$. Its popularity comes partly from the following easy completeness result [1].

Proposition 1. *A finitely branching abstract reduction system terminates if and only if there is a monotone embedding into $(\mathbb{N}, >)$.*

As in [1], observe that the technique is sound in the general case, but complete only for finitely branching systems. Indeed, the system with $A = \mathbb{N}^2$ and $\rightarrow$ defined by $(i + 1, j) \rightarrow (i, k)$, $(i, j + 1) \rightarrow (i, j)$, for all i, j, k, is terminating, whereas there is no monotone embedding from $(\mathbb{N}^2, \rightarrow)$ to $(\mathbb{N}, >)$ [1].

3   Stochastic Sequences and Markov Chains

Let us first come back to school [7, 10, 21]: a *σ-algebra* on a set Ω is a set of subsets of Ω which contains the empty-set, and is stable by countable union and complementation. In particular, the set of subsets is a natural σ-algebra for any countable set. A *measurable space* (Ω, σ) is a set with a σ-algebra on it. A *probability* is a function P from a σ-algebra to $[0, 1]$, which is countably additive, and such that $P(\Omega) = 1$. A triplet (Ω, σ, P) is called *a probability space*.

If (Ω, σ) and (Ω', σ') are measurable spaces, a function $f : \Omega \to \Omega'$ is *measurable* if for all W in σ', $f^{-1}(W) \in \sigma$. A *random variable* is a measurable function on some probability space. The *mean* of a random variable V taking values in the set $\mathbb{N}$ of integers is $E[V] = \sum_i iP(V = i)$. This value is always defined, even if it can be finite or infinite. Observe that such a random variable always satisfy the so-called *telescope formula* $E[X] = \sum_{n=0}^{\infty} P(X > n)$ [5]. For a random variable V taking values in $\mathbb{N} \cup \{+\infty\}$, the mean $E[V]$ can still always be defined: practically, it is infinite if $P(V = +\infty) > 0$ and equal to $E[V] = \sum_i iP(V = i)$ (which may still be infinite) otherwise.

Given $A, B \in \sigma$, when $P(B) > 0$, *the conditional probability of A given B* is by definition $P(A|B) = P(A \cap B)/P(B)$. The mean of random variable $V : \Omega \to \mathbb{N}$ conditioned by B is defined by $E[V|B] = \sum_i iP(V = i|B)$.

A *stochastic sequence on a set A* is a family $(X_i)_{i \in \mathbb{N}}$, of random variables defined on some fixed probability space (Ω, σ, P) with values on A. It is said to be *Markovian* if its conditional distribution function satisfies the so-called Markov property, that is for all n and $s \in A$,

$$P(X_n = s|X_0 = \pi_0, X_1 = \pi_1, \ldots, X_{n-1} = \pi_{n-1}) = P(X_n = s|X_{n-1} = \pi_{n-1}),$$

and *homogeneous* if furthermore this probability is independent of n.

The matrix $(p_{s,t}^i) = (P(X_{i+1} = t|X_i = s))$ is what is called a stochastic matrix (even when A is an infinite set) [5]. It has the nice property that columns sum to 1.

Giving a *Markov Chain* is of course equivalent to giving the sequence of its stochastic matrices. Given a *Homogeneous Markov Chain* corresponds to giving a unique stochastic matrix (at any rank, the matrix is the same).

4   Foster's Theorem

We are searching criteria in the spirit of Proposition 1. For that purpose, we now state the following result, that can be attributed to Foster [8]. It has strong connections with Martingale theory and can be seen as a consequence of very general results of (super) Martingale theory. However, it can be proved independently as in [5].

Theorem 1 (Foster's Theorem). *Given a homogeneous Markov chain over a countable space A whose matrix is $P = (p_{t,s})$, if there exists a measurable subset $C \subset A$, and some function $V : A \to \mathbb{R}$, with $\inf_{i \in A} V(i) > -\infty$ and such that the mean drift defined by $\Delta V(i) = \sum_{k \in A} p_{i,k}V(k) - V(i)$ satisfies for some $\epsilon > 0$ $\Delta V(i) \leq -\epsilon$ for all $i \notin C$, then almost surely one reaches C.*

Furthermore, the mean time to reach C from i is finite and less than $V(i)/\epsilon$.

Notice that the technique of using Foster's theorem in order to prove convergence to some set C has similarities with techniques used in self-stabilization as in [15, 23].

5   Probabilistic Abstract Reduction Systems

We are now ready to define probabilistic abstract reduction systems (PARS). We define PARS in a slightly modified way to [4]. The main motivation is that we want to allow non-deterministic systems.

In the same way that abstract reduction systems are also called *transition systems* in other contexts, PARS can be considered as *Markov Decision Processes* [20]. The only point is that, compared to usual definitions of Markov decision processes, we explicitly allow states to be terminal, and that we do not label transitions by actions.

The idea is that a PARS is given by some set A, and a relation that relate states to distributions on their successors.

Definition 1 (PARS). *Given some denumerable set S, we note $Dist(S)$ for the set of probability distributions on S: $\mu \in Dist(S)$ is a function $S \to [0,1]$ that satisfies $\sum_{i \in S} \mu(i) = 1$.*

A probabilistic abstract reduction system (PARS) is a pair $\mathcal{A} = (A, \to)$ consisting of a countable set A and a relation $\to \subset A \times Dist(A)$.

A PARS is said deterministic if, for all a, there is at most one μ with $a \to \mu$.

A state $a \in A$ with no μ such that $a \to \mu$ is said terminal.

We now need to explain how such systems evolve: a *history* (of length $n + 1$) is a finite sequence $a_0 a_1 \cdots a_n$ of elements of the state space A. It is non-terminal if a_n is. A *policy* ϕ, that can also be called a *strategy*, is a function that maps non-terminal histories to distributions in such a way that $\phi(a_0 a_1 \cdots a_n) = \mu$ is always one (of the possibly many) distribution μ with $a_n \to \mu$. A history is said *realizable*, if for all $i < n$, if μ_i denotes $\phi(a_0 a_1 \cdots a_i)$, one has $\mu_i(a_{i+1}) > 0$.

A *derivation of $\mathcal{A}$* is then a stochastic sequence where the non-deterministic choices are given by some policy ϕ, and the probabilistic choices are governed by the corresponding distributions.

Formally:

Definition 2 (Derivations). *A derivation π of $\mathcal{A}$ over policy ϕ is a stochastic sequence $\pi = (\pi_i)_{i \in \mathbb{N}}$ on $A \cup \{\bot\}$ such that for all n,*

$$P(\pi_{n+1} = \bot | \pi_n = \bot) = 1,$$

$$P(\pi_{n+1} = \bot | \pi_n = s) = 1 \text{ if } s \in A \text{ is terminal,}$$

$$P(\pi_{n+1} = \bot | \pi_n = s) = 0 \text{ if } s \in A \text{ is non-terminal,}$$

and for all $t \in A$.

$$P(\pi_{n+1} = t | \pi_n = a_n, \pi_{n-1} = a_{n-1}, \ldots, \pi_0 = a_0) = \mu(t)$$

whenever $a_0 a_1 \cdots a_n$ is a realizable non-terminal history and $\mu = \phi(a_0 a_1 \cdots a_n)$.

Several observations are in order.

Remark 1. Deterministic probabilistic abstract reduction systems correspond to probabilistic abstract reduction systems considered in [4].

Remark 2. The derivations are homogeneous and Markovian when the policy ϕ is *Markovian*, i.e. when the value of $\phi(a_0 a_1 \dots a_n)$ depends only on the value of a_n. In particular, this holds for deterministic systems.

6   Termination of a Probabilistic Abstract Reduction System

If a derivation is such that $\pi_n = \perp$ for some n, then $\pi_{n'} = \perp$ almost surely for all $n' \geq n$. Such a derivation is said to be *terminating*. In other words, a non-terminating derivation is such that $\pi_n \in A$ ($\pi_n \neq \perp$) for all n.

Definition 3 (Almost Sure Termination). *A PARS $\mathcal{A} = (A, \to)$ will be said almost surely (a.s) terminating iff for any policy ϕ, the probability that a derivation $\pi = (\pi_i)_{i \in \mathbb{N}}$ under policy ϕ terminates is 1: i.e. for all ϕ, $P(\exists n | \pi_n = \perp) = 1$.*

This can be restated as follows: given some policy ϕ, and some state a, consider the random variable $\tau[a, \phi]$ associated to a derivation π with $\pi_0 = a$, taking values in $\mathbb{N} \cup \{+\infty\}$, defined as $+\infty$ if $\pi_n \neq \perp$ for all n, and defined as $\tau[a, \phi] = \min\{n | \pi_n = \perp\}$ otherwise. Of course, $\tau[a, \phi]$ corresponds to the number of derivations from a under strategy ϕ before termination. $\tau[a, \phi]$ is easily proved to be a stopping time for all ϕ and a.

Previous definitions can then be stated as follows:

Proposition 2. *A PARS $\mathcal{A}$ is almost surely terminating iff for all strategies ϕ and all states a, $P(\tau[a, \phi] = +\infty) = 0$.*

As discussed in the introduction, this notion of termination is too weak. Even if $P(\tau[a, \phi] = \infty) = 0$, it might happen that the mean time before termination

$$T[a, \phi] = E[\tau(a, \phi)]$$

is not finite, and one may expect never to reach a terminal state.

That is why, we suggest to introduce the notion of positive almost sure termination. Note that the choice of the name "positive" is inspired by the distinction between positive recurrence and null recurrence in Markov chains theory [5].

Definition 4 (Positive Almost Sure Termination). *A PARS $\mathcal{A} = (A, \to)$ will be said positively almost surely (+a.s.) terminating if for all policies ϕ, for all states $a \in A$, $T[a, \phi]$ is finite.*

By the discussion in Section 3 on random variables taking values in $\mathbb{N} \cup \{+\infty\}$, we know that if $P(\tau(a, \phi) = \infty) > 0$ then necessarily $E[\tau(a, \phi)]$ is infinite. That means:

Proposition 3. *A positively almost surely terminating PARS is almost surely terminating.*

Remark 3. The previous notions subsume classical ones. As one may expect, non-probabilistic systems are special cases of probabilistic systems: an abstract reduction system is a probabilistic abstract reduction system where all the distributions are Dirac distributions. I.e. all the distributions μ have value 1 on a single point, and value 0 everywhere else. Strategies for abstract reduction systems do indeed correspond to strategies for corresponding probabilistic abstract reduction systems. Terminating derivations for abstract reduction systems do indeed correspond to terminating derivations for corresponding probabilistic abstract reduction systems. An abstract reduction system is terminating iff the corresponding probabilistic abstract reduction system is ((positively) almost surely) terminating. Note that positive almost sure termination corresponds to almost sure termination and to termination for those systems.

7   Proving Positive Almost Sure Termination

We are now going to discuss techniques for proving positive almost sure termination of a probabilistic abstract reduction system. We propose a technique that subsumes the technique of Proposition 1.

One must understand that it is not at all a coincidence, but more or less unavoidable: a deep consequence of remark 3 is that any technique for proving positive almost surely termination of probabilistic abstract reduction systems must also work for abstract reduction systems, and hence necessarily subsumes a technique for non-probabilistic abstract reduction systems.

First, we prove soundness of our technique

Theorem 2 (Soundness). *A PARS $\mathcal{A} = (A, \rightarrow)$ is +a.s. terminating if there exist some function $V : A \rightarrow \mathbb{R}$, with $\inf_{i \in A} V(i) > -\infty$, and some $\epsilon > 0$, such that, for all states $a \in A$, for all μ with $a \rightarrow \mu$, the drift in a according to μ defined by*

$$\Delta_\mu V(a) = \sum_i \mu(i) V(i) - V(a)$$

satisfies

$$\Delta_\mu V(a) \leq -\epsilon.$$

Proof. We would like to use Theorem 1. However, we can not work directly on the PARS, since even if we fix a strategy, a PARS is not necessarily an homogeneous Markov chain (the fixed policy can be non-Markovian).

The solution is to fix a policy ϕ and to work on an homogeneous Markov chain $\mathcal{M}_\phi$ defined on another state space: the state space of $\mathcal{M}_\phi$ is defined as the set of all realizable histories of $\mathcal{A}$.

The matrix of Markov chain $\mathcal{M}_\phi$ is then defined such that

- for all t, $p_{h,ht} = \mu(t)$ where $\mu = \phi(h)$ if $h = a_0 a_1 \cdots a_n$ is a realizable non-terminal history, where ht stands for history $a_0 a_1 \cdots a_n t$,
- $p_{h,h} = 1$ if h is a realizable terminal history,
- and every other entry of the matrix is 0.

By construction, $\mathcal{M}_\phi$ is an homogeneous Markov chain. Now clearly, a trajectory of PARS $\mathcal{A}$ starting from a reaches a terminal state under policy ϕ iff the corresponding trajectory of $\mathcal{M}_\phi$ of same length starting from a leads to a terminal history. Furthermore, the probabilities of corresponding derivations are preserved.

Consider now function $W : S_\phi \to \mathbb{R}$ defined by

$$W(a_0 a_1 \cdots a_n) = V(a_n)$$

for all realizable histories $a_0 a_1 \cdots a_n$.

We have

$$\Delta W(h) = \Delta_\mu V(h) \le -\epsilon$$

for any non-terminal realizable history h, where $\mu = \phi(h)$.

We can then apply Theorem 1 on $\mathcal{M}_\phi$, with C equal to the set of terminal realizable histories to conclude that the derivations starting from a in $\mathcal{M}_\phi$ reach terminal realizable histories in a time whose mean is less than $W(a)/\epsilon = V(a)/\epsilon$.

Hence, all the derivations starting from a in $\mathcal{A}$ under policy ϕ reach terminal states in a time whose mean is also less than $V(a)/\epsilon$. This holds for all a and ϕ.

We now prove that the technique is complete for finitely branching systems.

Definition 5. *A probabilistic abstract reduction system $\mathcal{A} = (A, \to)$ is finitely branching if for all a, there is at most a finite number of distributions μ with $a \to \mu$.*

Theorem 3 (Completeness for finitely branching systems). *If a finitely branching probabilistic abstract reduction system $\mathcal{A} = (A, \to)$ is $+a.s.$ terminating then there exist some function $V : A \to \mathbb{R}$, with $\inf_{i \in A} V(i) > -\infty$, and some $\epsilon > 0$, such that, for all states $a \in A$, for all μ with $a \to \mu$, the drift in a according to μ defined by*

$$\Delta_\mu V(a) = \sum_i \mu(i) V(i) - V(a)$$

satisfies

$$\Delta_\mu V(a) \le -\epsilon.$$

Proof. By hypothesis, for all states a, and policy ϕ, we have $T[a, \phi] < +\infty$. When h is a realizable history, and ϕ is a policy, we write $T[h, \phi]$ for the mean time before reaching $\perp$ after history h.

Note that for any policy ϕ, when h is a realizable non-terminal history, we have

$$T[h, \phi] = 1 + \sum_{x \in A} \phi(h)(x) T[hx, \phi] \qquad (1)$$

If policy ϕ is Markovian, we have $T[hx, \phi] = T[x, \phi]$, and hence

$$T[h, \phi] = 1 + \sum_{x \in A} \phi(h)(x) T[x, \phi]. \qquad (2)$$

The idea is to consider the "worst" strategy Φ. This strategy can be built as follows: in any realizable non-terminal history $h = a_0 \ldots a_n$, Φ maps h to the distribution μ with $a_n \to \mu$ that maximizes $\sup_\phi \sum_{x \in A} \mu(x)T[hx, \phi]$.

Since to any strategy ϕ on can associate a strategy ϕ' with

$$T[hx, \phi] = T[x, \phi']$$

(take $\phi'(h') = \phi(hh')$ for any realizable non-terminal history h'),

$$\sup_\phi \sum_{x \in A} \mu(x)T[hx, \phi] = \sup_\phi \sum_{x \in A} \mu(x)T[x, \phi],$$

and hence Φ is Markovian.

We claim that this is indeed the worst strategy, i.e.

$$\sup_\phi T[h, \phi] \leq T[h, \Phi] \tag{3}$$

for all realizable non-terminal histories h.

This follows from the following arguments: for any integer i, let Φ_i be the set of strategies that coincide with Φ on all histories of length less than i. Using repeatedly Equation 1, one gets for all integers i,

$$\sup_\phi T[h, \phi] \leq \sup_{\phi \in \Phi_i} T[h, \phi]$$

for all realizable non-terminal histories h of length less than i.

Now, since $T[h, \Phi]$ is the limit of $\sup_{\phi \in \Phi_i} T[h, \phi]$ when i goes to infinity, Equation 3 holds.

Now, in any non-terminal a, with $a \to \mu$,

$$\begin{aligned}
\sum_{x \in A} \mu(x)T[x, \Phi] &\leq \sup_\phi \sum_{x \in A} \mu(x)T[x, \phi] \\
&\leq \sup_\phi \sum_{x \in A} \Phi(a)(x)T[x, \phi] \\
&\leq \sum_{x \in A} \Phi(a)(x)T[x, \Phi].
\end{aligned} \tag{4}$$

where first inequality comes from the fact that Φ is a particular strategy, the second from the definition of $\Phi(a)(x)$, and the third from the fact that the sup of a sum is always less that the sum of the sups.

We are done: indeed, if we take $V(a) = T[a, \Phi]$ for all states a, and $\epsilon = 1$, we know that V is non-negative, and for any μ with $a \to \mu$, we have

$$\begin{aligned}
\Delta_\mu V(a) &= \sum_{x \in A} \mu(x)V(x) - V(a) \\
&= \sum_{x \in A} \mu(x)T[x, \Phi] - T[a, \Phi] \\
&= -1 + (\sum_{x \in A} \mu(x)T[x, \Phi] - \sum_{x \in A} \Phi(a)(x)T[x, \Phi]) \\
&\leq -1
\end{aligned}$$

where third equality comes from Equation 2, and last inequality from Equation 4.

Remark 4. Note that the restriction to finitely branching systems in the previous theorem is mandatory: this can be seen as a consequence of Remark 3. Indeed, consider the counter-example after Proposition 1, considered as a probabilistic abstract reduction system. If there were a function V and some $\epsilon > 0$ as in the conclusion of previous theorem, adding a constant if necessary, and multiplying by $1/\epsilon$ if necessary, we can assume V non-negative, and $\epsilon = 1$. Now, in any non-terminal state a with $a \to \mu$, since we should have $\Delta_\mu V(a) \le -1$, and since μ is a Dirac that is 0 except on some point x where it has value 1, we must have for that x, $V(x) \le V(a) - 1$. Now, if $k = V(1,1)$, consider the strategy going from $(1,1)$ to $(0,k)$, $(0,k-1)$, ..., $(0,0)$. V must decrease of at least 1 at each transition. That leads to a contradiction, since starting from k, one can not do it $k+1$ times keeping V non-negative.

8 Probabilistic Rewrite Systems

We are now introducing the notion of probabilistic rewrite system. Our motivation is to get something that covers classical (i.e. non-probabilistic) rewrite systems, and also Markov chains over finite spaces. Doing so, we can claim that all examples that have been modeled in literature using finite Markov chains (for e.g. in model-checking contexts [14]) can be modeled in this framework.

Definition 6 (Probabilistic Rewrite system). *Given a signature Σ and a set of variables X, the set of terms over Σ and X is denoted by $T(\Sigma, X)$.*

A probabilistic rewrite rule is an element of $T(\Sigma, X) \times Dist(T(\Sigma, X))$. A probabilistic rewrite system is a finite set $\mathcal{R}$ of probabilistic rewrite rules.

To a probabilistic rewrite system is associated a probabilistic abstract reduction system $(T(\Sigma, X), \to_\mathcal{R})$ over the set of terms $T(\Sigma, X)$ where $\to_\mathcal{R}$ is defined as follows: When $t \in T(\Sigma, X)$ is a term, let $Pos(t)$ be the set of its positions. For $\rho \in Pos(t)$, let $t|_\rho$ be the subterm of t at position ρ, and let $t[s]_\rho$ denote the replacement of the subterm at position ρ in t by s. The set of all substitutions is denoted by Sub.

Definition 7 (Reduction relation). *To a probabilistic rewrite system $\mathcal{R}$ is associated the following PARS $(T(\Sigma, X), \to)$ over terms:*

$$t \to_\mathcal{R} \mu$$

iff there is a rule $(g, M) \in \mathcal{R}$, some position $p \in Pos(t)$, some substitution $\sigma \in Sub$, such that $t|_p = \sigma(g)$, and, for all t',

$$\mu(t') = \sum_{t' = t[\sigma(d)]_p} M(d).$$

For example, a probabilistic rewrite rule can be $f(x, y) \mapsto g(a) : 1/2 | y : 1/2$, where $g(a) : 1/2 | y : 1/2$ denotes the distribution with value $1/2$ on $g(a)$ and value $1/2$ on y. Then $f(b, c)$ rewrites to $g(a)$ with probability $1/2$, and to c with probability $1/2$. Now, $f(b, g(a))$ rewrites to $g(a)$ with probability 1.

9   Termination of a Probabilistic Rewrite System

We now provide an equivalent of the result that says that a rewrite system is terminating iff there is a reduction order monotone on each rewrite rule [1, 13].

Theorem 4. *A probabilistic rewrite system $\mathcal{R}$ is positively almost surely terminating if and only if there exists some function $V : T(\Sigma, X) \to \mathbb{R}$, with $\inf_{i \in A} V(i) > -\infty$, and some $\epsilon > 0$, such that*

1. *"the drift of each rule is less than $-\epsilon$": for each probabilistic rewrite rule $g \to M \in \mathcal{R}$, the drift*

$$\Delta_M V(g) = \sum_d M(d)(V(d) - V(g))$$

   satisfies

$$\Delta_M V(g) \leq -\epsilon.$$

2. *"drift being less than $-\epsilon$ is preserved by substitutions":*
   for each term $s \in T(\Sigma, X)$, for all μ with $s \to \mu$, for all substitutions $\sigma \in Sub$, if $\Delta_\mu V(s) \leq -\epsilon$ then the drift

$$\Delta_{\sigma(\mu)} V(\sigma(s)) = \sum_{s'} \mu(s')(V(\sigma(s')) - V(\sigma(s)))$$

   satisfies

$$\Delta_{\sigma(\mu)} V(\sigma(s)) \leq -\epsilon$$

3. *"drift being less than $-\epsilon$ is preserved by contexts": for each term $s_1, ..., s_n, s \in T(\Sigma, X)$, for all μ with $s \to \mu$, for all function symbols f, if $\Delta_\mu V(s) \leq -\epsilon$, then the drift*

$$\Delta_{f(s_1,...,\mu,...,s_n)} V(f(s_1, \ldots, s, \ldots, s_n)) = \sum_{s'} \mu(s')(V(f(s_1, \ldots, s', \ldots, s_n)) - V(f(s_1, \ldots, s, \ldots, s_n)))$$

   satisfies

$$\Delta_{f(s_1,...,\mu,...,s_n)} V(f(s_1, \ldots, s, \ldots, s_n)) \leq -\epsilon.$$

Proof. If $\mathcal{R}$ is positively almost surely terminating, then by Theorem 3, there exists some function $V : T(\Sigma, X) \to \mathbb{R}$, with $\inf_{i \in A} V(i) > -\infty$, and some $\epsilon > 0$, such that, for all states $a \in T(\Sigma, X)$, for all μ with $a \to \mu$, $\Delta_\mu V(a) \leq -\epsilon$.

In particular, for $a = g$, we have $a \to \mu$, where $\mu(t') = M(t')$, and hence

$$\begin{aligned}
\Delta_\mu V(a) &= \sum_{t'} \mu(t')V(t') - V(a) \\
&= \sum_d M(d)(V(\sigma(d)) - V(a)) \\
&= \Delta_M V(g) \\
&\leq -\epsilon.
\end{aligned}$$

Now, when $s \to \mu'$, for $a = \sigma(s)$, we have $a \to \mu$, where $\mu(\sigma(s')) = \mu'(s')$, and hence

$$
\begin{aligned}
\Delta_\mu V(a) &= \sum_{t'} \mu(t')V(t') - V(a) \\
&= \sum_{t'=\sigma(s')} \mu'(s')V(\sigma(s')) - V(a) \\
&= \Delta_{\sigma(\mu')}V(\sigma(s)) \\
&\le -\epsilon.
\end{aligned}
$$

In a same way, when $s \to \mu'$, for $a = f(s_1, \ldots, s, \ldots, s_n)$, we have $a \to \mu$ where $\mu(f(s_1, \ldots, s', \ldots, s_n)) = \mu'(s')$, and hence

$$
\begin{aligned}
\Delta_\mu V(a) &= \sum_{t'} \mu(t')V(t') - V(a) \\
&= \sum_{s'} \mu'(s')V(f(s_1, \ldots, s', \ldots, s_n)) - V(f(s_1, \ldots, s, \ldots, s_n)) \\
&= \Delta_{f(s_1,\ldots,\mu',\ldots,s_n)}V(f(s_1, \ldots, s, \ldots, s_n)) \\
&\le -\epsilon.
\end{aligned}
$$

This proves that conditions 1, 2 and 3 are necessary.

Conversely, assume that conditions 1, 2 and 3, hold. We have $t \to \mu$ iff there is a rule $(g, M) \in \mathcal{R}$, some position $p \in Pos(t)$, some substitution $\sigma \in Sub$, such that $t|_p = \sigma(g)$, and, for all t', $\mu(t') = \sum_{t'=t[\sigma(d)]_p} M(d)$.

Since a derivation $t \to \mu$ is necessarily via some rule (g, M), from Theorem 2, we only need to prove that for all rules (g, M) and term t, if $t \to \mu$ via (g, M) then $\Delta_\mu V(t) \le -\epsilon$.

This is proved by induction on the length of p. If p is of length 0, then $t = \sigma(g)$. By condition 1, we know that $\Delta_M V(g) \le -\epsilon$. By condition 2, since $g \to M$, and $\Delta_M V(g) \le -\epsilon$, we have $\Delta_\mu V(t) = \Delta_{\sigma(M)}V(\sigma(g)) \le -\epsilon$, where the equality is established as in the third paragraph above.

If $p = p_1 p_2 \ldots p_k$ is of length $k > 0$, then t can be written as $f(s_1, \ldots, s, \ldots, s_n)$ and $s \to \mu'$ via (g, M). By induction hypothesis, $\Delta_{\mu'}V(s) \le -\epsilon$. By condition 3, $\Delta_\mu V(t) = \Delta_{f(s_1,\ldots,\mu',\ldots,s_n)}V(f(s_1, \ldots, s, \ldots, s_n)) \le -\epsilon$, where the equality is established as in the fourth paragraph above.

Sufficient conditions for 1, 2 and 3 can be established. Indeed:

Definition 8 (Context preservation of a function). *A function* $V : T(\Sigma, X) \to \mathbb{R}$ *is context preserving if for all* $t, t', s_1, \ldots, s_n$ *and function symbol* f,

$$
V(f(s_1, \ldots, t, \ldots, s_n)) - V(f(s_1, \ldots, t', \ldots, s_n)) = V(t) - V(t').
$$

Definition 9 (Substitution decrease on a rule). *A function* $V : T(\Sigma, X) \to \mathbb{R}$ *is substitution decreasing on a probabilistic rewrite rule* (g, M), *if for all substitution* $\sigma \in Sub$, *if we denote*

$$
\Delta_{\sigma(M)}V(\sigma(g)) = \sum_d M(d)(V(\sigma(d)) - V(\sigma(g)))
$$

and $\Delta_M V(g) = \sum_d M(d)(V(d) - V(g))$ *as before, we have*

$$
\Delta_{\sigma(M)}V(\sigma(g)) \le \Delta_M V(g).
$$

Theorem 5. *A probabilistic rewrite system $\mathcal{R}$ is positively almost surely terminating if there exists some function $V : T(\Sigma, X) \to \mathbb{R}$, with $\inf_{i \in A} V(i) > -\infty$, and some $\epsilon > 0$, such that the drift of each rule is less or equal to $-\epsilon$, V is context preserving, and V is substitution decreasing on every rule.*

Proof. Condition 1 holds by hypothesis.

Since V is context preserving, for all $f, s, s_1, \ldots, s_n$ and μ, we have

$$\Delta_{f(s_1,\ldots,\mu,\ldots,s_n)} V(f(s_1,\ldots,s,\ldots,s_n)) = \Delta_\mu V(s)$$

and so, condition 3 holds.

Now, given conditions 1 and 3, the proof of indirect sense of Theorem 4, only require that $\Delta_{\sigma(M)} V(\sigma(g)) = \sum_d M(d)(V(\sigma(d)) - V(\sigma(g))) \leq -\epsilon$ for each probabilistic rule (g, M) and substitution σ. Now, this holds, since V is substitution decreasing and so $\Delta_{\sigma(M)} V(\sigma(g)) \leq \Delta_M V(g) \leq -\epsilon$, by condition 1.

We are now discussing some examples:

Example 1. The probabilistic rewrite system restricted to the unique rule

$$a \to a : 1/2 | b : 1/2$$

is +a.s. terminating. Indeed, consider $V(a) = 10$, $V(b) = 2$, and observe that $1/2 \times 10 + 1/2 \times 2 - 10 < 0$.

Example 2. The probabilistic rewrite system

$$f(x) \to x : p_1 | f(f(x)) : 1 - p_1$$
$$f(x) \to x : p_2 | f(f(x)) : 1 - p_2$$

is +a.s. terminating if $p_1 > 1/2$ and $p_2 > 1/2$. Indeed, consider V that returns the size of a term. V is easily shown context preserving. V is also easily shown substitution decreasing on both rules. Now the drift of each rule is given by $-1 \times p_i + 1 \times (1 - p_i) = 1 - 2p_i \leq \min(1 - 2p_1, 1 - 2p_2) < 0$.

Example 3. The probabilistic rewrite system

$$f(x) \qquad \to f(f(x)) : p_{1_1} | g(f(x)) : p_{1_2} | x : p_{1_3}$$
$$f(h(f(x), x)) \to h(g(f(f(x))), f(x)) : p_{2_1} | g(f(x)) : p_{2_2} | f(g(f(f(x)))) : p_{2_3}$$

It is +a.s. terminating if $p_{1_1} + p_{1_3} < p_{1_3}$ and $p_{2_1} < p_{2_2}$. Indeed, consider same function V, which is clearly context preserving. An easy computation shows that the drift of the first rule (g_1, M_1) is $\Delta_{M_1} V(g_1) = p_{1_1} + p_{1_2} - p_{1_3}$. For the second rule (g_2, M_2), we have $\Delta_{M_2} V(g_2) = 2p_{2_1} - 2p_{2_2}$. Hence, both are negative. Now V is substitution decreasing on both rules: given some substitution $\sigma \in Sub$, if we denote $n = V(\sigma(x))$, some easy computations show that we have $\Delta_{\sigma(M_1)} V(\sigma(g_1)) = p_{1_1} + p_{1_2} - p_{1_3} = \Delta_{M_1} V(g_1)$ and $\Delta_{\sigma(M_2)} V(\sigma(g_2)) = (p_{2_1} - 1)n + 2p_{2_1} - p_{2_2} + p_{2_3} \leq 2p_{2_1} - 2p_{2_2} = \Delta_{M_2} V(g_2)$ since $n \geq 1$ and $(p_{2_1} - 1) = -p_{2_2} - p_{2_3} < 0$.

10   Conclusion and Perspective

In this paper we presented non-deterministic probabilistic abstract reduction systems, probabilistic rewrite systems, and we gave necessary and sufficient conditions for proving positive and almost sure termination of these systems. We also provided tractable sufficient conditions and application examples.

We believe that our notion of probabilistic rewrite system is very powerful since it covers all systems that can be encoded by classical rewrite systems and finite Markov chains. In particular, we already explored it to model a telecommunication protocol.

Next step include understanding whether there could be valid and interesting results generalizing techniques based on polynomial orders, or even on semantical methods.

Acknowledgment

We would like to thanks Thierry Heuillard for pointing out the fact that for protocols, and in almost all other contexts, only positive almost sure termination is interesting. His remarks and suggestions motivated a great part of this work. We would also like to thanks Claude Kirchner for many very fruitful discussions on this work.

References

1. Franz Baader and Tobias Nipkow. *Term Rewriting and All That*. Cambridge University Press, 1998.
2. Gianfranco Balbo. Introduction to Stochastic Petri nets. *Lecture Notes in Computer Science*, 2090:84, Springer-Verlag, 2001.
3. Olivier Bournez and Mathieu Hoyrup. Rewriting Logic and Probabilities. In Robert Nieuwenhuis, editor, *RTA 2003*, volume 2706 of *Lecture Notes in Computer Science*, pages 61–75. Springer-Verlag, 2003.
4. Olivier Bournez and Claude Kirchner. Probabilistic Rewrite Strategies: Applications to ELAN. In Sophie Tison, editor, *RTA 2002*, volume 2378 of *Lecture Notes in Computer Science*, pages 252–266. Springer-Verlag, 2002.
5. Pierre Brémaud. *Markov Chains, Gibbs Fields, Monte Carlo Simulation, and Queues*. Springer-Verlag, New York, 2001.
6. L. de Alfaro. *Formal Verification of Probabilistic Systems*. PhD thesis, Stanford University, 1997.
7. W. Feller. *An Introduction to Probability Theory and its Applications, volume 1*. Wiley, 1968.
8. F. G. Foster. On the Stochastic Matrices Associated with Certain Queuing Processes. *The Annals of Mathematical Statistics*, 24:355–360, 1953.
9. Thom Frühwirth, Alexandra Di Pierro, and Herbert Wiklicky. Toward Probabilistic Constraint Handling Rules. In Slim Abdennadher and Thom Frühwirth, editors, *RCoRP'01*, 2001.
10. G. Grimmett. *Probability Theory*. Cambridge University Press, 1993.

11. H. Hansson. *Time and Probability in Formal Design of Distributed Systems.* Series in Real-Time Safety Critical Systems. Elsevier, 1994.

12. Claire Jones. *Probabilistic Non-determinism.* PhD thesis, University of Edinburgh, 1990.

13. Jan Willem Klop. Term Rewriting Systems. In S. Abramsky, D. M. Gabbay, and T. S. E. Maibaum, editors, *Handbook of Logic in Computer Science*, volume 2, chapter 1, pages 1–117. Oxford University Press, Oxford, 1992.

14. Marta Kwiatkowska, Gethin Norman, and David Parker. PRISM: Probabilistic Symbolic Model Checker. *Lecture Notes in Computer Science*, 2324:200, Soringer-Verlag 2002.

15. N. Lynch. *Distributed Algorithms.* Morgan Kaufmann Publishers, Inc, 1997.

16. Narciso Martí-Oliet and José Meseguer. Rewriting Logic: Roadmap and Bibliography. *Theoretical Computer Science*, 285(2):121–154, 2002.

17. J. Meseguer. Conditional Rewriting Logic as a Unified Model of Concurrency. *Theoretical Computer Science*, 96(1):73–155, 1992.

18. Kumar Nirman, Koushik Sen, Jose Meseguer, and Gul Agha. A Rewriting Based Model for Probabilistic Distributed Object Systems. In *FMOODS'03*, volume 2884 of *Lecture Notes in Computer Science*, pages 32–46, Springer-Verlag 2003.

19. Panangaden. Does Combining Probability and Non-Determinism Makes Sense? *Bulletin of the EATCS*, 2001.

20. M.L. Puternam. *Markov Decision Processes - Discrete Stochastic Dynamic Programming.* Wiley Series in Probability and Mathematical Statistics. John Wiley & Sons, 1994.

21. W. Rudin. *Real and Complex Analysis, 3rd edition.* McGraw Hills, USA, 1987.

22. William H. Sanders and John F. Meyer. Stochastic Activity Networks: Formal Definitions and Concepts. *Lecture Notes in Computer Science*, 2090:315, Springer-Verlag 2001.

23. M. Schneider. Self-stabilization. *ACM Computing Surveys*, 25:45–67, 1993.

24. R. Segala and N. Lynch. Probabilistic Simulations for Probabilistic Processes. *Lecture Notes in Computer Science*, 836:481, Sringer-Verlag 1994.

25. Rob van Glabbeek, Scott A. Smolka, Bernhard Steffen, and Chris M. N. Tofts. Reactive, Generative, and Stratified Models of Probabilistic Processes. In *LICS'90*, pages 130–141, 1990. IEEE Computer Society Press.

26. M. Y. Vardi. Automatic Verification of Probabilistic Concurrent Finite-State Programs. In *FOCS'85*, pages 327–338, 1985.

Termination of Single-Threaded One-Rule Semi-Thue Systems

Wojciech Moczydłowski[1],* and Alfons Geser[2],**

[1] Cornell University, Dept. of Computer Science,
5162 Upson Hall, Ithaca, NY 14853, USA
wojtek@cs.cornell.edu
[2] National Institute for Aerospace (NIA),
144 Research Drive, Hampton, VA 23666, USA
geser@nianet.org

Abstract. This paper is a contribution to the long standing open problem of uniform termination of Semi-Thue Systems that consist of one rule $s \to t$. McNaughton previously showed that rules incapable of (1) deleting t completely from both sides, (2) deleting t completely from the left, and (3) deleting t completely from the right, have a decidable uniform termination problem. We use a novel approach to show that Premise (2) or, symmetrically, Premise (3), is inessential. Our approach is based on derivations in which every pair of successive steps has an overlap. We call such derivations single-threaded.

Key Words and Phrases: string rewriting, semi-Thue system, uniform termination, termination, one-rule, single-rule, single-threaded, well-behaved

1 Introduction

The decidability of the uniform termination problem of one-rule Semi-Thue Systems (1STS) has been open for 14 years. A systematic exploration of the problem was started by Kurth [5].

This problem is both a test case for the strength of termination proof methods and a trigger for their development. Remarkable progress has been made by investigating the consumption and introduction patterns in derivations [4, 7, 8].

McNaughton's notion of a well-behaved derivation is based on the idea that some rules act as if there was an invisible barrier ("inhibitor") somewhere at their right hand side. This inhibitor cannot be removed, so derivations cannot exhibit global communication through the string. McNaughton shows that it is decidable whether a rule is well-behaved, i.e. admits only well-behaved derivations. Moreover he shows that uniform termination is decidable for well-behaved rules.

---

* Partly supported by KBN Grant 7 T11C 028 20.

** Partly supported by the National Aeronautics and Space Administration under NASA Contract No. NAS1-97046.

J. Giesl (Ed.): RTA 2005, LNCS 3467, pp. 338–352, 2005.

In a well-behaved derivation the contractum introduced by any step during a derivation cannot be consumed completely. The contractum can be consumed partially from the left or from the right. We want to study non-well-behaved derivations and hence call a derivation:

- *both-sides-digestible* (BD) if the remainder of some step after partial consumption from the left and partial consumption from the right is consumed later completely;
- *left-digestible* (LD) if the remainder of some step after partial consumption from the left (without any partial consumption from the right) is consumed later completely;
- *right-digestible* (RD) if the remainder of some step after partial consumption from the right (without any partial consumption from the left) is consumed later completely.

We study the following question:

- A 1STS is obviously well-behaved iff it satisfies none of these properties. Can we decide uniform termination also if some of them are true?

An interesting special case is given when the left hand side of the rule has no self-overlap. For this self-overlap free (SOF) case, Kobayashi et al. [4] introduce derivation patterns that are less restrictive than well-behavedness and they call derivations which satisfy them *tame*, *gentle* and *simple*. They show that a gentle 1STS can be transformed to another Semi-Thue System which may have more rules. The two systems have equivalent uniform termination problems. Typically, the transformed system is more amenable to the classic termination criteria. Kobayashi et al. call the properties $\neg LD$, $\neg RD$, and their conjunction "left very gentle", "right very gentle", and "very gentle", respectively. They show that very gentle 1STSs are gentle and that the image of a simple 1STS is a context-free grammar whence its uniform termination problem is decidable. Other examples can often be solved by a transformation and a subsequent ad hoc argument. Beyond the SOF, simple systems no decidability result is available yet.

In a straightforward way the notions of tame, gentle, and simple 1STSs are generalized to non-SOF 1STSs [2]. These properties form a hierarchy:

$$\text{very gentle} \Rightarrow \text{gentle} \Rightarrow \text{tame}$$
$$\Uparrow \qquad\qquad \Uparrow$$
$$\text{well-behaved} \Rightarrow \text{simple} \Rightarrow \neg BD$$

It is easily verified that a 1STS is simple iff it is tame and $\neg BD$. We establish the following result:

- Uniform termination is decidable for 1STSs that satisfy $\neg BD \wedge (\neg LD \vee \neg RD)$.

We reduce the uniform termination problem of 1STSs that satisfy $\neg BD \wedge \neg RD$ to the uniform halting problem of pushdown automata which is decidable [12]. For this purpose we show that each non-terminating such 1STS has an infinite derivation where each step overlaps with the previous one. We call such derivations *single-threaded*. In this case the left and right contexts of the redex can be represented as the contents of two stacks. By $\neg RD$, the left stack is size bounded.

This class of 1STSs includes the following examples which are not covered by Kobayashi et al.: examples that are simple and non-SOF; examples that are non-simple (thus non-tame), non-SOF. On the other hand, Kobayashi et al. also cover the SOF, simple, left-digestible, right-digestible 1STSs, a class which however may be void.

Our examples are not covered by any existing automated termination criteria, except *inverse match-boundedness* [3]. Inverse match-boundedness covers all well-behaved 1STSs, but it is unknown what other classes of 1STSs it also covers.

This work is a thoroughly revised and extended version of the first author's master's thesis [9] and a Technical Report [10].

The paper is organized as follows: In Section 2, we introduce concepts important in our framework, such as chain graph and mother-in-law. In Section 3, we introduce the notion of single-threaded derivation and we derive the decidability result of uniform termination. In Section 4 we give examples of the systems to which our results apply.

2 Preliminaries

We assume familiarity of the reader with semi-Thue systems (string rewriting) [1].

A string u is called a *factor* of v, in symbols $u \sqsubseteq v$, if $v = xuy$ for some $x, y \in \Sigma^*$; a *prefix* if $v = uy$ for some $y \in \Sigma^*$; a *suffix* if $v = xu$ for some $x \in \Sigma^*$. The prefix or suffix u of v is called *proper* if $u \neq v$. The set of all proper suffixes of the word u is denoted by $\mathrm{Suf}(u)$.

The set of *overlaps* of a string u with a string v is defined by

$$\mathrm{OVL}(u, v) = \{w \in \Sigma^+ \mid u = u'w, v = wv', u'v' \neq \varepsilon, u', v' \in \Sigma^*\}$$

The *length* of a string u is denoted by $|u|$.

A *Semi-Thue System* R is a finite set of *rules* $(s, t) \in \Sigma^* \times \Sigma^*$, also written $s \to t$. The one-step *rewrite relation* $\to \subseteq \Sigma^* \times \Sigma^*$ is defined by $usv \to utv$ if $u, v \in \Sigma^*$ and $(s, t) \in R$. The factors s and t are also called the *redex* and the *contractum*, respectively. Occasionally we underline the redex and overline the contractum, as in the following two rewrite steps for the example system $ab \to ba$: $a\underline{ab} \to \underline{ab}\overline{a} \to \overline{ba}a$. A sequence of rewrite steps is called a *derivation*. We write $\mathcal{D} : w_0 \to w_1 \to \ldots$ to denote a derivation named $\mathcal{D}$ with rewrite steps $w_0 \to w_1 \to \ldots$. A system R is called *terminating* if there is no infinite derivation $w_0 \to w_1 \to \ldots$.

We focus on one-rule Semi-Thue Systems (1STS) $\{s \to t\}$, also written $s \to t$. As $s \to t$ is non-terminating if $s \sqsubseteq t$, and terminating if $|s| \geq |t|$ and $s \neq t$, we assume throughout the paper that $s \not\sqsubseteq t$ and $|s| < |t|$. A 1STS $s \to t$ is called *self-overlap free* (*SOF*), if $\mathrm{OVL}(s, s) = \emptyset$. If $\mathrm{OVL}(t, s) = \emptyset$ or $\mathrm{OVL}(s, t) = \emptyset$, then $s \to t$ terminates [5, Criterion D]. If $\mathrm{OVL}(t,s) \cap \mathrm{OVL}(s,t) \neq \emptyset$ ("bordered rule") then the uniform termination problem of $s \to t$ is reducible to that of a non-bordered rule [2, Theorem 6.21]. We henceforth assume that $\mathrm{OVL}(t,s)$ and $\mathrm{OVL}(s,t)$ are disjoint and non-empty.

Definition 1 ([4]). *If $\alpha \in \mathrm{OVL}(t,s)$ then let s_α and t_α be defined by $s = \alpha s_\alpha$ and $t = t_\alpha \alpha$. If $\beta \in \mathrm{OVL}(s,t)$ then let s_β and t_β be defined by $s = s_\beta \beta$ and $t = \beta t_\beta$.*

By $\mathrm{OVL}(t,s) \cap \mathrm{OVL}(s,t) = \emptyset$, there can be no confusion between s_α and s_β or between t_α and t_β.

2.1   Positions

By $[m,n]$ we mean the set of integer numbers between, and including, m and n. We flip the square bracket next to m or n to indicate that m or n, respectively, shall be excluded. Positions in a string w are integer numbers in $[0, |w|]$. We call 0 and $|w|$ the (left and right, respectively) *boundary positions* of w, and the other positions the *inner positions* of w. The inner positions represent the spaces between letters.

Let a (finite or infinite) derivation $\mathcal{D} : w_0 \to w_1 \to \dots$ be presupposed. We denote positions in $\mathcal{D}$ by pairs (i,p) where p is a position in w_i. The position $(i-1, p)$ *corresponds* to the position (i, q), in symbols $(i-1, p) \hookrightarrow_{res} (i, q)$, if there are $x, y \in \Sigma^*$ such that $w_{i-1} = xsy$, $w_i = xty$, and either $0 \le q = p \le |x|$ or $|xs| \le p \le |xsy|$ and $q = p - |s| + |t|$.

If to a given $(i-1, p)$ a q exists such that $(i-1, p) \hookrightarrow_{res} (i, q)$, then q is unique. If no such q exists, i.e., if $|x| < p < |xs|$, then p is said to be *consumed* at step i. Likewise if to a given (i, q) a p exists such that $(i-1, p) \hookrightarrow_{res} (i, q)$, then p is unique. If no such p exists, i.e., $|x| < q < |xt|$, then q is said to be *introduced* at step i.

The *redex position*, $R(i)$, of the i-th rewrite step in $\mathcal{D}$ is defined by $R(i) = |x|$ if $w_{i-1} = xsy$ and $w_i = xty$ for some $x, y \in \Sigma^*$.

The set of positions consumed in step i is $]R(i), R(i)+|s|[$. The set of positions introduced in step i is $]R(i), R(i)+|t|[$.

The equivalence closure of $\hookrightarrow_{res}$, denoted by $\sim_{res}$, allows us to identify a position in w_i with its corresponding position in w_j. If $(i, p) \sim_{res} (j, q)$ then the position p in w_i and the position q in w_j are called *residuals* (of each other). We will conveniently speak about a position p in string w_i when we mean the residual of p.

Example 1. As a running example we use the system $aabbab \to abbaabba$. Consider the following derivation $\mathcal{D}$:

$$w_0 = aab\underline{baabb}abbb \to \underline{aabbabb} * aabbabb \to \overline{abbaabbab} * \underline{aabbabb}$$
$$\to abb\underline{aabbab} * \overline{abbaabbab} \to abbabbaabba * abb\underline{aabbab}$$
$$\to abbabbaabba * \overline{abbabbaabba} \to abbabbaabbabbaabbabaabba = w_6$$

The set of positions consumed in the first step of $\mathcal{D}$ is $[5, 9]$. The set of positions introduced in the first step of $\mathcal{D}$ is $[5, 11]$. The position marked by $*$ in any word in $\mathcal{D}$ is a residual of the position marked by $*$ in any other word. According to our convention, we may say that the position $*$ introduced in the first step of $\mathcal{D}$ is consumed by the last step of $\mathcal{D}$.

Definition 2 ([8]). *A step i is called* digestible, *in symbols $D(i)$, if all contractum positions in w_i are later consumed. The derivation is called* well-behaved *if no step in it is digestible. The 1STS $s \to t$ is called* well-behaved *if all its derivations are well-behaved.*

Note that according to our definitions, the inner positions of the contractum are exactly the introduced positions.

Example 2. The first step in the derivation $\mathcal{D}$ from Example 1 is digestible.

Theorem 1 ([8]). *It is decidable whether an arbitrary 1STS is well-behaved. Uniform termination is decidable for the class of well-behaved 1STS.*

Definition 3 ([9, Definition 5.2]). *For each $j \geq i$ let $\mathrm{Rem}(i, j)$ (for "remainder") denote the set of all residuals in w_j of the set of contractum positions from step i. Step $j \geq i$ is said to* consume from the left *the remainder of step i if $\mathrm{Rem}(i, j) \neq \emptyset$ and*

$$\min \mathrm{Rem}(i, j-1) \in]R(j), R(j) + |s|[.$$

Step $j > i$ is said to consume from the right *the remainder of step i if $\mathrm{Rem}(i, j) \neq \emptyset$ and*

$$\max \mathrm{Rem}(i, j-1) \in]R(j), R(j) + |s|[.$$

Intuitively, step j consumes from the left (right) the remainder of step i if it consumes the leftmost (rightmost) position, but not every position, from the remainder at step $j - 1$.

Example 3. The second step in the derivation $\mathcal{D}$ from Example 1 consumes from the left the remainder of step 1, whereas the third step consumes it from the right.

Definition 4 ([9, Definition 5.5]). *We say that step i is*

- both-sides-digestible, *in symbols $BD(i)$, if $D(i)$ holds and some steps $j > i$ consume from the left the remainder of step i, and some steps $j > i$ consume from the right the remainder of step i;*
- left-digestible, *in symbols $LD(i)$, if $D(i)$ holds and all steps $j > i$ that partially consume the remainder of step i do so from the left (i.e., no steps $j > i$ consume from the right the remainder of step i);*
- right-digestible, *in symbols $RD(i)$, if $D(i)$ holds and all steps $j > i$ that partially consume the remainder of step i do so from the right (i.e., no steps $j > i$ consume from the left the remainder of step i).*

The conditions are mutually exclusive for given i. A derivation is said to satisfy BD, LD, or RD, if some of its steps i satisfy $BD(i)$, $LD(i)$, or $RD(i)$, respectively. A 1STS $s \to t$ satisfies BD, LD, or RD, if some of its derivations satisfy BD, LD, or RD, respectively. We define (both-sides, left, right)-indigestibility for steps, derivations and systems, denoting them by $\neg BD$, $\neg LD$, $\neg RD$, by negating the respective conditions. Note that by definition a 1STS is well-behaved if and only if it satisfies $\neg BD \wedge \neg LD \wedge \neg RD$.

Example 4. The condition $BD(1)$ holds for the derivation from Example 1.

Theorem 2 ([6]). *The Conditions LD and RD are decidable for 1STSs.*

Proof. Conditions LD and RD are equivalent to McNaughton's conditions $C2$ and $C3$, respectively [6, Theorem 6.1]. This shows up in cases I and II in his proof.   □

If $\neg LD \wedge \neg RD$ holds then BD is equivalent to McNaugton's Condition $C1$. However, Condition BD is not equivalent to $C1$ in the general case.

Example 5. The system from Example 1 satisfies $\neg RD$ and $\neg C1$. However, it satisfies BD as the derivation $\mathcal{D}$ exhibits.

2.2   Chain Graphs

The notion of chain graph gives one the means to reason in detail about the relation between steps in a derivation.

Definition 5. *Let $\mathcal{D} : w_0 \to w_1 \to \ldots$. Let $w_i = xsy$ for some i, x, y. The factor s in w_i is called* live *if:*

- *there is a step $j \geq i$ such that $(i, |x|) \sim_{res} (i-1, R(j))$, i.e., at step j the redex $|x|$ from w_i is reduced;*
- *$(i, p) \sim_{res} (j-1, p')$ for all $|x| \leq p \leq |xs|$; i.e., no position of s is consumed until s is rewritten.*

Informally speaking, a live factor is finally reduced and it is not touched before then. Note that the live factor in w_i need not be reduced in the very next step $w_i \to w_{i+1}$. Since the residuals of overlapping redexes overlap, live factors do not overlap.

Definition 6 ([5, Definition 4.25]). *The chain graph of a (finite or infinite) derivation $\mathcal{D} : w_0 \to w_1 \to \ldots$ is a directed graph (V, E). The vertices in V are the positions of live factors. The edges in $E = E_0 \cup E_1$ are defined as follows:*

- *if $(i-1, p) \hookrightarrow_{res} (i, q)$ and $(i-1, p), (i, q) \in V$ then $((i-1, p), (i, q)) \in E_0$;*
- *if $(i-1, R(i)), (i, q) \in V$, and some of the positions $(i, q), \ldots, (i, q + |s|)$ are introduced by step i, then $((i-1, R(i)), (i, q)) \in E_1$.*

We define selector functions $\mathsf{src}, \mathsf{tgt}, \mathsf{level} : E \to \mathbb{N}$ for the source, the target, and the level of an edge $k \in E$ by $\mathsf{src}(k) = p$, $\mathsf{tgt}(k) = q$, $\mathsf{level}(k) = i$ if $k = ((i-1, p), (i, q))$.

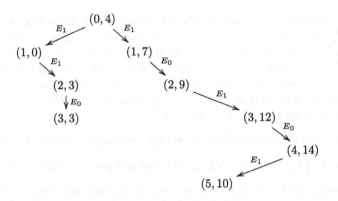

Fig. 1. The chain graph of the derivation from Example 1.

The chain graph is a forest of finitely many trees, $T_1, \ldots, T_K$, rooted at the positions $p_1 < \cdots < p_K$ of the live redexes in w_0.

Example 6. Figure 1 shows the chain graph of the system in Example 1. The lowest edge has the source vertex $(4, 14)$, the target vertex $(5, 10)$, it is in E_1, and its level is 5.

Definition 7. *Edges from the set E_1 will also be called* active. *An active edge k is called a* left edge *if* $\mathsf{src}(k) > \mathsf{tgt}(k)$; *and a* right edge *if* $\mathsf{src}(k) + |s| < \mathsf{tgt}(k) + |t|$. *We will call the active edges on the same level* rivals.

By $s \not\sqsubseteq t$ and $|s| < |t|$, every active edge is a left or a right edge. There can be at most one left edge and at most one right edge at each level.

Lemma 1. *If k is a left edge at level i then $w_{i-1} = zs_\beta sy$ and $w_i = zst_\beta y$ for some $\beta \in \mathrm{OVL}(s, t)$ and $z, y \in \Sigma^*$. Moreover $\mathsf{src}(k) = |zs_\beta|$ and $\mathsf{tgt}(k) = |z|$. If k is a right edge at level i then $w_{i-1} = xss_\alpha v$ and $w_i = xt_\alpha sv$ for some $\alpha \in \mathrm{OVL}(t, s)$ and $x, v \in \Sigma^*$. Moreover $\mathsf{src}(k) = |x|$ and $\mathsf{tgt}(k) = |xt_\alpha|$.*

Proof. Straightforward from the definitions. □

Lemma 2. *If $(i, p) \sim_{res} (j, q)$ and $(i, p') \sim_{res} (j, q')$ and $p < p'$ then $q < q'$.*

Proof. By induction on $|j - i|$, with the inductive step done by case analysis on $R(i) \le p$, $p < R(i) < p'$, and $p' \le R(i)$. □

2.3   Family Members

New tools developed in this section will enable us to speak in more detail about infinite derivations.

Definition 8 ([9, Definition 7.1]). *Let k be a right edge at level i. Then s_α and $(i-1, |xs|)$ in Lemma 1 are called the* husband *and its* position, *respectively. Likewise for a left edge, s_β and $(i-1, |z|)$ are called the* husband *and its* position, *respectively.*

Intuitively, a husband is a non-empty factor that is supplemented to a live redex by the next rewrite step. The husband positions of k are the residuals of the positions of the live redex created by k.

Example 7. In the chain graph of the derivation from Example 1, the husbands of the edges at level 1 are $aabb$ at position $(0, 0)$ and b at position $(0, 10)$.

Definition 9. *[[9, Definition 7.3]] Let p be a position in the husband h of an active edge k. Then we call the vertex $(i - 1, R(i))$ the* mother-in-law *of p if p is introduced in step i. A* mother-in-law *of the active edge k is the mother-in-law of one of the positions in the husband of k. The step that rewrites the target redex of k is called the* marriage consumption *step of k.*

Example 8. The vertex $(0, 4)$ in the chain graph of the derivation $\mathcal{D}$ from Example 1 is the mother-in-law of the position $*$ in w_4. The string $aabb$ at position $(4, 10)$ is the husband of the edge going from $(4, 14)$ to $(5, 10)$, and the vertex $(3, 3)$ is its mother-in-law.

Note that a mother-in-law need not be the source vertex of an edge. In other words, the rewrite step $w_{i-1} \rightarrow w_i$ need not create a live redex.

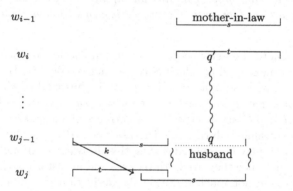

Fig. 2. Husband and mother-in-law.

3 Uniform Termination
of One-Rule Single-Threaded Systems

In this section we define single-threaded derivations, show how single-threadedness can be derived, and use single-threadedness for decidability of uniform termination in a special case.

3.1 Single-Threadedness and Independence

Definition 10. *A path in the chain graph of a derivation is called* single-threaded *if every edge on it is active. A derivation is called* single-threaded *if its chain graph is a single-threaded path. A 1STS is called* single-threaded *if it admits an infinite single-threaded derivation.*

Theorem 3 ([7, Theorem 7.4],[9, Theorem 3]). *Every non-terminating, well-behaved 1STS is single-threaded.*

McNaughton's xy-sequence corresponds to a single-threaded path.

Definition 11 ([9, Definitions 8.2 and 8.3]). *An active edge k is called* independent *if all its mothers-in-law are ancestors of k. A mother-in-law that is not an ancestor of k is called* alien *to k. A path is called* independent *if every active edge on it is independent.*

In other words, an edge is dependent iff it has an alien mother-in-law. In contrast, an independent path does not need any other paths to proceed with its reductions.

Example 9. The left path in Figure 1 is independent. However, the right path is not – the mother-in-law $(3,3)$ of the boundary position $(4,10)$ of the husband is a vertex in the left path and is hence alien to the edge $((4,14),(5,10))$.

Lemma 3. *If there is an infinite derivation whose chain graph contains an infinite independent path whose first i edges are active, then there is also an infinite derivation whose chain graph contains an infinite independent path whose first $i+1$ edges are active. Moreover, the two derivations coincide up to, and including, step i.*

Proof. Let k denote the active edge at level i in the independent path S. Let $j > i$ denote the next level at which S has an active edge, k'. By symmetry we may assume that k' is a right edge. Let h be the husband of k', i.e., there are $x, y, g \in \Sigma^*$ such that $w_{j-1} = xshy \to xthy = xgsy = w_j$. Since $k' \in S$ and all edges between k and k' are inactive, the occurrence of s is preserved, i.e., none of its positions is consumed, during the derivation $w_i \to^* w_{j-1}$. Only the parts left or right to it in w_i may be touched during this derivation. All mothers-in-law of h are above level i since they are both redexes and ancestor nodes of k. Hence h is present in w_i and not touched during the derivation $w_i \to^* w_{j-1}$ either. Let $x' \to^* x$ and $y' \to^* y$ render the changes that happened during the derivation $w_i \to^* w_{j-1}$. Then the derivation can be rearranged thus:

$$
\begin{array}{ccc}
w_i = x'shy' & \xrightarrow{\;*\;} & w_{j-1} = xshy \\
\downarrow & & \downarrow \\
w'_{i+1} = x'gsy' & \xrightarrow{\;*\;} & w'_j = w_j = xgsy
\end{array}
$$

We will show that the chain graph of the new derivation $\mathcal{D}' : w_0 \to w_1 \to \cdots \to w_i \to w'_{i+1} \to \cdots \to w'_{j-1} \to w'_j = w_j \to w_{j+1} \to \ldots$ has an infinite independent path with the first $i+1$ edges active. In steps $w'_{i+1} \to \cdots \to w'_{j-1} \to w'_j$ we execute reductions left and right from gs in the same order as they were executed in $\mathcal{D}$.

First note that the inactive edge at level $i+1$ having source $(i,|x'|)$ in the chain graph of $\mathcal{D}$ is replaced by the active edge $((i,|x'|),(i+1,|x'g|))$ in the chain graph of $\mathcal{D}'$. Let us denote this active edge by K.

Let S consist of vertices

$$v_0, \ldots, v_{i-1}, (i, |x'|), \ldots, (j-1, |x|), (j, |xg|), v_{j+1}, \ldots$$

and respective edges between them. The path S' consisting of the vertices:

$$v_0, \ldots, v_{i-1}, (i, |x'|), (i+1, |x'g|), \ldots, (j, |xg|), v_{j+1}, \ldots$$

in the chain graph of $\mathcal{D}'$ has by its construction first $i+1$ edges active. It suffices to show that it is an infinite independent path.

Suppose that there is a dependent edge $l \in S'$. There are 3 possible cases:

- $\mathsf{level}(l) \le i$. Since $\mathcal{D}'$ up to step i is the same as $\mathcal{D}$ and hence the respective parts of their chain graphs are the same, l is dependent also in the chain graph of $\mathcal{D}$, a contradiction.
- $\mathsf{level}(l) \in]i, j]$. Then $l = K$, since K is the only active edge in those levels. Hence one of the positions in the husband h of K is introduced before level $i+1$ by an alien mother-in-law m. But before level $i+1$ the derivations and their chain graphs are the same, hence m is also an alien mother-in-law of k', a contradiction.
- $\mathsf{level}(l) > j$. Then l was present in the original chain graph as well since all reduction steps later than j are the same. Let $m = (j'-1, R(j'))$ be an alien mother-in-law of l. We have 3 possible cases:
  - $j' > j$. Since the steps after j and hence their chain graphs are the same, m is alien to l in S as well.
  - $j' \in]i, j]$. Obviously, $j' \ne i+1$, because m is alien. Let p be the position in the husband of l introduced by the reduction corresponding to m. Then p has a residual p' in w'_j. We can either have $p' < |x|$ or $p' > |xgs|$, since other positions stay untouched during the derivation $w'_{i+1} \to^* w'_j$. Therefore p' is either an inner position of x or of y. To fix our attention, suppose that it is an inner position of y. By Lemma 2, we have $w'_{j'-1} = x''gsy_1sy_2$ for some $x'', y_1, y_2 \in \Sigma^*$, where $x' \to^* x'' \to^* x$ and $y' \to^* y_1sy_2 \to^* y$. Let $|x''gsy_1| < p'' < |x''gsy_1t|$ be the residual of p introduced in step j'. Consider the corresponding reduction step in $\mathcal{D}$: $w_{j'} = x''shy_1sy_2 \to x''shy_1ty_2 = w_{j'+1}$. The position $p''' = |h| - |g| + p''$ in $w_{j'+1}$ is introduced in this reduction. One shows that (j', p''') is a residual of (j, p') in $\mathcal{D}$. Since $(j', |x''|) \in S$, the mother-in-law $(j', |x''shy_1|)$ is alien. Hence S is not independent, a contradiction.
  - $j' \le i$. By definition of mother-in-law, the step $w'_{j'-1} \to w'_{j'}$ in $\mathcal{D}'$ introduces a residual of $(\mathsf{level}(\ell)-1, p)$. The same step in $\mathcal{D}$ introduces a residual of $(\mathsf{level}(\ell)-1, p)$ in $\mathcal{D}$, because the derivation $w_i \to^* w_{j-1} \to w_j$ touches exactly the same positions as the derivation $w_i \to w'_{i+1} \to^* w_j$. So m is an alien mother-in-law of ℓ also in the chain graph of $\mathcal{D}$, a contradiction.

So S' contains no dependent edge, which finishes the proof.   □

Example 10. Consider the first three steps of $\mathcal{D}$ from Example 1. Let B denote the right branch in its chain graph. The edges on B come from sets E_1, E_0, E_1. Pushing up the second active edge from B, results in the following derivation:

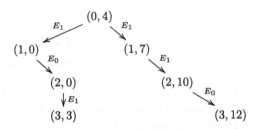

Fig. 3. The chain graph from Example 10.

$$aab\underline{baabbabb}bb \rightarrow aabb\overline{abb} * \underline{aabbabb} \rightarrow$$

$$\rightarrow \underline{aabbabb} * \overline{abbaabbab} \rightarrow \overline{abbaabbab} * abbaabbab$$

Its chain graph is shown in Figure 3. Note that the right path starts with two active edges.

Lemma 4 ([9, Lemma 15]). *If the chain graph of an infinite derivation contains an infinite independent path then there is a derivation that contains an infinite, single-threaded path starting from level 0.*

Proof. First we drop enough initial steps from the derivation, so that the independent path starts from level 0. Then we construct the i-the step of the target derivation and the i-th level of its chain graph by applying Lemma 3 i times. □

Lemma 5. *A derivation whose chain graph contains an infinite single-threaded path starting from level 0 is a single-threaded derivation.*

Proof. Let an infinite derivation be given that contains an infinite, single-threaded path, S. As every edge on the path is active, there cannot be, besides the path, another redex that is rewritten during the derivation. By definition of chain graph there is, therefore, no inactive edge in the chain graph. By the same token, the active edges have no rivals. So there are no edges outside S. □

The concepts and lemmas introduced so far can be used to prove:

Theorem 4. *Every both-sides-indigestible, non-terminating 1STS is single-threaded.*

3.2   Simulation by a Pushdown Automaton

We show in this section that the single-threaded derivations of a right-indigestible 1STS can be rendered by a pushdown automaton, whence the uniform termination problem for single-threaded, right-indigestible 1STSs is decidable.

Lemma 6 ([9, Proposition 34]). *Let $w_0 \rightarrow w_1 \rightarrow \ldots$ be an infinite single-threaded derivation and let $D(i)$ hold for some $i > 0$. If k_i is a left edge then $LD(i)$ holds. If k_i is a right edge then $RD(i)$ holds.*

Proof. To fix our attention, suppose that k_i is a left edge. Hence reduction of the target redex consumes positions introduced in the i-th step from the left. By induction we can show that no step consumes positions from the right. □

During the remainder of this section we assume that the given 1STS $s \to t$ is single-threaded and satisfies $\neg RD$.

Lemma 7 ([9, Lemma 35]). *Let k be a right edge at level i in the chain graph of a single-threaded derivation. Then no position $p \leq \mathsf{src}(k)$ in w_{i-1} is consumed later.*

Proof. By contradiction. Suppose that there is a right edge k at level i and the position $p \leq \mathsf{src}(k)$ is consumed later. Since the derivation is single-threaded, we can show by induction that all positions between $\mathsf{src}(k)$ and $\mathsf{tgt}(k)$ in w_i are also consumed; hence $D(i)$ holds. By Lemma 6, we get $RD(i)$, a contradiction.   □

Definition 12 ([9, Definition 10.5]). *To a 1STS $s \to t$, we assign a generalized pushdown automaton [11] $\mathcal{A}$ whose transitions will correspond to rewrite steps in a given derivation. The input alphabet and the stack alphabet are Σ each. The state of the automaton is encoded as the contents of a stack of size strictly bounded by $|t|$. So a configuration is a pair $(x, y) \in \Sigma^{<|t|} \times \Sigma^*$. The automaton has the transition relation $\vdash \subseteq (\Sigma^{<|t|} \times \Sigma^*) \times (\Sigma^{<|t|} \times \Sigma^*)$ defined by:*

$$\begin{cases} (x, y) \vdash (x', t_\beta y) & \text{if } x = x's_\beta,\ \beta \in \mathrm{OVL}(s, t),\ x \in \Sigma^{<|t|}, x', y \in \Sigma^* \\ (x, y) \vdash (t_\alpha, y') & \text{if } y = s_\alpha y',\ \alpha \in \mathrm{OVL}(t, s),\ x \in \Sigma^{<|t|},\ y', y \in \Sigma^* \end{cases}$$

The transition relation $\vdash$ is well-defined by $|x'| < |x| < |t|$ and $|t_\alpha| < |t|$. A finite or infinite sequence of transitions is called a computation.

Lemma 8 ([9, Lemma 37]). *If $\mathcal{A}$ admits an infinite computation then there is an infinite derivation.*

Proof. One shows that for all $x, x' \in \Sigma^{<|t|}$ and $y, y' \in \Sigma^*$, if $(x, y) \vdash (x', y')$ then $xsy \to x'sy'$ or $xsy \to xx'sy'$.   □

Definition 13. *We say that $\mathcal{A}$ is* put on *the derivation $w_0 \to w_1 \to \ldots$, if its configuration is set to (x, y), and $|x| < |t|$, where x and y are the left and right contexts of the first rewrite step, $w_0 \to w_1$.*

Lemma 9. *The automaton $\mathcal{A}$, put on an infinite, single-threaded derivation, admits an infinite computation.*

Proof. We prove that $\mathcal{A}$ admits one transition and thereafter it is put on an infinite, single-threaded derivation again. By applying this argument i times, we can construct the i-th transition of the automaton for any $i > 0$.

To prove the claim, let an infinite, single-threaded derivation $\mathcal{D} : w_0 \to w_1 \to \ldots$ be given, and let $w_0 = xsy$ and $R(1) = |x|$ for some $x, y \in \Sigma^*$. If $R(1) > R(2)$ (k_1 is a left edge), then $x = x's_\beta$ and $w_1 = x'st_\beta y$ for some $x' \in \Sigma^*$ and $\beta \in \mathrm{OVL}(s, t)$. The automaton can make a transition $(x, y) \vdash (x', t_\beta y)$, and is so put on the remaining derivation $w_1 \to w_2 \to \ldots$. If $R(1) < R(2)$ (k_1 is a right edge), then $y = s_\alpha y'$ and $w_1 = xt_\alpha sy'$ for some $y' \in \Sigma^*$ and $\alpha \in \mathrm{OVL}(t, s)$. By Lemma 7, the prefix x remains unaffected by the derivation $w_1 \to w_2 \to \ldots$. Now for all $i > 0$ let w_i' be defined by $w_i = xw_i'$. Then $w_1' \to w_2' \to \ldots$ is again an infinite, single-threaded derivation.   □

Lemma 10. *Let S be a path in the chain graph of an infinite, single-threaded derivation. If S contains infinitely many active edges then it contains infinitely many left edges and infinitely many right edges.*

Proof. Suppose that there are only finitely many left edges on S. Then there is some N such that k_n is a right edge, or an inactive edge, for all $n > N$. Let $a_n = |w_n| - \text{tgt}(k_n)$. Obviously $a_n \geq 0$ for all $n > N$. On the other hand, the subsequence of all a_n, $n > N$ for which k_n is a right edge strictly decreases. This gives a contradiction.   □

Lemma 11. *If $s \to t$ is a right-indigestible, single-threaded 1STS then $\mathcal{A}$ admits an infinite computation.*

Proof. Let $s \to t$ admit the infinite, single-threaded derivation $w_0 \to w_1 \to \ldots$. In order to work with Lemma 9, we need to ensure $|x| < |t|$ for the left context of the first rewrite step. This is not the case for an arbitrary derivation, but a suitable derivation can be derived as follows.

By Lemma 10, the single-threaded path of $w_0 \to w_1 \to \ldots$ contains a right edge, at level i say. Then the derivation $w_{i-1} \to w_i \to \ldots$ starts with a right edge: we have $w_{i-1} = xss_\alpha y'$ and $w_i = xt_\alpha sy'$ for some $x, y' \in \Sigma^*$ and $\alpha \in \text{OVL}(t, s)$. By Lemma 7 the prefix x remains unaffected by the derivation $w_{i-1} \to w_i \to \ldots$. Now for all $j \geq i - 1$ let w_j' be defined by $w_j = xw_j'$. Then $\mathcal{D} : w_1' \to w_2' \to \ldots$ is again an infinite, single-threaded derivation. Moreover $|t_\alpha| < |t|$ holds for the left context t_α of its first rewrite step. By Lemma 9, the automaton $\mathcal{A}$ put on $\mathcal{D}$ admits an infinite computation.   □

Example 11. Consider the well-behaved system $abcd \to cdcdbabab$ taken from [7], and the infinite derivation:

$$abcdcd \to cdcdbababcd \to cdcdbabcdcdbabab \to cdcdbcdcdbababcdbabab \to \ldots$$

The corresponding computation is:

$$(\varepsilon, cd) \vdash (cdcdbab, \varepsilon) \vdash (cdcdb, cdbabab) \vdash (cdcdbab, babab) \vdash \ldots$$

Definition 14. *The uniform halting problem of pushdown automata is the following problem: "Given a pushdown automaton $(\Sigma, Z, Q, \vdash, q_0, z_0)$ – is there $(x, y) \in Q \times Z^*$ that initiates an infinite computation?"*

Theorem 5. *The uniform termination problem is decidable for the class of 1STS $s \to t$ that satisfy $\neg BD \wedge (\neg RD \vee \neg LD)$.*

Proof. By symmetry we may assume $\neg RD$. By Lemmas 8 and 11, the uniform termination problem is reduced to the uniform halting problem of pushdown automata which is decidable [12].   □

4 Applications

There is a decidable sufficient criterion for both-sides-indigestibility of 1STSs. First BD can be characterized by the existence of two peculiar single-threaded derivations, then one can develop a simple test for non-existence of such derivations, based on the sets of suffixes of s and t that can be consumed and introduced, respectively. The question whether BD is decidable remains open.

We give several examples of systems to which this criterion and our theorems apply.

Example 12. The 1STS $R = \{caabca \to aabccaabc\}$ is both-sides-indigestible, satisfies $\neg LD$ and RD and is not tame.

Example 13 ([2]). The SOF 1STS $aaabbab \to abbaaabba$ satisfies $\neg BD \wedge \neg LD \wedge RD$ and is tame and terminating.

Example 14. The non-SOF 1STS $R = \{babbabb \to abbabbbba\}$ satisfies $\neg BD \wedge LD \wedge \neg RD$. It is non-tame and non-terminating.

Kobayashi et al. [4, page 603] find no instances for the case $SOF \wedge \neg BD \wedge RD \wedge LD$. Non-SOF systems satisfying $\neg BD \wedge RD \wedge LD$ however do exist:

Example 15. For every $n \geq 3$, the 1STS $R = \{ba(ab)^n \to (ab)^{n+2}a\}$ is not both-sides-digestible. However R is both left-digestible ($(ab)^{n+1}a$ suffix of $(ba)^{n+1}$) and right-digestible (aba prefix of $(ab)^n$).

5 Conclusion

We have shown that one-rule Semi-Thue Systems (1STSs) that satisfy $\neg BD \wedge (\neg RD \vee \neg LD)$ have a decidable uniform termination problem, for their non-terminating members admit infinite single-threaded derivations, which can be simulated by pushdown automata. The uniform termination problem for 1STSs that satisfy $\neg BD \wedge RD \wedge LD$ is open.

Acknowledgements

We are grateful to Robert McNaughton for his encouraging us to do this investigation and to the anonymous referees for their constructive comments. The first of the authors is grateful to Paweł Urzyczyn for his infinite patience.

References

1. Ronald Book and Friedrich Otto. *String-rewriting systems.* Texts and Monographs in Computer Science. Springer, New York, 1993.
2. Alfons Geser. *Is termination decidable for string rewriting with only one rule?* Habilitation thesis, Wilhelm-Schickard-Institut, Universität Tübingen, Germany, January 2002. 201 pages.

3. Alfons Geser, Dieter Hofbauer, and Johannes Waldmann. Termination proofs for string rewriting systems via inverse match-bounds. *J. Automated Reasoning*, 2005. In print.
4. Yuji Kobayashi, Masashi Katsura, and Kayoko Shikishima-Tsuji. Termination and derivational complexity of confluent one-rule string rewriting systems. *Theoretical Computer Science*, 262(1/2):583–632, 2001.
5. Winfried Kurth. *Termination und Konfluenz von Semi-Thue-Systemen mit nur einer Regel*. Dissertation, Technische Universität Clausthal, Germany, 1990.
6. Robert McNaughton. The uniform halting problem for one-rule Semi-Thue Systems. Technical Report 94-18, Dept. of Computer Science, Rensselaer Polytechnic Institute, Troy, NY, August 1994. See also "Correction to 'The Uniform Halting Problem for One-rule Semi-Thue Systems'", unpublished paper, August, 1996.
7. Robert McNaughton. Well-behaved derivations in one-rule Semi-Thue Systems. Technical Report 95-15, Dept. of Computer Science, Rensselaer Polytechnic Institute, Troy, NY, November 1995. See also "Correction by the author to 'Well-behaved derivations in one-rule Semi-Thue Systems'", unpublished paper, July, 1996.
8. Robert McNaughton. Semi-Thue Systems with an Inhibitor. *J. Automated Reasoning*, 26:409–431, 1997.
9. Wojciech Moczydłowski. Jednoregułowe systemy przepisywania słów. Masters thesis, Warsaw University, Poland, 2002.
10. Wojciech Moczydłowski and Alfons Geser. Termination of single-threaded one-rule Semi-Thue systems. Technical Report TR 02-08 (273), Warsaw University, December 2002. Available electronically at URL
   `research.nianet.org/~geser/papers/single.html`.
11. Arto Salomaa. *Theory of Automata*, volume 100 of *Intl. Series of Monographs in Pure and Applied Mathematics*. Pergamon Press, 1969.
12. J. V. Tucker. Computing in algebraic systems. In F. F. Drake and S. S. Wainer, editors, *Recursion Theory, its Generalisations and Applications*, volume 45 of *London Mathematical Society Lecture Note Series*, pages 215–235. Cambridge University Press, 1980.

On Tree Automata that Certify Termination of Left-Linear Term Rewriting Systems

Alfons Geser[1,*], Dieter Hofbauer[2], Johannes Waldmann[3], and Hans Zantema[4]

[1] National Institute of Aerospace, 144 Research Drive,
Hampton, Virginia 23666, USA
geser@nianet.org
[2] Mühlengasse 16, D-34125 Kassel, Germany
dieter@theory.informatik.uni-kassel.de
[3] Hochschule für Technik, Wirtschaft und Kultur (FH) Leipzig,
Fb IMN, PF 30 11 66, D-04251 Leipzig, Germany
waldmann@imn.htwk-leipzig.de
[4] Faculteit Wiskunde en Informatica, Technische Universiteit Eindhoven,
Postbus 513, 5600 MB Eindhoven, The Netherlands
h.zantema@tue.nl

Abstract. We present a new method for proving termination of term rewriting systems automatically. It is a generalization of the match bound method for string rewriting. To prove that a term rewriting system terminates on a given regular language of terms, we first construct an enriched system over a new signature that simulates the original derivations. The enriched system is an infinite system over an infinite signature, but it is locally terminating: every restriction of the enriched system to a finite signature is terminating. We then construct iteratively a finite tree automaton that accepts the enriched given regular language and is closed under rewriting modulo the enriched system. If this procedure stops, then the enriched system is compact: every enriched derivation involves only a finite signature. Therefore, the original system terminates. We present three methods to construct the enrichment: top heights, roof heights, and match heights. Top and roof heights work for left-linear systems, while match heights give a powerful method for linear systems. For linear systems, the method is strengthened further by a forward closure construction. Using these methods, we give examples for automated termination proofs that cannot be obtained by standard methods.

1 Introduction

We present a new method for proving automatically that a term rewriting system (TRS) terminates on each term from a given regular term language. Our method consists of two steps. In the first step, we switch to an *enrichment* of the given TRS, i.e., a rewriting system over a different signature that simulates the original derivations. We consider enriched systems over infinite signatures that are *locally*

* Partly supported by the National Aeronautics and Space Administration under NASA Contract No. NAS1-97046 while this author was in residence at the NIA.

terminating: every restriction to a finite signature is terminating. In the second step, we compute a *compatible* finite tree automaton for this enrichment, i.e., a tree automaton that contains the enriched given regular tree language and is closed under rewriting modulo the enriched system. The existence of such a compatible automaton ensures that the enriched TRS is *compact*, i.e., every infinite derivation involves only a finite signature. By local termination of the enrichment, the automaton certifies termination of the original system.

We have previously applied this method to string rewriting [7]. The string rewriting version is implemented in the tools TORPA [19], Matchbox [18] and AProVE [12]. In the present paper, we describe how to extend it to term rewriting. Non-linearities in the TRS complicate both the termination arguments and the automata constructions. The algorithms we present are implemented in Matchbox.

The enrichments that we consider are variants of the original TRS in which the symbols are labelled by natural numbers. An enrichment is more powerful than another if the labels in the right-hand sides are smaller. We introduce three enrichments with increasing power: top heights, roof heights and match heights. For match heights, linearity of the TRS is required for the desired theorem to hold. So for linear TRSs the best results are obtained by choosing the enrichment based on match heights, and for non-right-linear TRSs the best results are obtained by choosing the enrichment based on roof heights.

For linear systems, uniform termination can be concluded from termination on a restricted set of initial terms: the set of right-hand sides of forward closures. We use our method both to compute this set, and to prove termination on it, at the same time. This turns out to be more powerful than applying the method directly for the original system and the set of all terms. To our knowledge, this is the first method that computes finite representations of infinite sets of right-hand sides of forward closures on TRSs.

The paper is organized as follows. In Section 3 we define enrichments, give three instances, and compare them. In Section 4 we define compatible tree automata and in Section 5 we discuss how to construct them. Section 6 presents the simulation of forward closures by rewriting, while Section 7 shows how to implement this with an automata construction.

A preliminary version of this paper has been presented at the 7th International Workshop on Termination, Aachen 2004 [8].

2   Preliminaries

For a relation ρ on a set T and $t \in T$ write $\mathrm{SN}(t, \rho)$ if there is no infinite sequence $t_0, t_1, \ldots$ over T where $t = t_0$ and $t_i \ \rho \ t_{i+1}$ for every $i \geq 0$. Define $\mathrm{SN}(S, \rho)$ for $S \subseteq T$ by $\forall s \in S : \mathrm{SN}(s, \rho)$; then ρ is *terminating* (or: *strongly normalizing*) on S. Let $\mathrm{SN}(\rho)$ stand for $\mathrm{SN}(T, \rho)$. The reflexive closure of ρ is $\rho^=$, the composition of two relations $\rho \subseteq A \times B$ and $\sigma \subseteq B \times C$ is $\rho \circ \sigma = \{(a, c) \mid \exists b \in B : (a, b) \in \rho, (b, c) \in \sigma\}$.

For standard notations on term rewriting see [1, 20], for instance. Throughout we fix a signature Σ, a set of variables X, and consider term rewriting systems

$R \subseteq \mathcal{T}_\Sigma(X) \times \mathcal{T}_\Sigma(X)$. Unless otherwise stated, signatures and rewriting systems are finite. The set of left- and right-hand sides of R are denoted by $\text{lhs}(R)$ and $\text{rhs}(R)$ respectively. Since our topic is termination, we assume $\text{lhs}(R) \cap X = \emptyset$, and $X(r) \subseteq X(\ell)$ for rules $\ell \to r$. Let $X(t) \subseteq X$ denote the set of variables that occur in $t \in \mathcal{T}_\Sigma(X)$, and let $X(T) = \bigcup_{t \in T} X(t)$ for $T \subseteq \mathcal{T}_\Sigma(X)$. For a mapping $h : \mathcal{T}_\Sigma(X) \to \mathcal{T}_\Gamma(X)$ define the term rewriting system $h(R) = \{h(\ell) \to h(r) \mid \ell \to r \in R\}$ over signature Γ. For the symbol at position p in term t we write $t(p)$. For $Y \subseteq \Sigma \cup X$ let $\text{Pos}_Y(t)$ be the set of positions p such that $t(p) \in Y$. We use $<$ for the prefix ordering on positions. The set of *descendants* modulo R of a tree language $L \subseteq \mathcal{T}_\Sigma$ is $\to_R^*(L) = \{s \in \mathcal{T}_\Sigma \mid \exists t \in L : t \to_R^* s\}$.

The domain and the range of a substitution $\alpha : X \to \mathcal{T}_\Sigma(X)$ are $\text{dom}(\alpha) = \{x \in X \mid x\alpha \neq x\}$ and $\text{ran}(\alpha) = \{x\alpha \mid x \in \text{dom}(\alpha)\}$. For $Y \subseteq \text{dom}(\alpha)$ let $\alpha|_Y$ be the substitution with domain Y where $\alpha|_Y : x \mapsto x\alpha$ for $x \in Y$, $\alpha|_Y : x \mapsto x$ otherwise. For substitutions α and α' we write $\alpha \to_R \alpha'$ if $\text{dom}(\alpha) = \text{dom}(\alpha')$, and $x\alpha \to_R x\alpha'$ for some $x \in \text{dom}(\alpha)$ and $y\alpha = y\alpha'$ for every $y \neq x$.

A *tree automaton* $A = (Q, \Sigma, F, T)$ over a signature Σ consists of a set Q of constant symbols, disjoint from Σ, called *states*; a set $F \subseteq Q$ of *final states*; and a ground rewriting system T over $\Sigma \cup Q$ with rules (*transitions*) of the form $q_0 \to q$ or $f(q_1, \ldots, q_n) \to q$ for n-ary $f \in \Sigma$, $n \geq 0$, and $q_0, \ldots, q_n, q \in Q$. The automaton is *finite* if T is finite, and it is *deterministic* if T is non-overlapping. The *language* accepted by A is $\mathcal{L}(A) = \{t \in \mathcal{T}_\Sigma \mid \exists q \in F : t \to_T^* q\}$. For more on tree languages we refer to [2, 5].

3   Enrichments of Rewriting Systems

Definition 1. *A TRS R' over a signature Σ' is an enrichment of a TRS R over a signature Σ if there is a mapping* $\text{base} : \mathcal{T}_{\Sigma'} \to \mathcal{T}_\Sigma$ *such that every R-derivation step can be lifted to an R'-derivation step: for each step $s \to_R t$ and each $s' \in \text{base}^{-1}(s)$ there is some $t' \in \text{base}^{-1}(t)$ with $s' \to_{R'} t'$.*

We use enrichments to propagate termination properties:

Proposition 1. *Let R and R' be TRSs over Σ and Σ' resp., let $L \subseteq \mathcal{T}_\Sigma$ and $L' \subseteq \mathcal{T}_{\Sigma'}$. If R' is an enrichment of R via* $\text{base} : \mathcal{T}_{\Sigma'} \to \mathcal{T}_\Sigma$, *and $L \subseteq \text{base}(L')$, then termination of R' on L' implies termination of R on L.*

Suitable enrichments will satisfy the following property:

Definition 2. *A finite or infinite TRS R over a finite or infinite signature Σ is called* locally terminating *if every restriction of R to a finite signature $\Gamma \subseteq \Sigma$ is terminating: $R \cap (\mathcal{T}_\Gamma(X) \times \mathcal{T}_\Gamma(X))$ is terminating on $\mathcal{T}_\Gamma$.*

In the following, we will present three enrichments that are locally terminating, one of them under a suitable linearity restriction. We choose the enriched signature $\Sigma' = \Sigma \times \mathbb{N}$, and call the numbers *heights*. We often write f_h for (f, h). Define the mappings $\text{base} : \Sigma' \to \Sigma$, $\text{height} : \Sigma' \to \mathbb{N}$, and $\text{lift}_h : \Sigma \to \Sigma'$ by

$$\text{base} : (f, h) \mapsto f, \qquad \text{height} : (f, h) \mapsto h, \qquad \text{lift}_h : f \mapsto (f, h),$$

which are extended pointwise to term morphisms. E.g., $\mathrm{lift}_2(f(x,a)) = f_2(x,a_2)$ where a is a constant symbol, and x is a variable. We will use one fixed ordering on Σ', called the *height ordering*, given by $(f,h) < (f',h')$ iff $h > h'$. This ordering is well-founded when restricted to finite sets.

The enrichments label symbols in the right-hand side of a rule with the successor of the minimum of the heights of all symbols at a specified subset of positions in the left-hand side:

Definition 3. *For a term rewriting system R over Σ, and a function f that maps a rewriting rule $(\ell \to r)$ to a nonempty subset of $\mathrm{Pos}_\Sigma(\ell)$, we define the f-cover of R to be the term rewriting system over $\Sigma \times \mathbb{N}$ given by*

$$\mathrm{cover}_f(R) = \{\ell' \to \mathrm{lift}_h(r) \mid (\ell \to r) \in R,\ \mathrm{base}(\ell') = \ell,$$
$$h = 1 + \min\{\mathrm{height}(\ell'(p)) \mid p \in f(\ell,r)\}\}.$$

Note that $\mathrm{cover}_f(R)$ is indeed an enrichment of R.

To present the enrichments, we need one auxiliary definition:

Definition 4. *A position $p \in \mathrm{Pos}_\Sigma(t)$ is a roof position in $t \in \mathcal{T}_\Sigma(X)$ for a set of variables $Y \subseteq X$ if for each $y \in Y$ there is $q \in \mathrm{Pos}_{\{y\}}(t)$ such that $p < q$. Let $\mathrm{RPos}_Y(t)$ denote the set of all roof positions in t for Y.*

E.g., term $t = f(f(x, g(y)), a)$ has $\mathrm{RPos}_{\{y\}}(t) = \{\epsilon, 1, 12\}$ and $\mathrm{RPos}_{\{x\}}(t) = \mathrm{RPos}_{\{x,y\}}(t) = \{\epsilon, 1\}$, so position 12 of g is not a roof position for $\{x\}$ or $\{x,y\}$. Also for $s = f(f(x, g(y)), x)$ we get $\mathrm{RPos}_{\{x\}}(s) = \mathrm{RPos}_{\{x,y\}}(s) = \{\epsilon, 1\}$.

Now we define the enrichments that we will use in the rest of this paper.

Definition 5. – *The* top *enrichment* $\mathrm{top}(R)$ *is* $\mathrm{cover}_f(R)$ *for* $f(\ell,r) = \{\epsilon\}$.
  – *The* roof *enrichment* $\mathrm{roof}(R)$ *is* $\mathrm{cover}_f(R)$ *for* $f(\ell,r) = \mathrm{RPos}_{X(r)}(\ell)$.
  – *The* match *enrichment* $\mathrm{match}(R)$ *is* $\mathrm{cover}_f(R)$ *for* $f(\ell,r) = \mathrm{Pos}_\Sigma(\ell)$.

Example 1. Take $R = \{s(x) + 0 \to s(x)\}$. Then $\mathrm{top}(R)$ contains, among others, the rule $s_1(x) +_2 0_0 \to s_3(x)$, since 2 is the height of the top symbol $+_2$. The system $\mathrm{roof}(R)$ contains the rule $s_1(x) +_2 0_0 \to s_2(x)$, since 1 is the minimal height of a roof symbol (and 0_0 is not in roof position). Finally, $\mathrm{match}(R)$ contains the rule $s_1(x) +_2 0_0 \to s_1(x)$, since 0_0 has minimal height.

Lemma 1. *For a term rewriting system R, both the systems $\mathrm{top}(R)$ and $\mathrm{roof}(R)$ are locally terminating.*

Proof. $\mathrm{top}(R)$ and $\mathrm{roof}(R)$ are ordered by the recursive path ordering induced by the height ordering on Σ', which is well-founded for finite signatures. □

Lemma 2. *For a right-linear term rewriting system R, the system $\mathrm{match}(R)$ is locally terminating.*

Proof. To each term in a ground $\mathrm{match}(R)$-derivation assign the multiset of its symbols. By right-linearity, this sequence of multisets is decreasing with respect to the height ordering on Σ'. □

Remark 1. Right-linearity is essential, as shown by the non-terminating system $\{f_1(a_0, x) \to f_1(x, x)\} \subseteq \text{match}(\{f(a, x) \to f(x, x)\})$.

Definition 6. *For* $e \in \{\text{top}, \text{roof}, \text{match}\}$, *a term rewriting system* R *over* Σ *is called* e-*bounded by* $c \in \mathbb{N}$ *for a language* L *over* Σ *if the maximal height occurring in* $\to_{e(R)}^* (\text{lift}_0(L))$ *is at most* c.

Definition 7. *A finite or infinite term rewriting system* R *over a finite or infinite signature* Σ *is said to be* compact *for a language* $L \subseteq \mathcal{T}_\Sigma$ *if there exists a finite subset* $\Gamma \subseteq \Sigma$ *such that* $\to_R^* (L) \subseteq \mathcal{T}_\Gamma$.

Lemma 3. *If a finite or infinite term rewriting system* R *is locally terminating and compact for* $L \subseteq \mathcal{T}_\Sigma$, *then* R *is terminating on* L.

Obviously $e(R)$ is compact for every e-bounded TRS R. Together with Lemmas 1 and 2 we get:

Proposition 2. $-$ *If* R *is top-bounded for* L, *then* R *is terminating on* L.
$-$ *If* R *is roof-bounded for* L, *then* R *is terminating on* L.
$-$ *If* R *is right-linear and match-bounded for* L, *then* R *is terminating on* L.

Remark 2. All the enrichments discussed here are obtained as covers (Definition 3). This has two implications: since we take the minimum, each enrichment is monotonic (pointwise domination of heights is preserved by parallel derivations), and since the respective sets of positions are comparable by set inclusion the enrichments are comparable as well: for corresponding derivations, match heights are lower or equal to roof heights, and these are lower or equal to top heights. So we prefer roof-heights to top-heights in general, and we will use match-heights for right-linear systems.

Remark 3. Results on derivation lengths carry over from $\text{cover}_f(R)$ to R. For instance, for right-linear systems R, every restriction of $\text{match}(R)$ to a finite signature has linear derivational complexity, so the same complexity holds for every match-bounded right-linear system R. In contrast, for top-bounded R we can have (single) exponential complexity, as for the system $\{f(x) \to g(x, x)\}$ which is top-bounded by 1.

Remark 4. The correspondence between R and its enrichment R' is a *rewrite labelling* (with lift_0 as the initial labelling function) as defined by van Oostrom and de Vrijer [17], Section 8.4. They mention an earlier example of a labelling with the property that "bounded reductions are finite": the *Hyland-Wadsworth labelling* of a rewriting system R is defined just like $\text{match}(R)$, with the only difference of taking max instead of min in Definition 3.

Remark 5. In the string rewriting case, which can be seen as a particular form of linear term rewriting, all non-variable positions are roof positions, therefore $\text{match}(R)$ and $\text{roof}(R)$ coincide. Match-boundedness and top-boundedness differ, as the example $\{ab \to a\}$ shows, which is match-bounded by 1, but not top-bounded.

4   Compatible Tree Automata

Definition 8. *We call a tree automaton $A = (Q, \Sigma, F, T)$ compatible with a term rewriting system R over Σ and a language L over Σ if $L \subseteq \mathcal{L}(A)$, and for each rule $(\ell \to r) \in R$, for each state $q \in Q$, and for each substitution $\sigma : X(\ell) \to Q$, we have that $\ell\sigma \to_T^* q$ implies $r\sigma \to_T^* q$.*

Remark 6. We can decide compatibility of A with R and L in case A and R are finite, and L is given by a finite tree automaton, by just enumerating all cases.

A compatible automaton is closed under left-linear rewriting.

Lemma 4. *If A is compatible with R and L, and R is left-linear, then $\to_R^*(L) \subseteq \mathcal{L}(A)$.*

Proof. We show that R-derivations are covered "step by step" in A: if $t_1 \in \mathcal{L}(A)$ and $t_1 \to_R t_2$, then $t_2 \in \mathcal{L}(A)$. Let $t_1 = t_1[\ell\sigma]_p \to_R t_1[r\sigma]_p = t_2$ for some rule $\ell \to r$, position p, and substitution $\sigma : X(\ell) \to T_\Sigma$. Since $t_1 \in \mathcal{L}(A)$, there is a state q, a final state $\bar{q}$, and a substitution $\rho : X(\ell) \to Q$ such that $t_1 = t_1[\ell\sigma]_p \to_T^* t_1[\ell\rho]_p \to_T^* t_1[q]_p \to_T^* \bar{q}$. Note that ρ exists as R is left-linear. From $\ell\rho \to_T^* q$ and compatibility of A with R we get $r\rho \to_T^* q$. This implies $t_2 = t_1[r\sigma]_p \to_T^* t_1[r\rho]_p \to_T^* t_1[q]_p \to_T^* \bar{q}$, thus $t_2 \in \mathcal{L}(A)$. $\square$

The requirement of left-linearity in Lemma 4 cannot be dropped, as the following example shows.

Example 2. We take an automaton A with states $Q = \{1, 2, 3\}$ and transitions $a \to 1$, $a \to 2$, $f(1,2) \to 3 \in F$. Then $\mathcal{L}(A) = \{f(a,a)\}$. This automaton is compatible with the rewriting system $R = \{f(x,x) \to b\}$ since there are *no* rule $(\ell \to r) \in R$, state q and substitution $\sigma : X(\ell) \to Q$ with $\ell\sigma \to_T^* q$. On the other hand, A is not closed under rewriting, as $\to_R^*(\mathcal{L}(A)) = \{f(a,a), b\}$.

The premise "R is left-linear" in Lemma 4 may be exchanged with "A is deterministic". We don't follow on this branch in the present paper.

By Lemma 4 we get

Lemma 5. *If R is left-linear, and there is some finite automaton A that is compatible with R and L, then R is compact for L.*

5   Constructing Compatible Automata

The following obvious procedure yields an automaton $A = (\Sigma, Q, F, T)$ that is compatible with a rewriting system R and a regular tree language L whenever the procedure terminates:

> Start with an automaton A_0 that accepts L;
> $A := A_0$;
> while A is not compatible
>   choose $q \in Q$, $(\ell \to r) \in R$, $\sigma : X(\ell) \to Q$
>   such that $\ell\sigma \to_T^* q$ and $r\sigma \not\to_T^* q$;

add new states and transitions to A
   yielding a new automaton A' with transitions T'
   such that $r\sigma \to_{T'}^* q$;
$A := A'$;

The interesting issue is the *strategy*: exactly how new states and transitions are chosen. The straightforward strategy is to add a new state for each proper subterm of $r\sigma$ that is not in Q, and fill in the corresponding transitions.

Example 3. For the automaton A with transitions $\{a \to 0, b \to 1, f(0,1) \to 1\}$, and the rewriting system $R = \{\ell \to r\} = \{f(x,y) \to g(h(y),x)\}$ we have $\ell\sigma \to_T^* q$ for $q = 1$ and $\sigma = \{x \mapsto 0, y \mapsto 1\}$. Transitions and states have to be added such that $r\sigma = g(h(1),0) \to_T^* 1 = q$. We add one new state 2, corresponding to the subterm $h(1)$ of $r\sigma$, and transitions $\{h(1) \to 2, g(2,0) \to 1\}$.

The straightforward strategy is the basic idea behind automata closure constructions for various syntactically restricted classes of rewriting systems, e.g., ground, (generalized) (semi)-monadic, finite path overlapping systems. In each case, the syntactic restriction ensures that only finitely many states and transitions will be added.

We cannot generally avoid the addition of states. Therefore the completion procedure for tree automata need not stop. Indeed there are rewriting systems R, as in Example 5, for which the set of descendants is not regular. In such a case, we try to over-approximate the set of descendants by a compatible tree automaton. Genet [6] gets such an approximation by limiting the number of states that are added to the automaton during completion.

We follow a more simplistic approach here that works well for match-bounded string rewriting [10, 19]. We avoid generating some of the additional states as follows. If $r\sigma \not\to_T^* q$, we look for a context $D[\square]$, a context $C[\square, \ldots, \square]$, and terms $t_1, \ldots, t_n \in T_\Sigma(Q)$ such that $D[C[t_1, \ldots, t_n]] = r\sigma$. Suppose that $D[q_0] \to_T^* q$ for some state q_0, and $t_i \to_T^* q_i$ for states q_i, $1 \le i \le n$. Then we add a fresh state for each non-leaf, non-root subterm of $C[\square, \ldots, \square]$, and transitions such that $C[q_1, \ldots, q_n] \to_{T'}^* q_0$. In this way, we re-use states that occur in the derivations $D[q_0] \to_T^* q$ and $t_i \to_T^* q_i$.

This is a non-deterministic procedure. Our implementation chooses in each step one such context $C[\square, \ldots, \square]$ that requires the least number of new states.

For instance, take $R = \{\ell \to r\} = \{b(a(x)) \to c(b(x))\}$, and let A be a two-state automaton with transitions $T = \{e \to 0, a(0) \to 0, b(0) \to 1\}$, state 1 being final. Here, $\mathcal{L}(A) = b(a^*(e))$. Now we have $\ell\{x \mapsto 0\} = b(a(0)) \to_T 1$, but $r\{x \mapsto 0\} = c(b(0)) \not\to_T 1$. Here, $c(b(0)) = D[C[t_1]]$ for $D[\square] = \square$, $C[\square] = c(\square)$, and $t_1 = b(0)$. We have $t_1 \to_T^* 1$, so we add no new state (as $C[\square]$ has no non-trivial subterms), but the transition $c(1) \to 1$. The new automaton is now compatible with R and $\mathcal{L}(A)$, and it accepts $\to_R^*(\mathcal{L}(A)) = c^*(b(a^*(e)))$.

Note that a compatible automaton obtained this way may be an over-approximation of the set of descendants:

Example 4. Let $R = \{a(c) \to b(c)\}$, and $L = \{a(c), a(d)\}$ accepted by the automaton with states $\{0, 1\}$, state 1 being final, and transitions $\{c \to 0, d \to 0,$

$a(0) \to 1\}$. The rewrite rule matches in state 1, so we have to ensure $b(c) \to^* 1$. This could be done by adding a new state 2 and transitions $c \to 2$, $b(2) \to 1$. As $c \to^* 0$, we might want to avoid state 2 and instead add the single transition $b(0) \to 1$. But then $b(d) \to^* 1$ as well, so the automaton now accepts $\{a(c), a(d), b(c), b(d)\}$, which is a proper superset of $\to_R^*(L) = \{a(c), a(d), b(c)\}$.

For string rewriting, match-boundedness implies preservation of regularity of languages [7]. As the following example shows, the corresponding property does not hold for term rewriting.

Example 5. The system $R = \{g(f(x, y)) \to f(h(x), h(y))\}$ is top-bounded by 1. However, the language $\to_R^*(L) \cap f(h^*(a), h^*(a)) = \{f(h^n(a), h^n(a)) \mid n \geq 0\}$ is not regular for the regular language $L = g^*(f(a, a))$, so $\to_R^*(L)$ is not regular either.

So contrary to the string rewriting case, there is no *exact* construction for the sets of descendants of a regular language modulo top-bounded term rewriting. Note that the same holds for roof- and match-bounded rewriting, since these heights are majorized by top heights.

We conclude this section with a few examples that illustrate our approach. In order to visualize tree automata, a transition $f_h(q_1, \ldots, q_n) \to q$ is graphically represented as the hyperedge in the illustration at the right. Squares contain function symbols with height annotations as subscripts, where the argument ordering is indicated by numbers at the incoming arrows. Circles denote states, and double circles denote final states.

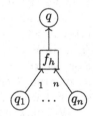

Example 6. For $R = \{f(x, f(a, a)) \to f(f(x, a), x)\}$ over $\{f, a\}$ we present the construction that proves that R is top-bounded by 3. We have to find an automaton A that is compatible with $\text{top}(R)$ and $\text{lift}_0(L)$. We start with the automaton $A_0 = (\Sigma \times \mathbb{N}, \{0\}, \{0\}, T_0)$ where $T_0 = \{a_0 \to 0, f_0(0, 0) \to 0\}$, which accepts $\text{lift}_0(\mathcal{T}_\Sigma)$. Now we have a derivation

$$f_0(0, f_0(a_0, a_0)) \to_{T_0}^* 0,$$

starting with a redex of the rule $f_0(x, f_0(a_0, a_0)) \to f_1(f_1(x, a_1), x)$ from $\text{top}(R)$. The automaton A_0 is not compatible, for $f_1(f_1(0, a_1), 0) \not\to_{T_0}^* 0$. There are no states we could re-use, so our first step follows the straightforward strategy: to add the new states 1 and 2, corresponding to the subterms $f_1(0, a_1)$ and a_1, respectively, and the rules

$$a_1 \to 2, \quad f_1(0, 2) \to 1, \quad f_1(1, 0) \to 0.$$

This way we get another automaton, $A_1 = (\Sigma \times \mathbb{N}, \{0, 1, 2\}, \{0\}, T_1)$, where $T_1 = T_0 \cup \{a_1 \to 2, f_1(0, 2) \to 1, f_1(1, 0) \to 0\}$, such that $f_1(f_1(0, a_1), 0) \to_{T_1} f_1(f_1(0, 2), 0) \to_{T_1} f_1(1, 0) \to_{T_1} 0$. For the new automaton, we again look for violations of compatibility: We have the redex match

$$f_1(1, f_0(a_0, a_0)) \to_{T_1}^* 0$$

with the rule $f_1(x, f_0(a_0, a_0)) \rightarrow f_2(f_2(x, a_2), x)$ in top(R). So A_1 is not compatible, for $f_2(f_2(1, a_2), 1) \not\rightarrow^*_{T_1} 0$. According to the straightforward strategy, we add states 3 and 4, corresponding to the subterms $f_2(1, a_2)$ and a_2, respectively, and transitions to A_1. We get $A_2 = (\Sigma \times \mathbb{N}, \{0, \ldots, 4\}, \{0\}, T_2)$ where

$$T_2 = T_1 \cup \{a_2 \rightarrow 4, f_2(1, 4) \rightarrow 3, f_2(3, 1) \rightarrow 0\},$$

and $f_2(f_2(1, a_2), 1) \rightarrow^*_{T_2} 0$ as wanted. For A_2 again there is a redex

$$f_2(3, f_1(a_0, a_1)) \rightarrow^*_{T_2} 0,$$

but $f_3(f_3(3, a_3), 3) \not\rightarrow^*_{T_2} 0$. We add states 5 and 6 for $f_3(3, a_3)$ and a_3, and transitions

$$a_3 \rightarrow 6, \quad f_3(3, 6) \rightarrow 5, \quad f_3(5, 3) \rightarrow 0.$$

The resulting automaton is displayed at the right; it is compatible with top(R) and lift$_0(T_\Sigma)$. By Remark 6 we can check compatibility with top(R) restricted to the signature $\Sigma \times \{0, 1, 2, 3, 4\}$, being a finite system. Since heights ≥ 4 do not occur, it is even compatible with the infinite system top(R). So R is top-bounded by 3 as claimed, and thus terminating.

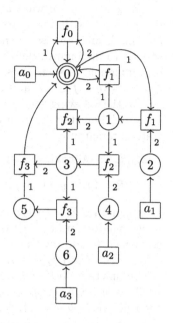

Example 7. For $R = \{f(f(x, a), a) \rightarrow f(x, f(x, a))\}$ we can easily show that it is not top-bounded, as we have the derivation

$$t_{n+2} = f(f(t_n, a), a) \rightarrow_R f(t_n, f(t_n, a)) = f(t_n, t_{n+1})$$

where $t_0 = a$ and $t_{n+1} = f(t_n, a)$. Thus by induction, $t_{n+2} \rightarrow^*_R f(t_n, f(t_{n-1}, \ldots f(a, a) \ldots))$. This derivation reaches top height $n + 1$. However, R is roof-bounded by 1, as the compatible automaton to the right reveals.

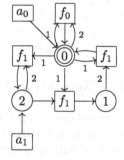

Example 8. Let $R = \{f(a, f(x, a)) \rightarrow f(a, f(f(a, a), x))\}$. As before, we start with the one-state automaton with transitions $\{a_0 \rightarrow 0, f_0(0, 0) \rightarrow 0\}$, accepting lift$_0(T_\Sigma)$. There is a redex match $f_0(a_0, f(0, a_0)) \rightarrow f_0(0, 0) \rightarrow 0$ so we have to ensure that $f_1(a_1, f_1(f_1(a_1, a_1), 0)) \rightarrow^* 0$. This is done by adding states $\{1, 2, 3, 4, 5\}$ and transitions $\{a_1 \rightarrow 1, a_1 \rightarrow 4, a_1 \rightarrow 5, f_1(4, 5) \rightarrow 3, f_1(3, 0) \rightarrow 2, f_1(1, 2) \rightarrow 0\}$. This produces another redex $f_1(a_1, f(3, a_0)) \rightarrow f_1(1, 2) \rightarrow 0$ which requires $f_1(a_1, f_1(f_1(a_1, a_1), 3)) \rightarrow^* 0$. Note that a_0 in the redex has minimal height, and thus the labels in the contractum are 1. This requirement can be fulfilled by adding the transition $f_1(3, 3) \rightarrow 2$, since then

$$f_1(a_1, f_1(f_1(a_1, a_1), 3)) \rightarrow^* f_1(1, f_1(f_1(4, 5), 3)) \rightarrow f_1(1, f_1(3, 3)) \rightarrow f_1(1, 2) \rightarrow 0.$$

Note that this is a state re-use corresponding to the choice $r\sigma = D[C[t_1, t_2]]$ with $t_1 = f_1(a_1, a_1) \to 3 = q_1$, $t_2 = 3 = q_2$, $C = f_1(\Box, \Box)$, $D = f_1(a_1, \Box)$, $q_0 = 2$, $D[q_0] \to 0 = q$. The resulting automaton is compatible with match(R) and $\text{lift}_0(\mathcal{T}_\Sigma)$, thus R is match-bounded by 1, and therefore terminating.

In general, if we want a compatible automaton for $L = \mathcal{T}_\Sigma$, then we may simply start with the automaton for L that has just one state q, which is also final, and for each symbol $f \in \Sigma$, a transition $f(q, \dots, q) \to q$. Doing so fails to distinguish between symbols, and this might cause non-termination of completion. In such cases it is better to "split" the automaton. We then take $Q = F = \{q_f \mid f \in \Sigma\}$, and transitions $\{f(q_1, \dots, q_n) \to q_f \mid f \in \Sigma, q_i \in Q\}$.

For the following example a termination proof via top heights can be obtained by completion starting with the split automaton, but not starting with the one-state automaton.

Example 9. (AProVE-forward_instantiation2, [15]) Consider the system

$$R = \{f(x, y, z) \to g(x, y, z),\ g(d, e, x) \to f(x, x, x),\ a \to b,\ a \to c\}.$$

over signature $\Sigma = \{a, b, c, d, e, f, g\}$. Completion does not stop if we start with an automaton for $\text{lift}_0(\mathcal{T}_\Sigma)$ with only one state q: We find a redex $f_0(q, q, q) \to q$, so we have to add the transition $g_1(q, q, q) \to q$. Since $g_1(d, e, q) \to_R^2 g_1(q, q, q) \to q$, we have to add $f_2(q, q, q) \to q$, and so forth, creating symbols $g_3, f_4, g_5, \dots$.

Completion does succeed if we start with a split automaton. It has 7 states in $Q = \{q_a, q_b, q_c, q_d, q_e, q_f, q_g\}$, and all symbols are labelled by 0. Because of the rules $\{a \to b, a \to c\}$, we add transitions $b_1 \to q_a, c_1 \to q_a$. Due to the rule $f(x, y, z) \to g(x, y, z)$ we add 7^3 transitions $\{g_1(q_x, q_y, q_z) \to q_f \mid q_x, q_y, q_z \in Q\}$, and due to rule $g(d, e, x) \to f(x, x, x)$ we add $\{f_2(q, q, q) \to q_g \mid q \in Q\}$. Rule $f(x, y, z) \to g(x, y, z)$ entails the transitions $\{g_3(q, q, q) \to q_f \mid q \in Q\}$. The result is a compatible automaton, so 3 is the top bound, and R is terminating.

6   Simulating Forward Closures by Rewriting

Forward closures [14] can be used to characterize uniform termination by termination on a restricted set of terms: For a right-linear [4] or non-overlapping [11] term rewriting system R, termination is equivalent to termination on the set RFC(R) of right-hand sides of forward closures.

Following [11], we inductively define the set RFC(R) as the least subset of $\mathcal{T}_\Sigma(X)$ that contains rhs(R), is closed under renaming of variables, and satisfies the condition:

- if $t \in$ RFC(R), $p \in \text{Pos}_\Sigma(t)$, $(\ell \to r) \in R$, ℓ variable-disjoint with t, μ a most general unifier of $t|p$ and ℓ, then $(t[r]_p)\mu \in$ RFC(R).

Next, we will show how to simulate the construction of RFC(R) by ordinary rewriting. Note that since we simulate unification by matching, we cannot cope with non-linearity in left- or right-hand sides. So for the rest of this section we consider *linear* rewrite rules only, i.e., rules with linear left- and right-hand sides.

Let $C \in T_\Sigma(X)$ be linear, and let $\alpha : X \to T_\Sigma(X)$ be a substitution with $\mathrm{dom}(\alpha) \subseteq X(C)$. We say that (C, α) is a *factorization* of $t \in T_\Sigma(X)$ if $t = C\alpha$, and we call C the *context* of the factorization. The factorization is *non-trivial* if $C \notin X$, $\mathrm{dom}(\alpha) \neq \emptyset$, and $x\alpha \notin X$ for every $x \in \mathrm{dom}(\alpha)$. In order to mark the border between the context and the substitution of a factorization, we use the constant $\sharp$, not contained in Σ. Let the substitution $\sigma_\sharp : X \to T_{\Sigma \cup \{\sharp\}}$ be defined by $\sigma_\sharp : x \mapsto \sharp$ for $x \in X$.

Definition 9. *For a linear TRS R define the TRS $R_\sharp$ over $\Sigma \cup \{\sharp\}$ by*

$$R_\sharp = R \cup \{C\sigma'\alpha \to r\sigma'' \mid \ell \to r \in R, \ (C, \alpha) \ a \ non\text{-}trivial \ factorization \ of \ \ell,$$
$$\sigma' = \sigma_\sharp|_{\mathrm{dom}(\alpha)} \ and \ \sigma'' = \sigma_\sharp|_{X(\mathrm{ran}(\alpha))}\}.$$

That is, $R_\sharp$ consists of all rules obtained in the following way. If $\ell \to r$ is a rule in R and (C, α) is a non-trivial factorization of ℓ, then $\ell' \to r'$ is a rule in $R_\sharp$, where ℓ' is obtained from C by replacing every variable in $\mathrm{dom}(\alpha)$ by $\sharp$, and r' is obtained from r by replacing every variable that occurs in $\mathrm{ran}(\alpha)$ by $\sharp$. Note that $R_\sharp$ is linear, and that $X(r) \subseteq X(\ell)$ for each rule $\ell \to r$ in $R_\sharp$.

Example 10. For $R = \{g(f(h(x), h(y))) \to f(y, x)\}$, the system $R_\sharp$ consists of R together with the rules

$$g(\sharp) \to f(\sharp, \sharp), \qquad\qquad g(f(\sharp, \sharp)) \to f(\sharp, \sharp),$$
$$g(f(\sharp, h(y))) \to f(y, \sharp), \qquad\qquad g(f(h(x), \sharp)) \to f(\sharp, x).$$

For $R = \{g(x) \to x\}$ we get $R_\sharp = R$, for $R = \{g(g(x)) \to x\}$ we have $R_\sharp = R \cup \{g(\sharp) \to \sharp\}$, and for $R = \{g(g(x)) \to a\}$ we obtain $R_\sharp = R \cup \{g(\sharp) \to a\}$.

Replacing variables by $\sharp$, we can now characterize $\mathrm{RFC}(R)$ as the set of descendants modulo $R_\sharp$ of $\mathrm{rhs}(R)$, provided R is linear:

Lemma 6. *If R is linear then $\mathrm{RFC}(R)\sigma_\sharp = \to^*_{R_\sharp}(\mathrm{rhs}(R)\sigma_\sharp)$.*

Abbreviate $\mathrm{RFC}(R)\sigma_\sharp$ by $\mathrm{RFC}_\sharp(R)$. A self-contained proof of the following theorem can be found in [9].

Theorem 1. *Let R be a linear term rewriting system. Then*

$$\mathrm{SN}(\to_R) \ if \ and \ only \ if \ \mathrm{SN}(\mathrm{RFC}_\sharp(R), \to_R).$$

Corollary 1. *If a linear term rewriting system R is match-bounded for $\mathrm{RFC}_\sharp(R)$, then R is terminating.*

Proof. If R is match-bounded for $\mathrm{RFC}_\sharp(R)$ then R is terminating on $\mathrm{RFC}_\sharp(R)$ by Proposition 2, thus terminating by Theorem 1. $\square$

Remark 7. In general we do not have $\mathrm{SN}(\to_R)$ iff $\mathrm{SN}(\mathrm{rhs}(R)\sigma_\sharp, \to_{R_\sharp})$. As a counter-example consider the terminating system $R = \{g(g(x)) \to g(x)\}$. Here, $R_\sharp = R \cup \{g(\sharp) \to g(\sharp)\}$ is non-terminating on $\mathrm{rhs}(R)\sigma_\sharp = \{g(\sharp)\}$.

Theorem 1 cannot be generalized to left-linear and non-overlapping systems:

Example 11. For $R = \{f(a, x) \to f(x, x)\}$ we get $R_\sharp = R \cup \{f(\sharp, x) \to f(x, x)\}$. Obviously, R is terminating on $\mathrm{RFC}_\sharp(R) = \to^*_{R_\sharp}(\{f(\sharp, \sharp)\}) = \{f(\sharp, \sharp)\}$, but not terminating.

7 Compatible Automata and Forward Closures

According to Corollary 1, termination of R can be shown by verifying that R is match-bounded for $\mathrm{RFC}_\sharp(R)$. Literally following the definition, we would first construct an automaton A with $\mathrm{RFC}_\sharp(R) \subseteq \mathcal{L}(A)$, that is, A should be compatible with $R_\sharp$ and $\mathrm{rhs}(R)\sigma_\sharp$. Then we find an automaton A' that is compatible with some enrichment R' of R, and a suitable language L' with $\mathcal{L}(A) \subseteq \mathrm{base}(L')$. Since R is linear, we want to use $R' = \mathrm{match}(R)$ and $L' = \mathrm{lift}_0(\mathcal{L}(A))$.

We can merge these two automata constructions into one. To do so, we need an additional, trivial enrichment that completely disregards heights in left-hand sides and assigns height 0 everywhere in the right-hand sides.

Definition 10. *For a term rewriting system R over Σ, the enrichment $\mathrm{zero}(R)$ over $\Sigma \times \mathbb{N}$ is defined by*

$$\mathrm{zero}(R) = \{\ell' \to \mathrm{lift}_0(r) \mid (\ell \to r) \in R,\ \mathrm{base}(\ell') = \ell\}.$$

Lemma 7. *For TRSs R and S and a language L over Σ, if a finite tree automaton A over $\Sigma \times \mathbb{N}$ is compatible with $\mathrm{match}(R) \cup \mathrm{zero}(S)$ and $\mathrm{lift}_0(L)$, then R is match-bounded for $\to^*_{R \cup S}(L)$.*

Proof. We will show that R is match-bounded for $\to^*_{R \cup S}(L)$ by c, where c is the maximal height occurring in $\mathcal{L}(A)$, which exists since A is assumed to be finite. Consider a derivation $t_1 \to^*_R t_2$ with $t_1 \in \to^*_{R \cup S}(L)$, and its canonical lifting $t'_1 \to^*_{\mathrm{match}(R)} t'_2$. We have to show that the maximal height in t'_2 is $\leq c$.

Define the relation $\leq$ on $\mathcal{T}_{\Sigma \times \mathbb{N}}$ by $s \leq t$ if $\mathrm{base}(s) = \mathrm{base}(t)$ and for each position p in s, $\mathrm{height}(s(p)) \leq \mathrm{height}(t(p))$.

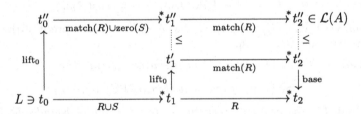

Since $t_1 \in \to^*_{R \cup S}(L)$, there is a term $t_0 \in L$ such that $t_0 \to^*_{R \cup S} t_1$. For the canonical lifting $t''_0 \to^*_{\mathrm{match}(R) \cup \mathrm{zero}(S)} t''_1$ of the derivation $t_0 \to^*_{R \cup S} t_1$ we have $\mathrm{base}(t''_1) = t_1 = \mathrm{base}(t'_1)$, therefore $t'_1 \leq t''_1$. Now there are two canonical liftings of the derivation $t_1 \to^*_R t_2$, both starting in terms with the same base: $t'_1 \to^*_{\mathrm{match}(R)} t'_2$ and $t''_1 \to^*_{\mathrm{match}(R)} t''_2$. From $t'_1 \leq t''_1$ we obtain $t'_2 \leq t''_2$ by monotonicity, cf. Remark 2. We have $t''_2 \in \mathcal{L}(A)$ by compatibility, since $t''_0 \in \mathrm{lift}_0(L)$ and $t''_0 \to^*_{\mathrm{match}(R) \cup \mathrm{zero}(S)} t''_2$. Therefore the maximal height in t''_2 is $\leq c$, and by $t'_2 \leq t''_2$ the same is true for t'_2. $\square$

Choosing $S = R_\sharp \setminus R$ and $L = \mathrm{rhs}(R)\sigma_\sharp$, in combination with Corollary 1 and Lemma 6 this will be used as follows.

Corollary 2. *For a linear TRS R, if some finite tree automaton is compatible with* $\mathrm{match}(R) \cup \mathrm{zero}(R_\sharp \setminus R)$ *and* $\mathrm{lift}_0(\mathrm{rhs}(R)\sigma_\sharp)$, *then R is terminating.*

Example 12. The system $R = \{f(f(a,x),a) \rightarrow f(a, f(x,a))\}$ is not match-bounded. This can be seen as follows. Writing $R_h x$ for $f_h(a_h, x)$, and $L_h x$ for $f_h(x, a_h)$, we have $\mathrm{match}(R)$-derivations $L_0^n R_0^n a_0 \rightarrow^* R_1 R_2 \ldots R_n L_n \ldots L_2 L_1 a_0$ for each $n \geq 0$, exceeding any given bound. On the other hand, the system *is* match-bounded for $\mathrm{RFC}_\sharp(R)$: We have

$$R_\sharp = \left\{ \begin{array}{ll} f(\sharp, \sharp) \rightarrow f(a, f(\sharp, a)), & f(f(\sharp, x), \sharp) \rightarrow f(a, f(x, a)), \\ f(f(a, x), \sharp) \rightarrow f(a, f(x, a)), & f(\sharp, a) \rightarrow f(a, f(\sharp, a)), \\ f(f(\sharp, x), a) \rightarrow f(a, f(x, a)) \end{array} \right.$$

and $\mathrm{rhs}(R)\sigma_\sharp = \{f(a, f(\sharp, a))\}$. This can be accepted by an automaton with states $Q = \{0, 1, 2, 3, 4\}$, $F = \{1\}$, and transitions $\{a_0 \rightarrow 2, \sharp_0 \rightarrow 0, a_0 \rightarrow 4, f_0(0, 4) \rightarrow 3, f_0(2, 3) \rightarrow 1\}$. There is only one redex match for $\mathrm{match}(R) \cup \mathrm{zero}(R_\sharp \setminus R)$, namely $f_0(\sharp_0, a_0) \rightarrow f_0(0, 4) \rightarrow 3$. So we need to ensure that $f_0(a_0, f_0(\sharp_0, a_0)) \rightarrow^* 3$, This can be achieved by adding the single transition $f_0(4, 3) \rightarrow 3$, because then $f_0(a_0, f_0(\sharp_0, a_0)) \rightarrow f_0(4, f_0(0, 4)) \rightarrow f_0(4, 3) \rightarrow 3$. The resulting automaton is compatible with $\mathrm{match}(R) \cup \mathrm{zero}(R_\sharp \setminus R)$. In particular, no rule of $\mathrm{match}(R)$ matches in the automaton. So the automaton certifies that R is match-bounded for $\mathrm{RFC}_\sharp(R)$ by 0, and thus the system is terminating. Termination of R can also be proved by standard methods.

Example 13. Take $R = \{f(a, f(a, x)) \rightarrow f(a, f(x, f(f(a, a), a)))\}$. Here, the set of descendants of $\mathrm{rhs}(R)\sigma_\sharp$ modulo $R_\sharp$ is actually finite: $\{f(a, f(\sharp, f(f(a, a), a))), f(a, f(f(f(a, a), a), f(f(a, a), a)))\}$. Since $\mathrm{match}(R)$ does not match at all, it is match-bounded for $\mathrm{RFC}_\sharp(R)$ by 0.

8 Conclusion

In this paper, we presented a new automated method for termination proofs in term rewriting: constructing compatible tree automata for systems enriched by height annotations. We offered three enrichment schemes – top, roof, and match – which are increasingly more powerful. We demonstrated that match-bounds on the set of right-hand sides of forward closures can be even more powerful for linear TRSs. In contrast to string rewriting, match-bounded systems do not preserve regular languages for term rewriting.

The power of standard methods, like path orderings and interpretations, markedly decreases for small signatures. The fewer symbols there are, the fewer orderings and statuses there are to choose from. To improve this situation, people develop methods that encode additional information into the signature. This can be semantic information (as in semantic labelling, e.g.), or syntactic information (as in dependency pairs, e.g.). Our method belongs to the latter category, for the construction of compatible automata can be seen as a detailed analysis of overlap patterns. An earlier use of tree automata for analyzing rewrite patterns is Middeldorp's estimation of dependency graphs [16].

The algorithms described in this paper have been implemented in the program Matchbox. It is a highly configurable testbed for string and term rewriting with height annotations. The string rewriting version has been described in [18]. Our program is freely available (Haskell source, GNU/Linux executable, CGI interface) via http://141.57.11.163/matchbox/.

An earlier version of Matchbox took part in the Termination Competition during WST04. It solved part of the problems from the contest data base. Since Matchbox does not implement any of the standard methods for automated termination, this illustrates the power of our approach.

The present paper provides known (Example 9) and new (Examples 8, 13) termination problem instances that Matchbox can solve but that other available automated provers cannot. We checked with CiME [3], AProVE [12], TTT [13].

Acknowledgements

We thank the anonymous referees for carefully reading the paper and suggesting improvements. Thanks to Vincent van Oostrom for pointing out the relation to bisimulation and rewriting labellings.

References

1. F. Baader and T. Nipkow. *Term Rewriting and All That*. Cambridge University Press, 1998.
2. H. Comon, M. Dauchet, R. Gilleron, F. Jacquemard, D. Lugiez, S. Tison, and M. Tommasi. *Tree automata techniques and applications*. 1997–2001. Available at http://www.grappa.univ-lille3.fr/tata/.
3. E. Contejean, C. Marché, B. Monate, and X. Urbain. Proving Termination of Rewriting with CiME, In A. Rubio (Ed.), *Proc. 6th Int. Workshop on Termination WST-03*, Technical Report DSIC II/15/03, pp. 71–73, Universidad Politécnica de Valencia, Spain
4. N. Dershowitz. Termination of linear rewriting systems. In S. Even and O. Kariv (Eds.), *Proc. 8th Int. Coll. Automata, Languages and Programming ICALP-81*, Lecture Notes in Comput. Sci. Vol. 115, pp. 448–458. Springer-Verlag, 1981.
5. F. Gécseg and M. Steinby. *Tree Languages*. In G. Rozenberg and A. Salomaa (Eds.), *Handbook of Formal Languages*, Vol. 3, pp. 1–68. Springer-Verlag, 1997.
6. T. Genet. Decidable approximations of sets of descendants and sets of normal forms. In T. Nipkow (Ed.), *Proc. 9th Int. Conf. Rewriting Techniques and Applications RTA-98*, Lecture Notes in Comput. Sci. Vol. 1379, pp. 151–165. Springer-Verlag, 1998.
7. A. Geser, D. Hofbauer and J. Waldmann. Match-bounded string rewriting systems. *Appl. Algebra Engrg. Comm. Comput.* 15(3-4):149–171, 2004.
8. A. Geser, D. Hofbauer, J. Waldmann, and H. Zantema. Tree automata that certify termination of term rewriting systems. In M. Codish and A. Middeldorp (Eds.), *Proc. 7th Int. Workshop on Termination WST-04*, Aachener Informatik Berichte AIB-2004-07, RWTH Aachen, Gemany, pp. 14–17, 2004.
9. A. Geser, D. Hofbauer, J. Waldmann, and H. Zantema. Tree automata that certify termination of left-linear term rewriting systems. 2005. Full version available at http://www.imn.htwk-leipzig.de/~waldmann/pub/rta05/

10. A. Geser, D. Hofbauer, J. Waldmann, and H. Zantema. Finding finite automata that certify termination of string rewriting. In M. Domaratzki, A. Okhotin, K. Salomaa, and S. Yu (Eds.), *Proc. 9th Int. Conf. Implementation and Application of Automata CIAA-04*, *Lecture Notes in Comput. Sci.* Vol. 3317, pp. 134–145. Springer-Verlag, 2004.

11. O. Geupel. Overlap closures and termination of term rewriting systems. Technical Report MIP-8922, Universität Passau, Germany, 1989.

12. J. Giesl, R. Thiemann, P. Schneider-Kamp, and S. Falke. Automated termination proofs with AProVE. In V. van Oostrom (Ed.), *Proc. 15th Int. Conf. Rewriting Techniques and Applications RTA-04*, *Lecture Notes in Comp. Sci.* Vol. 3091, pp. 210–220. Springer-Verlag, 2004.

13. N. Hirokawa and A. Middeldorp. Tsukuba Termination Tool. In R. Nieuwenhuis (Ed.), *Proc. 14th Int. Conf. Rewriting Techniques and Applications RTA-03*, *Lecture Notes in Comp. Sci.* Vol. 2706, pp. 311–320. Springer-Verlag, 2003.

14. D.S. Lankford and D.R. Musser. A finite termination criterion. Unpublished draft, Information Sciences Institute, University of Southern California, Marina-del-Rey, CA, 1978.

15. C. Marché and A. Rubio (Eds.) Termination Problems Data Base. 2004. http://www.lri.fr/~marche/wst2004-competition/tpdb.html

16. A. Middeldorp. Approximations for strategies and termination. In *Proc. 2nd Int. Workshop on Reduction Strategies in Rewriting and Programming*, *Electron. Notes Theor. Comput. Sci.* 70(6), 2002.

17. V. van Oostrom and R. de Vrijer. Equivalences of Reductions. In Terese, *Term Rewriting Systems*, pp. 301–474. Cambridge Univ. Press, 2003.

18. J. Waldmann. Matchbox: a tool for match-bounded string rewriting, In V. van Oostrom (Ed.), *Proc. 15th Int. Conf. Rewriting Techniques and Applications RTA-04*, *Lecture Notes in Comp. Sci.* Vol. 3091, pp. 85–94. Springer-Verlag, 2004.

19. H. Zantema. TORPA: Termination of rewriting proved automatically. In V. van Oostrom (Ed.), *Proc. 15th Int. Conf. Rewriting Techniques and Applications RTA-04*, *Lecture Notes in Comp. Sci.* Vol. 3091, pp. 95–104. Springer-Verlag, 2004. Updated version accepted for *J. Automat. Reason.*.

20. H. Zantema. Termination. In Terese, *Term Rewriting Systems*, pp. 181–259. Cambridge Univ. Press, 2003.

Twenty Years Later*

Jean-Pierre Jouannaud**

LIX, École Polytechnique,
91400 Palaiseau, France
jouannaud@lix.polytechnique.fr
http://www.lix.polytechnique.fr/Labo/jouannaud

Abstract. The first RTA conference took place in Dijon, in 1985. This
year, 2005, it takes place in Nara. Nara and Dijon share a glorious past
but can be considered as being "Sleeping Beauties", after the title of a
book by the Nobel price novelist yasunari Kawabata.

Is RTA sleeping on its glorious past? Back in the late 80s, many of us
feared that this would soon be the case, that research in rewrite systems
was deepening the gap with everyday's computer science practice, and
that we should develop rewrite-based powerful provers that would make
a difference with the state of art and help address real applications such
as software verification.

More than ten years later this has not really happened in the way we
thought it would. What has happened is that many research areas, such
as programming languages, constraint solving, first-order provers, proof
assistants, security theory, and verification have all been fertilized by
ideas coming from term rewriting. In return, our field has been renewed
by new problems and techniques coming from outside our small community.

I am convinced that this will continue, and that new subject areas will
join the journey. There are at least two reasons. To quote a celebrated
sentence that I have read in many papers: *Equations are ubiquitous in
computer science.* This is the first reason: we all like to use equations
for modeling problems. The second is that we have developed extremely
powerful, sophisticated tools to reason with equations. Many computer
scientists do not know these tools. It is our responsibility to preach for
their use by showing all we can do with them.

1   Introduction

My goal is to illustrate several aspects of the contributions of rewriting theory
to problems originating from programming and theorem proving, two closely
related fields that benefit from a term rewriting perspective. I will concentrate on
ordered paramodulation, a very old problem which is still progressing, rule-based
programming, tree automata, and proof assistants. I will not refrain quoting my
self.

---

* *Twenty years later* is the title of a novel by the french writer Alexandre Dumas.
** Project LogiCal, Pôle Commun de Recherche en Informatique du Plateau de Saclay,
   CNRS, École Polytechnique, INRIA, Université Paris-Sud.

J. Giesl (Ed.): RTA 2005, LNCS 3467, pp. 368–375, 2005.

2   Basic Ordered Resolution and Paramodulation

Ordered completion is a generalization of ground completion, because it yields a convergent rewrite system from an arbitrary set of equations, therefore providing with a uniform technique to reason about the word problem in an arbitrary equational theory, which is based on the existence of normal forms. The main difference with the case of ground completion is that the rewrite system may be infinite, and therefore no decidability result can be obtained in general. This is indeed the main strength of this method, that decidability is not required.

Ordered completion was first successfully addressed by Peterson [50], who addressed the case of a finite rewrite ordering, before the general case was solved by Hsiang and Rusinowitch who developed *transfinite semantic trees* for that purpose [31]. The use of transfinite semantic trees was made necessary by the possible use of transfinite orderings in ordered completion.

A more general, earlier framework was elaborated by Robinson and Wos [54], combining resolution and paramodulation: they were the first to propose the replacement of the axioms for equality by specific inference rules in order to reduce the search space. The idea of restricting the set of inferences further by systematically using normal forms generated by a rewrite ordering as it is the case with ordered completion is due to Lankford [42]. Lankford, however, did not have the tools to solve the problem in its full generality. This was done much later in a series of papers pioneered by Bachmair and Ganzinger, who came up with a novel, model theoretic method based on the idea of forcing [2–5]. It is interesting to notice that Goubault succeeded recently to improve over Bachmair and Ganzinger by using finite semantic trees: the very simple but beautiful idea, which Goubault himself ascribes to Rusinowitch, consists in applying a compactness argument before to construct the finite semantic tree of a given unsatisfiable set of clauses [28].

The last 15 years saw another achievement, with constraints taking over unification in deduction calculi [40]. This is an important phenomenon: after logic programming, constraints are making their way everywhere, in term rewriting theory where the difficult problem of local confluence of order-sorted rewrite rules was reduced to satisfiability of membership constraints [20], in automated deduction seen as a generalization of logic programming, but also in model checking where it allows to go smoothly from finite to some decidable infinite systems [25], or in functional programming where it yields a more elegant and powerful tool to express type inference algorithms [1]. In particular, constraints have been used very successfully to block inferences that were made inside substitutions inherited from previous inferences [6, 48]. This restriction of deduction calculi is dubbed *basic* after Hullot's pionneering work on *basic narrowing* [35].

Examples of use of ordered completion to *modularity* problems include: modular unification algorithms [13], modular confluence properties [36], and the study of CCC, a calculus of constructions embedding the congruence closure algorithm into the conversion rule [11]. Examples of use of constrained deduction to *decidability* results include: decidability of set constraints [7] and decidability of standard theories [47].

3   Rule-Based Programming

This topic is probably best examplified with MAUDE, a language developed by José Meseguer and his collaborators at SRI first, and now at the state university of Illinois at Urbana-Champaign [45]. Related efforts were conducted in parallel in France and in Japan, by Claude Kirchner and his group, who developed the language ELAN [41], and by Kokichi Futatsugi and his collaborators, who developed the language CAFE [46]. All three languages owe their origin to the OBJ-family of languages, a project started in California by Joseph Goguen in the early 80s, following his earlier work on Clear [16].

The main novelty of MAUDE was to consider that both functional and concurrent programming could be addressed uniformly by rewriting, depending whether confluence was satisfied or not. Elan goes even further by internalizing rewriting in the so-called ρ-calculus via a specific binding construct generalizing the λ-calculus [8], while Cafe insists on the use of co-algebrsa [23]. The use of rewriting as a functional model was of course well accepted [22], while the use of rewriting for non-functional programming had been advocated before for particular applications, especially unification, and more generally constraint solving [37].

There are even more familiar programming languages that use rule based constructs: this is the case of the Ocaml family of languages, where the case construct bases selection on pattern matching.

Another language based on rewriting is Isabelle, implementing Nipkow's higher-order rewrite systems [44]. Isabelle is targeting applications in which programs operate on data structures with binders, like program transformations. What makes this work apparently different from a language like MAUDE is that it uses higher-order pattern matching instead of plain matching. But this singularity is not really relevant: MAUDE uses pattern matching modulo associativity and commutativity, and a close look to Nipkow's higher-order rewrite systems shows that the problems are exactly those of rewriting modulo [39].

4   Tree Automata

At the first RTA, there was not a single paper using tree automata. There were of course papers in formal language theory using word automata. But no tree automaton. However, there were many informal talks about an almost published paper by David Plaisted, who solved the problem of inductive reducibility [51][1].

It is easy to see that tree automata are equivalent to OBJ's order-sorted signatures, and they were actually introduced in the 60s in a related context,

---

[1] Actually, the proof was wrong. I had found a counterexample to a *simple lemma* stated without proof, and the whole proof could not be repaired. When Emmanuel Kounalis and myself explained the problem to David Plaisted, he succeeded to found a new, completely different proof at the blackboard in ten minutes. This was really impressing: he understood our counterexample much better than ourselves. This new proof contained a complex argument that was much later understood as a pumping lemma on tree automata with equality tests [17].

see also [14]. But they had been almost completely forgotten. In some sense, RTA'85 was their second birth. They were later used in many different context, with the strong push coming from the rewriting community: set constraints [26], higher-order matching [33], strong sequentiality [19], Presbuger arithmetic [12], AC-inductive-reducibility [43], inductive theorem proving [15] and more. The theory of tree-automata and its many applications is studied in depth in [32].

5   Proof Assistants

Many will agree with me when saying that Isabelle, Coq and PVS are three among the most important proof assistants. Isabelle is based on Nipkow's higher-order rewriting [44]. PVS is based on Shostak's decision procedure for a combination of convex theories, whose ideas are clearly based on rewriting [52, 53]. Originally based on the Calculus of Constructions [21], then on the Calculus of Inductive Constructions [49], Coq is now rapidly moving towards a heavy use of rewriting, for defining inductive types on the one hand [10], and for specifying the conversion rule on the other hand [9]. None of these proof-assistants was available in 1985. At that time, most people in our community believed in the future of first-order provers, rather than higher-order ones. The situation is now reversed: many believe in the superiority of higher-order languages for modeling purposes: first-order provers are often seen as supplementing tactics for higher-order provers. And first-order decidability results are accordingly seen as a particular way to automate the higher-order prover in these cases. An even stronger argument is the existence of the Curry-Howard isomorphism which allows to see intuitionistic logic as a kind of abstract machine for implementing formal proofs. On the other hand, first-order provers have been successful for solving very particular problems such as crypto-attacks, for which a blind search appears adequate [24].

My own perspective on this question is that the coming years will see a new generation of proof assistants, in which (higher-order) rewriting superseeds the lambda calculus. Isabelle is the first prover of this kind, but has lost many of the important features of Curry-Howard based calculi. I anticipate both approaches to merge in the coming years, and some work has been done already [9, 11]. An other merge is coming as well: dependent types are making their way in programming languages [55], while modules and functors have been successfully added to the calculus of inductive constructions [18] as well as a compiler for reductions [29, 30]. This move towards harmony will make its way through in the coming years. I do not see a good reason in the present dichotomy between programming languages and proof assistants.

6   Conclusion

I tried to sketch what important unexpected developments based on rewriting took place in the past. I will try now to give my idea about the future.

First, I think that we need to continue investigating the fundamental properties of term rewriting formats: type preservation, termination and Church-Rosser properties are equally important. Any progress there means a progress with the applications. Another important fundamental question is the relationship between term rewriting and tree automata. We need to investigate these questions in various contexts, first-order, higher-order, and modulo. And we need to continue our work on abstract rewriting, in the light of Huet's work for the first-order case [34], and what was later done for the modulo [38] and higher-order cases [27]. The higher-order case, especially, needs more work, since the only abstract property investigated there was the finite development theorem.

Second I think we need to continue investigating the efficient implementation techniques of rewriting systems. Much has been done for the first-order case with Maude and Elan, and with the work of Ganzinger's group and of Nieuwenhuis's group with the SATURATE and SPASS systems, but this is not the end of the road. Since type-checking in proof-assistants like Coq relies on rewriting techniques, compilation techniques must be throroughly studied which combine first- and higher-order pattern matching.

Third, I think that we need to understand which other areas of computer science may benefit from our work. A recent interesting example was provided by security protocols: since rewrite rules can be seen as a specification language, security protocols can be modeled by rules. Using this approach, Rusinowitch showed that finding an attack to a cryptographic protocol could be achieved by using narrowing. Comon and others also showed that rewriting was a good tool for modeling security protocols since it allowed to smoothly integrate properties of the cryptographic primitives which were naturally expressed as equations.

Last, but not least, I think that we need to integrate the different existing kinds of rewriting, plain rewriting based on plain pattern matching, rewriting modulo based on plain pattern matching, rewriting modulo based on pattern matching modulo, normalized rewriting, normal rewriting, higher-order rewriting based on plain pattern matching, higher-order rewriting based on higher-order pattern matching, higher-order rewriting based on higher-order pattern matching modulo, into a single coherent framework in order to better understand how to design an abstract machine to implement them all, and make them available to users. This question is of course directly related to my view on the future of proof assistants that I sketched in the previous section. It is also related to the need of an abstract investigation of the fundamental properties of term rewriting formats

Acknowledgments

To all those I met during these 20 years and helped me understand that their problems were more important than mine.

References

1. A. Aiken and E. Wimmers. Type inclusion constraints and type inference. In *Proc. 7th ACM Conference on Functional programming and Computer Archtecture*, pages 31–41, Copenhaguen, Denmark, 1993.

2. Leo Bachmair and Harald Ganzinger. Completion of first-order clause with equality by strict superposition. In *Proc. 2nd Int. Workshop on Conditional and Typed Rewriting Systems, Montreal, LNCS 516*, 1990.

3. Leo Bachmair and Harald Ganzinger. On restrictions of ordered paramodulation with simplification. In *Proc. 10th Int. Conf. on Automated Deduction, Kaiserslautern, LNCS 449*, 1990.

4. Leo Bachmair and Harald Ganzinger. Rewrite-based equational theorem proving with selection and simplification. Technical Report MPI-I-91-208, Max-Planck-Institut für Informatik, Saarbrücken, September 1991. to appear in Journal of Logic and Computation.

5. Leo Bachmair and Harald Ganzinger. Non-clausal resolution and superposition with selection and redundancy criteria. In *Proc. of the LPAR'92*, 1992. Lecture Notes in Computer Science.

6. Leo Bachmair, Harald Ganzinger, Christopher Lynch, and Wayne Snyder. Basic paramodulation and superposition. In Deepak Kapur, editor, *Proc. 11th Int. Conf. on Automated Deduction, Saratoga Springs, NY, LNAI 607*. Springer-Verlag, June 1992.

7. Leo Bachmair, Harald Ganzinger, and Uwe Waldmann. Set constraints are the monadic class. In *Proceedings of the Eigth Annual IEEE Symposium on Logic in Computer Science*. IEEE Comp. Soc. Press, 1993.

8. Gilles Barthe, Horatiu Cirstea, Claude Kirchner, and Luigi Liquori. Pure patern type systems. In *Conference Record of the 30th Symposium on Principles of Programming Languages*, New-Orleans, USA, January 2003. ACM.

9. Frédéric Blanqui. Definitions by rewriting in the Calculus of Constructions. *Mathematical Structures in Computer Science*, 15(1):37–92, 2005.

10. Frédéric Blanqui. Inductive types in the Calculus of Algebraic Constructions. *Fundamenta Informaticae*, to appear.

11. Frédéric Blanqui, Jean-Pierre Jouannaud, and Pierre-Yves Strub. A calculus of congruent constructions. Technical report, École Polytechnique, 2005. submitted.

12. Alexandre Boudet and Hubert Comon. Diophantine equations, Presburger arithmetic and finite automata. In H. Kirchner, editor, *Proc. Coll. on Trees in Algebra and Programming (CAAP'96)*, Lecture Notes in Computer Science, pages 30–43, 1996.

13. Alexandre Boudet, Jean-Pierre Jouannaud, and Manfred Schmidt-Schauß. Unification in Boolean rings and Abelian groups. *Journal of Symbolic Computation*, 8:449–477, 1989.

14. A. Bouhoula, J.-P. Jouannaud, and J. Meseguer. Specification and proof in membership equational logic. In Michel Bidoit and Max Dauchet, editors, *Theory and Practice of Software Development*, volume 1214 of *Lecture Notes in Computer Science*, pages 67–92, Lille, France, April 1997. Springer-Verlag.

15. Adel Bouhoula and Jean-Pierre Jouannaud. Automata-driven automated induction. In *Twelfth Annual IEEE Symposium on Logic in Computer Science*, pages 14–25, Warsaw,Poland, June 1997. IEEE Comp. Soc. Press.

16. R. M. Burstall and J. A. Goguen. The semantics of CLEAR, a specification language. In *Proc. Winter School on Abstract Software Specifications, Copenhagen, LNCS 86*, 1979.

17. Anne-Cécile Caron, Jean-Luc Coquidé, and Max Dauchet. Encompassment properties and automata with constraints. In Claude Kirchner, editor, *5th International Conference on Rewriting Techniques and Applications*, volume 690 of *Lecture Notes in Computer Science*, pages 328–342, Montreal, Canada, June 1993. Springer-Verlag.

374   Jean-Pierre Jouannaud

18. Jacek Chrzaszcz. Modules in cow are and will be correct. In M. Coppo S. Berardi and F. Damiani, editors, *Proceedings TYPES'03*, volume 3085 of *Lecture Notes in Computer Science*, pages 135–150, Torino, Italy, 2003. Springer-Verlag.
19. Hubert Comon. Sequentiality, second-order monadic logic and tree automata. In Dexter Kozen, editor, *Tenth Annual IEEE Symposium on Logic in Computer Science*, pages 508–517, San Diego, CA, June 1995. IEEE Comp. Soc. Press.
20. Hubert Comon and Catherine Delor. Equational formulae with membership constraints. *Information and Computation*, 112(2):167–216, August 1994.
21. Thierry Coquand and Gérard Huet. The calculus of constructions. *Information and Computation*, 76:95–120, February 1988.
22. Nachum Dershowitz. Equations as programming language. In *Proceedings of the Fourth Jerusalem Conference on Information Technology*, pages 114–124, Jerusalem, Israel, May 1984. IEEE Computer Society.
23. Razvan Diaconescu and Kokichi Futatsugi. Cafeobj-report: The language, proof techniques and methodologies for object-oriented algebraic specification. In *AMAST series in Computing*, volume 6. World Scientific, 1998.
24. Michaël Rusinowitch et alii. The aviss security protocols analysis tool – system description. In *Proceedings of Computer-Aided Verification 02*, 2003.
25. Laurent Fribourg and Morcos Veloso Peixoto. Automates concurrents à contraintes. *Technique et Science Informatiques*, 13(6), 1994.
26. Rémy Gilleron, Sophie Tison, and Marc Tommasi. Solving systems of set constraints using tree automata. In *stacs93*, 1993.
27. G. Gonthier, J.-J. Lévy, and P.-A. Mellies. An abstract standardisation theorem. In *Proc. 7th IEEE Symp. on Logic in Computer Science*, Santa Cruz, CA, 1992.
28. Jean Goubault-Larrecq. Résolution ordonnée avec sélection et classes décidables de la logique du premier ordre, 2004. available from the web.
29. Benjamin Gregoire. *Compilation de termes de preuves. Un mariage entre Coq et OCaml*. PhD thesis, École Polytechnique, Palaiseau, France, 2003.
30. Olivier Hermant. A rewriting abstract machine for coq, 2004.
31. Jieh Hsiang and Michaël Rusinowitch. On word problems in equational theories. In Thomas Ottmann, editor, *14th International Colloquium on Automata, Languages and Programming*, volume 267 of *Lecture Notes in Computer Science*, pages 54–71, Karlsruhe, Germany, July 1987. Springer-Verlag.
32. Denis Lugiez Hubert Comon, Max Dauchet and Sophie Tison, editors. *Tree Automata techniques and Applications*. http://www.grappa.univ-lille3.fr/tata/, Lille, France, 2002.
33. Hubert Comon and Yann Jürsski. Higher-order matching and tree automata. In M. Nielsen and W. Thomas, editors, *Proc. 11th Computer Science Logic*, volume 1414 of *Lecture Notes in Computer Science*, Aarhus, Denmark, August 1997. Springer-Verlag.
34. Gérard Huet. Confluent reductions: abstract properties and applications to term rewriting systems. *Journal of the ACM*, 27(4):797–821, October 1980.
35. J.-M. Hullot. Canonical forms and unification. In W. Bibel and R. Kowalski, editors, *5th International Conference on Automated Deduction*, volume 87 of *Lecture Notes in Computer Science*, Les Arcs, France, July 1980. Springer-Verlag.
36. Jean-Pierre Jouannaud. Modular associative commutative confluence. Technical report, École Polytechnique, 2005.
37. Jean-Pierre Jouannaud and Claude Kirchner. Solving equations in abstract algebras: A rule-based survey of unification. In Jean-Louis Lassez and Gordon Plotkin, editors, *Computational Logic: Essays in Honor of Alan Robinson*. MIT-Press, 1991.

38. Jean-Pierre Jouannaud and Emmanuel Kounalis. Automatic proofs by induction in equational theories without constructors. In *Logic in Computer Science*, June 1986.

39. Jean-Pierre Jouannaud, Albert Rubio, and Femke Van Raamsdonk. Higher-order rewriting with types and arities. Technical report, École Polytechnique, 2005. submitted.

40. Claude Kirchner, Helene Kirchner, and Michaël Rusinowitch. Deduction with symbolic constraints. *Revue Française d'Intelligence Artificielle*, 4(3):9–52, 1990. Special issue on automatic deduction.

41. Claude Kirchner and Piere Moreau. Non deterministic computations in elan. In J.L. Fiadeiro, editor, *Proceedings 13th workshop on abstract data types*, volume 1589 of *Lecture Notes in Computer Science*, Lisbon, Portugal, October 1999. Springer-Verlag.

42. Dallas S. Lankford. Canonical inference. Memo ATP-32, University of Texas at Austin, March 1975.

43. D. Lugiez and J.-L. Moysset. Complement problems and tree automata in AC-like theories. In *Proc. Symp. on Theoretical Aspects of Computer Science*, Würzburg, 1993. also available as techincal report CRIN 92-R-175.

44. Richard Mayr and Tobias Nipkow. Higher-order rewrite systems and their confluence. *Theoretical Computer Science*, 192(1):3–29, February 1998.

45. José Meseguer. A logical theory of concurrent objects and its realization in the maude language. To appear in G. Agha, P.Wegner, and A.Yoneezawa (editors), Research Directions in Object-Based Concurrency, 1992.

46. A.T. Nakagawa and K. Futatsugi. An overview of cafe specification environment. In *Proc. of the 1st IEEE International Conference on Formal Engineering Methods*, pages 170–181. IEEE Computer Society Press, 1997.

47. Robert Nieuwenhuis. Basic paramodulation and decidable theories. In Amy Felty, editor, *Eleventh Annual IEEE Symposium on Logic in Computer Science*, New-Brunswick, CA, June 1996. IEEE Comp. Soc. Press.

48. Robert Nieuwenhuis and Albert Rubio. Completion of first-order clauses by basic superposition with ordering constraints. Tech. report, Dept. L.S.I., Univ. Polit. Catalunya, 1991. To appear in Proc. 11th Conf. on Automated Deduction, Saratoga Springs, 1992.

49. Christine Paulin-Mohring. Inductive definitions in the system COQ. In *Typed Lambda Calculi and Applications*, pages 328–345. Springer-Verlag, 1993. LNCS 664.

50. Gerald E. Peterson. A technique for establishing completeness results in theorem proving with equality. *SIAM Journal on Computing*, 12(1):82–100, February 1983.

51. David A. Plaisted. Semantic confluence tests and completion methods. *Information and Control*, 65(2-3):182–215, May/June 1985.

52. R. E. Shostak. An efficient decision procedure for arithmetic with function symbols. *J. of the Association for Computing Machinery*, 26(2):351–360, April 1979.

53. R.E. Shostak. Deciding combinations of theories. Technical Report CSL 132, SRI International, February 1982.

54. L. Wos, G. Robinson, D. Carson, and L. Shalla. The concept of demodulation in theorem proving. *Journal of the ACM*, 14:698–709, 1967.

55. Hongwei Xi and Franck Pfenning. Dependent types in practical programming. In *Conference Record of the 21st Symposium on Principles of Programming Languages*, San Antonio, Texas, 1998. ACM.

Open. Closed. Open.

Nachum Dershowitz

School of Computer Science, Tel Aviv University
Ramat Aviv 69978, Israel
nachum.dershowitz@cs.tau.ac.il

פתוח סגור פתוח.	Open closed open.
זה כל האדם.	That is Man.

—Yehudah Amichai (Israel)
Open Closed Open (1998)

Abstract. As a window into the subject, we recount some of the history (and geography) of two mature, challenging, partially open, partially closed problems in the theory of rewriting (numbers 13 and 21 from the original *RTA List of Open Problems*). One problem deals with (criteria for left-linear) confluence and the other with termination (of one linear or string rule), the two paradigmatic properties of interest for rewrite systems of any flavor. Both problems were formulated a relatively long time ago, have seen considerable progress, but remain open. We also venture to contemplate the future evolution and impact of these investigations.

1 Introduction

Twenty years later, and we're still hitting on a keyboard.

—Michael Capellas, Chairman and CEO of Compaq (USA),
Twentieth Anniversary of the PC,
Tech Museum of Innovation (August 2001)

Rewriting – in the sense of systematically replacing symbolic terms – is as old as algebra. Diophantus of Alexandria[1] (Egypt) in his famous (ca. 3rd c. CE) book, *Arithmetica*[2], reduced determinate and indeterminate equations to a form he knew how to solve. The use of rewriting nowadays in automated deductive engines derives from this ancient nascence of symbolic computation.

The formal study of rewriting and its properties began in 1910 with a paper by Axel Thue (Norway) [89]. Significantly, most early models of computation

---

[1] After whom Diophantine equations are named.

[2] It was in his copy of *this* book that Pierre de Fermat (France) wrote this frustratingly famous marginal note:

Cubem autem in duos cubos, aut quadratoquadratum in duos quadratoquadratos, et generaliter nullam in infinitum ultra quadratum potestatem in duos ejusdem nominis fas est dividere: cujus rei demonstrationem mirabilem sane detexi. Hanc marginis exiguitas non caparet.

J. Giesl (Ed.): RTA 2005, LNCS 3467, pp. 376–393, 2005.

were based on notions of rewriting strings or terms: Thue systems [90]; Andrei Markov's (Russia) normal algorithms [66]; and Alonzo Church's (USA) lambda calculus [10]. This all led to the continued study of rewriting in the context of programming language semantics.

As a window into the history of rewriting, I have chosen two problems (numbers 13 and 21) from the original *RTA List of Open Problems* [19]. The first relates to confluence and the second, to termination.

Confluence is perhaps better known as the (equivalent) "Church-Rosser property," after a 1936 paper by Church and Barkley Rosser (USA) [11]. Properties of this property were studied shortly thereafter by Maxwell Newman (England) [73], of "Newman's Lemma" fame, and remain central in Combinatory Logic and Lambda Calculus (the immediate ancestors of the study of rewriting). Thue's paper foreshadowed the use of these and other concepts in solving word problems; see the review by Magnus Steinby (Finland) and Wolfgang Thomas (Germany) [85]. Already in 1967, Saul Gorn (USA) [37] discussed the Church-Rosser property for the use of definitions in symbolic computation.

Termination ("uniform termination" or "strong normalization") is important in automated deduction applications, to guarantee that simplification of formulæ does not itself go on forever. Simplification is often essential for reasonable performance of theorem provers. Formal proofs of termination are as old as Euclid's (Egypt) algorithm for greatest common divisor.

Gorn also did early work on proofs of termination of symbolic computation. In the abstract to his 1973 paper [36], he wrote[3]:

> This paper ... explores such questions as (1) What different interpretations can be given to the expression "the intent of the process"? (2) Does the process, or should the process end? In either event, how do we prove it? (3) If the process does end, how do we prove that it does what was intended? This question may be meaningful even if the process does not end. (4) Is there a whole class of processes that stand or fall together? Can we adapt our proof of conclusiveness to cover the whole class? (5) Do the processes of the class yield the same or different results, and whichever it is, how do we prove it?

The RTA list open problems, whence the examples herein are drawn, was created by Jan Willem Klop (The Netherlands), Jean-Pierre Jouannaud (France), and myself (USA, at the time) on the occasion of the fourth *Rewriting Techniques and Applications* conference, held in 1991 (in Italy) and chaired by the

---

[3] Gorn is indirectly responsible for my interest in the subject of termination of rewriting: He discussed the issue with Bob Floyd (USA), who posed a question on the subject on a 1967 qualifying exam in computer science at Carnegie-Mellon University. Zohar Manna (USA) solved the problem and went on to write a dissertation on termination. Later, Zohar showed me a 1970 paper [65] of his with Steve Ness (USA) on termination of rewriting, notes of his discussions with Steve and with Amir Pnueli (Israel) on completeness of homomorphism-based methods, as well as the dissertation of another CMU student, Renato Iturriaga (USA, at the time), thereby sparking my unquenched interest.

late Ron Book (USA)[4]. Its 44 problems were compiled thanks to the contributions of many researchers who responded to messages on Pierre Lescanne's (France) rewriting mailing list[5], and from various older lists. Updated lists subsequently appeared in the proceedings of RTA '93 (Canada) [20] – which added 33 more problems, RTA '95 (Germany) [21] – 10 more, and RTA '98 (Japan) [23].

Since October 1997, the list of open problems has been maintained as a web service at

> http://www.lsv.ens-cachan.fr/~treinen/rtaloop

This effort is spearheaded by Ralf Treinen (France). Currently, the list comprises 103 problems[6], at least 28 of which have – gratifyingly – been solved to date, and many more have enjoyed significant progress.

2   Left-Linear Confluence

> E[lementary Problem] #1541:
> Find the maximum[7] and minimum[8] numbers of
> "Friday the 13th's" that can occur in a year.
>
> —George Clark Bush (Canada)
> *The American Mathematical Monthly* (1988)

The thirteenth problem in the original list of open problems is:

Problem #13: Give decidable (sufficient) criteria for left-linear rewriting systems to be Church-Rosser.

This problem was suggested for inclusion by Jean-Jacques Lévy (France)[9].

As already mentioned, the Church-Rosser property, $\leftrightarrow^* \subseteq \rightarrow^* {}^*\!\leftarrow$ (convergence implies joinability), had been thoroughly investigated in the context of lambda calculi and combinatory logic, and shown equivalent to the diamond confluence property ${}^*\!\leftarrow\rightarrow^* \subseteq \rightarrow^* {}^*\!\leftarrow$ (meetability implies joinability) by Max Newman (UK) in 1942 [73][10].

---

[4] For a summary of Ron's contributions to the theory of Thue systems, a.k.a. string rewriting, see Bob McNaughton's (USA) [63]. Book and Friedrich Otto (Germany) co-authored a monograph on the subject [7].

[5] Pierre has been caretaker of this mailing list since he founded it in 1988.

[6] One more than Harvey Friedman's (USA) list of hard problems in mathematical logic in *J. Symbolic Logic* **40**(2), pp. 113–129 (1975).

[7] Three, as shown by Charles Heuer, *AMM* **70**(7), p. 759. The editors of *AMM* mistakenly asserted that there can be four if any 12-month period counts as a "year." Their retraction appeared in *AMM* **98**(7), p. 649.

[8] One or none, depending on what is meant by a "year" (*AMM, ibid.*).

[9] Jean-Jacques is well-known for his work on optimal strategies in the lambda calculus and for his joint work with Gérard Huet (France) on sequentiality of rewriting [47] – work that had remained in technical-report form for some 12 years.

[10] I can't help preferring $\rightarrow^*$ over $\twoheadrightarrow$ for the reflexive-transitive closure.

In 1973, Barry Rosen (USA) [81] provided a proof (albeit for the variable-free case) that shows that when a term-rewriting system is orthogonal[11], confluence is guaranteed. In other words, when left-linear systems are also non-ambiguous (no left-hand side unifies – after renaming apart – with another left-hand side, or with any non-variable proper subterm of any left-hand side), the system is confluent. This is for much the same reason as combinatory logic is Church-Rosser, and is usually proved by recourse to an intermediate relation, such as parallel rewriting $\rightarrow^{\|}$ (for "rewriters," this means contracting redexes at disjoint, "parallel" positions), or complete developments $\rightarrow^{\perp}$ (in the sense of contracting all residuals)[12].

Newman had also shown that termination plus local confluence yield the (global) confluence property[13]. Huet, in his influential 1980 paper [44], referred to the Church-Rosser property as "confluence," and provided a beautiful proof of this "Diamond Lemma," based on Noetherian (well-founded) induction[14]. Steve Kleene (USA) had given (according to Roger Hindley (UK) [41]) a simple counterexample to confluence sans termination: $\bullet \leftarrow \circ \longleftrightarrow \circ \rightarrow \bullet$. But the rewrite system for this graph has more than one rule with identical left side.

In the late 1960s, Don Knuth (USA), with a student, Peter Bendix (USA), wrote a seminal paper [52] in which they showed that confluence of critical pairs is sufficient (and necessary) for confluence of a terminating, but not necessarily left-linear, system[15]. Using the notation $s \leftarrow\!\!\bowtie\!\!\rightarrow t$ for the critical-pair relation $s = u[r]\mu \leftarrow u[l]\mu = g\mu \rightarrow d\mu = t$ (for rules $l \rightarrow r$ and $g \rightarrow d$ and most general unifier μ of l with a non-variable subterm of g in context u), this amounts to $\leftarrow\!\!\bowtie\!\!\rightarrow \subseteq \rightarrow^* {}^*\!\!\leftarrow$ (joinability, or resolvability, of critical pairs)[16]. In his paper,

---

[11] I take some pride in having coined this term to replace its predecessors, "regular" and "non-ambiguous linear."

[12] My new notation for multi-steps at orthogonal positions.

[13] Marc Bezem (Norway) and Jan Willem Klop collect four proofs of this fact in the textbook [88] which they, and Roel de Vrijer (The Netherlands) edited: Newman's, Huet's, one based on decreasing diagrams, and one Jan Willem and I used, based on a multiset ordering of terms. For a discussion of its mechanization, see the column by Bezem and Thierry Coquand (Sweden) [5]. Unaware of Newman's lemma, several others after him proved weaker versions.

[14] This – the most general form of mathematical induction – is named after the great twentieth century algebraist, Emily Noether (Germany and the USA).

[15] Knuth is a great-great-grand-student of Thue. When I was a student, Knuth gave me an offprint of this paper (dated 1969 – the conference at which it had been presented took place in 1968), since I was working on termination methods and the paper included what is now called the "Knuth-Bendix ordering."

[16] Rather than argue æsthetics, as to which way a critical pair ought to be oriented, in those cases where it matters, we use this explicit notation. Critical pairs had been presaged in a paper by Trevor Evans (USA) [24], which served as starting point for Knuth's investigations.[17] Knuth also reinvented (syntactic) unification, as used by Alan Robinson (USA) in his resolution proof procedure [80], for the purpose of calculating critical pairs, since the goal is to obtain as generic a pair as necessary to encompass all critical peaks between two rules. Bendix implemented Knuth's algorithm in Fortran.

Huet also observed that Knuth's proof of his Critical Pair Lemma does not require termination; in other words, that a system is locally confluent if, and only if, its critical pairs resolve.

Huet provided an unambiguous (critical-pair-free; hence, locally confluent) example of the necessity of left-linearity for (global) confluence of non-terminating systems: $f(x,x) \to a$, $c \to g(c)$, $f(x,g(x)) \to b$. Klop gave a similar one (with only one non-left-linear rule, but two non-terminating ones) in his foundational study [51]: $f(x,x) \to a$, $c \to g(c)$, $g(x) \to f(x,g(x))$. Six years later, my student, Sivakumar (USA, at the time) constructed the following (weakly) normalizing (every term has a normal form) and unambiguous example of non-confluence: $f(x,x) \to g(x)$, $f(x,g(x)) \to b$, $h(c,y) \to f(h(y,c),h(y,y))$.

So, confluence is decidable for (finite) terminating systems, by the Critical Pair criterion. It is, however, undecidable for non-terminating systems, since the uniform word problem is, in general, undecidable, even for string (semi-Thue) rewrite systems (see below).

The question that now begged asking was how – notwithstanding the above – one might establish the confluence of ambiguous (overlapping) non-terminating systems. Indeed, functional programmers love to write interpreters and to use streams, but still desire unique normal forms. Though they are usually content with left-linear rule patterns, it is quite natural to code nondeterministically, with ambiguous left sides[18].

Accordingly, Huet proved that term-rewriting systems that are linear (that is, both left- and right-linear) are confluent if, but not only if, the two sides of every critical pair reduce in at most one step to a term reachable from the other side. In symbols: $\leftarrow\!\bowtie\!\to\ \subseteq\ (\to^=\ ^*\!\leftarrow) \cap (\to^*\ ^=\!\leftarrow)$ implies confluence. Huet also included a counterexample of Lévy's, showing the necessity for right-linearity. This criterion, however, is not very useful, since right-linearity is usually an impractical constraint, except in the string-rewriting setting (see the next section).

In any case, one cannot hope for a decidable necessary and sufficient critical-pair criterion in the general non-terminating linear case.

It was always clear that trivial critical pairs (of what are called "weakly orthogonal" systems: $\leftarrow\!\bowtie\!\to\ \subseteq\ =$) do no harm – vis-à-vis confluence, at least. Huet also proved that, without regard to right-linearity, left-linear systems are confluent if, but not only if, $\leftarrow\!\bowtie\!\to\ \subseteq\ \to^\|$, a property he dubbed "parallel closed." But his proof only works when the resolving parallel step applies to the reduct of the lower diverging step (on the open side of the symbol $\bowtie$).

Several years subsequent, in 1988, Yoshihito Toyama (Japan) [92] relaxed this condition to allow a resolution of the weak form $\to^\|\ ^\|\!\leftarrow$, but only for critical pairs generated from two rules overlapping at their roots, a situation that we will capture with a symmetric symbol: $\leftarrow\!\!\bowtie\!\!\to$. More precisely, Toyama's sufficient condition is: $\leftarrow\!\bowtie\!\to\ \subseteq\ \to^\| \cup ([\to^\|\ ^*\!\leftarrow] \cap [\to^*\ ^\|\!\leftarrow] \cap [\leftarrow\!\!\bowtie\!\!\to])$. In other words, root overlay pairs need only satisfy the weaker requirement $[\to^\|\ ^*\!\leftarrow] \cap [\to^*\ ^\|\!\leftarrow]$.

---

[18] Whether non-terminating systems are necessary in the more general framework of logic programming is a question; compare my arguments in [16].

These results and many others are usually based on strong versions of local confluence, for which all one-step divergences can be resolved by some variant of rewriting for which both terms resolve in at most one step[19]. But Huet's work left open various alternative conditions on critical pairs[20]:

Problem #13a: Is $\leftarrow\!\!\times\!\!\rightarrow\ \subseteq\ ^{\parallel}\!\!\leftarrow$ also enough for confluence?

Problem #13b: If yes, then maybe some critical pairs may resolve with a step in this direction ($^{\parallel}\!\!\leftarrow$), and others the other way around ($\rightarrow^{\parallel}$)? In other words: Is $\leftarrow\!\!\times\!\!\rightarrow\ \subseteq\ \leftrightarrow^{\parallel}$ enough (where the intent is the symmetric closure of $\rightarrow^{\parallel}$)?

Problem #13c: If not, then what about a stronger condition, namely, $\leftarrow\!\!\times\!\!\rightarrow\ \subseteq\ ^=\!\!\leftarrow$?

Problem #13d: If yes, then one could ask whether $\leftarrow\!\!\times\!\!\rightarrow\ \subseteq\ \leftrightarrow^=$ suffices?

A positive answer to any of these would provide a new criterion for confluence, and would suggest a Knuth-Bendix–like completion procedure for potentially non-confluent systems, adding equations to ensure that the condition is satisfied. Of course, for non-right-linear systems, a resultant critical pair may be non-linear on both sides, and, hence, unorientable. On the other hand, if these conditions are insufficient, counterexamples will have to be (besides left-linear) non-right-linear, non-terminating, and overlapping. To date, none of these conjectures has succumbed to a counterexample.

In 1991, Rolf Socher-Ambrosius (Germany) [84] wrote a short report on Problem #13a, in which an arbitrary ordering of rules induces a multiset-ordering condition on the rules used to resolve critical pairs.

In 1996, Bernhard Gramlich (Germany, at the time) [39] suggested expanding the overlaps being considered to include "parallel critical pairs[21]." Parallel rewriting is a standard tool for proving confluence of orthogonal systems, since it satisfies the diamond property $^{\parallel}\!\!\leftarrow\!\!\longrightarrow^{\parallel}\ \subseteq\ \rightarrow^{\parallel}\ ^{\parallel}\!\!\leftarrow$. The idea is to handle critical overlaps of such parallel steps, by requiring $\leftarrow\!\!\times\!\!\rightarrow\ \subseteq\ (\rightarrow^{\parallel}\ ^*\!\!\leftarrow)\cap\ (\rightarrow^*\cup([\rightarrow^*\,^{\parallel}\!\!\leftarrow]\cap[\leftarrow\!\!\bowtie\!\!\rightarrow]))$ for all ordinary critical pairs, plus $^{\parallel}\!\!\leftarrow\!\!\times\!\!\rightarrow\ \subseteq\ \rightarrow^*\cup([\rightarrow^{\parallel}\,^*\!\!\leftarrow]\cap[\rightarrow^*\,^{\parallel}\!\!\leftarrow]\cap[\leftarrow\!\!\bowtie\!\!\rightarrow])$ for all parallel pairs.

---

[19] This is an opportunity to apologize for sowing confusion by defining "strong confluence" in [18] as the "subcommutative" property, namely, $\leftarrow\!\!\longrightarrow\ \subseteq\ \rightarrow^=\ ^=\!\!\leftarrow$, whereas Huet used the term for his weaker condition $\leftarrow\!\!\longrightarrow\ \subseteq\ (\rightarrow^=\,^*\!\!\leftarrow)\cap(\rightarrow^*\,^=\!\!\leftarrow)$.

[20] Problem #13b was not posed explicitly in [19], but was included, for example, by Bernhard Gramlich (Germany, at the time) in [39].

[21] Gramlich generously attributes this extension of the notion of critical pairs to what underlies what are known as "critical pair criteria," as in the works of Franz Winkler and Bruno Buchberger (Austria) [93], Wolfgang Küchlin (USA, at the time) [54], Deepak Kapur (USA), Dave Musser (USA), and P. Narendran (USA) [49], and my student, Leo Bachmair (USA), and myself [1]. Around the same time, Dave Plaisted (USA) and Andrea Sattler-Klein (Germany) [79] also employed parallel critical pairs, but for other purposes.

Actually, parallel critical pairs and a related result were already present in an unpublished 1981 report in Japanese by Toyama [91]. There, the condition was the weaker inclusion $\leftarrow\!\!\bowtie\!\!\rightarrow\ \subseteq\ (\rightarrow^{\|}{}^{*}\!\leftarrow)\cap(\rightarrow^{*}{}^{\|}\!\leftarrow)$ for ordinary critical pairs, plus $^{\|}\!\leftarrow\!\!\bowtie\!\!\rightarrow\ \subseteq\ \rightarrow^{*}{}^{\|}\!\leftarrow$ for all parallel overlaps – the latter, however, subject to the extra requirement that all variables that appear in the contractum(s) of the resolving parallel step were also within the critical parallel redexes.

The next step transpired almost immediately, when Vincent van Oostrom – in discussions with Gramlich – realized that whatever can be said for parallel rewriting can also be said for developments[22]. Accordingly, he defined a development-closed criterion, improving on Toyama's 1988 weakening of Huet's 1980 parallel-closed condition, by replacing $\rightarrow^{\|}$ with $\rightarrow^{\perp}$ [76]. Specifically, a system is Church-Rosser if $\leftarrow\!\!\bowtie\!\!\rightarrow\ \subseteq\ \rightarrow^{\perp}\ \cup\ ([\rightarrow^{\perp}{}^{*}\!\leftarrow]\cap[\rightarrow^{*}{}^{\perp}\!\leftarrow]\cap[\leftarrow\!\!\bowtie\!\!\rightarrow])$. In the special case where the only overlaps are at the root, the condition is $\leftarrow\!\!\bowtie\!\!\rightarrow\ \subseteq\ ([\rightarrow^{\perp}{}^{*}\!\leftarrow]\cap[\rightarrow^{*}{}^{\perp}\!\leftarrow])$, which is satisfied when $\leftarrow\!\!\bowtie\!\!\rightarrow\ \subseteq\ (\rightarrow^{\perp}{}^{\perp}\!\leftarrow)$.

This led Aart Middeldorp (Japan, at the time) to raise the following question:

Problem #13-1: What if the critical pair reduces by an incomplete development, that is, if $^{\perp}\!\leftarrow\!\!\bowtie\!\!\rightarrow\ \subseteq\ \rightarrow^{\varnothing}$, where $\rightarrow^{\varnothing}$ signifies that only some of the redexes of a complete development $\rightarrow^{\perp}$ are contracted?

van Oostrom thinks the critical-pair theorem still holds, despite the fact that the invariant used in his proof for complete developments fails[23].

Plus, we have yet another unanswered question:

Problem #13-2: Is $^{\perp}\!\leftarrow\!\!\bowtie\!\!\rightarrow\ \subseteq\ ^{\perp}\!\leftarrow$ enough for confluence?

One can go further, by considering overlaps between developments. This condition, based on what I will call "orthogonal" critical pairs (but not define)[24], was presented by Satoshi Okui (Japan) at RTA '98 [74]. The conditions are: $\leftarrow\!\!\bowtie\!\!\rightarrow\ \subseteq\ (\rightarrow^{*}{}^{\perp}\!\leftarrow)\cap(\rightarrow^{\perp}{}^{*}\!\leftarrow)$ for ordinary critical pairs, plus $^{\perp}\!\leftarrow\!\!\bowtie\!\!\!\bowtie\!\!\rightarrow\ \subseteq\ \rightarrow^{*}{}^{\perp}\!\leftarrow$ for all orthogonal pairs. Independently, van Oostrom had obtained the same result – again, in the higher-order context. Whereas parallel reduction is a problem in the higher-order case, complete developments work nicely for both first-order and higher-order rewriting[25]. So, Okui and van Oostrom teamed up, and now have an unpublished generalization to the higher-order case.

---

[22] van Oostrom was motivated by attempts of Tobias Nipkow (Germany) and Richard Mayr (Germany, at the time) to extend Huet's condition to handle Nipkow's "higher-order pattern" rewrite systems. Some advantages of reasoning with orthogonal steps in Church-Rosser arguments had been pointed out by Masako Takahashi (Japan) in 1995 [87]. I heard Masako present her ideas at the *Toyohashi Symposium on Theoretical Computer Science* in 1990.

[23] van Oostrom: "I recall that in 1995 I came 'close' to solving it in the plane to Japan, but then we arrived, and I've never worked on it since."

[24] Instead of "simultaneous" or "multi-step" critical pairs.

[25] Vincent presented his ideas at a 1995 seminar in Munich (Germany), where he was holding a postdoctoral position with Tobias Nipkow at the time. He applied it to $\beta\eta\Omega$-reduction – see Henk Barendregt's (The Netherlands) book [3, p. 388] – with eight orthogonal-critical pairs.

Significant progress on Problem #13a was made by Michio Oyamaguchi (Japan) and Yoshikatsu Ohta (Japan) in 1997 [77, 78]. Let $\rightarrow^{\#}$ stand for $\rightarrow^{\parallel} \cup \; ^{\Lambda}\!\leftarrow$, where $^{\Lambda}\!\leftarrow$ signifies a root-step. They require $\leftarrow \bowtie \rightarrow \; \subseteq \; ^{\#}\!\leftarrow \cup \left([\rightarrow^{\#*}\!\leftarrow] \cap [\rightarrow^{*\,\#}\!\leftarrow] \cap [\leftarrow\bowtie\rightarrow] \right)$, but with an additional side condition on the parallel steps. The proof involves a beautiful invariant in terms of "outside in" sequences of $\rightarrow^{\parallel}$.

Lastly, five years ago, Toshimasa Matsumoto (Japan) [69] devised a new condition on the parallel resolution of ordinary critical pairs, based on Okui's work, but the extent of its applicability is unknown.

Perhaps critical pair criteria (see fn. 21), Nicolaas de Bruijn's (The Netherlands) and van Oostrom's decreasing diagrams [8, 75], and/or abstract semantic notions of criticality, as in Claude Kirchner's (France), Maria Paola Bonacina's (Italy), and my recent work [6, 22], can contribute to a fuller understanding of this fundamental problem.

3   One-Rule Termination

> If you leave it in existence and forget about it,
> all your future rewrite commands
> will be needlessly slow.
>
> —GNU Emacs Calc 2.02 Manual

Another problem on the original list was:

Problem #21a: Is termination of one (left- and right-) linear rule decidable?

This problem was contributed by Max Dauchet (France), who had recently (at RTA '89) shown that left-linearity alone is insufficient for decidability. This was the culmination of a series of efforts to delineate the borders of decidability.

Richard Lipton (USA) and Lawrence Snyder (USA) had claimed in a footnote to a 1997 paper [60] that three rules suffice for undecidability of termination. As they had not responded to a request for a proof, Huet and Dallas Lankford (USA) set out, in an unpublished report [45], to find one[26]. They used a string-rewriting simulation of Turing machines, similar to that used by Ann Yasuhara (USA) in her book on Recursion Theory [95]. Thus termination of string-rewriting systems was provenly undecidable – for an unbounded number of rules.

In the summer of 1980, visiting Lévy and Huet at INRIA, I managed to encode Turing machines in two rules, one of which was non-linear. Dauchet went one giant step further, and found a way of showing undecidability for only one non-linear rule [12, 13]. Pierre Lescanne (France) in 1994 [59] redid this more naturally, by reducing the Post Correspondence Problem to this case. So the

---

[26] Dallas Lankford was an early player in the field, along with Mike Ballantyne (USA). Dallas was probably the first to realize, in 1975 [57], that a process like Knuth-Bendix completion, which uses oriented equations, could replace paramodulation as a means of handling equality within resolution theorem provers.

question (still unanswered) was (and is) whether termination of one linear rule is decidable.

In a recent paper [33], Alfons Geser (USA), Aart Middeldorp (Austria), Enno Ohlebusch (Germany), and Hans Zantema (The Netherlands) leave the following question unanswered:

Problem #21-1: Is termination decidable for one (not necessarily linear) normalizing rule?

Geser (Germany, at the time) constructed a (overlapping) string rule that is normalizing but neither leftmost terminating nor rightmost terminating, and one that is rightmost terminating but non-terminating [27].

Most common term-rewriting termination proofs use simplification orderings, making terms always bigger than their subterms[27]. Aart Middeldorp (Japan, at the time) and Bernhard Gramlich (France, at the time) used Dauchet's trick and showed that it is also undecidable whether there exists a simplification ordering that proves termination of a single term-rewriting rule [71] (correcting a claim in [48]).

This negative answer suggested yet another problem:

Problem #87: Is it decidable whether a single term-rewriting rule can be proved terminating by a monotonic ordering that is total on ground terms?

Such orderings are important in deduction engines; see, for example, the work on unfailing completion of my former student, Jieh Hsiang (USA, at the time), with Michaël Rusinowitch (France) [43], and of Leo Bachmair (USA), Dave Plaisted (USA), and myself [2]. Zantema, who posed this one-rule problem, already knew that it is undecidable for more rules [96]. A negative solution to this question was given two years later by Geser, Middeldorp, Ohlebusch, and Zantema [32].

Now, one might think that a one-rule system is nonterminating only if it is looping in the sense of deriving a term from one of its subcontexts. But, it turns out that there is a non-looping, non-terminating one-rule term system, as well as such a two-rule string system [34]. This raises the following question:

Problem #95: Is there a one-rule string-rewriting system that is non-terminating but also non-looping?

A loop would be a string derivation of the form $s \to^+ usv$. Bob McNaughton (USA) [62] has conjectured that no such rule exists.

This all brings us around to a perhaps less ambitious, but long-standing open problem for the much simpler case of string rewriting:

Problem #21b: Is termination of one string rule decidable?

This had been mentioned in my survey with Jean-Pierre Jouannuad (France) [18], and was included in the second edition of our open problem list, in 1993.

---

[27] What I called "simplification orderings" in [14] are (in the fixed-arity case) the "divisibility orders" of Graham Higman (UK) [40].

Length-decreasing rules (however many) are obviously terminating. In 1991, Anne-Cécile Caron (France) had shown that termination is undecidable for multi-rule non-length-increasing string systems [9]. But a single length-preserving rule is only nonterminating when both sides are identical. In the latter case, one may still enquire about the length of derivations, the subject of a 1985 paper by Yves Métivier (France) [70], and of yet another problem in our original list:

> **Problem #20b: What is the best bound on the length of a derivation for a one-rule length-preserving string-rewriting (semi-Thue) system? Is it quadratic in the size of the initial term, as conjectured in [70], or of order n^k (for rules of length k and input of length n) as proved there?**

Métivier had provided a lower bound of $n^2/4$, easily reached by the derivation from $b^{n/2}a^{n/2}$ for the rule $ba \to ab$. His conjecture that this was also the upper bound for a binary alphabet was proved a few years later by Alain Bertrand (France) [4]. In that paper, Bertrand floated a new combinatorial conjecture relating to the positions of the letters in the input word[28].

String systems are confluent when no suffix of a left side is also a prefix, since that makes them orthogonal. For right-linear systems, in general, and string systems, in particular, termination of all forward closures (a subset of derivations in which only created redexes are contracted) is valid evidence of termination [15], an idea that grew out of an unpublished preliminary note [58] by Lankford and Dave Musser (USA). Moreover, when there are no left-side overlaps, the specific string-rewriting strategy (leftmost, rightmost, etc.) does not affect termination, and (weak) normalization implies termination [17, 38].

As an example of a difficult, though non-overlapping, length-increasing rule, Zantema suggested $bbaa \to aaabbb$, a problem that itself engendered a spate of interesting work by my student, Charles Hoot (USA) [17], Elias Tahhan [86] (France, at the time), Geser [25], and others. A complete classification of termination for a rule of the form $b^i a^j \to a^k b^\ell$ was presented by Geser and Zantema at RTA '95 (see [97]), which, in turn, was subsumed by the later work of Géraud Sénizergues (France) [82] and of Yuji Kobayashi, Masashi Katsura, and Kayoko Shikishima-Tsuji (all from Japan) [53], for $b^i a^j \to r$, where $r \in \{a, b\}^*$.

Geser picked up the gauntlet, obtaining partial results for single string-rewrite rules, culminating in his dissertation [29]. Rules with only one overlap had already been solved by Winfried Kurth (Germany) in his thesis [55], who also proved decidability of existence of loops of lengths 1–3 for one-rule string systems, and showed decidability for lone rules with right sides no more than six letters long [56]. Building on ideas of McNaughton [61, 62], Geser showed decidability for up to nine letters [26][29]. More recently, Geser [30] proved that termi-

---

[28] From inception and until recently, our on-line list stated: "Rumor has it that the conjecture has been shown true."

[29] Geser: "My termination sieve had a bug that I only detected after finishing my habilitation thesis in 2002. As a consequence of this bug, eight additional rules remain that cannot be solved by the methods in this paper."

nation is decidable for one-rule systems that have precisely one overlap between a prefix of the left and a suffix of the right and vice-versa. For fewer overlaps, this was already known. In [83], Shikishima-Tsuji, Katsura, and Kobayashi reduced the termination problem for a *confluent* overlapping rule to the non-overlapping case.

A grid rule is one in which some letter appears equally often on both sides (or diminishes). Grid rules cover all systems amenable to a total simplification ordering. Geser showed that termination is equivalent to the non-existence of loops of length one or two, which is decidable [28].

Dieter Hofbauer (Germany) and Johannes Waldmann (Germany) showed recently that string systems admitting a termination proof by the set extension (like the multiset extension, but for sets) of a symbol precedence preserve regular languages [42]. A string system is said to be match-bounded if only a finite section of a system annotated with symbol numbers can be used in any (labelled) derivation. Geser, Hofbauer, and Waldmann showed, in a series of papers, that match-bounded string systems are terminating; match-boundedness of right sides of forward closures is a stronger termination criterion; and inverse match-bounded string systems have a termination problems; see [31]. Decidability of match-boundedness is open.

Single-threaded derivations, where each pair of successive rewrites overlap, were introduced by Wojciech Moczydłowski (Poland; now in the USA) in his Masters thesis. He showed that one-rule string systems that are cannot consume all of a contractum from the right, nor from both sides, have a decidable termination problem. The second condition entails that the systems are either terminating or single-threaded, whence they can be simulated by a two-stack pushdown automaton; the first implies that one stack's size is bounded; hence, the problem is decidable. See his joint paper with Geser [72] in these proceedings.

In sum, the jury is still out on Problem #21b, one string-rule termination. Plaisted conjectured its decidability long ago; Kurth believes it is in general undecidable; McNaughton conjectures that at least the confluent case is decidable.

Turning again to the Church-Rosser property: The critical-pair test of Knuth and Bendix gave us a decision procedure for confluence of terminating systems, which, for non-terminating systems, remains undecidable. Confluence for one string rule is decidable, by the work of Celia Wrathall (USA) [94], but undecidable, even for just twelve string rules, as per Yuri Matiyasevich[30] (Russia) [67] (see [88, p. 151]), a bound that has been pared down to five by Matiyasevich and Sénizergues [68][31]. Accordingly, the 1993 list also included the following question:

Problem #21c: Is confluence of one linear rule decidable?

There are a number of cases for which decidability has been shown regardless of the number of linear rules. Most recently: Guillem Godoy (Spain), Ashish

---

[30] Of "Hilbert's Tenth Problem (Diophantine equations) is undecidable" fame.

[31] Derivability (accessibility) is undecidable for three string rules [68]. It is, however, decidable for one; see Bob McNaughton's (USA) [64].

Tiwari (USA), and Rakesh Verma (USA) have shown decidability when variables do not appear deeper than immediately below the outer function symbol [35].

To conclude, the questions raised in this section are interesting and important for demarcating the boundaries of decidability of termination and confluence. Their resolution, however, especially in the string case, seems combinatorial in nature, though some automata-based and residual-theory techniques are now entering the picture. The methods have ramifications for other decidability and complexity questions relating to semigroups and monoids (see, for example, the work of Katsura and Kobayashi with Friedrich Otto (Germany) [50]), topics of increasing interest in this bio-informatical era.

Acknowledgements

I thank Alfons Geser (USA), Bernhard Gramlich (Austria), Yoshihito Toyama (Japan), and Vincent van Oostrom (The Netherlands) for updates, references, and suggestions. I apologize in advance for any and all sins of omission, consequences of the need to converge and terminate.

> We use a deep mathematical result
> (namely, a minor modification of Kolmogorov's
> solution to Hilbert's 13th problem)
> to explain why fundamental physical equations are of second order.
>
> —Takeshi Yamakawa (Japan) and Vladik Kreinovich (USA),
> *International Journal of Theoretical Physics* (1999)

References

1. Leo Bachmair and Nachum Dershowitz. Critical pair criteria for completion. *J. Symbolic Computation*, 6(1):1–18, 1988.

2. Leo Bachmair, Nachum Dershowitz, and David A. Plaisted. Completion without failure. In H. Aït-Kaci and M. Nivat, editors, *Resolution of Equations in Algebraic Structures*, volume 2: Rewriting Techniques, chapter 1, pages 1–30. Academic Press, New York, 1989.

3. Henk Barendregt. *The Lambda Calculus, its Syntax and Semantics*. North-Holland, Amsterdam, second edition, 1984.

4. Alain Bertrand. Sur une conjecture d'Yves Métivier. *Theoretical Computer Science*, 123(1):21–30, 1994.

5. Marc Bezem and Thierry Coquand. Newman's Lemma – a case study in proof automation and geometric logic (Logic in Computer Science Column). *Bulletin of the EATCS*, 79:86–100, 2003.

6. Maria Paola Bonacina and Nachum Dershowitz. Abstract canonical inference. *ACM Transactions on Computational Logic*. To appear.

7. Ronald V. Book and Friedrich Otto. *String-Rewriting Systems*. Springer-Verlag, New York, 1993.

8. Nicolaas Govert de Bruijn. A note on weak diamond properties. Memorandum 1978-08, Department of Mathematics, Eindhoven University of Technology, Eindhoven, The Netherlands, August 1978.

9. Anne-Cécile Caron. Linear bounded automata and rewrite systems: Influence of initial configurations on decision properties. In *Proceedings of the International Joint Conference on Theory and Practice of Software Development, volume 1: Colloquium on Trees in Algebra and Programming (Brighton, U.K.)*, volume 493 of *Lecture Notes in Computer Science*, pages 74–89, Berlin, April 1991. Springer-Verlag.

10. Alonzo Church. *The Calculi of Lambda Conversion*, volume 6 of *Ann. Mathematics Studies*. Princeton University Press, Princeton, NJ, 1941.

11. Alonzo Church and J. Barkley Rosser. Some properties of conversion. *Transactions of the American Mathematical Society*, 39:472–482, 1936.

12. Max Dauchet. Simulation of Turing machines by a left-linear rewrite rule. In Nachum Dershowitz, editor, *Rewriting Techniques and Applications*, volume 355 of *Lecture Notes in Computer Science*, pages 109–120, Chapel Hill, NC, USA, April 1989. Springer-Verlag.

13. Max Dauchet. Simulation of Turing machines by a regular rewrite rule. *Theoretical Computer Science*, 103(2):409–420, 1992.

14. Nachum Dershowitz. A note on simplification orderings. *Information Processing Letters*, 9(5):212–215, November 1979.

15. Nachum Dershowitz. Termination of linear rewriting systems. In *Proceedings of the Eighth International Colloquium on Automata, Languages and Programming (Acre, Israel)*, volume 115 of *Lecture Notes in Computer Science*, pages 448–458, Berlin, July 1981. European Association of Theoretical Computer Science, Springer-Verlag.

16. Nachum Dershowitz. Goal solving as operational semantics. In *Proceedings of the International Logic Programming Symposium (Portland, OR)*, pages 3–17, Cambridge, MA, December 1995. MIT Press.

17. Nachum Dershowitz and Charles Hoot. Natural termination. *Theoretical Computer Science*, 142(2):179–207, May 1995.

18. Nachum Dershowitz and Jean-Pierre Jouannaud. Rewrite systems. In J. van Leeuwen, editor, *Handbook of Theoretical Computer Science*, volume B: Formal Methods and Semantics, chapter 6, pages 243–320. North-Holland, Amsterdam, 1990.

19. Nachum Dershowitz, Jean-Pierre Jouannaud, and Jan Willem Klop. Open problems in rewriting. In Ronald. V. Book, editor, *4th International Conference on Rewriting Techniques and Applications*, volume 488 of *Lecture Notes in Computer Science*, pages 445–456, Como, Italy, April 1991. Springer-Verlag.

20. Nachum Dershowitz, Jean-Pierre Jouannaud, and Jan Willem Klop. More problems in rewriting. In C. Kirchner, editor, *Proceedings of the Fifth International Conference on Rewriting Techniques and Applications, (Montreal, Canada, June 16–18, 1993)*, volume 690 of *Lecture Notes in Computer Science*, pages 468–487. Springer, 1993.

21. Nachum Dershowitz, Jean-Pierre Jouannaud, and Jan Willem Klop. Problems in rewriting III. In Jieh Hsiang, editor, *6th International Conference on Rewriting Techniques and Applications*, volume 914 of *Lecture Notes in Computer Science*, pages 457–471, Kaiserslautern, Germany, April 1995. Springer-Verlag.

22. Nachum Dershowitz and Claude Kirchner. Abstract canonical presentations. *Theoretical Computer Science*. To appear.

23. Nachum Dershowitz and Ralf Treinen. An on-line problem database. In Tobias Nipkow, editor, *Proceedings of the Ninth International Conference on Rewriting Techniques and Applications (Tsukuba, Japan)*, volume 1379 of *Lecture Notes in Computer Science*, pages 332–342. Springer, March 1998.

24. Trevor Evans. On multiplicative systems defined by generators and relations, I. *Proceedings of the Cambridge Philosophical Society*, 47:637–649, 1951.

25. Alfons Geser. A solution to Zantema's problem. Technical Report MIP-9314, Fakultät für Mathematik und Informatik, Universität Passau, 1993.

26. Alfons Geser. Termination of one-rule string rewriting systems $\ell \to r$ where $|r| \le$ 9. Technical report, Wilhelm-Schickard-Institut, Universität Tübingen, Germany, January 1998.

27. Alfons Geser. Note on normalizing, non-terminating one-rule string rewriting systems. *Theoretical Computer Science*, 243:489–498, 2000.

28. Alfons Geser. Decidability of termination of grid string rewriting rules. *SIAM J. Comput.*, 31(4):1156–1168, 2002.

29. Alfons Geser. *Is Termination Decidable for String Rewriting with Only One Rule?* Habilitation thesis, Universität Tübingen, Germany, January 2002.

30. Alfons Geser. Termination of string rewriting rules that have one pair of overlaps. In Robert Nieuwenhuis, editor, *14th International Conference on Rewriting Techniques*, volume 2706 of *Lecture Notes in Computer Science*, pages 410–423, Valencia, Spain, June 2003. Springer-Verlag.

31. Alfons Geser, Dieter Hofbauer, and Johannes Waldmann. Termination proofs for string rewriting systems via inverse match-bounds. *J. Automat. Reason.* To appear.

32. Alfons Geser, Aart Middeldorp, Enno Ohlebusch, and Hans Zantema. Relative undecidability in the termination hierarchy of single rewrite rules. In Michel Bidoit and Max Dauchet, editors, *Theory and Practice of Software Development*, volume 1214 of *Lecture Notes in Computer Science*, pages 237–248, Lille, France, April 1997. Springer-Verlag.

33. Alfons Geser, Aart Middeldorp, Enno Ohlebusch, and Hans Zantema. Relative undecidability in term rewriting, Part I: The termination hierarchy. *Inform. and Computation*, 178(1):101–131, 2002.

34. Alfons Geser and Hans Zantema. Non-looping string rewriting. *Theoret. Informatics Appl.*, 33(3):279–301, 1999.

35. Guillem Godoy, Ashish Tiwari, and Rakesh Verma. On the confluence of linear shallow term rewrite systems. In *Proc. of the 20th Annual Symposium on Theoretical Aspects of Computer Science (Berlin, Germany)*, volume 2607 of *Lecture Notes in Computer Science*, pages 85–96, January 2003.

36. S. Gorn. On the conclusive validation of symbol manipulation processes (how do you know it has to work?). *J. of the Franklin Institute*, 296(6):499–518, December 1973.

37. Saul Gorn. Handling the growth by definition of mechanical languages. In *Proceedings of the Spring Joint Computer Conference*, pages 213–224, Philadelphia, PA, Spring 1967.

38. Bernhard Gramlich. Relating innermost, weak, uniform and modular termination of term rewriting systems. SEKI-Report SR-93-09, Fachbereich Informatik, Universität Kaiserslautern, Kaiserslautern, West Germany, 1993.

39. Bernhard Gramlich. Confluence without termination via parallel critical pairs. In *Colloquium on Trees in Algebra and Programming*, pages 211–225, 1996.

40. Graham Higman. Ordering by divisibility in abstract algebras. *Proceedings of the London Mathematical Society (3)*, 2(7):326–336, September 1952.

41. J. Roger Hindley. *The Church-Rosser Property and a Result in Combinatory Logic*. PhD thesis, University of Newcastle-upon-Tyne, 1964.

42. Dieter Hofbauer and Johannes Waldmann. Deleting string rewriting systems preserve regularity. *Theoretical Computer Science*. To appear.

43. Jieh Hsiang and Michaël Rusinowitch. On word problems in equational theories. In T. Ottmann, editor, *Proceedings of the Fourteenth EATCS International Conference on Automata, Languages and Programming (Karlsruhe, West Germany)*, volume 267 of *Lecture Notes in Computer Science*, pages 54–71, Berlin, July 1987. Springer-Verlag.

44. Gérard Huet. Confluent reductions: Abstract properties and applications to term rewriting systems. *J. of the Association for Computing Machinery*, 27(4):797–821, October 1980.

45. Gérard Huet and Dallas S. Lankford. On the uniform halting problem for term rewriting systems. Rapport laboria 283, Institut de Recherche en Informatique et en Automatique, Le Chesnay, France, March 1978.

46. Gérard Huet and Jean-Jacques Lévy. Call by need computations in non-ambiguous linear term rewriting systems. Rapport Laboria 359, Institut National de Recherche en Informatique et en Automatique, Le Chesnay, France, August 1979.

47. Gérard Huet and Jean-Jacques Lévy. Computations in orthogonal rewriting systems, I and II. In Jean-Louis Lassez and Gordon Plotkin, editors, *Computational Logic: Essays in Honor of Alan Robinson*, pages 395–443. The MIT Press, Cambridge, MA, 1991. This is a revision of [46].

48. Jean-Pierre Jouannaud and Hélène Kirchner. Construction d'un plus petit ordre de simplification. *RAIRO Theoretical Informatics*, 18(3):191–207, 1984.

49. Deepak Kapur, David R. Musser, and Paliath Narendran. Only prime superpositions need be considered for the Knuth-Bendix procedure. *J. Symbolic Computation*, 4:19–36, August 1988.

50. Masashi Katsura, Yuji Kobayashi, and Friedrich Otto. Undecidable properties of monoids with word problem solvable in linear time. Part II– Cross sections and homological and homotopical finiteness conditions. *Theoretical Computer Science*, 301(1–3):79–101, May 2003.

51. Jan Willem Klop. *Combinatory Reduction Systems*, volume 127 of *Mathematical Centre Tracts*. Mathematisch Centrum, Amsterdam, 1980.

52. Donald E. Knuth and P. B. Bendix. Simple word problems in universal algebras. In J. Leech, editor, *Computational Problems in Abstract Algebra*, pages 263–297. Pergamon Press, Oxford, U. K., 1970. Reprinted in *Automation of Reasoning 2*, Springer-Verlag, Berlin, pp. 342–376 (1983).

53. Yuji Kobayashi, Masashi Katsura, and Kayoko Shikishima-Tsuji. Termination and derivational complexity of confluent one-rule string rewriting systems. *Theoretical Computer Science*, 262(1/2):583–632, 2001.

54. Wolfgang Küchlin. A confluence criterion based on the generalized Newman Lemma. In B. F. Caviness, editor, *Proceedings of the European Conference on Computer Algebra (Linz, Austria)*, volume 204 of *Lecture Notes in Computer Science*, pages 390–399, Berlin, 1985. Springer-Verlag.

55. Winfried Kurth. *Termination und Konfluenz von Semi-Thue-Systems mit nur einer Regel*. PhD thesis, Technische Universitat Clausthal, Clausthal, Germany, 1990. In German.

56. Winfried Kurth. One-rule semi-Thue systems with loops of length one, two, or three. *Inform. Theéor. Appl.*, 30:355–268, 1996.

57. Dallas S. Lankford. Canonical inference. Memo ATP-32, Automatic Theorem Proving Project, University of Texas, Austin, TX, December 1975.

58. Dallas S. Lankford and David R. Musser. A finite termination criterion. May 1978.

59. Pierre Lescanne. On termination of one rule rewrite systems. *Theoretical Computer Science*, 132(1–2):395–401, 1994.

60. Richard Lipton and Lawrence Snyder. On the halting of tree replacement systems. In *Proceedings of the Conference on Theoretical Computer Science*, pages 43–46, Waterloo, Canada, August 1977.

61. Robert McNaughton. The uniform halting problem for one-rule semi-Thue systems. Technical Report 94-18, Dept. of Computer Science, Rensselaer Polytechnic Institute, Troy, NY, August 1994. Correction, Aug. 1996.

62. Robert McNaughton. Well-behaved derivations in one-rule semi-Thue systems. Technical Report 95-15, Dept. of Computer Science, Rensselaer Polytechnic Institute, Troy, NY, November 1995. Correction, July 1996.

63. Robert McNaughton. Contributions of Ronald V. Book to the theory of string-rewriting systems. *Theor. Comput. Sci.*, 207(1):13–23, 1998.

64. Robert McNaughton. Semi-Thue systems with an inhibitor. *Journal of Automated Reasoning*, 26(4):409–431, May 2001.

65. Zohar Manna and Steven Ness. On the termination of Markov algorithms. In *Proceedings of the Third Hawaii International Conference on System Science*, pages 789–792, Honolulu, HI, January 1970.

66. Andrei A. Markov. *Theory of Algorithms*. Moscow, 1954.

67. J. V. Matiyasevich. Simple examples of undecidable associative calculi. *Soviet Mathematics (Dokladi)*, 8(2):555–557, 1967.

68. Yuri Matiyasevich and Géraud Sénizergues. Decision problems for semi-Thue systems with a few rules. In Ed Clarke, editor, *Proceedings of the Eleventh Annual IEEE Symp. on Logic in Computer Science*, pages 523–531. IEEE Computer Society Press, July 1996.

69. Toshimasa Matsumoto. On confluence conditions of left-linear term rewriting systems. Master's thesis, School of Information Science, Japan Advanced Institute of Science and Technology, February 2000. In Japanese. English abstract is available at http://www.jaist.ac.jp/library/thesis/is-master-2000/paper/tmatsumo/abstract.pdf.

70. Yves Métivier. Calcul de longueurs de chaînes de réécriture dans le monoïde libre. *Theoretical Computer Science*, 35(1):71–87, January 1985.

71. Aart Middeldorp and Bernhard Gramlich. Simple termination is difficult. *Applied Algebra on Engineering, Communication and Computer Science*, 6(2):115–128, 1995.

72. Wojciech Moczydłowski, Jr. and Alfons Geser. Termination of single-threaded one-rule semi-Thue systems. In *Proc. of the Sixteenth International Conf. on Rewriting Techniques and Applications (Nara, Japan)*, Lecture Notes in Computer Science. Springer, April 2005.

73. Maxwell Herman Alexander Newman. On theories with a combinatorial definition of 'equivalence'. *Annals of Mathematics*, 43(2):223–243, 1942.

74. Satoshi Okui. Simultaneous critical pairs and Church-Rosser property. In Tobias Nipkow, editor, *9th International Conference on Rewriting Techniques and Applications*, volume 1379 of *Lecture Notes in Computer Science*, pages 2–16, Tsukuba, Japan, April 1998. Springer-Verlag.

75. Vincent van Oostrom. Confluence by decreasing diagrams. *Theoretical Computer Science*, 126(2):259–280, 1994.

76. Vincent van Oostrom. Development closed critical pairs. In *Proceedings of the Second International Workshop on Higher-Order Algebra, Logic, and Term Rewriting (Paderborn, Germany; September 1995)*, volume 1074 of *Lecture Notes in Computer Science*, pages 185–200. Springer, 1996.

77. Michio Oyamaguchi and Yoshikatsu Ohta. A new parallel closed condition for Church-Rosser of left-linear term rewriting systems. In Hubert Comon, editor, *8th International Conference on Rewriting Techniques and Applications*, volume 1232 of *Lecture Notes in Computer Science*, pages 187–201, Barcelona, Spain, June 1997. Springer-Verlag.

78. Michio Oyamaguchi and Yoshikatsu Ohta. On the open problems concerning Church-Rosser of left-linear term rewriting systems. *Trans. of IEICE*, E87-D(2):290–298, 2004.

79. David A. Plaisted and Andrea Sattler-Klein. Proof lengths for equational completion. *Information and Computation*, 125(2):154–170, 1996.

80. J. Alan Robinson. A machine-oriented logic based on the resolution principle. *J. of the Association for Computing Machinery*, 12(1):23–41, January 1965.

81. Barry Rosen. Tree-manipulating systems and Church-Rosser theorems. *J. of the Association for Computing Machinery*, 20(1):160–187, January 1973.

82. Geraud Sénizergues. On the termination problem for one-rule semi-Thue systems. In *Proceedings of the Seventh International Conference on Rewriting Techniques and Applications (New Brunswick, NJ)*, volume 1103 of *Lecture Notes in Computer Science*, pages 302–316. Springer-Verlag, July 1996.

83. Kayoko Shikishima-Tsuji, Masashi Katsura, and Yuji Kobayashi. On termination of confluent one-rule string-rewriting systems. *Inf. Process. Lett.*, 61(2):91–96, 1997.

84. Rolf Socher-Ambrosius. On the Church-Rosser property in left-linear systems. Technical Report TR 91/17, SUNY at Stony Brook, 1991.

85. Magnus Steinby and Wolfgang Thomas. Trees and term rewriting in 1910: On a paper by Axel Thue. *Bulletin of the European Association for Theoretical Computer Science*, 72:256–269, 2000.

86. Elias Tahhan Bittar. Complexité linéaire du problème de Zantema. *Comptes Rendus de l'Académie des Sciences, Paris*, 323:1201–1206, 1996.

87. Masako Takahashi. Parallel reductions in λ-calculus. *Information and Computation*, 118(1):120–127, April 1995.

88. "Terese" (M. Bezem, J. W. Klop and R. de Vrijer, eds.). *Term Rewriting Systems*. Cambridge University Press, 2002.

89. Axel Thue. Die Lösung eines Spezialfalles eines generellen logischen Problems. *Kra. Videnskabs-Selskabets Skrifter I. Mat. Nat. Kl.*, (8), 1910.

90. Axel Thue. Probleme über Veranderungen von Zeichenreihen nach gegeben Regeln. *Skr. Vid. Kristianaia I. Mat. Naturv. Klasse*, 10/34, 1914.

91. Yoshihito Toyama. On the Church-Rosser property of term rewriting systems. ECL Technical Report 17672, NTT, December 1981. (In Japanese.) English précis, "On parallel critical pairs" (Nov. 1995).

92. Yoshihito Toyama. Commutativity of term rewriting systems. In K. Fuchi and L. Kott, editors, *Programming of Future Generation Computers II*, pages 393–407. North-Holland, 1988.

93. Franz Winkler and Bruno Buchberger. A criterion for eliminating unnecessary reductions in the Knuth-Bendix algorithm. In *Proceedings of the Colloquium on Algebra, Combinatorics and Logic in Computer Science*, Györ, Hungary, September 1983.

94. Celia Wrathall. Confluence of one-rule Thue systems. In *Proceedings of the First International Workshop on Word Equations and Related Topics (Tubingen)*, volume 572 of *Lecture Notes in Computer Science*, pages 237–246, Berlin, 1990. Springer-Verlag.

95. Ann Yasuhara. *Recursive Function Theory and Logic.* Academic Press, 1971.
96. Hans Zantema. Total termination of term rewriting is undecidable. *Journal of Symbolic Computation,* 20(1):43–60, 1995.
97. Hans Zantema and Alfons Geser. A complete characterization of termination of $0^p 1^q \rightarrow 1^r 0^s$. *Applicable Algebra in Engineering, Communication, and Computing,* 11(1):1–25, 2000.

A Tutorial Example of the Semantic Approach to Foundational Proof-Carrying Code

Amy P. Felty

School of Information Science and Technology
University of Ottawa, Canada
afelty@site.uottawa.ca

Abstract. Proof-carrying code provides a mechanism for insuring that a host, or code consumer, can safely run code delivered by a code producer. The host specifies a safety policy as a set of axioms and inference rules. In addition to a compiled program, the code producer delivers a formal proof of safety expressed in terms of those rules that can be easily checked. Foundational proof-carrying code (FPCC) provides increased security and greater flexibility in the construction of proofs of safety. Proofs of safety are constructed from the smallest possible set of axioms and inference rules. For example, typing rules are not included. In our semantic approach to FPCC, we encode a semantics of types from first principles and the typing rules are proved as lemmas. In addition, we start from a semantic definition of machine instructions and safety is defined directly from this semantics. Since FPCC starts from basic axioms and low-level definitions, it is necessary to build up a library of lemmas and definitions so that reasoning about particular programs can be carried out at a higher level, and ideally, also be automated. We describe a high-level organization that involves Hoare-style reasoning about machine code programs. This organization is presented using a detailed example. The example, as well as illustrating the above mentioned approach to organizing proofs, is designed to provide a tutorial introduction to a variety of facets of our FPCC approach. For example, it illustrates how to prove safety of programs that traverse input data structures as well as allocate new ones.

1 Introduction

In our first presentation of the semantic approach to foundational proof-carrying code (FPCC) [2], we encoded a semantics of types and proved typing rules as lemmas from the basic definitions. We also gave a direct encoding of machine semantics from which we built several layers of definitions so that reasoning about programs was similar to reasoning using Hoare-style program verification rules. This work extended the original proof-carry code (PCC) work [13] which stated typing rules as axioms and generated a safety theorem using a verification condition generator (VCG). Both the axioms and the VCG were parts of the system that had to be trusted.

In FPCC, much progress has been made in a variety of directions since our original work. Type systems that are currently handled are more sophisticated

J. Giesl (Ed.): RTA 2005, LNCS 3467, pp. 394–406, 2005.

and include contravariant recursive types [3] and mutable references [1]. Also, larger machine instructions sets have been encoded [9]. In addition, foundational versions of typed assembly languages (TAL) [10] have been developed for use in FPCC systems (e.g. [6, 14, 15]). Also, an alternative syntactic approach has been explored [8].

Although we presented an example in our first account [2], it was not large enough to illustrate the structure of proofs of safety in general, or demonstrate the style of reasoning that is used to build such proofs. This paper attempts to fill this gap. Although the example is larger, we keep the semantics simple. We only require the simple semantics of types and the same simple set of machine instructions as in our first account. Because any FPCC system is built in layers so that reasoning about particular programs is done at a fairly high level, this example could be carried over fairly directly to current FPCC systems which use machine instruction sets for real machines and more complex type systems. Like our previous work, we adopt the semantic approach to FPCC here. A more detailed comparison to syntactic approaches is future work.

The example presented here is a machine language program which reverses a list of integers. This example is complex enough to require recursive data types. It takes a list as input, and the computation includes traversing this input list as well as building a new one. The latter operation requires allocating new memory along the way. In addition, the program uses most of the instructions available in our simple instruction set.

After presenting the example program in Sect. 2, we present the typing lemmas in Sect. 3, followed by the encoding of machine instruction semantics. We present the machine semantics in two steps. As a first step, we prove the safety of our example program with respect to a set of Hoare-style program verification rules for machine instructions given in Section 4. Using such rules is fairly similar to the use of a VCG in the original PCC framework [13], but provides a slightly higher level of security. In original PCC, the safety proof is a proof of the formula output by the VCG; the VCG program must be trusted. Here, the proof steps which apply the Hoare-style rules are encoded as part of the safety proof. We must trust these rules because they are a part of our basic safety policy, but this should be simpler than trusting a VCG program. Roughly, using the Hoare rules corresponds to recording the primitive steps of the VCG in the proof so that they can be later checked. After presenting these rules, we discuss the safety proof of our example program.

Sect. 5 presents the second step in encoding machine instruction semantics. Here, we follow the approach of Appel and Michael [9]. We start with a direct encoding of machine instructions as a step relation relating one machine state to another, and we prove a theorems stating that safety follows from "progress" and "preservation" lemmas. We do not derive the Hoare rules of Sect 4, but we build up a library of lemmas which provides reasoning similar in style to using such rules. Describing both approaches here allows us to compare them. In particular, our example, discussed again in Sect. 6 provides enough detail to illustrate how the two styles of reasoning correspond. It would be interesting to take this work a step further and derive Hoare-style rules from the direct step-relation encoding.

Hamid and Shao [7], in fact, derive a version of Hoare-style rules in the context of reasoning using TAL in a syntactic FPCC system. Perhaps their approach could be carried over to our setting.

The proof discussed in Sect. 6 has been fully formalized in Coq [5]. We began by adopting and modifying some of the basic definitions in the Coq libraries used in Hamid et. al.'s syntactic approach to FPCC [8]. Most of the proof was done interactively, but we discuss its automation in Sect. 7, where we also discuss other issues and related work.

2   Example

We assume a representation of integer lists where the empty list uses one memory location and is just a tag whose value is 0. If the list is non-empty, then three consecutive memory locations are used. The first contains the tag value 1. The second contains an integer, and the third contains a pointer to the rest of the list. We assume there are 32 registers, denoted r_0 to r_{31}. We introduce our set of machine instructions by directly presenting the reverse program in Fig. 1. We

100	ST $m(r_8 + 0) := r_0$	store 0 at $m(r_8)$
101	ADDC $r_2 := r_8 + 0$	store r_8's value in r_2
102	ADDC $r_8 := r_8 + 1$	increase r_8 by 1
103	LD $r_5 := m(r_1 + 0)$	load tag of list r_1 into r_5
104	BEQ $(r_5 = r_0)$ 114	jump to point after loop end
105	LD $r_3 := m(r_1 + 1)$	load head of list r_1 into r_3
106	LD $r_1 := m(r_1 + 2)$	load tail of list r_1 into r_1
107	ADDC $r_4 := (r_0 + 1)$	r_4 gets value 1
108	ST $m(r_8 + 0) := r_4$	store this value in $m(r_8)$
109	ST $m(r_8 + 1) := r_3$	store head in $m(r_8) + 1$
110	ST $m(r_8 + 2) := r_2$	store r_2 (new tail) in $m(r_8) + 2$
111	ADDC $r_2 := r_8 + 0$	store r_8's current value in r_2
112	ADDC $r_8 := r_8 + 3$	update allocation pointer r_8 by 3
113	BEQ $(r_0 = r_0)$ 103	jump back to loop start
114	ADDC $r_1 := (r_2 + 0)$	r_1 gets value of r_2
115	JMP r_7	return

Fig. 1. A Program for Reversing a List.

assume that register r_0 has value 0 and that input register r_1 contains a list of integers. We also assume that there is a set of consecutive memory locations (unbounded) that are unallocated, and the first location in this set is given by the value of an *allocation pointer* whose value is stored in r_8. The program allocates new memory and increases the value of the allocation pointer as needed. The first 3 lines of the program perform the initialization steps; an empty list is stored at the memory location pointed to by the allocation pointer. Register r_2 stores the reversed list as it is built, and is initialized to point to the new empty list. Lines 103-113 contain the main loop of the program. First, the tag of the next location in the input list is loaded into r_5 and checked. If it is 0, then the program jumps to the point after the loop (line 114), puts the result in r_1, and jumps to some designated return point stored in r_7. Otherwise the body of the loop is executed. In this case, the next 3 memory locations starting at the allocation pointer are used to store the new list. The tail of the new list is assigned to the value of r_2, which is a pointer to the reversed list as constructed so far, and r_2 is updated to point to the new beginning of the reversed list. Finally, the allocation pointer is increased by 3, and control returns to the beginning of the loop. In addition to the instructions used in this program, our simple programming language also includes a MOV instruction, and another branching instruction BGT which compares two values and branches if the first is greater than the second.

To prove safety, the precondition of this program must include our assumptions $r_0 = 0$ and that r_1 contains a list of integers. We write this latter assumption as the typing judgment $(r_1 :_{m,r_8} intlist)$. Typing judgments depend on the contents of memory and the set of currently allocated locations; in particular all memory locations used to represent a list must be allocated. We leave the exact specification of this set unspecified here, but assume that it is a subset of all memory locations occurring before the allocation pointer r_8. To indicate this dependence, memory m and allocation pointer r_8 are given explicitly as subscripts to the typing judgment.

The precondition of this program must include additional information that is part of the loop invariant needed to prove safety of the program. For instance, the policy on readable memory locations is needed. We assume that all memory locations after a particular location $start$ are readable, expressed as the formula $Policy$ and defined as follows:

$$Policy := \forall w.(w \geq start \Rightarrow readable(w)).$$

We assume that the memory locations that we are permitted to write to are a subset of the readable locations. In particular, we assume they are all locations starting at the allocation pointer r_8 and that the allocation pointer r_8 is greater than $start$:

$$(r_8 \geq start) \wedge \forall w.(w \geq r_8 \Rightarrow writable(w)).$$

The loop invariant also includes $safe\_exit(r_7)$ which states that the return location is indeed safe. The complete precondition is stated as the precondition of the first line of code (line 100), defined as formula I_{100} in Fig. 2.

Fig. 2 also includes preconditions of some other lines of code in the program. In general, we write I_c to denote the precondition of line c of the code. In addition to the precondition of the entire program, we must have preconditions of all of the jump points, in this case lines 103 and 114. The precondition of line 103 is the loop invariant. When executing the loop body and when exiting it, we must know the type of register r_2, which stores the intermediate results, i.e., the reversed list as it is being constructed. This typing judgment appears in both I_{103} and I_{114}. Because line 103 is a load instruction, I_{103} contains the requirement that the load is from a readable location. Everything else in I_{103} comes from the precondition and remains invariant when executing the loop body. In this

I_{100} : $Policy \land (r_8 \geq start) \land \forall w.(w \geq r_8 \Rightarrow writable(w)) \land$
             $\land safe\_exit(r_7) \land r_0 = 0 \land (r_1 :_{m,r_8} intlist)$

I_{102} : $allocptr \; r_8 \; 1$

I_{103} : $Policy \land (r_8 - 1 \geq start) \land \forall w.(w \geq r_8 \Rightarrow writable(w)) \land$
             $safe\_exit(r_7) \land r_0 = 0 \land (r_1 :_{m,r_8} intlist) \land$
             $(r_2 :_{m,r_8} intlist) \land readable(r_1)$

I_{112} : $allocptr \; r_8 \; 3$

I_{114} : $safe\_exit(r_7) \land (r_1 :_{m,r_8} intlist) \land (r_2 :_{m,r_8} intlist)$

I_{115} : $safe\_exit(r_7) \land (r_1 :_{m,r_8} intlist)$

Fig. 2. Preconditions for Selected Lines of Code.

example, we also include a precondition for the last line of the program I_{115}. Much of the information provided in Fig. 2, including the typing information, can be generated automatically by a certifying compiler [13]. We call such compiler-generated formulas *hints* to distinguish from those we calculate later. To handle allocation correctly, we also need to know which lines in the program modify the allocation pointer. Lines I_{102} and I_{112} provide this information; in particular, the register serving as the allocation pointer and the amount it is increased at a given line is stated.

3 Types

We present the typing rules that are used in the proof of safety of the example program. We leave out the definitions and lemmas needed to prove these rules. We simply note that we require most of the definitions of types and type constructors and lemmas about them that were presented in our earlier work [2].

We define a *valid* type to be any type τ for which the following two rules hold.

$$\frac{w :_{m,A} \tau \qquad \neg A(v)}{w :_{m[v \mapsto u],A} \tau} \qquad\qquad \frac{w :_{m,A} \tau \qquad \forall x.A(x) \Rightarrow A'(x)}{w :_{m,A'} \tau}$$

In these rules A is an arbitrary *allocation predicate* specifying the set of allocated addresses. In our example, $A(w) := (start \leq w < r_8)$. The expression $m[v \mapsto u]$ denotes the memory m modified so that location v has value u. Integers and integer lists are both valid types [2].

The remaining typing rules we use in proving safety of our program are given in Fig. 3. They are stated in terms of lists of arbitrary type τ.

4 Machine Semantics as Hoare-Style Rules

Fig. 4 contains a set of Hoare-style rules for our machine instructions. Unlike the typing rules in the previous section, we take these rules as axioms. As mentioned earlier, although we must trust them, they provide more security than a VGC. The first rule is used to prove safety of a program with respect to a precondition *Pre*. We assume the program is a sequence of machine instructions ending with a

$$\frac{w :_{m,A} list(\tau) \qquad valid(\tau) \qquad m(w) \neq 0}{m(w+1) :_{m,A} \tau}$$

$$\frac{w :_{m,A} list(\tau) \qquad valid(\tau) \qquad m(w) \neq 0}{m(w+2) :_{m,A} list(\tau)}$$

$$\frac{w :_{m,A} list(\tau) \qquad valid(\tau)}{readable(w)}$$

$$\frac{w :_{m,A} list(\tau) \qquad valid(\tau) \qquad m(w) \neq 0}{readable(w+1)}$$

$$\frac{w :_{m,A} list(\tau) \qquad valid(\tau) \qquad m(w) \neq 0}{readable(w+2)}$$

$$\frac{valid(\tau) \qquad m(w) = 0 \qquad A(w) \qquad readable(w)}{m(w) :_{m,A} list(\tau)}$$

$$\frac{\begin{array}{ccc} & A(w) & readable(w) \\ m(w+1) :_{m,A} \tau \quad m(w) = 1 & A(w+1) & readable(w+1) \\ m(w+2) :_{m,A} list(\tau) \quad valid(\tau) & A(w+2) & readable(w+2) \end{array}}{m(w) :_{m,A} list(\tau)}$$

Fig. 3. Typing rules for integer lists.

$$\frac{Pre \Rightarrow I_1 \quad \{I_1\}S_1\{I_2\} \quad \dots \quad \{I_n\}S_n\{I_{n+1}\} \quad I_{n+1} \Rightarrow safe\_exit(r)}{safe(Pre, (S_1; \dots; S_n; \text{JMP } r))} \text{ safety}$$

$$\overline{\{I[r_s + c/r_d]\} \text{ ADDC } r_d := r_s + c \{I\}} \text{ addc} \qquad \overline{\{I[c/r_d]\} \text{ MOV } r_d := c \{I\}} \text{ mov}$$

$$\overline{\{I[r_{s_1} + r_{s_2}/r_d]\} \text{ ADD } r_d := r_{s_1} + r_{s_2} \{I\}} \text{ add} \qquad \overline{\{I_{m(r)}\} \text{ JMP } r \{I\}} \text{ jmp}$$

$$\overline{\{(r_{s_1} > r_{s_2} \to I_c) \wedge (\neg(r_{s_1} > r_{s_2}) \to I)\} \text{ BGT } (r_{s_1} > r_{s_2}) \, c \{I\}} \text{ bgt}$$

$$\overline{\{(r_{s_1} = r_{s_2} \to I_c) \wedge (r_{s_1} \neq r_{s_2} \to I)\} \text{ BEQ } (r_{s_1} = r_{s_2}) \, c \{I\}} \text{ beq}$$

$$\overline{\{I[m(r_s + c)/r_d] \wedge readable(r_s + c)\} \text{ LD } r_d := m(r_s + c) \{I\}} \text{ ld}$$

$$\overline{\{I[m[r_d + c \mapsto r_s]/m] \wedge writable(r_d + c)\} \text{ ST } m(r_d + c) := r_s \{I\}} \text{ st}$$

$$\frac{A \to A' \qquad \{A'\}S\{B'\} \qquad B' \to B}{\{A\}S\{B\}} \text{ Implied}$$

Fig. 4. Hoare-style rules for machine instructions.

JMP to a safe return point. Note that there is one rule for each machine instruction and that these rules are axioms.

A proof of safety is built by starting with the postcondition I_{n+1} and applying the rule corresponding to statement S_n to obtain I_n. Then I_n is used as the postcondition of statement S_{n-1} to compute I_{n-1}, etc. For any statement S_k $(1 \leq k \leq n)$, if there is an associated hint I_k, this hint is used as the postcondition of statement S_{k-1}. Let I'_k be the formula obtained by applying the axiom for statement S_k using postcondition I_{k+1}. At this point the Implied rule is used, resulting in proof obligation $I_k \Rightarrow I'_k$. The formula I_c in rules bgt and beq is the precondition of the statement at location c. Requiring hints for all jump points insures that such a formula always exists when applying the proof strategy just described.

The *allocptr* hints also generate proof obligations. The hint I_k tells us how to modify the postcondition of S_k. If I_{k+1} is the postcondition computed by applying the appropriate axiom, and I'_{k+1} is obtained from I_{k+1} because of the *allocptr* hint, the we have proof obligation $I'_{k+1} \Rightarrow I_{k+1}$. We will see how to use the *allocptr* hints to modify postconditions for our example program in Sect. 5.

Finally, we also have the proof obligations that appear as the first and last premise in the safety rule.

We can modify the safety rule so that the program includes a postcondition *Post* and the final premise states $I_n \Rightarrow (safe\_exit(r) \wedge Post)$. We use this version of the safety rule in our proof, so that in addition to safety, we prove that the output reversed list does indeed have type *intlist*.

5   Encoding Machine Semantics Directly

We define the type *Reg* to be the type of the set of 32 registers r_0 to r_{31}. *Word* is defined to be the set of natural numbers. For simplicity, we do not build in fixed-size words, though this can and has been done in various PCC systems (for example [9]). We write *Mem* to represent the function type $(Word \rightarrow Word)$. In particular, memory is modelled as a function from machine addresses to machine values. Similarly register banks are functions from registers to values; *RegFile* denotes the function type $(Reg \rightarrow Word)$.

We define a machine state to be a triple of the form (R, M, pc) where R is a register bank (of type *RegFile*), M is a memory (of type *Mem*), and pc is a *Word*. We define a step relation that relates two machine states, one before execution and one after execution of a particular instruction. We write $(R, M, pc \mapsto R', M', pc')$ to denote this relation, and $(R, M, pc \mapsto^* R', M', pc')$ to denote the reflexive transitive closure of this operation.

Machine instructions are encoded as 32-bit machine integers. These integers are decoded into machine instructions by extracting information from specific bits. The step relation is defined by extracting the instruction at line pc in M, decoding it, and changing the machine state according to the semantics of the particular instruction.

We leave out the details, which can be found in our earlier work [2]. We note that what we have described so far is the part of our formalization in Coq where we have adopted and modified some basic definitions from Hamid et. al. [8].

Following Michael and Appel [9], we define *safe*, *Progress*, and *Preservation* predicates as follows, and prove the safe* rule below.

$$safe(R, M, pc) := \forall R', M', pc'.[(R, M, pc \mapsto^* R', M', pc') \Rightarrow$$
$$\exists R'', M'', pc''.(R', M', pc' \mapsto R'', M'', pc'')]$$
$$Progress(Inv) := \forall R, M, pc.[Inv(R, M, pc) \Rightarrow$$
$$\exists R', M', pc'.(R, M, pc \mapsto R', M', pc')]$$
$$Preservation(Inv) := \forall R, M, pc, R', M', pc'.Inv(R, M, pc) \Rightarrow$$
$$(R, M, pc \mapsto R', M', pc') \Rightarrow Inv(R', M', pc')$$

$$\frac{Inv(R, M, pc) \qquad Progress(Inv) \qquad Preservation(Inv)}{safe(R, M, pc)}\ \textbf{safe}^*$$

The $safe$ predicate expresses the fact that execution of safe programs don't get stuck, for example, trying to execute a load from a non-readable location or a store to a non-writable location. Note that $safe$ is now a predicate on a machine state. The code is in M and pc points to the first instruction. In the definitions of $Progress$ and $Preservation$, Inv is a predicate which takes a machine state (R, M, pc) as an argument.

Finally, we define $safe\_exit$ as follows:

$$safe\_exit(w) := \forall R, M, pc.(pc = w \Rightarrow safe(R, M, pc)).$$

6 Example Revisited

It can be seen from the formalization discussed in the previous section that we start from a fairly low-level encoding of the machine semantics and end with the high-level derived rule **safe***. Reasoning using this rule corresponds closely to reasoning using the Hoare-style rules of Sect. 4. In the new setting, a proof of safety starts by applying **safe***. To do so, we need a predicate Inv which expresses a program invariant. Inv will have one clause for every line of the program stating what is true at the point when that line is executed. For the lines for which we already have hints, we use those hints fairly directly. We modify them to become predicates over a register bank and memory. If R is the function representing a register bank, we abbreviate $R(r_i)$ as R_i. We write M for memory functions. Using this encoding, we modify the formulas given in Fig. 2 to obtain the predicates in Fig. 5.

We must actually make one more modification. We must add information to the invariant so that we can show that this program does not include self-modifying code. This information is not needed in the proof using the Hoare-style rules of Sect. 4. In these rules, there is an implicit separation of code from

$$I_{100}(R, M) := Policy \wedge (R_8 \geq start) \wedge \forall w.(w \geq R_8 \Rightarrow writable(w)) \wedge$$
$$\wedge safe\_exit(R_7) \wedge R_0 = 0 \wedge (R_1 :_{M,R_8} intlist)$$
$$I_{103}(R, M) := Policy \wedge (R_8 - 1 \geq start) \wedge \forall w.(w \geq R_8 \Rightarrow writable(w)) \wedge$$
$$safe\_exit(R_7) \wedge R_0 = 0 \wedge (R_1 :_{M,R_8} intlist) \wedge$$
$$(R_2 :_{M,R_8} intlist) \wedge readable(R_1)$$
$$I_{114}(R, M) := safe\_exit(R_7) \wedge (R_1 :_{M,R_8} intlist) \wedge (R_2 :_{M,R_8} intlist)$$
$$I_{115}(R, M) := safe\_exit(R_7) \wedge (R_1 :_{M,R_8} intlist)$$

Fig. 5. Clauses of the invariant that come from hints.

data in memory because there is no connection between the statement part of judgments and the memory. In our new encoding, we define the program using a predicate $listrev(M)$ which states that decoding the instruction at line 100 gives instruction 100 as defined in Fig. 1, and similarly for all the lines of code in the program. If there was overlap between the code and data parts of memory, we would not be able to prove safety. To make explicit that there is no overlap, we add the formula $start > 1000$ to I_{100} and I_{103}. It then becomes part of the loop invariant, and since all of the store instructions are inside the loop, we can prove that $listrev(M)$ remains invariant even while M is changing. The formula $start > 1000$ does not appear in Fig. 5, and we continue to leave it out of the invariant clauses that we present below. Although it is important to the proof, it is not important to the rest of the presentation, and it is easy to prove that it remains a constant at each step.

The full predicate Inv has the following form:

$$Inv(R, M, pc) := [(listrev\ M)\ \wedge$$
$$(pc = 100 \wedge I_{100}(R, M)) \vee \cdots (pc = 115 \wedge I_{115}(R, M))] \vee$$
$$safe(R, M, pc)$$

The second clause of Inv's top-level disjunction is used when the program counter gets the value of r_7. The definition of $safe\_exit$ is used directly to prove this case. The clauses for lines of code that are not defined in Fig. 5 can be automatically calculated by simply applying the rules in Sect. 4. As we have stated, we do not prove the Hoare rules as lemmas from our new encoding of machine semantics. Instead, we apply them by hand to get Inv. It would be easy and much better to write a program to automatically generate them. Note that such a program would not be part of the trusted code; if an invariant is incorrect, it would not be possible to prove the program safe. To illustrate, some of these remaining clauses of Inv are given in Fig. 6. Given a memory function M, we write $M[a_1 \mapsto w_1, \ldots, a_n \mapsto w_n]$ to denote a new function which is the same as M except that for $i = 1, \ldots, n$, the new function maps address a_i to value w_i. We write $readable(\{w_1, \ldots, w_n\})$ to abbreviate $readable(w_1) \wedge \cdots \wedge readable(w_n)$, and similarly for $writable$.

First, consider I_{113} which is the precondition for the statement (BEQ $(r_0 = r_0)$ 103). We applied the beq rule to obtain I_{113} from I_{114}. Note that the precondition in this rule has a true and a false case. We only need the true case here, so we can take as a precondition simply I_c, which in our case is I_{103}. In addition, we need to consider the fact that the statement at line 113 follows a line which increased the allocation pointer r_8 by 3. We must modify I_{103} to account for this increase. Our signal to do so comes from the hint I_{112} in Fig. 2, which we repeat (for documentation purposes only) in Fig. 6 just before the definition of I_{112}. In particular, we must subtract 3 from the expression $R_8 - 1$ in I_{103} to obtain $(R_8 - 4 \geq start)$ in I_{113}.

To obtain I_{112}, we simply apply the addc rule to I_{113}, replacing R_8 with $R_8 + 3$. Most invariant clauses are obtained from such simple rule applications. I_{111} is also obtained by a simple application of addc.

$I_{101}(R, M) :=$ $Policy \wedge (R_8 \geq start) \wedge \forall w.(w \geq R_8 + 1 \Rightarrow writable(w)) \wedge$
$\quad\quad safe\_exit(R_7) \wedge R_0 = 0 \wedge (R_1 :_{M,R_8} intlist) \wedge$
$\quad\quad (R_8 :_{M,R_8} intlist) \wedge readable(R_1)$

$I_{102}(R, M) :=$ $allocptr\ R_8\ 1 :$
$\quad\quad Policy \wedge (R_8 \geq start) \wedge \forall w.(w \geq R_8 + 1 \Rightarrow writable(w)) \wedge$
$\quad\quad safe\_exit(R_7) \wedge R_0 = 0 \wedge (R_1 :_{M,R_8} intlist) \wedge$
$\quad\quad (R_2 :_{M,R_8} intlist) \wedge readable(R_1)$

$I_{104}(R, M) :=$ $(R_5 = R_0 \Rightarrow I_{114}) \wedge (R_5 \neq R_0 \Rightarrow I_{105})$

$I_{105}(R, M) :=$ $Policy \wedge (R_8 - 1 \geq start) \wedge \forall w.(w \geq R_8 + 3 \Rightarrow writable(w)) \wedge$
$\quad\quad \wedge safe\_exit(R_7) \wedge R_0 = 0 \wedge$
$\quad\quad (R_1 :_{M[R_8 \mapsto R_0+1, R_8+1 \mapsto M(R_1+1), R_8+2 \mapsto R_2], R_8} intlist) \wedge$
$\quad\quad (R_8 :_{M[R_8 \mapsto R_0+1, R_8+1 \mapsto M(R_1+1), R_8+2 \mapsto R_2], R_8} intlist) \wedge$
$\quad\quad readable(\{M(R_1 + 2), R_1 + 2, R_1 + 1\}) \wedge$
$\quad\quad writable(\{R_8, R_8 + 1, R_8 + 2\})$

$\quad\quad\vdots$

$I_{110}(R, M) :=$ $Policy \wedge (R_8 - 1 \geq start) \wedge \forall w.(w \geq R_8 + 3 \Rightarrow writable(w)) \wedge$
$\quad\quad \wedge safe\_exit(R_7) \wedge R_0 = 0 \wedge (R_1 :_{M[R_8+2 \mapsto R_2], R_8} intlist) \wedge$
$\quad\quad (R_8 :_{M[R_8+2 \mapsto R_2], R_8} intlist) \wedge readable(R_1) \wedge writable(R_8 + 2)$

$I_{111}(R, M) :=$ $Policy \wedge (R_8 - 1 \geq start) \wedge \forall w.(w \geq R_8 + 3 \Rightarrow writable(w)) \wedge$
$\quad\quad \wedge safe\_exit(R_7) \wedge R_0 = 0 \wedge (R_1 :_{M,R_8} intlist) \wedge$
$\quad\quad (R_8 :_{M,R_8} intlist) \wedge readable(R_1)$

$I_{112}(R, M) :=$ $allocptr\ R_8\ 3 :$
$\quad\quad Policy \wedge (R_8 - 1 \geq start) \wedge \forall w.(w \geq R_8 + 3 \Rightarrow writable(w)) \wedge$
$\quad\quad \wedge safe\_exit(R_7) \wedge R_0 = 0 \wedge (R_1 :_{M,R_8} intlist) \wedge$
$\quad\quad (R_2 :_{M,R_8} intlist) \wedge readable(R_1)$

$I_{113}(R, M) :=$ $Policy \wedge (R_8 - 4 \geq start) \wedge \forall w.(w \geq R_8 \Rightarrow writable(w)) \wedge$
$\quad\quad \wedge safe\_exit(R_7) \wedge R_0 = 0 \wedge (R_1 :_{M,R_8} intlist) \wedge$
$\quad\quad (R_2 :_{M,R_8} intlist) \wedge readable(R_1)$

Fig. 6. More clauses of the invariant.

Next, consider I_{110} which is the precondition for (ST $m(r_8 + 2) := r_2$). We obtain I_{110} by first replacing the memory expression M which appears as a subscript to the typing judgments by $M[R_8 + 2 \mapsto R_2]$, and then adding a new *writable* conjunct.

Working backward, I_{109} back through I_{105} are obtained by straightforward applications of the appropriate rules. We omit the details, showing only the last in the series, I_{105}. We then obtain I_{104} by applying the beq rule. This brings us back to I_{103} which was already given in Fig. 5. Note that at the point I_{103} was generated as a hint, the allocation pointer had to be taken into account; in this case, the increment by 1 at line 102 means we decremented R_8 by 1 to obtain $(R_8 - 1 \geq start)$ in I_{103}. Finally, I_{102} and I_{101} are also obtained by straightforward rule applications.

We show that our example program is safe whenever precondition I_{100} holds for the initial register bank and memory. We must add the fact that the program counter starts at line 100 and that $listrev(M_0)$ holds. Thus, the safety theorem is stated:

$$\forall R_0, M_0, pc_0.(pc_0 = 100 \wedge listrev(M) \wedge I_{100}(R_0, M_0)) \Rightarrow safe(R_0, M_0, pc_0).$$

To prove this theorem, we apply the **safe*** rule, which means we must show that $Inv(R_0, M_0, pc_0)$, $Progress(Inv)$, and $Preservation(Inv)$ hold under the assumptions $pc_0 = 100$, $listrev(M_0)$, and $I_{100}(R_0, M_0)$. $Inv(R_0, M_0, pc_0)$ is a disjunction, and we prove the first disjunct, and the proof is immediate. Since $pc_0 = 100$, showing $Inv(R_0, M_0, pc_0)$ reduces to showing that $listrev(M_0)$ and that $I_{100}(R_0, M_0)$. To show progress, we must show that no matter which line we are at in the program, there is a next step. This is straightforward, and includes proving *readable* and *writable* subgoals for load and store instructions. These follow immediately from the fact that the preconditions of all such instructions contain the necessary *readable* and *writable* facts.

$I'_{100}(R, M) := allocptr\ R_8\ 1 :$
$\qquad Policy \wedge (R_8 \geq start) \wedge \forall w.(w \geq R_8 + 1 \Rightarrow writable(w)) \wedge$
$\qquad safe\_exit(R_7) \wedge R_0 = 0 \wedge (R_1 :_{M[R_8 \mapsto R_0], R_8} intlist) \wedge$
$\qquad (R_8 :_{M[R_8 \mapsto R_0], R_8} intlist) \wedge readable(R_1) \wedge writable(R_8)$
$I'_{103}(R, M) := (M(R_1) = R_0 \Rightarrow I_{114}) \wedge (M(R_1) \neq R_0 \Rightarrow I_{105}) \wedge readable(R_1)$
$I'_{114}(R, M) := safe\_exit(R_7) \wedge (R_2 :_{M, R_8} intlist)$

Fig. 7. Clauses that form proof obligations.

Proving $Preservation(Inv)$ is where Hoare-style reasoning takes place. We have a case for each line of the program; for $pc = 100, \ldots, 114$, under the assumption that $Inv_i(R, M)$ and $(R, M, i \mapsto R', M', pc')$ hold, we show that $Inv_{i+1}(R', M', pc')$ holds. For the cases where we calculated Inv_i from Inv_{i+1} by a straightforward application of one of the Hoare-style axioms, the proof is immediate. The step relation encodes the same information as the corresponding Hoare rule, so all the work was done when we applied the rule by hand to determine the right Inv_i to include in Inv. More reasoning is needed for the cases when Inv_i comes from a hint. The subgoals we must prove correspond to the proof obligations that were described earlier. Consider the formulas in Fig. 7. Formula I'_k is obtained from formula I_{k+1} by an application of the Hoare axiom for the statement at line k. Note that there is one such clause in Fig. 7 for every line of code for which we started out with a hint. The proof obligations we are left with are to show that $I_{100} \Rightarrow I'_{100}$, $I_{103} \Rightarrow I'_{103}$, and $I_{114} \Rightarrow I'_{114}$. The third one is straightforward. The second one is the most complex. The first and second together require all of the typing rules in Fig. 3. This reasoning corresponds to applications of the Implied rule in the proof using the Hoare-style rules.

7   Discussion

The complete proof is approximately 3000 lines of Coq script. Roughly half of that is the foundational part and the other half is the proof of safety of the example program. The latter part could be fully automated. In fact, in our first prototype system, we used the typing rules in Sect. 3 and the Hoare-style rules in Sect 4 as axioms. Thus the system was not yet foundational, but instead concentrated on handling allocation of data structures correctly. This prototype was implemented in λProlog [11, 12], and proofs of safety of a variety of examples, including the list reverse program presented here, were constructed fully automatically. Since the typing rules have since been derived, and since reasoning using the safe* rule corresponds to reasoning using the Hoare-style rules, the proof we generated automatically is similar to the proof done by hand in Coq. In fact, our motivation for doing the Coq proof was to study the similarities and differences in the two styles of reasoning to gain an understanding of how to automate proofs using only the foundational rules. Most of the proof search involves determining which typing rules to apply and fairly straightforward reasoning about arithmetic equalities and inequalities, which can easily be handled by a system with simple but efficient rewriting capabilities. Proving that $listrev(M)$ is an invariant, which was not part of our original automated proof, involves simple but numerous subgoals which followed from simple arithmetic rules.

Our example program is one representative from a large class of programs that could be proved safe with the same kind of automated proof search. Although we did not include the basic definitions, our *intlist* type was defined using a library of definitions for a wide variety of type constructors. Any programs manipulating data structures built from such type constructors fit into this class.

PCC systems that use foundational versions of TAL go even further in the direction of easily automated safety proofs. They essentially reduce such proofs to type checking. Safety in such a setting is limited to what is expressible in TAL. Chang et. al. [4] argue that because there exist a variety of code verification strategies, it is better to use a verifier that is best suited to the code verification strategy. Most examples of safety policies have been simple. In fact, our example does not use a safety policy any more sophisticated than what can be expressed in TAL. But when extending such policies to include more complex properties, other strategies besides TAL may become important. Our approach to automating proofs should provide more flexibility in handling a variety of strategies. This is another subject of future work.

Acknowledgments

The author acknowledges the support of the Natural Sciences and Engineering Research Council of Canada.

References

1. Amal J. Ahmed, Andrew W. Appel, and Roberto Virga. A stratified semantics of general references embeddable in higher-order logic. In *Seventeenth Annual IEEE Symposium on Logic in Computer Science*, pages 75–86, July 2002.
2. Andrew W. Appel and Amy P. Felty. A semantic model of types and machine instructions for proof-carrying code. In *27th ACM SIGPLAN-SIGACT Symposium on Principles of Programming Languages*, pages 243–253, 2000.

3. Andrew W. Appel and David McAllester. An indexed model of recursive types for foundational proof-carrying code. *ACM Transactions on Programming Languages and Systems*, 13(5):657–683, September 2001.
4. Bor-Yuh Evan Chang, Adam Chlipala, George C. Necula, and Robert R. Schneck. The open verifier framework for foundational verifiers. In *ACM SIGPLAN Workshop on Types in Language Design and Implementation*, January 2005.
5. Coq Development Team. The Coq Proof Assistant reference manual: Version 7.4. Technical report, INRIA, 2003.
6. Karl Crary and Susmit Sarkar. Toward a foundational typed assembly language. In *Thirtieth ACM SIGPLAN-SIGACT Symposium on Principles of Programming Languages*, pages 198–212, 2003.
7. Nadeem A. Hamid and Zhong Shao. Interfacing hoare logic and type systems for foundational proof-carrying code. In *Seventeenth International Conference on the Applications of Higher Order Logic Theorem Proving*, volume 3223 of *Lecture Notes in Computer Science*, pages 118–135. Springer-Verlag, September 2004.
8. Nadeem A. Hamid, Zhong Shao, Valery Trifonov, Stefan Monnier, and Zhaozhong Ni. A syntactic approach to foundational proof-carrying code (extended version). *Journal of Automated Reasoning*, 31(3–4):191–229, 2003.
9. Neophytos G. Michael and Andrew W. Appel. Machine instruction syntax and semantics in higher order logic. In *Seventeenth International Conference on Automated Deduction*, Lecture Notes in Computer Science, pages 7–24. Springer-Verlag, June 2000.
10. Greg Morrisett, David Walker, Karl Crary, and Neal Glew. From System F to typed assembly language. *ACM Transactions on Programming Languages and Systems*, 21(3):528–569, May 1999.
11. Gopalan Nadathur and Dale Miller. An overview of λProlog. In K. Bowen and R. Kowalski, editors, *Fifth International Conference and Symposium on Logic Programming*. MIT Press, 1988.
12. Gopalan Nadathur and Dustin. J. Mitchell. System description: Teyjus — a compiler and abstract machine based implementation of λProlog. In *The Sixteenth International Conference on Automated Deduction*, pages 287–291. Springer-Verlag, July 1999.
13. George Necula. Proof-carrying code. In *24th ACM SIGPLAN-SIGACT Symposium on Principles of Programming Languages*, pages 106–119. ACM Press, January 1997.
14. Kedar N. Swadi and Andrew W. Appel. Foundational semantics for TAL syntactic rules via typed machine language. http://www.cs.princeton.edu/~kswadi/papers/tml.ps, March 2002.
15. Gang Tan, Andrew W. Appel, Kedar N. Swadi, and Dinghao Wu. Construction of a semantic model for a typed assembly language. In *Fifth International Conference on Verification, Model Checking and Abstract Interpretation*, volume 2937 of *Lecture Notes in Computer Science*, pages 30–43. Springer-Verlag, January 2004.

Extending the Explicit Substitution Paradigm

Delia Kesner[1] and Stéphane Lengrand[1,2]

[1] PPS, Université Paris 7, France
{kesner,lengrand}@pps.jussieu.fr
[2] School of Computer Science, University of St Andrews, UK

Abstract. We present a simple term language with explicit operators for erasure, duplication and substitution enjoying a sound and complete correspondence with the intuitionistic fragment of Linear Logic's Proof Nets. We establish the good operational behaviour of the language by means of some fundamental properties such as confluence, preservation of strong normalisation, strong normalisation of well-typed terms and step by step simulation. This formalism is the first term calculus with explicit substitutions having full composition and preserving strong normalisation.

1 Introduction

The *Curry-Howard* paradigm, according to which the terms/types/reduction of a term language respectively correspond to the proofs/propositions/normalisation of a logical system, has already shown its numerous merits in the computer science community. Such a correspondence gives a double reading of proofs as programs and programs as proofs, so that insight into one aspect helps the understanding of the other.

A typical example of the Curry-Howard correspondence is obtained by taking the simply typed *l*-calculus [11] as term language and Natural Deduction for Intuitionistic Logic as logical system. But both formalisms can be decomposed in the following sense: on the one hand the evaluation rule of *l*-calculus, known as β-reduction, can be decomposed into more elementary operations: the various tasks needed to implement β can be achieved by manipulation of various *explicit operators*, such as *erasure, duplication* and *substitution*. On the other hand *Linear Logic* [24] decomposes the intuitionistic logical connectives into more elementary connectives, such as the linear arrow and the exponentials, thus providing a more refined use of resources than that of Intuitionistic Logic.

We show that there is a deep connection between these two elementary decompositions. In order to relate them, we must bridge the conceptual gap between the formalism of a term syntax and that of *Proof Nets* [24] that we use to denote proofs in Linear Logic. Visually convenient to manipulate, the latter retains from the structure of a proof the part that is logically relevant, thus giving geometric insight into proof transformations. However, it is quite cumbersome in proof formalisations. On the other hand, term notation is more convenient to formalise and carry detailed proofs of properties, and also when one wants to implement them via some proof-assistant [12, 31].

J. Giesl (Ed.): RTA 2005, LNCS 3467, pp. 407–422, 2005.

Several works [16, 17] have already explored the relation between these two approaches, but none of them has pushed the formalism far enough to obtain a computational counterpart to Proof Nets that is sound and complete with respect to the underlying logical model.

We present a calculus with explicit operators for erasure, duplication and substitution, called λlxr, which can be seen as a functional computational counterpart to Proof Nets. The major features of this calculus are

- Simple syntax and natural semantics via reduction rules and equations;
- Sound and complete correspondence with the Proof Nets model, where the equations and reductions of terms have a natural correspondence with those of Proof Nets;
- Full composition of substitutions;
- Nice properties such as confluence, preservation of strong normalisation, strong normalisation for well-typed terms, and step by step simulation of β-reduction.

Explicit Operators and Proof Nets. Much work on explicit substitutions has been done in the last 10 years, for example [1, 5, 8, 32]. In particular, an unexpected result was given by Melliès [40] who has shown that there are β-strongly normalisable terms in l-calculus that are not strongly normalisable when evaluated by the reduction rules of an explicit version of the l-calculus, such as for example $\lambda\sigma$ [1] or $\lambda\sigma_{\Uparrow}$ [27]. In other words, $\lambda\sigma$ and $\lambda\sigma_{\Uparrow}$ do not enjoy the property known as Preservation of Strong Normalisation (PSN) [5].

This phenomenon shows a defect in the design of these calculi with explicit substitutions because they are supposed to implement their underlying language without losing its good properties. However, there are many ways to avoid Melliès' counter-example in order to recover the PSN property. One of them is to simply forbid the substitution operators to cross lambda-abstractions [22, 39]; another consists of avoiding composition of substitutions [5]; another one imposes a simple strategy on the calculus with explicit substitutions to mimic exactly the calculus without explicit substitutions [25]. The first solution leads to *weak* lambda calculi, not able to express *strong* beta-equality, which is used for example in implementations of proof-assistants [12, 31]. The second solution is drastic as composition of substitutions is needed in implementations of HO unification [20] or functional abstract machines [28]. The last one does not take benefit of the power of explicit operators because substitutions are neither controlled nor delayed.

In order to cope with this problem David and Guillaume [14] defined a calculus with *explicit labels*, called λ_{ws}, which allows *controlled* composition of substitutions without losing PSN. These labels are obtained by considering a *weakening* rule in the logical system that specifies the typing rules of λ_{ws}, and then by annotating in the term language the formula introduced by this rule via a label. But the λ_{ws}-calculus has a complicated syntax and its named version [17] is even more unreadable. On the positive side we should mention that λ_{ws}-calculus has very nice properties as it is confluent (or Church-Rosser) and enjoys PSN. Also, it can be shown [18] that there is a simple translation from λ_{ws}

into the Proof Nets of Linear Logic that preserves reduction. This translation gives at the same time an elegant proof of strong normalisation for well-typed λ_{ws}-terms. Moreover, the translation reveals a natural semantics for composition of explicit substitutions, and also suggests that explicit erasure and duplication can be added to the calculus without losing termination. These are the main ideas constituting the starting point of the calculus called λlxr that we present in this paper.

Explicit operators of typed λlxr have thus a nice logical interpretation: substitution is *cut*, duplication is *contraction*, erasure is *weakening*. From the point of view of implementation, this can be read as the facts that substitution can be delayed, and that duplication and erasure can be controlled.

Instead of translating a term syntax into Proof Nets, we *extract* a term calculus from Proof Nets, thus defining a simple and natural syntax involving not only reduction rules but also *equations*. Every term equation of λlxr can be seen as a computational counterpart to an equality between Proof Nets and, vice-versa, every Proof Net equality can be naturally read back as an equality between λlxr-terms.

It is then not surprising that we obtain a full correspondence between typed λlxr and the Intuitionistic fragment of Linear Logic's Proof Nets in the sense that the interpretation is not only *sound* but also *complete* (in contrast to the translation from λ_{ws} to Proof Nets, which was only sound).

Weakening and Garbage Collection. The *erasure/weakening operator* has an interesting computational behaviour that we illustrate via an example. Let us denote by $W_{\_}(\_)$ the weakening operator, so that a λlxr-term whose variable x is used to weaken the term N is written $W_x(N)$, that is, we explicitly annotate that the variable x does not appear free in the term N. Then, when evaluating the application of a term $\lambda x.W_x(N)$ to another term L, an explicit substitution $\langle x = L \rangle$ is created and the computation will continue with $W_x(N)\langle x = L \rangle$. Then, the weakening operator will be used to prevent the substitution $\langle x = L \rangle$ from going into the term N, thus making more efficient the propagation of a substitution with respect to the original term.

Another interesting feature of our system is that weakening operators are always *pulled out* to the top-level during λlxr-reduction. Moreover, free variables are *never* lost during computation because they get marked as weakening operators. Indeed, if t β-reduces to t', then its λlxr-interpretation reduces to that of t' where weakening operators are added at the top level to keep track of the variables that are lost during the β-reduction. Thus for example, when simulating the β-reduction steps $(lx.ly.x)Nz \longrightarrow^*_\beta N$, the lost variable z will appear in the result of the computation by means of a weakening operator at the top level, i.e. as $W_z(\overline{N})$ (where $\overline{N}$ is the interpretation of N in λlxr), thus preparing the situation for an efficient garbage collection on z.

The explicit weakening operator can thus be seen as a tool for an efficient implementation of *garbage collection*. This feature is not present in l-calculus, so one can think that l-calculus is better or simpler, but implementation of functional programming shows that garbage collection exists and must be taken into account.

It is worth noticing that the labels of the λ_{ws}-calculus cannot be pulled out to the top-level as in λlxr. Also, free variables may be lost during λ_{ws}-computation. Thus, garbage collection within λ_{ws} does not offer the advantages existing in λlxr.

Composition. From a rewriting point of view this calculus can be viewed as the first formalism that is confluent (or Church-Rosser) and strongly normalising on typed terms, *simulates β-reduction step by step*, and has PSN as well as *full composition*. By simulation of β-reduction step by step we mean that every β-reduction step in l-calculus induces a non-empty λlxr-reduction sequence. By full composition we mean that we can compute the application of an explicit substitution to a term, no matter which substitution remains non-evaluated within that term. In particular, in a term $N\langle y = P\rangle\langle x = L\rangle$, the external substitution is not blocked by the internal one and can be further evaluated without ever requiring any preliminary evaluation of $N\langle y = P\rangle$. In other words, the application of the substitution $\langle x = L\rangle$ to the term N can be evaluated independently from that of $\langle y = P\rangle$. A more technical explanation of the concept of full composition appears in Section 2.

Related Work. Besides the λ_{ws}-calculus [14] and its encoding in linear logic [17] already mentioned, other computational meanings of logic via the use of explicit operators have already been proposed.

Herbelin [30] proposes a term language with applicative terms and explicit substitutions which corresponds to the Gentzen-style sequent calculus LJT. A similar approach to intuitionistic logic is also studied in [49]. In a very different spirit, [10] relates the *pattern matching operator* in functional programming to the cut elimination process in sequent calculus for intuitionistic logic.

Abramsky [2] gives computational interpretations for intuitionistic and classical Linear Logic which are based on sequents rather than Proof Nets. As a consequence, no equalities between terms reflect the fact that some proofs of the sequent calculus approach get identified when expressed as Proof Nets. Many other term calculi based on sequents rather than Proof Nets have been proposed for Linear Logic, as for example [6, 23, 45, 50].

A related approach was independently developed by V. van Oostrom (available in course notes written in Dutch [48]), where explicit operators for contraction and weakening are added to the λ-calculus to present optimal reduction in a framework with implicit substitutions. We show here how the same operators allow a fine control of composition when using explicit substitutions, although the proofs of some fundamental properties, such as PSN and confluence, become harder. A complete overview on optimal sharing in functional programming languages, and its connection with linear logic can be found in [4].

Another approach is taken in [21], where a calculus with explicit operators is defined in order to study the notion of "closed reduction" in l-calculus. Although reduction rules take enormous advantage of the fact that some subterms are closed (i.e. without free variables), which greatly simplifies the definition of reduction, no deep relation with Proof Nets is exploited and no equalities appear at the level of terms.

Our completeness proof is inspired by [37], where polarised Proof Nets are proposed as a sound and complete model of the $\lambda\mu$ calculus [42]. Finally, a revised version of the calculus λ_{ws} with names is developed in [43].

The paper is organised as follows. Section 2 presents the syntax and operational semantics of the λlxr-calculus. Section 3 defines the model of the calculus and establishes soundness and completeness. Section 4 shows the relation between l-calculus and λlxr-calculus by giving mutual translations from one to the other. In Section 5 we state the main operational properties of λlxr. Finally we conclude and give some ideas for further work. For lack of space, we cannot give full proofs in this extended abstract; we refer the reader to [33] for further details.

2   The Calculus λlxr

The syntax for raw terms, given by the following grammar, is extremely simple[1] and can be just viewed as an extension of that of lx [8].

$$t ::= x \mid lx.t \mid t\ t \mid t\langle x = t\rangle \mid W_x(t) \mid C_x^{y,z}(t)$$

The term x is called a *variable*, $lx.t$ an *abstraction*, $t\ u$ an *application*, $t\langle x = u\rangle$ a *substitution*, $W_x(t)$ a *weakening* and $C_x^{y,z}(t)$ a *contraction*. The last three constructors are called *explicit operators*.

The terms $lx.t$ and $t\langle x = u\rangle$ define *binders* for the variable x (said to be *bound*) whose scope is t. The term $C_x^{y,z}(t)$ defines a *binder* for y and z (also said to be *bound*) whose scope is t, whereas x is *free* in the terms x, $C_x^{y,z}(t)$ and $W_x(t)$. We write $\mathcal{V}(t)$ to denote the set of variables of the term t and $\mathcal{FV}(t)$ to denote the subset of $\mathcal{V}(t)$ which contains only the free ones. As usual we shall consider α-conversion to guarantee that no variable is free and bound in a term at the same time and that bound variables have all different names.

We say that a term is *linear* if it satisfies the following: in every subterm, every variable has at most one free occurrence, and every binder binds a variable that does have a free occurrence (and hence only one). For instance, the terms $W_x(x)$ and $lx.xx$ are not linear. However, the latter can be represented in the λlxr-calculus by the linear term $lx.C_x^{y,z}(yz)$. More generally, every l-term can be translated to a linear λlxr-term (c.f. Section 4).

We use $\Phi, \Delta, \Sigma, \Pi, \ldots$ to denote finite *lists* of variables (with no repetition). We use the notation $W_{x_1,\ldots,x_n}(t)$ for $W_{x_1}(\ldots W_{x_n}(t))$, and $C_{x_1,\ldots,x_n}^{(y_1,\ldots,y_n),(z_1,\ldots,z_n)}(t)$ for $C_{x_1}^{y_1,z_1}(\ldots C_{x_n}^{y_n,z_n}(t))$. For any term t we define a *renaming operation* $R_{y_1,\ldots,y_n}^{x_1,\ldots,x_n}(t)$ as the result of *simultaneously* substituting y_i for every *free* occurrence x_i in t where $i \in 1\ldots n$. Thus for instance $R_{x',y'}^{x,y}(C_w^{y,z}(x(yz))) = C_w^{y,z}(x'(y'z))$.

We now introduce a congruence $\equiv$ (i.e. a symmetric, reflexive, transitive relation closed under any context) on terms which brings the typed version of the calculus closer to the Proof Nets modulo as defined in [15]. The relation $\equiv$ is the

---

[1] In contrast to l_{ws} with names [17, 18], where terms affected by substitutions have a complex format $t[x, u, \Gamma, \Delta]$

smallest congruence that includes the axioms in Figure 1 expressing associativity and commutativity for contraction (equations $A, C1_c, C2_c$), commutativity for weakening (equation C_w) and commutativity for independent substitutions (equation S), which is also called *parallel composition*. Terms up to rule S could be represented using sets of substitutions instead of atomic ones. Finally, contraction and substitution are treated at the same level using axiom $Cont2$.

$$
\begin{array}{lll}
C_w^{x,v}(C_x^{z,y}(t)) & \equiv_A & C_w^{x,y}(C_x^{z,v}(t)) & \text{if } x \neq y, v \\
C_x^{y,z}(t) & \equiv_{C1_c} & C_x^{z,y}(t) \\
C_{x'}^{y',z'}(C_x^{y,z}(t)) & \equiv_{C2_c} & C_x^{y,z}(C_{x'}^{y',z'}(t)) & \text{if } x \neq y', z' \ \& \ x' \neq y, z \\
W_x(W_y(t)) & \equiv_{C_w} & W_y(W_x(t)) \\
t\langle x = u\rangle\langle y = v\rangle & \equiv_S & t\langle y = v\rangle\langle x = u\rangle & \text{if } y \notin \mathcal{FV}(u) \ \& \ x \notin \mathcal{FV}(v) \ \& \ x \neq y \\
C_w^{y,z}(t)\langle x = u\rangle & \equiv_{Cont2} & C_w^{y,z}(t\langle x = u\rangle) & \text{if } x \neq w \ \& \ y, z \notin \mathcal{FV}(u)
\end{array}
$$

Fig. 1. Congruence axioms for λlxr-terms.

It can easily be proven that the congruence relation defined by the previous rules preserves free variables and linearity. Since we shall deal with rewriting modulo the congruence $\equiv$, it is worth noticing that $\equiv$ is decidable. More than that, each congruence class contains finitely many terms.

The congruence $\equiv$ enables us to write "$W_S(u)$", or "$C_\Phi^{\Delta,\Pi}(t)$ where $\Phi := S$" without ordering the variables in S. Besides, we shall sometimes not specify what the lists Δ and Π are, assuming them to be two *disjoint* lists of *fresh* variables.

The reduction relation of the calculus, denoted $\longrightarrow_{\lambda\mathsf{lxr}}$, is the relation generated by the reduction rules in Figure 2 modulo the congruence relation in Figure 1. In order to avoid variable capture, rules Abs and $CAbs$ respectively need the side-conditions $(y \notin \mathcal{FV}(u))$ and $(x \neq y, z)$, which can always be satisfied by α-conversion, so that their nature is different from that of the other side-conditions. The rules should be understood in the prospect of applying them to linear terms. Indeed, it can be shown that if t is linear and $t \longrightarrow_{\lambda\mathsf{lxr}} t'$, then t' is linear and $\mathcal{FV}(t) = \mathcal{FV}(t')$. The last statement is achieved in particular by the weakening operator, and coincides with the property called "interface preserving" [36] in interaction nets. The fact that linearity is preserved is a essential requirement of the system, so that we can henceforth consider linear terms only.

It is worth noticing that weakening and contraction can naturally be viewed, respectively, as explicit *erasure* and *duplication* operators. The former may be pulled out to the top level by using rules $WAbs, WApp1, WApp2, WSubs$ in order to eliminate void substitutions as soon as possible, while the latter may be pushed in by using rules $CAbs, CApp1, CApp2, CSubs$ in order to delay the duplication of substitutions as much as possible.

For any reduction relation $\longrightarrow_j$, we denote by $\longrightarrow^+_j$ the transitive closure and by $\longrightarrow^*_j$ the reflexive and transitive closure.

Owing to the linearity constraints previously imposed, the $Comp$ rule is equivalent to the following rule $Comp_{bad}$:

$$
t\langle y = v\rangle\langle x = u\rangle \longrightarrow t\langle y = v\langle x = u\rangle\rangle \qquad x \notin \mathcal{FV}(t)
$$

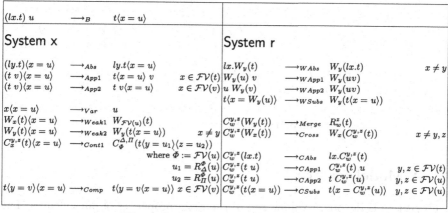

Fig. 2. Reduction rules for λlxr-terms.

However, if the linearity constraints are not taken into account, the $Comp_{bad}$ rule causes failure of the PSN property [7].

Also, when linearity constraints are not considered, four cases may occur when composing two substitutions as in $t\langle y = v\rangle\langle x = u\rangle$: either (1) $x \in \mathcal{FV}(t) \cap \mathcal{FV}(v)$, or (2) $x \in \mathcal{FV}(t) \setminus \mathcal{FV}(v)$, or (3) $x \in \mathcal{FV}(v) \setminus \mathcal{FV}(t)$, or (4) $x \notin \mathcal{FV}(t) \cup \mathcal{FV}(v)$.

Composition is said to be *partial* in calculi like λ_{ws} [14] because only cases (1) and (3) are considered by the reduction rules. Because of the linearity constraints of λlxr, cases (1) and (4) have to be dealt with by the introduction of a contraction for case (1) and a weakening for case (4). Those operators will interact with external substitutions by the use of rules $(Weak1)$ and $(Cont1)$, respectively. Case (3) is treated by rule $(Comp)$, and case (2) by the congruence rule $\equiv_S$. We say in this case that composition is *full* as all cases (1)-(4) are treated. Thus, λlxr turns out to be the first term calculus with explicit substitutions having full composition and preserving β-strong normalisation (c.f. Theorem 7).

We now define a typing system for λlxr *Types* are defined by means of the following syntax, where σ ranges over a countable set of atomic types.

$$A ::= \sigma \mid A \to A$$

An *environment* is a *set* of decorated variables of the form $x : A$, where A is any type. A *judgement* is a triplet of the form $\Gamma \vdash t : A$, where Γ is an environment, t is a λlxr-term, and A is a type.

We say that t is *well-typed* if there is an environment Γ and a type A such that the judgement $\Gamma \vdash t : A$ is derivable from the set of *typing rules* in Figure 3.

Remark that $\Gamma \vdash t : A$ always implies that the domain of Γ is exactly $\mathcal{FV}(t)$. Also, when writing $\Gamma, x : B$, the variable x is supposed to be fresh w.r.t Γ.

As expected, Subject Reduction holds:

Theorem 1. *If* $\Gamma \vdash s : A$ *and* $s \longrightarrow_{\lambda lxr} s'$, *then* $\Gamma \vdash s' : A$.

$$\frac{}{x : A \vdash x : A} \qquad \frac{\Gamma, x : A \vdash t : B}{\Gamma \vdash lx.t : A \to B} \qquad \frac{\Gamma, x : A, y : A \vdash M : B}{\Gamma, z : A \vdash C_z^{x,y}(M) : B}$$

$$\frac{\Gamma, x : B \vdash t : A \quad \Delta \vdash M : B}{\Gamma, \Delta \vdash t \langle x = M \rangle : A} \qquad \frac{\Gamma \vdash t : A \to B \quad \Delta \vdash v : A}{\Gamma, \Delta \vdash (t\,v) : B} \qquad \frac{\Gamma \vdash t : A}{\Gamma, x : B \vdash W_x(t) : A}$$

Fig. 3. Typing Rules for λlxr-terms.

3   A Model for λlxr

This section is devoted to show two of the main properties of our calculus. The first one concerns strong normalisation of well-typed terms, which is achieved by translating well-typed λlxr-terms to MELL Proof Nets. The second one shows that the translation from λlxr to Proof Nets is sound and complete w.r.t the appropriate equivalence relations on terms and proof nets respectively.

We briefly recall here the traditional notion of Proof Nets of Linear Logic and some of its basic properties. We refer the interested reader to [24] for more details.

Let $\mathcal{A}$ be a set of *atomic formulae* equipped with an involutive function $\perp : \mathcal{A} \to \mathcal{A}$, called *linear negation*. The set of formulae of the multiplicative exponential fragment of linear logic (called MELL) is defined as follows:

$$\mathcal{F} ::= \mathcal{A} \mid \mathcal{F} \otimes \mathcal{F} \mid \mathcal{F} \,\otimes\, \mathcal{F} \mid !\mathcal{F} \mid ?\mathcal{F}$$

The formula $\mathcal{F} \,\otimes\, \mathcal{G}$ denotes a "non-economic" version of the classical disjunction, whereas $?\mathcal{F}$ and $!\mathcal{F}$ are used to indicate where contraction or weakening can take place. We extend the notion of *linear negation* to formulae as follows:

$$(?A)^\perp = !(A^\perp) \qquad (A \otimes B)^\perp = A^\perp \,\otimes\, B^\perp$$
$$(!A)^\perp = ?(A^\perp) \qquad (A \,\otimes\, B)^\perp = A^\perp \otimes B^\perp$$

The set of Proof Nets is denoted PN (we refer the reader to [24] or [33] for a formal definition).   Proof Nets are the *computational objects* behind Linear Logic, where the notion of reduction (called also "cut elimination") corresponds exactly to the cut-elimination procedure on sequent derivations. The traditional reduction system for MELL consists in *cut elimination rules*, we refer the reader to [24] or [33] for a formal definition.

Unfortunately, the original notion of reduction on PN is not well adapted to simulate either the β rule of λ-calculus, or the rules dealing with propagation of substitution in explicit substitution calculi: too many inessential details about the order of application of the rules are still present, and in order to get rid of them, one is naturally led to define an equivalence relation on PN, as is done in [15], defined by the axioms A and B in Figure 4. Equivalence A turns contraction into an associative operator. Equivalence B abstracts away the relative order of application of the rules of box-formation and contraction. Finally,

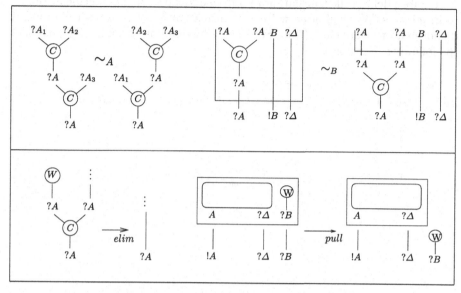

Fig. 4. Axioms and extra reduction rules for MELL Proof Nets.

besides these equivalence relations defined in [15], we shall also need the two extra reduction rules in Figure 4: *elim* is used to remove unneeded weakening links when simulating explicit substitutions and *pull* allows weakening links to go outside boxes.

Notation: Henceforth, we shall call R the system made of rules *Ax-cut*, *℘-⊗*, *w-b*, *d-b*, *c-b*, *b-b* and *elim* and *pull*. We shall write $\sim_E$ for the congruence (reflexive, symmetric, transitive, closed by contexts) relation on Proof Nets generated by axioms A, B. We shall write R/E for the system made of the rules in R and the axioms in A, B. System R/E is actually defining a notion of *reduction modulo an equivalence relation*, so we write $r \longrightarrow_{R/E} s$ if and only if there exist r' and s' such that $r \sim_E r' \longrightarrow_R s' \sim_E s$.

In order to prove the properties of the translation from typed λlxr to Proof Nets, we shall use the following result:

Theorem 2. *The reduction relation* $\longrightarrow_{R/E}$ *terminates.*

Proof. This result is proved in [43] for which we refer the interested reader for full details. The proof uses postponement of rule *pull* w.r.t to the rest of the system for which termination is stated in [15]. For the sake of completeness, we include the proof in English in the full version of our paper [33].

We now present the natural interpretation of typed λlxr-terms as Proof Nets. For that, we use the translation of types introduced in [13] given by:

$$A^* \quad\quad = A \quad\quad\quad \text{for atomic types}$$
$$(A \to B)^* = ?((A^*)^{\perp}) \,℘\, B^* \quad \text{otherwise}$$

Figure 5 defines the translation $T(\_)$ from derivable typing judgements of λlxr to Proof Nets. We shall often write $T(t)$ instead of $T(\Gamma \vdash t : A)$ when Γ and A do not matter or are clear (from Subject Reduction, for example). $T(\_)$ satisfies the following properties:

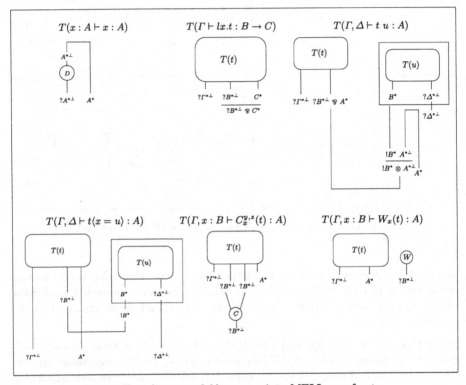

Fig. 5. Encoding typed λlxr-terms into MELL proof-nets.

Lemma 1 (Simulation of λlxr-reduction).

- If $t \equiv t'$, then $T(t) \sim_E T(t')$.
- If $t \longrightarrow_B t'$, then $T(t) \longrightarrow^+_{R/E} T(t')$.
- If $t \longrightarrow_{xr} t'$, then $T(t) \longrightarrow^*_{R/E} T(t')$.

As a consequence we obtain one of the main important properties of λlxr:

Theorem 3 (Strong Normalisation).
The relation $\longrightarrow_{\lambda lxr}$ *is strongly normalising on well-typed λlxr-terms.*

Proof. An infinite λlxr-reduction sequence would have infinitely many B-steps. Indeed, system xr can be proven terminating [33]. But this would lead by Lemma 1 to an infinite R/E-reduction sequence which is impossible by Theorem 2.

In order to show Theorem 3 another technique [43] using Preservation of Strong Normalisation (c.f. Section 5) together with the strong normalisation of typed λ-calculus could be used. Direct proofs using for instance reducibility by perpetuality [9, 38] seem much more difficult to adapt to our case. We remark however that those techniques are no simpler than ours, as many auxiliary properties about the computational behaviour of the calculus need to be establish.

The relevant computational steps of the cut-elimination procedure in Proof Nets are those modifying the box structure. Hence, we are interested in characterising those terms that are translated by $T(\_)$ into Proof Nets that have identical box structures. Let TB be the reduction relation on PN generated by the rules that do not modify the box structure, namely *pull*, *elim*, *Ax-cut* and $\otimes$-$\otimes$, modulo the congruence $\sim_E$.

Proposition 1. *The reduction relation TB is confluent and terminating. Hence, the normal form of a proof net r w.r.t this reduction relation, written $TB(r)$, exists and is unique up to the congruence $\sim_E$.*

Hence, "having the same box structure" can be expressed by the equivalence relation $\approx$ defined as: $r \approx r'$ if and only if $TB(r) \sim_E TB(r')$.

We now define for the terms a congruence $\cong$ obtained by adding to $\equiv$ the following rules turned into equalities:

$$\{B, Abs, App1, Weak2, WAbs, WApp1, WApp2, Cross, Merge, CAbs, CApp1, CApp2\}$$

Remark that $WSubs$ and $CSubs$ are included in $\cong$.

The following property relates two $\cong$-convertible terms w.r.t. their semantic translations into Proof Nets and is one of the interesting results about λlxr.

Theorem 4 (Soundness and Completeness). *The interpretation $T(\_)$ is sound and complete, i.e. given two λlxr-terms t_1, t_2 we have $t_1 \cong t_2$ iff $T(t_1) \approx T(t_2)$.*

The proof is inspired by that of a similar characterisation, given in [37], for $l\mu$-terms with respect to Polarized Proof Nets, where equality in the term syntax is an extension of the σ-equivalence on l-terms defined in [44]. Yet, the latter needs to consider specific permutations of β-redexs to achieve the characterisation, instead of simply turning some reduction rules into equivalence rules.

4   Recovering the *l*-Calculus

We show in this section the relation between λlxr-terms and l-terms. More precisely, we show that the linearity constraints and the use of explicit resource operators in λlxr are sufficient to decompose the β-reduction step into smaller steps. We shall use the notation $\Gamma \vdash_l t : A$ to denote typing judgements and typing derivability in l-calculus.

We now describe how to encode a l-term (possibly not linear) into a λlxr one, up to the congruence $\equiv$.

Definition 1. *The encoding of l-terms is defined by induction as follows:*

$$
\begin{aligned}
\mathcal{A}(x) &:= x \\
\mathcal{A}(lx.t) &:= lx.\mathcal{A}(t) && \text{if } x \in \mathcal{FV}(t) \\
\mathcal{A}(lx.t) &:= lx.W_x(\mathcal{A}(t)) && \text{if } x \notin \mathcal{FV}(t) \\
\mathcal{A}(tu) &:= C_\Phi^{\Delta,\Pi}(R_\Delta^\Phi(\mathcal{A}(t))\; R_\Pi^\Phi(\mathcal{A}(u))) \text{ where } \Phi := \mathcal{FV}(t) \cap \mathcal{FV}(u)
\end{aligned}
$$

Note that $\mathcal{A}(tu) = \mathcal{A}(t)\mathcal{A}(u)$ in the particular case $\mathcal{FV}(t) \cap \mathcal{FV}(u) = \emptyset$.

It is worth noticing that $\mathcal{A}$ commutes with renaming (i.e. $\mathcal{A}(R_\Delta^\Phi(t)) = R_\Delta^\Phi(\mathcal{A}(t))$) and that $\mathcal{A}$ preserves free variables (i.e. $\mathcal{FV}(t) = \mathcal{FV}(\mathcal{A}(t))$). As a consequence, the encoding of a λ-term is a linear λlxr-term. For instance, given $t = lx.ly.y(zz)$, we have $\mathcal{A}(t) = lx.W_x(ly.(y\; C_z^{z_1,z_2}(z_1\; z_2)))$.

Notice that a β-reduction step may not preserve the set of free variables whereas any reduction in λlxr does. Indeed, we have $t = (\lambda x.y)\; z \longrightarrow_\beta y$, but

$$
\mathcal{A}(t) = (lx.W_x(y))\; z \longrightarrow^*_{\lambda\text{lxr}} W_z(y) = W_z(\mathcal{A}(y))
$$

It is hence the following statement that we prove by induction on terms.

Theorem 5. *If $t \longrightarrow_\beta t'$, then $\mathcal{A}(t) \longrightarrow^+_{\lambda\text{lxr}} W_{\mathcal{FV}(t)\setminus\mathcal{FV}(t')}(\mathcal{A}(t'))$.*

As for the types, a straightforward induction on typing derivations allows us to show the soundness of the translation $\mathcal{A}$:

Lemma 2 ($\mathcal{A}$ preserves types). *If $\Gamma \vdash_l t : A$, then $\Gamma \vdash W_{\Gamma\setminus\mathcal{FV}(t)}(\mathcal{A}(t)) : A$.*

We now show how to encode a λlxr-term into a l-term.

Definition 2. *Let t be a λlxr-term. We define the function $\mathcal{B}(t)$ by induction on the structure of t as follows:*

$$
\begin{aligned}
\mathcal{B}(x) &= x & \mathcal{B}(t\langle x = u\rangle) &= \mathcal{B}(t)\{x\backslash\mathcal{B}(u)\} \\
\mathcal{B}(lx.t) &= lx.\mathcal{B}(t) & \mathcal{B}(t\; u) &= \mathcal{B}(t)\mathcal{B}(u) \\
\mathcal{B}(W_x(t)) &= \mathcal{B}(t) & \mathcal{B}(C_x^{y,z}(t)) &= \mathcal{B}(t)\{y\backslash x\}\{z\backslash x\}
\end{aligned}
$$

The translation $\mathcal{B}$ commutes with renaming (i.e. $\mathcal{B}(R_\Delta^\Phi(t)) = R_\Delta^\Phi(\mathcal{B}(t))$) and does not add new free variables (i.e. $\mathcal{FV}(\mathcal{B}(t)) \subseteq \mathcal{FV}(t)$). Now the following simulation result can be proven:

Theorem 6. *If $M \longrightarrow_{\lambda\text{lxr}} N$, then $\mathcal{B}(M) \longrightarrow^*_\beta \mathcal{B}(N)$.*

Remark that congruent terms are mapped to the same l-term, so that it makes sense to consider $\mathcal{B}(\mathcal{A}(\_))$, which is in fact the identity: $t =_\alpha \mathcal{B}(\mathcal{A}(t))$.

A straightforward induction on typing derivations allows us to show:

Lemma 3 ($\mathcal{B}$ preserves types). *If $\Gamma \vdash t : A$, then $\Gamma \vdash_l \mathcal{B}(t) : A$.*

5   Operational Properties

In Sections 2, 3 and 4 we have already established the properties of subject reduction, strong normalisation of well-typed λlxr-terms and simulation of β-reduction step by step. But a calculus which is defined in order to implement l-calculus is also expected to preserve fundamental properties such as confluence and preservation of strong normalisation. We state in this section both properties which hold not only for well-typed terms but for all (linear) terms.

The original notion of PSN [5] has to be properly reformulated in our context as follows: every strongly normalisable l-term *is encoded* into a strongly normalisable λlxr-term. We establish PSN of λlxr by simulating reductions in λlxr by reductions in the λI-calculus of Church-Klop [35] with its associated reduction relations β,π. We refer the reader to [47, 51] for a survey on different techniques based on the λI-calculus to infer normalisation properties. Our proof technique can be summarised as follows (full details are given in [33]).

1. Define a relation $\mathcal{I}$ between linear λlxr-terms and lI-terms.
2. Show that $t\,\mathcal{I}\,T$ and $t \longrightarrow_{\mathsf{xr}} t'$ imply $t'\,\mathcal{I}\,T$,
3. Show that $t\,\mathcal{I}\,T$ and $t \longrightarrow_B t'$ imply $\exists T'$ such that $t'\,\mathcal{I}\,T'$ and $T\longrightarrow^+_{\beta\pi} T'$.
4. Deduce from 1,2,3 that if $t\,\mathcal{I}\,T$ and $T \in SN_{\beta\pi}$, then $t \in SN_{\lambda\mathsf{lxr}}$.
5. Define an encoding $i() : l \mapsto \lambda I$ such that if $t \in SN_\beta$ then $i(t) \in WN_{\beta\pi}$.
6. Show that $\mathcal{A}(t)\,\mathcal{I}\,i(t)$, where $\mathcal{A}(t)$ is the encoding given in Section 4.

Theorem 7 (PSN). *For any l-term t, if $t \in SN_\beta$, then $\mathcal{A}(t) \in SN_{\lambda\mathsf{lxr}}$.*

Proof. If $t \in SN_\beta$, then by the above point 5, $i(t) \in WN_{\beta\pi}$. A well-known result of Nederpelt [41] states that $WN_{\beta\pi} \subseteq SN_{\beta\pi}$, so $i(t) \in SN_{\beta\pi}$ and by points 6 and 4 we have $\mathcal{A}(t) \in SN_{\lambda\mathsf{lxr}}$.

We now use both simulations presented in Section 4 to derive the confluence property via a generalisation of the Interpretation Method [26]. We refer the reader to [33] for more details.

Theorem 8. *The system λlxr is confluent.*

6   Conclusion and Further Work

This paper extends the explicit substitution paradigm by showing how the Proof Nets of Linear Logic can be suitable as a logical model of a language with explicit operators for erasure, duplication and substitution.

Our term language is expressed by a simple syntax, and enjoys natural operational semantics via a well-established notion of reduction modulo a set of equations. Soundness and completeness of λlxr are shown with respect to its Proof Nets model.

In contrast to other calculi in the literature, λlxr has full composition and enjoys PSN. Moreover, λlxr enjoys all the nice properties that one expects such as confluence, strong normalisation of well-typed terms and step by step simulation of β-reduction. All these properties are shown by considering the complex notion of reduction modulo an equivalence which we have associated to λlxr-terms.

We claim that weakening operators are well-adapted to implement efficient garbage collection. Indeed, free variables are never lost and weakening operators are pulled out to the top-level during computation.

Our soundness and completeness proofs illustrate that the only rules with computational relevance in λlxr are $\{App2, Comp, Var, Weak1, Cont1\}$, just as the interesting rules in Proof Nets are only those concerning boxes. More precisely, $App2$ and $Comp$ in λlxr correspond to b-b in PN, Var to d-b, $Weak1$ to w-b and $Cont1$ to c-b.

It is worth mentioning the calculus obtained by turning the equation $Cont2$ into a reduction rule (from left to right) and by eliminating reduction rules $WSubs$ and $CSubs$ enjoys exactly the same properties as the calculus presented in this paper, namely Theorems 1,3,5,6,7,8. However, they seem to be necessary for the confluence on open terms (ongoing work).

We think that many interesting points raised in this work deserve further development. The first one concerns the study of reduction strategies well-adapted to handle explicit operators of substitution, erasure and duplication. This may take into account the notion of weak reduction used to implement functional programming [39].

Proof techniques used in the literature to show PSN of calculi with explicit substitutions (zoom-in [3], minimality [5], semantic RPO [7], PSN by standardisation [34], or intersection types [19]) are not all easy to adapt/extend to λlxr and other formalisms. We believe that the proof technique used here is really flexible.

But using the PSN result, we believe that we can characterise very neatly the strongly normalising terms of λlxr as the terms typable with intersection types, as it it the case in l-calculus as well as in the explicit substitution calculus lx [38].

First-order term syntax for λlxr via de Bruijn indices, or other special notation to avoid α-conversion as for example explicit scoping [29] or also director strings [46], would make implementation easier and bring the term language even closer to the Proof Nets model which has no notion of binding.

Acknowledgements

We are very grateful to R. Dyckhoff, J. Forest, B. Guillaume, O. Laurent, P. Lescanne, J. Mc Kinna and V. van Oostrom for valuable comments and suggestions.

References

1. M. Abadi, L. Cardelli, P. L. Curien, and J.-J. Lévy. Explicit substitutions. *JFP*, 4(1):375–416, 1991.
2. S. Abramsky. Computational interpretations of linear logic. *TCS*, 111:3–57, 1993.

3. A. Arbiser, E. Bonelli, and A. Ríos. Perpetuality in a lambda calculus with explicit substitutions and composition. *WAIT, JAIIO*, 2000.
4. A. Asperti and S Guerrini. *The Optimal Implementation of Functional Programming Languages*, volume 45 of *Cambridge Tracts in Theoretical Computer Science*. Cambridge University Press, 1998.
5. Z.-E.-A. Benaissa, D. Briaud, P. Lescanne, and J. Rouyer-Degli. λv, a calculus of explicit substitutions which preserves strong normalisation. *JFP*, 6(5):699–722, 1996.
6. N. Benton, G. Bierman, V. de Paiva, and M. Hyland. A term calculus for intuitionistic linear logic. *TLCA, LNCS* 664, pages 75–90, 1993.
7. R. Bloo and H. Geuvers. Explicit substitution: on the edge of strong normalization. *TCS*, 211(1-2):375–395, 1999.
8. R. Bloo and K. Rose. Preservation of strong normalization in named lambda calculi with explicit substitution and garbage collection. In *Computing Science in the Netherlands*, pages 62–72, Netherlands Computer Science Research Foundation, 1995.
9. E. Bonelli. Perpetuality in a named lambda calculus with explicit substitutions, *MSCS*, 11(1):409–450, 2001.
10. S. Cerrito and D. Kesner. Pattern matching as cut elimination. *LICS*, pages 98–108, 1999.
11. A. Church. The calculi of lambda conversion, Princeton University Press, 1941.
12. The Coq Proof Assistant. http://coq.inria.fr/.
13. V. Danos, J.-B. Joinet, and H. Schellinx. Sequent calculi for second order logic. *Advances in Linear Logic*, Cambridge University Press, 1995.
14. R. David and B. Guillaume. A λ-calculus with explicit weakening and explicit substitution. *MSCS*, 11:169–206, 2001.
15. R. Di Cosmo and S. Guerrini. Strong normalization of proof nets modulo structural congruences. *RTA, LNCS* 1631, pages 75–89, 1999.
16. R. Di Cosmo and D. Kesner. Strong normalization of explicit substitutions via cut elimination in proof nets. *LICS*, pages 35–46, 1997.
17. R. Di Cosmo, D. Kesner, and E. Polonovski. Proof nets and explicit substitutions. *FOSSACS, LNCS* 1784, pages 63–81, 2000.
18. R. Di Cosmo, D. Kesner, and E. Polonovski. Proof nets and explicit substitutions. *MSCS*, 13(3):409–450, 2003.
19. D. Dougherty and P. Lescanne. Reductions, Intersection Types and Explicit Substitutions. *TLCA, LNCS* 2044, pages 121-135, 2001.
20. G. Dowek, T. Hardin, and C. Kirchner. Higher-order unification via explicit substitutions. *LICS*, 1995.
21. M. Fernández and I. Mackie. Closed reductions in the lambda calculus. *CSL, LNCS* 1683, 1999.
22. J. Forest. A weak calculus with explicit operators for pattern matching and substitution. *RTA, LNCS* 2378, pages 174–191, 2002.
23. N. Ghani, V. de Paiva, and E. Ritter. Linear explicit substitutions. *IGPL*, 8(1):7–31, 2000.
24. J.-Y. Girard. Linear logic. *TCS*, 50(1):1–101, 1987.
25. J. Goubault-Larrecq. A proof of weak termination of typed lambda sigma-calculi. *TYPES, LNCS* 1512, pages 134–151, 1996.
26. T. Hardin. Résultats de confluence pour les règles fortes de la logique combinatoire catégorique et liens avec les lambda-calculs. PhD Thesis, Université Paris 7, 1987.
27. T. Hardin and J.-J. Lévy. A confluent calculus of substitutions. In *France-Japan Artificial Intelligence and Computer Science Symposium*, 1989.

28. T. Hardin, L. Maranget, and B. Pagano. Functional back-ends within the lambda-sigma calculus. *ICFP*, 1996.
29. D. Hendriks and V. van Oostrom. Adbmal. *CADE, LNAI* 2741, pages 136–150, 2003.
30. H. Herbelin. A *l*-calculus structure isomorphic to sequent calculus structure. *CSL, LNCS 933*, 1994.
31. The HOL system. `http://www.dcs.gla.ac.uk/~tfm/fmt/hol.html`.
32. F. Kamareddine and A. Ríos. A λ-calculus à la de Bruijn with explicit substitutions. *PLILP, LNCS* 982, pages 45–62, 1995.
33. D. Kesner and S. Lengrand. An Explicit Operator Calculus as the Syntactic Counterpart to a Proof-Net Model. Available at
`http://www.pps.jussieu.fr/~kesner/papers`, 2004.
34. Z. Khasidashvili, M. Ogawa, V. van Oostrom. Uniform Normalization Beyond Orthogonality. *RTA, LNCS* 2051, pages 122–136, 2001.
35. J.-W. Klop. *Combinatory Reduction Systems*, PhD Thesis, volume 127 of *Mathematical Centre Tracts*. CWI, Amsterdam, 1980.
36. Y. Lafont. Interaction Nets. *POPL*, pages 95–108, 1990.
37. O. Laurent. Polarized proof-nets and lambda-mu calculus. *TCS*, 1(290):161–188, 2003.
38. S. Lengrand. P. Lescanne, D. Dougherty, M. Dezani-Ciancaglini and S. van Bakel. Intersection types for explicit substitutions. *I & C*, 189(1):17–42, 2004.
39. J.-J. Lévy and L. Maranget. Explicit substitutions and programming languages. *FSTTCS, LNCS* 1738, pages 181–200, 1999.
40. P.-A. Melliès. Typed λ-calculi with explicit substitutions may not terminate. *TLCA, LNCS* 902, pages 328–334, 1995.
41. R. Nederpelt. Strong Normalization in a Typed Lambda Calculus with Lambda Structured Types. PhD Thesis, *Eindhoven University of Technology*, 1973.
42. M. Parigot. λμ-calculus: an algorithmic interpretation of classical natural deduction. *LPAR, LNCS* 624, pages 190–201, 1992.
43. E. Polonovski. Substitutions explicites et preuves de normalisation. PhD thesis, Université Paris 7, 2004.
44. L. Regnier. Une équivalence sur les lambda-termes. *TCS*, 2(126):281–292, 1994.
45. S. Ronchi della Rocca and L. Roversi. Lambda calculus and intuitionistic linear logic. *Studia Logica*, 59(3), 1997.
46. F.-R. Sinot and M. Fernández and I. Mackie. Efficient Reductions with Director Strings. *RTA, LNCS* 2706, pages 46–60, 2003.
47. M.H. Sorensen. Strong Normalization From Weak Normalization in Typed Lambda-Calculi. *I&C*, 37:35–71, 1997.
48. V. van Oostrom. Net-calculus. Course Notes available on `http://www.phil.uu.nl/~oostrom/typcomp/00-01/net.ps`, 2001.
49. R. Vestergaard and J. Wells. Cut Rules and Explicit Substitutions. *MSCS*, 11(1), 2001.
50. P. Wadler. A syntax for linear logic. *MFPS, LNCS* 802, pages 513–529, 1993.
51. H. Xi. Weak and Strong Beta Normalisations in Typed Lambda-Calculi. *TLCA, LNCS* 1210, pages 390–404, 1997.

Arithmetic as a Theory Modulo

Gilles Dowek and Benjamin Werner

Projet LogiCal
Pôle Commun de Recherche en Informatique du Plateau de Saclay
École polytechnique, INRIA, CNRS and Université de Paris-Sud
LIX, École polytechnique, 91128 Palaiseau Cedex, France
{Gilles.Dowek,Benjamin.Werner}@polytechnique.fr

Abstract. We present constructive arithmetic in Deduction modulo with rewrite rules only.

In natural deduction and in sequent calculus, the cut elimination theorem and the analysis of the structure of cut free proofs is the key to many results about predicate logic with no axioms: analyticity and non-provability results, completeness results for proof search algorithms, decidability results for fragments, constructivity results for the intuitionistic case...

Unfortunately, the properties of cut free proofs do not extend in the presence of axioms and the cut elimination theorem is not as powerful in this case as it is in pure logic. This motivates the extension of the notion of cut for various axiomatic theories such as arithmetic, Church's simple type theory, set theory and others. In general, we can say that a new axiom will necessitate a specific extension of the notion of cut: there still is no notion of cut general enough to be applied to any axiomatic theory. Deduction modulo [2, 3] is one attempt, among others, towards this aim.

In deduction modulo, a theory is not a set of axioms but a set of axioms combined with a set of rewrite rules. For instance, the axiom $\forall x \; x + 0 = x$ can be replaced by the rewrite rule $x + 0 \longrightarrow x$. The point is that replacing the axiom by the rewrite rule introduces short-cuts in the corresponding proofs, which avoid axiomatic cuts. When the set of rewrite rules is empty, one is simply back to regular predicate logic. On the other hand, when the set of axioms is empty we have theories expressed by rewrite rules only. For such theories, cut free proofs are similar to cut free proofs in pure logic, in particular they end with an introduction rule. Thus, when a theory can be expressed in deduction modulo with rewrite rules only and, in addition, cuts can be eliminated modulo these rewrite rules, the theory has most of the properties of pure logic. This leads to the question of which theories can be expressed with rewrite rules only in such a way that cut-elimination holds.

It is known that several theories can be expressed in such a setting, for instance all equational theories, type theory, set theory, etc... But arithmetic was an important example of a theory that lacked such a presentation. The goal of this paper is to show that arithmetic can indeed be presented in deduction modulo without axioms in such a way that cut elimination holds. The cut elimination result is built using the generic tools introduced in [3].

J. Giesl (Ed.): RTA 2005, LNCS 3467, pp. 423–437, 2005.

When considering arithmetic, it is customary to keep the cut-elimination argument predicative. We show that these generic tools also make it possible to build a predicative proof.

It should be noticed that second-order arithmetic can be embedded in simple type theory with the axiom of infinity and thus that it can be expressed in deduction modulo. Our presentation of first-order arithmetic in deduction modulo uses many ideas coming from second-order arithmetic. However, our presentation of arithmetic has exactly the power of first-order arithmetic.

1   Deduction Modulo

1.1   Identifying Propositions

In deduction modulo, the notions of language, term and proposition are those of predicate logic. But, a theory is formed with a set of axioms Γ *and a congruence* $\equiv$ defined on propositions. Such a congruence may be defined by a rewrite system on terms and on propositions (as propositions contain binders — quantifiers — these rewrite systems are in fact *combinatory reduction systems* [10]). Then, the deduction rules take this congruence into account. For instance, the *modus ponens* is not stated as usual

$$\frac{A \Rightarrow B \quad A}{B}$$

as the first premise need not be exactly $A \Rightarrow B$ but may be only congruent to this proposition, hence it is stated

$$\frac{C \quad A}{B} \text{ if } C \equiv A \Rightarrow B$$

All the rules of intuitionistic natural deduction may be stated in a similar way (see Figure 1).

For example, we can define a congruence with the following rewrite system

$$0 + y \to y \qquad\qquad S(x) + y \to S(x + y)$$
$$0 \times y \to 0 \qquad\qquad S(x) \times y \to x \times y + y$$

In the theory formed with a set of axioms Γ containing the axiom $\forall x \ x = x$ and this congruence, we can prove, in natural deduction modulo, that the number 4 is even

$$\frac{\dfrac{\dfrac{}{\Gamma \vdash_{\equiv} \forall x \ x = x} \text{ axiom}}{\dfrac{\Gamma \vdash_{\equiv} 2 \times 2 = 4}{\Gamma \vdash_{\equiv} \exists x \ 2 \times x = 4}} \langle x, x = x, 4 \rangle \text{ } \forall\text{-elim}}{} \langle x, 2 \times x = 4, 2 \rangle \text{ } \exists\text{-intro}$$

Substituting the variable x by the term 2 in the proposition $2 \times x = 4$ yields the proposition $2 \times 2 = 4$, that is congruent to $4 = 4$. The transformation of one proposition into the other, that requires several proof steps in usual formulations of natural deduction, is dropped from the proof in deduction modulo.

$$\frac{}{\Gamma \vdash_\equiv B} \text{ axiom if } A \in \Gamma \text{ and } A \equiv B$$

$$\frac{\Gamma, A \vdash_\equiv B}{\Gamma \vdash_\equiv C} \Rightarrow\text{-intro if } C \equiv (A \Rightarrow B)$$

$$\frac{\Gamma \vdash_\equiv C \quad \Gamma \vdash_\equiv A}{\Gamma \vdash_< \equiv B} \Rightarrow\text{-elim if } C \equiv (A \Rightarrow B)$$

$$\frac{\Gamma \vdash_\equiv A \quad \Gamma \vdash_\equiv B}{\Gamma \vdash_\equiv C} \wedge\text{-intro if } C \equiv (A \wedge B)$$

$$\frac{\Gamma \vdash_\equiv C}{\Gamma \vdash_\equiv A} \wedge\text{-elim if } C \equiv (A \wedge B)$$

$$\frac{\Gamma \vdash_\equiv C}{\Gamma \vdash_\equiv B} \wedge\text{-elim if } C \equiv (A \wedge B)$$

$$\frac{\Gamma \vdash_\equiv A}{\Gamma \vdash_\equiv C} \vee\text{-intro if } C \equiv (A \vee B)$$

$$\frac{\Gamma \vdash_\equiv B}{\Gamma \vdash_\equiv C} \vee\text{-intro if } C \equiv (A \vee B)$$

$$\frac{\Gamma \vdash_\equiv D \quad \Gamma, A \vdash_\equiv C \quad \Gamma, B \vdash_\equiv C}{\Gamma \vdash_\equiv C} \vee\text{-elim if } D \equiv (A \vee B)$$

$$\frac{}{\Gamma \vdash_\equiv A} \top\text{-intro if } A \equiv \top$$

$$\frac{\Gamma \vdash_\equiv B}{\Gamma \vdash_\equiv A} \bot\text{-elim if } B \equiv \bot$$

$$\frac{\Gamma \vdash_\equiv A}{\Gamma \vdash_\equiv B} \langle x, A \rangle \ \forall\text{-intro if } B \equiv (\forall x \ A) \text{ and } x \notin FV(\Gamma)$$

$$\frac{\Gamma \vdash_\equiv B}{\Gamma \vdash_\equiv C} \langle x, A, t \rangle \ \forall\text{-elim if } B \equiv (\forall x \ A) \text{ and } C \equiv (t/x)A$$

$$\frac{\Gamma \vdash_\equiv C}{\Gamma \vdash_\equiv B} \langle x, A, t \rangle \ \exists\text{-intro if } B \equiv (\exists x \ A) \text{ and } C \equiv (t/x)A$$

$$\frac{\Gamma \vdash_\equiv C \quad \Gamma, A \vdash_\equiv B}{\Gamma \vdash_\equiv B} \langle x, A \rangle \ \exists\text{-elim if } C \equiv (\exists x \ A) \text{ and } x \notin FV(\Gamma B)$$

Fig. 1. Natural deduction modulo.

In this example, the rewrite rules apply to terms only. Deduction modulo permits also to consider rules rewriting atomic propositions to arbitrary ones. For instance, in the theory of integral domains, we can take the rule

$$x \times y = 0 \rightarrow x = 0 \vee y = 0$$

that rewrites an atomic proposition to a disjunction.

Notice that, in the proof above, we do not need the axioms of addition and multiplication. Indeed, these axioms are now redundant: since the terms $0 + y$ and y are congruent, the axiom $\forall y \ 0+y = y$ is congruent to the axiom of equality $\forall y \ y = y$. Hence, it can be dropped. Thus, rewrite rules replace axioms.

This equivalence between rewrite rules and axioms is expressed by the the *equivalence lemma* that for every congruence $\equiv$, we can find a theory $\mathcal{T}$ such that $\Gamma \vdash_\equiv A$ is provable in deduction modulo if and only if $\mathcal{T}, \Gamma \vdash A$ is provable in ordinary predicate logic [2]. Hence, deduction modulo is not a true extension of predicate logic, but rather an alternative formulation of predicate logic. Of course, the provable propositions are the same in both cases, but the proofs are very different.

1.2   Model of a Theory Modulo

A *model* of a congruence $\equiv$ is a model such that if $A \equiv B$ then for all assignments, A and B have the same denotation. A *model* of a theory modulo $\Gamma, \equiv$ is a model of the theory Γ and of the congruence $\equiv$. Unsurprisingly, the completeness theorem extends to classical deduction modulo [6] and a proposition is provable in the theory $\Gamma, \equiv$ if and only if it is valid in all the models of $\Gamma, \equiv$.

1.3   Normalization in Deduction Modulo

Replacing axioms by rewrite rules in a theory changes the structure of proofs and in particular some theories may have the normalization property when expressed with axioms and not when expressed with rewrite rules. For instance, from the normalization theorem for predicate logic, we get that any proposition that is provable with the axiom $A \Leftrightarrow (B \wedge (A \Rightarrow \perp))$ has a normal proof. But if we transform this axiom into the rule $A \rightarrow B \wedge (A \Rightarrow \perp)$ (Crabbé's rule [1]) the proposition $B \Rightarrow \perp$ has a proof, but no normal proof.

We have proved a *normalization theorem*: proofs normalize in a theory modulo if this theory bears a *pre-model* [3]. A pre-model is a many-valued model whose truth values are reducibility candidates, i.e. sets of proof-terms. Hence we first define proof-terms, then reducibility candidates and finally pre-models.

Definition 1 (Proof-term). *We write $t, u \ldots$ for terms of the language. Proof-terms denoted by $\pi, \sigma \ldots$ and are inductively defined as follows.*

$$
\begin{aligned}
\pi ::=\ & \alpha & &\mid I \\
& \mid \lambda\alpha\ \pi \mid (\pi\ \pi') & &\mid \delta_{\perp}(\pi) \\
& \mid \langle \pi, \pi' \rangle \mid fst(\pi) \mid snd(\pi) & &\mid \lambda x\ \pi \mid (\pi\ t) \\
& \mid i(\pi) \mid j(\pi) \mid \delta(\pi_1, \alpha\pi_2, \beta\pi_3) & &\mid \langle t, \pi \rangle \mid \delta_{\exists}(\pi, x\alpha\pi')
\end{aligned}
$$

Each proof-term construction corresponds to an intuitionistic natural deduction rule: terms of the form α express proofs built with the axiom rule, terms of the form $\lambda\alpha\ \pi$ and $(\pi\ \pi')$ express proofs built with the introduction and elimination rules of the implication, terms of the form $\langle \pi, \pi' \rangle$ and $fst(\pi), snd(\pi)$ express proofs built with the introduction and elimination rules of the conjunction, terms of the form $i(\pi), j(\pi)$ and $\delta(\pi_1, \alpha\pi_2, \beta\pi_3)$ express proofs built with the introduction and elimination rules of the disjunction, the term I expresses the proof built with the introduction rule of the truth, terms of the form $\delta_{\perp}(\pi)$ express proofs built with the elimination rule of the contradiction, terms of the form $\lambda x\ \pi$ and $(\pi\ t)$ express proofs built with the introduction and elimination rules of the universal quantifier and terms of the form $\langle t, \pi \rangle$ and $\delta_{\exists}(\pi, x\alpha\pi')$ express proofs built with the introduction and elimination rules of the existential quantifier.

Definition 2 (Reduction). Reduction *on proof-terms is defined as the contextual closure of the following rules that eliminate cuts step by step.*

$$(\lambda\alpha\ \pi_1\ \pi_2) \triangleright (\pi_2/\alpha)\pi_1 \qquad (\lambda x\ \pi\ t) \triangleright (t/x)\pi$$
$$fst(\langle\pi_1,\pi_2\rangle) \triangleright \pi_1 \qquad snd(\langle\pi_1,\pi_2\rangle) \triangleright \pi_2$$
$$\delta(i(\pi_1),\alpha\pi_2,\beta\pi_3) \triangleright (\pi_1/\alpha)\pi_2 \qquad \delta(j(\pi_1),\alpha\pi_2,\beta\pi_3) \triangleright (\pi_1/\beta)\pi_3$$
$$\delta_\exists(\langle t,\pi_1\rangle,\alpha x\pi_2) \triangleright (t/x,\pi_1/\alpha)\pi_2$$

We write $\triangleright^*$ for the reflexive-transitive closure of the relation $\triangleright$.

In the following, the techniques are usual for normalization proofs by reducibility. The setting, however, is different.

Definition 3 (Reducibility candidates). *A proof-term is said to be* neutral *if it is a proof variable or an elimination (i.e. of the form $(\pi\ \pi')$, $fst(\pi)$, $snd(\pi)$, $\delta(\pi_1,\alpha\pi_2,\beta\pi_3)$, $\delta_\perp(\pi)$, $(\pi\ t)$, $\delta_\exists(\pi,x\alpha\pi')$), but not an introduction. A set R of proof-terms is a* reducibility candidate *if*

- *whenever $\pi \in R$, then π is strongly normalizable,*
- *whenever $\pi \in R$ and $\pi \triangleright^* \pi'$ then $\pi' \in R$,*
- *whenever π is neutral and if for every π' such that $\pi \triangleright^1 \pi'$, $\pi' \in R$ then $\pi \in R$.*

We write $\mathcal{CR}$ for the set of all reducibility candidates.

Definition 4. *Let $\mathcal{SN}$ be the set of all strongly normalizable proof-terms and $\perp\!\!\!\perp$ be the set of all strongly normalizing proof-terms whose normal form is neutral.*

It is easy to check that both $\mathcal{SN}$ and $\perp\!\!\!\perp$ are reducibility candidates. Furthermore, they are respectively the maximal and minimal reducibility candidate with respect to inclusion.

Definition 5 (Pre-model). *A pre-model $\mathcal{M}$ for a many-sorted language $\mathcal{L}$ is given by:*

- *for every sort s a set M_s,*
- *for every function symbol f of rank $\langle s_1,\dots,s_n,s_{n+1}\rangle$ a mapping $\hat{f}$ from $M_{s_1} \times \dots \times M_{s_n}$ to $M_{s_{n+1}}$,*
- *for every predicate symbol P of rank $\langle s_1,\dots,s_n\rangle$ a mapping $\hat{P}$ from $M_{s_1} \times \dots \times M_{s_n}$ to $\mathcal{CR}$.*

Definition 6 (Denotation in a pre-model). *Let $\mathcal{M}$ be a pre-model, ϕ an assignment mapping any variable x of sort s to an element of M_s and let t be a term of sort s. We define the object $[\![t]\!]_\phi \in M_s$ by induction over the structure of t.*

- $[\![x]\!]_\phi = \phi(x),$
- $[\![f(t_1,\dots,t_n)]\!]_\phi = \hat{f}([\![t_1]\!]_\phi,\dots,[\![t_n]\!]_\phi).$

Let A be a proposition and ϕ a well-sorted assignment as above. We define the reducibility candidate $[\![A]\!]_\phi$ by induction over the structure of A.

If A is an atomic proposition $P(t_1, \ldots, t_n)$ then $[\![A]\!]_\phi = \hat{P}([\![t_1]\!]_\phi, \ldots, [\![t_n]\!]_\phi)$.

When A is a non-atomic proposition, its interpretation is defined by the following, usual, equations:

$$[\![B \Rightarrow C]\!]_\phi = \{\pi \in \mathcal{SN} | \pi \triangleright^* \lambda\alpha\ \pi' \Rightarrow \forall\sigma \in [\![B]\!]_\phi\ (\sigma/\alpha)\pi' \in [\![C]\!]_\phi\}$$

$$[\![B \vee C]\!]_\phi = \{\pi \in \mathcal{SN} \mid \pi \triangleright^* i(\pi_1) \Rightarrow \pi_1 \in [\![B]\!]_\phi \wedge \pi \triangleright^* j(\pi_2) \Rightarrow \pi_2 \in [\![C]\!]_\phi\}$$

$$[\![B \wedge C]\!]_\phi = \{\pi \in \mathcal{SN} | \pi \triangleright^* \langle \pi_1, \pi_2 \rangle \Rightarrow (\pi_1 \in [\![B]\!]_\phi \wedge \pi_2 \in [\![C]\!]_\phi)\}$$

$$[\![\top]\!]_\phi = \mathcal{SN}$$

$$[\![\bot]\!]_\phi = \mathcal{SN}$$

$$[\![\exists x\ B]\!]_\phi = \{\pi \in \mathcal{SN} | \pi \triangleright^* \langle t, \pi' \rangle \Rightarrow \exists X \in M_s\ \pi' \in [\![B]\!]_{\phi, X/x}\}$$

$$[\![\forall x\ B]\!]_\phi = \{\pi \in \mathcal{SN} | \pi \triangleright^* \lambda x\ \pi' \Rightarrow \forall X \in M_s \forall t \in \mathcal{T}\ (t/x)\pi' \in [\![B]\!]_{\phi, X/x}\}$$

where $\mathcal{T}$ is th set of terms of the language.

Definition 7. *A pre-model is said to be a* pre-model of a congruence $\equiv$ *if when $A \equiv B$ then for every assignment ϕ, $[\![A]\!]_\phi = [\![B]\!]_\phi$.*

Theorem 1 (Normalization). *[3] If a congruence $\equiv$ has a pre-model all proofs modulo $\equiv$ strongly normalize.*

In this article we will be able to shorten some proofs using the following remark; it simply states that the previous definition can also be reformulated in a more conventional way.

Proposition 1. *A proof term σ belongs to $[\![A \Rightarrow B]\!]_\phi$ if and only if for any proof term $\pi \in [\![A]\!]_\phi$, $(\sigma\ \pi) \in [\![B]\!]_\phi$.*

A proof term σ belongs to $[\![\forall x_s A]\!]_\phi$ if and only if for any term t of the language and any element X of M_s, $(\sigma\ t) \in [\![A]\!]_{\phi, X/x}$.

2   An Alternative Presentation of Arithmetic

Heyting arithmetic is usually presented as a theory in predicate logic with the axioms of Definition 8 below. Before we give a presentation of arithmetic in deduction modulo, we shall give an alternative presentation HA_{Class} of arithmetic in predicate logic in Definition 9 below and prove the equivalence with HA. This equivalence is proved in several steps using two intermediate theories. Let us first recall the usual presentation of arithmetic.

2.1   Heyting Arithmetic

Definition 8 (HA). *The language of the theory HA is formed with the symbols $0, S, +, \times$ and $=$. The axioms are the axioms of equality corresponding to these symbols and the propositions*

$$\forall x \ \forall y \ (S(x) = S(y) \Rightarrow x = y)$$

$$\forall x \ \neg(0 = S(x))$$

$$((0/x)P \Rightarrow \forall y \ ((y/x)P \Rightarrow (S(y)/x)P) \Rightarrow \forall n \ (n/x)P)$$

$$\forall y \ (0 + y = y) \qquad \forall x \ \forall y \ (S(x) + y = S(x + y))$$

$$\forall y \ (0 \times y = 0) \qquad \forall x \ \forall y \ (S(x) \times y = x \times y + y)$$

2.2 A Symbol for Predecessor

The first step is to add a predecessor symbol to arithmetic and the axioms

$$Pred(0) = 0 \qquad Pred(S(x)) = x$$

$$\forall x \forall y \ (x = y \Rightarrow Pred(x) = Pred(y))$$

We prove that the theory obtained this way, called HA_{Pred} is a conservative extension of HA. This is a consequence of Skolem's theorem for constructive logic (see, for instance, [5]). But notice that in order to obtain the third axiom above, it is not sufficient to skolemize the theorem

$$\forall x \exists y \ ((x = 0 \wedge y = 0) \vee x = S(y))$$

but we need to skolemize the theorem

$$\forall x \exists y \ ((x = 0 \Rightarrow y = 0) \wedge \forall z \ (x = S(z) \Rightarrow y = z))$$

2.3 A Symbol for Natural Numbers

The second step is to introduce a theory HA_N where the universe of discourse is not restricted to the natural numbers and where we have a predicate symbol N to characterize the natural numbers. The language of this theory is formed with the symbols 0, S, $+$, $\times$, $=$, $Pred$, $Null$ and N. The axioms are the axioms of equality (including those related to $Pred$, $Null$ and N) and the propositions

$$(0/x)P \Rightarrow \forall y \ (N(y) \Rightarrow (y/x)P \Rightarrow (S(y)/x)P) \Rightarrow \forall n \ (N(n) \Rightarrow (n/x)P)$$

$$N(0) \qquad \qquad \forall x \ (N(x) \Rightarrow N(S(x)))$$
$$Pred(0) = 0 \qquad \qquad \forall x \ (Pred(S(x)) = x)$$
$$Null(0) \qquad \qquad \forall x \ (\neg Null(S(x)))$$
$$\forall y \ (0 + y = y) \qquad \forall x \ \forall y \ (S(x) + y = S(x + y))$$
$$\forall y \ (0 \times y = 0) \qquad \forall x \ \forall y \ (S(x) \times y = x \times y + y)$$

In the induction scheme, all the symbols, including $Pred$, $Null$ and N may occur in the proposition P

Because of the introduction of the predicate N, we must define a translation from the language of HA_{Pred} to the language of HA_N.

- $|P| = P$, if P is atomic, $|\top| = \top$, $|\bot| = \bot$, $|A \wedge B| = |A| \wedge |B|$, $|A \vee B| = |A| \vee |B|$, $|A \Rightarrow B| = |A| \Rightarrow |B|$,
- $|\forall x\ A| = \forall x\ (N(x) \Rightarrow |A|)$, $|\exists x\ A| = \exists x\ (N(x) \wedge |A|)$.

Then we prove that HA_N is a conservative extension of HA_{Pred} in the sense that if A is a closed proposition formed in the language of HA_{Pred} then A is provable in HA_{Pred} if and only if $|A|$ is provable in HA_N. Proving that HA_N is an extension of HA_{Pred} is relatively easy as it just requires to prove that if a proposition A is an axiom of HA_{Pred} then $|A|$ is provable in HA_N and an induction over proof structure. Proving that the extension is conservative is achieved using the completeness theorem by verifying that all constructive models of HA_{Pred} extend to models of HA_N.

2.4   A Sort for Classes of Numbers

Finally, we introduce a second sort for classes of natural numbers and use these number classes to express equality and the induction scheme.

Definition 9 (HA_{Class}).
  The theory HA_{Class} is a many sorted theory with two sorts ι and κ. The language contains a constant 0 of sort ι, function symbols S and $Pred$ of rank $\langle \iota, \iota \rangle$ and $+$ and $\times$ of rank $\langle \iota, \iota, \iota \rangle$, predicate symbols $=$ of rank $\langle \iota, \iota \rangle$, $Null$ and N of rank $\langle \iota \rangle$ and $\in$ of rank $\langle \iota, \kappa \rangle$ and for each proposition P in the language 0, S, $Pred$, $+$, $\times$, $=$, $Null$ and N and whose free variables are among $x, y_1, \ldots, y_n$ of sort ι, a function symbol $f_{x,y_1,\ldots,y_n,P}$ of rank $\langle \iota, \ldots, \iota, \kappa \rangle$. The symbol $f_{x,y_1,\ldots,y_n,P}$ is written f_P when the variables $x, y_1, \ldots, y_n$ are clear from the context. The axioms are

$$\forall y \forall z\ (y = z \Leftrightarrow \forall p\ (y \in p \Rightarrow z \in p))$$

$$\forall n\ (N(n) \Leftrightarrow \forall p\ (0 \in p \Rightarrow \forall y\ (N(y) \Rightarrow y \in p \Rightarrow S(y) \in p) \Rightarrow n \in p))$$

$$\forall x \forall y_1 \ldots \forall y_n\ (x \in f_{x,y_1,\ldots,y_n}P(y_1,\ldots,y_n) \Leftrightarrow P)$$

$Pred(0) = 0$	$\forall x\ (Pred(S(x)) = x)$
$Null(0)$	$\forall x\ (\neg Null(S(x)))$
$\forall y\ (0 + y = y)$	$\forall x \forall y\ (S(x) + y = S(x + y))$
$\forall y\ (0 \times y = 0)$	$\forall y\ (S(x) \times y = x \times y + y)$

The theory HA_{Class} is a conservative extension of HA_N. Again, proving that is is an extension is relatively simple, while proving that the extension is conservative requires to prove that prove that all constructive models of HA_N extend to models of HA_{Class}.

The conclusion is the equivalence between HA and HA_{Class}.

Proposition 2. *Let A be a closed proposition in the language of HA. Then A is provable in HA if and only if $|A|$ is provable in HA_{Class}.*

3   Arithmetic in Deduction Modulo

Definition 10 (The theory HA$_{\longrightarrow}$)).
The language of the theory HA$_{\longrightarrow}$ is the same as that of the theory HA_{Class}. This theory has no axioms and the rewrite rules

$$y = z \longrightarrow \forall p \ (y \in p \Rightarrow z \in p)$$

$$N(n) \longrightarrow \forall p \ (0 \in p \Rightarrow \forall y \ (N(y) \Rightarrow y \in p \Rightarrow S(y) \in p) \Rightarrow n \in p)$$

$$x \in f_{x,y_1,\dots,y_n,P}(y_1,\dots,y_n) \longrightarrow P$$

$$
\begin{array}{ll}
Pred(0) \longrightarrow 0 & Pred(S(x)) \longrightarrow x \\
Null(0) \longrightarrow \top & Null(S(x)) \longrightarrow \bot \\
0 + y \longrightarrow y & S(x) + y \longrightarrow S(x + y) \\
0 \times y \longrightarrow 0 & S(x) \times y \longrightarrow x \times y + y
\end{array}
$$

Proposition 3. *The theory HA$_{\longrightarrow}$ is a conservative extension of HA.*

Proof. It is equivalent to HA_{Class}.

Remark 1. The variant of HA$_{\longrightarrow}$ where the rule

$$N(n) \longrightarrow \forall p \ (0 \in p \Rightarrow \forall y \ (N(y) \Rightarrow y \in p \Rightarrow S(y) \in p) \Rightarrow n \in p)$$

is replaced by

$$N(n) \longrightarrow \forall p \ (0 \in p \Rightarrow \forall y \ (y \in p \Rightarrow S(y) \in p) \Rightarrow n \in p)$$

is also a conservative extension of HA.

   We favor the first formulation that allows more natural induction proofs (see Section 6).

4   Cut Elimination

In this section, we build a pre-model to show that HA$_{\longrightarrow}$ has the cut elimination property.

Proposition 4. *The theory HA$_{\longrightarrow}$ has the cut elimination property.*

Proof. We build a pre-model as follows. We take $M_\iota = \mathbb{N}$, $M_\kappa = \mathcal{CR}^{\mathbb{N}}$. The denotations of 0, S, $+$, $\times$, *Pred* are obvious. We take $\hat{Null}(n) = \mathcal{SN}$. The denotation of $\in$ is the function mapping n and f to $f(n)$. Then we can define the denotation of

$$\forall p \ (y \in p \Rightarrow z \in p)$$

and the denotation of $=$ accordingly.

   To define the denotation of N, for each function f of $\mathcal{CR}^{\mathbb{N}}$ we can define an interpretation $\mathcal{M}_f$ of the language of the proposition

$$\forall p \ (0 \in p \Rightarrow \forall y \ (N(y) \Rightarrow y \in p \Rightarrow S(y) \in p) \Rightarrow n \in p)$$

where the symbol N is interpreted by the function f. We define the function Φ from $\mathcal{CR}^{\mathbb{N}}$ to $\mathcal{CR}^{\mathbb{N}}$ mapping f to the function mapping the natural number x to the candidate

$$[\![\forall p\ (0 \in p \Rightarrow \forall y\ (N(y) \Rightarrow y \in p \Rightarrow S(y) \in p) \Rightarrow n \in p)]\!]_{x/n}^{\mathcal{M}_f}$$

The order on $\mathcal{CR}^{\mathbb{N}}$ defined by $f \subseteq g$ if for all n, $f(n) \subseteq g(n)$ is a complete order and the function Φ is monotonous as the occurrence of N is positive in

$$\forall p\ (0 \in p \Rightarrow \forall y\ (N(y) \Rightarrow y \in p \Rightarrow S(y) \in p) \Rightarrow n \in p)$$

Hence it has a fixpoint g. We interpret the symbol N by the function g.

Finally, the denotation of the symbols of the form f_P is defined in the obvious way.

This pre-model is a pre-model of each rule of $HA_{\longrightarrow}$ by construction.

Remark 2. Building a premodel for the variant of $HA_{\longrightarrow}$ with the rule

$$N(x) \longrightarrow \forall p\ (0 \in p \Rightarrow \forall y\ (y \in p \Rightarrow S(y) \in p) \Rightarrow n \in p)$$

is even simpler, we do not need to use the fixpoint theorem and we just define the denotation of the proposition $N(n)$ as the denotation of

$$\forall p\ (0 \in p \Rightarrow \forall y\ (y \in p \Rightarrow S(y) \in p) \Rightarrow n \in p)$$

5   A Predicative Cut Elimination Proof

The normalization proof of the previous section is essentially obtained by mapping arithmetic into second order arithmetic and then applying the usual normalization proof of second order arithmetic.

This proof is impredicative, indeed, to define the reducibility candidates interpreting the propositions $t = u$ and $N(t)$ we use a quantification over the set M_κ of functions mapping natural number to reducibility candidates.

We shall now see that it is possible to build also a predicative proof. In this proof, we restrict the set M_κ to contain only some functions from natural numbers to candidates, typically definable functions. Then these functions can be replaced by indices, for instance, a natural number and the quantification over functions from natural numbers to candidates can be replaced by a simple quantification over natural numbers. The difficulty here is that to define the denotation of $=$ and N we must use quantification on elements of the set M_κ. To define this set we need to define the notion of definable functions and as the symbols $=$ and N occur in the language, to define this notion we need to use the denotation of the symbols $=$ and N. To break this circularity, we give another definition of the interpretation of $t = u$ and $N(t)$ that does not use quantification over the elements of M_κ. Then the rewrite rules are not valid by construction anymore and we have to check their validity *a posteriori*.

Thus, we shall start by constructing the reducibility candidates E and E' used for interpreting equality and P_n used for interpreting the symbol N.

5.1   The Construction of Some Candidates

Definition 11. *Let A be a set of strongly normalizing terms. The set $[A]$ is inductively defined by*

- *if $\pi \in A$ then $\pi \in [A]$,*
- *if $\pi \in [A]$ and $\pi \vartriangleright^* \pi'$ then $\pi' \in [A]$,*
- *if π is an elimination and all its one step reducts are in $[A]$ then $\pi \in [A]$.*

It is routine to check that if A is a set of strongly normalizing proof-terms, then $[A]$ is the smallest reducibility candidate containing A.

The smallest reducibility candidate $\perp\!\!\!\perp$ can be defined by $\perp\!\!\!\perp = [\emptyset]$. For each strongly normalizing proof-term σ we define C_σ, the smallest reducibility candidate containing σ, by $C_\sigma = [\{\sigma\}]$.

Definition 12.

$$E = \{\pi \in \mathcal{SN} \mid \forall t \; \forall \sigma \in \mathcal{SN} \; (\pi \; t \; \sigma) \in C_\sigma\}$$
$$E' = \{\pi \in \mathcal{SN} \mid \forall t \; \forall \sigma \in \mathcal{SN} \; (\pi \; t \; \sigma) \in \perp\!\!\!\perp\}$$

Let $P = (P_i)_{i \in \mathbb{N}}$ and $Q = (Q_i)_{i \in \mathbb{N}}$ be two sequences of reducibility candidates, recall that the order defined by $P \subseteq Q$ if for all n, $P_n \subseteq Q_n$ is a complete order.

Definition 13. *Let σ_0 and σ_S be two proof terms and P be a sequence of reducibility candidates, we define the family of candidates $C_n^{\sigma_0,\sigma_S,P}$ by induction on n.*

$$C_0^{\sigma_0,\sigma_S,P} = [\{\pi \mid \pi = \sigma_0 \wedge \pi \in \mathcal{SN}\}]$$

$$C_{n+1}^{\sigma_0,\sigma_S,P} = [\{(\sigma_S \; t \; \rho \; \pi) \in \mathcal{SN} \mid \rho \in P_n \wedge \pi \in C_n^{\sigma_0,\sigma_S,P}\}]$$

It is easy to check that if $P \subseteq Q$ then $C_n^{\sigma_0,\sigma_S,P} \subseteq C_n^{\sigma_0,\sigma_S,Q}$.

Definition 14 (P-Peano pair). *A pair of proof-terms $\langle \sigma_0, \sigma_S \rangle$ is called a P-Peano pair if*

- *σ_0 is $\mathcal{SN}$,*
- *σ_S is $\mathcal{SN}$ and for every term t, for every natural number n, every proof-term $\rho \in P_n$, and for every proof-term π in $C_n^{\sigma_0,\sigma_S,P}$, the term $(\sigma_S \; t \; \rho \; \pi)$ is $\mathcal{SN}$.*

It is easy to check that if $P \subseteq Q$ then ($\langle \sigma_0, \sigma_S \rangle$ is a P-Peano pair $\Leftarrow \langle \sigma_0, \sigma_S \rangle$ is a Q-Peano pair).

Finally we define a family of candidates $\Phi(P)$.

Definition 15.

$$(\Phi(P))_n = \{\pi \in \mathcal{SN} \mid \forall t \forall \sigma_0 \forall \sigma_S \; \langle \sigma_0, \sigma_S \rangle \text{ is a } P\text{-Peano pair}$$
$$\Rightarrow (\pi \; t \; \sigma_0 \; \sigma_S) \in C_n^{\sigma_0,\sigma_S,P}\}.$$

It is easy to check that is $P \subseteq Q$ then $\Phi(P) \subseteq \Phi(Q)$, i.e. that the function Φ is monotonous.

As this function is monotonous, it has a least fixpoint. Let $(P_i)_{i \in \mathbb{N}}$ be the least fixpoint of Φ. By definition

$$P_n = \{\pi \in \mathcal{SN} \mid \forall t \forall \sigma_0 \forall \sigma_S \; \langle \sigma_0, \sigma_S \rangle \text{ is a } P\text{-Peano pair} \Rightarrow (\pi \; t \; \sigma_0 \; \sigma_S) \in C_n^{\sigma_0,\sigma_S,P}\}.$$

5.2   A Pre-model

As in the impredicative construction, we take $M_\iota = \mathbb{N}$, we interpret the symbols 0, S, $+$, $\times$, *Pred* in the obvious way and we take $\hat{Null}(n) = \mathcal{SN}$. Then, we define the interpretation of the symbols $=$ and N as follows.

$$\hat{=}(n,n) = E$$
$$\hat{=}(n,m) = E' \text{ if } n \neq m$$
$$\hat{N}(n) = P_n$$

Before we define the set M_κ and the interpretation of the symbol $\in$, we introduce a notion of definable function from the set of natural numbers to the set of candidates.

Definition 16 (Definable function). *A function f from $\mathbb{N}$ to $\mathcal{CR}$ is said to be definable if there exists a proposition P in the language of $HA_{\longrightarrow}$ without the symbol $\in$ and an assignment ϕ such that for all n $f(n) = [\![P]\!]_{\phi,n/x}$.*

We then define the set M_κ, as the (countable) set of functions from $\mathbb{N}$ to $\mathcal{CR}$ containing

- definable functions,
- constant functions taking the value C_σ for some proof-term σ,
- and functions mapping k to $C_k^{\sigma_0, \sigma_S, P}$ for some proof-terms σ_0 and σ_S.

Finally, we complete the construction of the pre-model by defining the denotation of $\in$ as the obvious application function and the denotation of the symbols of the form f_P accordingly. The validity of all the rewrite rules of $HA_{\longrightarrow}$ is routine, except that of the rules

$$y = z \longrightarrow \forall p \ (y \in p \Rightarrow z \in p)$$

$$N(x) \longrightarrow \forall p \ (0 \in p \Rightarrow (\forall y \ (N(y) \Rightarrow y \in p \Rightarrow S(y) \in p)) \Rightarrow x \in p)$$

each of them requiring a lemma.

Remark 3. **(Making the proof predicative).** The pre-model construction as presented above is not obviously predicative since to define the reducibility candidate associated to proposition $\forall p \ A$ we use quantification over M_κ that is a set of functions mapping natural numbers to reducibility candidates. However as the set M_κ is countable, it is not difficult to associate a natural number to each of its elements and to define a function U that maps each number to the associated function. Then we can replace M_κ by $\mathbb{N}$ and define the interpretation of $\in$ as the function mapping n and m to $U(m)(n)$. The construction obtained this way is predicative. For instance, it could be formalized in Martin-Löf's Type Theory with one universe.

Remark 4. For the variant of HA$_{\longrightarrow}$ with the rule

$$N(n) \longrightarrow \forall p \ (0 \in p \Rightarrow \forall y \ (y \in p \Rightarrow S(y) \in p) \Rightarrow n \in p)$$

the proof is simpler as we do not need to apply the fixpoint theorem. If σ_0 and σ_S be two proof terms, we define the family of candidates $C_n^{\sigma_0,\sigma_S,P}$ by induction on n without the parameter P. Peano pairs and the family P_n can be defined directly and the rest of the proof is similar.

6   The System T

More traditional cut elimination proofs for arithmetic use the normalization of Gödel system T. We show here that the normalization of system T also can be obtained as a corollary of the normalization theorem of [3] although the system T contains a specific rewrite rule on proofs and [3] allows only specific rewrite rules on terms and propositions but uses fixed rewrite rules on proofs.

Consider the symbol $nat = f_{N(x)}$ and $\rightarrow= f_{x \in y \Rightarrow x \in z}$. In HA$_{\longrightarrow}$, we have

$$x \in nat \longrightarrow N(x)$$

$$x \in (y \rightarrow z) \longrightarrow x \in y \Rightarrow x \in z$$

and of course

$$N(n) \longrightarrow \forall p \ (0 \in p \Rightarrow \forall y \ (N(y) \Rightarrow y \in p \Rightarrow S(y) \in p) \Rightarrow n \in p)$$

We can drop the first rule, replacing all propositions of the form $N(x)$ the proposition $x \in nat$ and we get this way the rewrite system with two rules

$$n \in nat \longrightarrow \forall p \ (0 \in p \Rightarrow \forall y \ (y \in nat \Rightarrow y \in p \Rightarrow S(y) \in p) \Rightarrow n \in p)$$

$$x \in (y \rightarrow z) \longrightarrow x \in y \Rightarrow x \in z$$

In this system, we get rid of all terms of type ι. We get the following theory

Definition 17 (The theory $\mathcal{T}$).

$$\varepsilon(nat) \longrightarrow \forall p \ (\varepsilon(p) \Rightarrow (\varepsilon(nat) \Rightarrow \varepsilon(p) \Rightarrow \varepsilon(p)) \Rightarrow \varepsilon(p))$$

$$\varepsilon(y \rightarrow z) \longrightarrow \varepsilon(y) \Rightarrow \varepsilon(z)$$

Proposition 5. *The theory $\mathcal{T}$ has the cut elimination property.*

The proof is structurally similar to the one of Section 5.

Definition 18 (The system T). *The system T is the extension of simply typed lambda-calculus with a constant 0, a unary function symbol S and a ternary function symbol Rec^A for each type A and the rules*

$$Rec(a, f, 0) \longrightarrow a$$

$$Rec(a, f, S(b)) \longrightarrow (f \ b \ Rec(a, f, b))$$

Proof normalization for the theory $\mathcal{T}$ implies normalization for the system T. Indeed, types of the system T are terms of the theory $\mathcal{T}$ and terms of type A in the system T can be translated into proofs of $\varepsilon(A)$ in the theory $\mathcal{T}$ (Parigot's numbers [11]):

- $|x| = x$, $|u\ v| = |u|\ |v|$, $|\lambda x : A\ u| = \lambda x : \varepsilon(A)\ |u|$,
- $|0| = \lambda p\ \lambda x : \varepsilon(p)\ \lambda f : \varepsilon(nat) \Rightarrow \varepsilon(p) \Rightarrow \varepsilon(p)\ x$,
- $|S(n)| = \lambda p\ \lambda x : \varepsilon(p)\ \lambda f : \varepsilon(nat) \Rightarrow \varepsilon(p) \Rightarrow \varepsilon(p)\ (f\ |n|\ (|n|\ p\ x\ f))$,
- $|Rec^A(x, f, n)| = (|n|\ A\ x\ f)$.

It is routine to check that if $t \longrightarrow^1 u$ in the system T then $|t| \longrightarrow^+ |u|$ in the theory $\mathcal{T}$. For instance:

$$|Rec^A(x, f, 0)| = (|0|\ A\ x\ f) = (\lambda p\ \lambda x : \varepsilon(p)\ \lambda f : \varepsilon(nat) \Rightarrow \varepsilon(p) \Rightarrow \varepsilon(p)\ x)\ A\ x\ f$$
$$\longrightarrow^+ x.$$

Here we reap the benefit of having chosen the rule

$$N(n) \longrightarrow \forall p\ (0 \in p \Rightarrow \forall y\ (N(y) \Rightarrow y \in p \Rightarrow S(y) \in p) \Rightarrow n \in p)$$

and not

$$N(n) \longrightarrow \forall p\ (0 \in p \Rightarrow \forall y\ (y \in p \Rightarrow S(y) \in p) \Rightarrow n \in p)$$

that would have given us only the termination of the variant of system T where the recursor is replaced by an iterator.

References

1. M. Crabbé. Non-normalisation de la théorie de Zermelo. Manuscript (1974).
2. G. Dowek, Th. Hardin, and C. Kirchner. Theorem proving modulo. *Journal of Automated Reasoning*, 31 (2003) pp. 33-72.
3. G. Dowek and B. Werner. Proof normalization modulo, *The Journal of Symbolic Logic*, 68, 4 (2003) pp. 1289-1316.
4. G. Dowek and B. Werner. Arithmetic as a theory modulo. Manuscript (2004).
5. G. Dowek and B. Werner. A constructive proof of Skolem theorem for constructive logic, Manuscript (2004).
6. G. Dowek. *La part du Calcul*. Habilitation thesis, Université de Paris 7 (1999).
7. J.Y. Girard. Interprétation fonctionnelle et élimination des coupures dans l'arithmétique d'ordre supérieur, *Thèse d'État*, Université de Paris 7 (1972).
8. J.Y. Girard, Y. Lafont and P. Taylor. *Proofs and Types*, Cambridge University Press (1989).
9. K. Gödel. Über eine bisher noch nicht benüzte Erweiterung des finiten Standpunktes, *Dialectica*, 12 (1958) pp. 280-287. Reproduced in S. Feferman *et al.* (eds.), *Collected Works*, vol. II, Oxford University Press (1990) pp. 241-251.
10. J.-W. Klop, V. van Oostrom and F. van Raamsdonk. Combinatory reduction systems: introduction and survey. *Theoretical Computer Science*, 121, (1993) pp. 279-308.

11. M. Parigot. Programming with proofs: A second order type theory. *European Symposium on Programming*, H. Ganzinger (ed.), Lecture Notes in Computer Science, 300, (1988) pp. 145-159.

12. D. Prawitz, Natural deduction. A proof-theoretical study. Almqvist & Wiksell (1965).

13. H. Rasiowa and R. Sikorski, *The mathematics of metamathematics*, Polish Scientific Publishers (1963).

Infinitary Combinatory Reduction Systems
(Extended Abstract)

Jeroen Ketema[1] and Jakob Grue Simonsen[2]

[1] Department of Computer Science, Vrije Universiteit Amsterdam
De Boelelaan 1081a, 1081 HV Amsterdam, The Netherlands
jketema@cs.vu.nl
[2] Department of Computer Science, University of Copenhagen (DIKU)
Universitetsparken 1, DK-2100 Copenhagen Ø, Denmark
simonsen@diku.dk

Abstract. We define infinitary combinatory reduction systems (iCRSs). This provides the first extension of infinitary rewriting to higher-order rewriting. We lift two well-known results from infinitary term rewriting systems and infinitary λ-calculus to iCRSs:
1. every reduction sequence in a fully-extended left-linear iCRS is compressible to a reduction sequence of length at most ω, and
2. every complete development of the same set of redexes in an orthogonal iCRS ends in the same term.

1 Introduction

One of the main reasons for the initial research in infinitary rewriting was to have a model of lazy or stream-based programming languages easily accessible to people familiar with term rewriting. Two notions of infinitary rewriting were developed: infinitary (first-order) term rewriting systems (iTRSs) [1–3] and infinitary λ-calculus (iλc) [3, 4]. However, the standard notion of rewriting employed to model higher-order programs is *higher-order* rewriting, and thus goes beyond λ-calculus. The absence of a general notion of infinitary higher-order rewriting thus constitutes a gap in the arsenal of the rewriting theorist bent on modelling lazy or stream-based languages.

In the present paper we aim to plug this gap by investigating infinitary higher-order rewriting.

As for iTRSs and iλc some finitary system needs to be chosen as a starting point. We choose the notion of higher-order rewriting most familiar to the authors, namely combinatory reduction systems (CRSs) [3, 5, 6].

The definition of infinitary combinatory reduction systems (iCRSs) consists of a combination of the usual four-stage definition of CRSs and the corresponding four-stage definition of iTRSs and iλc:

CRSs	iTRSs/iλc
1a. Meta-terms	
1b. Terms	1. Infinite terms
2. Substitutions	2. Substitutions
3. Rewrite rules	3. Rewrite rules
4. Rewrite relation	4. Rewrite relation

J. Giesl (Ed.): RTA 2005, LNCS 3467, pp. 438–452, 2005.

Given the definition of iCRSs, we seek to answer two of the most pertinent questions asked for any notion of infinitary rewriting:

1. Are reduction sequences compressible to reduction sequences of length at most ω?
2. Do complete developments of the same set of redexes end in the same term?

For iTRSs these questions have positive answers under assumption of respectively left-linearity and orthogonality. For iλc the same holds as long as the η-rule is not introduced. Apart from the definition of iCRSs, the main contribution of this paper is that similar positive answers can be given in the case of iCRSs.

The remainder of this paper is organised as follows. In Sect. 2 we give some preliminary definitions, and in Sect. 3, we define infinite (meta-)terms and substitutions. Thereafter, in Sect. 4 we define infinitary rewriting and prove compression, and in Sect. 5 we investigate complete developments. Finally, in Sect. 6 we give directions for further research.

2 Preliminaries

Prior knowledge of CRSs [6] and infinitary rewriting [3] is not required, but will greatly improve the reader's understanding of the text.

Throughout the paper we assume a signature Σ, each element of which has finite arity. We also assume a countably infinite set of variables, and, for each finite arity, a countably infinite set of meta-variables. Countably infinite sets are sufficient, given that we can employ 'Hilbert hotel'-style renaming. We denote the first infinite ordinal by ω, and arbitrary ordinals by $\alpha, \beta, \gamma, \ldots$.

The set of *finite meta-terms* is defined as follows:

1. each variable x is a finite meta-term,
2. if x is a variable and s is a finite meta-term, then $[x]s$ is a finite meta-term,
3. if Z is a meta-variable of arity n and $s_1, \ldots, s_n$ are finite meta-terms, then $Z(s_1, \ldots, s_n)$ is a finite meta-term,
4. if $f \in \Sigma$ has arity n and $s_1, \ldots, s_n$ are finite meta-terms, then $f(s_1, \ldots, s_n)$ is a finite meta-term.

A finite meta-term of the form $[x]s$ is called an *abstraction*. Each occurrence of the variable x in s is *bound* in $[x]s$. If s is a finite meta-term, we denote by $root(s)$ the root symbol of s.

The *set of positions* of a finite meta-term s, denoted $\mathcal{P}os(s)$, is the set of finite strings over $\mathbb{N}$, with ϵ the empty string, such that:

- if $s = x$ for some variable x, then $\mathcal{P}os(s) = \{\epsilon\}$,
- if $s = [x]t$, then $\mathcal{P}os(s) = \{\epsilon\} \cup \{0 \cdot p \mid p \in \mathcal{P}os(t)\}$,
- if $s = Z(t_1, \ldots, t_n)$, then $\mathcal{P}os(s) = \{\epsilon\} \cup \{i \cdot p \mid 1 \leq i \leq n, p \in \mathcal{P}os(t_i)\}$,
- if $s = f(t_1, \ldots, t_n)$, then $\mathcal{P}os(s) = \{\epsilon\} \cup \{i \cdot p \mid 1 \leq i \leq n, p \in \mathcal{P}os(t_i)\}$.

Given $p, q \in \mathcal{P}os(s)$, we say that p is a *prefix* of q, denoted $p \leq q$, if there exists an $r \in \mathcal{P}os(s)$ such that $p \cdot r = q$. If $r \neq \epsilon$, then we say that the prefix is *strict* and we write $p < q$. Moreover, if neither $p < q$ nor $q < p$, then we say that p and q are *parallel*, which we denote $p \parallel q$. We denote by $s|_p$ the subterm of s at position p.

3   (Meta-)Terms and Substitutions

In iTRSs and iλc, terms are defined by means of introducing a metric on the set of finite terms and subsequently taking the completion of the metric. That is, taking the least set of objects containing the set finite terms such that every Cauchy sequence converges [2, 4, 7]. Intuitively, in such a metric, two terms s and t are close to each other if the first 'conflict' between them occurs 'deep' according to some depth measure. In iTRSs, a conflict is a position p such that $root(s|_p) \neq root(t|_p)$. In iλc, a conflict is defined similarly, but also takes into account α-equivalence. The metric, denoted $d(s,t)$, is defined as 0 when no conflict occurs between s and t and otherwise as 2^{-k}, where k denotes the minimal depth such that a conflict occurs between s and t.

To define terms and meta-terms for iCRSs, we first define the notions of a conflict and α-equivalence for finite meta-terms. In the definition we denote by $s[x \rightarrow y]$ the replacement in s of the occurrences of the free variable x by the variable y.

Definition 3.1. *Let s and t be finite meta-terms. A conflict of s and t is a position $p \in \mathcal{P}os(s) \cap \mathcal{P}os(t)$ such that:*

1. *if $p = \epsilon$, then $root(s) \neq root(t)$,*
2. *if $p = i \cdot q$ for $i \geq 1$, then $root(s) = root(t)$ and q is a conflict of $s|_i$ and $t|_i$,*
3. *if $p = 0 \cdot q$, then $s = [x_1]s'$ and $t = [x_2]t'$ and q is a conflict of $s'[x_1 \rightarrow y]$ and $t'[x_2 \rightarrow y]$, where y does not occur in either s' or t'.*

The finite meta-terms s and t are α-equivalent if no conflict exists [4].

We next define the depth measure D.

Definition 3.2. *Let s be a meta-term and $p \in \mathcal{P}os(s)$. Define:*

$$D(s, \epsilon) = 0$$
$$D(Z(t_1, \ldots, t_n), i \cdot p') = D(t_i, p')$$
$$D([x]t, 0 \cdot p') = 1 + D(t, p')$$
$$D(f(t_1, \ldots, t_n), i \cdot p') = 1 + D(t_i, p')$$

Note that meta-variables are not counted by D. Changing the second clause to $D(Z(t_1, \ldots, t_n), i \cdot p') = 1 + D(t_i, p')$ yields the 'usual' depth measure, which counts the number of symbols in a position.

The measure D is employed in the definition of the metric, which is defined precisely as in the case of iTRSs and iλc.

Definition 3.3. *Let s and t be meta-terms. The metric d is defined as:*

$$d(s,t) = \begin{cases} 0 & \text{if s and t are α-equivalent} \\ 2^{-k} & \text{otherwise,} \end{cases}$$

where k is the minimal depth with respect to the measure D such that a conflict occurs between s and t.

Following precisely the definition of terms in the case of iTRSs and iλc, we define the meta-terms.

Definition 3.4. *The set of* meta-terms *over a signature Σ is the metric completion of the set of finite meta-terms with respect to the metric d.*

Note that, by definition of metric completion, the set of finite meta-terms is a subset of the set of meta-terms.

The notions of a set of positions and a subterm of a finite meta-term carry over directly to the meta-terms, we use the same notation in both cases.

The metric completion allows precisely those meta-terms such that the depth measure D increases to infinity along all infinite paths in the meta-term. Thus, by the definition of D and d, no meta-term has a subterm s such that there exists an infinite string p over $\mathbb{N}$ with the property that each finite prefix q of p is a position of s with $root(s|_q)$ a meta-variable. Informally, *no meta-term has an infinite chain of meta-variables.*

Examples of candidate 'meta-terms' that are disallowed by the definition of meta-term are:

$$Z(Z(\dots(Z(\dots))))$$
$$Z_1(Z_2(\dots(Z_n(\dots))))$$

A construction that *is* allowed is an infinite number of *finite* chains of meta-variables 'guarded' by abstractions or function symbols. For example, the following is allowed:

$$[x_1]Z_1([x_2]Z_2(\dots([x_n]Z_n(\dots))))$$

If we had wanted to include 'meta-terms' with infinite chains of meta-variables we should have used the usual depth measure on finite meta-terms instead of the measure D.

We explain the reason for the exclusion of meta-terms with infinite chains of meta-variables after the definition of substitutions. The idea of the exclusion of certain meta-terms comes from iλc where it is possible to define subsets of the set of infinite λ-terms by slightly changing the notion of the depth measure on which the metric is based [4]. It is, for example, possible to define a subset in which no λ-terms with infinite chains of λ-abstractions occur, i.e., subterms of the form $\lambda x_1.\lambda x_2 \dots \lambda x_n \dots$ are disallowed.

The terms can now be defined as in the finite case [3, 5, 6]. The only difference is that meta-terms now occur in the definition instead of finite meta-terms.

Definition 3.5. *The set of* terms *is the largest subset of the set of meta-terms, such that no meta-variables occur in the meta-terms.*

Note that the definition of meta-terms, as defined by the measure D, only restricts meta-terms containing meta-variables, not meta-terms *without* meta-variables. Hence, the set of terms is independent of the use of either D in Definition 3.3 or the usual depth measure. As a consequence, both the set of (infinite) first-order terms and the set of (infinite) λ-terms are easily shown to be included in the set of terms.

We next define substitutions. The required definitions are the same as in the case of CRSs [3, 6], except that coinduction is employed instead of induction. This is identical to what is done in the case of iTRSs and iλc with respect to the finite systems they are based on. In the definitions we use $\boldsymbol{x}$ and $\boldsymbol{t}$ as a short-hands for respectively the sequences $x_1, \ldots, x_n$ and $t_1, \ldots, t_n$ with $n \geq 0$. We assume n fixed in the next two definitions.

Definition 3.6. *A substitution of the terms $\boldsymbol{t}$ for distinct variables $\boldsymbol{x}$ in a term s, denoted $s[\boldsymbol{x} := \boldsymbol{t}]$, is coinductively defined as:*

1. $x_i[\boldsymbol{x} := \boldsymbol{t}] = t_i$,
2. $y[\boldsymbol{x} := \boldsymbol{t}] = y$ *if y does not occur in $\boldsymbol{x}$,*
3. $([y]s')[\boldsymbol{x} := \boldsymbol{t}] = [y](s'[\boldsymbol{x} := \boldsymbol{t}])$,
4. $f(s_1, \ldots, s_m)[\boldsymbol{x} := \boldsymbol{t}] = f(s_1[\boldsymbol{x} := \boldsymbol{t}], \ldots, s_m[\boldsymbol{x} := \boldsymbol{t}])$.

The above definition implicitly takes into account the variable convention [8] in the third clause to avoid the binding of free variables by the abstraction.

Definition 3.7. *An n-ary substitute is a mapping denoted $\underline{\lambda}x_1, \ldots, x_n.s$ or $\underline{\lambda}\boldsymbol{x}.s$, with s a term, such that:*

$$(\underline{\lambda}\boldsymbol{x}.s)(t_1, \ldots, t_n) = s[\boldsymbol{x} := \boldsymbol{t}]. \tag{1}$$

Reading Eq. (1) from left to right gives rise to the rewrite rule

$$(\underline{\lambda}\boldsymbol{x}.s)(t_1, \ldots, t_n) \rightarrow s[\boldsymbol{x} := \boldsymbol{t}].$$

This rule can be seen a *parallel β-rule*. That is, a variant of the β-rule from iλc which substitutes for multiple variables simultaneously. The root of $(\underline{\lambda}\boldsymbol{x}.s)$ is called the $\underline{\lambda}$-abstraction and the root of the left-hand side of the parallel β-rule is called the $\underline{\lambda}$-application.

Definition 3.8. *A valuation $\bar{\sigma}$ is an extension of a function σ which assigns n-ary substitutes to n-ary meta-variables. It is coinductively defined as:*

1. $\bar{\sigma}(x) = x$,
2. $\bar{\sigma}([x]s) = [x](\bar{\sigma}(s))$,
3. $\bar{\sigma}(Z(s_1, \ldots, s_m)) = \sigma(Z)(\bar{\sigma}(s_1), \ldots, \bar{\sigma}(s_m))$,
4. $\bar{\sigma}(f(s_1, \ldots, s_m)) = f(\bar{\sigma}(s_1), \ldots, \bar{\sigma}(s_m))$.

Similar to Definition 3.6, the above definition implicitly takes into account the variable convention in the second clause to avoid the binding of free variables by the abstraction.

Thus, applying a substitution means applying a valuation and proceeds in two steps: In the first step each subterm of the form $Z(t_1, \ldots, t_n)$ is replaced by a subterm of the form $(\underline{\lambda}\boldsymbol{x}.s)(t_1, \ldots, t_n)$. In the second step Eq. (1) is applied to each subterm of the form $(\underline{\lambda}\boldsymbol{x}.s)(t_1, \ldots, t_n)$ as introduced in the first step.

In the light of the rewrite rule introduced just below Definition 3.7 the second step can be viewed as a complete development of the parallel β-redexes introduced in the first step. This is obviously a complete development in a variant of iλc. The variant has the parallel β-rule and a signature containing the $\underline{\lambda}$-application, the $\underline{\lambda}$-abstraction, the abstractions, the meta-variables, and the elements of Σ.

As in the finite case [5, Remark II.1.10.1], we need to prove that the application of a valuation to a meta-term yields a unique term.

Proposition 3.9. *Let s be a meta-term and $\bar{\sigma}$ a valuation. There exists a unique term that is the result of applying $\bar{\sigma}$ to s.*

Proof (Sketch). That the first step in applying $\bar{\sigma}$ to s has a unique result is an immediate consequence of being defined coinductively. We denote the result of the first step by s_σ. The set of parallel β-redexes in s_σ is denoted $\mathcal{U}$.

To prove that the second step also has a unique result we employ the rewriting terminology as introduced above. Although omitted, the definitions of a development and a complete development can be easily derived from the iλc definitions.

Note that to repeatedly rewrite the root of s_σ by means of the parallel β-redex, the root must look like

$$(\underline{\lambda}x.x_i)(t_1, \ldots, t_n),$$

with $1 \leq i \leq n$ and t_i again such a redex. This is only possible if there exists in s_σ an infinite chain of such redexes which starts at the root. However, this requires an infinite chain of meta-variables to be present in s, which is not allowed by the definition of meta-terms. Thus, the root can only be rewritten finitely often in a development. Applying the same reasoning to the roots of the subterms, gives that a complete development is obtained by reducing the redexes in $\mathcal{U}$ in an outside-in fashion. As all parallel β-redexes occur in $\mathcal{U}$ and as no $\underline{\lambda}$-applications and $\underline{\lambda}$-abstractions occur in s the result of the complete development, which we denote $\bar{\sigma}(s)$, is necessarily a term.

To show that each complete development ends in $\bar{\sigma}(s)$, note that we can view each parallel β-redex $(\underline{\lambda}x_1, \ldots, x_n.s)(t_1, \ldots, t_n)$ as a sequence of β-redexes:

$$(\lambda x_1(\ldots((\lambda x_n.s)t_n)\ldots))t_1.$$

This means that each complete development in our variant of iλc corresponds to a complete development in iλc extended with some function symbols. As each complete development in iλc ends in the same term, a result independent of added function symbols, the complete developments of the second step must also end in the same term. Hence, $\bar{\sigma}(s)$ is unique. $\square$

Let us now see why we excluded 'meta-terms' with infinite chains of meta-variables from Definition 3.4. Consider the 'meta-term'

$$Z(Z(\ldots(Z(\ldots)))).$$

Applying the valuation that assigns to Z the substitute $\underline{\lambda}x.x$ yields:

$$(\underline{\lambda}x.x)((\underline{\lambda}x.x)(\ldots((\underline{\lambda}x.x)(\ldots))))$$

which has no complete development, as no matter how many parallel β-redexes are contracted, it reduces only to itself and not to a term. This is inadequate, as rewrite steps in iCRSs need to relate terms to terms.

The previous problem does not depend on only a single meta-variable being present in the 'meta-term'. The same behaviour can occur with different meta-variables of different arities. In that case, we can define a valuation that assigns $\underline{\lambda}\boldsymbol{x}.y$ to each meta-variable Z in the 'meta-term' with y in $\boldsymbol{x}$ such that y corresponds to an argument of Z which is a chain of meta-variables.

The above 'meta-term' still has the nice property that it exhibits confluence with respect to the parallel β-rule. Unfortunately, there are 'meta-terms' that do not have this property. Consider a signature with constants a and b and also consider the 'meta-term'

$$Z(a, Z(b, Z(a, Z(b, Z(\ldots)))))\,.$$

Applying the valuation that assigns to Z the substitute $\underline{\lambda}xy.y$ yields the '$\underline{\lambda}$-term' of Fig. 1. It reduces by means of two different developments to the $\underline{\lambda}$-terms of Fig. 2 and Fig. 3. These last two $\underline{\lambda}$-terms have no common reduct with respect to parallel β-reduction. They reduce only to themselves. Note that this problem also occurs in iλc [4, Sect. 4].

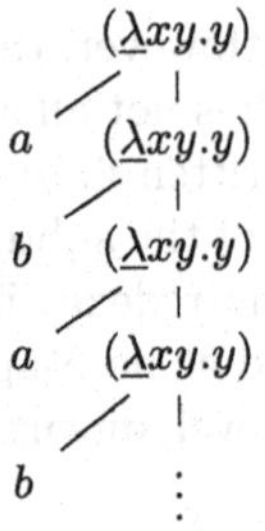

Fig. 1. 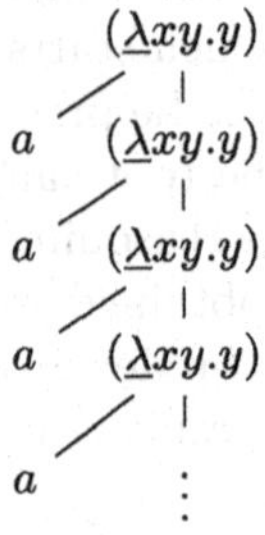   Fig. 2. 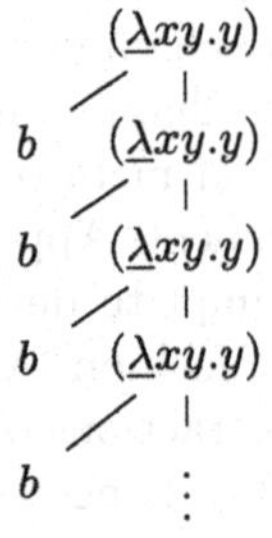   Fig. 3.

Concluding, when we allow 'meta-terms' with infinite chains of meta-variables we have two problems. First, substitution in such a 'meta-term' does not always yield a term. Second, substitution may yield distinct results, none of which are terms. We can overcome these problems by not allowing infinite chains of meta-variables to occur in meta-terms, as shown in Proposition 3.9.

4   Infinitary Rewriting

We continue to combine the definitions of iTRSs and iλc and those of CRSs. We start with a definition that comes directly from CRS theory.

Definition 4.1. *A finite meta-term is a* pattern *if each of its meta-variables has distinct bound variables as its arguments. Moreover, a meta-term is* closed *if all its variables occur bound.*

We next define rewrite rules and iCRSs. In analogy to the rewrite rules of iTRSs, the definition is identical to the one in the finitary case, but without the finiteness restriction on the right-hand sides of the rewrite rules [1, 2].

Definition 4.2. *A* rewrite rule *is a pair* (l, r), *denoted* $l \to r$, *where* l *is a finite meta-term and* r *is a meta-term, such that:*

1. l *is a pattern and of the form* $f(s_1, \ldots, s_n)$ *with* $f \in \Sigma$ *of arity* n,
2. *all meta-variables that occur in* r *also occur in* l, *and*
3. l *and* r *are closed.*

An infinitary combinatory reduction system (iCRS) *is a pair* $C = (\Sigma, R)$ *with* Σ *a signature and* R *a set of rewrite rules.*

As the rewrite rules of iTRSs and iλc only have finite chains of meta-variables when their rules are considered as rewrite rules in the above sense, it follows easily that iTRSs and iλc are iCRSs.

A context is a term over $\Sigma \cup \{\Box\}$ where $\Box$ is a fresh constant. One-hole contexts are defined in the usual way. We now define redexes and rewrite steps.

Definition 4.3. *Let* $l \to r$ *be a rewrite rule. Given a valuation* $\bar{\sigma}$, *the term* $\bar{\sigma}(l)$ *is called a* $l \to r$-redex. *If* $s = C[\bar{\sigma}(l)]$ *for some context* $C[\Box]$ *with* $\bar{\sigma}(l)$ *a* $l \to r$-*redex and* p *the position of the hole in* $C[\Box]$, *then an* $l \to r$-*redex, or simply a* redex, *occurs* at position p *and* depth $D(s, p)$ *in* s. *A* rewrite step *is a pair* (s, t), *denoted* $s \to t$, *such that a* $l \to r$-*redex occurs in* $s = C[\bar{\sigma}(l)]$ *and such that* $t = C[\bar{\sigma}(r)]$.

We can now define what a transfinite reduction sequence is. The definition copies the definition from iTRSs and iλc verbatim [2, 4].

Definition 4.4. *A* transfinite reduction sequence *of ordinal length* α *is a sequence of terms* $(s_\beta)_{\beta < \alpha+1}$ *such that* $s_\beta \to s_{\beta+1}$ *for all* $\beta < \alpha$. *For each rewrite step* $s_\beta \to s_{\beta+1}$, *let* d_β *denote the depth of the contracted redex. The reduction sequence is* weakly convergent *or* Cauchy convergent *if for every ordinal* $\gamma \le \alpha$ *the distance between* t_β *and* t_γ *tends to 0 as* β *approaches* γ *from below. The reduction sequence is* strongly convergent *if it is weakly convergent and if* d_β *tends to infinity as* β *approaches* γ *from below.*

Notation 4.5. By $s \twoheadrightarrow^\alpha t$, respectively $s \twoheadrightarrow^{\le \alpha} t$, we denote a *strongly convergent* transfinite reduction sequence of ordinal length α, respectively of ordinal length less than or equal to α. By $s \twoheadrightarrow t$ we denote a *strongly convergent* transfinite reduction sequence of arbitrary ordinal length and by $s \to^* t$ we denote a reduction sequence of finite length.

As in [2–4], we prefer to reason about strongly converging reduction sequences. This ensures that we can restrict our attention to reduction sequences of length at most ω by the so-called *compression property*. To prove the property we need the following lemma and definitions.

Lemma 4.6. *If $s \twoheadrightarrow t$, then the number of steps contracting redexes at depths less than $d \in \mathbb{N}$ is finite for any d.*

Proof. This is exactly the proof of [2, Lemma 3.5]. □

Definition 4.7. *A rewrite rule $l \to r$ is left-linear, if each meta-variable occurs at most once in l. Moreover, an iCRS is left-linear if all its rewrite rules are left-linear.*

Definition 4.8. *A pattern is fully-extended [9, 10], if, for each of its meta-variables Z, and each abstraction $[x]$ having Z in its scope, x is an argument of Z. Moreover, an iCRS is fully-extended if the left-hand sides of all rewrite rules are fully-extended.*

Left-linearity and fully-extendedness ensure no redex is created by either making two subterms equal in an infinite number of steps or by erasing some variable in an infinite number of steps.

Theorem 4.9 (Compression). *For every fully-extended, left-linear iCRS, if $s \twoheadrightarrow^\alpha t$, then $s \twoheadrightarrow^{\leq \omega} t$.*

Proof (Sketch). Let $s \twoheadrightarrow^\alpha t$, and proceed by ordinal induction on α. By [3, Theorem 12.7.1] it suffices to show that the theorem holds for $\alpha = \omega + 1$: The cases where α is 0, a limit ordinal, or a successor ordinal greater than $\omega + 1$ do not depend on the definition of rewriting.

For $\alpha = \omega+1$ it follows by Lemma 4.6 that we can write $s \twoheadrightarrow^\alpha t$ as $s \to^* s' \twoheadrightarrow^\omega s'' \to t$, such that all rewrite steps in $s' \twoheadrightarrow^\omega s''$ occur below the meta-variable positions of the redex contracted in the step of $s'' \to t$. By fully-extendedness and left-linearity it follows that a redex of which the redex contracted in $s'' \to t$ is a residual occurs in s'. Hence, we can contract the redex in s', which yields a term t'.

The result now follows if we can construct a strongly convergent reduction sequence $t' \twoheadrightarrow^{\leq\omega} t$. To construct such a reduction sequence, assume $t_0 = t'$ and construct for each $d > 0$ a reduction sequence $t_{d-1} \to^* t_d$ where all rewrite steps occur at depths greater or equal to $d - 1$, and where $d(t_d, t) \leq 2^{-d}$. That the construction of these reduction sequences is possible follows by a proof that is similar to the proof of compression for iλc [4]. Using the fact that only finite chains of meta-variables occur in meta-terms is essential to the proof. By the requirements on the constructed reduction sequences, it follows that $t_0 \to^* t_1 \to^* \ldots \to^* t_{d-1} \to^* t_d \to^* \ldots t$ is a strongly convergent reduction sequence of length at most ω. As $s \to^* t'$, we then have that $s \twoheadrightarrow^{\leq\omega} t$, as required. □

The previous theorem does not hold in general for iCRSs that are not left-linear or fully-extended. For left-linearity, this follows from the iTRS counterexample in [2]. For fully-extendedness, this follows from the infinitary $\lambda\beta\eta$-calculus in which reduction sequences occur that are not compressible to reduction sequences of length at most ω [3, 4]. The η-rule is not fully-extended.

5   Developments

In this section we prove that each complete development of the same set of redexes in an orthogonal iCRS ends in the same term. As all the left-hand sides of the rewrite rules in iCRSs are finite, the definition of orthogonality carries over immediately from CRSs.

Definition 5.1. *Let* $R = \{l_i \to r_i \mid i \in I\}$ *be a set of rewrite rules.*

1. *R is* non-overlapping *if it holds that:*
   - *each* $l_i \to r_i$*-redex that occurs at a position p in an* $l_j \to r_j$*-redex with* $i \neq j$ *occurs such that there exists a position $q \leq p$ with $q \in \mathcal{P}os(l_j)$ and* $root(l_j|_p)$ *a meta-variable,*
   - *likewise for $p \neq \epsilon$ and $i = j$.*
2. *R is* orthogonal *if it is left-linear and non-overlapping.*
3. *An iCRS is* orthogonal *if its set or rewrite rules is orthogonal.*

In the remainder of this section we assume an orthogonal iCRS, a term s, and a set $\mathcal{U}$ of redexes in s.

5.1   Descendants and Residuals

Before we can consider developments, we need to define descendants and residuals. The definition of descendant across a rewrite step $\bar{\sigma}(l) \to \bar{\sigma}(r)$ follows the definition of substitution, and is thus defined in two steps. The first step defines descendants in $\bar{\sigma}(r)$ where only the valuation is applied and not Eq. (1). The second step defines descendants across application of Eq. (1).

Given that the second step of the substitution is just a complete development in a variant of iλc, the second step in the definition of descendants is just a variant of descendants in iλc [3, 4]. For this reason, the step is not made explicit here.

We next give a definition of the first step. In the definition we denote by 0 the position of the subterm on the left-hand side of a λ-application and also the position of the body of a λ-abstraction. By $1, \ldots, n$ we denote the positions of the subterms on the right-hand side of the λ-application. This means that $(\underline{\lambda}x.s)(t_1, \ldots, t_n)|_0 = (\underline{\lambda}x.s)$, $\underline{\lambda}x.s|_0 = s$, and $Z(t_1, \ldots, t_n)|_i = (\underline{\lambda}x.s)(t_1, \ldots, t_n)|_i = t_i$ for $1 \leq i \leq n$. We denote by $\bar{\sigma}(l) \to r_\sigma$ the rewrite step $\bar{\sigma}(l) \to \bar{\sigma}(r)$ when only the first step of the substitution applied to r.

Definition 5.2. *Let $l \to r$ be a rewrite rule, $\bar{\sigma}$ a valuation, and $p \in \mathcal{P}os(\bar{\sigma}(l))$. Suppose $u : \bar{\sigma}(l) \to r_\sigma$. The set $p/^1u$ is defined as follows:*

- *if a position $q \in \mathcal{P}os(l)$ exists such that $p = q \cdot q'$ and $root(l|_q) = Z$, then define $p/^1u = \{p' \cdot 0 \cdot 0 \cdot q' \mid p' \in P\}$ with $P = \{p' \mid root(r|_{p'}) = Z\}$,*
- *if no such position exists, then define $p/^1u = \emptyset$.*

Note that $\mathcal{P}os(r) \subseteq \mathcal{P}os(r_\sigma)$ by the notation of positions in subterms of the form $(\underline{\lambda}x.s)(t_1, \ldots, t_n)$. From this it follows that $P \subseteq \mathcal{P}os(r_\sigma)$.

We can now give a complete definition of a descendant across a rewrite step.

Definition 5.3. *Let* $u : C[\bar{\sigma}(l)] \to C[\bar{\sigma}(r)]$ *be a rewrite step, such that p is the position of the hole in $C[\Box]$, and let $q \in \mathcal{P}os(C[\bar{\sigma}(l)])$. The set of descendants of q across u, denoted q/u, is defined as $q/u = \{q\}$ in case $p \parallel q$ or $p < q$. In case $q = p \cdot q'$, it is defined as $q/u = \{p \cdot q'' \mid p'' \in Q\}$, where Q is the set of descendants of $q'/^1 u'$ with $u' : \bar{\sigma}(l) \to r_\sigma$ across complete development of the parallel β-redexes in r_σ.*

Descendants across a reduction sequence are defined as for iTRSs and iλc.

Definition 5.4. *Let $s_0 \twoheadrightarrow^\alpha s_\alpha$ and let $P \subseteq \mathcal{P}os(s_0)$. The set of descendants of P across $s_0 \twoheadrightarrow^\alpha s_\alpha$, denoted $P/(s_0 \twoheadrightarrow^\alpha s_\alpha)$, is defined as follows:*

- *if $\alpha = 0$, then $P/(s_0 \twoheadrightarrow^\alpha s_\alpha) = P$,*
- *if $\alpha = 1$, then $P/(s_0 \to s_1) = \bigcup_{p \in P} p/(s_0 \to s_1)$,*
- *if $\alpha = \beta + 1$, then $P/(s_0 \twoheadrightarrow^{\beta+1} s_{\beta+1}) = (P/(s_0 \twoheadrightarrow^\beta s_\beta))/(s_\beta \to s_{\beta+1})$,*
- *if α is a limit ordinal, then $p \in P/(s_0 \twoheadrightarrow^\alpha s_\alpha)$ iff $p \in P/(s_0 \twoheadrightarrow^\beta s_\beta)$ for all large enough $\beta < \alpha$.*

By orthogonality, if there exists a redex at a position p using a rewrite rule $l \to r$ that is not contracted in rewrite step and if p has descendants across the step, then there exists a redex at each descendant of p also employing the rule $l \to r$. Hence, there exists a well-defined notion of *residual* by strongly convergent reduction sequences. We overload the notation $\cdot/\cdot$ to denote both the descendant and the residual relation.

5.2 Complete Developments

We now define developments. Recall that we assume we are working in an orthogonal iCRS and that $\mathcal{U}$ is a set of redexes in a term s.

Definition 5.5. *A development of $\mathcal{U}$ is a strongly convergent reduction sequence such that each step contracts a residual of a redex in $\mathcal{U}$. A development $s \twoheadrightarrow t$ is complete if $\mathcal{U}/(s \twoheadrightarrow t) = \emptyset$.*

To prove that each complete development of the same set of redexes ends in the same term, we extend the technique of the Finite Jumps Developments Theorem [3] to orthogonal iCRSs. The theorem employs notions of paths and path projections. In essence, paths and path projections are 'walks' through terms starting at the root and proceeding to greater and greater depths. An important property of paths and path projections is that when a walk encounters a redex to be contracted in a development, a 'jump' is made to the right-hand side of the employed rewrite rule. It continues there until a meta-variable is encountered, at which point a jump back to the original term occurs.

In the following definition, we denote by p_u the position of the redex u in s.

Definition 5.6. *A path of s with respect to $\mathcal{U}$ is a sequence of nodes and edges. Each node is labelled either (s, p) with $p \in \mathcal{P}os(s)$ or (r, p, q) with r a right-hand side of a rewrite rule, $p \in \mathcal{P}os(r)$, and $q = p_u$ with $u \in \mathcal{U}$. Each directed edge is either unlabelled or labelled with an element of $\mathbb{N}$.*

Every path starts with a node labelled (s, ϵ). If a node n of a path is labelled (s, p) and if it has an outgoing edge to a node n', then:

1. *if the subterm at p is not a redex in $\mathcal{U}$, then for some $i \in \mathcal{P}os(s|_p) \cap \mathbb{N}$ the node n' is labelled $(s, p \cdot i)$ and the edge from n to n' is labelled i,*
2. *if the subterm at p is a redex $u \in \mathcal{U}$ with $l \to r$ the employed rewrite rule, then the node n' is labelled (r, ϵ, p_u) and the edge from n to n' is unlabelled,*
3. *if $s|_p$ is a variable x bound by an abstraction $[x]$ occurring in the left-hand side of the rule $l \to r$ of a redex $u \in \mathcal{U}$, then the node n' is labelled $(r, p' \cdot i, p_u)$ and the edge from n to n' is unlabelled, such that (r, p', p_u) was the last node before n with p_u, $root(r|_{p'}) = Z$, the unique position of Z in l is q, and $l|_{q \cdot i} = x$.*

If a node n of a path is labelled (r, p, p_u) and if it has an outgoing edge to a node n', then:

1. *if $root(r|_p)$ is not a meta-variable, then for some $i \in \mathcal{P}os(r|_p) \cap \mathbb{N}$ the node n' is labelled $(r, p \cdot i, p_u)$ and the edge from n to n' is labelled i,*
2. *if $root(r|_p)$ is a meta-variable Z, then the node n' is labelled $(s, q \cdot q')$ and the edge from n to n' is unlabelled, such that $l \to r$ is the rewrite rule employed in u, q is the position of u in s, and q' is the unique position of Z in l.*

We say that a path is *maximal* if it is not a proper prefix of another path. We write a path P as a (possibly infinite) sequence of alternating nodes and edges $P = n_1 e_1 n_2 \ldots$.

Definition 5.7. *Let $P = n_1 e_1 n_2 \ldots$ be a path of s with respect to $\mathcal{U}$. The path projection of P is a sequence of alternating nodes and edges $\phi(P) = \phi(n_1)\phi(e_1)\phi(n_2) \ldots$ such that for each node n in P:*

1. *if n is labelled (t, p), then $\phi(n)$ is unlabelled if $root(t|_p)$ is a redex in $\mathcal{U}$ or a variable bound by some redex in $\mathcal{U}$ and it is labelled $root(t|_p)$ otherwise,*
2. *if n is labelled (r, p, q), then $\phi(n)$ is unlabelled if $root(r|_p)$ is a meta-variable and it is labelled $root(r|_p)$ otherwise.*

For each edge e, if e is labelled i, then $\phi(e)$ has the same label, and if e is unlabelled, then $\phi(e)$ is labelled ϵ.

Example 5.8. Consider the iCRS with the following rewrite rule $l \to r$:

$$f([x]Z(x), Z') \to Z(g(Z(Z'))).$$

Also, consider the terms $s = f([x]g(x), a)$ and $t = g(g(g(a)))$, the meta-term $r = Z(g(Z(Z')))$, and the set $\mathcal{U}$ containing the only redex in s. Obviously, $s \to t$ is a complete development.

The term s has one maximal path with respect to $\mathcal{U}$:

$$(s, \epsilon) \to (r, \epsilon, \epsilon) \to (s, 10) \to_1 (s, 101) \to (r, 1, \epsilon) \to_1 (r, 11, \epsilon)$$
$$\to (s, 10) \to_1 (s, 101) \to (r, 111, \epsilon) \to (s, 2)$$

The term t has one maximal path with respect to $\mathcal{U}/\mathcal{U} = \emptyset$:

$$(t, \epsilon) \to_1 (t, 1) \to_1 (t, 11) \to_1 (t, 111).$$

The path projections of the maximal paths are respectively

$$\cdot \rightarrow_\epsilon \cdot \rightarrow_\epsilon g \rightarrow_1 \cdot \rightarrow_\epsilon g \rightarrow_1 \cdot \rightarrow_\epsilon g \rightarrow_1 \cdot \rightarrow_\epsilon \cdot \rightarrow_\epsilon a$$

and

$$g \rightarrow_1 g \rightarrow_1 g \rightarrow_1 a .$$

Let $\mathcal{P}(s,\mathcal{U})$ denote the set of path projections of maximal paths of s with respect to $\mathcal{U}$. The following result can be witnessed in the above example.

Lemma 5.9. *Let $u \in \mathcal{U}$ and let $s \rightarrow t$ be the rewrite step contracting u. There is a surjection from $\mathcal{P}(s,\mathcal{U})$ to $\mathcal{P}(t,\mathcal{U}/u)$. Given a path projection $\phi(P) \in \mathcal{P}(s,\mathcal{U})$, its image under the surjection is acquired from $\phi(P)$ by deleting finite sequences of unlabelled nodes and ϵ-labelled edges from $\phi(P)$.*

Proof (Sketch). By straightforwardly, but very tediously, tracing through the construction of paths, it is evident that the set of maximal paths of t with respect to $\mathcal{U}/u$ can be obtained from the set of maximal paths of s with respect to $\mathcal{U}$ by replacing or deleting nodes of the form (r, p, p_u). If a maximal path of t is obtained from a maximal path of s in this way, then they have identical path projections, except that sequences of ϵ-labelled edges and unlabelled nodes may have been deleted (due to the contraction of u). This establishes the desired surjection. It is easy to see that the sequences of deleted edges can only be infinite if there is an infinite chain of meta-variables in the right-hand side of the rule of u, which is impossible by definition of meta-terms. □

We next define a property for sets $\mathcal{P}(s,\mathcal{U})$: the finite jumps property. We also define some terminology to relate a term to a set $\mathcal{P}(s,\mathcal{U})$.

Definition 5.10. *If no path projection occurring in $\mathcal{P}(s,\mathcal{U})$ contains an infinite sequences of unlabelled nodes and ϵ-labelled edges, then we say that $\mathcal{U}$ has the finite jumps property. Moreover, we say that a term t matches $\mathcal{P}(s,\mathcal{U})$, if, for all $\phi(P) \in \mathcal{P}(s,\mathcal{U})$, and for all prefixes of $\phi(P)$ ending in a node n labelled f, we have that $root(t|_p) = f$, where p is the concatenation of the edge labels in the prefix (starting at the first node of $\phi(P)$ and ending at $\phi(n)$).*

We have the following.

Proposition 5.11. *If $\mathcal{U}$ has the finite jumps property, then there exists a unique term, denoted $\mathcal{T}(s,\mathcal{U})$, that matches $\mathcal{P}(s,\mathcal{U})$.*

Proof. The proof is identical to the proof of Proposition 12.5.8 in [3]. □

We can now finally prove the Finite Jumps Developments Theorem:

Theorem 5.12 (Finite Jumps Developments Theorem). *If $\mathcal{U}$ has the finite jumps property, then:*

1. *every complete development of $\mathcal{U}$ ends in $\mathcal{T}(s,\mathcal{U})$,*
2. *for any $p \in \mathcal{P}os(s)$, the set of descendants of p by a complete development of $\mathcal{U}$ is independent of the complete development,*
3. *for any redex u of s, the set of residuals of u by a complete development of $\mathcal{U}$ is independent of the complete development, and*
4. *$\mathcal{U}$ has a complete development.*

Proof (Sketch). The proof is identical to the proof of Proposition 12.5.9 in [3], except that Lemma 5.9 is employed instead of tracing. □

With the Finite Jumps Developments Theorem in hand, we can now precisely characterise the sets of redexes having complete developments. This characterisation seems to be new.

Lemma 5.13. *The set $\mathcal{U}$ has a complete development if and only if $\mathcal{U}$ has the finite jumps property.*

Proof. To prove that the finite jumps property follows if $\mathcal{U}$ has a complete development, suppose $\mathcal{U}$ does not have the finite jumps property. In this case there is a path projection which ends in an infinite sequence of unlabelled nodes and ϵ-labelled edges.

By Lemma 5.9 we have for each step $s \rightarrow t$ contracting a redex in $\mathcal{U}$ that there is a surjection from $\mathcal{P}(s,\mathcal{U})$ to $\mathcal{P}(t,\mathcal{U}/u)$ which deletes only finite sequences of unlabelled nodes and ϵ-labelled edges. Hence, for all path projections we have that the nodes and edges left after the contraction of a redex in $\mathcal{U}$ either stay at the same distance from the first node of the path projection in which they occur or move closer to the first node. But then it follows immediately by ordinal induction that a path projection with an infinite sequence of unlabelled nodes and ϵ-labelled edges is present after each development. In particular, such an infinite sequence is present after the complete development. However, by definition of paths and path projections this means that a descendant of a redex in $\mathcal{U}$ is present in the final term of the complete development. But this contradicts the fact that no descendants of redexes in $\mathcal{U}$ exist in the final term of a complete development. Hence, $\mathcal{U}$ has the finite jumps property.

That $\mathcal{U}$ has a complete development if it has the finite jumps property is an immediate consequence of Theorem 5.12(4). □

The result we were aiming at now follows easily.

Theorem 5.14. *If $\mathcal{U}$ has a complete development then all complete developments of $\mathcal{U}$ end in the same term.*

Proof. By Lemma 5.13, if $\mathcal{U}$ has a complete development then it has the finite jumps property. But then each complete development of $\mathcal{U}$ ends in the same final term by Theorem 5.12(1). □

452 Jeroen Ketema and Jakob Grue Simonsen

6 Further Directions

We have defined and proved the first results for iCRSs, but a number of questions that have been answered for iTRSs and iλc remain open: Does there exist a notion of meaningless terms [11] that allows for the construction of Böhm-like trees? Can we prove a partial confluence property [2, 3, 11] showing infinitary confluence up to equivalence of meaningless terms?

Furthermore, can the treatment of iCRS in this paper be extended to the other formats of higher-order rewriting? The fact that CRSs have a clean separation of abstractions (in terms and rewrite rules) and substitutions which is not present in some of the other forms of higher-order rewriting [3] may constitute a stumbling block in this respect.

Finally, it is as yet unclear how to relax the requirement that no infinite chains of meta-variables are allowed in meta-terms while still retaining a meaningful notion of substitution.

References

1. Dershowitz, N., Kaplan, S., Plaisted, D.A.: Rewrite, rewrite, rewrite, rewrite, rewrite, ... TCS **83** (1991) 71–96
2. Kennaway, R., Klop, J.W., Sleep, R., de Vries, F.J.: Transfinite reductions in orthogonal term rewriting systems. I&C **119** (1995) 18–38
3. Terese: Term Rewriting Systems. Cambridge University Press (2003)
4. Kennaway, J.R., Klop, J.W., Sleep, M., de Vries, F.J.: Infinitary lambda calculus. TCS **175** (1997) 93–125
5. Klop, J.W.: Combinatory Reduction Systems. PhD thesis, Rijksuniversiteit Utrecht (1980)
6. Klop, J.W., van Oostrom, V., van Raamsdonk, F.: Combinatory reduction systems: introduction and survey. TCS **121** (1993) 279–308
7. Arnold, A., Nivat, M.: The metric space of infinite trees. Algebraic and topological properties. Fundamenta Informaticae **3** (1980) 445–476
8. Barendregt, H.P.: The Lambda Calculus: Its Syntax and Semantics. Second edn. Elsevier Science (1985)
9. Hanus, M., Prehofer, C.: Higher-order narrowing with definitional trees. In Ganzinger, H., ed.: Proc. of the 7th Int. Conf. on Rewriting Techniques and Applications (RTA'96). Volume 1103 of LNCS., Springer-Verlag (1996) 138–152
10. van Oostrom, V.: Higher-order families. In Ganzinger, H., ed.: Proc. of the 7th Int. Conf. on Rewriting Techniques and Applications (RTA '96). Volume 1103 of LNCS., Springer-Verlag (1996) 392–407
11. Kennaway, R., van Oostrom, V., de Vries, F.J.: Meaningless terms in rewriting. The Journal of Functional and Logic Programming **1** (1999)

Proof-Producing Congruence Closure

Robert Nieuwenhuis* and Albert Oliveras**

Technical University of Catalonia, Jordi Girona 1, 08034 Barcelona, Spain
www.lsi.upc.es/{~roberto,~oliveras}

Abstract. Many applications of congruence closure nowadays require the ability of recovering, among the thousands of input equations, the small subset that caused the equivalence of a given pair of terms. For this purpose, here we introduce an incremental congruence closure algorithm that has an additional *Explain* operation.

First, two variations of union-find data structures with *Explain* are introduced. Then, these are applied inside a congruence closure algorithm with *Explain*, where a k-step proof can be recovered in almost optimal time (quasi-linear in k), without increasing the overall $O(n \log n)$ runtime of the fastest known congruence closure algorithms.

This non-trivial (ground) equational reasoning result has been quite intensively sought after (see, e.g., [SD99, dMRS04, KS04]), and moreover has important applications to verification.

1 Introduction

Union-find data structures maintain the *equivalence* relation induced by a given sequence of *Union* operations between pairs of elements. Similarly, *congruence closure* algorithms maintain a *congruence* relation given by a sequence of pairs of *terms* (i.e., equations) without variables. The difference between equivalence closure and congruence closure is that the congruence relation, in addition to reflexivity, symmetry and transitivity, also satisfies the *monotonicity* axioms saying, for all f, that $f(x_1 \ldots x_n) = f(y_1 \ldots y_n)$ whenever $x_i = y_i$ for all i in $1 \ldots n$.

Example 1. The equation $a=b$ belongs to the congruence generated by the three equations: $b=d$, $f(b)=d$, and $f(d)=a$.

This is equivalent to saying that $a=b$ is a logical consequence (in first-order logic with equality) of these three ground equations. □

Congruence closure is closely related to ground Knuth-Bendix completion; in fact, as usual, our congruence closure algorithm applied to an input set of ground equations E builds a convergent term rewrite system for E (possibly with some new symbols).

Decision procedures based on congruence closure are used in numerous deduction and verification systems, where the generation of *explanations* is highly

---

* Both authors partially supported by Spanish Min. of Educ. and Science by the LogicTools project (TIN2004-03382).
** Supported by FPU grant AP2002-3533 from the Spanish MECD Ministry.

J. Giesl (Ed.): RTA 2005, LNCS 3467, pp. 453–468, 2005.

desirable if not required. For instance, this is crucial in the so-called *lazy* approaches to decision procedures for Boolean formulae over theory atoms. In these procedures, the Boolean formulae frequently include equality atoms; see, e.g., CVC-Lite, at `verify.stanford.edu/CVCL,` and [dMR02, ABC+02, BDS02, FJOS03]. All these approaches are *lazy* in the sense that initially each equality atom is simply abstracted by considering it as a distinct propositional variable, and the resulting propositional formula is sent to a SAT solver. If the SAT solver reaches a (partial) model that is not a congruence, an additional propositional clause (a *lemma*) precluding that model is added; this is iterated (*many* times) until the SAT solver finds a congruence model or all assignments have been explored.

Example 2. Assume that, in such a lazy approach, the model being built by the SAT solver is fed into the congruence closure algorithm as a (long!) sequence of atoms that, in particular, includes $b = d$, $f(b) = d$, and $f(d) = a$. Then, if additionally $a \neq b$ is given, it is no longer a congruence (see Example 1).

At that point, the congruence closure algorithm has to generate as a lemma the clause $b{=}d \wedge f(b){=}d \wedge f(d){=}a \longrightarrow a{=}b$, because the first three atoms are the explanation of $a{=}b$. It is hence crucial in these applications to efficiently recover this small explanation among the (thousands of) originally input equations. □

Another recent approach for the flexible generation of decision procedures given in [GHN+04] also heavily relies on incremental congruence closure with intermixed *Explain* operations. The basic idea is similar to the $CLP(X)$ scheme for constraint logic programming: to provide a clean and efficient integration of specialized theory solvers within the Davis-Putnam-Logemann-Loveland procedure [DP60, DLL62]. A general engine DPLL(X) is used, where X can be instantiated with a solver for a given theory T, thus producing a system DPLL(T). Each time the DPLL(T) procedure produces a conflict, explanations need to be generated by the theory solver for building the *conflict graph* that is used for *non-chronological backtracking* in modern SAT solvers like Chaff [MMZ+01]. The fact that this approach currently outperforms previous techniques on logics with equality is largely due to the efficient incremental algorithm for congruence closure with explanations described here (see [GHN+04] for details about the DPLL(T) approach and experiments on benchmarks from a large variety of verification problems).

Since in such an incremental setting many *Explain* operations occur during a single congruence closure procedure, it is crucial to efficiently recover these explanations, even at the expense of making the congruence closure algorithm slightly slower in practice. If the congruence closure procedure deals with input equations of size n and *Explain*, say, were linear in n, the cost of *Explain* would enormously dominate the $O(n \log n)$ runtime of the overall congruence closure algorithm. Here we present, to our knowledge, the first congruence closure algorithm able to produce these explanations in an efficient way.

Section 2 of this paper is on union-find data structures with *Explain*. Indeed, already for union-find data structures the problem requires some thinking, since the information about the original input unions is, in general, lost in the compact representations of the equivalence relation. We first very briefly introduce some basic notions about union-find data structures and define the *Explain* operation. A first solution that supports optimal *Union* and *Find* operations (as in Tarjan's well-known algorithm with path compression [Tar75]) and recovers the k-step proof in time $O(k \log n)$ is given in Subsection 2.1. In Subsection 2.2 we describe another union-find data structure that has optimal $O(k)$ *Explain* operations and optimal *Find*, at the expense of a slightly more costly *Union*, which has an amortized time bound of $O(\log n)$.

Section 3 is the core of this paper, where the latter union-find data structure is applied inside an incremental congruence closure algorithm. Its complexity is analyzed in Subsection 3.3, where we show that the use of this more costly union-find algorithm (needed for bookkeeping the explanations) does not increase the overall $O(n \log n)$ runtime of the fastest known congruence closure algorithms.

The *Explain* operation is given in Subsection 3.4, and analyzed in detail in Subsection 3.5, showing that it is almost optimal, running in $O(k \, \alpha(k, k))$ time for a k-step explanation, where $\alpha(k, k)$ (related to the inverse of Ackermann's function) is in practice never larger than 4. Subsection 3.6 discusses quality issues of explanations, gives extensive experimental results, and introduces several extensions with practical impact of our explanation algorithms.

2   Union-Find with Proofs

For the sake of self-containedness of this paper and for introducing some notation, here we first shortly explain the classical union-find data structure (see, e.g., [CLR90] for details). A binary *relation* over a set $\mathcal{E}$ is a subset of $\mathcal{E} \times \mathcal{E}$. It is an *equivalence relation* if it is reflexive, symmetric and transitive. The *equivalence closure of a relation* U is the smallest equivalence relation containing U.

The union-find data type maintains the equivalence closure of a relation $U = \{ \, (e_1, e_1') \ldots (e_p, e_p') \, \}$ given incrementally (on-line) as a sequence of operations $Union(e_1, e_1') \ldots Union(e_p, e_p')$. Each equivalence class is identified by its *representative*, which is a certain element of the class. After initializing the data type with the singleton classes $\{e_1\}, \{e_2\}, \ldots, \{e_n\}$, it supports the operations:

- *Union(e, e')*: merges the classes containing e and e' into a new class. We will assume that e and e' were not in the same class prior to the operation, or, equivalently, that *redundant* unions are ignored.
- *Find(e)*: returns the current representative of the class containing e.

A very well-known implementation of this data type is a set of trees, i.e., a forest, where each tree represents one class. Each node in a tree is (labelled with) some element e_i, and the root of a tree is the representative of that class. Then, $Find(e)$ amounts to returning the root r_e of its tree, and each $Union(e, e')$ first finds the two roots by doing $Find(e)$ and $Find(e')$, and then adds the tree rooted with r_e as an additional child of $r_{e'}$ (or vice versa). This is implemented efficiently by an array A of n integers where $A[i] = j$ if the parent of e_i is e_j, and $A[i] = -1$ if e_i is a root (i.e., a representative). This way, the cost of both operations depends only on the depth of the trees. This depth can be kept logarithmic in n, by adding, in each *Union* operation, the tree with fewer nodes as an additional child of the larger one's root. Then, each time an element increases its depth by one, the size of its class is at least doubled, which cannot happen more than $\log n$ times. Note that the size of each tree can be kept as a negative number at its root. Thus, both operations can be done in $O(\log n)$.

Example 3. Below we show a (numbered) sequence of 12 *Union* operations and the tree and array representations of the resulting two classes. Each edge in the trees is labelled with the union that caused it.

$$\underbrace{(1,8)}_{1}, \underbrace{(7,2)}_{2}, \underbrace{(3,13)}_{3}, \underbrace{(7,1)}_{4}, \underbrace{(6,7)}_{5}, \underbrace{(9,5)}_{6}, \underbrace{(9,3)}_{7}, \underbrace{(14,11)}_{8}, \underbrace{(10,4)}_{9}, \underbrace{(12,9)}_{10}, \underbrace{(4,11)}_{11}, \underbrace{(10,7)}_{12}$$

$$
\begin{array}{ccc}
e_8 \xleftarrow{\ 12\ } e_{11} & & e_{13} \\
\nearrow\ \uparrow^4\ \nwarrow^5 \quad \uparrow^{11}\ \nwarrow^8 & & \nearrow^3\ \uparrow^7\ \nwarrow^{10} \\
e_1\quad e_2\quad e_6 \qquad e_4 \qquad e_{14} & & e_3\quad e_5\quad e_{12} \\
\uparrow^2 \qquad\qquad \uparrow^9 & & \uparrow^6 \\
e_7 \qquad\qquad e_{10} & & e_9
\end{array}
$$

8	8	13	11	13	8	2	−9	5	4	8	13	−5	11
1	*2*	*3*	*4*	*5*	*6*	*7*	*8*	*9*	*10*	*11*	*12*	*13*	*14*

$\square$

An optimization known as *path compression* aims at further decreasing the trees' depth: at each *Find(e)*, for every element e' on the path from e to r_e a direct shortcut to r_e is created, that is, all such e' become children of r_e; this comes at the expense of (roughly) duplicating the work at each *Find*. It turns out (see [Tar75]) that a sequence of $n-1$ *Union* operations (i.e., in the end there is only one class), intermixed with $m \geq n$ *Find* operations, is processed in $\Theta(m\,\alpha(m,n))$ time by the algorithm with path compression, where $\alpha(m,n)$ is a *very* slowly growing function. This $\Theta(m\,\alpha(m,n))$ bound is optimal [Tar79].

Our purpose here is to extend the data structure in order to support the following operation, that is able to explain at any point of the computation "why" two given elements e and e' are equivalent at that moment:

– *Explain(e, e')*: if a sequence U of unions of pairs $(e_1, e_1')\ldots(e_p, e_p')$ has taken place, it returns a minimal subset E of U such that (e, e') belongs to the equivalence relation generated by E and it returns $\bot$ if no such E exists.

Example 4 (Example 3 continued). On the previous example, *Explain(e_1, e_4)* returns the explanation $\{(e_7, e_1), (e_{10}, e_7), (e_{10}, e_4)\}$. $\square$

Proposition 1. *The subset E returned by Explain is unique if it exists.*

The previous property is easy to see by considering the undirected graph which has as edges the pairs in the sequence U of unions. Since U includes no redundant unions, this graph (which we will re-visit in Section 2.2) has no cycles. Therefore, *Explain(e, e')* consists exactly of the edges on the unique path between e and e'.

2.1   Union-Find with an $O(k \log n)$ *Explain* Operation

In this section a data structure will be developed that supports optimal *Union* and *Find* operations and where *Explain* takes time $O(k \log n)$ for a k-step explanation. The starting point will be the classical union-find forest implementation, without path compression.

Example 5 (Example 3 continued). Consider again $Explain(e_1, e_4)$ on:

$$\underbrace{(1,8)}_{1}, \underbrace{(7,2)}_{2}, \underbrace{(3,13)}_{3}, \underbrace{(7,1)}_{4}, \underbrace{(6,7)}_{5}, \underbrace{(9,5)}_{6}, \underbrace{(9,3)}_{7}, \underbrace{(14,11)}_{8}, \underbrace{(10,4)}_{9}, \underbrace{(12,9)}_{10}, \underbrace{(4,11)}_{11}, \underbrace{(10,7)}_{12}$$

The key observation is the following. Among the three unions (unions #1, #11, and #12) corresponding to the paths from e_1 and from e_4 to their nearest common ancestor e_8, only the one that occurs *last* (union #12) in the sequence U is sure to belong to $Explain(e_1, e_4)$.   □

In the example we have seen how to find one union (a, b) in $Explain(e, e')$. The remaining unions can be found with two recursive calls $Explain(e, a)$ and $Explain(b, e')$, which, as we will see, gives an algorithm for *Explain* of cost $O(k \log n)$. Note however that the recursive calls could also be $Explain(e, b)$ and $Explain(a, e')$; in order to know which one of the two situations applies, it is also necessary to distinguish at each edge in which direction the union was applied, i.e., each edge has an *oriented* associated union. In a given sequence U of unions of pairs $(e_1, e_1') \ldots (e_p, e_p')$, a union (e_i, e_i') will be called *newer* than a union (e_j, e_j') whenever $i > j$. Now, another technical detail is that the pair (a, b) found as in the example is in fact the newest union in $Explain(e, e')$; therefore, the newest union in both recursive calls will be strictly older than (a, b) and hence an infinite recursion cannot occur. Below all this is formalized.

Lemma 1. *Consider a union-find forest data structure without path compression. For each pair of constants (e, e') with nearest common ancestor c, the newest associated union (a, b) of the paths from e to c and from e' to c belongs to $Explain(e, e')$. Moreover, (a, b) is the newest union in $Explain(e, e')$.*

The previous lemma can be used in a data structure where the unions are kept as a numbered sequence of (oriented) pairs, and where the array representation now also contains the associated unions corresponding to each edge. Since path compression is necessary in order to have optimal *Find* and *Union* operations, we will also keep, with each element, the parents on the data structure with path compression.

Example 6 (Example 3 continued). For our example, the data structures are:

$$\underbrace{(1,8)}_{1}, \underbrace{(7,2)}_{2}, \underbrace{(3,13)}_{3}, \underbrace{(7,1)}_{4}, \underbrace{(6,7)}_{5}, \underbrace{(9,5)}_{6}, \underbrace{(9,3)}_{7}, \underbrace{(14,11)}_{8}, \underbrace{(10,4)}_{9}, \underbrace{(12,9)}_{10}, \underbrace{(4,11)}_{11}, \underbrace{(10,7)}_{12}$$

	1	2	3	4	5	6	7	8	9	10	11	12	13	14
associated union:	1	4	3	11	7	5	2	−1	6	9	12	10	−1	8
parent w/ path c. :	8	8	13	11	13	8	8	−9	13	11	8	13	−5	11
parent wo/ path c.:	8	8	13	11	13	8	2	−9	5	4	8	13	−5	11

$\Box$

Theorem 1. *The previous data structure performs a sequence of $m \geq n$ finds and $n-1$ intermixed unions in time $\Theta(m\,\alpha(m,n))$. Moreover, any $Explain(e,e')$ is supported in $O(k \log n)$ where k is the size of the proof.*

Proof. If always *Find* and *Union* are computed first on the compressed trees, in the same way as it is done in [Tar75], the remaining work at each *Union* (updating the non-compressed parents and the associated union) only needs constant time. Hence, this extra work will not affect the $\Theta(m\,\alpha(m,n))$ runtime. For $Explain(e,e')$, by Lemma 1 we can identify one pair of the proof in time $O(\log n)$ from the non-compressed tree which has depth $O(\log n)$. This work will only be repeated k times, and hence, the total complexity for $Explain(e,e')$ is $O(k \log n)$. $\Box$

2.2   Union-Find with an $O(k)$ *Explain* Operation

Here we develop a data structure in which *Explain* can be answered in optimal time $O(k)$ for a k-step proof, at the expense of slightly more costly *Unions*, which have an amortized time bound of $O(\log n)$.

The main idea is to consider again, as we did for Proposition 1, the graph which has as edges the pairs in the sequence U of unions. As said, since U includes no redundant unions, this graph has no cycles, i.e., it is a forest, and therefore $Explain(e,e')$ consists exactly of the edges on the unique path between e and e'. Of course this forest can be maintained with only constant work at each *Union*, and hence the only problem is how to efficiently find this unique path for a given *Explain* operation.

For this purpose we will choose a root for each tree and direct all its edges towards that root. With this structure being invariant, $Explain(e,e')$ will amount to returning the edges in the paths from e and e' to their common ancestor, which is computable in time $O(k)$, k being the length of the proof. This concrete structure, which in the following will be called *proof forest*, can be kept invariant as follows. At each $Union(e,e')$, assume, w.l.o.g., that the tree of e has no more elements than the one of e', and do:

1. Reverse all edges on the path between e and the root of its tree.
2. Add an edge $e \rightarrow e'$.

It is not difficult to see that this preserves the aforementioned tree structure, as well as the invariant that the path between two nodes is found by computing their nearest common ancestor. Moreover, each time an edge is reversed, the size of its tree is at least doubled. Therefore we have the following:

Lemma 2. *In a sequence of $n-1$ Union operations, each edge in the proof forest is reoriented at most $O(\log n)$ times.*

Example 7. (Example 3 revisited).
Assume that again the following sequence of unions takes place:

$$\underbrace{(1,8)}_{1}, \underbrace{(7,2)}_{2}, \underbrace{(3,13)}_{3}, \underbrace{(7,1)}_{4}, \underbrace{(6,7)}_{5}, \underbrace{(9,5)}_{6}, \underbrace{(9,3)}_{7}, \underbrace{(14,11)}_{8}, \underbrace{(10,4)}_{9}, \underbrace{(12,9)}_{10}, \underbrace{(4,11)}_{11}, \underbrace{(10,7)}_{12}$$

Then the proof forest could be as follows (but note that it is not unique):

$$8 \rightarrow 1 \;\; \rightarrow 7 \leftarrow 2 \qquad\qquad 12 \rightarrow 9 \rightarrow 3 \rightarrow 13$$
$$\nearrow \uparrow \qquad\qquad\qquad\qquad \uparrow$$
$$14 \rightarrow 11 \rightarrow 4 \rightarrow 10 \quad 6 \qquad\qquad 5$$

$\square$

The algorithm we propose is to use the standard union-find with path compression and maintain at the same time the proof forest, which can be represented by an array of pointers (integers) to parents, as it is done in the union-find data structure itself. Altogether, the only operation whose cost will be increased is *Union*.

Theorem 2. *In a sequence of $m \geq n$ finds and $n-1$ intermixed unions, the previous data structure performes each Union in an amortized time bound of $O(\log n)$. Moreover, any Explain(e, e') operation is supported in $O(k)$ where k is the size of the proof.*

Proof. For every call to *Union* the only extra work to be done is the reorientation of the appropriate edges. Since we will have a maximum of $n-1$ edges and each edge will be reoriented at most $O(\log n)$ times, this extra work will be $O(n \log n)$ in the whole sequence, hence giving an amortized time bound of $O(\log n)$ for each *Union*. Note that the *Find* operations are still as efficient as in [Tar75].

As explained above, *Explain(e, e')* will consist of the edges in the paths from e and e' to their common ancestor, which is computable in time $O(k)$ with the invariant structure of the proof forest. $\square$

3 Incremental Congruence Closure with *Explain*

Let $\mathcal{F}$ be a set of (fixed-arity) function symbols and let $T(\mathcal{F})$ be the set of terms without variables built over $\mathcal{F}$: all constants (0-ary symbols) are terms, and $f(t_1, \ldots, t_n)$ is a term whenever f is a non-constant n-ary symbol and $t_1, \ldots, t_n$ are terms. A binary relation $=$ over $T(\mathcal{F})$ is a *congruence relation* if it is reflexive, symmetric, transitive and monotonic; the latter property states, for every non-constant function symbol f, that $f(x_1 \ldots x_n) = f(y_1 \ldots y_n)$ whenever $x_i = y_i$ for all i in $1 \ldots n$. The *congruence closure of a relation U* is the smallest congruence relation containing U. Well-known algorithms for computing the congruence closure of a given set of equations between terms without variables were already given in the early 1980s, such as the $O(n \log n)$ DST algorithm by Downey, Sethi, and Tarjan, [DST80], the Nelson-Oppen one of [NO80] and Shostak's algorithm [Sho78].

Here we will define an *incremental* congruence closure algorithm: we consider a sequence of n *Merge(s, t)* operations, for terms s and t, intermixed with *AreCongruent?(s, t)* operations asking whether s and t are currently congruent, and *Explain(s, t)* operations for recovering the original *Merge* operations causing s and t to be congruent.

3.1   Initial Assumptions and Operations

We will use as a starting point the (non-incremental) congruence closure algorithm of [NO03], which is essentially a simplification of the DST algorithm. DST needs an initial transformation to directed acyclic graphs of outdegree 2, which in [NO03] is replaced by another one, at the formula representation level. This is done by *Currifying*, as in the implementation of functional languages; as a result, there will be only one binary "apply" function symbol (denoted here by an f) and constants. For example, Currifying $g(a, h(b), b)$ gives $f(f(f(g, a), f(h, b)), b)$.

Furthermore, as in the abstract congruence closure approaches (see [Kap97], [BT00]), new constant symbols c are introduced for giving names to non-constant proper subterms t; such t are then replaced everywhere by c, and the equation $t=c$ is added. Then, in combination with Currification, one can obtain the same efficiency as in more sophisticated DAG implementations by appropriately indexing the new constants such as c, which play the role of the pointers to the (shared) subterms such as t in the DAG approaches. For example, the equation $f(f(f(g, a), f(h, b)), b)=b$ is flattened by replacing it by the four equations $f(g, a)=c$, $f(h, b)=d$, $f(c, d)=e$, and $f(e, b)=b$.

These two (structure-preserving) transformations can be done in linear time, and, *in all practical applications we are aware of*, also *once and for all*, instead of at each call to the congruence closure procedure. The transformations could even be done back-and-forth at each operation without increasing the asymptotic complexity bounds (although then the k in the complexity of *Explain* becomes the proof *size*, rather than its number of steps, since each step can involve large terms).

Hence, along this section, we will assume that the equations input to *Merge* are of the form $f(a, b)=c$, or of the form $a=b$, where a, b and c always denote constants. This makes the algorithm surprisingly simple and clean and more efficiently implementable than algorithms for arbitrary terms (in fact, its non-incremental version of [NO03] is about 50 times faster than other recent implementations such as [TV01] on the benchmarks of [TV01]). Due to its simplicity, our algorithm is also easier to extend; e.g., in [NO03] we considered congruence closure with *integer offsets*.

In the remainder of this paper we consider an abstract data type for incremental congruence closure with the following operations:

- *Merge*(t, c) : where t is a flat term of the form $f(a, b)$ or a constant a.
- *AreCongruent?*(s, t) : returns *"yes"* if s and t currently belong to the same congruence class and *"no"* otherwise.
- *Explain*(s, t): assume a sequence M of merges $(s_1, t_1) \ldots (s_p, t_p)$ has occurred, and that (s, t) is in the congruence closure of M; then *Explain*(s, t) returns a subset $E = (s_{i_1}, t_{i_1}) \ldots (s_{i_k}, t_{i_k})$ of M, with $1 \le i_1 < \ldots < i_k \le p$, such that exactly at the i_k-th merge s and t became congruent, due to the merge operations in E.

Theorem 3. *A sequence of n Merge operations can be processed in $O(n \log n)$ time, and hence each one of them in $O(\log n)$ amortized time. Furthermore, each question AreCongruent?(s, t) can be answered in $O(|s| + |t|)$, i.e., in constant time if s and t are constants. For the Explain(s, t) operation between constants, a k-step proof can be found in time $O(k \, \alpha(k, k))$, to which, for arbitrary terms s and t, an additional cost $O(|s| + |t|)$ has to be added.*

3.2 Implementation of *Merge*

The underlying union-find data structure used here applies *eager path compression*, that is, for each constant symbol c_i, its representative can always be returned in constant time by accessing an array *Representative*$[i]$. In order to maintain this *Representative* table, there will be additional *Class Lists* containing for each representative the constants in its class.

The basic data structures for the congruence closure algorithm are:

1. *Pending*: a list whose elements are input equations $a{=}b$, or pairs of input equations $(f(a_1, a_2){=}a, f(b_1, b_2){=}b)$ where a_i and b_i are already congruent for $i = 1, 2$. In both cases, when inserting such an element in *Pending*, what is pending is the merge of the constants a and b.
2. The *Use lists*: for each representative a, *UseList*(a) is a list of input equations $f(b_1, b_2){=}b$ such that a is the representative of b_1 or of b_2 (or of both).
3. The *Lookup table*: for all pairs of representatives (b, c), *Lookup*(b, c) is some input equation $f(a_1, a_2){=}a$ such that b and c are the current respective representatives of a_1 and a_2 iff such an equation exists. Otherwise, *Lookup*(b, c) is $\perp$. A useful additional invariant is that, for representatives b and c, $f(a_1, a_2){=} a$ is in *UseList*(b) and in *UseList*(c) iff *Lookup*(b, c) is $f(a_1, a_2) = a$.
4. The *Proof forest* data structure, as presented in the previous section.

Here we present the algorithms in a way as similar as possible to the nonincremental one of [NO03], and separate the treatment of the proof forest from the congruence closure algorithm itself. The data structures are initialized as expected: all *Use Lists*, *Pending*, and the *Proof forest* are empty, and *Lookup*(a, b) is $\perp$ for all pairs (a, b). Each *ClassList*(a) is initialized to contain only a and each *Representative*(a) is initialized with a. Note that *Lookup* could also be stored in a hash table (since a 2-dimensional array will be almost empty), and that the non-constant time initializations can also be avoided[1]. In the following algorithms, a' always denotes $Representative(a)$ for each constant a.

1. **Procedure** $Merge(s{=}t)$
2.   **If** s and t are constants a and b **Then** {
3.   add $a{=}b$ to $Pending$
4. $Propagate()$ }
5.   **Else** /* $s{=}t$ is of the form $f(a_1, a_2){=}a$ */
6.   **If** $Lookup(a_1', a_2')$ is some $f(b_1, b_2){=}b$ **Then** {
7.   add ($f(a_1, a_2){=}a, f(b_1, b_2){=}b$) to $Pending$
8. $Propagate()$ }
9.   **Else** {
10.   set $Lookup(a_1', a_2')$ to $f(a_1, a_2){=}a$
11.   add $f(a_1, a_2){=}a$ to $UseList(a_1')$ and to $UseList(a_2')$ }

12. **Procedure** $Propagate()$
13.   **While** $Pending$ is non-empty **Do** {
14.   Remove E of the form $a{=}b$ or $(f(a_1, a_2){=}a, f(b_1, b_2){=}b)$ from $Pending$
15.   **If** $a' \neq b'$ and, wlog., $|ClassList(a')| \leq |ClassList(b')|$ **Then** {
16. $old\_repr\_a := a'$
17.   Insert edge $a \rightarrow b$ labelled with E into the *proof forest*
18.   **For each** c in $ClassList(old\_repr\_a)$ **Do** {
19.   set $Representative(c)$ to b'
20.   move c from $ClassList(old\_repr\_a)$ to $ClassList(b')$ }
21.   **For each** $f(c_1, c_2){=}c$ in $UseList(old\_repr\_a)$ **Do**
22.   **If** $Lookup(c_1', c_2')$ is some $f(d_1, d_2){=}d$ **Then** {
23.   add $(f(c_1, c_2){=}c, f(d_1, d_2){=}d)$ to $Pending$
24.   remove $f(c_1, c_2){=}c$ from $UseList(old\_repr\_a)$ }
25.   **Else** {
26.   set $Lookup(c_1', c_2')$ to $f(c_1, c_2){=}c$
27.   move $f(c_1, c_2){=}c$ from $UseList(old\_repr\_a)$ to $UseList(b')$ }}}

Each iteration of the $Propagate()$ algorithm picks a pending union. If this union is not redundant, it is added to the proof forest (line 17) and (lines 19 and 20) to the union-find data structure. Lines 21–27 traverse the $UseList$ of the constant whose representative has changed and, checking the lookup table, detect new pairs of constants to be merged.

3.3 Complexity of *Merge* and *AreCongruent*?

As said, an amortized analysis is done over the whole sequence of n *Merge* operations. The procedure *Merge* itself has no loops. Concerning $Propagate()$, let m be the number of different constants (note that $m \leq 3n$). The loop at lines

---

[1] For example, $Representative[a]$ can be updated by storing in $Representative[a]$ an index k to an auxiliary array A, where $A[k]$ contains a and its representative, and with a counter max indicating that, for all $k < max$, $A[k]$ contains correct (i.e., initialized) information; re-initialization then simply amounts to setting max to 0 (in fact, this general idea can always avoid any n-dimensional array initialization).

19 and 20 is executed in total $O(m \log m)$ times, namely when some constant changes its representative, which for each one of the m constants happens at most $\log m$ times, because each time the size of its class is at least doubled. Line 17 inserts an edge between pair of constants in the proof forest. This is done at most $m-1$ times, and by Lemma 2 the total time for the $m-1$ insertions is $O(m \log m)$. In the loop at lines 21–27, each one of the at most n input equations of the form $f(c_1, c_2)=c$ is treated when c_1 or c_2 changes its representative (which, as before, cannot happen more than $\log m$ times). Altogether, we obtain an $O(n \log n)$ runtime. Re-using *UseList* and *ClassList* nodes, only linear space is required.

Note that the equivalence relation between all constants is dealt with by the *Representative* array, i.e., by union-find with "eager path compression". This makes the algorithm simpler and allows one to handle *AreCongruent?*(a, b) in constant time; asymptotically speaking, it causes no overhead to *Merge*. In practice, lazy path compression may globally perform better since strictly less "compressions" are done, although it has an overhead caused by additional checks (whether the representative has been reached or not, etc.). In any case, all asymptotic bounds given here carry over straightforwardly to the case of lazy path compression.

3.4 Implementation of *Explain*

As said, each edge $a-b$ in the proof forest is labelled with a single input equation $a=b$ or with a pair of input equations $(f(a_1, a_2)=a, f(b_1, b_2)=b)$. The way the proof forest (and the information associated to its edges) is represented is not described here; it can be done e.g., as in Subsection 2.2, by an array of pointers.

Example 8. Below we show a (numbered) sequence of 6 *Merge* operations and the state of the proof forest after processing them. Each edge of the proof forest is annotated with its corresponding input equation or pair of input equations:

$$\underbrace{f(g,h)=d,}_{1} \underbrace{c=d,}_{2} \underbrace{f(g,d)=a,}_{3} \underbrace{e=c,}_{4} \underbrace{e=b,}_{5} \underbrace{b=h,}_{6} \qquad a \xrightarrow{1,3} d \xleftarrow{2} c \xleftarrow{4} e \xleftarrow{5} b \xleftarrow{6} h$$

On an *Explain*(a, b) operation, the nearest common ancestor d is detected, and the merge operations on the paths $a-d$ (1,3) and $b-d$ (5,4,2) are output as part of the proof; but from 1 and 3 also recursively *Explain*(h, d) needs to be output. In order to obtain the desired complexity bound, it is necessary to avoid repeated visits to nodes like b, e, c, d in such recursive calls. After the merge operations in the path $b-d$ have been output, the constants b, e, c and d can be considered to be inside the same equivalence class C. Since the information in the edges in the path $b-d$ has already been output, in any future traversal one can jump from any element of C to d (here d is the *highest node* of C, the element of C that is closest to the root of its tree in the proof forest). Hence, in the recursive call to *Explain*(h, d), only the edge $b-h$ is traversed, since from b one can directly jump to d. $\qquad\square$

The data structures for avoiding such repeated visits and the way they are used is explained in the following two points:

The Additional Union-Find, and *HighestNode*. At each call to $Explain(s,t)$, an additional union-find data structure with path compression keeps track of the classes of constants that are already equivalent by the proof output so far. More precisely, apart from the $Find(a)$ operation, there is also a $HighestNode(a)$ operation, which returns the *highest node* among all nodes of the proof tree in the equivalence class of a; this highest node is simply stored at the node of $Find(a)$. Maintaining the *HighestNode* information will be easy: since only unions of the form $Union(a, parent(a))$ take place, the *HighestNode* of the new class is always the *HighestNode* of the second argument of the call, i.e. the *HighestNode* of $parent(a)$.

Finding the Nearest Common Ancestor in the Proof Forest. There is also a $NearestCommonAncestor(a,b)$ operation that retrieves the *highest node* of the class of the nearest common ancestor of a and b in the proof forest. When looking for it, as it happens in the *ExplainAlongPath* procedure below, one has to jump over whole classes of equivalent constants by means of the *HighestNode* operation in order to avoid traversing unnecessary edges.

Now we are ready to present the two procedures implementing *Explain*:

1. **Procedure** $Explain(c_1, c_2)$
2.   Set $PendingProofs$ to $\{c_1{=}c_2\}$
3.   **While** $PendingProofs$ is not empty **Do** {
4.   Remove an equation $a{=}b$ from $PendingProofs$
5. $c := NearestCommonAncestor(a,b)$
6. $ExplainAlongPath(a,c)$
7. $ExplainAlongPath(b,c)$ }

8. **Procedure** $ExplainAlongPath(a,c)$
9. $a := HighestNode(a)$
10.   **While** $a{\neq}c$ **Do** {
11. $b := parent(a)$
12.   **If** edge $a \rightarrow b$ is labelled with a single input merge $a{=}b$
13.   Output $a{=}b$
14.   **Else** {   /* edge labelled with $f(a_1, a_2){=}a$ and $f(b_1, b_2){=}b$ */
15.   Output $f(a_1, a_2){=}a$ and $f(b_1, b_2){=}b$
16.   Add $a_1{=}b_1$ and $a_2{=}b_2$ to $PendingProofs$ }
17. $Union(a, b)$
18. $a := HighestNode(b)$ }

3.5   Complexity of *Explain*

Let k be the number of steps in the final proof that is output. There are in total $O(k)$ iterations of the *ExplainAlongPath* loop since at each iteration either one (line 13) or two (line 15) such steps are output. In fact, for each call of the form $ExplainAlongPath(a,c)$, the number of iterations corresponds to the number of different equivalence classes along the path from a to c, and at each iteration, one union between classes takes place, as well as one call to *HighestNode* (i.e., one *Find*). Hence in total $O(k)$ such classes are merged along the whole proof. The total work done for searching nearest common ancestors (line 5 of procedure *Explain*) is also $O(k)$, because it can be done in time linear in the number of classes that are merged in the subsequent two calls *ExplainAlongPath(a,c)* and *ExplainAlongPath(b,c)*. Furthermore, for each iteration of *ExplainAlongPath*, at most two equalities are added to *PendingProofs*, and hence the loop of procedure *Explain* is executed $O(k)$ times. Altogether, the global runtime is dominated by the $O(k)$ unions of classes and the $O(k)$ calls to *Find*, which has a total cost of $O(k\,\alpha(k,k))$ in the union-find algorithm with path compression.

3.6 Quality of Explanations. Experiments

Example 9. After a given sequence of input equations E, there can be several explanations for an equation $s=t$. Consider the sequence of 7 input equations E:
$$a=b_1 \quad b_1=b_2 \quad b_2=b_3 \quad b_3=c \quad f(a_1,a_1)=a \quad f(c_1,c_1)=c \quad a_1=c_1$$
In our algorithm, *Explain*$(a=c)$ will return the first four equations, although the last three equations also form a correct explanation of $a=c$. □

Finding *short* explanations is good for most practical applications, and also finding the *oldest* explanation (i.e., the one contained in the shortest prefix of the sequence E) is desirable (roughly, because it allows one to do more powerful backjumping). Since our algorithm always returns the oldest explanation (see the definition of *Explain* before Theorem 3), from now on we will focus on length.

Unfortunately, trying to always find the *shortest* explanation (in number of steps) is too ambitious: given such an E, an equation $s=t$, and a natural number k, deciding whether an explanation of size smaller than k exists for $s=t$ is already an NP-hard problem[2]. Therefore, the usual criterion for quality of an explanation is its *irredundancy*: after removing any step, it is no longer a valid explanation. Surprisingly, the explanations found by our algorithm as presented in the previous subsection sometimes still contain redundant steps.

Example 10. After the sequence of input equations:
$$a_1=b_1 \quad a_1=c_1 \quad f(a_1,a_1)=a \quad f(b_1,b_1)=b \quad f(c_1,c_1)=c$$
the proof forest may consist of the two trees: $a \rightarrow b \leftarrow c$ and $b_1 \rightarrow a_1 \leftarrow c_1$. Now *Explain*$(a=c)$ will return all five equations. However, the two equations containing b_1 are redundant. □

We have run our algorithm as given in the previous subsection over a very large set of benchmarks (all the EUF examples mentioned in [GHN+04], available at the second author's home page). There, on average, explanations have 12.6 steps; redundant explanations are returned in 13.56 percent of the cases, having, on average, 40 steps of which 6 are redundant.

Fortunately, one can easily and efficiently post-process explanations in order to fully remove all redundant steps. On the one hand, it is not very hard to see

---

[2] Ashish Tiwari. Personal communication.

that one of our explanations can be redundant only if it contains at least three
equations of the same *structural class*, i.e., of the form $f(a_1, a_2)=a$, $f(b_1, b_2)=b$,
$f(c_1, c_2)=c$ where a_i, b_i and c_i have the same representative for $i = 1, 2$. This
can be checked in time linear in k, and is an extremely good filter: three such
equations occur only in 0.27 percent of the irredundant explanations.

The 13.8 percent of the explanations marked as "possibly redundant" by
this test can be post-processed as follows in time $O(k^2 \log k)$ in order to remove
all redundancies: while not all equations are marked as "necessary", pick an
unmarked one, remove it if the remaining equations are still a correct explanation
(checking this takes $O(k \log k)$ time), and otherwise mark it as "necessary".

Proof Forests with Structural Classes as Nodes. We have also implemented a variant of our proof forests where the nodes are these structural classes
and hence all edges are labelled with a single input equation between constants.
Now, instead of inserting edges labelled with $(f(a_1, a_2) = a, f(b_1, b_2) = b)$, one
merges the two nodes (classes) [...a...] and [...b...] into a single one.

Example 11. Consider again the input sequence of the previous example:
    $a_1=b_1$ $a_1=c_1$ $f(a_1, a_1)=a$ $f(b_1, b_1)=b$ $f(c_1, c_1)=c$
Now the proof forest will consist of the two trees: $[a, b, c]$ and $b_1 \rightarrow a_1 \leftarrow c_1$
and *Explain*$(a=c)$ will return only the structural equations involving a and c
and its corresponding recursive explanation that $a_1=c_1$. □

In such proof forests, the *Explain* operation is implemented in a very similar
way as before. For simplicity, in the previous subsection we have not mentioned
this improvement, but it is not hard to see that all results apply.

With this new approach, only 3.5 percent of the explanations are still redundant, having on average 34 steps, of which 6 are redundant. Using the test, now
postprocessing is needed only in 3.95 percent of the cases.

Example 12. Let's see why some redundancies can still appear. Consider:
    1. $f(a_1, a_1)=a$ 2. $f(b_1, b_1)=b$ 3. $f(c_1, c_1)=c$ 4. $f(d_1, d_1)=d$
    5. $a_1=b_1$ 6. $c_1=d_1$ 7. $a_1=c$ 8. $a_1=a$ 9. $d=d_1$

Then the proof tree may become: $[a, b] \rightarrow a_1 \leftarrow [c, d] \leftarrow d_1 \leftarrow c_1$
                                                            $\uparrow$
                                                            b_1

and *Explain*$(b=d_1)$ returns the set of all 9 input equations, of which #1 and
#8 are redundant. This redundancy is caused by the two equivalent classes $[a, b]$
and $[c, d]$. Indeed, it can be shown that if no two such equivalent non-singleton
structural classes exist, proofs will always be irredundant. But it seems too
expensive to maintain that property during the congruence closure procedure;
in particular, the difficulties arise when two such classes become equivalent (in
the example, after $d=d_1$) while they are already in the same tree, i.e., when they
are already equal by equations between constants. □

4   Related and Future Work

To our knowledge, this is the first congruence closure algorithm able to produce explanations in time that does not depend on the number of input equations n. Moreover, the congruence closure algorithm itself is not only simple, but it also runs in the best known time, namely $O(n \log n)$, and is indeed very fast in practice.

We believe that this kind of fundamental algorithmic developments are extremely useful, because we have seen several less adequate ad-hoc solutions being applied in modern deduction and verification tools. E.g., in [BDS02] where the CVC tool is described (verify.stanford.edu/CVC), a trial-and-error method for finding explanations is given. Another example of this phenomenon is SRI's "lemmas-on-demand" approach in the ICS tool: in [dMR02] it is mentioned that "Unfortunately, current domain-specific decision procedures lack such a conflict explanation facility. Therefore, we developed an algorithm that calls C-solver $O(k \times n)$ times, where k is given, for finding such an overapproximation". Several authors have attacked the specific problem of generating explanations in the context of union-find and congruence closure [SD99, KS04, dMRS04]). In particular, in the paper "Justifying Equality" [dMRS04], for union-find *Explain* is done in time $O(n \, \alpha(n))$, i.e., it depends on the number of unions that have taken place. For the (strict) generalization to congruence closure, this is indeed also the case (although no concrete bound is given in that paper), and the notion of *local irredundancy* achieved in [dMRS04] already holds for our basic algorithm of Section 3.

Concerning future work, we plan to study extensions, such as the version for congruence closure with *integer offsets* [NO03], to which the ideas given here for *Explain* can also be applied. It also remains to be studied whether irredudant proofs can be generated directly without any postprocessing (although this does not seem to lead to more practical efficiency).

References

[ABC+02] G. Audemard, P. Bertoli, A. Cimatti, A. Kornilowicz, and R. Sebastiani. A SAT based approach for solving formulas over boolean and linear mathematical propositions. In *CADE-18*, LNCS 2392, pages 195–210, 2002.

[BDS02] Clarke Barrett, David Dill, and Aaron Stump. Checking satisfiability of first-order formulas by incremental translation into sat. In *Procs. 14th Intl. Conf. on Computer Aided Verification (CAV)*, LNCS 2404, 2002.

[BT00] L. Bachmair and A. Tiwari. Abstract congruence closure and specializations. In *Conf. Autom. Deduction, CADE*, LNAI 1831, pages 64–78, 2000.

[CLR90] Thomas T. Cormen, Charles E. Leiserson, and Ronald L. Rivest. *Introduction to algorithms*. MIT Press, 1990.

[DLL62] Martin Davis, George Logemann, and Donald Loveland. A machine program for theorem-proving. *Comm. of the ACM*, 5(7):394–397, 1962.

[dMR02] Leonardo de Moura and Harald Rueß. Lemmas on demand for satisfiability solvers. In *Procs. 5th Int. Symp. on the Theory and Applications of Satisfiability Testing, SAT'02*, pages 244–251, 2002.

[dMRS04]   L. de Moura, H. Rueß, and N. Shankar. Justifying equality. In *Proc. of the Second Workshop on Pragmatics of Decision Procedures in Automated Reasoning*, Cork, Ireland, 2004.

[DP60]   Martin Davis and Hilary Putnam. A computing procedure for quantification theory. *Journal of the ACM*, 7:201–215, 1960.

[DST80]   Peter J. Downey, Ravi Sethi, and Robert E. Tarjan. Variations on the common subexpressions problem. *J. of the Association for Computing Machinery*, 27(4):758–771, 1980.

[FJOS03]   C. Flanagan, R. Joshi, X. Ou, and J. B. Saxe. Theorem proving using lazy proof explanation. In *Procs. 15th Int. Conf. on Computer Aided Verification (CAV)*, LNCS 2725, 2003.

[GHN$^+$04] Harald Ganzinger, George Hagen, Robert Nieuwenhuis, Albert Oliveras, and Cesare Tinelli. DPLL(T): Fast decision procedures. In R. Alur and D. Peled, editors, *Proceedings of the 16th International Conference on Computer Aided Verification, CAV'04 (Boston, Massachusetts)*, volume 3114 of *Lecture Notes in Computer Science*, pages 175–188. Springer, 2004.

[Kap97]   Deepak Kapur. Shostak's congruence closure as completion. In *Procs. 8th Int. Conf. on Rewriting Techniques and Applications*, LNCS 1232, 1997.

[KS04]   Robert Klapper and Aaron Stump. Validated proof-producing decision procedures. In *Proceedings of the Second Workshop on Pragmatics of Decision Procedures in Automated Reasoning*, Cork, Ireland, 2004.

[MMZ$^+$01] Matthew W. Moskewicz, Conor F. Madigan, Ying Zhao, Lintao Zhang, and Sharad Malik. Chaff: Engineering an Efficient SAT Solver. In *Proc. 38th Design Automation Conference (DAC'01)*, 2001.

[NO80]   Greg Nelson and Derek C. Oppen. Fast decision procedures bases on congruence closure. *Journal of the Association for Computing Machinery*, 27(2):356–364, April 1980.

[NO03]   Robert Nieuwenhuis and Albert Oliveras. Congruence closure with integer offsets. In *10h Int. Conf. Logic for Programming, Artif. Intell. and Reasoning (LPAR)*, LNAI 2850, pages 78–90, 2003.

[SD99]   Aaron Stump and David L. Dill. Generating proofs from a decision procedure. In A. Pnueli and P. Traverso, editors, *Proceedings of the FLoC Workshop on Run-Time Result Verification*, Trento, Italy, 1999.

[Sho78]   Robert E. Shostak. An algorithm for reasoning about equality. *Commun. ACM*, 21(7), 1978.

[Tar75]   Robert Endre Tarjan. Efficiency of a good but not linear set union algorithm. *Journal of the ACM (JACM)*, 22(2):215–225, April 1975.

[Tar79]   Robert Endre Tarjan. A class of algorithms that require nonlinear time to maintain disjoint sets. *J. Comput. and Sys. Sci.*, 18(2):110–127, 1979.

[TV01]   Ashish Tiwari and Laurent Vigneron. Implementation of Abstract Congruence Closure with randomly generated CC problem instances, 2001. At www.csl.sri.com/users/tiwari.

The Algebra of Equality Proofs

Aaron Stump and Li-Yang Tan

Dept. of Computer Science and Engineering
Washington University in St. Louis
St. Louis, Missouri, USA
http://cl.cse.wustl.edu/

Abstract. Proofs of equalities may be built from assumptions using proof rules for reflexivity, symmetry, and transitivity. Reflexivity is an axiom proving x=x for any x; symmetry is a 1-premise rule taking a proof of x=y and returning a proof of y=x; and transitivity is a 2-premise rule taking proofs of x=y and y=z, and returning a proof of x=z. Define an equivalence relation to hold between proofs iff they prove a theorem in common. The main theoretical result of the paper is that if all assumptions are independent, this equivalence relation is axiomatized by the standard axioms of group theory: reflexivity is the unit of the group, symmetry is the inverse, and transitivity is the multiplication. Using a standard completion of the group axioms, we obtain a rewrite system which puts equality proofs into canonical form. Proofs in this canonical form use the fewest possible assumptions, and a proof can be canonized in linear time using a simple strategy. This result is applied to obtain a simple extension of the union-find algorithm for ground equational reasoning which produces minimal proofs. The time complexity of the original union-find operations is preserved, and minimal proofs are produced in worst-case time $O(n^{log_2 3})$, where n is the number of expressions being equated. As a second application, the approach is used to achieve significant performance improvements for the CVC cooperating decision procedure.

1 Introduction

Ground equational reasoning plays an important role in many approaches to verification and automated reasoning [3, 8, 9, 12–14, 17]. Recently, there has been interest in producing minimal proofs from algorithms for ground equational reasoning [5, 10, 11]. Minimal proofs are of interest primarily for performance reasons: they can be exported to a fast SAT solver as conflict clauses, which greatly improve search space pruning [1].

In this paper, we approach the problem of minimal proofs by studying the algebra of equality proofs themselves (Section 2). It turns out that theorem equivalence of equality proofs with independent assumptions is completely characterized by the axioms for free groups (Sections 3, 4, and 5). This enables us to use a standard convergent rewrite system for free group terms to put equality proofs into canonical form (Section 5). If all assumptions are independent, this

J. Giesl (Ed.): RTA 2005, LNCS 3467, pp. 469–483, 2005.

form is minimal, in the sense that it uses the unique minimal set of assumptions needed to prove the equality. We analyze the number of steps required for canonization using the rewrite rules. A simple strategy yields canonization in time linear in the size of the equality proof (Section 6), although without a strategy canonization can take cubic time (Section 7). We then show how these results can be used to obtain minimal proofs of equations $x = y$ from a simple augmentation of the standard union-find algorithm for ground equational reasoning in $(O(n^{\log_2 3})$ time, where n is the size of the equivalence class for x and y (Section 8). Finally, major improvements in search space pruning and overall performance are obtained in the context of the CVC tool [16] using the algebra of equality proofs (Section 9).

2   Equality Proofs

Notation: If f is an n-ary function symbol and $A_1, \ldots, A_n$ are sets of terms, we write $f(A_1, \ldots, A_n)$ for $\{f(t_1, \ldots, t_n) \mid t_1 \in A_1, \ldots, t_n \in A_n\}$.

Let $\mathcal{U}$ be a set of *assumptions*, and let $\mathcal{P}$ be the set of *equality proofs* inductively defined as follows, where *Refl*, *Symm*, and *Trans* are function symbols of arities 0, 1, and 2, respectively:

$$\mathcal{P} ::= \mathcal{U} \mid \mathit{Refl} \mid \mathit{Symm}(\mathcal{P}) \mid \mathit{Trans}(\mathcal{P}, \mathcal{P})$$

Let $\mathcal{A}$ be a set of *atoms*, and let $\mathcal{E}$ be the set of all equations between atoms. Let :: be a total function from $\mathcal{U}$ to $\mathcal{E}$, specifying which equation is proved by each assumption. Extend this to a non-functional relation on all of $\mathcal{P} \times \mathcal{E}$, which says which theorems are proved by which equality proofs; this is done inductively by the universal closures of the following clauses:

$$\mathit{Refl} :: a = a$$
$$p :: a_1 = a_2 \;\Rightarrow\; \mathit{Symm}(p) :: a_2 = a_1$$
$$p_1 :: a_1 = a_2 \;\wedge\; p_2 :: a_2 = a_3 \;\Rightarrow\; \mathit{Trans}(p_1, p_2) :: a_1 = a_3$$

We say that the assumptions in $\mathcal{U}$ are *independent* iff no equation proved by an assumption is equationally entailed by the other assumptions. For example, assumptions of $a = b$, $b = c$, and $a = c$, respectively, are not independent.

Finally, define a relation $==$ of *theorem equivalence* on $\mathcal{P} \times \mathcal{P}$ by the universal closure of

$$p_1 == p_2 \;\Leftrightarrow\; (\exists e.\; p_1 :: e \;\wedge\; p_2 :: e)$$

Two equality proofs are in this relation iff they prove a theorem in common. For example, $\mathit{Trans}(\mathit{Refl}, \mathit{Refl}) == \mathit{Refl}$, because each proof proves $a = a$, for some $a \in \mathcal{A}$. It turns out that technical reasons prevent us from adopting the stronger notion where proofs are equivalent iff they prove exactly the same theorems. Not every equality proof proves a theorem. Write $\equiv$ for meta-equality on sets like $\mathcal{A}$, $\mathcal{E}$, and $\mathcal{P}$. Then, for example, if $p \in \mathcal{P}$ proves $a = b$ with $a \not\equiv b$, then $\mathit{Trans}(p, p)$ does not prove any theorem. The following lemmas help justify the use of the notion of proving a theorem in common as our notion of theorem equivalence. We define $\hat{\mathcal{P}}$ to be $\mathit{Dom}(::)$, so that $p \in \hat{\mathcal{P}}$ iff p proves a theorem.

Lemma 1 *If $p \in \hat{\mathcal{P}}$ does not contain any assumptions, then $\forall a \in \mathcal{A}. \; p :: a = a$. Further, if $a \not\equiv b$, then p does not prove $a = b$.*

Lemma 2 *If $p \in \hat{\mathcal{P}}$ contains an assumption, then p proves exactly one theorem.*

3 Algebraic Characterization of Theorem Equivalence

In this Section, we show that theorem equivalence on proofs which prove a theorem is axiomatized by the standard axioms of group theory, as long as the assumptions in $\mathcal{U}$ are independent. In more detail, for all $p_1, p_2 \in \hat{\mathcal{P}}$, $p_1 ==$ p_2 holds iff the equation $p_1 \cong p_2$ is provable using standard congruence rules for equality and the universal closures of the formulas given in Figure 1. For comparison, the group axioms are given in more customary form in Figure 2. *Refl* plays the role of the group unit, *Symm* is the inverse operator, and *Trans* is the multiplication.

$$
\begin{array}{lll}
[Associativity] & Trans(Trans(p_1, p_2), p_3) & \cong\; Trans(p_1, Trans(p_2, p_3)) \\
[Unit] & Trans(Refl, p) & \cong\; p \\
[Inverse] & Trans(Symm(p), p) & \cong\; Refl
\end{array}
$$

Fig. 1. Group Axioms for Equality Proofs.

$$
\begin{array}{lll}
[Associativity] & (x_1 * x_2) * x_3 & \cong\; x_1 * (x_2 * x_3) \\
[Unit] & 1 * x & \cong\; x \\
[Inverse] & x^{-1} * x & \cong\; 1
\end{array}
$$

Fig. 2. Group Axioms.

We observe first that strictly speaking, neither $\mathcal{P}$ nor $\hat{\mathcal{P}}$ forms a group with operators *Refl*, *Symm*, and *Trans* and equivalence relation $==$. For the case of $\mathcal{P}$, this is because if p is an equality proof which does not prove any theorem, then the left hand side (lhs) of the [*Inverse*] axiom of Figure 1 does not prove any theorem, but the rhs (*Refl*) does. Hence, the two proofs do not prove the same theorems, so the axiom is just false (if $==$ is taken for $\cong$) for domain $\mathcal{P}$. We prove below that all the axioms are sound with respect to $==$ for $\hat{\mathcal{P}}$, but $\hat{\mathcal{P}}$ is not closed in general under *Trans*. As noted above, if $p :: a = b$ with $a \not\equiv b$, then $Trans(p, p)$ is not in $\hat{\mathcal{P}}$. So, $\hat{\mathcal{P}}$ does not form a group under *Refl*, *Symm*, and *Trans*. Nevertheless, we have the following results, which are proved in Sections 4 and 5 below.

Theorem 1 (Soundness) *For all $p_1, p_2 \in \hat{\mathcal{P}}$, if $p_1 \cong p_2$, then $p_1 == p_2$.*

Theorem 2 (Completeness) *For all $p_1, p_2 \in \hat{\mathcal{P}}$, if $p_1 == p_2$, then $p_1 \cong p_2$.*

4   Proof of Soundness

Lemma 3 (Theorem Determinacy) *Suppose $p \in \hat{P}$, $p :: a_1 = a_2$ and $p ::$ $b_1 = b_2$. Then $a_1 \equiv b_1$ iff $a_2 \equiv b_2$.*

Proof. If p contains no assumption, then by Lemma 1, $a_1 \equiv a_2$ and $b_1 \equiv b_2$, and we have the result by transitivity of $\equiv$. If p contains an assumption, then by Lemma 2, p proves exactly one theorem, and so $a_1 \equiv b_1$ and $a_2 \equiv b_2$.

Proof (Soundness (Theorem 1)). First we observe that the congruence rules for $\cong$ are sound. If $p_1 == p_2$, then $Symm(p_1) == Symm(p_2)$. For the congruence rule for *Trans*, we reason as follows. Suppose one of p_1 or p_2 contains assumptions. WLOG, say it is p_1. Then by Lemma 2, both p_1 and $Trans(p_1, p)$ prove at most one theorem, for any p. If the latter proves no theorem, we contradict the hypothesis of Theorem 1 that the proofs in question prove a theorem. If $Trans(p_1, p)$ proves a theorem, then since p_2 proves a theorem in common with p_1, p_2 must prove the same theorem as p_1, and hence $Trans(p_2, p)$ proves a theorem in common with $Trans(p_1, p)$. If neither p_1 nor p_2 contains an assumption, then by Lemma 1, they prove all the same theorems, and hence $Trans(p_1, p)$ and $Trans(p_2, p)$ prove a theorem in common, assuming again that they prove any theorem at all. We now consider the group axioms for $\cong$.

Case [*Associativity*]: Suppose $Trans(Trans(p_1, p_2), p_3) :: a = d$. By the definition of ::, this implies that there is a c such that $Trans(p_1, p_2) :: a = c$ and $p_3 :: c = d$. The former fact implies again by the definition of :: that there is a b such that $p_1 :: a = b$ and $p_2 :: b = c$. Then clearly $Trans(p_1, Trans(p_2, p_3))$ proves $a = d$.

Case [*Unit*]: If $Trans(Refl, p)$ proves $a = b$, then p must prove $a = b$.

Case [*Inverse*]: If $Trans(Symm(p), p)$ proves $a = c$, then there must be a b such that $Symm(p) :: a = b$ and $p :: b = c$. The former consequence implies that $p :: b = a$. By Lemma 3, $a \equiv c$, so $Trans(Symm(p), p) :: a = a$. We also have, of course, $Refl :: a = a$. Note that this is the point at which defining theorem equivalence as proving exactly the same theorems breaks down. For we also have, e.g., $Refl :: b = b$, but by Lemma 1, if p contains an assumption, $Trans(Symm(p), p)$ cannot prove two theorems. And hence, since it proves $a = a$, it cannot prove $b = b$. So this axiom would not be sound with the stronger version of theorem equivalence.

5   Canonical Proofs and Completeness

The proof of our completeness theorem (Theorem 2) relies on the canonical forms for equality proofs. Recall that the rewrite rules of Figure 3 are a convergent completion of the group axioms of Figure 2, oriented from left to right [7]. These rules are given again in Figure 4, formulated for equality proofs.

$$(x_1 * x_2) * x_3 \rightarrow x_1 * (x_2 * x_3)$$
$$1 * x \qquad\qquad \rightarrow x$$
$$x * 1 \qquad\qquad \rightarrow x$$
$$x^{-1} * x \qquad\quad \rightarrow 1$$
$$x * x^{-1} \qquad\quad \rightarrow 1$$
$$1^{-1} \qquad\qquad\; \rightarrow 1$$
$$(x^{-1})^{-1} \qquad\; \rightarrow x$$
$$(x * y)^{-1} \qquad \rightarrow y^{-1} * x^{-1}$$
$$x^{-1} * (x * y) \rightarrow y$$
$$x * (x^{-1} * y) \rightarrow y$$

Fig. 3. Convergent System for Simplifying Group Terms.

$$Trans(Trans(p_1, p_2), p_3) \qquad\qquad \rightarrow Trans(p_1, Trans(p_2, p_3))$$
$$Trans(Refl, p) \qquad\qquad\qquad\quad\; \rightarrow p$$
$$Trans(p, Refl) \qquad\qquad\qquad\quad\; \rightarrow p$$
$$Trans(Symm(p), p) \qquad\qquad\quad \rightarrow Refl$$
$$Trans(p, Symm(p)) \qquad\qquad\quad \rightarrow Refl$$
$$Symm(Refl) \qquad\qquad\qquad\qquad \rightarrow Refl$$
$$Symm(Symm(p)) \qquad\qquad\qquad \rightarrow p$$
$$Symm(Trans(p_1, p_2)) \qquad\qquad \rightarrow Trans(Symm(p_2), Symm(p_1))$$
$$Trans(Symm(p_1), Trans(p_1, p_2)) \rightarrow p_2$$
$$Trans(p_1, Trans(Symm(p_1), p_2)) \rightarrow p_2$$

Fig. 4. Convergent System for Canonizing Proofs.

Theorem 3 (Canonical Proofs) *Suppose an equality proof p is in canonical form with respect to the rewrite system of Figure 4. Then p is either Refl or in the set $\mathcal{C}$ inductively defined by:*

$$\mathcal{C} ::= \mathcal{U} \;\mid\; Symm(\mathcal{U}) \;\mid\; Trans(\mathcal{U}, \mathcal{C}) \;\mid\; Trans(Symm(\mathcal{U}), \mathcal{C})$$

Furthermore, no assumption is used twice in p, and if $p \in \mathcal{C}$, then there is no $a \in \mathcal{A}$ such that $p :: a = a$.

Proof. The proof is by induction on the form of p. If p is *Refl* or in $\mathcal{U}$, it is clearly in $\mathcal{C}$ and satisfies the condition on assumptions. It does not prove any equation of the form $a = a$, by independence of assumptions. If $p \equiv Symm(p_1)$ for some p_1, then by IH, $p_1 \in \mathcal{C}$. We cannot have $p_1 \equiv Symm(p_2)$, since then p is not canonical. For the same reason, we cannot have $p_1 \equiv Trans(p_2, p_3)$ or $p_1 \equiv Refl$. The only possibility is that $p_1 \in \mathcal{U}$, which shows that $p \in \mathcal{C}$. This also implies that p does not prove any equations of the form $a = a$, since p_1 does not by independence of assumptions. The condition on assumptions is clearly satisfied, since p contains a single assumption.

Finally, if $p \equiv Trans(p_1, p_2)$ for some p_1, p_2, then by IH we may assume $p_1, p_2 \in \mathcal{C}$; p is not canonical if one of p_1 or p_2 is *Refl*. We also cannot have $p_1 \equiv Trans(p_3, p_4)$, since p would not be canonical. Similar considerations show that the only possibilities are $p_1 \equiv u \in \mathcal{U}$ or $p_1 \equiv Symm(u) \in Symm(\mathcal{U})$. This shows $p \in \mathcal{C}$. To show that the condition on assumptions is satisfied by p, we have by IH that it is satisfied by p_2. It now suffices to show that u cannot occur in p_2. Suppose $p_1 :: a = b$ for some $a, b \in \mathcal{A}$. Since p_1 contains an assumption, a and b are, in fact, the unique atoms such that $p_1 :: a = b$. Suppose u occurs in p_2. Since $p_2 \in \mathcal{C}$, this means that there is a sequence $L_1, \ldots, L_n$ of proofs in $\mathcal{U} \cup Symm(\mathcal{U})$ such that $p_2 \equiv Trans(L_1, Trans(\ldots, Trans(L_n, q)))$ for some $q \in \mathcal{C}$; where $L_n \equiv u$ or $L_n \equiv Symm(u)$. Suppose $L_n \equiv p_1$. If $n = 1$, then the only way $p \ (\equiv Trans(p_1, Trans(p_1, q)))$ can prove a theorem is if $a \equiv b$, which contradicts independence of assumptions. If $n > 1$, then $Trans(L_1, Trans(\ldots, Trans(L_{n-2}, L_{n-1}))) :: b = a$. This contradicts independence, since this latter term and L_n do not contain the same assumption by IH. Suppose now that $L_n \not\equiv p_1$, and hence, $L_n :: b = a$. Suppose $n > 1$. We have $Trans(L_1, Trans(\ldots, Trans(L_{n-2}, L_{n-1}))) :: b = b$. If $n = 2$ then $L_{n-1} :: b = b$, contradicting independence of assumptions. Otherwise, $Trans(L_1, Trans(\ldots, L_{n-2})) :: b = x$ for some x, and $L_{n-1} :: x = b$. This again contradicts independence of assumptions. A similar argument shows that p does not prove any equation of the form $b = b$. If $n = 1$, then $p \equiv Trans(p_1, Trans(L_1, q))$, where either $p_1 = Symm(L_1)$ or $L_1 = Symm(p_1)$. In either case, p is not in canonical form.

Completeness now follows easily:

Proof (Completeness (Theorem 2)). Assuming $p_1, p_2 \in \hat{\mathcal{P}}$ prove a theorem in common, we must show $p_1 \cong p_2$. By Soundness (Theorem 1), we may assume p_1 and p_2 are in the canonical form of Theorem 3. Suppose $p_1 \not\equiv p_2$. Then neither one can be *Refl*, since by Theorem 3, *Refl* is the only proof in that canonical form which proves an equation of the form $a = a$. So $p_1, p_2 \in \mathcal{C}$. Since p_1 and p_2 both prove a theorem and both contain an assumption, by Lemma 2, we may suppose they both prove just $a = b$ (where $a \not\equiv b$). Suppose $p_2 \in \mathcal{U} \cup Symm(\mathcal{U})$. Then independence of assumptions is violated, since $p_1 :: a = b$ but $p_1 \not\equiv p_2$. Suppose $p_2 \equiv Trans(L, Trans(L_1, \ldots L_n))$. Then $L :: a = x$ and $Trans(L_1, \ldots L_n) :: x = b$ for some x. Hence,

$$Symm(Trans(Trans(L_1, \ldots L_n), Symm(p_1))) :: a = x$$

which again contradicts independence of assumptions (since $L \in \mathcal{U} \cup Symm(\mathcal{U}$ also proves $a = x$).

6 A Linear-Time Strategy for Canonizing Proofs

In this Section, we show that if the rules of Figure 4 are applied with a leftmost outermost strategy, a given equality proof can be canonized in a number of steps linear in its size. Let **Collapse** be the set of all rewrite rules from the Figure which have either a variable or a constant on the rhs. We call the remaining two rules *RAssoc* and *InverseIn*, respectively:

$$Trans(Trans(p_1, p_2), p_3) \rightarrow Trans(p_1, Trans(p_2, p_3))$$
$$Symm(Trans(p_1, p_2)) \rightarrow Trans(Symm(p_2), Symm(p_1))$$

Define the *right-linear spine* of a proof p to be the longest position π of the form 1^* such that $p|_{\pi'}$ is a *Trans* expression, for every prefix π' of π. The internal nodes of a proof are those of its subexpressions that are *Trans* or *Symm* expressions. A node p' is on the right-linear spine π of a proof p if there is a prefix π' of π such $p|_{\pi'} \equiv p'$. Now the leftmost outermost strategy decreases the measure $(m_1(p), m_2(p), m_3(p))$ in the lexicographic combination of the usual arithmetic ordering, where the components are:

$m_1(p)$ The number of subterms occuring in some subterm of p of the form $Symm(p')$.

$m_2(p)$ The size (denoted $|p|$) of p.

$m_3(p)$ The number of internal nodes not on the right-linear spine of p.

Note that the whole measure is bounded above by $(|p|, |p|, |p|)$. We now show that every rule decreases this measure. *InverseIn* decreases m_1. This is because we are using an outermost strategy, and so if *InverseIn* applies to a subterm, that subterm is not itself contained in a *Symm* node at a shorter position in p. No other rule increases m_1, no matter what strategy is used. Note that when *InverseIn* decreases m_1, it causes m_2 to increase by 1. No other rule can increase m_2, and the **Collapse** rules all decrease it, without increasing m_3. This is again true no matter what strategy is used. The rule *RAssoc* decreases m_3. This is so again because we are using an outermost strategy. If *RAssoc* applies to a subterm of p, that subterm must actually occur on the right-linear spine of p. If not, it is either immediately beneath some *Symm* node, in which case *InverseIn* applies; or else it is the left child of some other *Trans* node. That other *Trans* node cannot be on the right-linear spine, since if it were, *RAssoc* would have been applied to it using our leftmost outermost strategy. An inductive argument based on similar reasoning shows that other *Trans* node cannot be off the right-linear spine. Hence, the original subterm we considered must have indeed been on the right-linear spine. And then the length of that spine is increased by one by applying *RAssoc*. We can then bound the number of steps to canonize p above by $4|p|$. This is so since each component is decreased linearly from an amount bounded by $|p|$, with only the linear increase in m_2 caused by applying *InverseIn* to account for additionally.

7   Canonizing Proofs Without a Strategy

In this Section, we analyze the complexity of canonization if no strategy is used. Canonization of term p can take time at least cubic in $|p|$. To see this, first observe that right associating a left-associated term of size n takes $O(n^2)$ time if a leftmost innermost strategy is used. This is because with an innermost strategy, each assumption starting with the third one from the left must be pushed past all the assumptions to its left. This takes $\Sigma_{i=3}^{n} i = O(n^2)$ time. Now consider the following example

$$Symm_1(\ldots(Symm_{\frac{n}{2}}(Trans(\ldots(Trans(a_1, a_2), a_3), \ldots), a_{\frac{n}{2}}))\ldots)$$

where $a_1, \ldots, a_{\frac{n}{2}}$ are assumptions. For each instance of the inverse operator $Symm$, it obviously takes (at least) linear time to distribute it inwards to the innermost positions. Since *InverseIn* swaps the positions of its arguments, each distribution of *Symm* into a proof in completely right associated form results in a completely left-associated form. The proof then has to be put into right associated form again. If we alternate pushing inverses in and right associating fully left associated terms, the overall time complexity is $O(n^3)$. The following theorem shows that this is, in fact, the worst case.

Theorem 4 (Analysis with no strategy) *It takes $O(|p|^3)$ steps to normalize a proof p using the rules of Figure 4 (without a fixed strategy).*

Proof. Let *TransPos(p)* be the set of positions in p at which there is a *Trans* operator. Let *lefts(π)* be the number of 0s in position π, and let *invs_above(π, p)* be the number of prefixes of π at which p has a *Symm* operator. Now define the following three measures:

$$m_1(p) = \Sigma_{\pi \in TransPos(p)} \ invs\_above(\pi, p),$$
$$m_2(p) = \Sigma_{\pi \in TransPos(p)} \ lefts(\pi),$$
$$m_3(p) = |p|$$

The claim is that each rule reduces the measure

$$m(p) = (m_1(p), m_2(p), m_3(p))$$

in the lexicographic combination of the usual less-than relation on natural numbers. **Collapse** rules clearly reduce m_3, and can readily be seen to preserve m_1 and m_2. *InverseIn* reduces m_1, and *RAssoc* maintains m_1 while reducing m_2.

We obtain a bound of $2|p|^3$ for the number of rewrite steps to normalize p. This is done using a more refined analysis of the changes to the measure m, presented by the table of Figure 5. Each row bounds the effect of a rule or rules on $m(p)$, for an arbitrary proof p, by showing the worst-case (slowest decrease) change on the measure when p is rewritten to some p'. In the worst case, *RAssoc* rewrites *Trans(Trans(p_1, p_2), p_3)* to *Trans(p_1, Trans(p_2, p_3))* where p_1 is an assumption. This is the worst case because p_1 then contributes nothing to $m_2(p)$, since its position is not in *TransPos(p)*. If p_1 contained a *Trans* node, then m_2 would decrease by more than 1. As it is, the only decrease to $m_2(p)$ is due to the fact that the *Trans* node at position 1 in p' is to the left of one fewer *Trans* nodes than the node at position 0 in p. The worst case shown by Figure 5 for *InverseIn* occurs when the rule is applied at the top position of *Symm(Trans(p_1, p_2))*, where p_1 is an assumption and p_2 is hence a term of size $|p| - 3$. In this case, m_2 increases by $|p| - 3$, because p_2 occurs to the left of one more *Trans* node in the resulting term than in the original term p.

To obtain the $2|p|^3$ bound, we next observe that $m_1(p)$ and $m_2(p)$ are both bounded by $|p|^2$ for any p. Each use of *InverseIn* results in a decrease by 1 in m_1 and an increase by $|p| - 3$ in m_2. To offset the latter increase, *RAssoc* will have to be used some number of times bounded by $|p|$. Hence, the overall number of steps is bounded by the sum of $|p|^2$ for the initial value of m_2; $|p|^3$ to reduce all the additions to m_2 caused by reducing m_1 ($|p|$ for each reduction to m_1, which

Rule(s)	Starting measure	Ending measure
Collapse	(m_1, m_2, m_3)	$(m_1, m_2, m_3 - 1)$
RAssoc	(m_1, m_2, m_3)	$(m_1, m_2 - 1, m_3)$
InverseIn	(m_1, m_2, m_3)	$(m_1 - 1, m_2 + |p| - 3, m_3 + 1)$

Fig. 5. How the rules change the measure $m(p)$ of a proof p.

is bounded by $|p|^2$); $|p|^2$ to offset the additions to m_3 incurred by *InverseIn*; and $|p|$ for the initial value of m_3. For any p with $|p| > 2$, this sum is bounded above by $2|p|^3$. (The only reducible term of size less than or equal to 2 is *Symm(Refl)*, which reduces to *Refl* in just one step.)

8   Minimal Proofs from Union-Find

This Section presents an approach to producing minimal proofs from the well-known union-find algorithm [4, Chapter 22]. Proofs are minimal in the sense that they use the unique minimal subset of independent assumptions from which a given equation can be derived. Recall that union-find maintains equivalence classes of atoms in balanced, lazily path-compressed trees. Each atom has an associated *find pointer* which points towards the root of its tree. Roots of trees have null find pointers.

We obtain minimal proofs from union-find by first instrumenting the code for union and find to maintain proofs (cf. [15, Chapter 5]). Unioning the equivalence classes for atoms x and y requires a proof that $x = y$. Here, such proofs are just assumptions from $\mathcal{U}$. Finding the representative y of x's equivalence class produces a proof that $x = y$. Each non-null find pointer has an associated proof. We maintain the invariant that if x's find pointer points to y, then the associated proof is a proof of $x = y$. This invariant is maintained as illustrated in Figures 6 and 7. Find pointers are denoted with solid arrows, and the associated proofs are written (using the compact group notation) next to them. In Figure 7, the dotted arrow is for an assumption given to union. The proof produced for a call to find for atom x is just the proof associated with x's find pointer after path compression has modified it to point directly to the root of x's tree. If x is the root of its tree, the proof is just *Refl*.

On top of proof-producing union and find, we define a check function, that checks whether or not atoms x and y are equal under the assumptions given to union. If they are equal, the function produces a minimal proof of $x = y$.

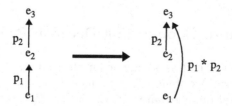

Fig. 6. Maintaining proofs during path compression.

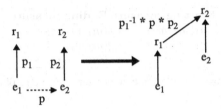

Fig. 7. Maintaining proofs during unions.

The implementation is as follows. We call find on x and y. If they have the same representative z, then find produces proofs p_1 and p_2 of $x = z$ and $y = z$, respectively. To compute the minimal proof that $x = y$, we simply canonize the proof $Trans(p_1, Symm(p_2))$ using the rules of Figure 4. Since we can canonize a proof using the strategy of Section 6 in time linear in the proof's size, it suffices to bound the size of $Trans(p_1, Symm(p_2))$ to bound the additional time needed for producing the minimal proof.

We bound the size of proofs returned by find as follows. Define the *proof to the root* for a non-root atom x in a union-find tree to be $Trans(p_1, Trans(p_2, \ldots, p_n))$, where $p_1, \ldots, p_n$ are the proofs associated with the find pointers on the path from x to the root of the tree. Clearly the size of the proof to the root for x is the size of the proof returned by find for x. Suppose that $T(n)$ is a bound on the size of the biggest proof to the root in a tree of size n maintained by union-find. Clearly the size of the biggest proof to the root is not affected by path compression. So consider the effect of doing a union. The worst case is when two trees of equal size n are merged. The size of the new tree is $2 * n + 1$. The proof associated with the new find pointer (see Figure 7) is clearly bounded by $2 * T(n) + 4$, since it consists of one $Symm$ node, two $Trans$ nodes, an assumption, and two proofs to the roots of the merged trees. The size of the biggest proof to the root in the resulting tree is hence $3 * T(n) + 4$. This is because all the paths in one of the merged trees have been augmented by a find pointer whose associated proof is of size $2 * T(n) + 4$. So $T(n)$ must satisfy $T(2 * n + 1) = 3 * T(n) + 4$. Textbook techniques yield a solution to this recurrence of $T(n) = O(n^{\log_2 3})$. In the worst case, when the size of the minimal proof is $O(n)$, this result is quite close to the result of $O(n \ log n)$ obtained in [10]. It must be noted, however, that in that work, minimal proofs of size k are obtained in $O(k \ log n)$ time, which is clearly better than the bound obtained here, if $k \ll n$ or n is very large. The advantage of the approach presented here is its simplicity: the cited work requires rather subtle additions to union-find to compute minimal proofs.

9   Application to Cooperating Decision Procedures

In this Section, the above ideas are applied in the context of the CVC ("Cooperating Validity Checker") system to obtain major performance improvements on benchmark formulas from hardware verification [16]. In review, CVC and similar tools like CVC Lite (CVC's successor) and ICS separate boolean reasoning from theory-specific reasoning [2, 6]. A fast propositional SAT solver tries to find an assignment to the propositional skeleton of the input formula, or possibly to an equisatisfiable CNF version. Either at each step as the assignment is generated or once it is found, cooperating decision procedures (DPs) are consulted about the consistency of the assignment. For example, a large formula might contain many equalities or other interpreted atomic formulas between ground terms. The SAT solver chooses an assignment to some of those formulas which makes the goal formula satisfiable, if the meanings of the interpreted predicates are ignored. The cooperating DPs then determine if that assignment is consistent with the meanings of the interpreted predicates. If not, a subset of the assignment is identified as inconsistent and returned to the SAT solver as a *conflict clause*. Conflict clauses are maintained during subsequent search for a satisfying assignment, and can greatly prune the search space [1]. Smaller conflict clauses are always more effective than their supersets at pruning the search space.

CVC leverages its infrastructure for generating proofs to track assumptions. The basic idea is that when the cooperating DPs discover that an assignment proposed by the SAT solver is inconsistent, a subset of the assumptions used in that assignment can be determined by inspecting an explicit proof of the contradiction. Such proofs are generated by CVC's DPs. In its fastest mode before the present work, the DPs generate not a full proof, but an *abstract proof* consisting just of the assumptions that would have appeared in the full proof [1]. This greatly reduces the time required to manipulate proofs and extract conflict clauses.

For the first experiments reported in this Section, CVC was modified to canonize equality proofs according to the linear-time strategy of Section 6. Note that this requires full proofs instead of abstract proofs. Each time CVC's DPs try to build an equality proof, that proof is put in canonical form. It turns out that an additional transformation on proofs is required to get significant benefits for CVC's equational reasoning. CVC's congruence closure algorithm rewrites asserted disequalities each time one of the sides is asserted equal to something else. The modified disequality is then asserted. The resulting proofs of contradictions turn out often to involve subproofs of the following form:

$$\frac{a=b \quad \dfrac{a=c \quad b=d}{(a=b) \Leftrightarrow (c=d)}SubstEquiv}{c=d}EquivMP$$

Such subproofs are rewritten to ones of the following form in order to take advantage of the canonization algorithm:

$$\frac{\dfrac{a=c}{c=a}Symm \quad \dfrac{a=b \quad b=d}{a=d}Trans}{c=d}Trans$$

Algebraically, this corresponds to adding the following rewrite rule to the rules of Figure 4:

$$EquivMP(p_1, SubstEq(p_2, p_3)) \;\rightarrow\; Trans(Symm(p_2), Trans(p_1, p_3))$$

Adding this rule to those of Figure 4 leads to no new critical pairs, and the resulting system is obviously still terminating. Hence, it remains convergent. Leftmost outermost application is readily seen to remain linear time.

Figure 8 compares the number of *decisions* (the number of times a value was chosen for an atomic formula in a propositional assignment) and wallclock time on 6 benchmark formulas from hardware verification, using the approach with abstract proofs and the approach with canonized full proofs. The sizes of the benchmark formulas themselves in ASCII text are also listed. We see that canonization reduces the number of decisions from anywhere from 15% to 65% on these benchmarks, but in all but one case (pp-dmem) requires more time overall. Profiling the largest benchmark (pp-regfile) reveals that 60% of the overall runtime is going to proof canonization. This is an unacceptably high price to pay for the search space pruning we are achieving. We address this problem, but first consider data on the canonization itself.

Benchmark	Size (KB)	dec. orig	time orig (s)	dec. canon	time canon (s)
dlx-regfile	70.9	2807	2.1	2430	5.6
dlx-dmem	71.0	1336	1.0	1025	1.8
pp-regfile	2480.0	115197	295.7	43610	336.9
pp-dmem	1842.2	25928	68.1	11991	58.2
pp-bloaddata	314.2	4060	1.7	3502	3.3
pp-TakenBranch	1842.3	15364	26.1	9616	33.2

Fig. 8. Comparison of original CVC and CVC with canonization of equality proofs on hardware verification benchmarks ("dec." stands for decisions).

Figure 9 breaks out the number of uses of the different rewrite rules during canonization. Note that the numbers for the **Collapse** rules do not count uses of the following rules, where it is never necessary to build the left hand side at all when canonizing equality proofs as they are being built:

$$1 * x \rightarrow x$$
$$x * 1 \rightarrow x$$
$$1^{-1} \rightarrow 1$$

We address the problem of spending too much time canonizing full equality proofs as follows. Instead of canonizing equality proofs and then extracting the assumptions from proofs of contradictions (with canonical equality subproofs), we extract assumptions from uncanonized proofs of contradictions in a way that incorporates the algebra of equality proofs. In more detail, for each equality subproof occurring in a proof of a contradiction, we compute the difference between the number of positive and the number of negative occurrences of each assumption in that subproof. An occurrence is positive if it beneath an even number of uses of *Symm*, and negative otherwise. During this computation, we queue up non-equality subproofs of the equality proof for later consideration. It is an easy lemma that the difference between the number of positive and the

Benchmark	Collapse	RAssoc	InverseIn	Total
dlx-regfile	87515	156463	85961	329939
dlx-dmem	25576	46584	22721	94881
pp-regfile	4620398	10391230	3933697	18945325
pp-dmem	964806	1885856	737122	3587784
pp-bloaddata	41329	61975	31831	135135
pp-TakenBranch	486355	834499	273832	1594686

Fig. 9. Number of rewrites by category for canonization of equality proofs when running modified CVC on the given benchmarks.

Benchmark	dec. orig	time orig (s)	dec. smart	time smart (s)
dlx-regfile	2807	2.1	2430	2.5
dlx-dmem	1336	1.0	1048	0.9
pp-regfile	115197	295.7	44547	121.9
pp-dmem	25928	68.1	11899	23.7
pp-bloaddata	4060	1.7	3461	2.1
pp-TakenBranch	15364	26.1	11928	24.6

Fig. 10. Comparison of original CVC and CVC with smart abstraction of assumptions from uncanonized proofs of contradictions.

number of negative occurrences of an assumption is 0 iff that assumption does not occur in the canonized version of the subproof (cf. Theorem 3).

With this "smart" abstraction of otherwise uncanonized proofs of contradictions, we obtain the favorable results in Figure 10. Wallclock times range from slightly slower for some of the smaller benchmarks to 60% faster in the case of the two toughest benchmarks. Profiling the pp-regfile benchmark reveals that smart abstraction now takes a much more acceptable 10% of the overall runtime.

The numbers of decisions used for the smart version are slightly different from the numbers of decisions for the canonizing version (Figure 8). Careful inspection of trace data from canonization shows that in some cases, a **Collapse** rule applies to eliminate an entire non-equality subproof. This sort of elimination will not be possible in general in the case of smart abstraction. And once different conflict clauses begin to be added, the behaviors of the two versions of CVC are highly likely to diverge.

10   Conclusion and Future Work

Theorem equivalence of equality proofs using independent assumptions is completely characterized by the standard axioms for free groups. Using a standard completion of the group axioms taken as rewrite rules, equality proofs can be put into canonical form. This form is minimal in the sense that the fewest possible assumptions are used. Canonization can be performed using a simple strategy in time linear in the size of the equality proof. Without a strategy, canonization can take cubic time in the proof's size. Using these results, the standard union-find algorithm for ground equational reasoning can be instrumented to produce minimal proofs of equations $x = y$ in additional time $O(n^{\log_2 3})$, where n is the size of the equivalence class of x and y. Using the algebra of equality proofs, major improvements were achieved in the performance of the CVC tool on hardware verification benchmark formulas. The approach is attractive, because rather than carefully modifying specific algorithms to produce minimal proofs (as in [5, 10]), we apply a simple general-purpose technique. This can result, for example, in improvements for other decision procedures, like arithmetic, that do equational reasoning.

The most exciting avenue for future work is to extend the algebra of equality proofs to an algebra of congruence proofs. In the case of unary function symbols, the congruence proof rule functions like a homomorphism: $Congr(Trans(p_1, p_2))$ $= Trans(Congr(p_1), Congr(p_2))$. If this observation can be generalized appropriately to higher arities, we may be able to canonize congruence proofs. This promises further speedups for tools like CVC, which rely heavily on congruence closure. Careful inspection of some of the conflict clauses generated reveal cases where out of large clauses (e.g., 21 literals) derived from proofs using congruence rules, only a very small number (e.g., 3) are needed for inconsistency.

The authors wish to thank the anonymous reviewers for their helpful comments, as well as Grigori Mints and David Dill for earlier feedback on the ideas.

References

1. C. Barrett, D. Dill, and A. Stump. Checking Satisfiability of First-Order Formulas by Incremental Translation to SAT. In *14th International Conference on Computer-Aided Verification*, 2002.
2. Clark Barrett and Sergey Berezin. CVC Lite: A new implementation of the cooperating validity checker. In *Proceedings of the 16th International Conference on Computer Aided Verification*, 2004.
3. Jerry R. Burch and David L. Dill. Automatic verification of pipelined microprocessor control. In David L. Dill, editor, *Conference on Computer-Aided Verification*, volume 818 of *Lecture Notes in Computer Science*, pages 68–80. Springer-Verlag, 1994.
4. T. Cormen, C. Leiserson, and R. Rivest. *Introduction to Algorithms*. MIT Press, 1992.
5. L. de Moura, H. Rueß, and N. Shankar. Justifying Equality. In S. Ranise and C. Tinelli, editors, *2nd International Workshop on Pragmatics of Decision Procedures in Automated Reasoning*, 2004.
6. J. Filliâtre, S. Owre, H. Rueß, and N. Shankar. ICS: integrated canonizer and solver. In G. Berry, H. Comon, and A. Finkel, editors, *13th International Conference on Computer-Aided Verification*, 2001.
7. D. Knuth and P. Bendix. Simple Word Problems in Universal Algebras. In J. Leech, editor, *Computational Problems in Abstract Algebra*, pages 263–297. Pergamon Press, 1970.
8. S. Lahiri, R. Bryant, A. Goel, and M. Talupur. Revisiting Positive Equality. In *Tools and Algorithms for the Construction and Analysis of Systems*, volume 2988 of *LNCS*, pages 1–15. Springer-Verlag, 2004.
9. G. Nelson and D. Oppen. Fast decision procedures based on congruence closure. *Journal of the Association for Computing Machinery*, 27(2):356–64, 1980.

10. R. Nieuwenhuis and A. Oliveras. Union-Find and Congruence Closure Algorithms that Produce Proofs. In S. Ranise and C. Tinelli, editors, *2nd International Workshop on Pragmatics of Decision Procedures in Automated Reasoning*, 2004. (short paper).

11. R. Nieuwenhuis and A. Oliveras. Proof-producing Congruence Closure. In J. Giesl, editor, *16th International Conference on Rewriting Techniques and Applications*, 2005. (under review).

12. A. Pnueli, Y. Rodeh, O. Shtrichman, and M. Siegel. Deciding Equality Formulas by Small Domains Instantiations. In *Proceedings of the 11th International Computer-Aided Verification Conference*, volume 1633 of *Lecture Notes in Computer Science*, pages 455–469. Springer-Verlag, 1999.

13. H. Ruess and N. Shankar. Deconstructing Shostak. In *16th IEEE Symposium on Logic in Computer Science*, 2001.

14. R. Shostak. Deciding combinations of theories. *Journal of the Association for Computing Machinery*, 31(1):1–12, 1984.

15. A. Stump. *Checking Validities and Proofs with CVC and flea*. PhD thesis, Stanford University, 2002. available from http://www.cs.wustl.edu/~stump/.

16. A. Stump, C. Barrett, and D. Dill. CVC: a Cooperating Validity Checker. In *14th International Conference on Computer-Aided Verification*, 2002.

17. M. Velev and R. Bryant. Superscalar Processor Verification Using Efficient Reductions of the Logic of Equality with Uninterpreted Functions. In L. Pierre and T. Kropf, editors, *Correct Hardware Design and Verification Methods*, volume 1703 of *Lecture Notes in Computer Science*, pages 37–53. Springer-Verlag, 1999.

On Computing Reachability Sets
of Process Rewrite Systems

Ahmed Bouajjani and Tayssir Touili

LIAFA, University of Paris 7, 2 place Jussieu, 75251 Paris cedex 5, France
{abou,touili}@liafa.jussieu.fr

Abstract. We consider the problem of symbolic reachability analysis of
a class of term rewrite systems called Process Rewrite Systems (PRS).
A PRS can be seen as the union of two mutually interdependent sets
of term rewrite rules: a prefix rewrite system (or, equivalently, a pushdown system), and a multiset rewrite system (or, equivalently, a Petri
net). These systems are natural models for multithreaded programs with
dynamic creation of concurrent processes and recursive procedure calls.
We propose a generic framework based on tree automata allowing to
combine (finite-state automata based) procedures for the reachability
analysis of pushdown systems with (linear arithmetics/semilinear sets
based) procedures for the analysis of Petri nets in order to analyze PRS
models. We provide a construction which is parametrized by such procedures and we show that it can be instantiated to (1) derive procedures
for constructing the (exact) reachability sets of significant classes of PRS,
(2) derive various approximate algorithms, or exact semi-algorithms, for
the reachability analysis of PRS obtained by using existing symbolic
reachability analysis techniques for Petri nets and counter automata.

1 Introduction

Software verification is one of the main challenges in computer-aided verification.
Among the difficulties to face when dealing with this problem, we can mention
the fact that (1) programs manipulate data ranging over infinite domains, and
the fact that (2) programs may have complex control structures. Concerning the
first point, abstraction techniques (such as predicate abstraction) can be used to
obtain abstract programs on finite data domains (see, e.g., [5]). As for the second
point, standard model-checking algorithms can be used if the control structure of
the program is finite. However, programs may have unbounded control structures
due to, e.g., (unbounded depth) recursive procedure calls, and dynamic creation
of concurrent processes (threads). Therefore, we need to extend the capabilities
of automatic verification to deal with infinite-state models which capture the
behaviors of such (abstract) programs.

Recently, model-checking techniques for infinite-state systems have been successfully used in this context. Pushdown systems have been proposed as a natural model for sequential programs with procedure calls [23, 26], whereas Petri
nets have been used to reason about multi-threaded programs without procedure calls (in this case, each thread is a finite-state system, but there may be
an arbitrary number of them running at the same time) [4, 19]. In both cases,

J. Giesl (Ed.): RTA 2005, LNCS 3467, pp. 484–499, 2005.

symbolic reachability analysis techniques are used to verify properties, basically safety properties, on these models.

In this paper, our aim is to define reachability analysis techniques for models which subsume pushdown systems and Petri nets, allowing to deal with multi-threaded programs with procedures calls. We consider models based on term rewrite systems called *Process Rewrite Systems* (PRS). These models can be seen as combinations of prefix and multiset rewrite systems.

The construction of the reachability sets of PRS is a hard problem, in particular because they subsume Petri nets which are known to be not semilinear in general (i.e., their reachability sets cannot be defined in Presburger arithmetics) [30]. However, there exist well-known classes of semilinear Petri nets (see, e.g., [25]), and moreover, several algorithms or semi-algorithms have been developed (and implemented) recently for computing exact or upperapproximate reachability sets of counter automata, using various representation structures for linear arithmetical constraints (polyhedra, automata, etc), and fixpoint acceleration techniques [2, 7, 8, 14, 17, 18, 28, 39].

Then, our approach is to design a framework where all existing semilinear sets-based symbolic analysis procedures for Petri nets (or counter automata) can be integrated with existing automata-based symbolic analysis algorithms for pushdown systems in order to derive procedures for the reachability analysis of PRS models.

Our main contribution is a *generic* procedure which constructs a tree automata-based representation of the reachability set of a given PRS by invoking procedures for the analysis of prefix and multiset rewrite systems. While the procedure for analyzing prefix rewrite systems can be considered as fixed (e.g., the one of [15]), the construction is parameterized by a procedure for the analysis of multiset rewrite systems using semilinear sets (or Presburger arithmetics) for the representation of sets of configurations (markings). This procedure can be:

- an *exact algorithm*, but applicable to some particular subclass of multiset rewrite systems (which is known to be semilinearity preserving), or to some particular subclass of semilinear sets (which is known to be closed under multiset rewriting),
- an *approximate algorithm*, or an *exact semi-algorithm* (for which termination is not guaranteed), but applicable in general to any multiset rewrite system and to any semilinear set of configurations.

The construction we propose allows to derive, for every class of multiset rewrite systems $\mathcal{C}$ for which we have an algorithm (resp. semi-algorithm) for exact (resp. upperapproximate) reachability analysis, an algorithm (resp. semi-algorithm) for exact (resp. upperapproximate) reachability analysis for the class of PRS obtained by combining $\mathcal{C}$ systems with prefix rewrite systems. We show that our construction can be instantiated in such a manner to derive:

- an algorithm for computing the exact forward and backward reachability sets for the synchronization-free PRS (equivalent to the so-called PAD systems) which subsume pushdown systems and synchronization-free Petri nets. As a corollary, we obtain an algorithm for *global model checking* of PAD vs the

EF fragment of the temporal logic CTL, i.e., an algorithm for computing the set of all configurations of a PAD system satisfying some given formula in the logic EF. These results extend all the existing ones concerning symbolic reachability analysis for subclasses of PRS, and solves for the first time the problem of global model checking of EF for PAD systems.
- exact/approximate procedures for the analysis of PRS based on various analysis procedures and invariant generation techniques for Petri nets and counter automata.

In order to characterize the reachability sets of PRS, we use a class of tree automata allowing to define (nonregular) sets of trees with unbounded width, which are closed under commutation of the children of some of their nodes (those corresponding to the parallel operator). This class of automata enjoys all closure and decision properties which are necessary in symbolic reachability analysis.

Related Work: Mayr has proved in [34] that the reachability problem between two fixed terms is decidable using a reduction to the reachability problem of Petri nets. The problem we consider here is the constructibility problem of the (potentially infinite) set of terms which are reachable from a given possibly infinite set of terms.

Symbolic reachability analysis based on (word) automata techniques has been used for model-checking pushdown systems in, e.g., [9, 29]. In [33], this approach is used for the analysis of PA processes, i.e., combination of context-free prefix and multiset rewrite rules (left-hand-sides of the rules are reduced to single symbols). In that work, the authors use finite bounded-width tree automata as symbolic representation structures.

Our construction allows to handle (in particular) the class of PAD systems (or synchronization-free PRS) which is strictly larger than both pushdown systems and PA. The class PAD allows for instance to take into account return values of procedures whereas PA cannot. In the case of a pushdown system, our construction will behave like the algorithms for this class of systems (it will compute the set of all reachable configurations). In the case of PA systems our construction will also compute the precise set of all reachable configurations, whereas the construction in [33] computes only a set of representatives of terms w.r.t. associativity and associativity-commutativity of the sequential and parallel composition. Indeed, the set of all reachable configurations of PA processes is not regular in general in the sense that it cannot be represented by a bounded-width tree automaton. In [12], we have extended the approach of [33] for constructing representatives of the reachability sets to the class of PAD. The construction we give in this paper is more general in the sense that it allows to compute the whole set of all backward/forward reachable configurations, and allows also to solve the problem of global model checking for PAD against the EF fragment of CTL. The decidability problem of PAD vs EF has been shown to be decidable by Mayr [34]. However, his proof is rather complex and concerns only the model checking problem for a single process term against a formula, whereas our approach allows to compute for the first time the whole satisfiability set of a PAD formula.

The application of symbolic reachability analysis of PRS to program analysis has been advocated in [23] for sequential recursive programs using pushdown systems. This approach is extended in [27] to parallel programs using PA systems. In [21] a discussion about the modeling power of PRS is given. In [4, 19], the model of Petri nets is proposed for the verification of multi-threaded programs without procedure calls (threads are finite-state communicating systems).

In [10, 11] we define a different framework for the analysis of concurrent programs with procedure calls based on models which are either communicating pushdown systems [10] or synchronized PA systems (PA with synchronization à-la CCS) [11] The verification problem for these model is undecidable, and therefore we propose an analysis approach based on computing (either finite or commutative) abstractions of the path languages. While the considered models in [10, 11] are more general than those we consider in this paper, their analysis is (by necessity) approximate.

In [35], another approach for dealing with concurrent programs with procedures is proposed based on combining the so-called summarization technique (control location reachability in pushdown systems) with a partial order-like approach. Basically, the authors show that under some restrictions on the occurrences of synchronizations, the program has a representative (modulo some action commutations) where the summarization technique can be applied. When these conditions are not satisfied, the technique may not terminate. It is not completely clear how our models are related to the class of programs for which the algorithm of [35] terminates. However, in principle our symbolic techniques are more general than summarization-based techniques since they allow to construct the whole infinite set of reachable configurations. Moreover, the approach we present in this paper allows to deal with dynamic creation of concurrent processes.

Finally, the tree automata we use in this paper are extensions of the hedge automata [13] recognizing sets of trees with unbounded width (i.e., tree languages closed under associativity). Our automata, called commutative hedge automata, are equivalent to other kinds of automata defined recently in the literature for instance in [16, 32, 37]. However, as far as we know, our use of commutative hedge automata in the context of the analysis of process rewrite systems is original.

2   Models

We introduce hereafter Process Rewrite Systems (PRS for short) [34]. Our presentation of PRS does not follow the standard one given in [34]. In fact, we adopt the view that PRS are sets of (mutually dependent) multiset and prefix rewrite rules. It can be proved that every (standard) PRS can be transformed into an equivalent one that has the form we consider here [34].

2.1   Process Terms

Let $Const = \{X, Y, \ldots\}$ be a set of process constants, and let T_p be the set of process terms t defined by the following syntax:

$$t ::= 0 \mid X \mid t \odot t \mid t \| t$$

where, intuitively, 0 corresponds to the idle process, and "$\odot$" (resp. "$\|$") represents the sequential composition (resp. parallel composition) operator. We use ω to denote in a generic way $\odot$ or $\|$. We denote by $\bar{\omega}$ the operator $\odot$ (resp. $\|$) if $\omega = \|$ (resp. $\omega = \odot$).

Process terms are considered modulo the following algebraic properties: associativity of "$\odot$", commutativity and associativity of "$\|$", and neutrality of 0 w.r.t. both "$\odot$" and "$\|$". Let $\simeq$ be the equivalence relation on T_p induced by these properties.

Process terms in *canonical form* are terms t defined by:

$$t ::= 0 \mid s \mid p$$
$$s ::= X \mid p_1 \odot p_2 \cdots \odot p_n, \ n \geq 2$$
$$p ::= X \mid s_1 \| s_2 \cdots \| s_n, \ n \geq 2$$

It can easily be seen that every term has an $\simeq$-equivalent term in canonical form. From now on, we work only with terms in canonical form.

A *seq-term* (resp. *paral-term*) is either 0, a constant X, or a term of the form $p_1 \odot \cdots \odot p_n$, called $\odot$-*rooted term* (resp. $s_1 \| \cdots \| s_n$, called $\|$-*rooted term*), for $n \geq 2$. A *flat seq-terms* (resp. *flat paral-terms*) is a term of the form $X_1 \odot \cdots \odot X_n$ (resp. $X_1 \| \cdots \| X_n$) for $n \geq 0$ (the case $n = 0$ corresponds to the term 0, and the case $n = 1$ corresponds to a process constants X).

2.2   Process Rewrite Systems

A PRS is a set of rewrite rules of the forms:

$$X_1 \odot \cdots \odot X_n \hookrightarrow Y_1 \odot \cdots \odot Y_m \tag{1}$$
$$X_1 \| \cdots \| X_n \hookrightarrow Y_1 \| \cdots \| Y_m \tag{2}$$

for $n, m \geq 0$. Rules of the form (1) (resp. (2)) are called $\odot$-*rules* (resp. $\|$-*rules*).

A PRS R induces a transition relation $\rightarrow_R$ over T_p defined as the smallest relation between process terms such that:

1. if $t_1 \hookrightarrow t_2$ is a rule in R, then $t_1 \rightarrow_R t_2$,
2. if $t_1 = t \omega t_2$, and $t \rightarrow_R t'$, then $t_1 \rightarrow_R t' \omega t_2$,
3. if $t_1 \simeq t_1'$, $t_1' \rightarrow_R t_2'$, and $t_2' \simeq t_2$, then $t_1 \rightarrow_R t_2$,

Let $Post_R(t) = \{t' \in T_p \mid t \rightarrow_R t'\}$, and $Pre_R(t) = \{t' \in T_p \mid t' \rightarrow_R t\}$. As usual, $Post_R^*(t)$ and $Pre_R^*(t)$ denote respectively the reflexive-transitive closures of $Post_R(t)$ and $Pre_R(t)$. We omit the subscript R when it is understood from the context. Also, we write sometimes $R(t)$ instead of $Post_R(t)$, and similarly $R^*(t)$ instead of $Post_R^*(t)$. These definitions and notations can be extended to sets of terms in the obvious manner. Given a system R, we denote by R^{-1} the system obtained by swapping the left-hand-sides and right-hand-sides of the rules of R. Notice that for every set of process terms $\mathcal{L}$, $Pre_R^*(\mathcal{L}) = Post_{R^{-1}}^*(\mathcal{L})$.

2.3 Subclasses of PRS and Program Modeling

PRS is a natural formal model for multithreaded programs with procedure calls (see, e.g., [21, 23, 27]). It subsumes several well-known classes of (infinite-state) models. We mention hereafter the ones which are relevant to this paper and mention their relevance in program modeling.

- **Prefix rewrite systems** are sets of $\odot$-rules. They are equivalent to **pushdown systems (PDS)**. Prefix rewrite systems are models for sequential programs with procedure (recursive) calls ranging over finite data domains (see, e.g., [23, 36]): values of global variables correspond to control states, local variables and program control points are modeled as stack symbols which are stored at each recursive call.
- **BPA processes** (or context-free processes) are prefix rewrite systems where all the left-hand-sides of the rules are process constants. They are equivalent to pushdown systems with a single control state. Therefore, they do not allow to take into account global variables.
- **Multiset rewrite systems** are sets of $\parallel$-rules. These systems are equivalent to **Petri nets (PN)**. They are natural models of multithreaded programs (with arbitrary number of parallel finite-state systems) (see, e.g., [18]).
- **BPP processes** (commutative context-free processes) are multiset rewrite systems where all left-hand-sides of the rules are process constants. They are equivalent to synchronization-free Petri nets.
- **PA processes** are PRSs where all the left-hand-sides of the rules are process constants (i.e., they are the nesting of BPA and BPP systems). PA processes are abstract models for programs with procedure calls and dynamic creation of (asynchronous) parallel processes [27].
- **PAD processes** are PRSs which are the nesting of BPP and prefix rewrite systems (i.e., PRS such that all the left-hand-sides of their rules are seq-terms). PAD systems subsume pushdown systems and PA processes. Contrary to PA processes, they allow to take into account return values of procedure calls.

Let $\mathcal{C}$ be a class of multiset rewrite systems. We denote by **PRS[$\mathcal{C}$]** the class of PRSs that are unions of prefix rewrite systems and multiset rewrite systems in $\mathcal{C}$. For instance, PRS[BPP] is precisely the class PAD.

Given a class $\mathcal{C}$ of PRS, we denote by co$\mathcal{C}$ the *dual* class of $\mathcal{C}$ which consists of all systems R such that R^{-1} is in $\mathcal{C}$. Clearly, PRS = coPRS, and the same holds for prefix and multiset rewrite systems.

3 Tree Automata-Based Symbolic Representations

We use a class of tree automata, called *commutative hedge automata*, for the representation and the manipulation of infinite sets of PRS process terms. The automata we consider extend (bottom-up) hedge automata recognizing sets of arbitrary-width trees [13]. They recognize sets of terms modulo associativity of

⊙ and associativity-commutativity of ∥. Our automata are very close to the ones defined in [16], and can be seen as particular cases of those introduced in [32, 37]. Let us give briefly the intuition behind the definition of these automata: Consider first a "classical" bottom-up tree automaton over fixed-width trees, say binary trees. To accept a tree in a bottom-up manner, the automaton has to find a labelling of the nodes of this tree by control states which (1) puts a final state at the root, and (2) is compatible with rules of the form: either (i) $a \rightarrow q$ allowing to label a leaf a with state q, or (ii) $f(q_1, q_2) \rightarrow q$ allowing to label a tree $f(t_1, t_2)$ with q provided that its subtrees t_1 and t_2 are labeled by q_1 and q_2. Now, assume that f represents an associative operator. Then, we consider that a node corresponding to such an operator can have an arbitrary number of sons. Therefore, we use labelling rules of the form $f(L) \rightarrow q$, where L is a regular language over the alphabet of control states, which allow to label a tree $f(t_1, \ldots, t_n)$ with q provided that each sub-tree t_i is labelled by q_i and the word $q_1 \cdots q_n$ is in the language L. Assume furthermore that f is associative *and* commutative. In such a case, the ordering between sons is not relevant. Therefore, we use rules of the form $f(\varphi) \rightarrow q$, where φ is an arithmetical constraint, allowing to label a tree $f(t_1, \ldots, t_n)$ with q provided that each subtree t_i is labelled by q_i and the number of occurrences of each control state in the word $q_1 \cdots q_n$ satisfies the constraint expressed by φ.

In the sequel, we give the definition of the general class of commutative hedge automata, and then we describe the particular automata which are used for the representation of sets of process terms.

3.1   Preliminaries

Presburger arithmetics is the first order logic of integers with addition and linear ordering. Given a formula φ, we denote by $FV(\varphi)$ the set of its free variables. Let $FV(\varphi) = \{x_1, \ldots, x_n\}$. Then, a vector $\boldsymbol{u} = (u_1, \ldots, u_n) \in \mathbb{Z}^n$ satisfies φ, written $\boldsymbol{u} \models \varphi$, if $\varphi(\boldsymbol{u}) = \varphi[x_i \leftarrow u_i]$ is true. Each formula φ defines a set of integer vectors $[\![\varphi]\!] = \{\boldsymbol{u} \in \mathbb{Z}^n \mid \boldsymbol{u} \models \varphi\}$. Presburger formulas define *semilinear sets* of integer vectors, i.e., finite union of sets of the form $\{\boldsymbol{x} \in \mathbb{Z}^n \mid \exists k_1, \ldots, k_m \in \mathbb{Z}, \boldsymbol{x} = \boldsymbol{v}_0 + k_1\boldsymbol{v}_1 \cdots + k_m\boldsymbol{v}_m\}$, where $\boldsymbol{v}_i \in \mathbb{Z}^n$, for $1 \leq i \leq m$.

Given a word w over an alphabet $\Sigma = \{a_1, \ldots, a_n\}$, the *Parikh image* of w, denoted $Parikh(w)$, is the vector $(|w|_{a_1}, \ldots, |w|_{a_n})$. This definition can be generalized to sets of words (languages) over Σ in the obvious manner.

As usual, a set of words is *regular* if it is definable by a finite-state automaton. The notion of regularity can be transfered straightforwardly to sets of flat seq-terms. Similarly, the notion of semilinearity can be transfered to sets of flat paral-term by associating with a term $X_1 \| \cdots \| X_n$ the vector $Parikh(X_1 \cdots X_n)$.

In the sequel, we will represent by γ a *constraint* which is either a regular language or a Presburger formula. We say that a word $w = a_1 a_2 \ldots a_n$ *satisfies* the constraints γ if $w \in \gamma$ (resp. $Parikh(w) \models \gamma$) when γ is a language (resp. a formula).

3.2 Commutative Hedge Automata

Let $\Sigma = \Sigma' \cup \Sigma_A$ be a finite alphabet, where Σ' is a ranked alphabet, and Σ_A is a finite set of associative operators. We assume that Σ' and Σ_A are disjoint. For $k \geq 0$, let Σ_k denote the set of elements of Σ' of rank k.

Σ-Terms: Let $\mathcal{X}$ be a fixed denumerable set of variables $\{x_1, x_2, \ldots\}$. The set $T_\Sigma[\mathcal{X}]$ of Σ-terms over $\mathcal{X}$ is the smallest set such that:

- $\Sigma_0 \cup \mathcal{X} \subseteq T_\Sigma[\mathcal{X}]$,
- for $k \geq 1$, if $f \in \Sigma_k$ and $t_1, \ldots, t_k \in T_\Sigma[\mathcal{X}]$, then $f(t_1, \ldots, t_k) \in T_\Sigma[\mathcal{X}]$,
- if $f \in \Sigma_A$, $t_1, \ldots, t_n \in T_\Sigma[\mathcal{X}]$ for some $n \geq 1$, and $root(t_i) \neq f$ for every $1 \leq i \leq n$, then $f(t_1, \ldots, t_n) \in T_\Sigma[\mathcal{X}]$, where $root(\sigma) = \sigma$ if $\sigma \in \Sigma_0 \cup \mathcal{X}$, and $root(g(u_1, \ldots, u_m)) = g$.

Terms without variables are called *ground terms*. Let T_Σ be the set of ground terms over Σ. A term t in $T_\Sigma[\mathcal{X}]$ is *linear* if each variable occurs at most once in t. A *context* C is a linear term of $T_\Sigma[\mathcal{X}]$. Let $t_1, \ldots, t_n$ be terms of T_Σ, then $C[t_1, \ldots, t_n]$ denotes the term obtained by replacing in the context C the occurrence of the variable x_i by the term t_i, for each $1 \leq i \leq n$.

Definition of CH-Automata. Let us consider that $\Sigma_A = \Sigma'_A \cup \Sigma'_{AC}$ where Σ'_{AC} is a set of associative and commutative operators. We assume that Σ'_A and Σ'_{AC} are disjoint. Then, a CH-automaton is a tuple $\mathcal{A} = (Q, \Sigma, F, \Delta)$ where:

- Q is a union of disjoint finite sets of states $Q' \cup \bigcup_{f \in \Sigma_A} Q_f$,
- $F \subseteq Q$ is a set of final states,
- Δ is a set of rules of the form:
  1. $a \to q$, where $q \in Q', a \in \Sigma_0$,
  2. $f(q_1, \ldots, q_k) \to q$, where $f \in \Sigma_k, q \in Q'$, and $q_i \in Q$,
  3. $q \to q'$, where $(q, q') \in Q' \times Q' \cup \bigcup_{f \in \Sigma_A} Q_f \times Q_f$,
  4. $f(L) \to q$, where $f \in \Sigma'_A$, $L \subseteq (Q \setminus Q_f)^*$, and $q \in Q_f$,
  5. $f(\varphi) \to q$, where $f \in \Sigma'_{AC}$, $q \in Q_f$, and φ is a Presburger formula such that $FV(\varphi) = \{x_q \mid q \in Q \setminus Q_f\}$.

We define a *move relation* $\to_\Delta$ between ground terms in $T_{\Sigma \cup Q}$ as follows: for every two terms t and t', we have $t \to_\Delta t'$ iff there exist a context C and a rule $r \in \Delta$ such that $t = C[s]$, $t' = C[s']$, and:

- $r = a \to q$, with $s = a$ and $s' = q$, or
- $r = q \to q'$, with $s = q$ and $s' = q'$, or
- $r = f(q_1, \ldots, q_k) \to q$, with $s = f(q_1, \ldots, q_k)$ and $s' = q$, or
- $r = f(L) \to q$, with $f \in \Sigma'_A$, $s = f(q_1, \ldots, q_n)$, $q_1 \cdots q_n \in L$, and $s' = q$, or
- $r = f(\varphi) \to q$, with $f \in \Sigma'_{AC}$, $s = f(q_1, \ldots, q_n)$, $Parikh(q_1 \cdots q_n) \models \varphi$, and $s' = q$.

Let $\xrightarrow{*}_\Delta$ denote the reflexive-transitive closure of $\to_\Delta$. A ground term $t \in T_\Sigma$ is accepted by a state q if $t \xrightarrow{*}_\Delta q$. Let $L_q = \{t \mid t \xrightarrow{*}_\Delta q\}$. A ground term t is accepted by the automaton $\mathcal{A}$ if there is some state q in F such that $t \xrightarrow{*}_\Delta q$. The CH-language of $\mathcal{A}$, denoted by $L(\mathcal{A})$, is the set of all ground terms accepted by $\mathcal{A}$.

By adapting the constructions for hedge automata [13], it is possible to prove the following fact (see [16, 32, 37, 38]):

Theorem 1. *The class of CH-automata is effectively closed under boolean operations. Moreover, the emptiness problem of CH-automata is decidable.*

3.3   CH-Automata for PRS Process Terms

We consider PRS process terms as trees and use CH-automata to represent sets of such trees. Indeed, the set T_p of PRS process terms can be seen as the set of Σ-terms T_Σ where $\Sigma_0 = \{0\} \cup Const$, $\Sigma'_A = \{\odot\}$, and $\Sigma'_{AC} = \{\|\}$.

Sets of process terms are recognized by CH-automata $\mathcal{A} = (Q, \Sigma, F, \Delta)$ such that (1) Q is the disjoint union $Q = Q' \cup Q_\odot \cup Q_\|$ where Q' is itself the disjoint union $Q' = Q_0 \cup Q_-$, and (2) the rules in Δ are of the form: (a) $X \to q$, where $q \in Q_-$, $X \in Const$, (b) $0 \to q$, where $q \in Q_0$, (c) $q \to q'$, where $(q, q') \in (Q_0)^2 \cup (Q_-)^2 \cup (Q_\odot)^2 \cup (Q_\|)^2$, (d) $\odot(L) \to q$, where $L \subseteq \big(Q \setminus (Q_\odot \cup Q_0)\big)^*$ and $q \in Q_\odot$, and (e) $\|(\varphi) \to q$, where $q \in Q_\|$, and φ is a Presburger formula such that $FV(\varphi) = \{x_q \mid q \in Q \setminus (Q_\| \cup Q_0)\}$. In other words, the states in $Q_\odot$ (resp. $Q_\|$) recognize trees whose root is $\odot$ (resp. $\|$). The states in Q_- recognize constants in $Const$, and the states in Q_0 recognize 0.

4   A Generic Construction of PRS Reachability Sets

We provide a construction of the reachability analysis of PRS which is *parameterized* by two algorithms $\Theta_\odot$ and $\Theta_\|$ such that (1) $\Theta_\odot$ is an algorithm for the reachability analysis of prefix rewrite systems based on regular languages, and (2) $\Theta_\|$ is an algorithm for the reachability analysis of multiset rewrite systems based on semilinear sets.

4.1   Preliminaries

A class of multiset rewrite systems (Petri nets) $\mathcal{C}$ is *effectively semilinear* if we have for it an algorithm $\Theta_\|$ which constructs, for every given system $M \in \mathcal{C}$ and every semilinear set of paral-terms S (markings), a set $\Theta_\|(M, S)$ which is semilinear and equal to $M^*(S)$.

Let us fix for the rest of this section an effectively semilinear class $\mathcal{C}$ of multiset rewrite systems and let $\Theta_\|$ be the algorithm we have for its symbolic reachability analysis. For a reason which will be clear in the next subsection, we assume that $\mathcal{C}$ is *1-rules closed*, i.e., for every system $M \in \mathcal{C}$, and every X, Y in $Const$, the system $M \cup \{X \to Y\}$ is also in the class $\mathcal{C}$.

Let us fix also an algorithm $\Theta_\odot$ for computing regular reachability sets of prefix rewrite systems (let us assume for instance that it corresponds to one of the algorithms in [9, 15, 22]).

Then, let us consider a PRS $R = R_\odot \cup R_\|$, where $R_\odot$ is a prefix rewrite system, and $R_\|$ is a multiset rewrite system in $\mathcal{C}$, and let $\mathcal{A} = (Q, \Sigma, F, \Delta)$ be a CH-automaton recognizing a set of initial configurations (process terms) $\mathcal{L}$. The rest of this section is devoted to the construction of a CH-automaton $\mathcal{A}^*[R, \Theta_\odot, \Theta_\|] = (\widetilde{Q}, \Sigma, \widetilde{F}, \widetilde{\Delta})$ which recognizes $R^*(\mathcal{L})$, where $\widetilde{Q}$ is the set of states, $\widetilde{F}$ is the set of final states, and $\widetilde{\Delta}$ the set of rules.

4.2   The Set of States

The set of states $\widetilde{Q}$ includes the set of states Q of $\mathcal{A}$ and contains new states q_X, which are assumed to accept precisely the singletons $\{X\}$ (i.e., $L_{q_X} = \{X\}$), for each $X \in Const$. Let Q_R be the set of states $\{q_X \mid X \in Const\}$. In addition, the set $\widetilde{Q}$ contains states which recognize the *successors* by R of terms in L_q for each $q \in Q \cup Q_R$ (these states will have this property at the end of the iterative construction of automaton described below). In order to ensure (during the construction) that the recognized trees are always in canonical form, we need to partition the sets of recognized trees according to their types (given by their root). We associate with each $q \in Q \cup Q_R$ different states $(q, -), (q, 0), (q, \odot)$, and $(q, \|)$ recognizing successors of terms in L_q which are respectively constants in $Const$, null (equal to 0), $\odot$-rooted terms, and $\|$-rooted terms.

Let $Q = Q_0 \cup Q_- \cup Q_\odot \cup Q_\|$. We consider that the set $\widetilde{Q}$ is equal to the union of the following sets: (1) $\widetilde{Q}_0 = Q_0 \cup \{(q, 0) : q \in Q \cup Q_R\}$, (2) $\widetilde{Q}_- = Q_- \cup Q_R \cup \{(q, -) : q \in Q \cup Q_R\}$, and (3) $\widetilde{Q}_\omega = Q_\omega \cup \{(q, \omega) : q \in Q \cup Q_R\}$, for $\omega \in \{\odot, \|\}$. Moreover, we consider that $\widetilde{F} = \{q, (q, -), (q, 0), (q, \odot), (q, \|) : q \in F\}$.

4.3   Nested Prefix/Multiset Rewriting

The construction of the automaton $\mathcal{A}^*[R, \Theta_\odot, \Theta_\|]$ is based on closure rules which add new transitions to those originally in the automaton $\mathcal{A}$ in order to recognize terms obtained by applying the transitive closure of R. The added transitions are defined by computing new constraints reflecting the effect of applying sequences of rewriting steps using the systems $R_\odot$ and $R_\|$. Intuitively, given a rule $w(\gamma) \to q$, we would like to add a rule $w(\gamma') \to q$ where γ' is obtained, roughly speaking, by applying Θ_w to γ. However, many problems appear since $\odot$ and $\|$-rooted terms are nested, and the applications of $R_\odot$ and $R_\|$ may interact at different levels of the term. The main issue is to deal with these interactions in such a manner that only a finite number of transitions needs to be added to the automaton $\mathcal{A}$. The crucial idea is to use, instead of the system R, an extended system R' (called its transitive normal form) where nested prefix and multiset rewriting have been taken into account. The computation of this system is a preliminary step of our construction. Then, the construction itself deals with the problems which come from closing the language of the given CH-automaton under the application of the system S.

Transitive Normal Form: The *transitive normal form* of R is the union of the multiset and prefix rewrite systems $R'_\|$ and $R'_\odot$ defined, for $\omega \in \{\odot, \|\}$, by $R'_\omega = R_\omega \cup \{X \to Y : Y \in R^*(X)\}$.

Lemma 1. *We have* $R^* = (R')^*$. *Moreover, for every flat seq-term (resp. paral-term) t, $R'_\odot{}^*(t)$ (resp. $R'_\|{}^*(t)$) is the set of all the flat seq-terms (resp. paral-terms) in $R^*(t)$.*

The systems R'_ω, for $\omega \in \{\odot, \|\}$, can be computed iteratively as the limits of the ascending chains of sets of rules $(R_i^\omega)_{i \geq 0}$ defined by $R_0^\omega = R_\omega$, and $R_{i+1}^\omega = R_i^\omega \cup \{X \to Y : Y \in \Theta_{\overline\omega}(R_i^{\overline\omega}, X)\}$, for $i \geq 0$. Notice that these chains are finite since there is a finite number of pairs (X, Y).

Rewrite System over the Alphabet of States: Rules in CH-automata (of the forms $\omega(\gamma) \to q$) involve constraints on sequences of *states*, whereas the systems $R'_\odot$ and $R'_\|$ are defined over the alphabet of process constants. Therefore, we define the systems $R''_\odot = \alpha(R'_\odot)$ and $R''_\| = \alpha(R'_\|)$ where α is the substitution such that $\alpha(X) = q_X$, for every $X \in Const$ (extended in the standard way to terms, rules, and sets of rules).

Successor Closure: The system S used in the construction is the union of the systems S_ω, for $\omega \in \{\odot, \|\}$, defined by $S_\omega = R''_\omega \cup \{q \to (q, -), (q, \overline\omega) : q \in Q \cup Q_R\}$.

The role of the additional rules is, roughly speaking, to close the set of accepted trees under the "succession relation": If a rule $\|(\varphi) \to q$ is added to the automaton, whenever it allows to recognize a tree $\|(t_1, t_2, \cdots, t_n)$ at state q, it should also recognize any tree of the form $\|(t'_1, t'_2, \cdots, t'_n)$ where each t'_i is a (constant or $\odot$-rooted) successor of t_i. For the case of the operator $\odot$, this closure concerns only the left-most tree due to prefix rewriting.

4.4   The Set of Transition Rules

The set $\widetilde\Delta$ is inductively defined as the smallest set of transition rules which (1) contains Δ, (2) contains the set of rules $X \to q_X$ for every $X \in Const$, and (3) is such that:

(β_1) **Initialization rules:**

For every state $q \in Q \cup Q_R$, (a) if $q \in Q_0$, then $0 \to (q, 0) \in \widetilde\Delta$, (b) if $q \in Q_- \cup Q_R$, then $q \to (q, -) \in \widetilde\Delta$, and (c) if $q \in Q_\omega$, then $q \to (q, \omega) \in \widetilde\Delta$.

These rules express that $L_q \subseteq L_{(q,0)}$ if $q \in Q_0$, $L_q \subseteq L_{(q,-)}$ if $q \in Q_- \cup Q_R$, and $L_q \subseteq L_{(q,\omega)}$ if $q \in Q_\omega$.

(β_2) **Simulation of the ϵ-rules:**

If $q \to q' \in \Delta$, then $\{(q, -) \to (q', -), (q, 0) \to (q', 0), (q, \omega) \to (q', \omega)\} \subseteq \widetilde\Delta$.

This rule expresses that if initially $L_q \subseteq L_{q'}$, then any successor of L_q is also a successor of $L_{q'}$.

(β_3) **Term flattening rules:**

(a) If $\omega(\gamma) \to (q, \omega) \in \tilde{\Delta}$, $(q', -) \in \gamma$, and $(q', \overline{w}) \in \gamma$, then $(q', 0) \to (q, 0) \in \tilde{\Delta}$, $(q', -) \to (q, -) \in \tilde{\Delta}$, and $(q', \overline{w}) \to (q, \overline{w}) \in \tilde{\Delta}$,

(b) If $\omega(\gamma) \to (q, \omega) \in \tilde{\Delta}$ and $0 \in \gamma$, then $0 \to (q, 0) \in \tilde{\Delta}$.

These rules express that if $\omega(t)$ is a successor of L_q, then t and all its successors are also successors of L_q.

(β_4) **Closure rules: successors of process constants and 0:**

(a) If $X \xrightarrow{*}_{\tilde{\Delta}} (q, -)$, then $\omega\big(\Theta_\omega(S_\omega, qx)\big) \to (q, \omega) \in \tilde{\Delta}$,

(b) If $0 \xrightarrow{*}_{\tilde{\Delta}} (q, 0)$, then $\omega\big(\Theta_\omega(S_\omega, 0)\big) \to (q, \omega) \in \tilde{\Delta}$.

The rule (a) says that if X is a successor of some term in L_q, then all its ω-successors obtained by applying the system R'_ω are also successors of L_q. The rule (b) says the same thing for successors of 0.

(β_5) **Closure rule: successors of ω-rooted terms:**

If $\omega(\gamma) \to p \in \Delta$, then $\omega\big(\Theta(S'_\omega, \sigma(\gamma))\big) \to (p, \omega) \in \tilde{\Delta}$, where:

$-$ σ is the substitution such that $\forall q \in Q \cup Q_R$, $\sigma(q) = \{q\} \cup \{qx \; : \; X \xrightarrow{*}_{\Delta} q\}$,

$-$ $S'_\omega = S_\omega \cup \{q \to qx \; : \; X \xrightarrow{*}_{\tilde{\Delta}} (q, -), q \in Q \cup Q_R\} \cup \{q \to 0 \; : \; 0 \xrightarrow{*}_{\tilde{\Delta}} (q, 0)\}$

This rule concerns the case of ω-rooted terms where rewritings have not occurred so far at their root (they have occurred starting from the level of the child of the root). The rule says the following: Let $t = \omega(t_1, \ldots, t_n)$ be a term initially accepted at p due to the fact that each of the t_i's can be labelled with a q_i with $q_1 \cdots q_n \models \gamma$. Let t' be a successor of t, and assume that it is equal to t where some of the t_i's have been rewritten to 0, and some others have been transformed into constants X. Then, any successor of t', obtained by applying the rewrite system R'_ω to the subsequence of all constants appearing among its first-level children, must also be a successor of t.

    The set of rules $\tilde{\Delta}$ can be constructed iteratively as the limit of an increasing sequence $\tilde{\Delta}_1 \subseteq \tilde{\Delta}_2 \subseteq \ldots$ of set of rules, obtained by applying iteratively the closure and flattening β-rules. It can be seen that this iterative procedure terminates. Indeed, the flattening rules (β_3) create rewrite rules of the form $(q, \sim) \to (q', \sim)$. Since there is a finite number of pairs (q, q'), (β_3) can only be applied a finite number of times. The same holds for the closure rules (β_4) and (β_5) since they can only be applied if there is a new pair (X, q) such that $X \xrightarrow{*}_{\tilde{\Delta}} (q, -)$, and there is a finite number of such pairs. We can prove the following fact (the proof can be found in the full version of the paper and in [38]):

Lemma 2. *For every process term t, and every $q \in Q \cup Q_R$ we have: (1) $t \xrightarrow{*}_{\tilde{\Delta}}$ $(q, 0)$ iff $t \in Post^*(L_q)$ and $t = 0$, (2) $t \xrightarrow{*}_{\tilde{\Delta}} (q, -)$ iff $t \in Post^*(L_q)$ and $t \in Const$, and (3) $t \xrightarrow{*}_{\tilde{\Delta}} (q, \omega)$ iff $t \in Post^*(L_q)$ and $root(t) = \omega$, for $\omega \in \{\odot, \|\}$.*

Theorem 2. *Let $\Theta_\odot$ be an algorithm for computing Post* images of regular sets by prefix rewrite systems. Let C be a 1-rules closed, effectively semilinear class of multiset rewrite systems, and let $\Theta_{\parallel}$ be an algorithm for computing Post* images of semilinear sets by systems in C. Then, for every system R in PRS[C], and every CH-automaton $\mathcal{A}$, we have $Post_R^*(L(\mathcal{A})) = L(\mathcal{A}^*[R, \Theta_\odot, \Theta_{\parallel}])$.*

5 Application: Reachability Analysis of PRS

5.1 Computing Reachability Sets of PAD Systems

As said in section 2, the class of PAD system is precisely the class PRS[BPP], and moreover, the class coPAD is equal to PRS[coBPP]. It has been shown that the reachability relation of BPP systems is semilinear and effectively constructible [20]. This implies that both classes BPP and coBPP are effectively semilinear. These classes are also 1-rules closed, by definition. Therefore, an immediate consequence of Theorem 2 is the following result:

Corollary 1. *For every PAD system R, and every CH-automaton $\mathcal{A}$, the sets $Post_R^*(L(\mathcal{A}))$ and $Pre_R^*(L(\mathcal{A}))$ are computable and effectively representable by CH-automata.*

5.2 Global Model Checking for EF

We consider the EF fragment of the temporal logic CTL: the set of formulas built over a finite set of atomic propositions, and closed under boolean operators and the EF operator (EFϕ means that there exists a computation path where ϕ is eventually true). We consider that atomic propositions are interpreted as CH-automata definable sets of process terms.

The global model checking problem is "given a PRS R and a formula ϕ, compute the set of all configurations (process terms) satisfying ϕ". Then, since EF corresponds to the *Pre** operator and CH-automata are closed under boolean operations (Theorem 1), we obtain as a consequence of Corollary 1 the fact that:

Corollary 2. *For every PAD system R, and for every formula ϕ in the EF fragment of CTL over CH-automata definable valuations of atomic propositions, the set of configurations of R satisfying ϕ is computable and effectively representable by a CH-automaton.*

5.3 Integrating Reachability Analysis Procedures for Pushdown Systems and Petri Nets

Our construction provides a framework for extending any procedure (exact or approximate) for symbolic reachability analysis of Petri nets using semilinear sets (or linear arithmetics) to a procedure for symbolic reachability analysis of PRS.

Indeed, we can compute upper approximations of reachability sets of PRS by instantiating the parameter $\Theta_{\|}$ of the construction by an algorithm which computes, for any given multiset rewrite system, a semilinear upper approximation of its reachability set. There are several such algorithms, leading to different analysis procedures for PRS with different precisions, such as the Karp-Miller algorithm for computing the coverability set, and procedures which generate invariants based on, e.g, flow constraints and trap constraints (see [24]).

Other possible instantiations can be obtained by using various existing *approximate* algorithms or *exact* semi-algorithms for symbolic reachability analysis of counter automata (multiset rewrite systems can of course be seen as particular cases of such general models). Many of such procedures have been developed in the last few years using either polyhedra-based representations or automata-based representations for linear constraints or Presburger arithmetics [2, 7, 8, 14, 17, 18, 28, 39]. These procedures are quite efficient thanks to the use of different fixpoint acceleration techniques (such as widening-based techniques [2, 7, 14, 17], or meta-transition techniques [8, 28]) allowing to force or to help termination of the reachability analysis.

6 Conclusion

We have defined automata-based techniques for computing reachability sets of PRS. These techniques provide a general framework for modeling and analyzing programs with dynamic creation of concurrent processes and recursive procedure calls. Indeed, such programs can be modeled quite naturally using term rewrite systems combining prefix and multiset rewrite systems. Then, our generic construction allows to use procedures for reachability analysis of both kinds of rewrite systems in order to derive various analysis procedures for their combination. This allows in particular to establish new analysis and model-checking algorithms for the class of PAD systems. Our results generalize and unify all existing results on the kind of models we consider based on process rewrite systems. Furthermore, our construction provides a theoretical basis for the construction of a tool for symbolic analysis of multithreaded programs modeled as PRS, based on the integration of existing, and quite efficient, tools for symbolic reachability analysis of pushdown systems (e.g., [36]) and tools for symbolic reachability analysis of counter automata (e.g., [1, 3, 6, 31]).

References

1. P. Ammirati, G. Delzanno, P. Ganty, G. Geeraerts, J.-F. Raskin, and L. Van Begin. Babylon: An integrated toolkit for the specification and verification of parameterized systems. In *SAVE'02*, 2002.
2. A. Annichini, E. Asarin, and A. Bouajjani. Symbolic Techniques for Parametric Reasoning about Counter and Clock Systems. In *CAV'00*. LNCS 1855, 2000.
3. A. Annichini, A. Bouajjani, and M. Sighireanu. TReX: A Tool for Reachability Analysis of Complex Systems. In *CAV'01*. LNCS 2102, 2001.
4. T. Ball, S. Chaki, and S. K. Rajamani. Parameterized verification of multithreaded software libraries. In *TACAS 2001, LNCS 2031*, 2001.

5. T. Ball, A. Podelski, and S. K. Rajamani. Boolean and cartesian abstractions for model checking c programs. In *TACAS 2001, LNCS 2031*, 2001.

6. S. Bardin, A. Finkel, J. Leroux, and L. Petrucci. FAST: Fast Acceleration of Symbolic Transition systems. In *CAV'03*. LNCS 2725, 2003.

7. B. Boigelot, A. Legay, and P. Wolper. Iterating transducers in the large. In *Proc. CAV'03*. LNCS 2725, 2003.

8. B. Boigelot and P. Wolper. Symbolic Verification with Periodic Sets. In *CAV'94*. LNCS 818, 1994.

9. A. Bouajjani, J. Esparza, and O. Maler. Reachability Analysis of Pushdown Automata: Application to Model Checking. In *CONCUR'97*. LNCS 1243, 1997.

10. A. Bouajjani, J. Esparza, and T. Touili. A generic approach to the static analysis of concurrent programs with procedures. In *Proc. of the 30th ACM Symp. on Principles of Programming Languages, (POPL'03)*, 2003.

11. A. Bouajjani, J. Esparza, and T. Touili. Reachability Analysis of Synchronized PA Systems. In *Proc. Infinity Workshop*, 2004.

12. Ahmed Bouajjani and Tayssir Touili. Reachability Analysis of Process Rewrite Systems. In *Proc. 23rd Intern. Conf. on Foundations of Software Technology and Theoretical Computer Science (FSTTCS'03)*. LNCS 2914, 2003.

13. A. Bruggemann-Klein, M. Murata, and D. Wood. Regular tree and regular hedge languages over unranked alphabets. Research report, 2001.

14. T. Bultan, R. Gerber, and C. League. Verifying Systems With Integer Constraints and Boolean Predicates: A Composite Approach. In *Proc. of the Intern. Symp. on Software Testing and Analysis*. ACM press, 1998.

15. D. Caucal. On the regular structure of prefix rewriting. *Theoret. Comput. Sci.*, 106:61–86, 1992.

16. T. Colcombet. Rewriting in the partial algebra of typed terms modulo ac. In *Electronic Notes in Theoretical Computer Science, volume 68. Elsevier Science Pub., Proc. Infinity Workshop*, 2002.

17. Patrick Cousot and Nicholas Halbwachs. Automatic discovery of linear restraints among variables of a program. In *POPL'78*. ACM, 1978.

18. G. Delzanno, L. Van Begin, and J.-F. Raskin. Attacking symbolic state explosion in parametrized verification. In *CAV'01*, 2001.

19. G. Delzanno, L. Van Begin, and J.-F. Raskin. Toward the automated verification of multithreaded java programs. In *TACAS 2002*, 2002.

20. J. Esparza. Petri nets, commutative context-free grammars, and basic parallel processes. In *Fundamentals of computation theory*, volume 965 of *LNCS*, 1995.

21. J. Esparza. Grammars as processes. In *Formal and Natural Computing*. LNCS 2300, 2002.

22. J. Esparza, D. Hansel, P. Rossmanith, and S. Schwoon. Efficient algorithm for model checking pushdown systems. In *CAV'00*, volume 1885 of *LNCS*, 2000.

23. J. Esparza and J. Knoop. An automata-theoretic approach to interprocedural data-flow analysis. In *FOSSACS'99*. LNCS 1578, 1999.

24. J. Esparza and S. Melzer. Verification of safety properties using integer programming: Beyond the state equation. *Formal Methods in System Design*, 16, 2000.

25. J. Esparza and M. Nielsen. Decidability issues for Petri nets - a survey. *Bulletin of the EATCS*, 52:85–107, 1994.

26. J. Esparza and S. Schwoon. A bdd-based model checker for recursive programs. In *In Proc. of CAV'01, number 2102 in Lecture Notes in Computer Science, pages 324-336. Springer-Verlag*, 2001.

27. Javier Esparza and Andreas Podelski. Efficient algorithms for pre * and post * on interprocedural parallel flow graphs. In *Symposium on Principles of Programming Languages*, pages 1–11, 2000.
28. A. Finkel and J. Leroux. How to compose Presburger-accelerations: Applications to broadcast protocols. In *FSTTCS'02*. LNCS 2556, 2002.
29. A. Finkel, B. Willems, and P. Wolper. A Direct Symbolic Approach to Model Checking Pushdown Systems. In *Infinity'97*, 1997.
30. J. Hopcroft and J.-J. Pansiot. On The Reachability Problem for 5-Dimensional Vector Addition Systems. *Theoret. Comput. Sci.*, 8, 1979.
31. The Liège Automata-based Symbolic Handler (LASH), 2001.   Available at http://www.montefiore.ulg.ac.be/~boigelot/research/lash/.
32. D. Lugiez. Counting and equality constraints for multitree automata. In *FoSSaCS 2003*, pages 328–342, 2003.
33. D. Lugiez and Ph. Schnoebelen. The regular viewpoint on PA-processes. In *Proc. 9th Int. Conf. Concurrency Theory (CONCUR'98), Nice, France, Sep. 1998*, volume 1466, pages 50–66. Springer, 1998.
34. R. Mayr. Decidability and Complexity of Model Checking Problems for Infinite-State Systems. Phd. thesis, Techn. Univ. of Munich, 1998.
35. S. Qadeer, S.K. Rajamani, and J. Rehof. Procedure Summaries for Model Checking Multithreaded Software. In *POPL'04*. ACM, 2004.
36. Stefan Schwoon. *Model-Checking Pushdown Systems*. PhD thesis, Technische Universität München, 2002.
37. H. Seidl, Th. Schwentick, and A. Muscholl. Numerical Document Queries. In *PODS'03*. ACM press, 2003.
38. Tayssir Touili. Analyse symbolique de systèmes infinis basée sur les automates: Application à la vérification de systèmes paramétrés et dynamiques. Phd. thesis, University of Paris 7, 2003.
39. P. Wolper and B. Boigelot. Verifying Systems with Infinite but Regular State Spaces. In *CAV'98*. LNCS 1427, 1998.

Automata and Logics
for Unranked and Unordered Trees

Iovka Boneva and Jean-Marc Talbot

Laboratoire d'Informatique Fondamentale de Lille – UMR CNRS/USTL 8022
INRIA Futurs – MOSTRARE Project

Abstract. In this paper, we consider the monadic second order logic (MSO) and two of its extensions, namely Counting MSO (CMSO) and Presburger MSO (PMSO), interpreted over unranked and unordered trees. We survey classes of tree automata introduced for the logics PMSO and CMSO as well as other related formalisms; we gather results from the literature and sometimes clarify or fill the remaining gaps between those various formalisms. Finally, we complete our study by adapting these classes of automata for capturing precisely the expressiveness of the logic MSO.

1 Introduction

Relationship between logics and tree automata for ranked trees has been established by Thatcher and Wright in their seminal paper [19]: they proved that languages of finite and ranked trees that are accepted by tree automata coincide with the models of monadic second-order logic (MSO) sentences when interpreted over (ranked) tree structures.

Recently, due to the development of semi-structured databases and in particular, of XML, there has been some new interest in unranked and ordered trees; for those trees, the number of children of some node is not a *priori* bounded and for instance, does not depend on the symbol labeling this position in the tree. Moreover, those trees are said to be ordered in the sense that there exists a total ordering on children of each node. The relationship between logics and automata has been carried over unranked and ordered trees [13],[1]: once again, languages that are definable by means of tree automata are exactly models of MSO sentences.

In this paper we consider unranked and unordered trees, ie trees that are unranked but without any ordering relation between children of the same node. As noticed by Courcelle in [4], the fact that there is no order between siblings drastically reduces the expressiveness of MSO: hence, for ordered unranked trees, properties such as "the root has an even number of children labeled with *b*" or such as "the number of nodes in the tree is a multiple of 5" can be expressed in MSO (where the ordering relation on sibling nodes is represented as an ordering relation or as some successor relation). It goes differently for unranked and unordered trees where those two latter properties can no longer be expressed in

J. Giesl (Ed.): RTA 2005, LNCS 3467, pp. 500–515, 2005.

MSO. Courcelle proposed in [4] to extend MSO with some constraints for counting modulo on cardinalities of sets. He showed that this logic, named Counting MSO (CMSO), can be related to tree automata by the notion of algebraic recognizability in the sense of [12]: a set of trees can be expressed by some CMSO sentence iff it is recognizable.

Recently, Seidl, Schwentick and Muscholl introduced Presburger monadic second order logic (PMSO) [18]: it extends MSO with a new kind of atomic formula x/ϕ; in such an atomic formula, x is a variable denoting a node of the tree and ϕ is a Presburger formula expressing arithmetical constraints on the cardinality of sets when restricted to the children of x. Seidl *et al.* also defined a notion of automata, called Presburger tree automata, and showed that tree languages accepted by Presburger tree automata are precisely models of PMSO sentences.

The objective of this paper is two folds: first, we gather results concerning formalisms that can express sets of unranked and unordered trees definable by PMSO and CMSO sentences. This survey permits to clarify or sometimes make explicit the relationship of different formalisms, in particular, various classes of tree automata (*eg* Presburger tree automata [18], ACU equational tree automata [15], [20] and equational tree languages [4] when considering the logic PMSO). Our second aim is to try to get a uniform view on tree languages that can be defined by the logics CMSO and PMSO, but also by MSO: in particular, for PMSO and CMSO, we try to adapt systematically (when possible) a formalism associated to some specific logic to the other one. Finally, we investigate the expressiveness of the logic MSO: considering formalisms used for describing CMSO and PMSO definable sets, we propose subclasses capturing precisely MSO over unranked and unordered trees.

This paper is organized as follows: Section 2 presents definitions for trees as graphs, an algebraic view of trees and recall Presburger formulas. In Section 3 we define the three logic formalisms MSO, CMSO and PMSO. Sections 4 and 5 survey PMSO- and CMSO-complete formalisms respectively, and in Sections 6 and 7 we present new characterizations of PMSO- and CMSO-definable sets of trees. Finally, in Section 8 we give characterizations of MSO-definable sets of trees.

2   Preliminaries

2.1   The Tree Model

We consider here edge-labeled[1] unranked and unordered trees (called simply trees in the rest of the paper).

Trees will be finite non-empty directed graphs with a distinguished node, the root of the tree, such that for any node, there exists exactly one path from the root to this node. Additionally, we suppose a mapping associating with each

---

[1] For simplicity, we assume that nodes are unlabeled. However, the results presented here could be extended to trees where both edges and nodes are labeled.

edge of the graph a label from a finite set Λ. Formally, a tree is given by a triple (V, E, λ) such that V is a finite non-empty set of nodes, $E \subseteq V \times V$ is a finite set of edges and λ is a mapping from E to Λ. Moreover, it satisfies that any node is reached by a unique path from the root: for any nodes $v_n, v'_{n'}$, for any two sequences $v_0, v_1, \ldots, v_n$ and $v'_0, v'_1, \ldots, v'_{n'}$ such that v_0, v'_0 both denote the root of the tree, $v_n, v'_{n'}$ are equal and (v_i, v_{i+1}), (v'_j, v'_{j+1}) belong to E for all $0 \le i \le n-1$, $0 \le j \le n'-1$, the two sequences are identical.

As usual, we consider two isomorphic trees as being equal. We denote Tree the set of all trees. We denote $\mathrm{root}(\tau)$ the root of the tree τ and for any node v, $\mathrm{children}(v)$ the set of nodes $\{v' \mid (v, v') \in E\}$.

2.2   An Algebraic View of Trees

We adopt the algebraic view of trees proposed in [4]. We consider the signature Σ given by the constant $\mathbf{0}$, the unary function symbols a for each a in Λ and the binary (infix) symbol $|$.

Let $\mathcal{T}$ be the Σ-algebra whose domain is the set of all finite edge-labeled trees. The constant $\mathbf{0}$ is interpreted in $\mathcal{T}$ as $\mathbf{0}^{\mathcal{T}}$ the tree having one single node and no edge (we consider only non-empty graphs). For any tree τ defined as (V, E, λ), the tree $a^{\mathcal{T}}(\tau)$ is given by $(V \cup \{r\}, E \cup \{r, \mathrm{root}(\tau)\}, \lambda')$ where r is a new node (not belonging to V) and λ' extends λ by letting $\lambda'((r, \mathrm{root}(\tau))) = a$. For trees τ, τ' defined as (V, E, λ), (V', E', λ') respectively, $\tau \mid^{\mathcal{T}} \tau'$ is the tree given by (V'', E'', λ'') where (assuming $V \cap V' = \varnothing$):

- $V'' = (V \cup V' \cup \{r\}) \smallsetminus \{\mathrm{root}(\tau), \mathrm{root}(\tau')\}$ (where $r \notin V \cup V'$)
- $E'' = \{(r, v) \mid v \in \mathrm{children}(\mathrm{root}(\tau)) \cup \mathrm{children}(\mathrm{root}(\tau'))\} \cup (E \smallsetminus \{(\mathrm{root}(\tau), v) \mid v \in V\}) \cup (E' \smallsetminus \{(\mathrm{root}(\tau'), v') \mid v' \in V\})$.
- λ'' is defined as λ and λ' for edges in E'' coming from E and E' respectively and by $\lambda''((r, v)) = \lambda((\mathrm{root}(\tau), v))$ if $v \in E$ and $\lambda''((r, v)) = \lambda((\mathrm{root}(\tau'), v))$ if $v \in E'$.

Informally, $a^{\mathcal{T}}(t)$ adds a new edge labeled by a from a new node (the new root) to the ancient root of t whereas $t \mid^{\mathcal{T}} t'$ is obtained from t and t' by merging their roots. Figure 1 illustrates algebraic operations on trees.

One can remark that the set of trees Tree is finitely generated by Σ, that is each tree in Tree can be obtained by combining the operators from the Σ-algebra $\mathcal{T}$.

Fig. 1. Algebraic operations over trees.

It should also be noticed that the operation $|^{\mathcal{T}}$ is associative and commutative over trees and that $\mathbf{0}^{\mathcal{T}}$ is its neutral element. Therefore, $(\mathsf{Tree}, |^{\mathcal{T}}, \mathbf{0}^{\mathcal{T}})$ is a commutative monoid.

We will also consider $\mathcal{C}$ the algebra of terms built over the signature Σ (ie the initial algebra over Σ). We will denote $h_{\mathcal{C}}$ the unique homomorphism from $\mathcal{C}$, the Σ-algebra of terms to $\mathcal{T}$, the algebra of trees.

2.3   Arithmetical Formulas

In this paper, we will have to consider different kinds of arithmetical formulas interpreted over $\mathbb{N}$ the set of natural numbers. Different logics will be defined depending on the atomic predicates that are allowed.

Let $\mathcal{U}$ be a set of natural variables and B be a set of atomic formulas whose free variables belong to $\mathcal{U}$. We define $\mathcal{F}_{\mathcal{U}}(B)$ as the least set of formulas such that (i) B is included in $\mathcal{F}_{\mathcal{U}}(B)$ and (ii) if ϕ, ϕ' are in $\mathcal{F}_{\mathcal{U}}(B)$ then $\phi \wedge \phi'$, $\neg\phi$ are in $\mathcal{F}_{\mathcal{U}}(B)$ as well.

For our purpose, we are going to consider only two kinds of atomic formulas: $p \le p'$ and $Div_k(p)$, where k is some fixed natural number different from zero and p is an arithmetical term defined as:

$$p ::= n \mid u \mid p + p \qquad (u \in \mathcal{U}, n \in \mathbb{N})$$

Formulas in $\mathcal{F}_{\mathcal{U}}(B)$ are interpreted over $(\mathbb{N}, \{+\}, \{\le, Div_k\})$ the structure of naturals where $+$ is interpreted as the addition function, $\le$ as the usual ordering over $\mathbb{N}$ and finally, Div_k is the unary predicate such that $Div_k(n)$ holds if n is divisible by k. The semantics for Boolean connectives is the usual one.

Let ϕ be a formula from $\mathcal{F}_{\mathcal{U}}(B)$. We say that a valuation μ mapping free variables of ϕ to naturals is a solution of ϕ if the structure $(\mathbb{N}, \{+\}, \{\le, Div_k\})$ is a model of ϕ under the valuation μ.

Formulas from $\mathcal{F}_{\mathcal{U}}(\{p \le p', Div_k(p)\})$ are called *Presburger formulas* and the ones from $\mathcal{F}_{\mathcal{U}}(\{p \le p'\})$ are called *ordering formulas*.

Strictly speaking, Presburger formulas usually allow also existential quantification $\exists u.\phi$. However, it is well-known that for any Presburger formula ϕ with quantification, there exists an equivalent (quantifier-free) formula ϕ' from $\mathcal{F}_{\mathcal{U}}(\{p \le p', Div_k(p)\})$ [2]. Note that this is not the case for ordering formulas for which adding existential quantification strictly increases their expressiveness[3].

An atomic formula from $\{p \le p', Div_k(p)\}$ is said to be *unary* if this formula contains only one variable (but possibly several occurrences of it). By extension, a formula ϕ from $\mathcal{F}_{\mathcal{U}}(\{p \le p', Div_k(p)\})$ is *unary* if it is built over unary atomic formulas. Note that a unary formula may contain several different variables but any of its atoms contains only one variable.

---

[2] The first-order theory of formulas built over $\{p \le p', Div_k(p)\}$ interpreted over natural numbers admits quantifier elimination.

[3] In presence of existential quantifications, Presburger and ordering formulas are equally expressive as $Div_k(p)$ can be written as $\exists y.p = \underbrace{y + \ldots + y}_{k}$.

We will denote $\mathcal{F}_{\mathcal{U}}^1(\{p \leq p', Div_k(p)\})$ (resp. $\mathcal{F}_{\mathcal{U}}^1(\{p \leq p'\})$) the set of unary Presburger formulas (resp. of unary ordering formulas).

2.4   Presburger-Definable Sets and Multiset Languages

Let $\mathbb{N}^l$ be the set of tuples of length l of naturals. A subset N of $\mathbb{N}^l$ is said to be *Presburger-definable* (resp. *ordering-definable*) if there exists a Presburger formula (resp. an ordering formula) ϕ whose free variables are $(x_1, \ldots, x_l)$ considered as totally ordered and such that for any tuple $(n_1, \ldots, n_l)$ from N, the valuation $\{x_1 \mapsto n_1, \ldots, x_l \mapsto n_l\}$ is a solution of ϕ.

Let $A = (a_1, \ldots, a_l)$ be a sequence of symbols. We denote $\mathbb{M}(A)$ the set of all multisets whose elements are in A. The Parikh mapping [17] is a mapping from $\mathbb{M}(A)$ to $\mathbb{N}^l$ defined as $\pi_A(m) = (|m|_{a_1}, \ldots, |m|_{a_l})$, where $|m|_{a_i}$ is the number of occurrences of a_i in the multiset m. Parikh mappings are extended as mappings from multiset languages to subsets of $\mathbb{N}^l$ as follows: for $M \subseteq \mathbb{M}(A)$, $\pi_A(M) = \{\pi_A(m) \mid m \in M\}$.

Denoting $\varnothing$ the empty multiset and $\uplus$ the multiset union,

Definition 1. *The family* $Rat(\mathbb{M}(A))$ *of rational multiset languages is the least subset of* $\mathbb{M}(A)$ *which contains any finite subset of* $\mathbb{M}(A)$ *and such that if* L, L' *belong to* $Rat(\mathbb{M}(A))$ *then* $L \cup L'$, $L \uplus L' = \{m \uplus m' \mid m \in L \text{ and } m' \in L'\}$, $L^* = \bigcup_{n \in \mathbb{N}} L^n$ *(where* $L^0 = \varnothing$ *and* $L^{i+1} = L^i \uplus L$ *for* $i > 0$*) belong to* $Rat(\mathbb{M}(A))$.

It is well-known that

Note 1. Let N be a subset of $\mathbb{N}^l$ and $A = (a_1, \ldots, a_l)$ be some alphabet. Then N is Presburger-definable iff $\pi_A^{-1}(N) \in Rat(\mathbb{M}(A))$.

Definition 2. *A multiset language* $L \in \mathbb{M}(A)$ *is recognizable if there exists a monoid morphism* h *from* $(L, \uplus, \{\varnothing\})$ *to a finite monoid* $(D, +, \iota)$ *and a finite subset* D' *of* D *such that* $L = h^{-1}(D')$.

We denote $Rec(\mathbb{M}(A))$ the set of recognizable multiset languages. It is well-known that the set of recognizable multisets is strictly included into the set of rational multisets, ie $Rec(\mathbb{M}(A)) \subsetneq Rat(\mathbb{M}(A))$.

3   MSO-Based Logics for Trees

We consider in this section monadic second-order logic (MSO) as well as two extensions of it. First, let us recall how trees can be viewed as logical structures over which logical formulas are interpreted.

Let σ be the signature $\{label_a \mid a \in \Lambda\}$ where the $label_a$'s are binary predicates. With a tree $\tau = (V, E, \lambda)$, we associate a finite σ-structure $\mathcal{S}^\tau = \langle V, \{label_a^\tau \mid a \in \Lambda\}\rangle$, such that $label_a^\tau(v, v')$ holds in $\mathcal{S}^\tau$ if $(v, v') \in E$ and $\lambda((v, v')) = a$.

We assume a countable set of first-order variables ranging over by $x, y, z, \ldots$ and a countable set of second-order variables ranging over by $X, Y, Z, \ldots$.

Definition 3. *The formulas of the logic MSO are defined by the following syntax:*

$$\psi ::= \mathsf{label}_a(x,y) \mid x \in X \mid \psi \vee \psi \mid \neg\psi \mid \exists x.\psi \mid \exists X.\psi$$

Let $\mathcal{S}$ be a σ-structure whose domain is V. Let ρ be a valuation mapping first-order variables to elements of V and second-order variables to subsets of V. The structure $\mathcal{S}$ is a model of a MSO formula ψ under the valuation ρ (defined for free variables of ψ) denoted $\mathcal{S} \models_\rho \psi$, if:

- ψ is $\mathsf{label}_a(x,y)$ and $\mathsf{label}_a(\rho(x),\rho(y))$ holds in $\mathcal{S}$;
- ψ is $x \in X$ and $\rho(x)$ belongs to $\rho(X)$;
- ψ is $\psi_1 \vee \psi_2$ (resp. $\neg\psi'$) and $\mathcal{S} \models_\rho \psi_1$ or $\mathcal{S} \models_\rho \psi_2$ (resp. $\mathcal{S} \not\models_\rho \psi'$) holds;
- ψ is $\exists x.\psi'$ and there exists an element v from V s.t. $\mathcal{S} \models_{\rho[x \to v]} \psi'$ holds.
- ψ is $\exists X.\psi'$ and there exists a subset V' of V s.t. $\mathcal{S} \models_{\rho[X \to V']} \psi'$ holds.

Overloading the notation, for a closed MSO formula ψ and a tree τ, we write $\tau \models \psi$ whenever $\mathcal{S}^\tau \models \psi$ for the σ-structure $\mathcal{S}^\tau$ associated with τ; moreover, we write $[\![\psi]\!]$ to denote the set of all trees τ such that $\tau \models \psi$. We say that a set of trees T is *MSO-definable* if there exists some closed MSO formula ψ such that $[\![\psi]\!] = T$.

The logic CMSO Courcelle defined in [4] the counting MSO logic (CMSO) as an extension of MSO. The syntax of CMSO[4] augments the one from MSO with an atomic formula $Mod^i_j(X)$ where X is a second-order variable and i, j are naturals such that $i \neq 0$ and $j < i$. The formula $Mod^i_j(X)$ holds for a σ-structure $\mathcal{S}$ and a mapping ρ associating with X a subset of the domain of $\mathcal{S}$ if the cardinality of $\rho(X)$ modulo i is equal to j.

The logic PMSO Seidl et al. introduced in [18] an extension of MSO called Presburger MSO (PMSO). This extension is defined by a new kind of atomic formulas of the form x/ϕ, ϕ being a Presburger formula from $\mathcal{F}_\mathcal{V}(\{p \leq p', Div_k(p)\})$, where $\mathcal{V}$ is the set of integer variables $\{\#X \mid X$ is a second-order variable$\}$.

The formula x/ϕ holds in some σ-structure $\mathcal{S}$ under a valuation ρ if the valuation μ mapping each variable $\#X$ from ϕ to the cardinality of the set $\rho(X) \cap \mathsf{children}(\rho(x))$ is a solution for ϕ [5].

CMSO-definable and *PMSO-definable* set of trees are defined on the same way that MSO-definable set of trees.

---

[4] Actually, the syntax of CMSO from [4] is richer than the one we consider here; there, the logic has two sorts for both individual and set variables, respectively a sort for nodes and a sort for edges. However, Courcelle showed in [5] that this two-sorted extension does not add expressive power when trees are considered.

[5] PMSO allows to express quite complex relationships between cardinalities of sets; however, those sets are always relative to some precise node. For arbitrary sets, the associated monadic second order logic would be undecidable [11].

4   A Survey on PMSO-Complete Formalisms

In this section, we present various formalisms which are able to express precisely PMSO definable sets of trees.

4.1   Presburger Tree Automata

In [18], Seidl *et al.* introduced Presburger tree automata which correspond to the logic PMSO. We define here an adaptation of these automata for edge-labeled trees. Later on we identify precisely subclasses of these automata for the logics MSO and CMSO. These automata are also very close to sheaves automata from [8],[7].

Definition 4. *A Presburger tree automaton (PTA) is given by a tuple (Λ, Q, F, δ) where Λ is a finite set of labels, Q is a finite set of states, δ is a transition mapping from $Q \times \Lambda$ to $\mathcal{F}_{\mathcal{U}}(\{p \leq p', Div_k(p)\})$ where $\mathcal{U}$ is $\{x_q \mid q \in Q\}$ and finally, $F \in \mathcal{F}_{\mathcal{U}}(\{p \leq p', Div_k(p)\})$ is the acceptance condition.*

A run r_A for a tree $\tau = (V, E, \lambda)$ and a PTA $A = (\Lambda, Q, F, \delta)$ is a mapping from E to Q such that for all edges (v, v') in E, $\mu_v \models \delta(r_A((v, v')), \lambda((v, v')))$ where μ_v is the valuation associating with each variable x_q the cardinality of the set $\{(v', v'') \mid (v', v'') \in E$ and $r_A((v', v'')) = q\}$.

Informally, a run labels edges with states from Q: the state labeling some edge $e = (v, v')$ depends on the label of the edge as well as on the multiplicity of the states labeling the edges originating from the node v' (ie edges of the form (v', v'') for some node v'').

A tree $\tau = (V, E, \lambda)$ is accepted by a Presburger tree automaton $A = (\Lambda, Q, F, \delta)$ if there exists a run r for τ and A such that $\mu_F \models F$ where μ_F is the valuation associating with each variable x_q the cardinality of the set $\{(\text{root}(\tau), v) \mid (\text{root}(\tau), v) \in E$ and $r_A((\text{root}(\tau), v)) = q\}$. For some PTA A, we denote $L(A)$ the set of all trees accepted by A.

Example 1. The Presburger tree automaton A_1 here after accepts precisely the set of trees of height 1 such that the root has as many a outgoing edges as b ones: $A_1 = (\{a, b\}, \{q_a, q_b\}, x_{q_a} = x_{q_b}, \delta)$ where δ is the transition mapping such that $\delta((q_a, a)) = \delta((q_b, b)) = x_{q_a} \leq 0 \wedge x_{q_b} \leq 0$ and $\delta((q_a, b)) = \delta((q_b, a)) = false$. The automaton A_2 accepts precisely the set of trees satisfying that each node has as many a outgoing edges as b ones: $A_2 = (\{a, b\}, \{q_a, q_b\}, x_{q_a} = x_{q_b}, \delta)$ where

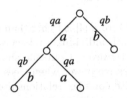

Fig. 2. Run of the automaton A_2.

δ is the transition mapping such that $\delta((q_a, a)) = \delta((q_b, b)) = (x_{q_a} = x_{q_b})$ and $\delta((q_a, b)) = \delta((q_b, a)) = $ *false*.

Theorem 1. *[18] For any set of trees T, T is PMSO-definable iff there exists some Presburger tree automaton A that accepts T.*

4.2 Rational-Multiset Tree Automata

Colcombet proposed in [2] rational-multiset tree automata. We give here a slightly rephrased definition of those automata.

Definition 5. *A rational-multiset automaton (RatMA) is a tuple (Λ, Q, F, δ) where Λ is a finite set of labels, Q is a finite set of states, δ is a transition mapping from $Q \times \Lambda$ to $Rat(\mathbb{M}(Q))$ and $F \in Rat(\mathbb{M}(Q))$ is the acceptance condition.*

A run r_A for a tree $\tau = (V, E, \lambda)$ and a RatMA $A = (\Lambda, Q, F, \delta)$ is a mapping from E to Q such that for all edges (v, v') in E, the multiset $\{r_A((v', v_1)), \dots, r_A((v', v_n))\}$ belongs to $\delta(r_A((v, v')), \lambda((v, v')))$, $v_1 \dots v_n$ being exactly the children of v'.

A tree $\tau = (V, E, \lambda)$ is accepted by a RatMA $A = (\Lambda, Q, F, \delta)$ if there exists a run r for τ and A such that $\{r_A((\text{root}(\tau), v_1)), \dots, r_A((\text{root}(\tau), v_n))\}$ belongs to F, $v_1 \dots v_n$ being exactly the children of $\text{root}(\tau)$. For some RatMA A, we denote $L(A)$ the set of all trees accepted by A.

Using Note 1, it is straightforward that

Proposition 1. *For any set of trees T, T is PMSO-definable iff there exists a rational-multiset automaton A that accepts T.*

4.3 ACU Equational Tree Automata

Let us consider the equational theory ACU stating that | is associative and commutative and that **0** is its neutral element. Formally,

$$\text{ACU} \begin{cases} x \mid \mathbf{0} = x \\ x \mid y = y \mid x \\ x \mid (y \mid z) = (x \mid y) \mid z \end{cases}$$

We write $t \simeq_{\text{ACU}} t'$ whenever the two Σ-terms t and t' are equal modulo ACU. It is well-known that even when a term language L is regular, its ACU-closure, that is the set of terms $\{t \mid t \simeq_{\text{ACU}} t' \text{ and } t' \in L\}$, may not be regular.

For dealing with languages obtained as closure of regular term languages by some equational theory, Ohsaki [15],[16] and Verma [20] have independently introduced so-called *equational tree automata*[6].

An ACU equational tree automaton A over the signature Σ is given by a tuple (Σ, Q, F, Δ) where Q is a finite set of states, $F \subseteq Q$ is the set of final

---

[6] For some equational theory, the classes of automata defined respectively in [15] and in [20] may differ. However, they do coincide for the ACU equational theory.

states and Δ is a finite set of transition rules of the form (q, q_1, q_2 being states from Q and a a unary symbol from Σ):

$$\mathbf{0} \to q \qquad a(q_1) \to q \qquad q_1 \mid q_2 \to q$$

A run for a Σ-term t in an ACU equational tree automaton $A = (\Sigma, Q, F, \Delta)$ is a sequence $t_1, \ldots, t_n$ of terms built over the signature $\Sigma \cup Q$ (where states from Q are considered as constants) such that $t_1 = t$, $t_n \in Q$ and for all $1 \leq i \leq n$, $t_i \simeq_{\mathsf{ACU}} t' \to_\Delta t'' \simeq_{\mathsf{ACU}} t_{i+1}$ for some terms t', t''. The relation $\to_\Delta$ is the ground rewriting relation induced by Δ. A run $t_1, \ldots, t_n$ is *accepting* if the state t_n belongs to F. A Σ-term t is accepted by some ACU equational tree automaton A if there exists an accepting run for t in A. Finally, the language accepted by an ACU equational tree automaton A over the signature Σ is the set of all Σ-terms having an accepting run in A.

Definition 6. *A set of Σ-terms is* ACU-*regular if it is accepted by an* ACU *equational tree automaton.*

Ohsaki showed in [15] that a language E is ACU-regular iff there exists a regular set of Σ-terms E' such that $E = \{t' \mid t \simeq_{\mathsf{ACU}} t' \text{ and } t \in E'\}$.

Lemma 1. *For any two Σ-terms t, t', if $t \simeq_{\mathsf{ACU}} t'$ then $h_C(t) = h_C(t')$*

Proof. By definition, $t \simeq_{\mathsf{ACU}} t'$ holds iff there exists a sequence of terms $t_1, \ldots, t_n$ such that $t = t_1$, $t' = t_n$ and for all $i \in \{1, \ldots, n-1\}$, there exists an equation $l = r$ or $r = l$ in the ACU theory satisfying that $t_i = C[\theta(l)]$ and $t_{i+1} = C[\theta(r)]$ for some context C and some substitution θ mapping variables from l, r to Σ-terms. The proof goes by trivial induction over the context C.

From Colcombet's work [2], it follows easily that

Proposition 2. *For any* ACU-*closed set of Σ-terms E, E is* ACU-*regular iff $h_C(E)$ is accepted by some rational-multiset automaton.*

As for any set of trees T, $h_C^{-1}(T)$ is always ACU-closed (see Lemma 1),

Corollary 1. *For any set of trees T, T is accepted by some rational-multiset automaton iff $h_C^{-1}(T)$ is* ACU-*regular.*

5   A Survey on CMSO-Complete Formalisms

We present here formalisms expressing precisely CMSO definable sets of trees.

5.1   Algebraic Recognizability

We focus first on the notion of *algebraic recognizability* in the sense of Mezei and Wright [12].

Definition 7 ([12]). *Let $\mathcal{M}$ be a Σ-algebra and B be a subset of the domain of $\mathcal{M}$. Then B is said to be $\mathcal{M}$-recognizable if there exists a finite Σ-algebra $\mathcal{A}$ with domain $dom(\mathcal{A})$, a homomorphism from $\mathcal{M}$ to $\mathcal{A}$ and a finite subset D of $dom(\mathcal{A})$ such that $B = h^{-1}(D)$.*

As a particular case, a tree language T is $\mathcal{T}$-recognizable if there exists a finite Σ-algebra $\mathcal{A}$ with domain $dom(\mathcal{A})$, a homomorphism from $\mathcal{T}$ to $\mathcal{A}$ and a finite subset D of $dom(\mathcal{A})$ such that $T = h^{-1}(D)$.

Theorem 2 ([4]). *For any set of trees T, T is CMSO-definable iff T is $\mathcal{T}$-recognizable.*

Starting from a sightly different algebra for trees, Niehren and Podelski defined in [14] a notion of (feature) tree automata for which accepted languages coincide with $\mathcal{T}$-recognizable sets of trees. Note that $\mathcal{T}$-recognizability can be defined alternatively as:

Proposition 3. *A tree language T is $\mathcal{T}$-recognizable iff there exists a finite Σ-algebra $\mathcal{A}$ with domain $dom(\mathcal{A})$ such that $(dom(\mathcal{A}), |^{\mathcal{A}}, 0^{\mathcal{A}})$ is a commutative monoid, h is a homomorphism from $\mathcal{T}$ to $\mathcal{A}$ and D is a finite subset of $dom(\mathcal{A})$ such that $T = h^{-1}(D)$.*

Proof. As T is $\mathcal{T}$-recognizable, there exists a finite Σ-algebra $\mathcal{A}$ with domain $dom(\mathcal{A})$, an homomorphism h from $\mathcal{T}$ to $\mathcal{A}$ and a finite subset D of $dom(\mathcal{A})$ such that $T = h^{-1}(D)$. Let us consider the sub-algebra $\mathcal{A}'$ of $\mathcal{A}$ whose domain is precisely $h(\mathsf{Tree})$. Obviously, T is $\mathcal{T}$-recognizable using the finite algebra $\mathcal{A}'$, the homomorphism h and the set $D \cap dom(\mathcal{A}')$. It is then easy to prove that $(dom(\mathcal{A}'), |^{\mathcal{A}'}, 0^{\mathcal{A}'})$ is a commutative monoid.

5.2   Recognizable-Multiset Tree Automata

In [6], Courcelle introduced a notion of tree automaton whose transitions are defined by means of recognizable sets of finite multisets. This notion can be rephrased in our settings as follows:

Definition 8. *A recognizable-multiset tree automaton is a rational-multiset tree automaton (Λ, Q, F, δ) such that $F \in Rec(\mathbb{M}(Q))$ and for all q in Q and a in Λ, $\delta(q, a) \in Rec(\mathbb{M}(Q))$.*

Theorem 3. *[6] For any set of trees T, T is CMSO-definable iff there exists some recognizable-multiset automaton A that accepts T.*

As recognizable sets of multisets are strictly included into rational sets of multisets, we have:

Corollary 2. *The PMSO logic is strictly more powerful than the CMSO logic over unranked and unordered trees.*

Courcelle proved in [4] that CMSO is strictly more expressive than MSO on unranked and unordered trees. So, this shows that MSO-CMSO-PMSO is a strict hierarchy for this kind of trees; this has to be contrasted with the case of ranked trees where it is known that MSO and CMSO have the same expressive power [4]. It is also not difficult to see that the extension to PMSO does not bring neither some new expressiveness for ranked trees. For unranked and ordered trees, it is quite simple to write an MSO formula for the atom $Mod_j^i(X)$, and thus, showing that MSO and CMSO have in that case the same expressiveness. But, PMSO is for unranked and ordered trees strictly more expressive than MSO [18].

6   New Characterizations for PMSO Definable Sets

We consider first sets of trees defined by means of a system of equations, namely, equational trees languages. Then as done in Section 5.1 for CMSO, we give a fully algebraic characterization of PMSO definable sets of trees.

6.1   Equational Tree Languages

Let $X_1, \ldots, X_n$ be a finite set of variables. We consider the signature $\Sigma \cup \{+\} \cup \{X_1, \ldots, X_n\}$ where $+$ is a binary symbol used in infix notation and $X_1, \ldots, X_n$ are considered as constants.

A system S of equations over the signature $\Sigma \cup \{+\}$ and the variables $X_1, \ldots, X_n$ is a set of equations of the form $X_i = s_i$ such that s_i is a term built over $\Sigma \cup \{+\} \cup \{X_1, \ldots, X_n\}$ and for each X_i, there exists precisely one equation in S.

For a Σ-algebra $\mathcal{M}$ and a set of variables $\{X_1, \ldots, X_n\}$, a $\mathcal{M}$-valuation $\mathcal{I}$ is a mapping associating with each variable X_i a subset of the domain of $\mathcal{M}$. A $\mathcal{M}$-valuation $\mathcal{I}$ is extended to terms built over the signature $\Sigma \cup \{+\}$ as follows:

- $\mathcal{I}(0) = \{0^{\mathcal{M}}\}$
- $\mathcal{I}(a(s)) = \{a^{\mathcal{M}}(t) \mid t \in \mathcal{I}(s)\}$
- $\mathcal{I}(s_1 \mid s_2) = \{t_1 \mid^{\mathcal{M}} t_2 \mid t_1 \in \mathcal{I}(s_1), t_2 \in \mathcal{I}(s_2)\}$
- $\mathcal{I}(s_1 + s_2) = \mathcal{I}(s_1) \cup \mathcal{I}(s_2)$

A $\mathcal{M}$-valuation $\mathcal{I}$ is a solution of a system of equations S for the Σ-algebra $\mathcal{M}$ if for all equations $X_i = s_i$ in S, it holds that $\mathcal{I}(X_i)$ is equal to $\mathcal{I}(s_i)$. Valuations (and thus, solutions) over the same set of variables are equipped with a natural partial ordering: $\mathcal{I}$ is smaller than $\mathcal{I}'$ if for all X_i, $\mathcal{I}(X_i) \subseteq \mathcal{I}(X_i')$. It is not difficult to prove that any system of equations S admits a least solution; we will denote $Least(S, \mathcal{M})$ the least $\mathcal{M}$-valuation which is a solution of S.

Definition 9 ([12]). *For a Σ-algebra $\mathcal{M}$, a subset L of the domain of $\mathcal{M}$ is equational if there exists a system of equations S (over the signature $\Sigma \cup \{+\}$) with some designated variable X such that $Least(S, \mathcal{M})(X) = L$.*

As a particular case for the Σ-algebra $\mathcal{T}$, a set of trees T is equational if there exists a system of equations $\mathcal{S}$ with some designated variable X such that $Least(\mathcal{S}, \mathcal{T})(X) = T$. We denote $\mathsf{Equat}(\mathcal{T})$ the set of equational tree languages.

Courcelle proved in [4] that CMSO-definable languages are equational but that the converse is not true: some languages are equational but not CMSO-definable.

We recall in the next two propositions some useful properties of equational languages.

Proposition 4 ([12]). *Let $\mathcal{M}, \mathcal{M}'$ be two Σ-algebras and h a homomorphism from $\mathcal{M}$ to $\mathcal{M}'$. For any system of equations $\mathcal{S}$, for any variable X from $\mathcal{S}$, it holds that $Least(\mathcal{S}, \mathcal{M}')(X) = h(Least(\mathcal{S}, \mathcal{M})(X))$.*

Proposition 5 ([12]). *For the Σ-algebra of terms $\mathcal{C}$, a language L is equational (ie $L \in \mathsf{Equat}(\mathcal{C})$) iff L is regular (ie accepted by some "classical" tree automaton).*

Theorem 4. *For any set of trees T, T is PMSO-definable iff $T \in \mathsf{Equat}(\mathcal{T})$.*

Proof. By proposition 5, a set S of Σ-terms is regular iff it is equational over $\mathcal{C}$. By Proposition 4, we have that $h_{\mathcal{C}}(S)$ is equational over $\mathcal{T}$. Conversely, if T is equational over $\mathcal{T}$ then still by Proposition 4, there exists S equational over $\mathcal{C}$ such that $T = h_{\mathcal{C}}(S)$. Then, by Proposition 1, it is sufficient to prove that $h_{\mathcal{C}}(S)$ is accepted by some rational-multiset automaton.

Let us denote $\mathsf{ACU}(S)$ the ACU-closure of S. By Proposition 2, $h_{\mathcal{C}}(\mathsf{ACU}(S))$ is accepted by some rational-multiset automaton. We conclude easily using that $h_{\mathcal{C}}(S) = h_{\mathcal{C}}(\mathsf{ACU}(S))$.

6.2 An Algebraic Characterization of PMSO Definability

We are going to define now an algebraic recognizability criteria for the logic PMSO. Recalling that $\mathcal{C}$ is the algebra of terms built over the signature Σ, it is obvious to see that the notion of $\mathcal{C}$-recognizability is the same as the one defined by "classical" tree automata [3] for ranked trees written over the signature Σ (ie for Σ-terms): the set of states is the domain of the finite Σ-algebra $\mathcal{A}$, the interpretation of the function symbols from Σ in $\mathcal{A}$ provides the transition rules (which are bottom-up deterministic) and D is the set of final states.

We define weak $\mathcal{T}$-recognizability for unranked and unordered trees as follows:

Definition 10. *A tree language T is weakly $\mathcal{T}$-recognizable iff there exists some $\mathcal{C}$-recognizable set of Σ-terms M such that $T = h_{\mathcal{C}}(M)$.*

Intuitively, we can consider Σ-terms as representatives for trees and $h_{\mathcal{C}}$ as the mapping associating with each Σ-term the tree it represents. However, $h_{\mathcal{C}}$ is not injective, ie a single tree may have several representatives (actually, countably many). The intuition of weak $\mathcal{T}$-recognizability is to consider recognizability for the representatives (*ie* the Σ-terms) instead of the trees themselves. This notion is therefore different from $\mathcal{T}$-recognizability as $\mathcal{T}$-recognizability requires all the representatives of some tree to be recognized (see Proposition 6).

Theorem 5. *A set of trees T is PMSO-definable iff T is weakly $\mathcal{T}$-recognizable.*

Sketch of proof. By definition, T is weakly $\mathcal{T}$-recognizable iff there exists some $\mathcal{C}$-recognizable set of Σ-terms M such that $T = h_{\mathcal{C}}(M)$. By Proposition 5, this is equivalent to the existence of some equational language M over the algebra $\mathcal{C}$ such that $T = h_{\mathcal{C}}(M)$. Using Proposition 4, this latter holds iff T is an equational language over the algebra $\mathcal{T}$. Finally, by Theorem 4, this amounts to have T PMSO-definable.

7 New Characterizations for CMSO Definable Sets

In this section we reformulate CMSO definability first in terms of $\mathcal{C}$-recognizability and then by a restricted subclass of Presburger tree automata.

7.1 CMSO-Definability and $\mathcal{C}$-Recognizability

Proposition 6. *For any set of trees T, T is CMSO-definable iff the set of Σ-terms $h_{\mathcal{C}}^{-1}(T)$ is $\mathcal{C}$-recognizable.*

Proof. Immediate from Theorem 2 and Proposition 4.4 from [6] stating that T is $\mathcal{T}$-recognizable iff $h_{\mathcal{C}}^{-1}(T)$ is $\mathcal{C}$-recognizable.

7.2 CMSO-Definability and Presburger Tree Automata

Definition 11. *A unary Presburger tree automaton is a PTA (Λ, Q, F, δ) such that $F \in \mathcal{F}_{\mathcal{U}}^1(\{p \le p', Div_k(p)\})$ and for all $q \in Q$ and all $a \in \Lambda$, $\delta(q, a)$ belongs to $\mathcal{F}_{\mathcal{U}}^1(\{p \le p', Div_k(p)\})$.*

Lemma 2. *Let N be a subset of $\mathbb{N}^l$ and $A = (a_1, \dots, a_l)$ be some alphabet. Then N is unary ordering-definable iff $\pi_A^{-1}(N) \in Rec(\mathbb{M}(A))$.*

Proof. Courcelle showed in [6] that $\pi_A^{-1}(N) \in Rec(\mathbb{M}(A))$ iff N is a finite union of Cartesian products of l ultimately periodic sets of naturals, *ie* N is a finite union of sets of the form $B_1 \times \dots \times B_l$ where for each i, $B_i = \{b + \alpha p \mid \alpha \in \mathbb{N}\}$ for some $b, p \in \mathbb{N}$. We just prove then that N is unary ordering-definable iff N is a finite union of Cartesian products of l ultimately periodic sets of naturals

Then, as for Presburger tree automata and PMSO, we have

Proposition 7. *For any set of trees T, T is CMSO-definable iff there exists some unary Presburger tree automaton A that accepts T.*

Proof. Straightforward using Theorem 3 and Lemma 2.

8 Some Characterizations for MSO Definable Sets

In this section, we investigate sets of trees definable by means of MSO sentences; Mainly, we are going to study how restrictions over formalisms used to characterize CMSO or PMSO can be put.

8.1 MSO-Definability and Presburger Tree Automata

Definition 12. *A* unary ordering tree automaton *is a PTA* (Λ, Q, F, δ) *such that* $F \in \mathcal{F}_{\mathcal{U}}^1(\{p \leq p'\})$ *and for all* $q \in Q$ *and all* $a \in \Lambda$, $\delta(q, a)$ *belongs to* $\mathcal{F}_{\mathcal{U}}^1(\{p \leq p'\})$.

Proposition 8. *For any set of trees* T, T *is MSO-definable iff there exists some unary ordering tree automaton* A *that accepts* T.

Sketch of proof. The proof is rather standard. We show first that the existence of an accepting run for a tree can be expressed by some MSO sentence. For the converse, we show closure of the unary ordering tree automaton under union, complementation (by computing first a deterministic and complete automaton) and relabeling morphism. Then, we build such an automaton inductively over the structure of the MSO formula.

8.2 MSO-Definability and Aperiodic-Recognizable Tree Automata

Definition 13. *A multiset language* $L \in \mathbb{M}(A)$ *is* aperiodically recognizable *if there exists a monoid morphism* h *from* $(L, \uplus, \{\varnothing\})$ *to a finite aperiodic[7] monoid* $(D, +, \iota)$ *and a finite subset* D' *of* D *such that* $L = h^{-1}(D')$.

We denote $ApRec(\mathbb{M}(A))$ the set of aperiodically recognizable multiset languages.

Definition 14. *An* aperiodic-recognizable multiset tree automaton *is a rational-multiset tree automaton* (Λ, Q, F, δ) *such that* $F \in ApRec(\mathbb{M}(Q))$ *and for all* q *in* Q *and* a *in* Λ, $\delta(q, a) \in ApRec(\mathbb{M}(Q))$.

Lemma 3. *Let* N *be a subset of* $\mathbb{N}^l$ *and* $A = (a_1, \dots, a_l)$ *be some alphabet,* N *is unary ordering-definable iff* $\pi_A^{-1}(N) \in ApRec(\mathbb{M}(A))$.

Sketch of proof. We prove first that N is unary ordering-definable iff N is a finite union of Cartesian products of l ultimately periodic sets of naturals with periods in $\{0, 1\}$, ie N is a finite union of sets of the form $B_1 \times \dots \times B_l$ where for each i, $B_i = \{b + \alpha p \mid \alpha \in \mathbb{N}\}$ for some $b \in \mathbb{N}$ and $p \in \{0, 1\}$. Then, we use a result from [9] stating that N is a finite union of Cartesian products of l ultimately periodic sets of naturals with periods in $\{0, 1\}$ iff N is a star-free subset of $\mathbb{N}^l$, ie N can be obtained from finite subsets of $\mathbb{N}^l$ using sum $+$ and Boolean operations

---

[7] We recall that a monoid $(S, .)$ is said to to be aperiodic if for all $s \in S$, there exists some natural n such that $s^n = s^{n+1}$ where $s^1 = s$ and $s^{k+1} = s^k . s$.

(union, intersection, complement). Finally, we can conclude using that $(\mathbb{N}^l, +)$ is isomorphic to $(\mathbb{M}(A), \uplus)$ and that over commutative monoids, star-free languages are precisely the recognizable and aperiodic ones [10].

Theorem 6. *For any set of trees T, T is MSO-definable iff there exists some aperiodic-recognizable multiset automaton A that accepts T.*

Proof. Straightforward from Proposition 8 and Lemma 3.

8.3   An Algebraic Characterization of MSO Definability

We relate here MSO definability and algebraic $\mathcal{T}$-recognizability.

Definition 15. *A tree language T is aperiodically $\mathcal{T}$-recognizable iff there exists a finite Σ-algebra $\mathcal{A}$ with domain $dom(\mathcal{A})$ such that $(dom(\mathcal{A}), |^{\mathcal{A}}, 0^{\mathcal{A}})$ is an aperiodic and commutative monoid, h is a homomorphism from $\mathcal{T}$ to $\mathcal{A}$ and D is a finite subset of $dom(\mathcal{A})$ such that $T = h^{-1}(D)$.*

Theorem 7. *For any set of trees T, T is MSO-definable iff T is aperiodically $\mathcal{T}$-recognizable.*

References

1. J. Carme, J. Niehren, and M. Tommasi. Querying Unranked Trees with Stepwise Tree Automata. In *International Conference on Rewriting Techniques and Applications*, volume 3091 of *LNCS*, pages 105–118. Springer, 2004.
2. T. Colcombet. Rewriting in the partial algebra of typed terms modulo AC. In *Electronic Notes in Theoretical Computer Science*, volume 68. Elsevier Science Publishers, 2002.
3. H. Comon, M. Dauchet, R. Gilleron, F. Jacquemard, D. Lugiez, S. Tison, and M. Tommasi. Tree Automata Techniques and Applications. Available on: http://www.grappa.univ-lille3.fr/tata, 1997. release October, 1rst 2002.
4. B. Courcelle. The Monadic Second-Order Logic of Graphs. I. Recognizable Sets of Finite Graphs. *IC*, 85(1):12–75, 1990.
5. B. Courcelle. The monadic second order logic of graphs VI: on several representations of graphs by relational structures. *Discrete Applied Mathematics*, 54(2-3):117–149, 1994.
6. B. Courcelle. Basic notions of universal algebra for language theory and graph grammars. *Theoretical Computer Science*, 163:1–54, 1996.
7. S. Dal-Zilio and D. Lugiez. XML Schema, Tree Logic and Sheaves Automata. In *Rewriting Techniques and Applications, 14th International Conference, RTA 2003*, volume 2706 of *LNCS*, pages 246–263. Springer, 2003.
8. S. Dal-Zilio, D. Lugiez, and C. Meyssonnier. A logic you can count on. In *31st Annual ACM SIGPLAN-SIGACT Symposium on Principles of Programming Languages*, 2004.
9. S. Gaubert and A. Giua. Petri net languages and infinite subsets of $\mathbb{N}^m$. *Journal of Computer System Sciences*, 59(3):373–391, 1999.

10. G. Guaiana, A. Restivo, and S. Salemi. On Aperiodic Trace Languages. In *STACS 91, 8th Annual Symposium on Theoretical Aspects of Computer*, volume 480 of *LNCS*, pages 76–88. Springer, 1991.

11. F. Klaedtke and H. Rueß. Monadic Second-Order Logics with Cardinalities. In *30th International Colloquium on Automata, Languages and Programming, ICALP 2003*, volume 2719 of *Lecture Notes in Computer Science*. Springer Verlag, 2003.

12. J. Mezei and J.B. Wright. Algebraic automata and context-free sets. *Information and Control*, 11(2-3):3–29, 1967.

13. F. Neven and T. Schwentick. Query automata over finite trees. *Theoretical Computer Science*, 275(1–2):633–674, March 2002.

14. J. Niehren and A. Podelski. Feature Automata and Recognizable Sets of Feature Trees. In *Theory and Practice of Software Development, International Joint Conference CAAP/FASE/TOOLS*, volume 668 of *LNCS*, pages 356–375. Springer, 1993.

15. H. Ohsaki. Beyond Regularity: Equational Tree Automata for Associative and Commutative Theories. In *Proceedings of 15th International Conference of the European Association for Computer Science Logic - CSL 2001*, volume 2142 of *LNCS*, pages 539–553. Springer, 2001.

16. H. Ohsaki and T. Takai. Decidability and Closure Properties of Equational Tree Languages. In *Proceedings of 13th International Conference on Rewriting Techniques and Applications*, volume 2378 of *LNCS*, pages 114–128. Springer, 2002.

17. R. J. Parikh. On context-free languages. *Journal of the ACM*, 13(4):570–581, 1966.

18. H. Seidl, T. Schwentick, and A. Muscholl. Numerical Document Queries. In *Twenty-Second ACM SIGACT-SIGMOD-SIGART Symposium on Principles of Database Systems*, pages 155–166. ACM, 2003.

19. J. W. Thatcher and J. B. Wright. Generalized finite automata with an application to a decision problem of second-order logic. *Mathematical System Theory*, 2:57–82, 1968.

20. K. N. Verma. Two-Way Equational Tree Automata for AC-like Theories: Decidability and Closure Properties. In *Proceedings of 14th International Conference on Rewriting Techniques and Applications*, volume 2706 of *LNCS*, pages 180–197. Springer, 2003.

Author Index

Lecture Notes in Computer Science

For information about Vols. 1–3356

please contact your bookseller or Springer

Vol. 3407: Z. Liu, K. Araki (Eds.), Theoretical Aspects of Computing - ICTAC 2004. XIV, 562 pages. 2005.

Vol. 3406: A. Gelbukh (Ed.), Computational Linguistics and Intelligent Text Processing. XVII, 829 pages. 2005.

Vol. 3404: V. Diekert, B. Durand (Eds.), STACS 2005. XVI, 706 pages. 2005.

Vol. 3403: B. Ganter, R. Godin (Eds.), Formal Concept Analysis. XI, 419 pages. 2005. (Subseries LNAI).

Vol. 3401: Z. Li, L.G. Vulkov, J. Waśniewski (Eds.), Numerical Analysis and Its Applications. XIII, 630 pages. 2005.

Vol. 3399: Y. Zhang, K. Tanaka, J.X. Yu, S. Wang, M. Li (Eds.), Web Technologies Research and Development - APWeb 2005. XXII, 1082 pages. 2005.

Vol. 3398: D.-K. Baik (Ed.), Systems Modeling and Simulation: Theory and Applications. XIV, 733 pages. 2005. (Subseries LNAI).

Vol. 3397: T.G. Kim (Ed.), Artificial Intelligence and Simulation. XV, 711 pages. 2005. (Subseries LNAI).

Vol. 3396: R.M. van Eijk, M.-P. Huget, F. Dignum (Eds.), Agent Communication. X, 261 pages. 2005. (Subseries LNAI).

Vol. 3395: J. Grabowski, B. Nielsen (Eds.), Formal Approaches to Software Testing. X, 225 pages. 2005.

Vol. 3394: D. Kudenko, D. Kazakov, E. Alonso (Eds.), Adaptive Agents and Multi-Agent Systems III. VIII, 313 pages. 2005. (Subseries LNAI).

Vol. 3393: H.-J. Kreowski, U. Montanari, F. Orejas, G. Rozenberg, G. Taentzer (Eds.), Formal Methods in Software and Systems Modeling. XXVII, 413 pages. 2005.

Vol. 3392: D. Seipel, M. Hanus, U. Geske, O. Bartenstein (Eds.), Applications of Declarative Programming and Knowledge Management. X, 309 pages. 2005. (Subseries LNAI).

Vol. 3391: C. Kim (Ed.), Information Networking. XVII, 936 pages. 2005.

Vol. 3390: R. Choren, A. Garcia, C. Lucena, A. Romanovsky (Eds.), Software Engineering for Multi-Agent Systems III. XII, 291 pages. 2005.

Vol. 3389: P. Van Roy (Ed.), Multiparadigm Programming in Mozart/OZ. XV, 329 pages. 2005.

Vol. 3388: J. Lagergren (Ed.), Comparative Genomics. VII, 133 pages. 2005. (Subseries LNBI).

Vol. 3387: J. Cardoso, A. Sheth (Eds.), Semantic Web Services and Web Process Composition. VIII, 147 pages. 2005.

Vol. 3386: S. Vaudenay (Ed.), Public Key Cryptography - PKC 2005. IX, 436 pages. 2005.

Vol. 3385: R. Cousot (Ed.), Verification, Model Checking, and Abstract Interpretation. XII, 483 pages. 2005.

Vol. 3383: J. Pach (Ed.), Graph Drawing. XII, 536 pages. 2005.

Vol. 3382: J. Odell, P. Giorgini, J.P. Müller (Eds.), Agent-Oriented Software Engineering V. X, 239 pages. 2005.

Vol. 3381: P. Vojtáš, M. Bieliková, B. Charron-Bost, O. Sýkora (Eds.), SOFSEM 2005: Theory and Practice of Computer Science. XV, 448 pages. 2005.

Vol. 3380: C. Priami, Transactions on Computational Systems Biology I. IX, 111 pages. 2005. (Subseries LNBI).

Vol. 3379: M. Hemmje, C. Niederee, T. Risse (Eds.), From Integrated Publication and Information Systems to Information and Knowledge Environments. XXIV, 321 pages. 2005.

Vol. 3378: J. Kilian (Ed.), Theory of Cryptography. XII, 621 pages. 2005.

Vol. 3377: B. Goethals, A. Siebes (Eds.), Knowledge Discovery in Inductive Databases. VII, 190 pages. 2005.

Vol. 3376: A. Menezes (Ed.), Topics in Cryptology – CT-RSA 2005. X, 385 pages. 2005.

Vol. 3375: M.A. Marsan, G. Bianchi, M. Listanti, M. Meo (Eds.), Quality of Service in Multiservice IP Networks. XIII, 656 pages. 2005.

Vol. 3374: D. Weyns, H.V.D. Parunak, F. Michel (Eds.), Environments for Multi-Agent Systems. X, 279 pages. 2005. (Subseries LNAI).

Vol. 3372: C. Bussler, V. Tannen, I. Fundulaki (Eds.), Semantic Web and Databases. X, 227 pages. 2005.

Vol. 3371: M.W. Barley, N. Kasabov (Eds.), Intelligent Agents and Multi-Agent Systems. X, 329 pages. 2005. (Subseries LNAI).

Vol. 3370: A. Konagaya, K. Satou (Eds.), Grid Computing in Life Science. X, 188 pages. 2005. (Subseries LNBI).

Vol. 3369: V.R. Benjamins, P. Casanovas, J. Breuker, A. Gangemi (Eds.), Law and the Semantic Web. XII, 249 pages. 2005. (Subseries LNAI).

Vol. 3368: L. Paletta, J.K. Tsotsos, E. Rome, G.W. Humphreys (Eds.), Attention and Performance in Computational Vision. VIII, 231 pages. 2005.

Vol. 3367: W.S. Ng, B.C. Ooi, A. Ouksel, C. Sartori (Eds.), Databases, Information Systems, and Peer-to-Peer Computing. X, 231 pages. 2005.

Vol. 3366: I. Rahwan, P. Moraitis, C. Reed (Eds.), Argumentation in Multi-Agent Systems. XII, 263 pages. 2005. (Subseries LNAI).

Vol. 3365: G. Mauri, G. Păun, M.J. Pérez-Jiménez, G. Rozenberg, A. Salomaa (Eds.), Membrane Computing. IX, 415 pages. 2005.

Vol. 3363: T. Eiter, L. Libkin (Eds.), Database Theory - ICDT 2005. XI, 413 pages. 2004.

Vol. 3362: G. Barthe, L. Burdy, M. Huisman, J.-L. Lanet, T. Muntean (Eds.), Construction and Analysis of Safe, Secure, and Interoperable Smart Devices. IX, 257 pages. 2005.

Vol. 3361: S. Bengio, H. Bourlard (Eds.), Machine Learning for Multimodal Interaction. XII, 362 pages. 2005.

Vol. 3360: S. Spaccapietra, E. Bertino, S. Jajodia, R. King, D. McLeod, M.E. Orlowska, L. Strous (Eds.), Journal on Data Semantics II. XI, 223 pages. 2005.

Vol. 3359: G. Grieser, Y. Tanaka (Eds.), Intuitive Human Interfaces for Organizing and Accessing Intellectual Assets. XIV, 257 pages. 2005. (Subseries LNAI).

Vol. 3358: J. Cao, L.T. Yang, M. Guo, F. Lau (Eds.), Parallel and Distributed Processing and Applications. XXIV, 1058 pages. 2004.

Vol. 3357: H. Handschuh, M.A. Hasan (Eds.), Selected Areas in Cryptography. XI, 354 pages. 2004.